AAPG MEMOIR 42

INTERPRETATION OF THREE-DIMENSIONAL SEISMIC DATA

FOURTH EDITION

By

ALISTAIR R. BROWN

CONSULTING RESERVOIR GEOPHYSICIST

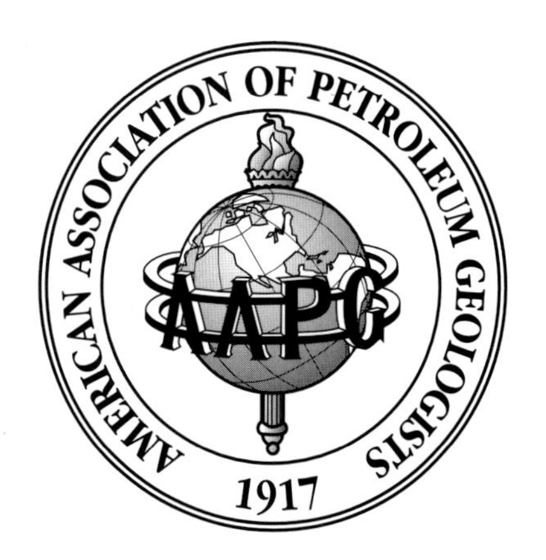

PUBLISHED BY
AMERICAN ASSOCIATION OF PETROLEUM GEOLOGISTS
TULSA, OKLAHOMA 74101, U.S.A.

Library of Congress Cataloging-in-Publication Data

Brown, Alistair
Interpretation of three-dimensional seismic data.
Fourth edition.

(AAPG memoir; 42)
Includes bibliographies and index.
1.Seismology-Methodology. 2. Seismic reflection method. 3. Petroleum-Geology-Methodology. I. Title.
II. Series.
QE539.B78 1986 551.2'2 86-22341
ISBN 0-89181-346-2

Association Editor: Kevin T. Biddle
Science Director: Richard Steinmetz
Publications Manager: Kenneth M. Wolgemuth
Special Projects Editor: Anne H. Thomas

PREFACE TO THE FOURTH EDITION

Consider everything to be geology until proved otherwise.

— *MILO BACKUS*

This is being written shortly after the 20th anniversary of the first commercial 3-D survey. Few could then have imagined how important and widespread the technology would become. Mature petroleum areas are now totally covered with 3-D, surveys being contiguous, overlapping or on top of each other. Speculative 3-D surveys are commonplace and have made more 3-D data available to more interpreters. Many speculative surveys are very large; one in the Gulf of Mexico covers 700 blocks or more than 16,000 km^2. Surveys over producing fields are being repeated for seismic monitoring of production, generally known as 4-D.

There are now many published stories of exploration and exploitation successes attributed to 3-D seismic data. Three-D reduces finding costs, reduces risk, and improves success rates. Royal Dutch/Shell reports that its exploration success outside North America increased from 33% in 1990 to 45% in 1993 based largely on 3-D. Their seismic expenditures are now 90% on 3-D surveys. Exxon considers "3-D seismic to be the single most important technology to ensure the effective and cost-efficient exploration and development of our oil and gas fields." Exxon reports that their success in the Gulf of Mexico in the period 1987-92 was 43% based on 2-D data and 70% based on 3-D data; in the same period in The Netherlands the numbers were 47% (2-D) and 70% (3-D). Mobil reports that in the South Texas Lower Wilcox trend their success based on 2-D was 70% but this rose to 84% based on 3-D. Amoco have concluded that "the average exploitation 3-D survey detects six previously unknown, high quality drill locations," and "adds $9.8 million of value" to a producing property. Petrobras reports that in the Campos Basin offshore Brazil their success rate has increased from 30% based on 2-D data to over 60% based on 3-D.

With this tremendous level of activity and euphoria, and with exploration and development problems becoming more difficult, the issue of the moment is to apply the technology appropriately. There is still a great amount of data underutilization. In an attempt to correct this, let us not impose too rigorous a geologic model on our interpretations; let us seek a full understanding of the seismic character, and allow the data to speak to us. "Consider everything to be geology until proved otherwise."

On the other hand our data has its shortcomings and interpreters benefit greatly from an understanding of geophysical principles and of the processes that the data has been through before it reaches the interpretation workstation. Reductions in acquisition costs have sometimes been over-zealous resulting in significant data irregularities which can only be partly fixed in data processing. There is no substitute for good signal-to-noise ratio. We cannot expect "to make a silk purse out of a sow's ear" and 3-D is certainly not a universal panacea. Reservoir evaluation or characterization using 3-D data is popular today and so it should be, but data quality imposes limitations. I know several projects where the results have been disappointing because the data just wasn't good enough. We must have realistic expectations.

The largest single development in 3-D interpretation techniques since the publication of the last edition has been the generation, display and use of seismic attributes. This Fourth Edition has a whole new chapter on the subject. In addition there are many new data examples and procedural diagrams distributed throughout the book in an attempt to bring the treatment of every aspect of 3-D interpretation up-to-date.

Alistair R. Brown
Dallas, Texas
April 1996

PREFACE TO THE THIRD EDITION

The 3-D seismic method is now mature. Few people would doubt this, and the huge number of geophysicists, geologists and engineers using it are testimony to the accepted power of 3-D seismic technology. Three-D seismic is used for exploration, for development and for production, and hardly a corner of the world is as yet untouched by the technology. Substantially more than 50% of all seismic activity in the Gulf of Mexico and the North Sea is now 3-D! The total land area of The Netherlands is now 30% covered by 3-D seismic data! Execution of 3-D surveys is a condition for the granting of some licenses. Some companies, or divisions of companies, have given up 2-D data collection altogether!

The new Foreword to this edition provides a striking accolade for 3-D seismic and its association with the interactive workstation. Workstations are today almost as numerous as 3-D surveys, and so they should be. But both of them are underutilized. The amount of information in modern 3-D seismic data is very great and the capability to extract it lies in the proper use of the computer-driven workstation. All too many of today's practitioners are applying traditional 2-D methods carried over from their experience of 2-D data. This is natural but inefficient, time-consuming and misdirected. The 3-D interpreter needs to understand and use the tools available to him in order to do justice to his investment in 3-D data. Oil company management needs to offer appropriate encouragement to geoscientists. The next phase of our technological evolution must be to make proper use of what we already have.

Another impediment to proper utilization of 3-D data is confused terminology. We find a plethora of terms referring to the same product. For example, a horizontal section or time slice is also referred to, unfortunately, as a Seiscrop, Seiscrop section, isotime (slice or section), horizontal time slice, time-slice map or seiscut. At one time companies saw a competitive advantage in special or trademarked names, but that time has passed. Everybody in the 3-D processing or display business can make a time slice. Interpreters of three-dimensional data need to make regular use of time slices as they are essential to a complete interpretation. Fancy names just encourage inexperienced 3-D interpreters to distance themselves from the product and develop the opinion that they are a phenomenon to be marvelled at rather than a section pregnant with geologic information. I believe that much of the confusing terminology has arisen because of a lack of distinction between the process and the product. We use the process of *amplitude extraction* to make the product of a **horizon slice**; we construct a section in the *trace* direction to make a **crossline**; we *reconstruct a cut* through the volume to make an **arbitrary line**. The interactive system vendors generate most of these capabilities for us and are concerned more about the procedure. Interpreters are concerned more about the utilization of the product. This book attempts to clarify these issues by using only the more accepted terms.

The Third Edition sees a further significant expansion in material with many new companies—oil companies, service companies, and interactive workstation vendors—contributing data examples. Examples from Europe play a more significant role than in previous editions and there are five new case histories.

Alistair R. Brown
Dallas, Texas
September 1991

PREFACE TO THE SECOND EDITION

Since publication of the first edition, 3-D seismic technology has continued its trend toward universal acceptance and maturity. Much of this has resulted from the emphasis on development and production prompted by the recent depression in exploration.

I have found a great demand for short courses on interpretation of three-dimensional seismic data, for which this book has served as the text, and this has fueled the need to update the content for a Second Edition. The expansion in text and figures is about 30%, including more case history examples. During the expansion my objective has been to extend the application and appeal of the book by broadening the field of contributing companies, of types of display, interactive system and color usage, and of the range of subsurface problems addressed with 3-D seismic data. Emphasis continues on the synergistic benefits of amplitude, phase, interactive approaches and color.

Alistair R. Brown
Dallas, Texas
June 1988

PREFACE TO THE FIRST EDITION

The whole is more than the sum of the parts.

— *ARISTOTLE*

Three-dimensional seismic data have spawned unique interpretation methodologies. This book is concerned with these methodologies but is not restricted to them. The theme is two-fold:

—How to use 3-D data in an optimum fashion, and

—How to extract the maximum amount of subsurface information from seismic data today.

I have assumed a basic understanding of seismic interpretation which in turn leans on the principles of geology and geophysics. Most readers will be seismic interpreters who want to extend their knowledge, who are freshly confronted with 3-D data, or who want to focus their attention on finer subsurface detail or reservoir properties.

Color is becoming a vital part of seismic interpretation and this is stressed by the proportion of color illustrations herein.

Alistair R. Brown
Dallas, Texas
January 1986

ACKNOWLEDGMENTS FOR THE SECOND, THIRD AND FOURTH EDITIONS

I really appreciate the help that so many people have provided. Most particularly I must thank the principal authors of the contributed material. Also, many individuals provided me with one, two or three figures and secured for me their release; in some cases this involved considerable effort because several companies were involved in group surveys. My classes of short course students have provided critical comment and discussion and these have prompted me to sharpen up the subject matter and to generate several new explanatory diagrams. To all of these helpful people—a big Thank-you.

ACKNOWLEDGMENTS FOR THE FIRST EDITION

I have found the writing and organization of this book daunting, challenging and rewarding. But it certainly has not been accomplished without the help of many friends and colleagues. First, I would like to thank Geophysical Service Inc. (GSI) and especially Bob Graebner for encouraging the project. Bob Sheriff, University of Houston, has been my mentor in helping me to discover what writing a book entails. Bob McBeath has been a constant help and source of technical advice; he also read all the manuscript. I am indebted to many companies who released data for publication, and also to the many individuals within those companies who provided their data and discussed its interpretation with me. In particular, Roger Wright and Bill Abriel, Chevron U.S.A., New Orleans, were outstandingly helpful. Colleagues within GSI who provided significant help were Mike Curtis, Keith Burkart, Tony Gerhardstein, Chuck Brede, Bob Howard, and Jennifer Young. Last but not least, my wife, Mary, remained sane while typing and editing the manuscript on a cantankerous word processor.

CONTENTS

ABOUT THE AUTHOR

Alistair Ross Brown was born and raised in Carlisle in the northernmost part of England. The first and middle names demonstrate Scots ancestry. He graduated in Physics from Oxford University in 1963, having attended The Queen's College. Later the necessary geology component was obtained at the Australian National University in Canberra, Australia. He married Mary, another Oxford graduate, in 1963 and they have three children. Now there are also two grandchildren.

Alistair's professional career in geophysics began in Australia where for seven years he was employed by the Bureau of Mineral Resources, and there gained experience in seismic data collection, processing, and interpretation. The Brown family returned to England in 1972 where Alistair worked for Geophysical Service International (GSI). He soon specialized in experimental seismic interpretation and was asked to interpret the first commercial 3-D seismic survey in 1975. Early experimental 3-D interpretation and display soon brought him to Dallas, the worldwide headquarters of GSI, and the family relocated there in 1978.

As 3-D surveys became more and more numerous during the 1980s, Alistair continued to investigate the best ways to interpret them. Interactive workstations emerged in the early part of the decade and he started using an early version in late 1980. After presenting several papers on aspects of 3-D interpretation in the late 1970s and early 1980s, Alistair started teaching the subject to oil company personnel. This led to his independence in 1987.

He is now a Consulting Reservoir Geophysicist specializing in the interpretation of 3-D seismic data and the effective use of interactive workstations. His courses and consultation are acclaimed worldwide and his time is dedicated to helping interpreters get more out of their 3-D seismic data.

Alistair is an active member of SEG, AAPG, and EAGE. He received SEG's Best Presentation Award in 1975; he was recognized by Texas Instruments as a Senior Member of Technical Staff in 1981; he has been a continuing education instructor of SEG and AAPG; he was an AAPG Distinguished Lecturer in 1988, an SEG Distinguished Lecturer in 1991 and the Petroleum Exploration Society of Australia Distinguished Lecturer in 1994.

FOREWORD

Previous editions of this book have contained only the brief laudation to 3-D seismic that appears in the opening lines of Chapter 1. The following was presented on September 25, 1990 in the course of the SEG Convention in San Francisco and is the most comprehensive accolade I have yet heard. I therefore reproduce it here verbatim.

—Alistair R. Brown

THE VALUE OF 3-D SEISMIC AND THE WORKSTATION

Robert E. Osborne, Chief Development Geologist (Retired), Chevron Corporation

I am glad to have the opportunity to acknowledge the splendid contribution that modern geophysics has made towards the development and further exploitation of our oil and gas fields, and to thank geophysicists for that. As a development geologist, with a responsibility for developing oil and gas fields, I have followed the growth of this new sector of activity called development geophysics, and have welcomed the dimension given to our projects by those efforts.

Development geophysics has emerged as a vital sector of your profession and now has become a vital sector of our industry as well. What is happening in development geophysics is impacting many other parts of our industry. I would like to focus my remarks on some aspects of this impact because I am not sure that all of our colleagues, and perhaps geophysicists as well, fully appreciate the new business environment that has been created by modern geophysics.

In my opinion, modern geophysics is more than just a simple technical improvement—it is a true technological *breakthrough*. It is not, as some might conclude, merely the result of 3-D seismic techniques. After all, 3-D seismic has been around for 15 years or so, and although it is admittedly a marvelous innovation, modern geophysics goes beyond 3-D seismic.

Modern geophysics combines 3-D data with the workstation. This is then combined with special geophysical groups that focus that technology toward the development and further exploitation of existing oil and gas fields, which has been shown to be the best area for the combined applications.

It is the workstation that has revolutionized geophysics. Just thinking about it boggles the mind. We can seismically examine a volume of rock, big or small, slice through it in all directions, flatten a specific event, examine specific geophysical variations, add some color to make it easier to study, identify faults or depositional relationships that can be traced in the subsurface, put some quantification to the variables, locate wells in the seismic data, complete with their log traces, generate usertracks (arbitrary lines) that can connect key wells, and perform a host of other data manipulations. These are truly powerful applications. That's why we can call it "modern geophysics." That is why it is a breakthrough.

Not long ago, geophysical organizations were assigned to exploration efforts, out there on the frontier, that were concerned with defining prospects with 2-D seismic lines, to be followed by high-risk wildcat wells.

Modern geophysical methods, however, have their best application in areas requiring precision geology, with attention to small geologic variations, in localities where large capital investments already made can absorb the additional cost of expensive geophysical techniques and where the small revelations of the geology go to the very heart of the problems to be resolved. Those conditions, of course, are found in an existing or developing oil or gas field.

The shift of geophysical attention from exploration for fields towards development of existing fields has affected the way we conduct our geophysical business. Modern geophysics, in my opinion, can alter the way we manage our whole industry.

Consider these statistics from Chevron's operations in just one province—the offshore Gulf of Mexico for the years 1980-1990. In that time period, Chevron shot 36 3-D surveys and spent over $75 million doing it. This does not count the several 3-D surveys shot by Chevron prior to 1980, nor the surveys shot by Gulf Oil Company prior to 1985 and inherited in that year by Chevron, nor does this include the 3-D surveys shot by Tenneco prior to Chevron's acquisition of those properties in 1988. However, if we include these acquisitions, Chevron now has over 100 3-D surveys from just this one province.

The money invested in obtaining and interpreting these geophysical surveys is awesome. It is also increasing. Chevron will spent two to three times as much for Gulf Coast 3-D geophysics in 1990 as it did in 1987.

The manpower devoted to this same effort is equally impressive. Development Geophysics, an organization Chevron created to handle all of this, has grown from just one person in 1980 to more than 22 in 1990, and this does not include many geologists who have rotated through this geophysical section and who are incorporating 3-D seismic work in their development geology projects.

These numbers are from just one province. Add to them Chevron's efforts in the North Sea, west and central Africa, Indonesia, Australia, Canada, and the rest of the United States, including Alaska, and the total effort spawned by these new geophysical techniques is, to use a previous expression, awesome.

Chevron now spends almost as much money for Development Geophysics as it does for Exploration Geophysics, and we have as many if not more earth scientists devoted to the management of our existing fields than we have in the exploration search for new fields, especially in the United States. This is caused partly by the property acquisitions Chevron made in the last decade, but it also reflects the increased efforts now focused on producing properties and their problems.

Chevron is but one company. If we multiply these numbers by 10 or even 20 times, we can begin to get a feel for the size of activity in modern development geophysics.

One might ask if all of the cost and all of this effort directed to modern geophysics in our existing fields is justifiable. Experience so far suggests that it is very justifiable. Compared with other investments in producing an oil or gas field, a 3-D seismic survey is not particularly expensive. Such surveys invariably reveal some critical aspect about a field not even suspected beforehand, and almost always lead to some opportunity created or some unfruitful expense avoided. Such information is very valuable and should easily justify the cost of its acquisition. Our files in Chevron are full of examples of new pools found, new reserves added, production declines arrested or reversed, bad wells avoided, reservoir anomalies explained, and of payouts of 3-D surveys achieved in a very short time.

While the economics of modern geophysics appear to be justified, at least for now, some peripheral considerations that are now emerging will affect how we will conduct our business in the future. Geophysics will both impact that conduct and be impacted by it. Here are some of the problems that I foresee.

Revelations about the geology of our fields where we have shot a 3-D survey have been so profound that we will shortly mistrust any analysis in a field that does not have a 3-D survey. I foresee programs, extended over a period of time, in which we will systematically run 3-D surveys on all important fields. It will be done to check for new fault blocks, deeper potential, or unusual geologic conditions that could be important; indeed, it will be done as a form of insurance for conditions that are not even suspected.

It does not take much more vision to recognize that 3-D seismic will become an important new exploration tool not only in mature basins away from or between fields, but also out on the remote frontier. Exploration strategies, especially in the water, will include 3-D surveys up front of the first wildcat well. Of course, this approach will cost more than our conventional approach, but those costs will be recovered in better exploratory well success ratios, in shortening the time from discovery to production, and by providing confidence for large future capital expenditures. There is going to be a lot of money pumped into 3-D geophysics in the coming years. That is, if it remains affordable and if we have time. And those are two big "ifs" that we will discuss more later.

Modern geophysics, as mentioned earlier, provides a chance for new life in our older fields. It should follow that our existing fields will receive a lot of geophysical attention in the future. However, there is a trend in most major companies to sell some of their producing properties in an effort to raise cash, or reduce costs, or achieve efficiencies by retreating to so-called "core" properties. While these retained properties could receive additional geophysical attention, those that are sold may not. Certainly, some of these properties could be candidates for a new round of further exploration using modern geophysics.

If, however, these properties are transferred to small companies that are less familiar with the power of modern geophysics or that are unable to afford these new techniques, then this group of existing fields will not receive any geophysical attention and will merely putter along until depletion. That would be unfortunate for both our industry and our country, because we will eventually need all of the oil we can get. However, this possibility cannot be dismissed.

There is a great paradox to all of this. We must remember that oil fields deplete, wells eventually must be plugged, and acreage once held by production eventually expires. It is possible that many of our fields, no matter who owns them, will experience these conditions before we can get around to applying the new technologies that could have extended their life. Like the patient who died before the vaccine was delivered to the bedside, so too could many of our fields be abandoned before we can marshall the right combination of new technical resources that could have extended their life. We must, it seems to me, develop a sense of urgency about using modern geophysics in certain areas or risk losing its contribution altogether.

Yet, necessity is the mother of invention. It may be that a new type of entrepreneur will emerge in our industry, a group who will purchase old properties just to apply these new 3-D geophysical technologies. Such groups could replace the old independent wildcatter, whose approach was to assemble acreage over a new prospect and drill an exploratory well. This new approach carries some risk, but perhaps no more so than the old way. I would not be surprised if such a new approach finds more new oil than that older traditional approach—that is, if such groups ever are formed and if they act quickly.

Geophysics has provided a new tool for exploring our existing fields more thoroughly, but in doing so the distinction between exploration geology and development geology has become more blurred. So, too, has the distinction between geology and geophysics, between exploration geophysics and development geophysics, between geology and engineering, between the formal research for new technology and the evolutionary developments of new technology in operating companies. It's hard to compartmentalize our respective disciplines anymore. Indeed, there is a need to integrate them more thoroughly. There is both good and bad about this convergence of disciplines; good because it removes barriers to understanding, promotes cooperation, and insures technical integration, but bad because it inhibits the ability to focus on necessary technical avenues. And, technology is getting so complicated that we need that ability to focus.

It's the age-old concern about generalists and specialists. It's hard enough to learn one discipline, let alone several, but more and more we are being forced to integrate disciplines by the nature of our work. This will require more training, broader experience paths, altered career development, mixed supervisory relationships, and probably some injured egos. The very concept of a career is itself going to be redefined. Where one fits into the big scheme of things will tax the wisdom and the patience of both supervisors and those that are supervised. Indeed, that distinction itself will change, in some way or another.

If there is confusion about this, we should not be surprised at some confusion in some of our companies about which organizational segments are expected to do which activity. Which unit is to explore, and which is to develop? Which sector is to push which technology, and which sector is to use it? Which is to do research, which to develop technology, which to implement it? And then we have the problem of who pays the bills. Is it the same group of people that decides on the deployment of technology and can judge its merits, or someone uninformed about technology and its benefits? Things are just not as straightforward as they used to be.

We are observing some restructuring of the internal organizations of many petroleum companies, each searching for the right combination of business effectiveness and technical mission. All business units in all companies are being pressed to reduce costs, to become more efficient, to generate profits. In this atmosphere it is possible that the favorable economics for modern geophysics I referred to earlier could be forgotten. Geophysical projects could be abandoned or postponed, and treated more as a cost burden than as an investment opportunity. We cannot rule out the possibility that technologies, especially expensive technologies like geophysics, might be used less because they cost too much, and might be controlled by people with little understanding of their merit.

Geophysics must become more economic if we expect it to continue to be widely used. That means cheaper costs, but it especially means shorter time frames for 3-D surveys—from planning, to shooting, to processing, to interpretation. Lengthy time spans awaiting 3-D results often can preclude shooting the survey in the first place, and that's bad. Time is often more critical than money. Managers of oil fields, especially those in the process of early development, are not noted for patience. I am not sure how you can reduce the time for 3-D seismic surveys but I hope that you find out soon.

Modern geophysics, we said before, is very well applied to our existing fields, but our fields are managed by producing departments. The search for new pools pursued at high risk puts producing directly into the exploration business and directly into the geophysics business as well.

Producing departments are managed by engineers using engineering thinking. Are producing departments able to understand modern geophysics, with its peculiar jargon, its analytical techniques, and its bewildering array of software acronyms? Have we educated our engineers about the power and perils of modern geophysics? Can they manage this technology? *Should* they manage this technology? Just how should we organize geophysics in this changing business environment? We had better pay attention to this problem as well, because how we organize any technology is basic to how it will eventually be utilized.

It is obvious that geophysicists are, and will continue to be, drawn into producing department activities. They will have to learn about producing's objectives, its needs, its style, and its culture. We in development geology have had a long association with production departments and are comfortable in that environment. We have influence upon production projects because we advance our geologic ideas in the context of production objectives. Geophysicists will need to make a similar adjustment. Geophysicists will need to consider themselves part of production's effort, as part of their team, rather than just as providers of a simple geophysical service. To be effective, one must belong.

Downhole geophysics will be another area of possible new activity. Cross-well tomography for monitoring EOR projects, or for finer-tuned reservoir descriptions, will be of great use once it is perfected. Even the drilling engineers now are embracing geophysical data and techniques. One thing is for sure: geophysicists and engineers are going to get to know each other a lot better in the future, and that will be a good thing.

Frontier exploration in the U.S., as it once used to be, is declining, but exploitation around oil fields will increase. The technical ability to do this kind of development and exploitation work, using large data sources and 3-D geophysics, associated with complex geology and multiple maps, working with and for engineers, will become a valuable skill.

Another trend will be the closer association of geophysics with a knowledge of depositional systems. The geophysical response to ancient stratigraphic time-rock units will become a big field of study. Geophysicists will have to become good stratigraphers and good sedimentologists, and vice versa. Stratigraphy, rather than structure, will become the dominant new area of geophysical investigation.

All of this means that geophysicists, and geologists who understand geophysics, and those who can relate geophysics back to geology, will have a broad spectrum of activity to consume their talent, regardless of whether we call it exploration, exploitation, or development. And, that talent will still be needed far into the future. Yet, such talent could also be in short supply for a variety of reasons we do not need to elaborate upon here. We will need to handle the problem of manpower availability very carefully because it is possible that we could develop a great technology without enough of the right people to use it.

The computer age has arrived in our business. The workstation, which is just a few years old, is emerging as the basic analytical tool for all earth scientists and all engineers. There will be a growing necessity to develop comprehensive data bases, computer systems, and interactive workstations that will integrate a wide variety of well, seismic, geologic, engineering, and production history data from many sources and disciplines. Such systems will be expensive to develop and difficult to maintain, but they will be essential for the future. When they arrive and our earth scientists and engineers become acquainted with their full power and effectiveness, we will see a radical change in the way by which we all do our work. I believe such systems are much closer to reality than we may realize. We are soon going to do our work differently. It will be done better, faster, and with more new ideas than before. The full computerization of the upstream will be the next major breakthrough of our business.

The impacts I have described would cause severe adjustments in our industry under normal circumstances, but now we have the traumatic events caused by the current crisis in the Middle East. It is too soon to tell if these shocks to the world's political and economic psyche are of short or long duration. Surely, our industry will not be the same again, but where it will go, or be driven, is still uncertain. Yet one thing does seem clear—that the fields we already have are suddenly more important than they were before, and that we will need to nurse them along further than we might have done otherwise.

The close and intense technical scrutiny to be given to our existing fields will require a strong sense of *teamwork* among people, with a de-emphasis on rigid boundaries either of organization or technical discipline. The necessary components of tomorrow's workplace will be Flexibility, Communication, Teamwork, Skill, and Synergism.

In conclusion, let me say that geophysics is sure to play a key role in our business in the future, just as it has in the past. Those of you who are engaged in development geophysics will lead the way in developing the skills necessary to harness this new geophysical era, because you represent "modern geophysics." Your efforts will have an impact far beyond your present perceptions of your job. You will show us how things can be done with modern geophysics. Any expansion of this into new exploration efforts, or into new business arrangements, will flow from an appreciation of that simple observation—modern geophysicists already see and know how to handle the future.

I would remind you, however, that it is not enough to know how to do something. You must inform others of what can be done and should be done with geophysics. You must educate your geologic and engineering colleagues in this new geophysical era. You must educate your managers as well, and their managers, too. You must take the path that says geophysics is an investment and not just a cost. You must, in short, become champions of the geophysical cause—not to perpetuate your careers (although that will happen), but because it is a responsible contribution to the health of our industry.

INTRODUCTION

History and Basic Ideas

The earth has always been three-dimensional and the petroleum reserves we seek to find or evaluate are contained in three-dimensional traps. The seismic method, however, in its attempt to image the subsurface has traditionally taken a two-dimensional approach. It was 1970 when Walton (1972) presented the concept of three-dimensional seismic surveys. In 1975, 3-D surveys were first performed on a normal contractual basis, and the following year Bone, Giles and Tegland (1976) presented the new technology to the world.

The essence of the 3-D method is areal data collection followed by the processing and interpretation of a closely-spaced data volume. Because a more detailed understanding of the subsurface emerges, 3-D surveys have been able to contribute significantly to the problems of field appraisal, development and production as well as to exploration. It is in these post-discovery phases that many of the successes of 3-D seismic surveys have been achieved and also where their greatest economic benefits have been enjoyed. The scope of 3-D seismic for field development was first reported by Tegland (1977).

In the late 1980s and early 1990s, the use of 3-D seismic surveys for exploration has increased significantly. This started in the mid-1980s with widely-spaced 3-D surveys called, for example, Exploration 3-D. This technology is briefly discussed in Chapter 3; its success was modest. Today, speculative 3-D surveys, properly sampled and covering huge areas, are available for purchase piecemeal in mature areas like the Gulf of Mexico. This, however, is not the only use for exploration. Some companies are acquiring 3-D surveys over prospects routinely, so that the vast majority of their seismic budgets are for 3-D operations.

In 20 years of 3-D survey experience (1975-95) many successes and benefits have been recorded. Five particular accolades are reproduced here; others are found in the case histories of Chapter 9 and implied at many other places throughout this book. There is a major symbiosis between modern 3-D seismic data and the interactive workstation. A significant accolade to the combined technology appears in the Foreword to this book.

> *"...there seems to be unanimous agreement that 3-D surveys result in clearer and more accurate pictures of geological detail and that their costs are more than repaid by the elimination of unnecessary development holes and by the increase in recoverable reserves through the discovery of isolated reservoir pools which otherwise might be missed."*
> (Sheriff and Geldart, 1983)

> *"The leverage seems excellent for 3-D seismic to pay for itself many times over in terms of*

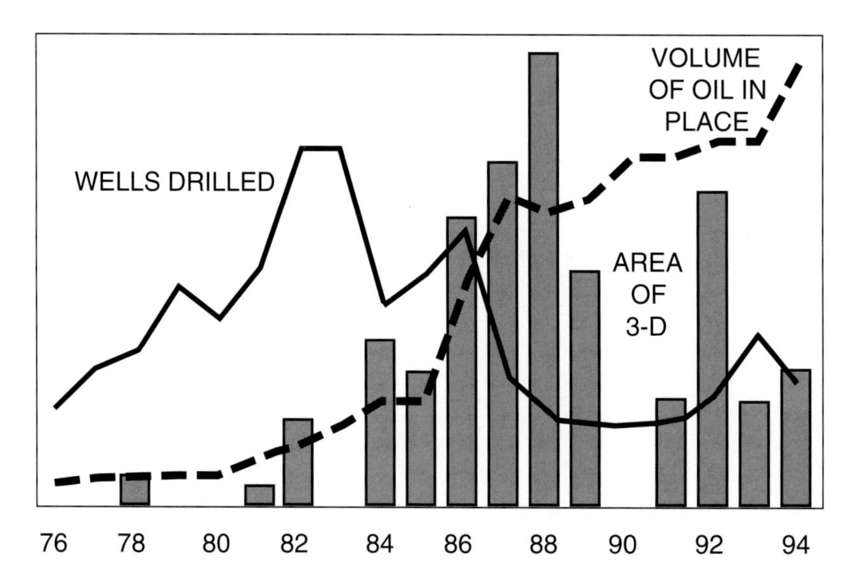

Fig. 1-1. Area covered by 3-D surveys, exploratory wells drilled and volume of oil in place for the period 1976 to 1994 in the Campos Basin offshore Brazil (from Martins et al, 1995). (Courtesy Petrobras.)

reducing the eventual number of development wells."
(West, 1979)

"...the 3-D data are of significantly higher quality than the 2-D data. Furthermore, the extremely dense grid of lines makes it possible to develop a more accurate and complete structural and stratigraphic interpretation...Based on this 3-D interpretation, four successful oil wells have been drilled. These are located in parts of the field that could not previously be mapped accurately on the basis of the 2-D seismic data because of their poor quality. This eastward extension has increased the estimate of reserves such that it was possible to declare the field commercial in late 1980."
(Saeland and Simpson, 1982)

"...3-D seismic surveying helped define wildcat locations, helped prove additional outpost locations, and assisted in defining untested fault blocks. Three-D seismic data helped find additional reserves and, most certainly, provided data for more effective reservoir drainage while being cost-effective...Gulf participated in 16 surveys that covered 26 blocks and has invested $15,000,000 in these data. The results show that a 3-D seismic program can be cost-effective since it can improve the success ratio of development drilling and can encourage acceleration of a development program, thereby improving the cash flow."
(Horvath, 1985)

"We acquired two offshore blocks which contained a total of seven competitor dry holes. Our exploration department drilled one more dry hole before making a discovery. At that point we conducted a 3-D survey while the platform was being prepared. When drilling commenced, guided by the 3-D data, we had 27 successful wells out of the next 28 drilled. In this erratic depositional environment, we believe that such an accomplishment would not have been possible without the 3-D seismic data."
(R. M. Wright, Chevron U.S.A. Inc., personal communication, May, 1988)

Sheriff (1992) addresses many benefits of 3-D seismic in *Reservoir Geophysics;* a few quotations from that volume follow:

3-D seismic is an extremely powerful delineation tool, and spectacularly cost-effective, particularly when well costs are high.

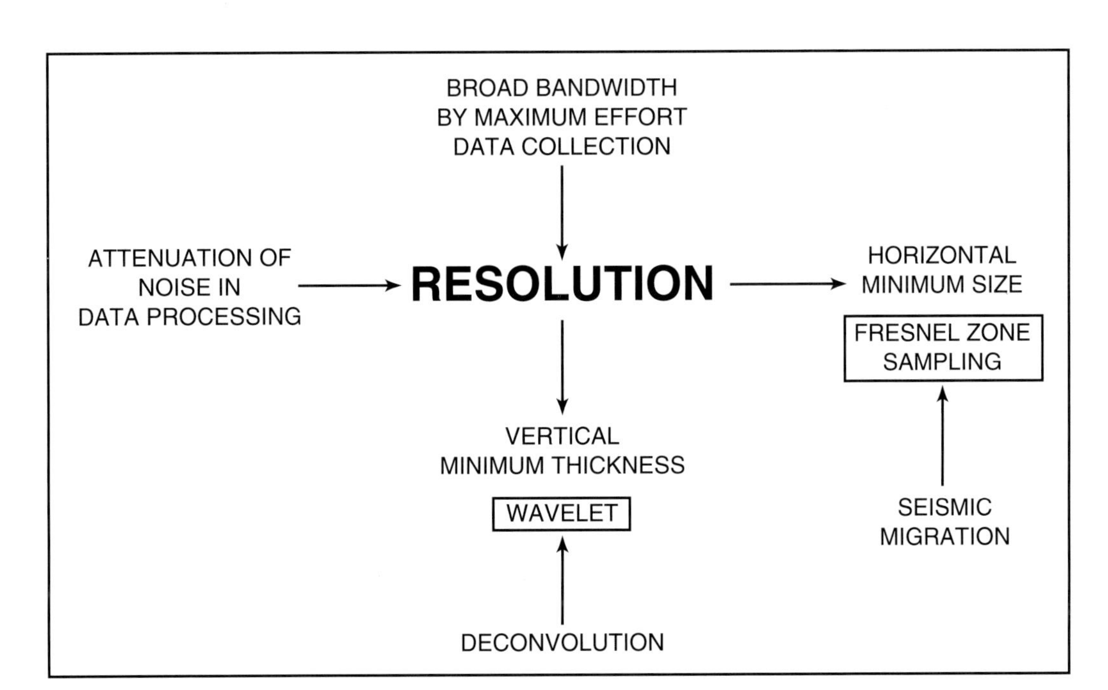

Fig. 1-2. Factors affecting horizontal and vertical seismic resolution.

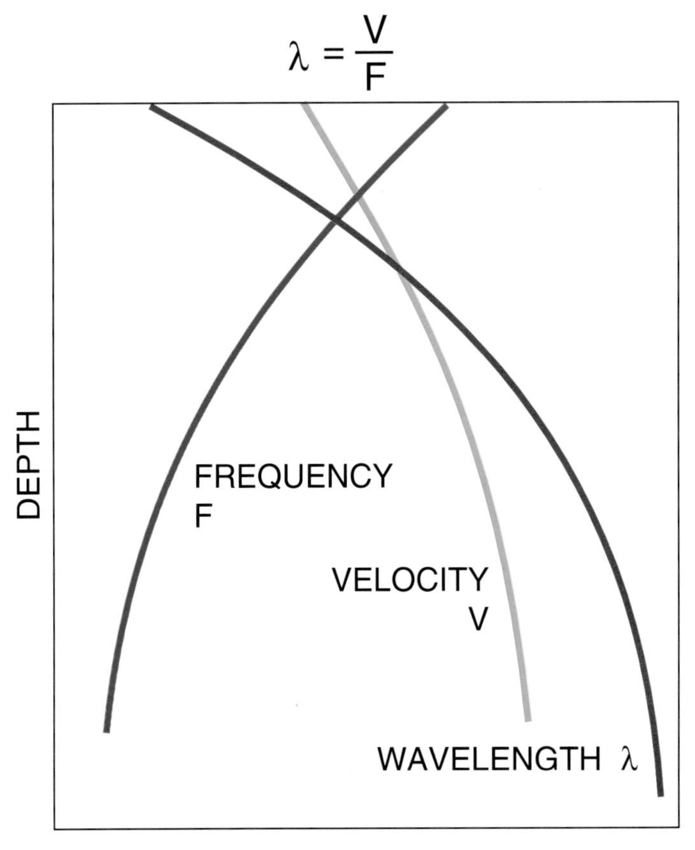

Fig. 1-3. Wavelength, the seismic measuring rod, increases significantly with depth making resolution poorer.

The success is directly attributable to the better structural interpretation made possible by the 3-D survey.

The greatest impact of 3-D surveys has been the ability to match platform size, number of well slots, and production facilities to the more accurately determined field reserves.

Martins et al (1995), working in the Campos Basin offshore Brazil, have tracked the amount of 3-D survey coverage in relation to the wells drilled and the oil reserves

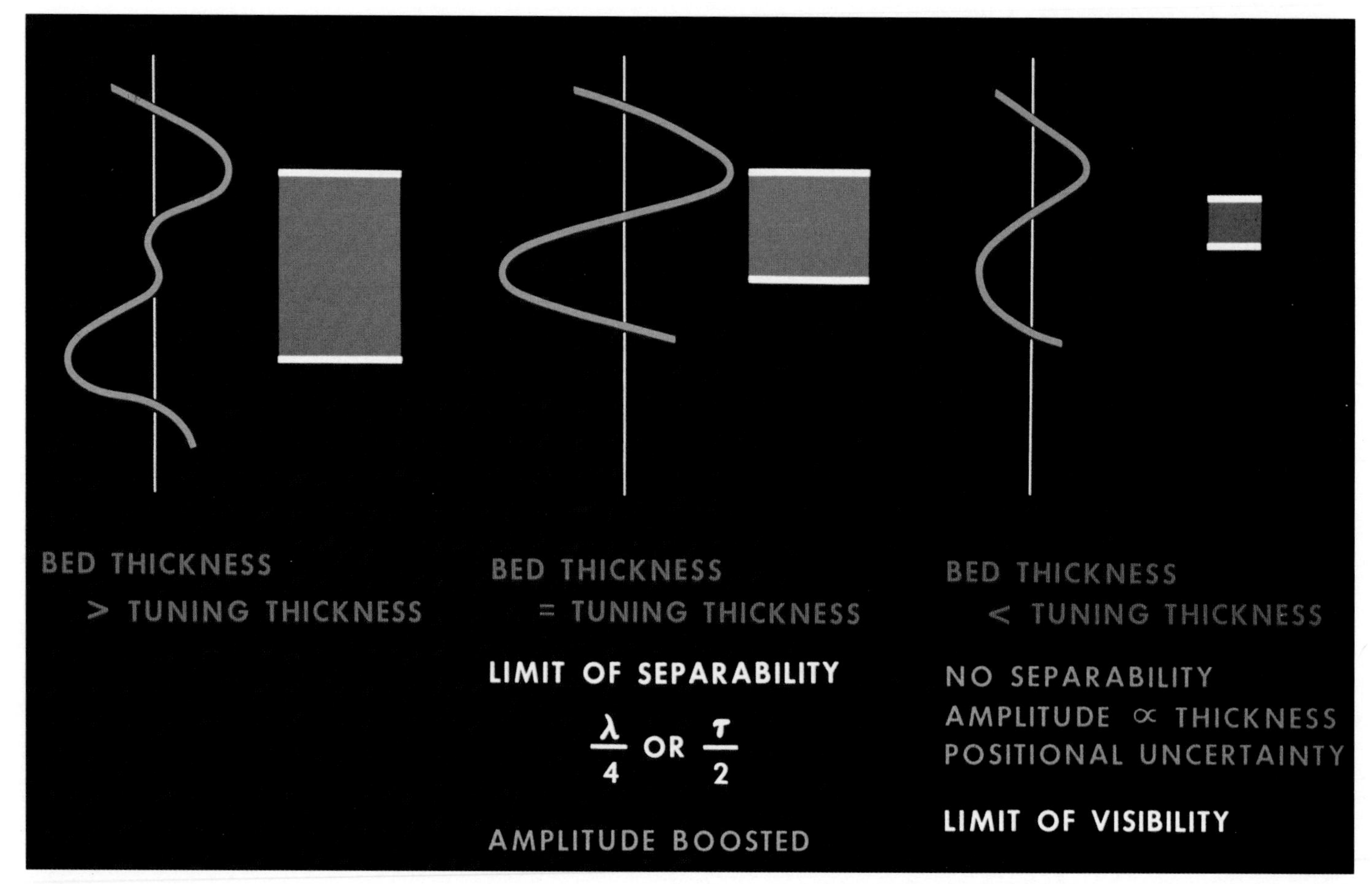

Fig. 1-4. Resolution of the reflections from the top and bottom of a bed is dependent on the interaction of closely spaced wavelets.

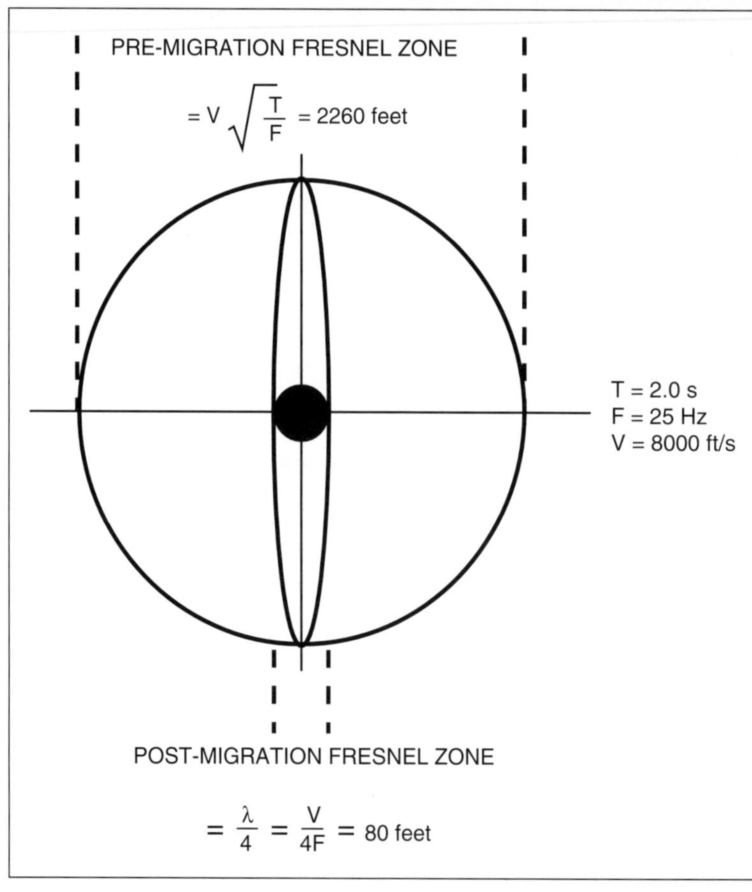

Fig. 1-5. Effect on Fresnel zone size and shape of 2-D and 3-D migration.

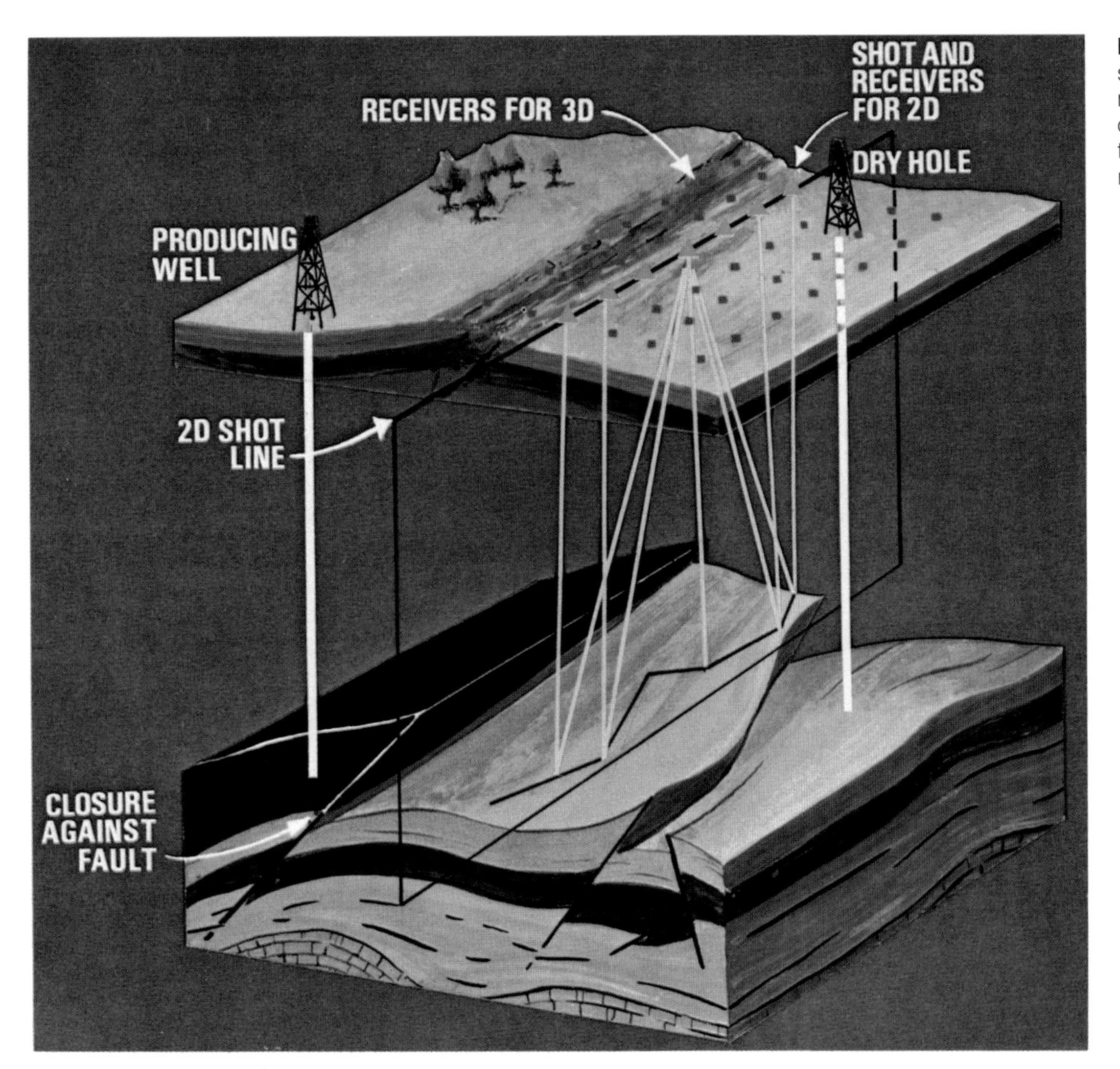

Fig. 1-6. Subsurface structure causes reflection points to lie outside the vertical plane through shots and receivers.

booked (Figure 1-1). This demonstrates very nicely that 3-D seismic is indeed replacing exploration wells!

Resolution

The fundamental objective of the 3-D seismic method is increased resolution. Resolution has both vertical and horizontal aspects and Sheriff (1985) discusses the subject qualitatively. The resolving power of seismic data is always measured in terms of the seismic wavelength, which is given by the quotient of velocity and frequency (Figure 1-3). Seismic velocity increases with depth because the rocks are older and more compacted. The predominant frequency decreases with depth because the higher frequencies in the seismic signal are more quickly attenuated. The result is that the wavelength increases significantly with depth, making resolution poorer.

Figure 1-2 summarizes resolution issues. Vertical resolution has two limits, both resulting from the interaction of the wavelets from adjacent reflecting interfaces. The **limit of separability** is equal to one-quarter of a wavelength (or half a period) and is simply the closest separation of two wavelets of a given bandwidth (Figure 1-4). For thinner intervals than this, the amplitude is progressively attenuated until the **limit of visibility** is reached, when the reflection signal becomes obscured by the background noise. These issues are discussed at greater length under the subject of Tuning in Chapter 6. Wavelet processing and the understanding of wavelet phase are highly significant for improving vertical resolution and will be considered from an interpretive standpoint in Chapter 2.

Fig. 1-7. Model of two anticlines and one fault with seismic data along Line 6 showing comparative effects of 2-D and 3-D migration (from French, 1974).

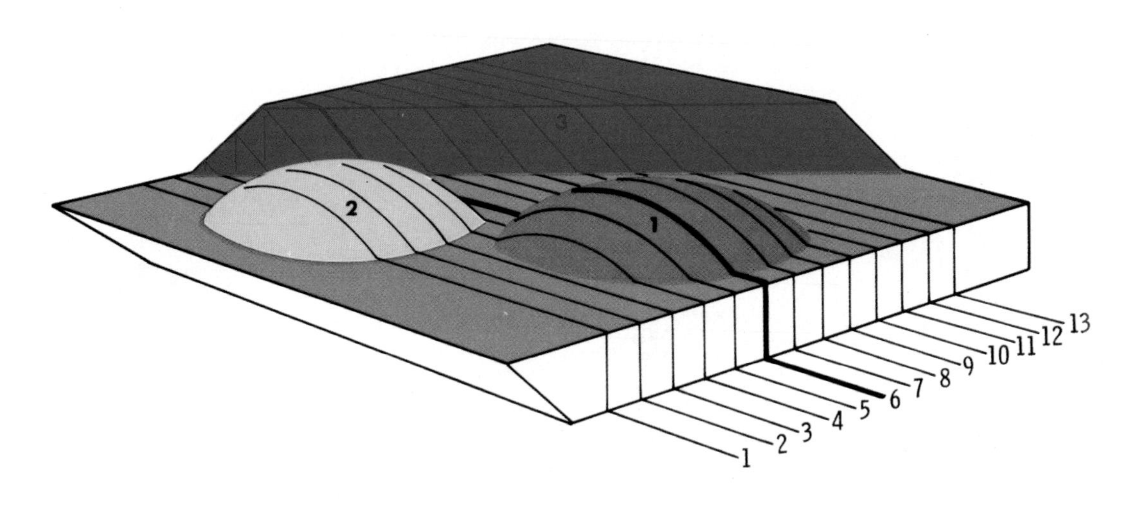

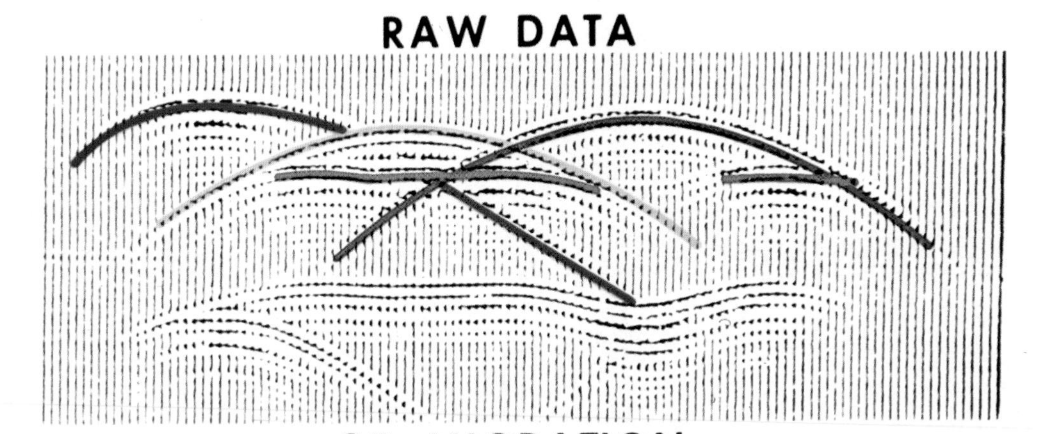

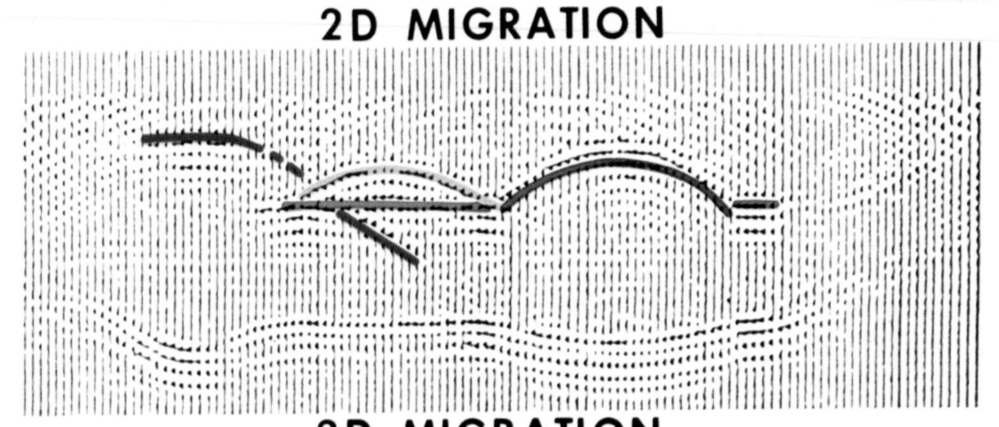

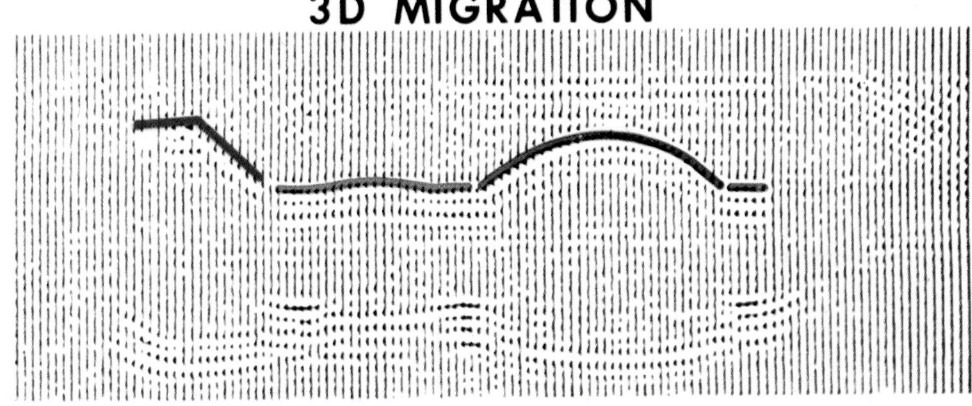

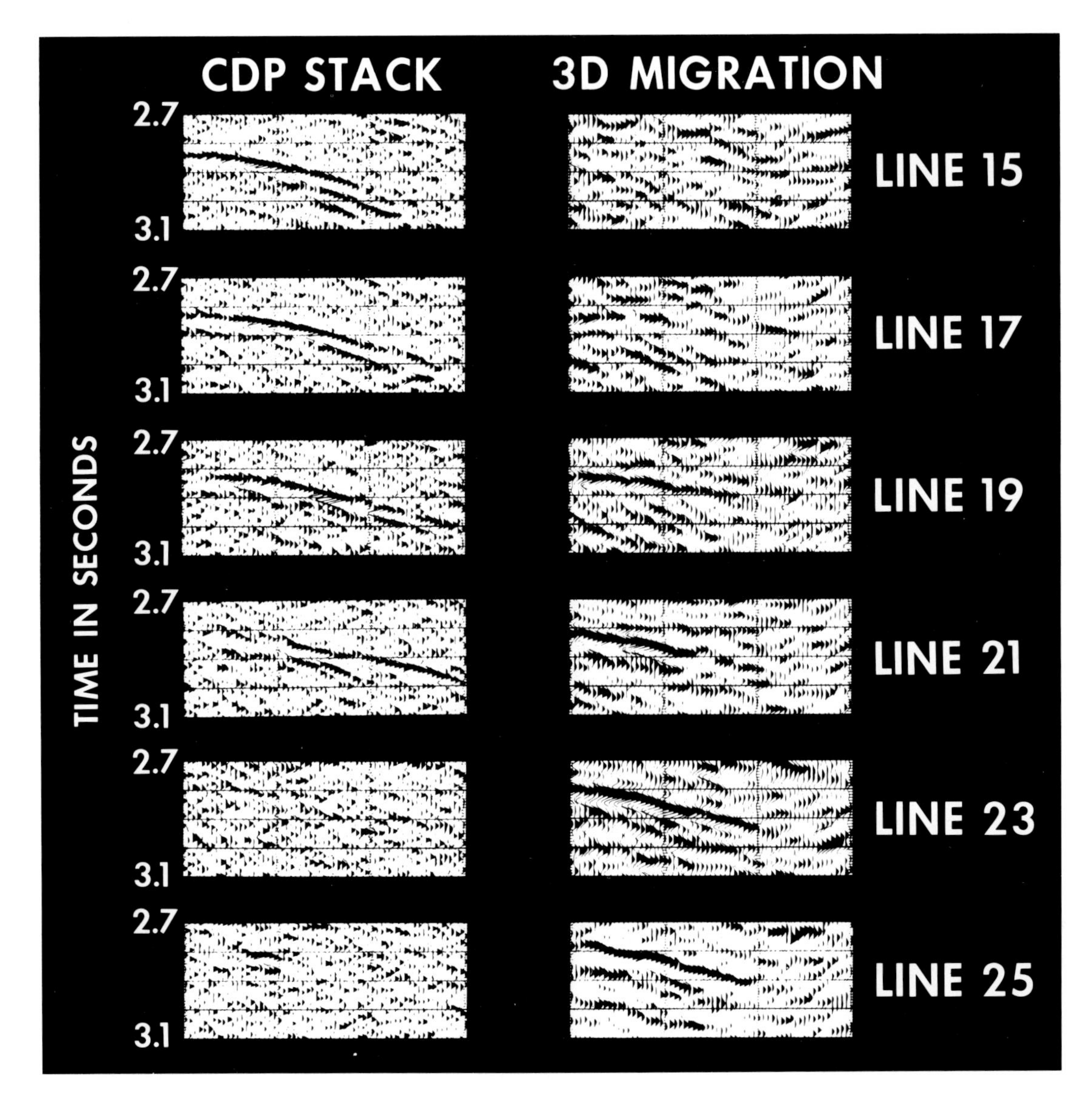

Fig. 1-8. Three-dimensional movement of a dipping reflection by 3-D migration. (Courtesy Geophysical Service Inc.)

Migration is the principal technique for improving horizontal resolution, and in doing so performs three distinct functions. The migration process (1) repositions reflections out-of-place because of dip, (2) focuses energy spread over a Fresnel zone, and (3) collapses diffraction patterns from points and edges. Seismic wavefronts travel in three dimensions and thus it is obvious that all the above are, in general, three-dimensional issues. If we treat them in two dimensions, we can only expect part of the potential improvement. In practice, 2-D lines are often located with strike and dip of major features in mind so that the effect of the third dimension can be minimized, but rarely eliminated. Figure 1-5 shows the focussing effect of migration in two and three dimensions. The Fresnel zone will be reduced to an ellipse perpendicular to the line for 2-D migration (Lindsey, 1989) and to a small circle by 3-D migration. The diameter of one-quarter of a wavelength indicated in Figure 1-5 is for perfect migration. In practice, the residual Fresnel zone may be about twice this size.

The accuracy of 3-D migration depends on the velocity field, signal-to-noise ratio, migration aperture and the approach used. Assuming the errors resulting from these factors are small, the data will be much more interpretable both structurally and stratigraphically. Intersecting events will be separated, the confusion of diffraction patterns will be gone, and dipping events will be moved to their correct subsurface positions. The collapsing of energy from diffractions and the focusing of energy spread over Fresnel zones will make amplitudes more accurate and more directly

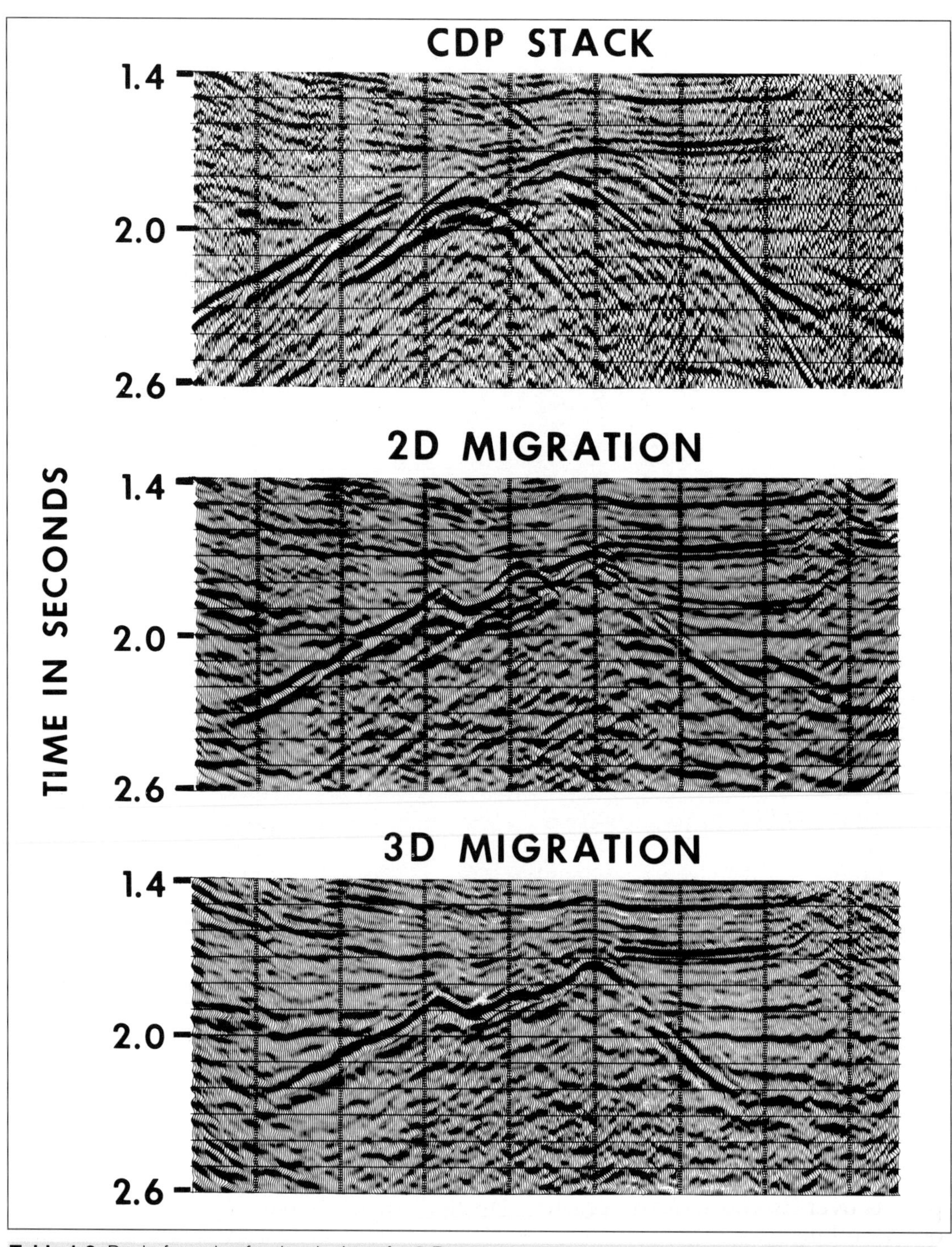

Fig. 1-9. Improved structural continuity of an unconformity reflection resulting from 2-D and 3-D migration.

Table 1-1. Alias frequency (in hertz) as a function of subsurface spacing (in meters) and dip (in degrees) for an RMS velocity of 2500 m/s.

	SUBSURFACE SPACING				
DIP	12.5	25	50	75	100
5	574	287	143	96	72
10	288	144	72	48	36
15	193	96	48	32	24
20	146	73	37	24	18
25	118	59	30	20	15

Table 1-2. Basic formulas for the design of a 3-D survey.

Maximum subsurface spacing (2 samples per wavelength) $= \dfrac{1}{2F_{max}\,DIP_{max}}$

Desirable subsurface spacing (3 samples per wavelength) $= \dfrac{1}{3F_{max}\,DIP_{max}}$

Migration distance (or half-aperture) $= \dfrac{TV^2DIP}{4}$

Fresnel zone radius $= \dfrac{V}{2}\sqrt{\dfrac{T}{F_{min}}}$

where T is seismic two-way travel time in seconds
DIP is measured in seconds per unit distance
F is seismic frequency
V is seismic velocity

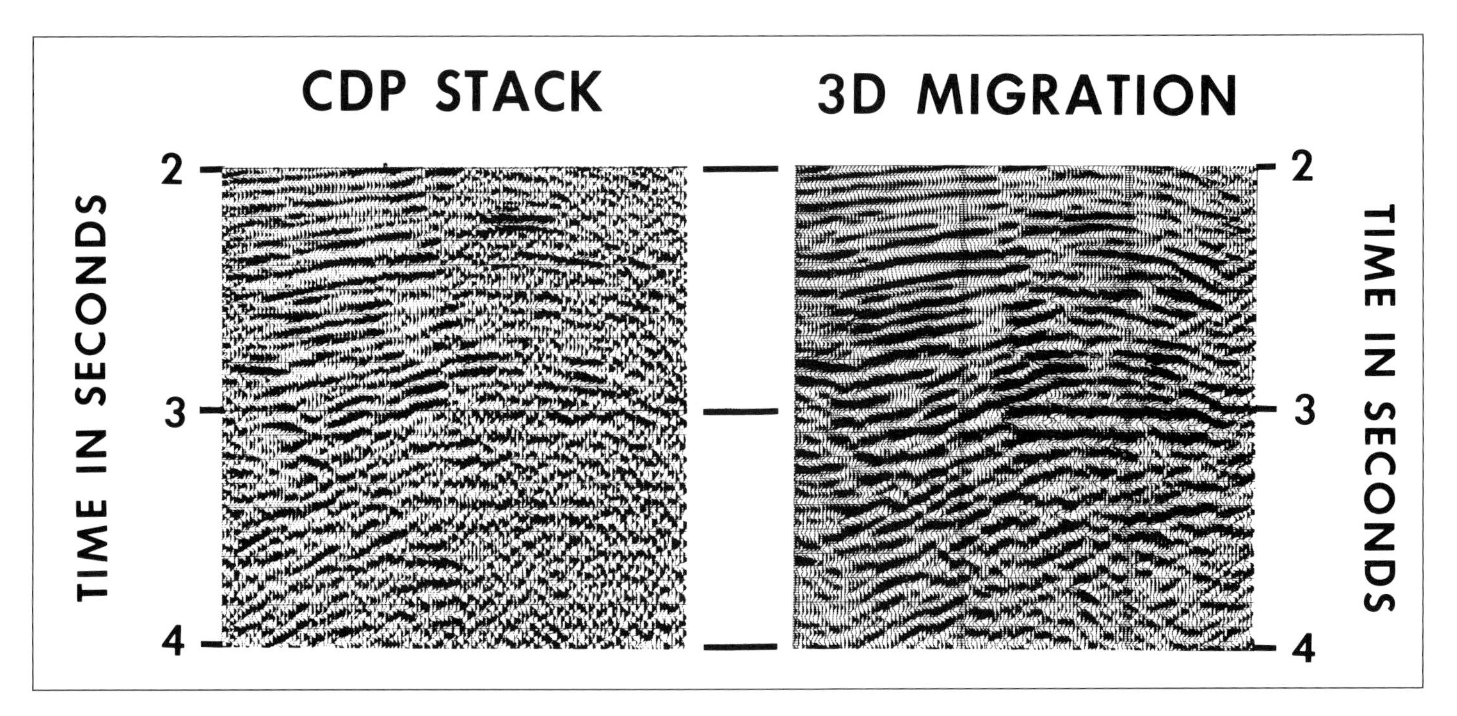

Fig. 1-10. Improved visibility of a flat spot reflection after removal of interfering events by 3-D migration.

interpretable in terms of reservoir properties. The determination of true velocity for accurate migration and depth conversion is a significant issue. It is desirable to collect data with a reasonable distribution of offsets and azimuths, so that the three-dimensional dip effects in the velocity field can be removed properly.

Examples of 3-D Data Improvement

The interpreter of a 2-D vertical section normally assumes that the data were recorded in one vertical plane below the line traversed by the shots and receivers. The extent to which this is not so depends on the complexity of the structure perpendicular to the line. Figure 1-6 demonstrates that, in the presence of moderate structural complexity, the points at depth from which normal reflections are obtained may lie along an irregular zig-zag track. Only by migrating along *and* perpendicular to the line direction is it possible to resolve where these reflection points belong in the subsurface.

French (1974) demonstrated the value of 3-D migration very clearly in model experiments. He collected seismic data over a model containing two anticlines and a fault scarp (Figure 1-7). Thirteen lines of data were collected although only the results for Line 6 are shown. The raw data have diffraction patterns for both anticlines and the fault so the section appears very confused. The situation is greatly improved with 2-D migration and anticline number 1 (shown in green) is correctly imaged, as Line 6 passed over its crest. However, anticline number 2 (shown in yellow) should not occur on Line 6 and the fault scarp has the wrong slope. The 3-D migration has correctly imaged the fault scarp and moved the yellow anticline away from Line 6 to where it belongs.

Figure 1-8 demonstrates this three-dimensional event movement on real data. The same panel is presented before and after 3-D migration for six lines. Here we can observe the movement of a discrete patch of reflectivity to the left and in the direction of higher line numbers.

Figure 1-9 shows improved continuity of an unconformity reflection. The 2-D migration has collapsed most of the diffraction patterns but some confusion remains. The crossline component of the 3-D migration removes energy not in the plane of this section and clarifies the shape of the unconformity surface in detail.

Figure 1-10 shows the effect of 3-D migration in enhancing the visibility of a fluid contact reflection by removing energy not belonging in the plane of this section.

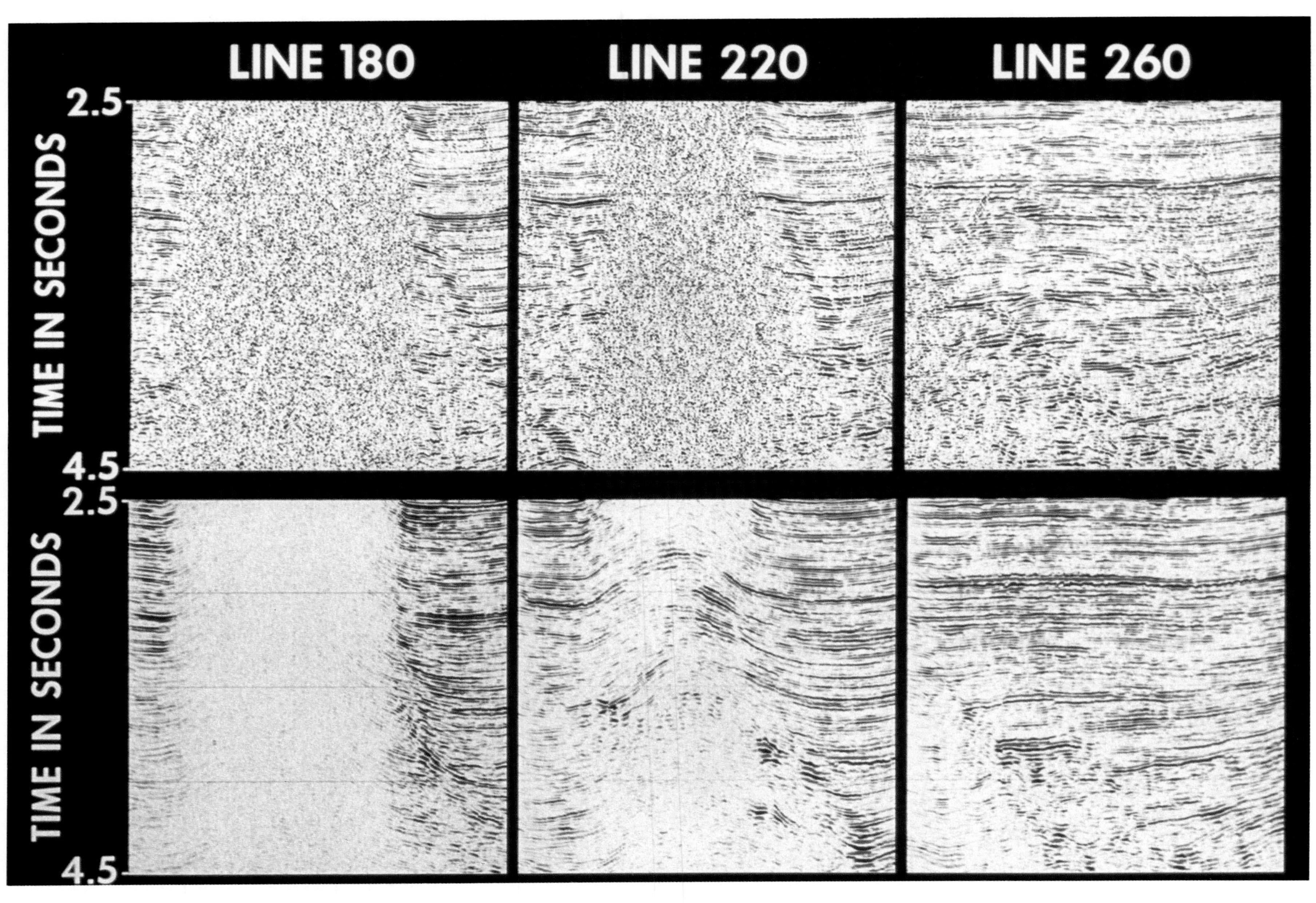

Fig. 1-11. Three vertical sections through or adjacent to a Gulf of Mexico salt dome before migration (top) and after migration (bottom), showing the repositioning of several reflections near the salt face. (Courtesy Hunt Oil Company.)

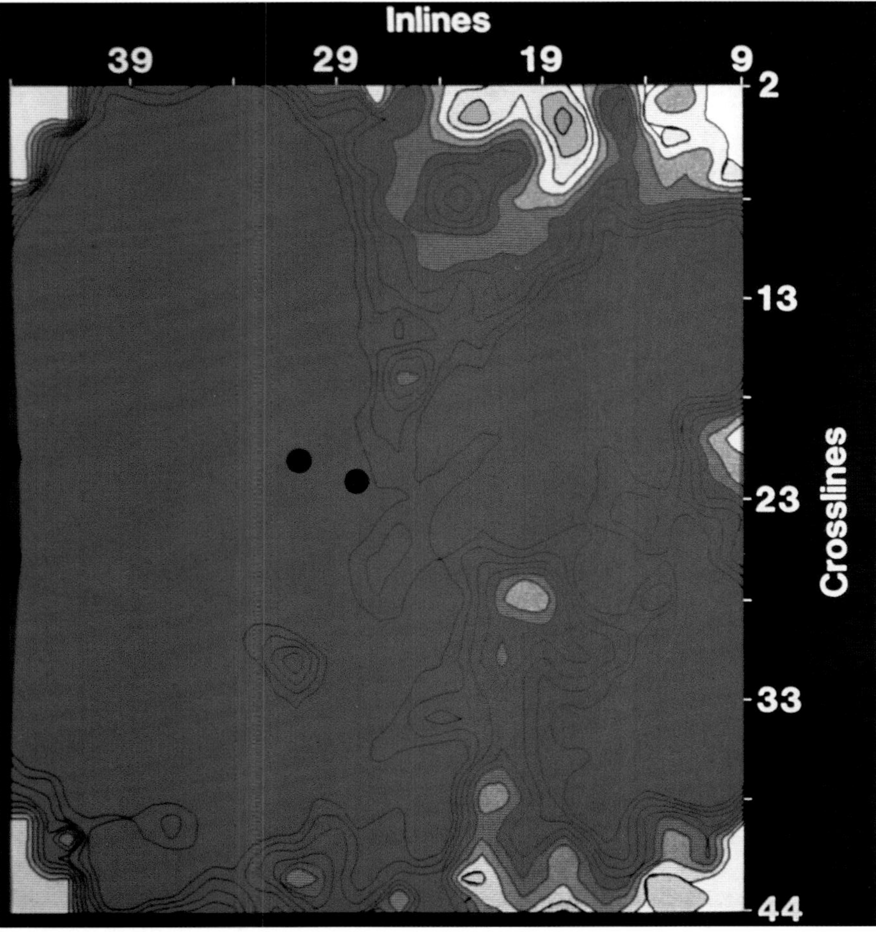

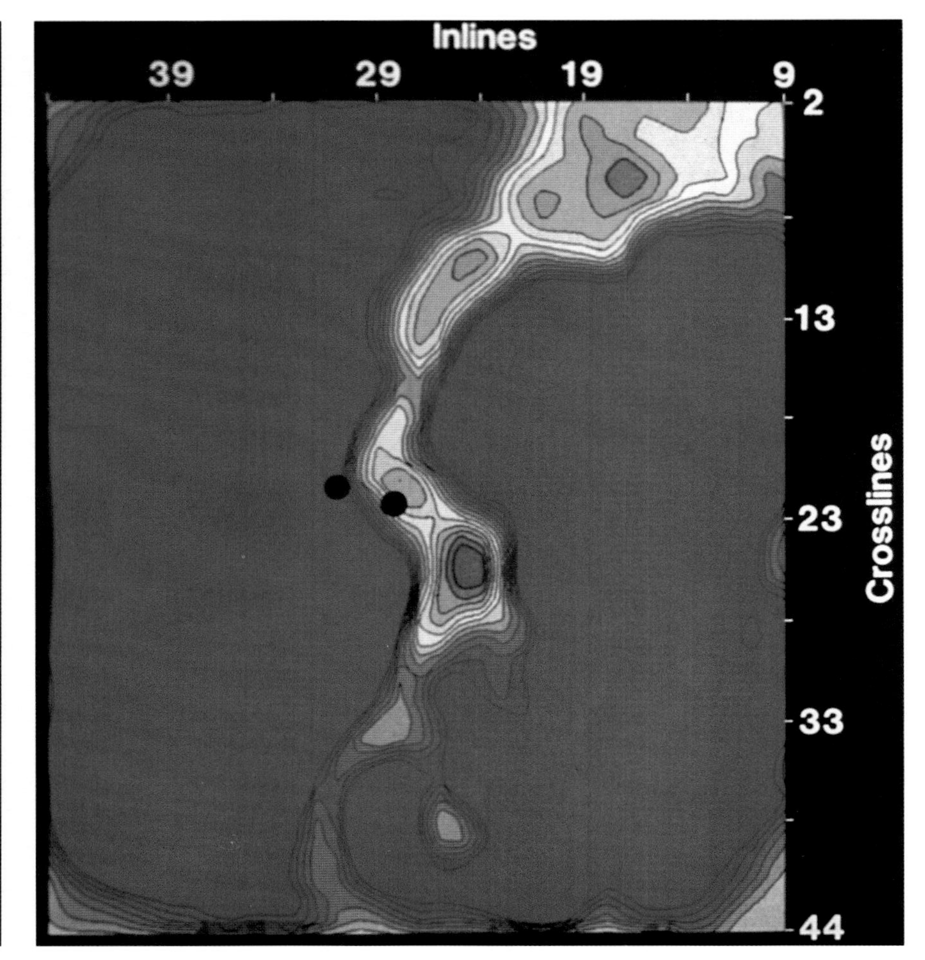

Fig. 1-12. Horizontal sections before migration (left) and after migration (right) showing the necessity of 3-D migration for the observation of shallow channels. (Courtesy Amoco Canada Petroleum Company Limited and N. E. Pullin.)

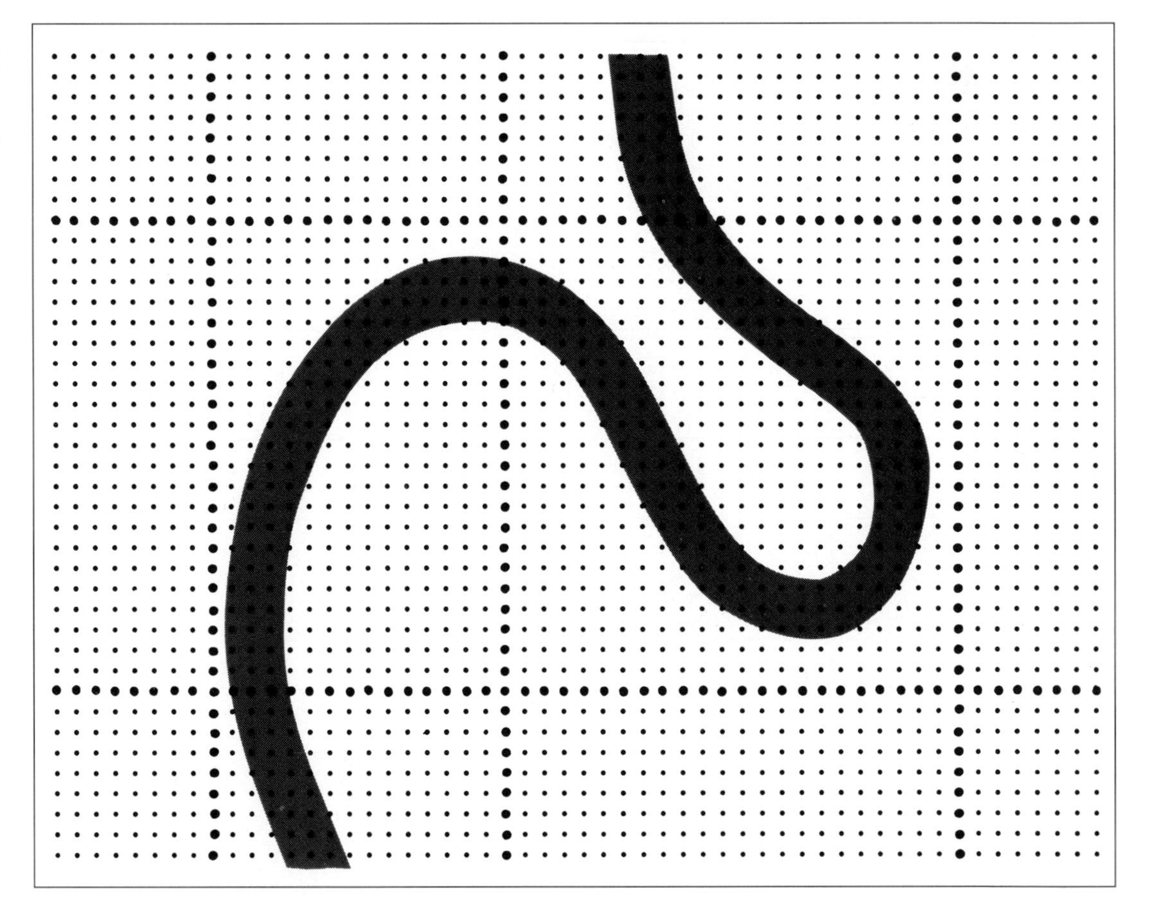

Fig. 1-13. Areal coverage of a 3-D survey compared to the coverage of a grid of five 2-D lines, and the ability of each to delineate a meandering channel.

Figure 1-11 shows portions of three lines passing through and close to a salt diapir. Line 180 shows steeply-dipping reflections at the edge of the salt mass brought into place by the 3-D migration. Line 220 shows an apparent anticline which is caused by reflections dipping up steeply toward the salt face in a plane perpendicular to that in the migrated (lower) portion of Figure 1-11. In this prospect, 3-D migration imaged reflections underneath a salt overhang and provided valuable detail about traps located there against the salt face (Blake, Jennings, Curtis, Phillipson, 1982).

When comparing sections before and after 3-D migration to appraise its effectiveness, it is important to bear in mind the way in which reflections have moved around. In the presence of dip perpendicular to the section under scrutiny the visible data before and after 3-D migration are different. It is unreasonable to compare detailed character and deduce what 3-D migration did. It is possible to compare a section before 3-D migration with the one from the same location after 3-D migration and find that a good quality reflection has disappeared. The migrated section is not consequently worse; the good reflection has simply moved to its correct location in the subsurface.

Figure 1-12 shows a horizontal section at a time of 224 ms from a very high resolution 3-D survey in Canada aimed at monitoring a steam injection process. The section on the left is from the 3-D volume before migration and the section on the right is from the volume after migration. The two black dots indicate wells. The striking visibility of a channel after migration results from the focusing of energy previously spread over the Fresnel zone. The fact that one well penetrates the channel and the other does not is significant: they are only 10 m apart.

Sampling Requirements

The sampling theorem requires that, for preservation of information, a waveform must be sampled such that there are at least two samples per cycle for the highest fre-

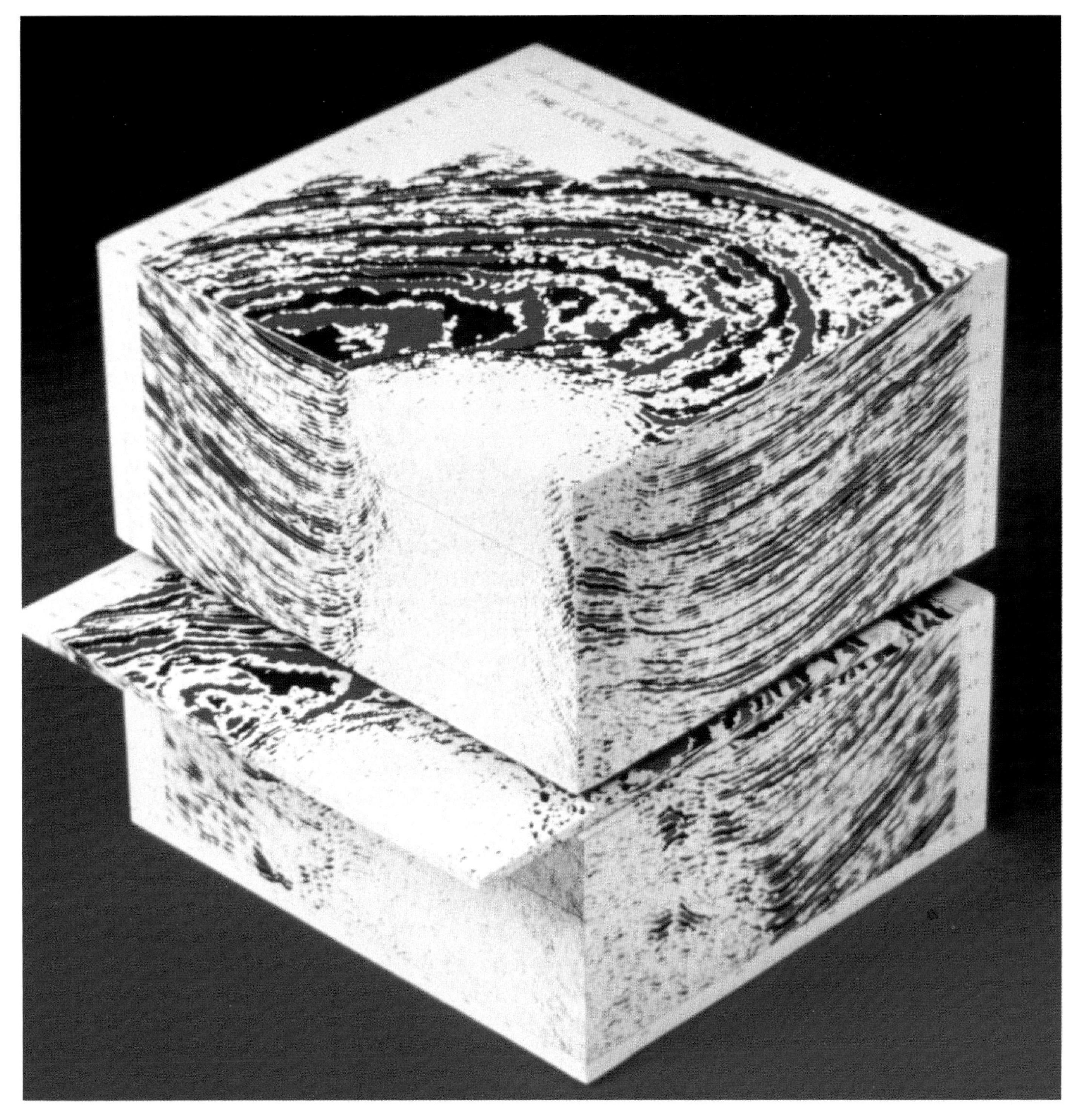

Fig. 1-14. 3-D data volume showing a Gulf of Mexico salt dome and associated rim syncline. (Courtesy Hunt Oil Company).

quency. Since the beginning of the digital era, we have been used to sampling a seismic trace in time. For example, 4 ms sampling is theoretically adequate for frequencies up to 125 Hz. In practice we normally require at least three samples per cycle for the highest frequency. With this safety margin, 4 ms sampling is adequate for frequencies up to 83 Hz.

In space, the sampling theorem translates to the requirement of at least two, and preferably three, samples per shortest wavelength in every direction. In a normal 2-D survey layout this will be satisfied by the depth point spacing along lines but not by the spacing between lines. Hence the restriction that widely-spaced 2-D lines can be processed individually on a 2-D basis but not together as a 3-D volume.

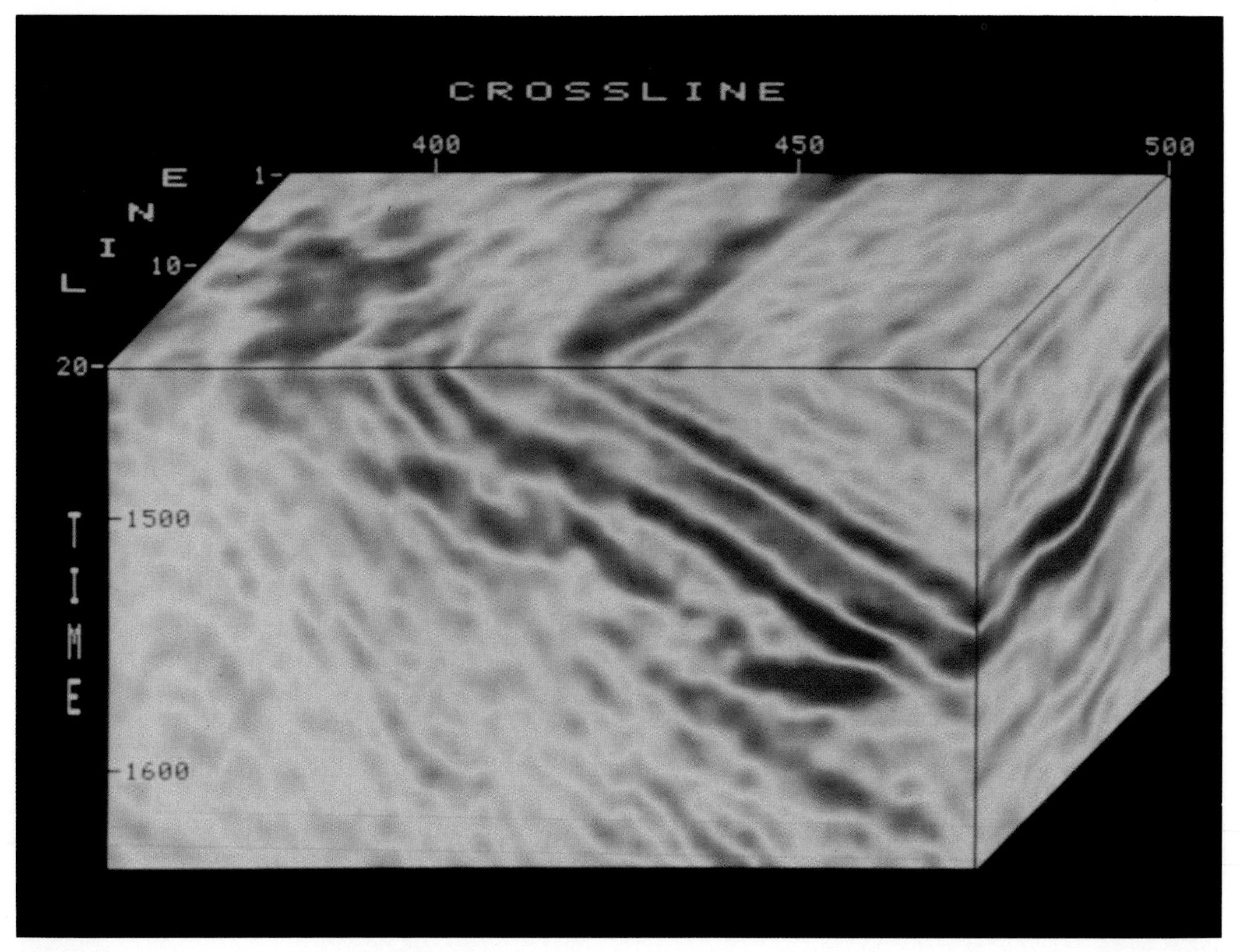

Fig. 1-15. 3-D data volume showing a bright spot from a Gulf of Mexico gas reservoir. (Courtesy Chevron U.S.A. Inc.)

If the sampling theorem is not satisfied the data are aliased. In the case of a dipping event, the spatial sampling of that event must be such that its principal alignment is obvious; if not, aliases occur and spurious dips result after multichannel processing. Table 1-1 shows the frequencies at which this aliasing occurs for various dips and subsurface spacings. Clearly, a 3-D survey must be designed such that aliasing during processing does not occur. Tables like the one presented can be used to establish the necessary spacing considering the dips and velocities present. In order to impose the safety margin of three samples, rather than two, per shortest wavelength, the frequency limit is normally considered to be around two-thirds of each number tabulated. The formulas in Table 1-2 provide a general method of establishing the spacings required. The first formula, based on two samples per shortest wavelength, gives the maximum spacing that can be used to image the structure. Given our ignorance of the subsurface structure at the time the 3-D survey is being designed, we should allow a significant safety margin by collecting at least three samples per shortest spatial wavelength. Table 1-2 also shows the two formulas needed to calculate the width of the extra strip around the periphery of the prospect over which data must be collected in order to ensure proper imaging in the area of interest. The calculation of migration distance should use the local value of dip measured perpendicular to the prospect boundary. The Fresnel zone radius needs to be considered for the proper focusing of amplitudes. The two strip widths thus calculated should be added together in defining the total survey area. In practice, the design solution also includes consideration of the distribution of subsurface interest and economics.

Fig. 1-16. Seismodel Display Unit, one of the approaches to true 3-D display. (Courtesy Geophysical Service Inc.)

Proper design of a 3-D survey is critical to its success, and sufficiently close spacing is always the first consideration. The formulas and table are addressing structural design issues. In areas of shallow dip where the survey objectives are stratigraphic, the selected spacing must be such that there are at least two samples within the lateral extent of any expected stratigraphic feature of interest, for example the width of a channel. Figure 1-13 demonstrates a typical comparison between the subsurface sampling of a 2-D and 3-D survey. The bold dots indicate the 2-D survey depth points which satisfy the sampling theorem along each line. The 3-D survey requires a similarly close spacing in both directions over the whole area. In addition to the opportunity for three-dimensional processing which the areal coverage provides, note the sampling and thus potential definition of a meandering stream channel. Sampling for

Fig. 1-17. Voxel-rendered view of data volume which, by making the voxels semi-transparent, permits the interpreter to look *into* the volume. (Courtesy CogniSeis Development.)

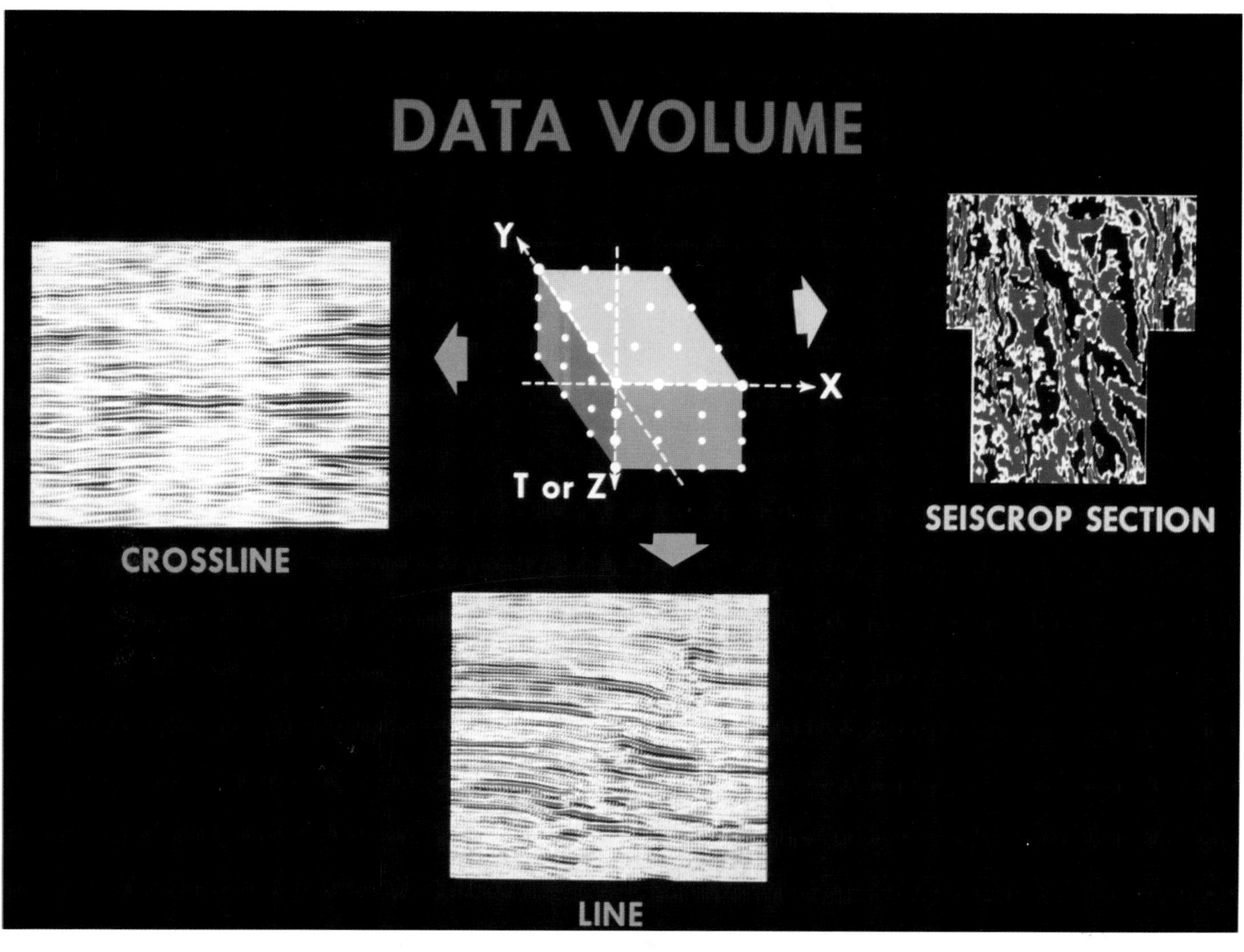

Fig. 1-18. Three sets of orthogonal slices through a data volume provide the basic equipment of the 3-D seismic interpreter.

stratigraphic features like this channel requires at least two but preferably three or four samples within the channel width. In practice, 3-D depth point spacing ranges between 6 and 50 m.

Volume Concept

Collection of closely-spaced seismic data over an area permits three-dimensional processing of the data as a volume. The volume concept is equally important to the seismic interpreter. With 3-D data, the interpreter is working directly with a volume rather than interpolating a volumetric interpretation from a widely-spaced grid of observations. The handling of this volume and what can be extracted from it are principal subjects of this book. One property of the volume pervades everything the 3-D interpreter does: The subsurface seismic wavefield is closely sampled in every direction, so that there is no grid loop around which the interpreter must tie, and no grid cell over which he must guess at the subsurface structure and stratigraphy. This is an opportunity which an interpreter must use to full advantage. Because the sampling requirements for interpretation are the same as for processing, all the processed data points contain unique information and thus should be used in the interpretation. (The one exception to this might be where dip varies over the prospect, and thus, so could design, but it is operationally expedient to maintain a constant spacing for the whole survey.) Thus, the interpreter of a 3-D volume should not decimate the data available to him but, given that he has time constraints imposed on him, he should use innovative approaches with horizontal sections, specially selected slices, and automatic spatial tracking, in order to comprehend all the information in the data. In this way the

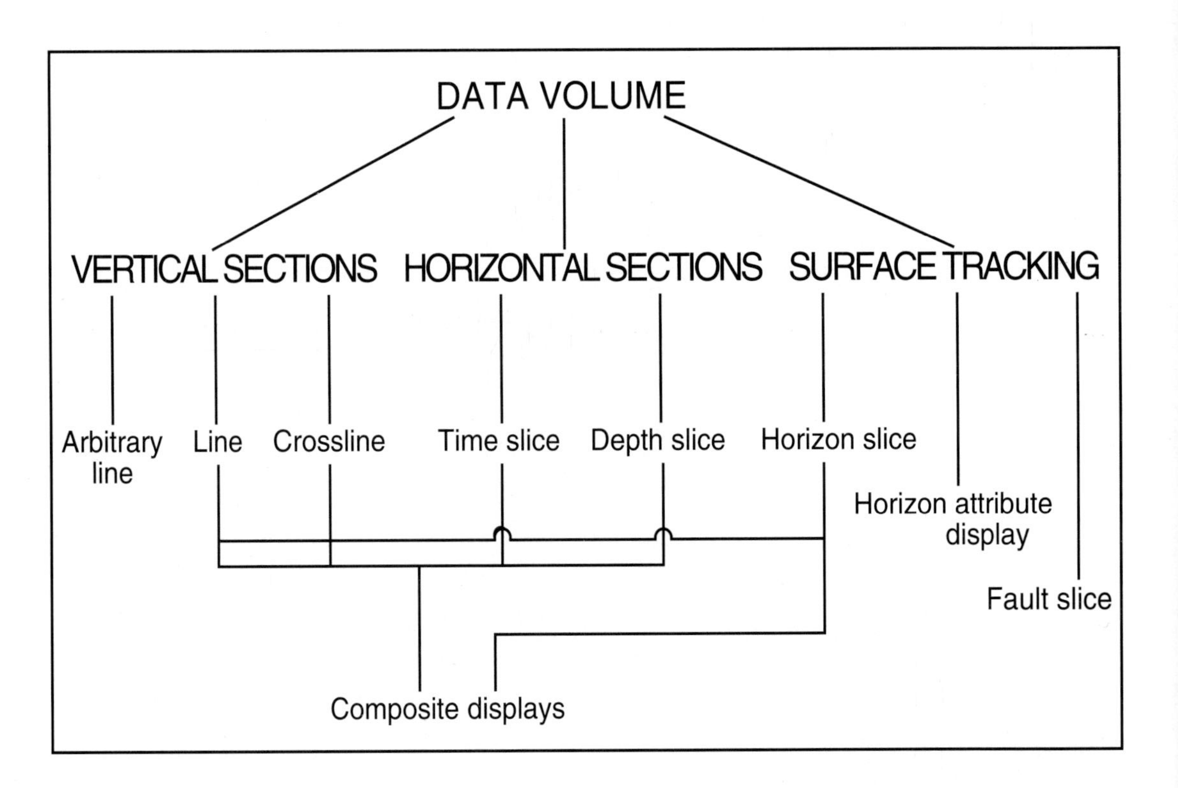

Fig. 1-19. Recognized and approved terms for display products from 3-D seismic data. All display seismic amplitude unless specified otherwise. Use of all other terms should be discouraged.

3-D seismic interpreter will generate a more accurate and detailed map or other product than his 2-D predecessor in the same area.

Figure 1-14 shows a view of a 3-D data volume through a salt dome. It demonstrates the volume concept well and the interpreter can use a display of this kind to help in appreciation of subsurface three-dimensionality. Figure 1-15 shows another cube, in this case generated interactively, which helps in the three-dimensional appreciation of a much more detailed subsurface objective. Neither of these displays, however, permits the interpreter to look *into* the volume of data.

True 3-D display of a volumetric image is a difficult problem. Nelson (1983) reviewed the applicable technologies but all fall short of what the seismic interpreter really needs. Most address very small volumes of data and also lack dynamic range. The author has personally experimented with holography and several seismic data holograms exist. However, the interpreter cannot interact with the image and the dynamic range is inadequate for most purposes.

An early approach to true 3-D display of a seismic data volume is the Seismodel Display Unit (Figure 1-16). Here individual vertical sections from the volume are printed onto transparent plastic plates which are positioned in accurately-engineered grooves in a metal box. A stack of these plates is then illuminated from behind and viewed from in front. The interpreter may look along a fault plane or down the axis of a structure. A plate may be removed, as shown in Figure 1-16, marked by the interpreter, and then returned to its place in the stack. The interpretation on this one section can then be seen in relationship to other adjacent sections and other plates can be marked accordingly. The principal shortcoming of the Seismodel Display Unit is that, in order to increase the transparency of each plate, the data have to be displayed with a very low gain and peaks only. Hence the dynamic range of the displayed data is low, restricting its use to structural interpretation.

True 3-D display has recently become a reality on computer workstations and Figure 1-17 shows an example. The portion of the volume being displayed is composed of voxels, or volume elements, and these are rendered with differing degrees of transparency so that the interpreter can really see into the volume. In Figure 1-17 there are four interpreted surfaces as well as the semi-transparent data. As with any volumetric display the dynamic range is reduced because of the quantity of data viewed. These

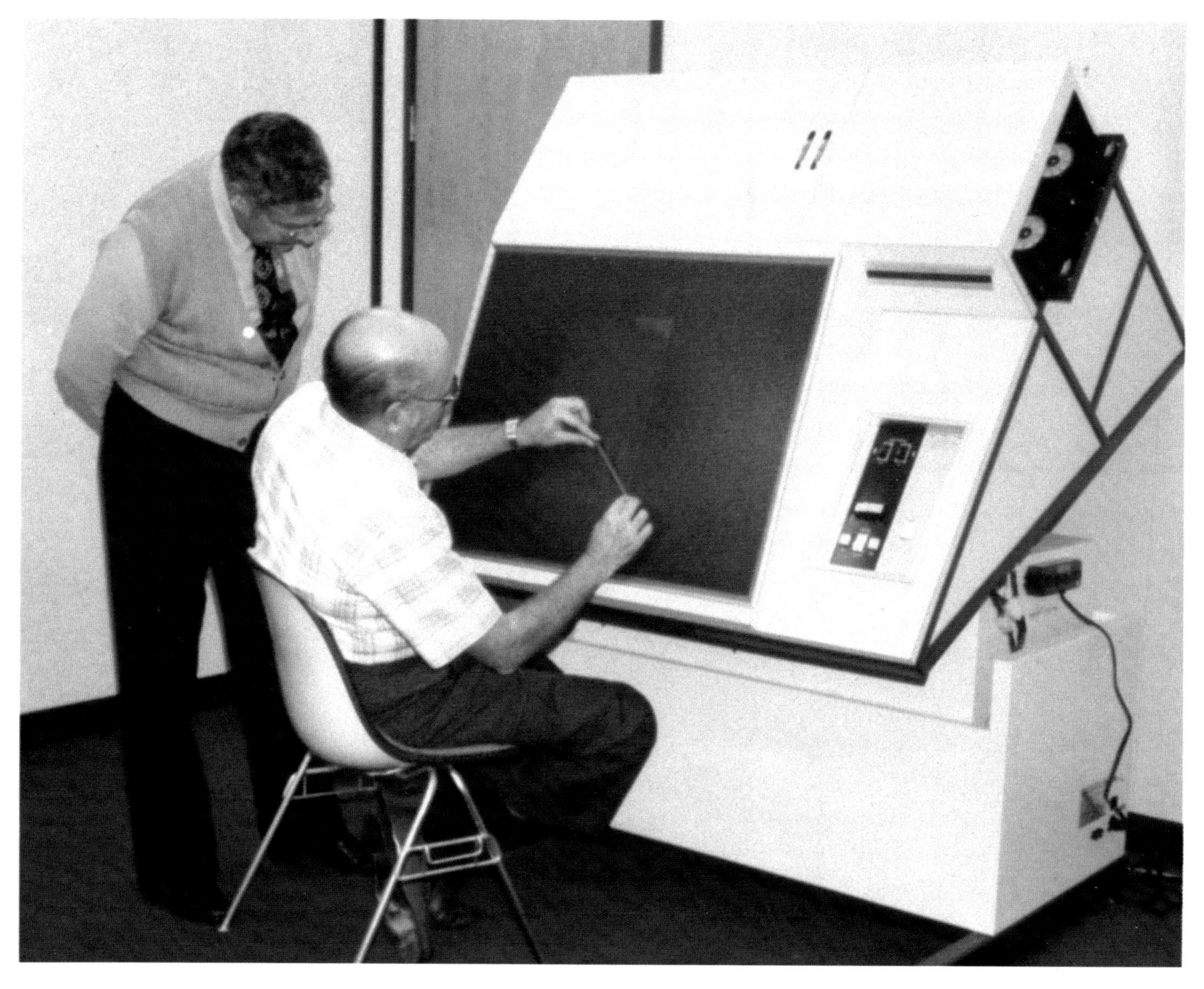

Fig. 1-20. Seiscrop Interpretation Table. (Courtesy Geophysical Service Inc.)

types of display are very useful for data visualization but they are not yet fully integrated into mainstream interpretation systems.

Slicing the Data Volume

The vast majority of 3-D interpretation is performed on slices through the data volume. There are no restrictions on the dynamic range for the display of any one slice, and therefore all the benefits of color, dual polarity, etc., can be exploited (see Chapter 2). The 3-D volume contains a regularly-spaced orthogonal array of data points defined by the acquisition geometry and probably adjusted during processing. The three principal directions of the array define three sets of orthogonal slices or sections through the data, as shown in Figure 1-18.

The vertical section in the direction of boat movement or cable lay-out is called a **line** (sometimes an **inline**). The vertical section perpendicular to this is called a **crossline**. The horizontal slice is called a **horizontal section, time slice, Seiscrop* section, or depth slice**. The terminology used for slices through 3-D data volumes has become somewhat confused. One of the objectives of this chapter is to clarify terms in common use today.

*Trademark of Geophysical Service Inc.

Fig. 1-21. Seismic Interactive Interpretation System in action.

Three sets of orthogonal slices through the data volume (as defined above) are regarded as the basic equipment of the 3-D interpreter. A complete interpretation will make use of some of each of them. However, many other slices through the volume are possible. A **diagonal line** may be extracted to tie two locations of interest, such as wells. A zig-zag sequence of diagonal line segments may be necessary to tie together several wells in a prospect. In the planning stages for a production platform, a diagonal line may be extracted through the platform location along the intended azimuth of a deviated well. All these are vertical sections and are referred to as **arbitrary lines**.

More complicated slices are possible for special applications. A slice along or parallel to a structurally interpreted horizon, and hence along one bedding plane, is a **horizon slice, horizon Seiscrop section,** or **amplitude map**. Slices of this kind have particular application for stratigraphic interpretation, which is explored in Chapter 4. **Fault slices** generated parallel to a fault face have various applications in structural and reservoir interpretation and will be discussed in Chapter 7. **Horizon attribute displays** are the subject of Chapter 8.

Figure 1-19 shows a hierarchy of approved terms for display products from 3-D seismic data. It shows, for example, the equivalence of horizontal and vertical sections, and the equivalence of time slices with lines and crosslines. In order to aid worldwide communication, use of other terms is discouraged.

Manipulating the Slices

Because 3-D interpretation is performed with data slices and because there is a very large number of slices for a typical data volume, several innovative approaches for manipulating the data have emerged. In the early days of 3-D development a

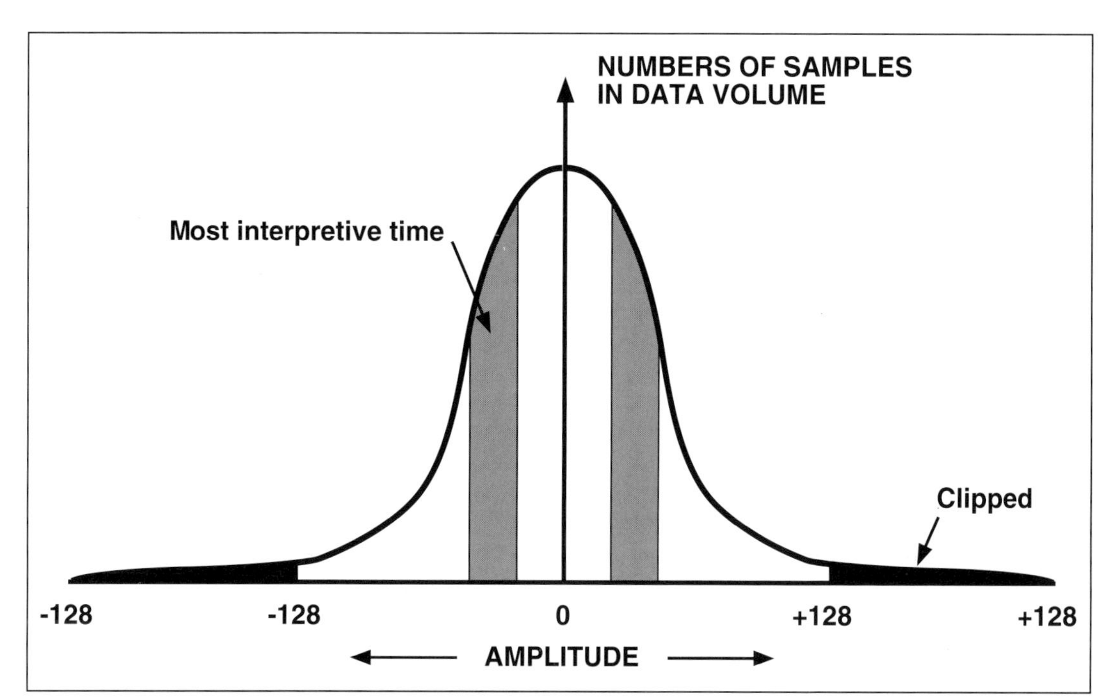

Fig. 1-22. Typical statistical distribution of amplitudes in a 3-D data volume. Plus or minus 128, the largest number which can be described by 8 bits, may be set to the largest amplitude, or alternatively to some smaller amplitude, thus causing data clipping.

Fig. 1-23. Test for and demonstration of data clipping.

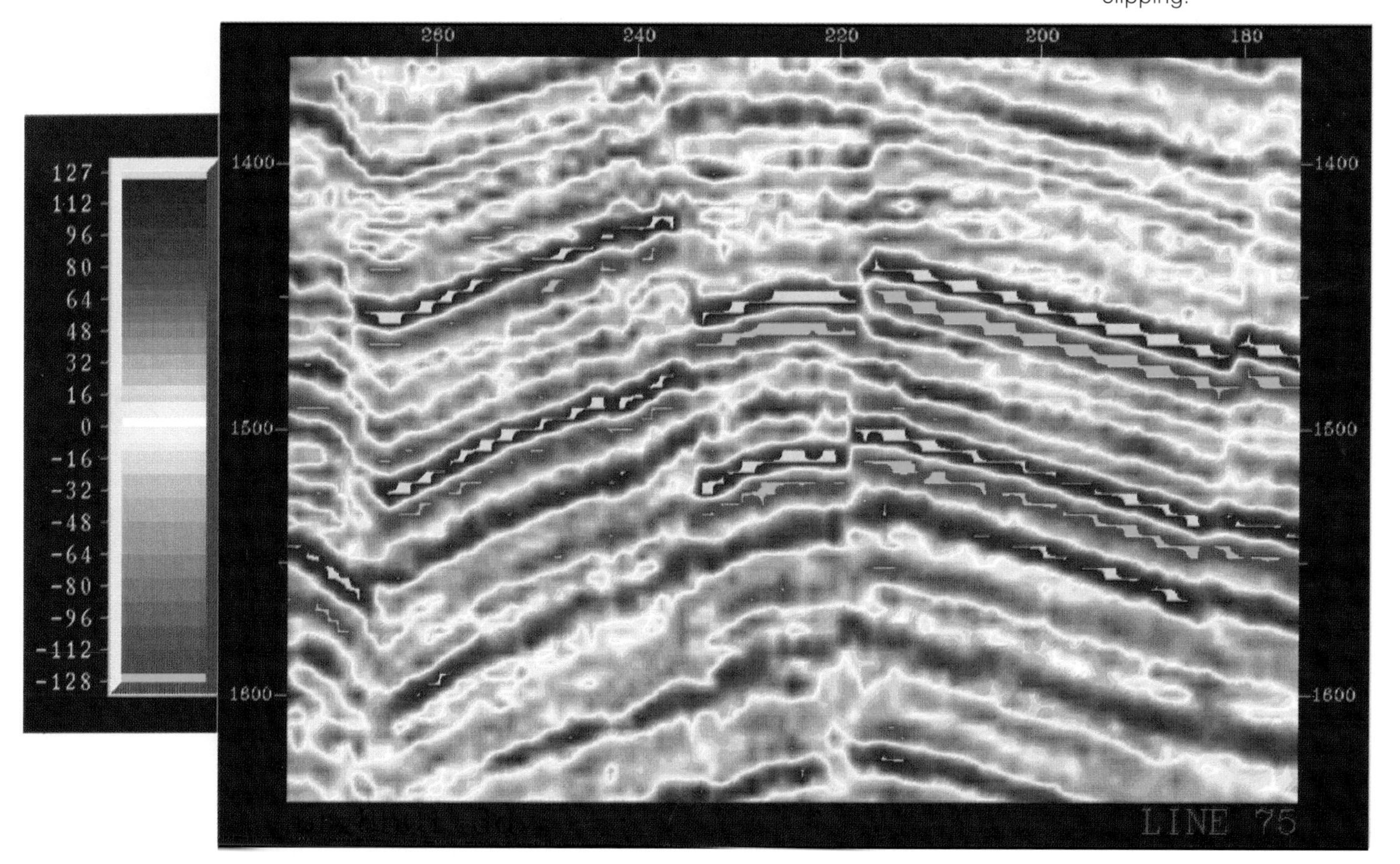

sequence of horizontal sections was displayed on film-strip and shown as a motion picture (Bone, Giles, Tegland, 1983). From this developed the Seiscrop Interpretation Table—initially a commercially-available piece of equipment incorporating a 16mm analytical movie projector. This machine was originally developed for coaches wanting to examine closely the actions of professional athletes.

The Seiscrop Interpretation Table then evolved into a custom-built device (Figure 1-20). The data, either horizontal or vertical sections, were projected from 35mm film-strip onto a large screen. The interpreter fixed a sheet of transparent paper over the screen for mapping and then adjusted the size of the data image, focus, frame advance, or movie speed by simple controls.

Today 3-D interpretation is performed interactively and there has been an explosion in workstation usage in recent years (see Foreword). The interpreter calls the data from disk and views them on the screen of a color monitor (Figure 1-21). The large amount of regularly-organized data in a 3-D volume gives the interactive approach enormous benefits. In fact, many interactive interpretation systems addressed 3-D data first as the easier problem, and then developed 2-D interpretation capabilities later.

Most of the interpretation discussed in this book resulted from use of an interactive workstation, and many of the data illustrations are actual screen photographs. Furthermore, the facilities of the system contributed in several significant ways to the success of many of the projects reported here. Hence it is appropriate to review the interpretive benefits of an interactive interpretation system.

(1) **Data management**—The interpreter needs little or no paper; the selected seismic data display is presented on the screen of a color monitor and the progressive results of interpretation are returned to the digital database.

(2) **Color**—Flexible color display provides the interpreter with maximum optical dynamic range adapted to the particular problem under study.

(3) **Image composition**—Data images can be composed on the screen so that the interpreter views what is needed, no more and no less, for the study of one particular issue. Slices through the data volume are designed by the user in order to customize the perspective to the problem.

(4) **Idea flow**—The rapid response of the system makes it easy to try new ideas. The interpreter can rapidly generate innovative map or section products in pursuit of a better interpretation.

(5) **Interpretation consistency**—The capability to review large quantities of data in different forms means that the resulting interpretation should be more consistent with all available evidence. This is normally considered the best measure of interpretation quality.

(6) **More information**—Traditional interpretive tasks performed interactively will save time; however, the extraction of more detailed subsurface information is more persuasive and far-reaching.

Dynamic Range and Data Loading

Interactive interpretation must commence with data loading and this is a critical first step. Should the data be loaded at 8, 16 or 32 bits? Is clipping of the highest amplitudes acceptable?

Data processing has always been performed using 32 bits to describe each amplitude value. This large word size ensures that significance is retained during all computations. The first interactive systems in the early 1980's were 32-bit machines but soon a demand for speed dictated that data be loaded using 8 bits only. The small word reduces response time and minimizes storage space for the survey data. Today interactive systems offer a choice of 8-bit, 16-bit or 32-bit dynamic range although color monitors normally display 8 bits only.

Figure 1-22 shows a typical statistical distribution of amplitudes in a data volume. There are a large number of very low amplitudes, a fairly large number of moderate amplitudes but a very small number of high amplitudes. Mainstream structural interpretation tends to work on moderate amplitude horizons. The high amplitude tails of the distribution are localized anomalies which, in tertiary clastic basins, are often the hydrocarbon bright spots. The interpreter avoids the low amplitudes as much as possible because they are the most subject to noise. Thus most interpretive time is devoted to the amplitudes lying in the stippled areas of Figure 1-22.

If interpretation is to be conducted using 8-bits only, scaling 32-bit amplitude numbers to 8-bit amplitude numbers must be done during data loading. If the maximum amplitude in the volume is set to ± 128, relative amplitudes are preserved within the precision of the 8 bits. However, this often severely limits the dynamic range available in the stippled, or heavily used, amplitude regions. Clipping of the highest amplitudes is a common reaction to this problem so that a smaller value is set to ±128. More dynamic range is then available for the mainstream structural interpretation but the highest amplitudes are destroyed and hence unavailable for stratigraphic or reservoir analysis. This can be very damaging in areas like the Gulf of Mexico.

A common and generally desirable solution today is to load the data using 16 bits for each amplitude value. In this way clipping is irrelevant and unnecessary as there is plenty of dynamic range for structural interpretation and bright spot studies.

An interesting comparison of 8-bit and 16-bit interpretation was conducted by Roberts and Hughes (1995). They concluded that there are always differences between interpretation products from 8-bit and 16-bit volumes but they are generally less than 5%. These are often tolerable but they stressed the need for *sensible clipping*. Figure 1-23 is a test for and demonstration of data clipping. Contrasting colors have been placed in the extremities of the otherwise-gradational color scheme. The large amounts of yellow and cyan demonstrate an anomalously high occupancy of those highest amplitudes, that is the data has been heavily clipped.

The author is opposed to data clipping as it places restrictions on interpretation activities. Generally the best solution is to use 16 bits and sometimes 32 bits. The total interpretation project today often involves a significant amount of post-interpretation computation. The larger number of bits helps ensure that numeric significance is maintained during these operations. Fortunately faster and cheaper hardware is now available which makes the use of 16 or 32 bits much less of a burden than it was in the past.

Synergism and Pragmatism in Interpretation

Seismic technology has, over the years, become increasingly complex. Whereas a party chief used to handle data collection, processing, and interpretation, experts are now generally restricted to each discipline. Data processing involves many highly sophisticated operations and is conducted in domains unfamiliar to the nonmathematically-minded interpreter. The ability of certain processes to transform data in adverse as well as beneficial ways is striking.

Today's seismic interpreter must understand in some detail what has been done to the data and must understand data processing well enough to ask meaningful questions of the processing staff. A summary of 3-D data collection and processing issues is included as Appendix A. Today's interpreter will also benefit greatly by using high technology aids, such as an interactive system. Critical to maximum effectiveness is an understanding of the advantages of color and how to work with horizontal sections, acoustic impedance sections, frequency sections, vertical seismic profiles, attribute displays, and the like.

Seismic interpretation today thus involves a wide range of seismic technologies. If the results of these are studied by the interpreter in concert, significant synergism can result. However, pragmatism retains its place. The interpreter must continue to take a broad view, to integrate geology and geophysics, and, to an increasing degree, engineering, and to make simplifying assumptions in order to get the job done. The progress of seismic interpretation depends on the continued coexistence of technological synergism and creative pragmatism.

References

Blake, B. A., J. B. Jennings, M. P. Curtis, and R. M. Phillipson, 1982, Three-dimensional seismic data reveals the finer structural details of a piercement salt dome: Offshore Technology Conference paper 4258, p. 403-406.

Bone, M. R., B. F. Giles, and E. R. Tegland, 1976, 3-D high resolution data collection, processing and display: Houston, Texas, presented at 46th Annual SEG Meeting.

Bone, M. R., B. F. Giles, and E. R. Tegland, 1983, Analysis of seismic data using horizontal cross-sections: Geophysics, v. 48, p. 1172-1178.

French, W. S., 1974, Two-dimensional and three-dimensional migration of model-experiment reflection profiles: Geophysics, v. 39, p. 265-277.

Horvath, P. S., 1985, The effectiveness of offshore three-dimensional seismic surveys—case histories: Geophysics, v. 50, p. 2411-2430.

Lindsey, J. P., 1989, The Fresnel zone and its interpretive significance: The Leading Edge, v. 8, no. 10, p. 33-39.

Martins, C. C., and C. A. da Costa, C. E Theodoro, L. R. Guardado, and V. F. Andrade, 1995, 3-D Seismic: A successful strategy in the Campos Basin: The Leading Edge, v. 14, p. 701–704.

Nelson, H. R., Jr., 1983, New technologies in exploration geophysics: Houston, Texas, Gulf Publishing Company, p. 187-206.

Roberts, G. A., and M. J. Hughes, 1995, Workstation interpretation—a comparison of data stored at 8 and 16 bits: Poster paper and EAGE Annual Convention, Glasgow.

Saeland, G. T., and G. S. Simpson, 1982, Interpretation of 3-D data in delineating a sub-unconformity trap in Block 34/10, Norwegian North Sea, *in* M. T. Halbouty, ed., The deliberate search for the subtle trap: AAPG Memoir 32, p. 217-236.

Sheriff, R. E., 1985, Aspects of seismic resolution, *in* O. R. Berg and D. Woolverton, eds., Seismic stratigraphy II: an integrated approach to hydrocarbon exploration: AAPG Memoir 39, p. 1-10.

Sheriff, R. E. (editor), 1992, Reservoir Geophysics, SEG Investigations in Geophysics no. 7.

Sheriff, R. E. and L. P. Geldart, 1983, Exploration seismology; v. 2, data-processing and interpretation: Cambridge University Press, p. 130.

Tegland, E. R., 1977, 3-D seismic techniques boost field development: Oil and Gas Journal, v. 75, no. 37, p. 79-82.

Walton, G. G., 1972, Three-dimensional seismic method: Geophysics, v. 37, p. 417-430.

West, J., 1979, Development near for Thailand field: Oil and Gas Journal, v. 77, no. 32, p. 74-76.

COLOR, CHARACTER AND ZERO-PHASENESS

Color Principles

"The total quantity of information recorded on a typical seismic line is enormous. It is virtually impossible to present all this information to the user in a comprehensible form." This quotation from Balch (1971) is even more true today than it was in 1971 and color has become an important contributor to the problem's solution. The human eye is very sensitive to color and the seismic interpreter can make use of this sensitivity in several ways. Taner and Sheriff (1977) and Lindseth (1979) were among the first to present color sections which demonstrated the additional information color can convey. Of equal importance is the increased optical dynamic range of a color section compared to its black and white variable area/wiggle trace equivalent. Both these properties are of great importance in stratigraphic interpretation.

Some understanding of color principles will help an interpreter maximize the use of color. It is helpful to visualize colors as a three-dimensional solid but there are three relevant sets of coordinates in terms of which the color solid can be expressed:

(1) the three additive primary colors—red, green, blue;
(2) the three subtractive primary colors—magenta, yellow, cyan; and,
(3) hue, saturation, density.

Figure 2-1 is a diagrammatic representation of a color cube showing the interrelationship of the above sets of coordinates. Figure 2-2 is a photograph of an actual color cube oriented to correspond to the diagram of Figure 2-1. Figure 2-3 is a photograph of the same cube from the opposite direction.

This cube was made using an Applicon color plotter, but the principles under discussion are independent of the plotting device. Any system which combines pigments employs the **subtractive primary colors**—magenta, yellow and cyan. Figure 2-2 and 2-3 show the *absence* of any color, which is *white*, at the top and progressively increasing quantities of magenta, yellow and cyan down the upper edges of the cube. These primaries, paired in equal quantities, give the **additive primary colors**—red, green and blue—at the three lower corners. All three subtractive primaries combined in equal quantities give black, seen at the bottom apex of the cube.

Any display system which combines light, such as a color monitor, follows the cube of Figures 2-2 and 2-3 from *bottom to top*. The *absence* of color is then *black*. Light of the three additive primary colors, red, green and blue, combine in pairs to make magenta, yellow and cyan and altogether to make white.

The cube photographs display only those colors on the surface of the cube. In fact, a much larger number of colors is inside. Down the vertical axis from white to black is the gray scale for which the **density** increases progressively (Figure 2-1). The **satura-**

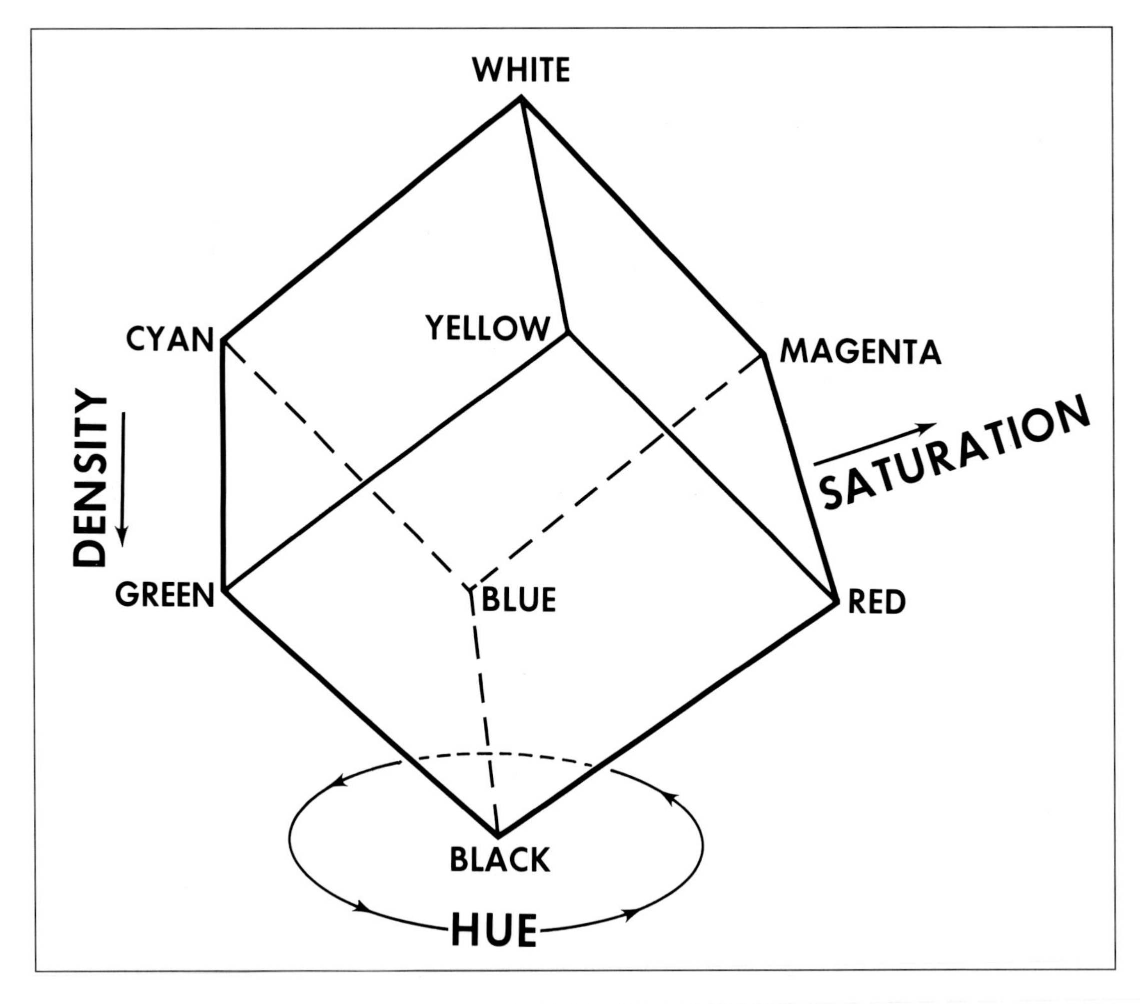

Fig. 2-1. Diagram of a color cube showing the relationship of the subtractive primary colors (magenta, yellow and cyan) to the additive primary colors (red, green and blue) to the color parameters (hue, saturation and density).

tion measures the distance from this central axis, ranging from zero on the axis to 100% on the surface of the cube. The **hue** is the rotational parameter measuring the spectral content of a color.

For the color cube illustrated in Figures 2-2 and 2-3 there are 17 levels (0-16) of each of the subtractive primaries—magenta, yellow and cyan. The total number of colors in the cube is thus 17 × 17 × 17 or 4,913, of which 1,538 are fully saturated colors on the surface. One way of studying the colors available inside the cube is to slice it along a chosen density level. Figure 2-4 shows density level 16, which has maximum strength magenta, yellow and cyan at the corners and gray of density 33% at the center. This display clearly demonstrates the significance of hue as the rotational parameter and saturation as the radial distance from the gray axis. The additive primaries, red, green and blue, lie on density level 32 with gray of density 67% at the center.

Figure 2-5 is a color chart used in an interactive interpretation system (Gerhardstein and Brown, 1984). It is based on the mixing of light and hence involves the additive primaries—red, green and blue. All the colors displayed in Figure 2-5 are fully saturated; that is, they lie only on the surface of the color cube. The right half of the chart is a projection of a color cube similar to that of Figures 2-2 and 2-3 when viewed from the top. The left half of the chart is a view of the same color cube from the bottom. Interactive workstations make the selection and building of logical, efficient color schemes easier if the selection chart is founded directly on the color cube, as in Figure 2-5.

Interpretive Value of Color

Today's interpreter uses color in two fundamentally different ways: with a *contrasting* or with a *gradational* color scheme. A map or a section displayed in *contrasting* col-

Fig. 2-2. Photograph of a color cube oriented the same as the diagram in Figure 2-1.

ors is normally accompanied by a legend, so the reader can identify the value of the displayed attribute at any point by reading the range of values associated with each color. Figure 2-6 is a structural contour map with a contour interval of 20 ms.

For an effective color display it is important that the range of values associated with each color, the number of colors used and their sequence, the contrast between adjacent colors, and the display scales are all carefully chosen. A color display must convey useful information and at the same time be aesthetically pleasing. For a map such as Figure 2-6 it is desirable to perceive equal visual contrast between adjacent colors, so that no one color boundary is more outstanding than another. A spectral sequence of colors was selected as the only really logical sequence available.

Figure 2-7 is a nomogram used for assessing visual color contrast. Visual contrast between two colors is, of course, somewhat subjective. Numerical color contrast is the sum of the absolute values of the differences in the amounts of the three primary colors. Zero density is white, maximum density (100%) is black, and density can have

Fig. 2-3. Photograph of the same color cube as in Figure 2-2 from the back.

either arbitrary or percentage units between these extremes. Figure 2-7 shows that, for a particular visual color contrast, numerical contrast should be approximately proportional to average density. In other words, a larger numerical contrast is needed between darker colors.

A *gradational* color scheme is used when the interpreter is looking for trends, shapes, patterns and continuity. Figure 2-8 includes a vertical section displayed with gradational blue for positive amplitudes (peaks) and gradational red for negative amplitudes (troughs). Absolute amplitude levels are unimportant but relative levels are very important. Much stratigraphic information is implied by the lateral variations in amplitude along each reflection. The blue and red give equal visual weight to peaks and troughs. If the display gain is properly set, only a few of the highest amplitudes reach the fully saturated color and the full range of gradational shades expresses the varying amplitudes in the data. This increased dynamic range gives the interpreter the best opportunity to judge the extent and the character of amplitude anomalies of interest.

Figure 2-8 also provides a comparison of gradational color and variable area/wig-

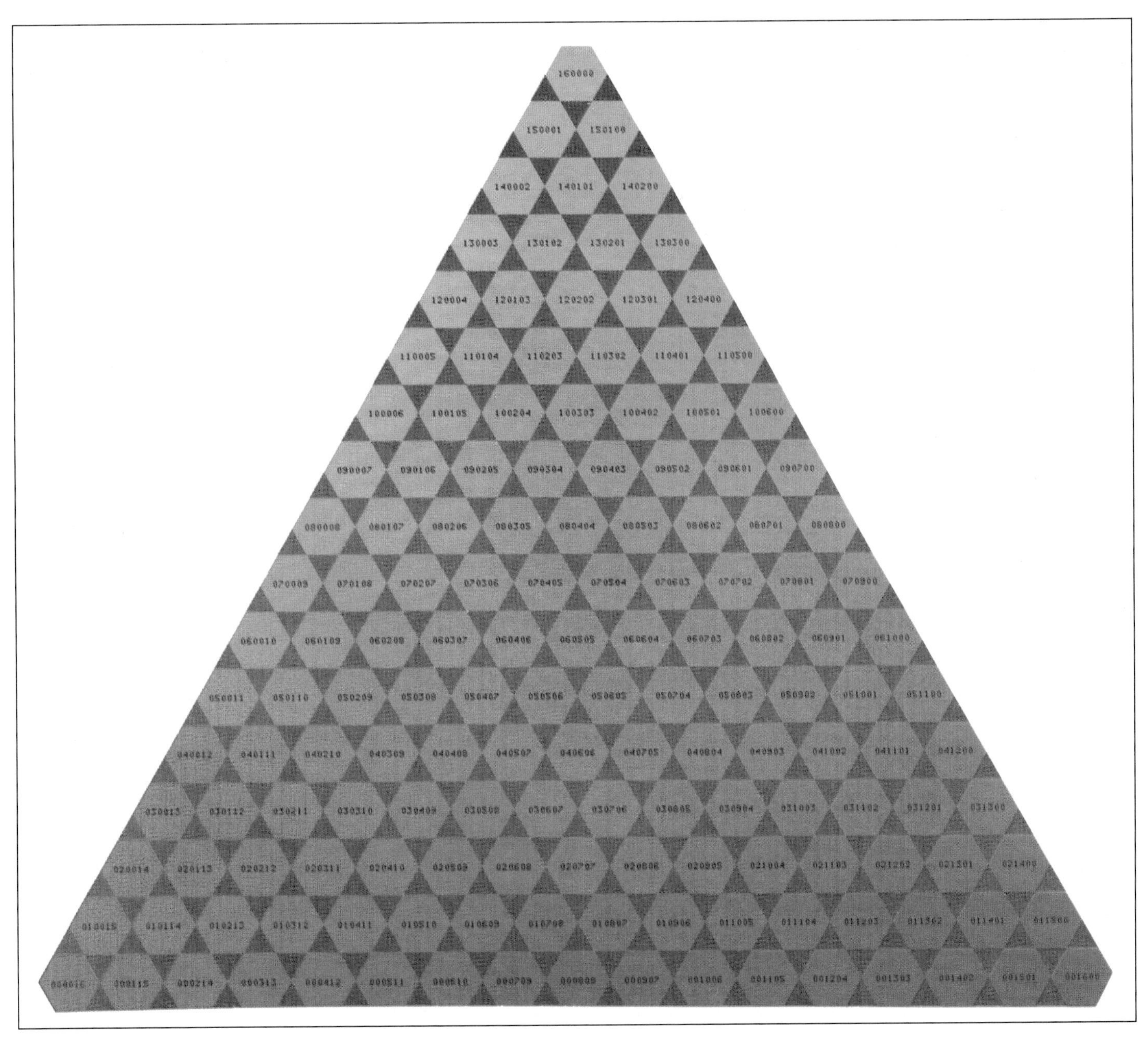

Fig. 2-4. Horizontal slice through the color cube at density level 33%, showing magenta, yellow and cyan at the corners and gray at the center.

gle trace for the same piece of data. The shortcomings in the variable area/wiggle trace display relative to the color section are: (1) the visual weights of peaks and troughs are very different, which makes comparison difficult and biases the interpreter's eye towards the peaks; (2) the peaks are saturated or clipped; and (3) the troughs, where they have significant amplitudes, are not visible beneath the depth points where they belong. The red flat spot reflection is clearly visible on the color section as are the relative amplitudes of peaks and troughs. At the extreme right of the section, coincident amplitude maxima in the peak and the trough indicate a tuning phenomenon (see Chapter 6).

Figure 2-9 provides a similar comparison between gradational color and variable area/wiggle trace. In addition, three different horizontal scales are used from which a further shortcoming of variable area/wiggle trace is apparent—that the dynamic range is dependent on horizontal scale.

The color schemes in Figures 2-8 and 2-9 are more explicitly called double-gradational schemes where the symmetry of blue and red about zero amplitude is important. The dynamic range of this type of scheme can be increased further by adding

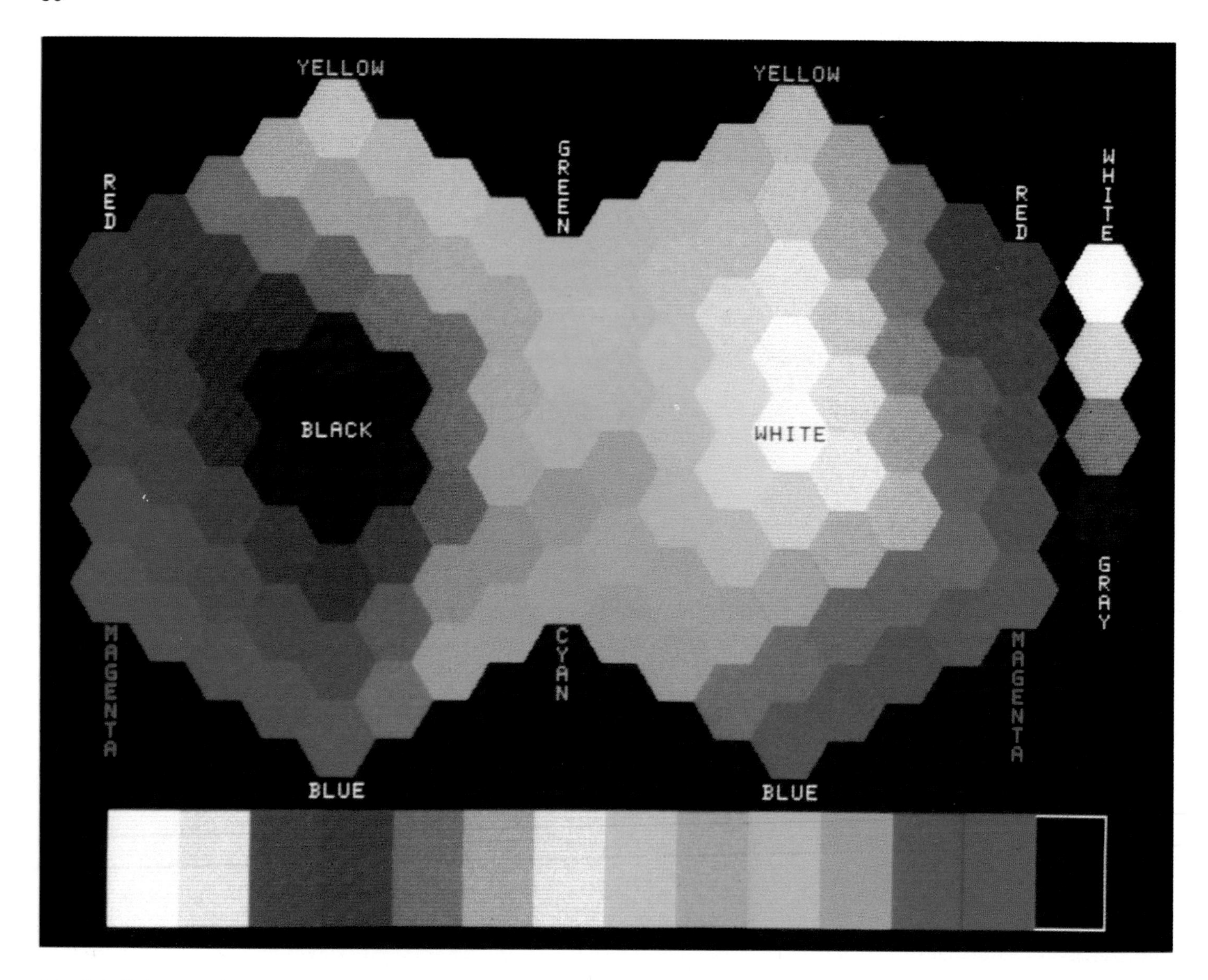

Fig. 2-5. Color selection chart from an interactive interpretation system. Note how its organization is based on the color cube.

extra gradations while still maintaining symmetry. In Figure 2-10, for example, gradational cyan has been added for the highest positive amplitudes and gradational yellow for the highest negative amplitudes. This is a very efficient color scheme for studying subtle amplitude anomalies, for example hydrocarbon indicators.

The need for equal visibility of peaks and troughs has long been recognized. Backus and Chen (1975) generated dual polarity variable area sections with the peaks in black and the troughs in red. Figure 2-11 is an example of this display from Galbraith and Brown (1982). Dual polarity variable area rectifies some of the shortcomings of variable area/wiggle trace but has less dynamic range than gradational color. An additional benefit of dual polarity variable area relative to variable area/wiggle trace is that there are effectively twice the number of event terminations at faults, making their recognition and detailed placement easier for the structural interpreter. This is apparent in Figure 2-11 and was considered an interpretive benefit by Galbraith and Brown.

A quite different display, but one that also has benefits for fault interpretation, is shown in Figure 2-12. Here a single gray scale from black for the largest peaks to white for the largest troughs enhances the visibility of low amplitude events. This also increases the number of event terminations visible at faults. The gradational blue and red, already discussed, is better for studying higher amplitudes but these need not be in conflict. Figure 5-26 (Chapter 5) shows gradational blue and red for the

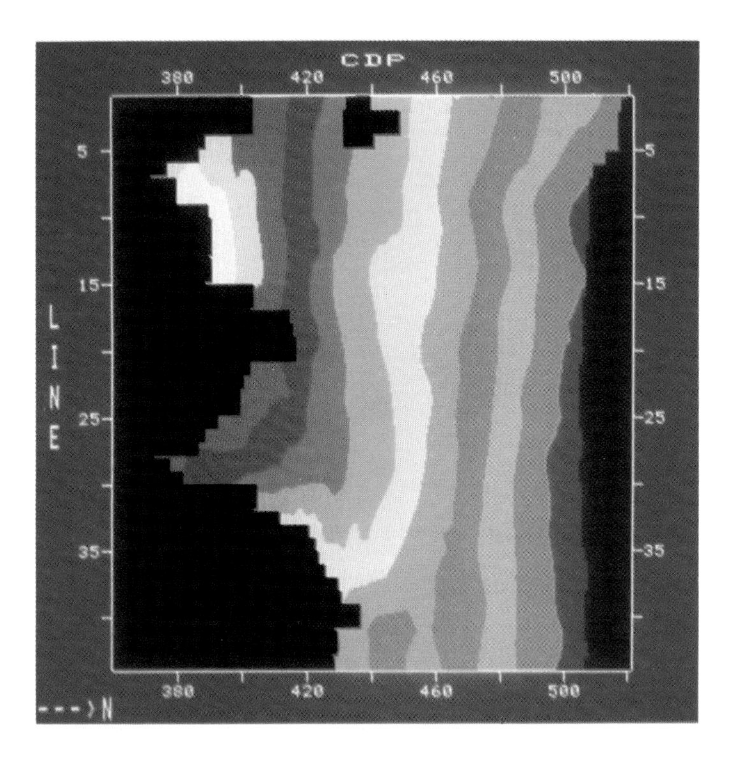

Fig. 2-6. Time structure map displayed in a contrasting spectral color scheme.

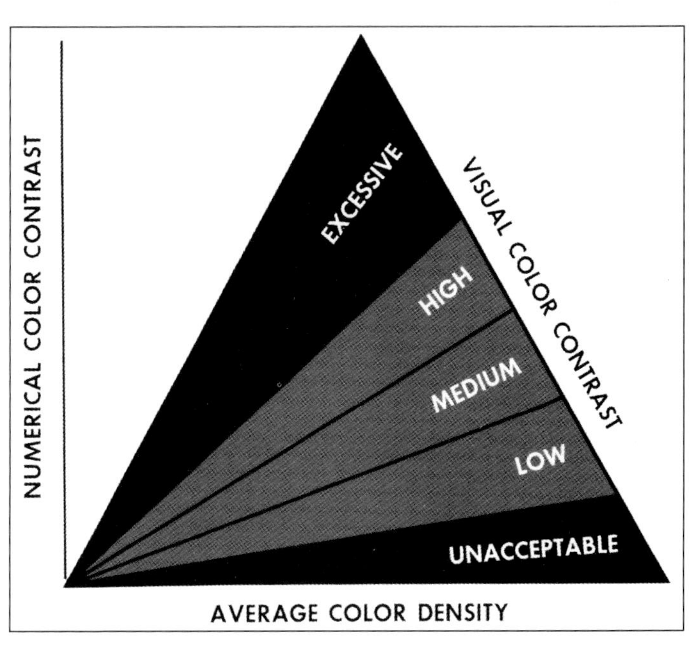

Fig. 2-7. Contrast-density nomogram used for establishing a color scheme with acceptable visual contrast between adjacent colors.

higher amplitudes and gradational gray for the lower amplitudes. This facilitates the study of bright spots in the presence of faulting.

An almost infinite number of color schemes can be applied to the same piece of data. Many of them are useful and interpreters' preferences vary. Care and thought are needed to develop a good scheme. Figure 2-13 shows four color schemes applied to the same vertical section segment. In the bottom left is the gradational blue and red, which is effectively the industry standard. Above this in the upper left is a scheme made with the same blue and red hues but many fewer intensity levels. It is sometimes useful to compare actual amplitude levels between two points. Here also the highest amplitudes have been distinguished with contrasting colors. This is generally not a good idea unless it has been possible to identify those highest amplitudes as having special meaning. For normal interpretation purposes, it is important to have an unbiased color scheme so that the interpreter can make his own interpretation of data phenomena. The upper right panel of Figure 2-13 shows the color scheme preferred by another interpreter. The lower right panel illustrates a color scheme selected for a different purpose and is clearly inappropriate here. It demonstrates the inappropriateness of a contrasting color scheme for data.

The key to success with colors is learning how to adapt the display to suit the type of interpretation underway. Hence flexible color display has become one of the key benefits of an interactive interpretation system. Each interpreter needs to use the color schemes with which he is most comfortable, but an innovative interpreter can beneficially scan through many schemes seeking the one which conveys, to him, the most information about the issue at hand.

Neidell and Beard (1985) are vociferous in their promotion of color display. In fact, in drawing conclusions concerning point bar and channel sands, they say, "Such interpretation would not be reasonable without the support of the color display and the stratigraphic visibility which it provides." Their detailed use of color is, however, slightly different from that discussed above in that stratigraphic acoustic impedance

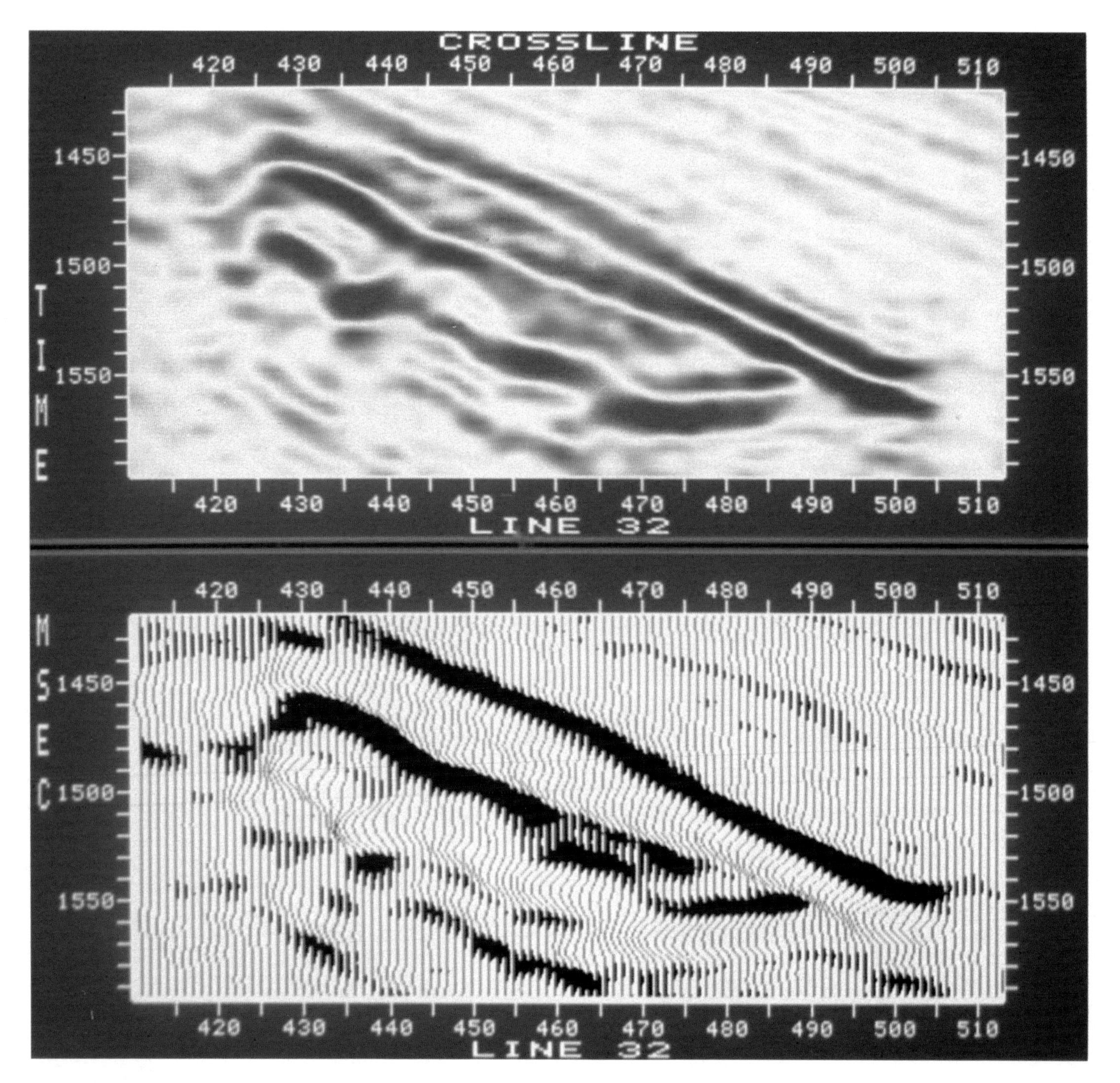

Fig. 2-8. Vertical seismic section displayed with gradational blue for peaks and gradational red for troughs compared to same section displayed in variable area/wiggle trace. (Courtesy Chevron U.S.A. Inc.)

sections are displayed with a very large number of mildly-contrasting colors.

A sun-shaded color scheme (Figure 2-14) is an aspect of display technology borrowed from satellite imagery. Here, an imaginary sun shining from the east enhances the visibility of lineations caused by subtle faults. This is discussed in Chapter 8.

The recognition of channels, bars and other depositional features on horizontal sections and horizon slices is becoming increasingly important for the stratigraphic interpreter. Here again the proper use of gradational color coded to amplitude helps the detectability of these features because of the eye's ability to integrate a wide range of densities. Figures 2-15 and 2-16 illustrate an inferred channel on a horizon slice (see Chapter 4) and the use and abuse of color for its detection. A well at about Line 55, Crossline 250, indicates that at least the lower part of the areal bright spot (Figure 2-15) is a sand-filled channel. How extensive is this channel? It seems probable that it extends to include the central zone between Lines 70 and 80 and between

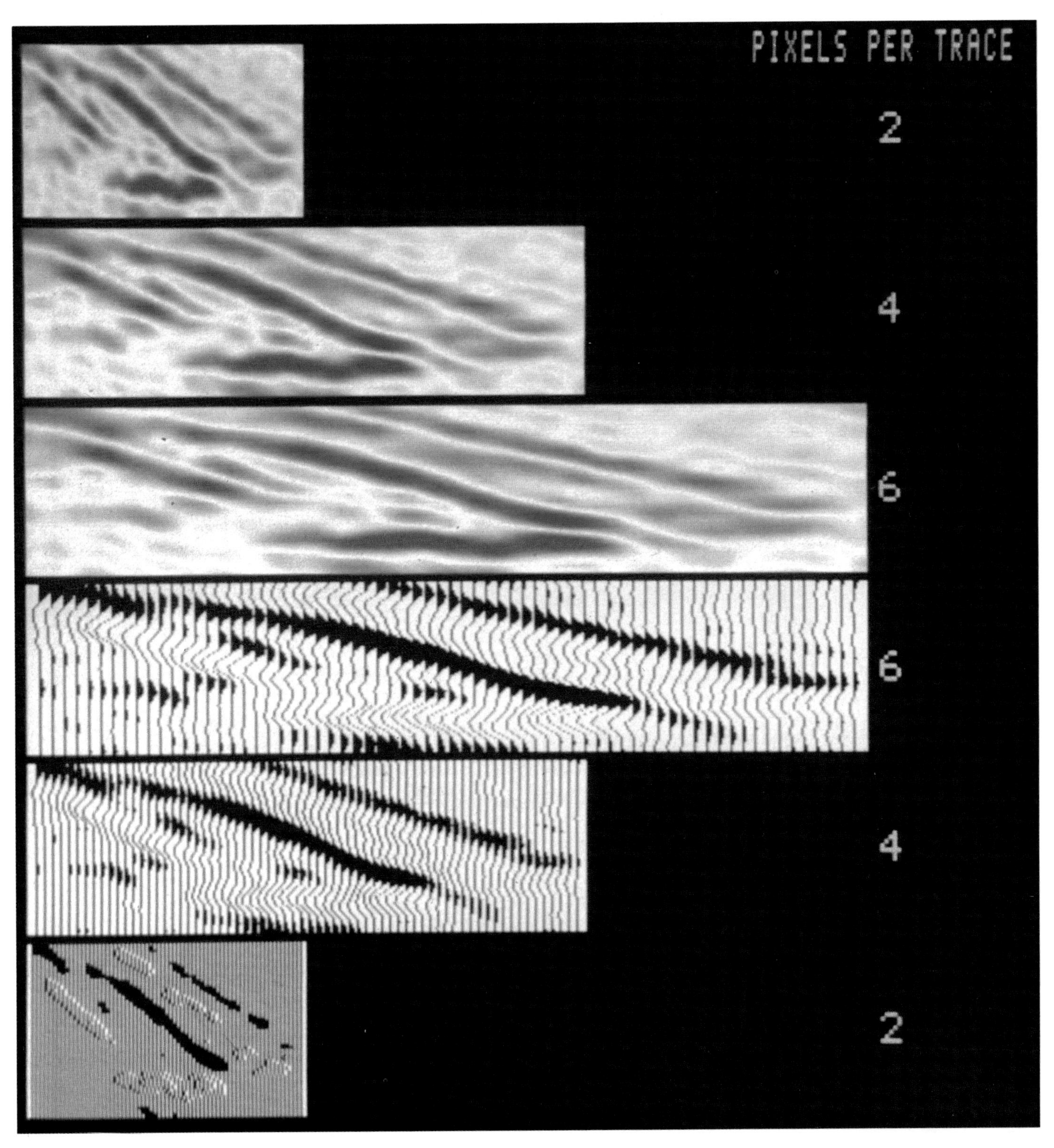

Fig. 2-9. Comparison of double-gradational blue and red with variable area/wiggle trace display illustrating respectively independence and dependence of dynamic range on horizontal scale.

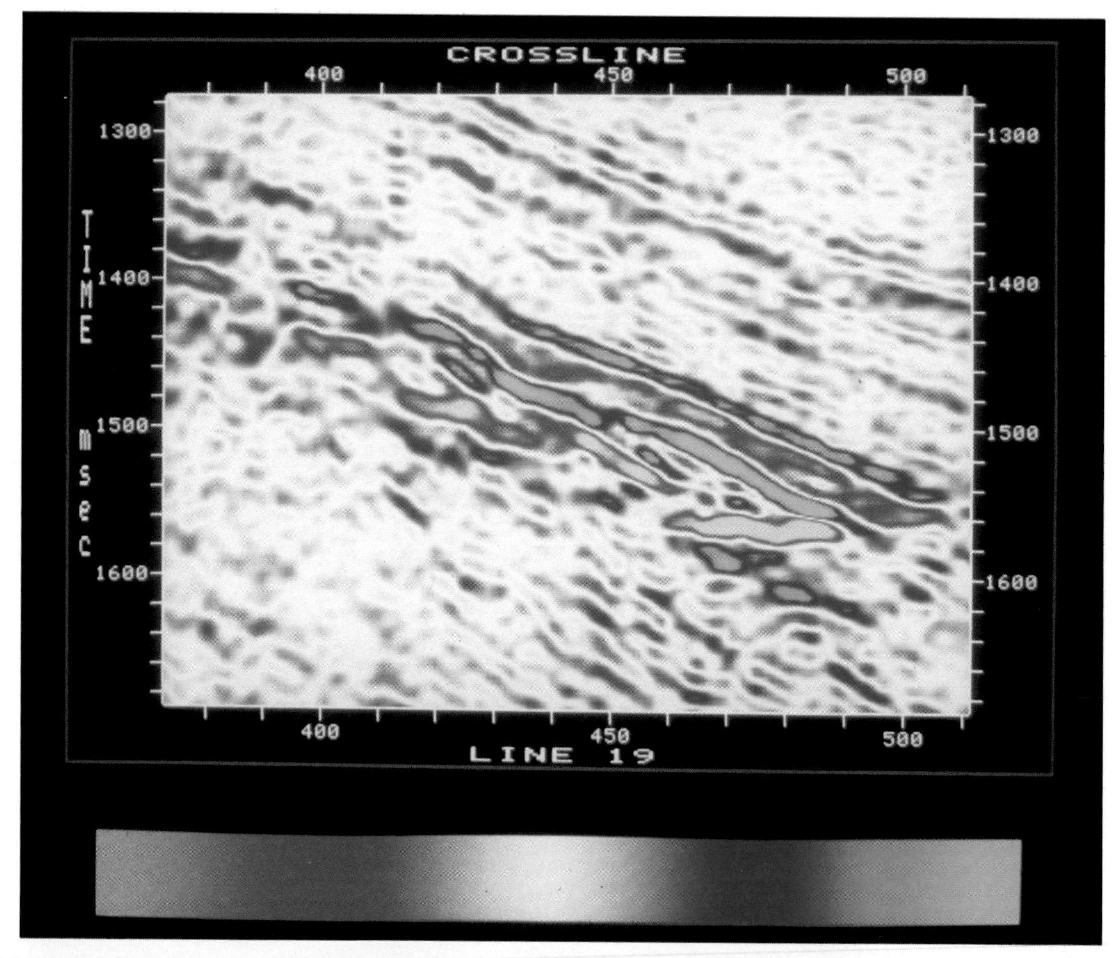

Fig. 2-10. Enhanced dynamic range gradational color scheme where cyan has been added for the highest positive amplitudes and yellow has been added for the highest negative amplitudes. (Courtesy Chevron U.S.A. Inc.)

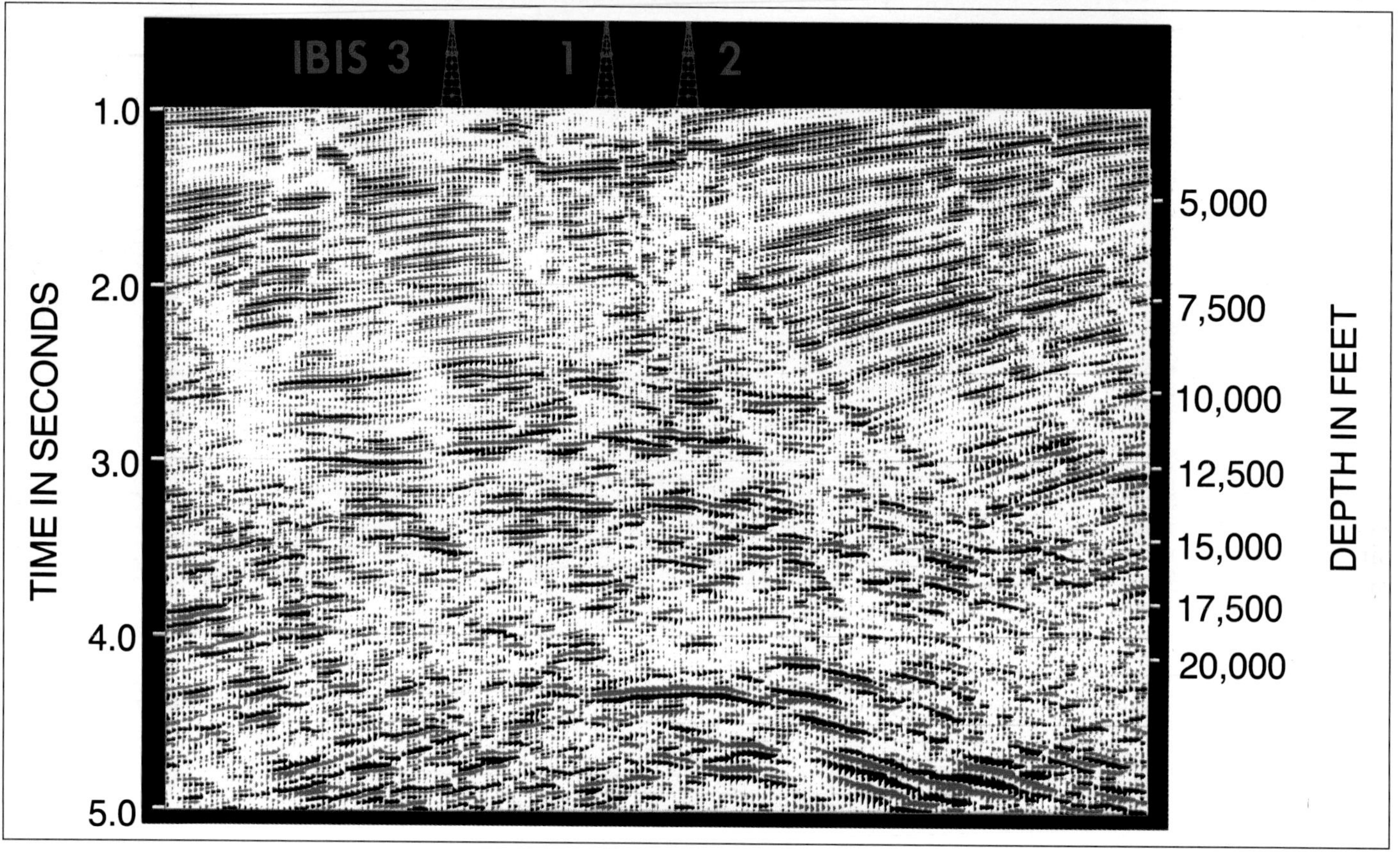

Fig. 2-11. Vertical section displayed in dual polarity variable area showing fault definition. (Courtesy Texaco Trinidad Inc.)

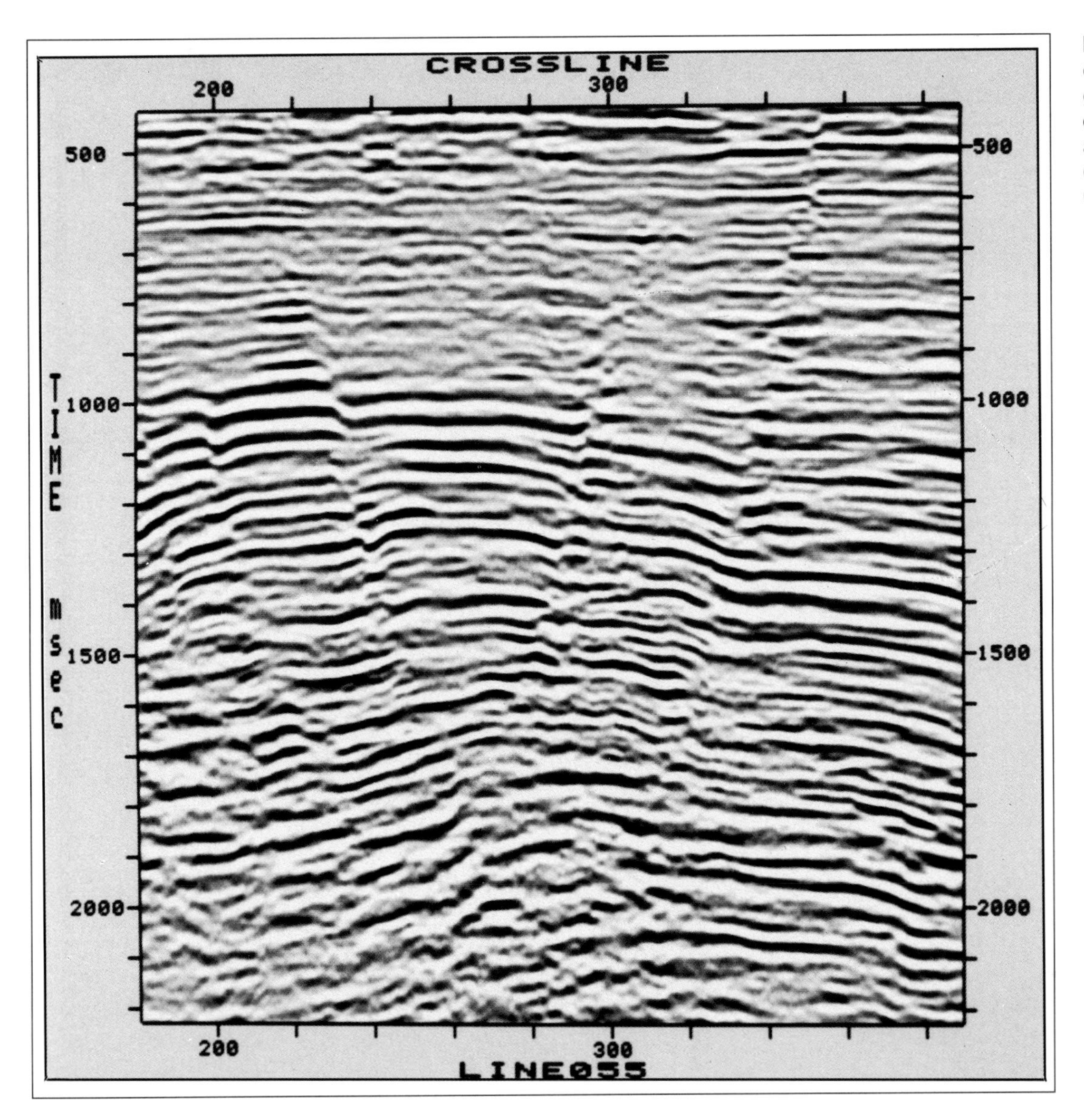

Fig. 2-12. Vertical section displayed with single gradational gray scale in order to enhance low amplitude events. (Courtesy Texas Pacific Oil Company Inc.)

Crosslines 180 and 270. However, after crossing two faults, a curvilinear feature can be seen continuing to the upper right to Line 122, Crossline 330. Is this a continuation of the channel system even though the amplitude is much reduced? We do not know the answer to this question, but we have been able to observe the continuity of this extensive curvilinear feature because of the use of gradational color.

Figure 2-16 shows the same section in contrasting colors and the detectability of the inferred channel is much reduced. In fact the eye tends to be drawn to the red and pink circular maxima at Crossline 250 between Lines 45 and 60 rather than the longer arcuate high amplitude trends. When applying a gradational color scheme to a horizon slice showing the spatial amplitude distribution of one trough, as in Figure 2-15, a single gradational color scheme should be used to match the one-sided range of amplitudes. For the horizon slice for a peak (for example Figure 4-28) the same principle applies but the color scheme should be inverted.

Assessment of Zero-Phaseness

Most interpreters today prefer zero-phase data. The reasons they give to support this preference include the following:

(1) the wavelet is symmetrical with the majority of the energy being concentrated in the central lobe;

(2) this wavelet shape minimizes ambiguity in associating observed waveforms

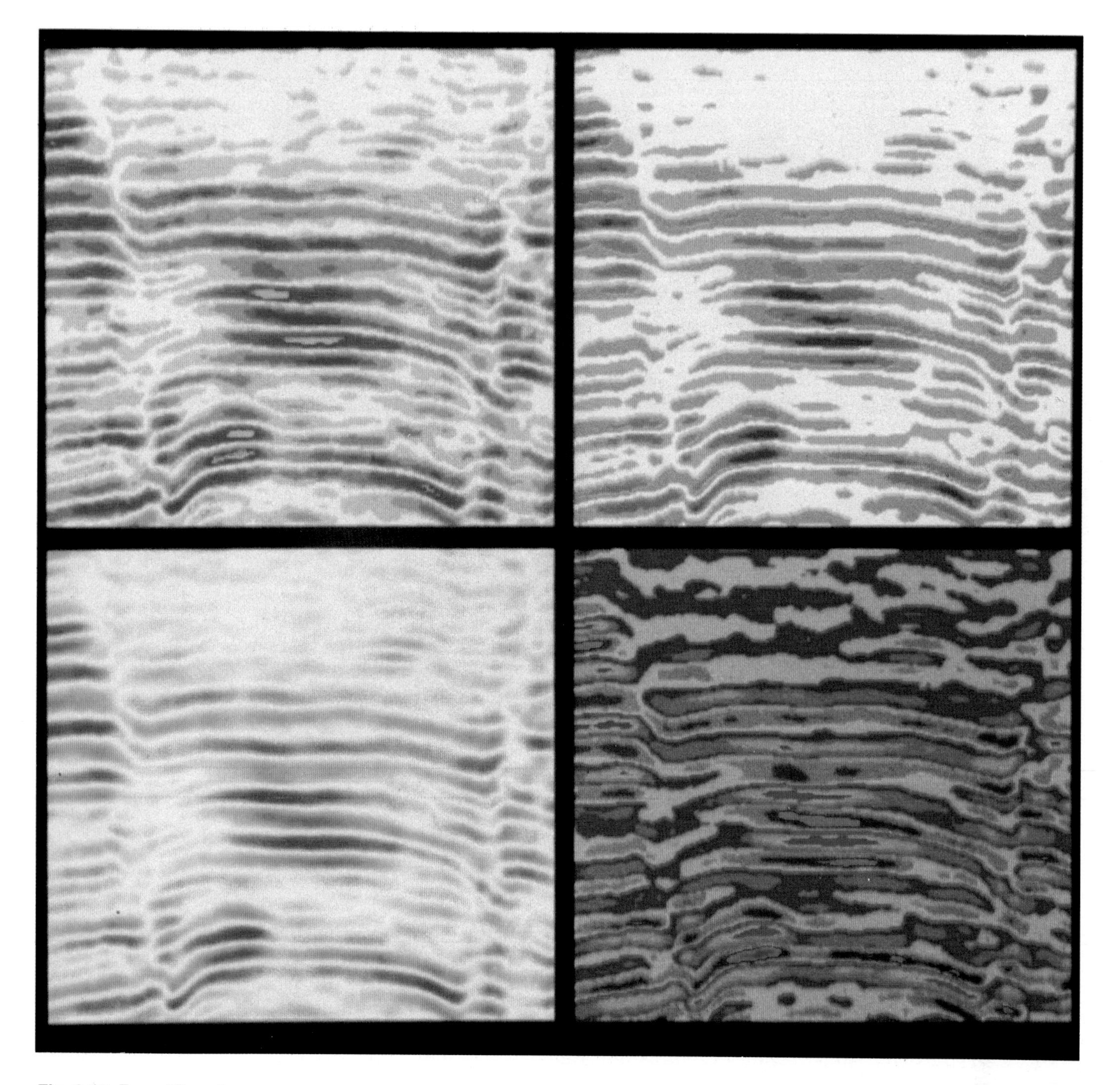

Fig. 2-13. Four different color schemes applied to the same vertical section segment. (Courtesy Texas Pacific Oil Company Inc.)

with subsurface interfaces;

(3) a horizon track drawn at the center of the wavelet coincides in time with the travel time to the subsurface interface causing the reflection;

(4) the maximum amplitude occurs at the center of the waveform and thus coincides with the time horizon; and,

(5) the resolution is better than for other wavelets with the same frequency content.

Much data processing research has been devoted to wavelet processing, which can be defined as the replacement of the source wavelet, the receiver response, and the filtering effects of the earth by a wavelet of known and desirable characteristics. Wood (1982) outlined the principles of wavelet processing and the properties of zero-phase wavelets, and Kallweit and Wood (1982) addressed the issues of resolution. Some relevant processing issues are discussed in Appendix A. Today's interpreter, particularly one who has a stratigraphic objective, wants to be able to assess whether the data pro-

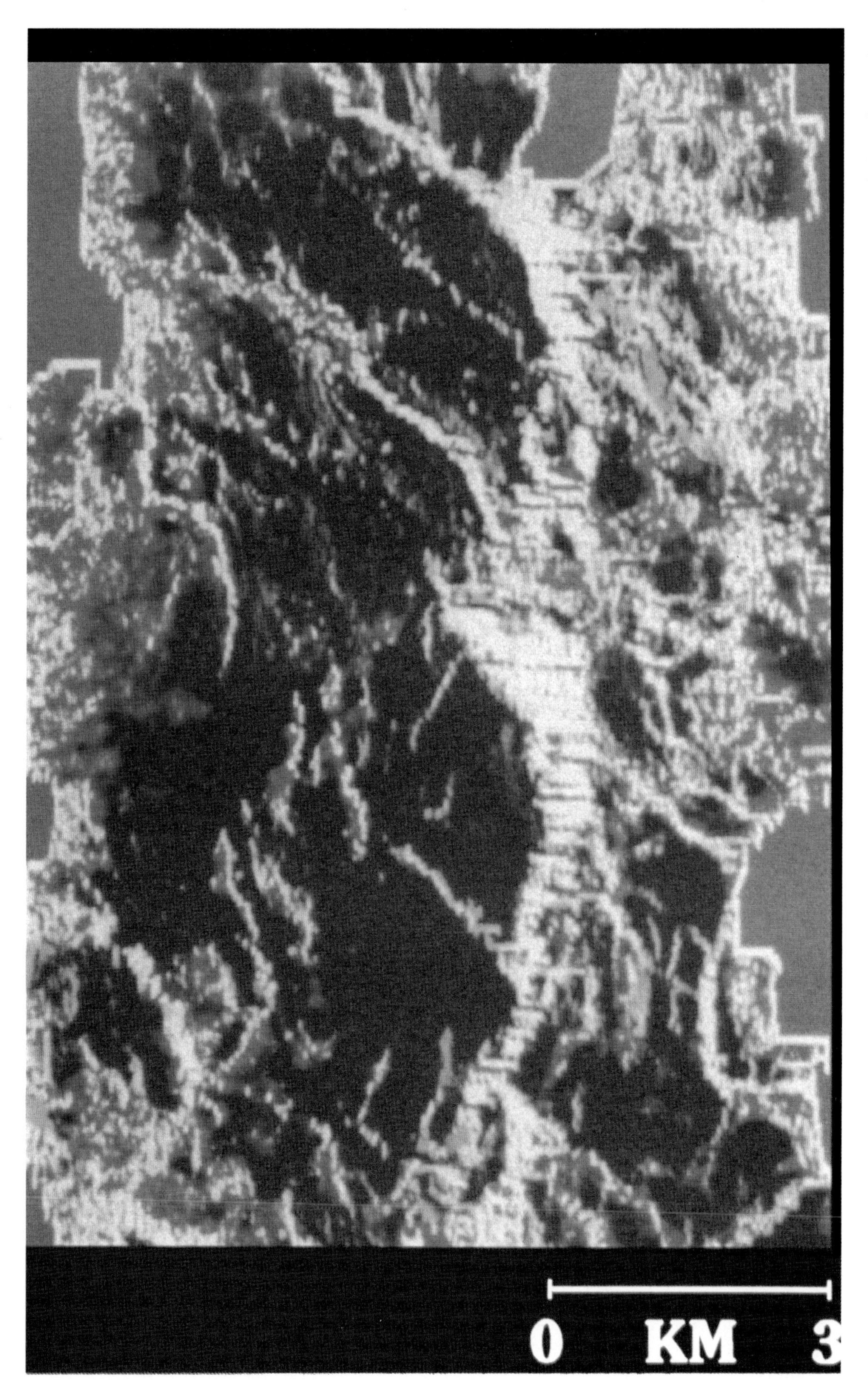

Fig. 2-14. Horizon slice from the North Sea displayed in a sun-shaded color scheme thus helping to reveal subtle fault lineations. (Courtesy ARCO Oil and Gas Company.)

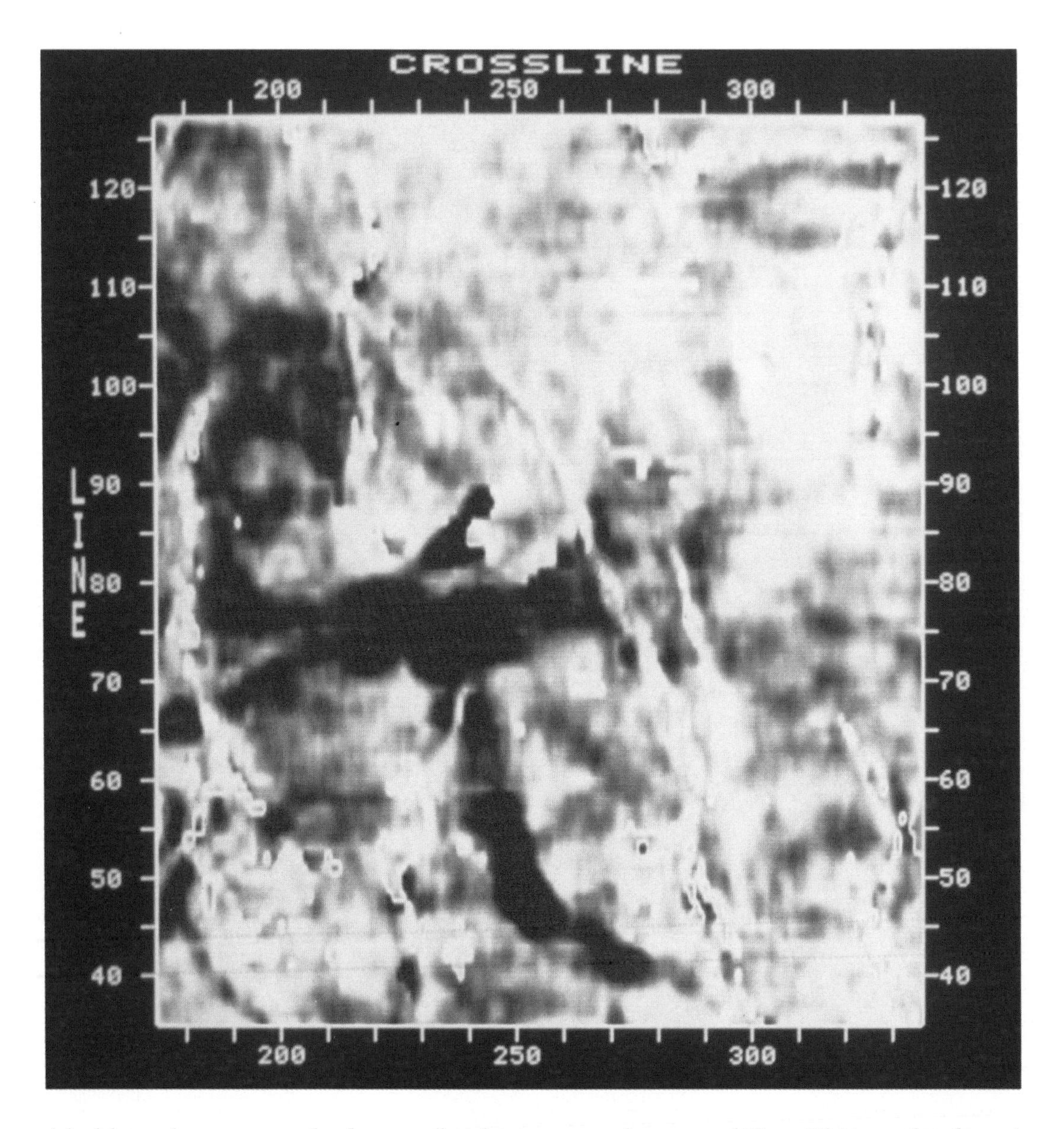

Fig. 2-15. Horizon slice showing an inferred channel system displayed with a gradational color scheme. (Courtesy Texas Pacific Oil Company Inc.)

vided have been properly deconvolved to a zero-phase condition. This can be done in many ways. Cross-correlation of a synthetic seismogram with the seismic trace at the well location is a good analytical technique. So is the extraction of a wavelet from the data and the study of its shape. But whatever is done, today's interpreter needs an increased awareness of zero-phaseness and the ability to recognize it, or other phases, in his or her data.

Understanding wavelet phase gives increased importance to the understanding of polarity and the adoption of a consistent polarity convention. For processed seismic data, polarity convention is confused, and in addition color display introduces the need for conventions in color usage. In presenting an interpretation using colored sections, the critical issue is to communicate the polarity and color usage for that data. It is less important what conventions are used because peaks and troughs are equally visible in color display. The author has developed a subjective appraisal of the polarity and color conventions in use today and these are diagrammed in Figure 2-17. The author's "normal" conventions are that a positive reflection coefficient arising from an increase in acoustic impedance is portrayed as a trough and is colored red. (Beware: not every display in this book uses these same conventions.)

The interpretive assessment of zero-phaseness requires high signal-to-noise ratio reflections and maximum dynamic range color display. But first zero-phaseness will be considered on model data. Figure 2-18 shows three zero-phase wavelets and their

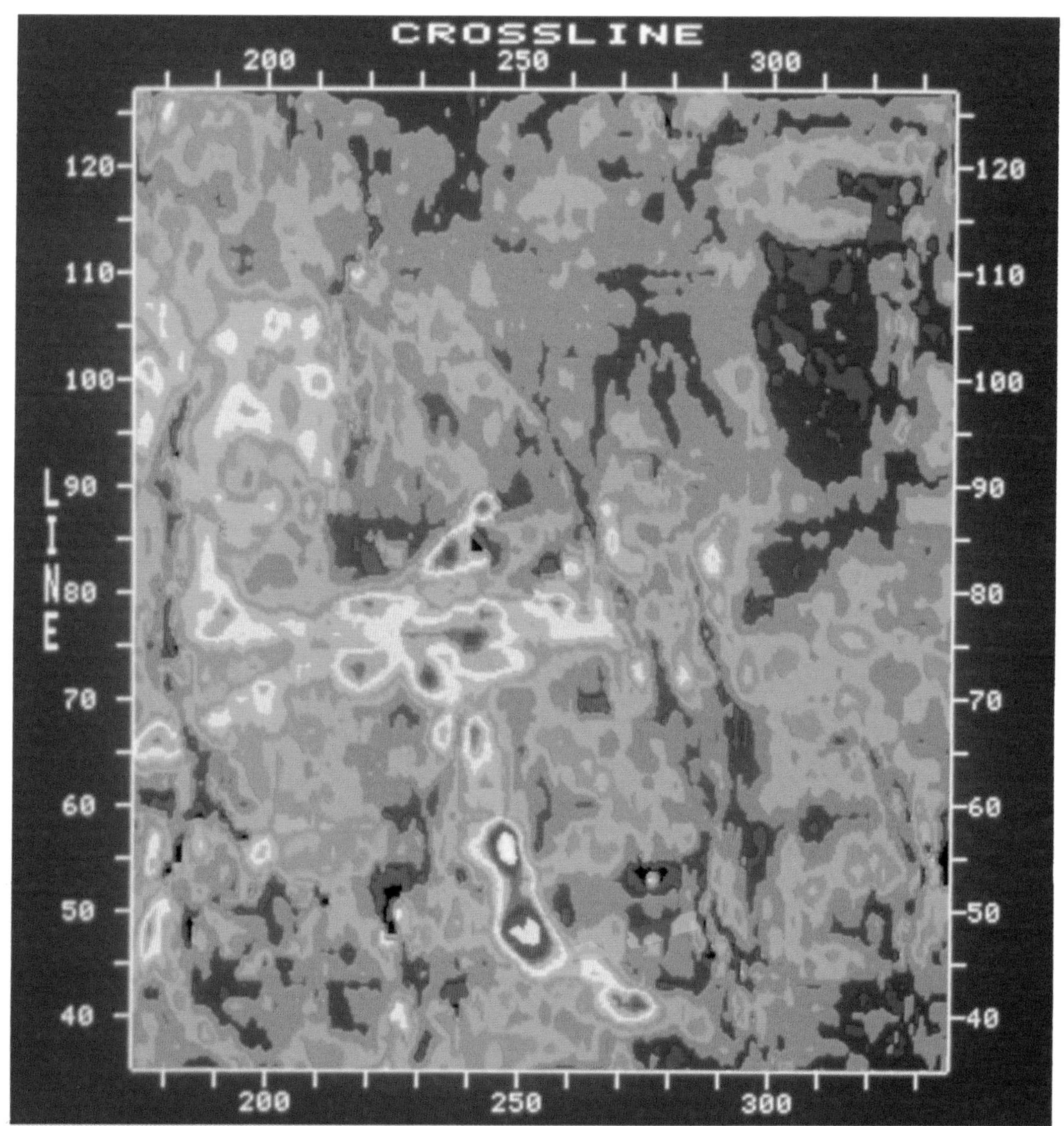

Fig. 2-16. Same horizon slice as in Figure 2-15 displayed with a contrasting color scheme, which reduces visibility of the channel system. (Courtesy Texas Pacific Oil Company Inc.)

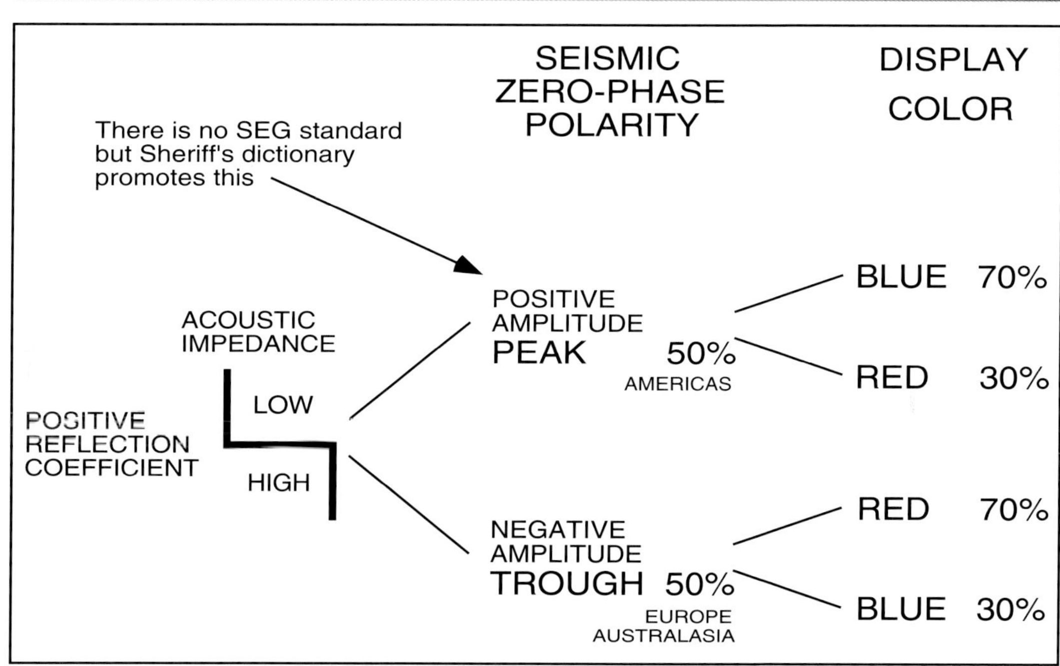

Fig. 2-17. Polarity and color conventions. *Low* and *high* refer to acoustic impedance and thus represent an impedance increase. The percentages refer to estimated worldwide industry usage. Continents refer to preferences in those areas. The colors *blue* and *red* refer to the extremities of a double-gradation color scheme broadly similar to that shown elsewhere in this book. Conventions most common in this book are that a positive reflection coefficient is red; thus a negative reflection coefficient is blue. See Figure 5-5.

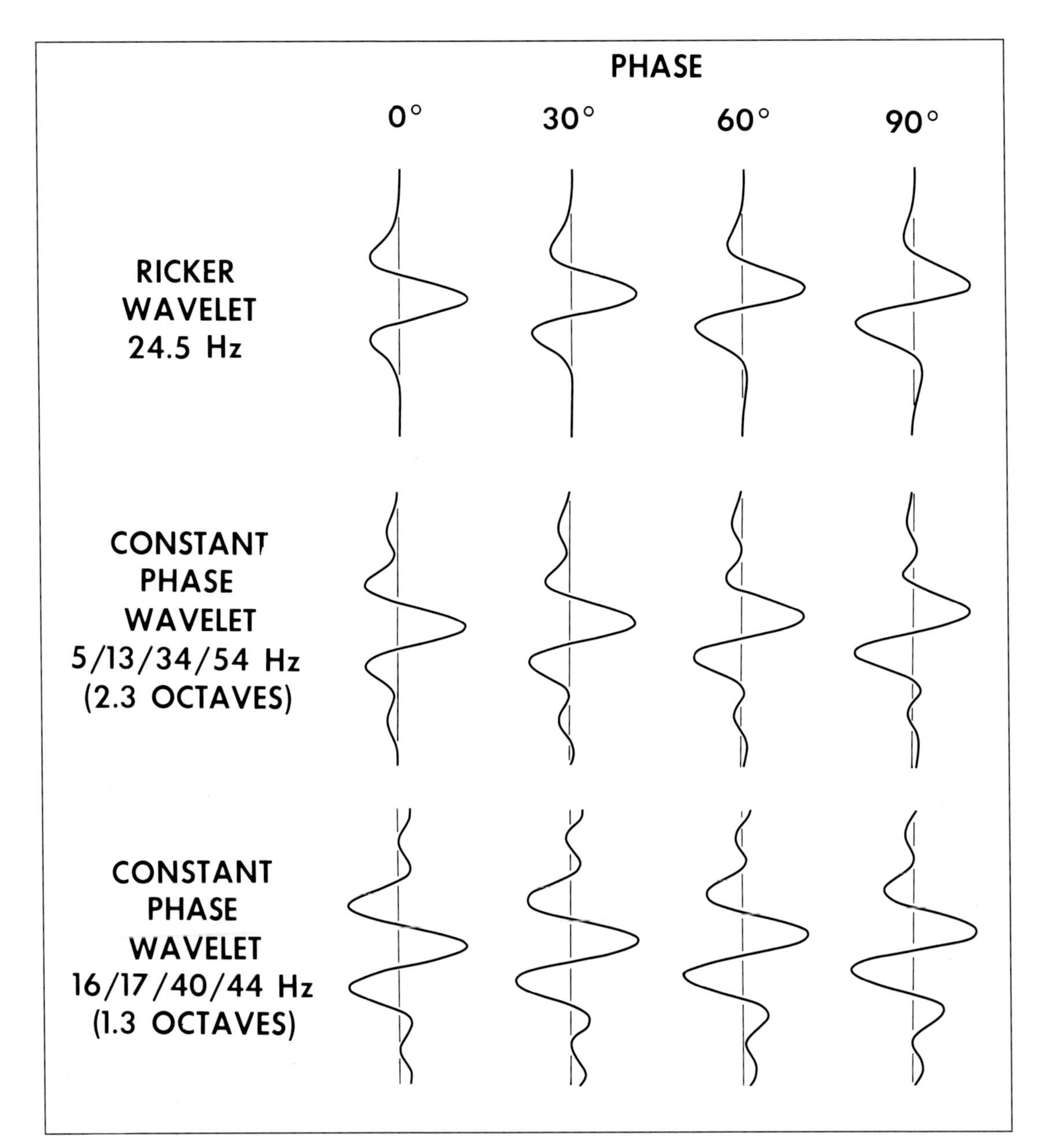

Fig. 2-18. Effect of phase shifting constant phase wavelets.

equivalents shifted by 30, 60, and 90 degrees. The first is a Ricker wavelet, the second is derived from a bandpass filter of 2.3 octaves with gentle slopes, and the third is derived from a bandpass filter of 1.3 octaves with steep slopes. The common property of these three wavelets is that the separation of central peak and first side lobe is the same for each—16 ms. The Ricker wavelet has no side lobes beyond the first. The 2.3 octave wavelet is a good wavelet extracted from actual processed data and has low side lobes. The 1.3 octave wavelet is a poor wavelet with relatively high side lobes.

The visual assessment of zero-phaseness amounts to a visual assessment of wavelet symmetry. In these model examples 30° of distortion is visible for all the wavelets but the higher side lobe levels of the narrower band wavelet make the distortion less pronounced. For the larger distortions, for example at 60°, the central peak and the larger side lobe are more easily confused for the narrower band wavelet, so in practice it may be difficult to decide whether the peak or the trough is the principal extremum. At a distortion of 90° the time horizon lies at the zero crossing between the largest amplitude peak and trough, and these are of equal size.

Figure 2-19 is a single trace example from real data where there was a known low velocity gas sand. The top of the low velocity zone should be a peak and the base a

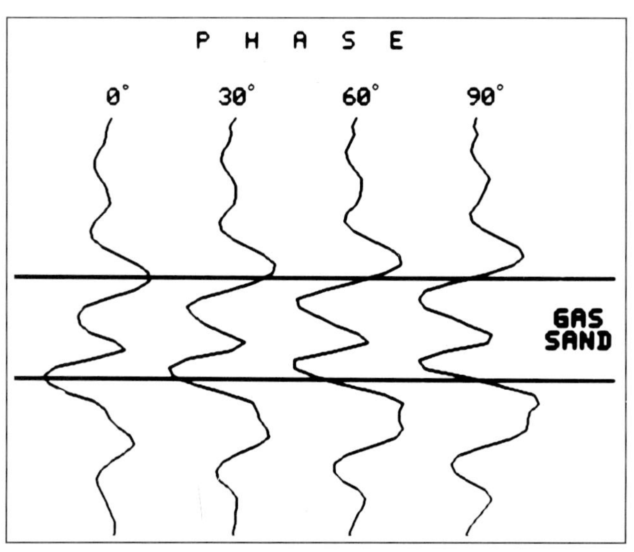

Fig. 2-19. Effect of phase shifting a real data trace showing reflections from the top and base of a gas sand. (Courtesy Chevron U.S.A. Inc.)

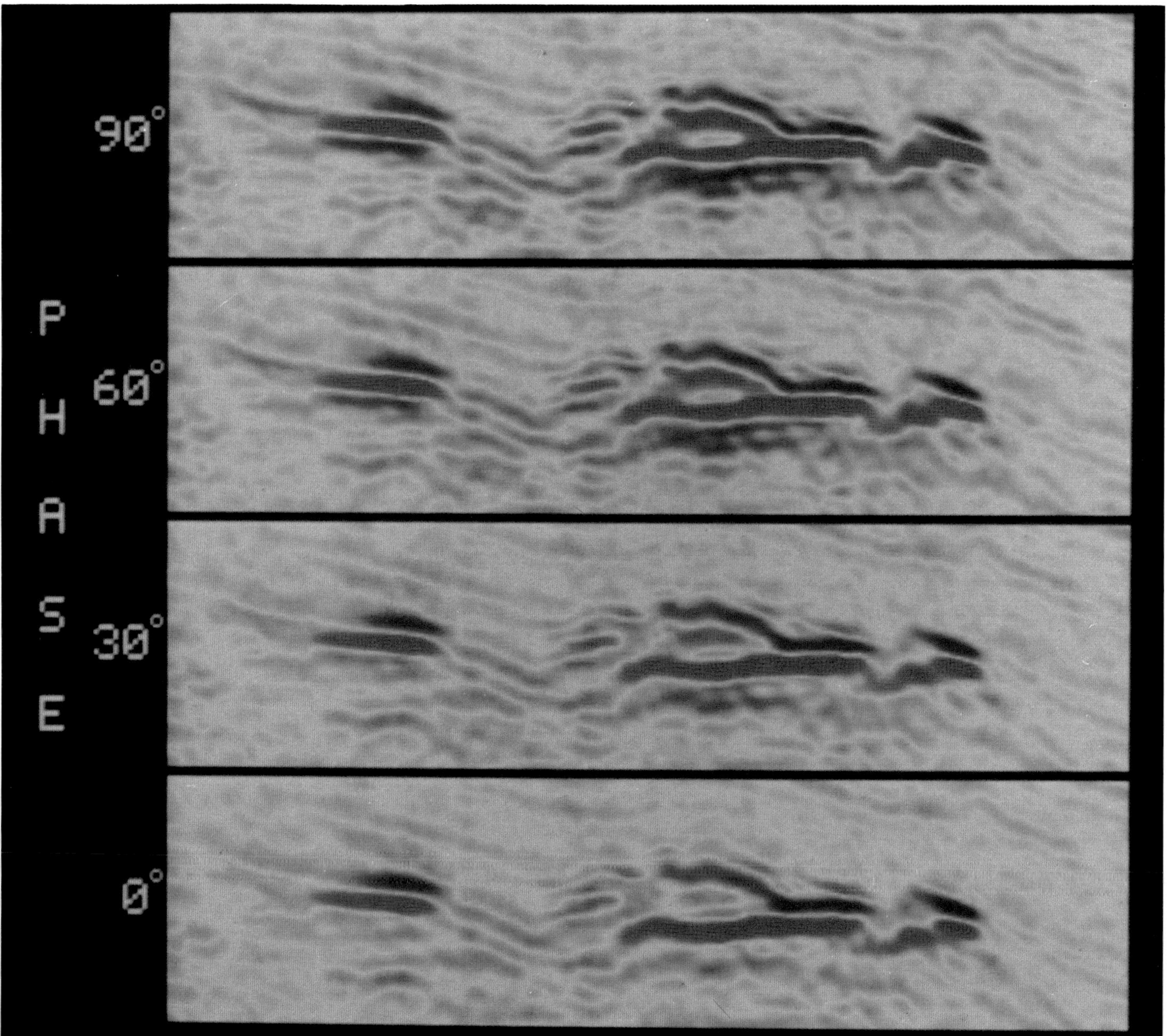

Fig. 2-20. Bright reflections from the top and base of a gas sand with constant phase shifts applied. (Courtesy Chevron U.S.A. Inc.)

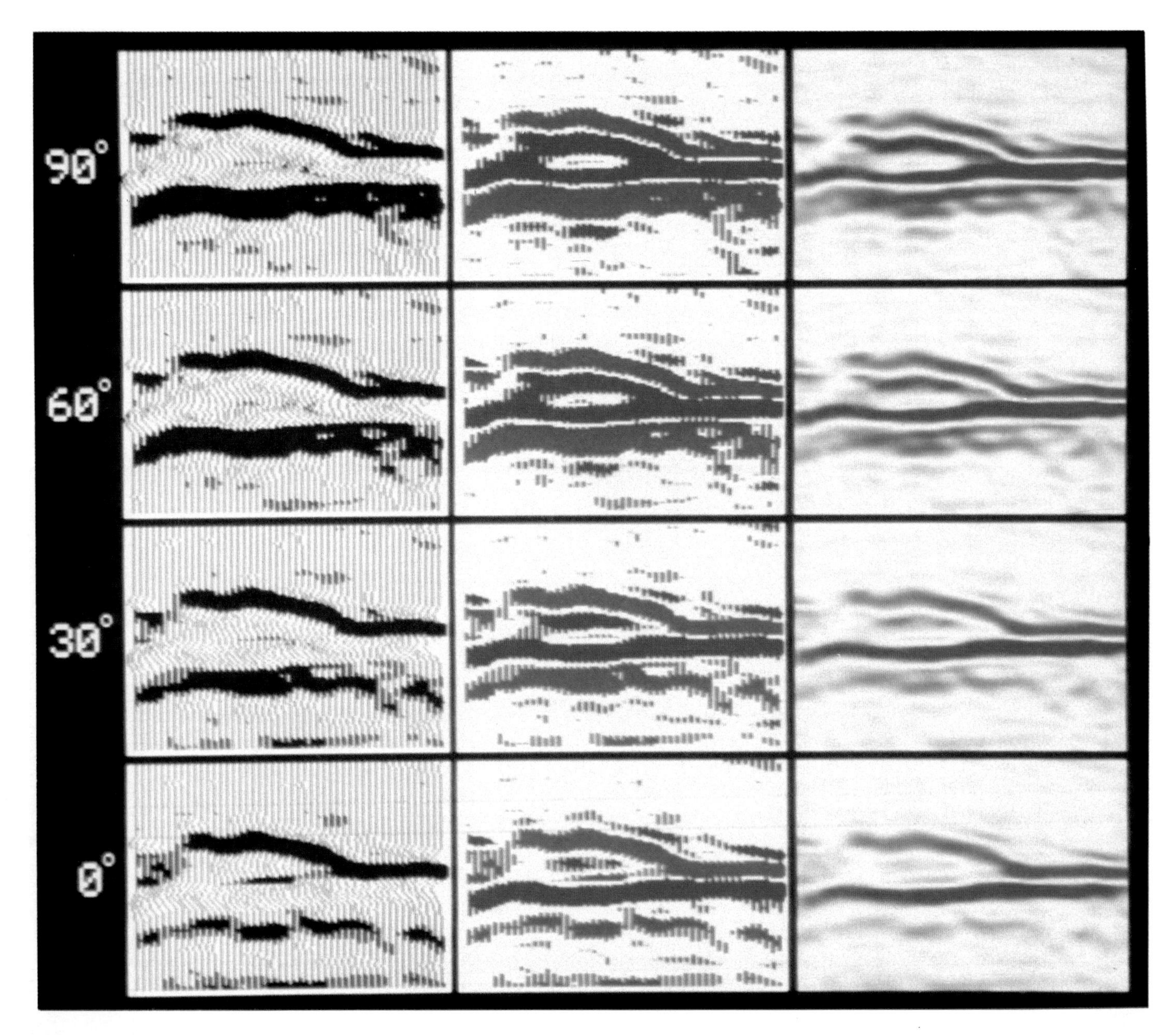

Fig. 2-21. Comparison between variable area/wiggle trace, dual polarity variable area and gradational color for the interpretive assessment of zero-phaseness. (Courtesy Chevron U.S.A. Inc.)

trough (according to the polarity convention considered normal in this book). The trace labelled 0° shows peak and trough each symmetrically placed over their corresponding interfaces. The phase distortions are again fairly evident when presented in this way.

In practice, interpreters must assess zero-phaseness on a section containing many traces in case one trace is unrepresentative. We select a high amplitude reflection, which, on the basis of a simple model, can be related to a single interface. The interpreter can then assume that the interference of events from adjacent parallel interfaces, multiples or noise is small. Figure 2-20 illustrates a bright spot from a gas reservoir where it is assumed that the above conditions hold. In the panel labelled 0° there is one blue event from the top of the reservoir and one red event from its base, and they have approximately the same amplitude. Side lobes are low and symmetrical as far as can be determined. This is the signature expected for the zero-phase response of a gas sand.

For the 90° case in Figure 2-20 the top of the gas sand has a signature of peak-over-trough and the base one of trough-over-peak. This confirms the modeling illustrated in Figure 2-18 and certainly shows a more complex character than the zero-phase sec-

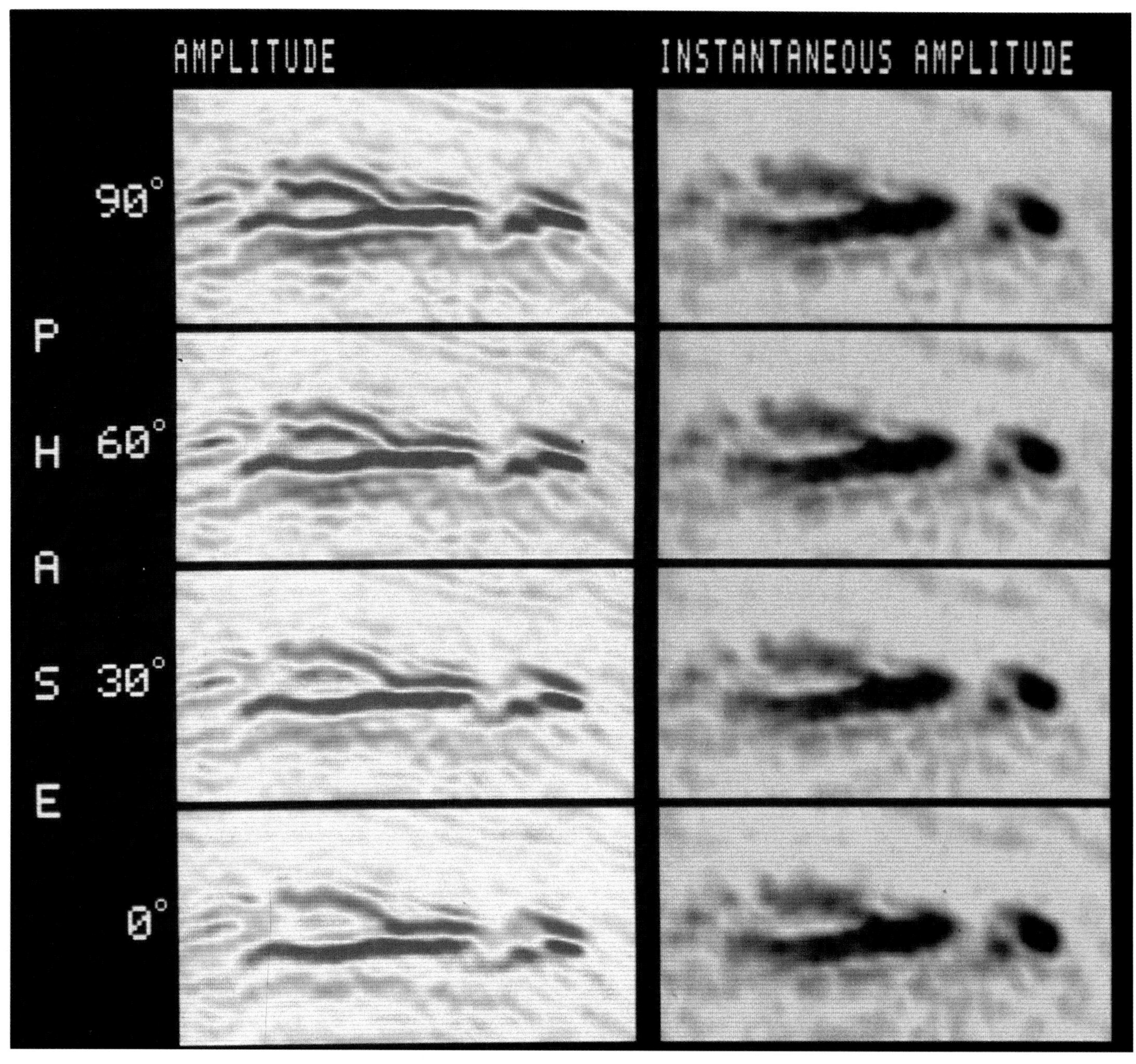

Fig. 2-22. Use of instantaneous amplitude or envelope amplitude to obscure the effects of phase distortion. (Courtesy Chevron U.S.A. Inc.)

tion. The intermediate levels of phase distortion show the progression from the 0° to 90° condition. Observation of these more complicated phase characteristics can be followed by experimental phase rotation of the data before interpretation of the data set commences in earnest.

The interpreter's ability to make this kind of assessment of zero-phaseness depends critically on the display used. Figure 2-21 presents the same data panel in the same four phase conditions for three different modes of display. Variable area/wiggle trace demonstrates how the visual imbalance between peaks and troughs makes the assessment of relative amplitudes extremely difficult. Dual polarity variable area has corrected the visual imbalance but demonstrates the limited dynamic range of variable area as the bright events are all saturated. Gradational color demonstrates the visual balance between peaks and troughs and also the improved dynamic range. Relative amplitudes of peaks, troughs and side lobes can now be assessed with maximum available clarity for fairly high trace density. One disadvantage, however, of gradational color display is the stringency imposed on the reproduction process. The illus-

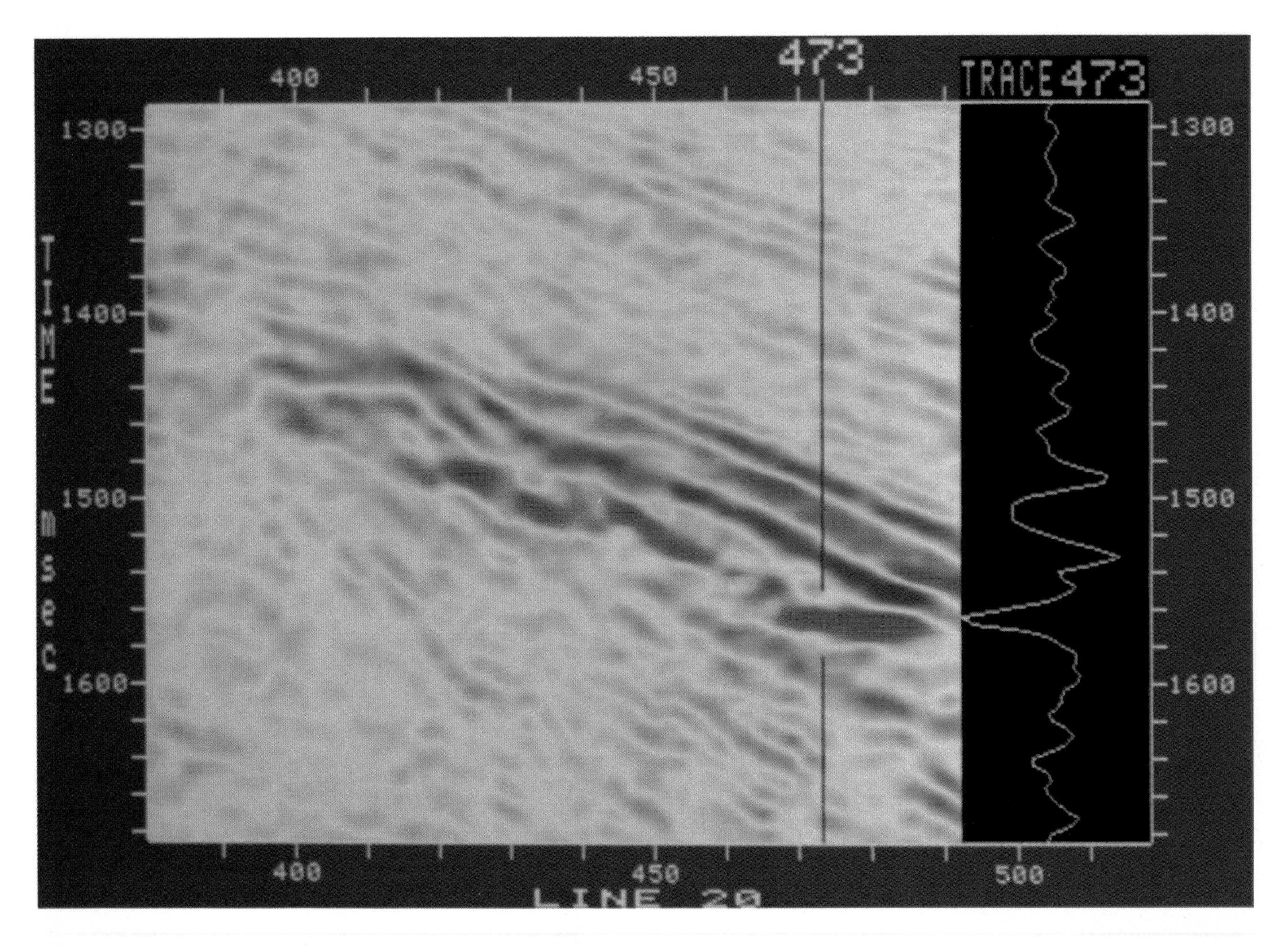

Fig. 2-23. Flat spot reflection displaying zero-phaseness, visible in gradational red for many traces and in wiggle format for one trace. (Courtesy Chevron U.S.A. Inc.)

tration that you, the reader, are studying is of reduced quality compared to the screen image of the color monitor on which the original assessment was made.

If the phase of the data is unknown and cannot be assessed, instantaneous amplitude (also known as envelope amplitude or reflection strength; Taner and Sheriff, 1977) provides a display in which amplitude can be studied independent of phase. Figure 2-22 shows identical instantaneous amplitude sections corresponding to the four regular amplitude sections with different phases.

Any high amplitude reflection which can be assumed to originate from a single interface is usable for assessing zero-phaseness when displayed in color. A fluid contact reflection, or flat spot, is normally an excellent candidate. If the structural horizons have moderate dip and the reservoir is fairly thick, the flat spot reflection will be well resolved and structurally unconformable. (The characteristics of fluid contact and other reservoir reflections are discussed more extensively in Chapter 5.) The flat spot in Figure 2-23 shows clearly one high-amplitude symmetrical red trough, indicating that the data are at least close to zero phase. The flat spot in Figure 2-24 shows a high-amplitude trough-over-peak (red-over-blue) character, indicating an approximately 90° phase condition.

Figure 2-25 illustrates diagrammatically the sources of seismic reflections that often have a sufficient signal-to-noise ratio to be useful for interpretive phase assessment. Top of salt is good but may not be smooth enough. The water bottom is good only in deep water when unaffected by the processing ramp. Basement, limestone, and coal are useful but the single interface assumption may not hold. Shallow gas is excellent.

Figure 2-26 shows reflections from the top and base of shallow gas. Both are double events indicating a phase of 90°. Figure 2-27 shows strong reflections from a Miocene

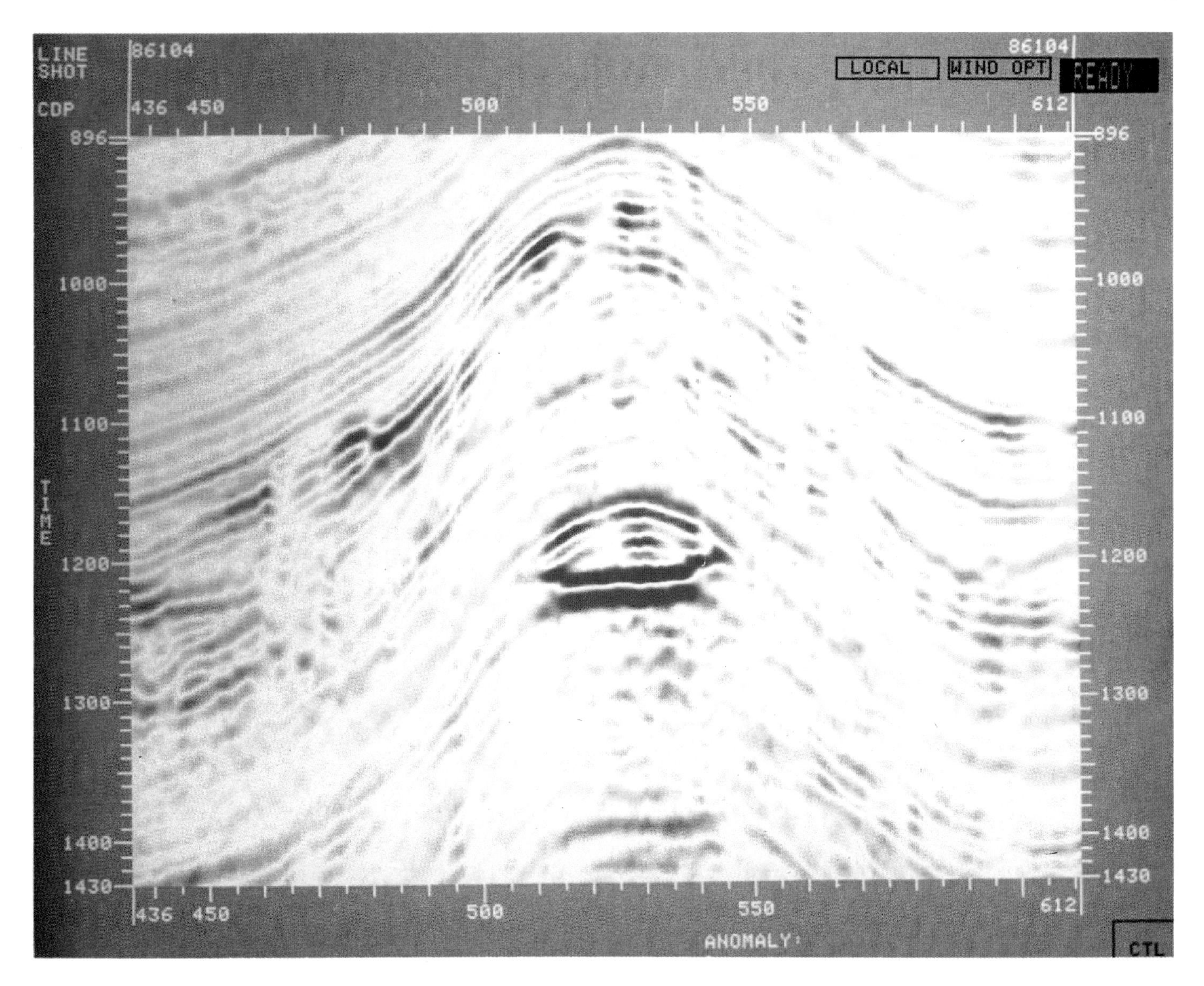

Fig. 2-24. Gulf of Mexico flat spot displaying a phase of approximately 90°. (Courtesy Geophysical Service Inc.)

gas reservoir. Here the reservoir is thin so that the reflection from the top and the reflection from the bottom overlap each other, thus giving reinforcement of the red/yellow trough in the center. This again is an indication of 90° phase data, a remarkably common phenomenon.

Figure 2-28 shows an outstanding basement reflection which is probably also from a single subsurface interface. The waveform of the reflection is clear, almost symmetrical, and spatially consistent. This indicates that the data are close to zero phase, at least around the time of 3 seconds. Figure 2-29 shows a strong water bottom reflection in deep water which is also a clear symmetrical waveform, again indicating zero-phaseness. On top of the seamount the phase assessment is ambiguous but at the red arrow the zero-phaseness is evident.

Psychological Impact of Color

Studies on the psychological impact of color have shown that hues of yellow, orange and red are advancing and attracting, while hues of green and blue are cooler and receding. The interpreter can take advantage of this in communicating his results. It would seem logical to display the structural highs, the isopach thicks and the bright spots in advancing colors in order to promote their prospectivity. Figure 2-6 is a structure map which demonstrates this point.

Figures 2-30, 2-31 and 2-32 are the same horizon slice displaying reflection ampli-

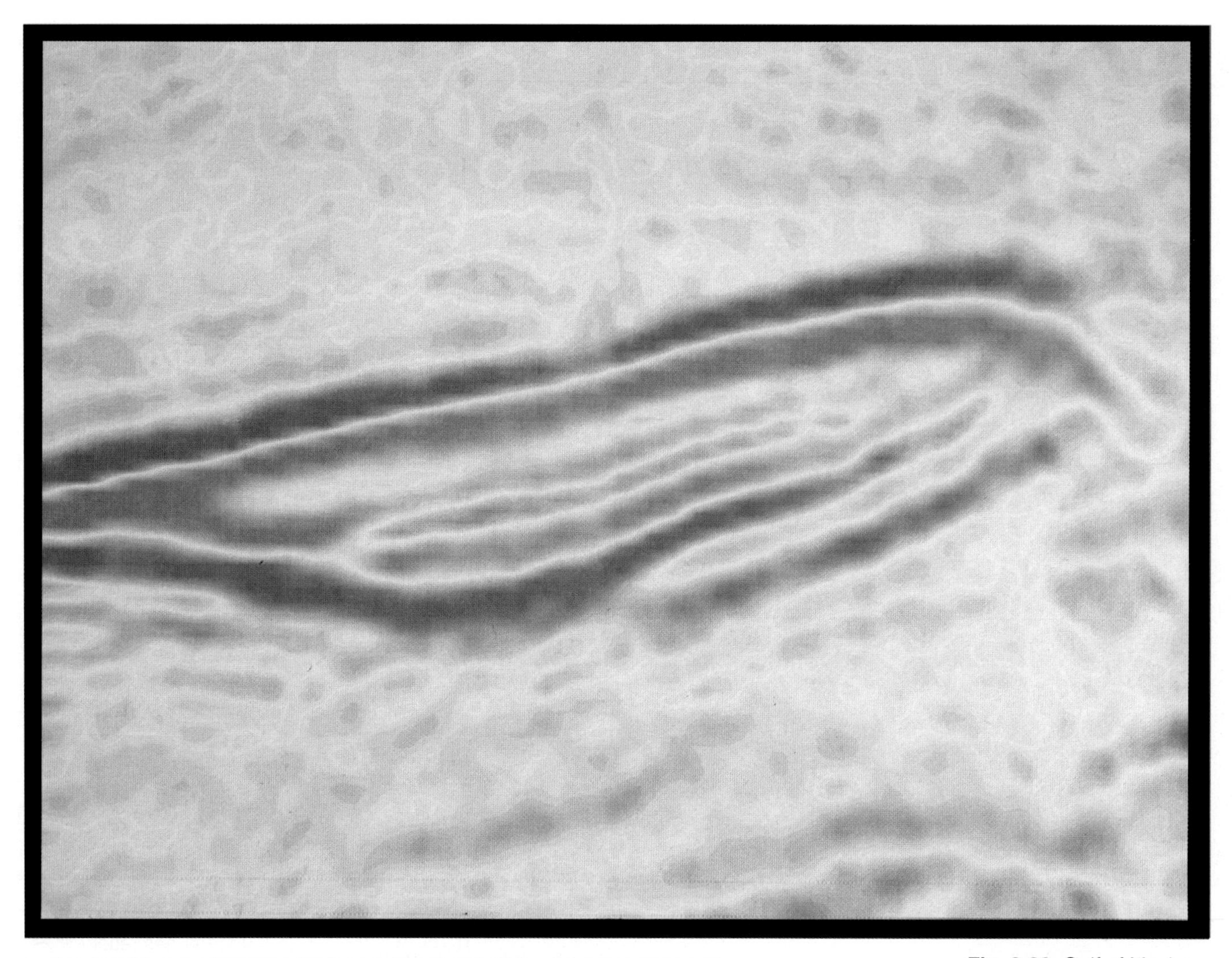

Fig. 2-26. Gulf of Mexico shallow gas reflections showing a phase of approximately 90°. (Courtesy Mobil Exploration & Producing U.S. Inc.)

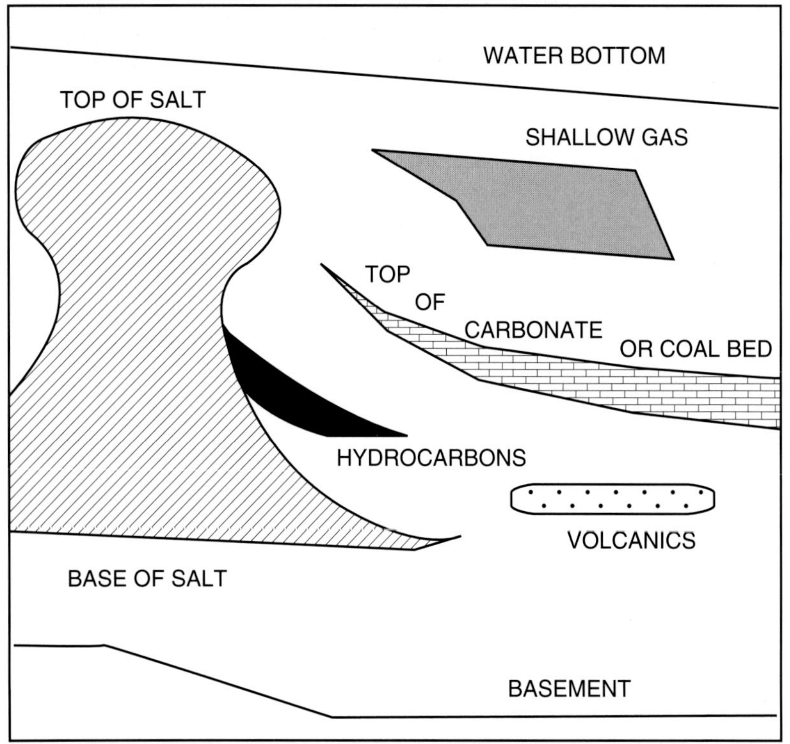

Fig. 2-25. Sub-surface features which can generate sufficiently high amplitude, single interface reflections to be useful for interpretive phase assessment.

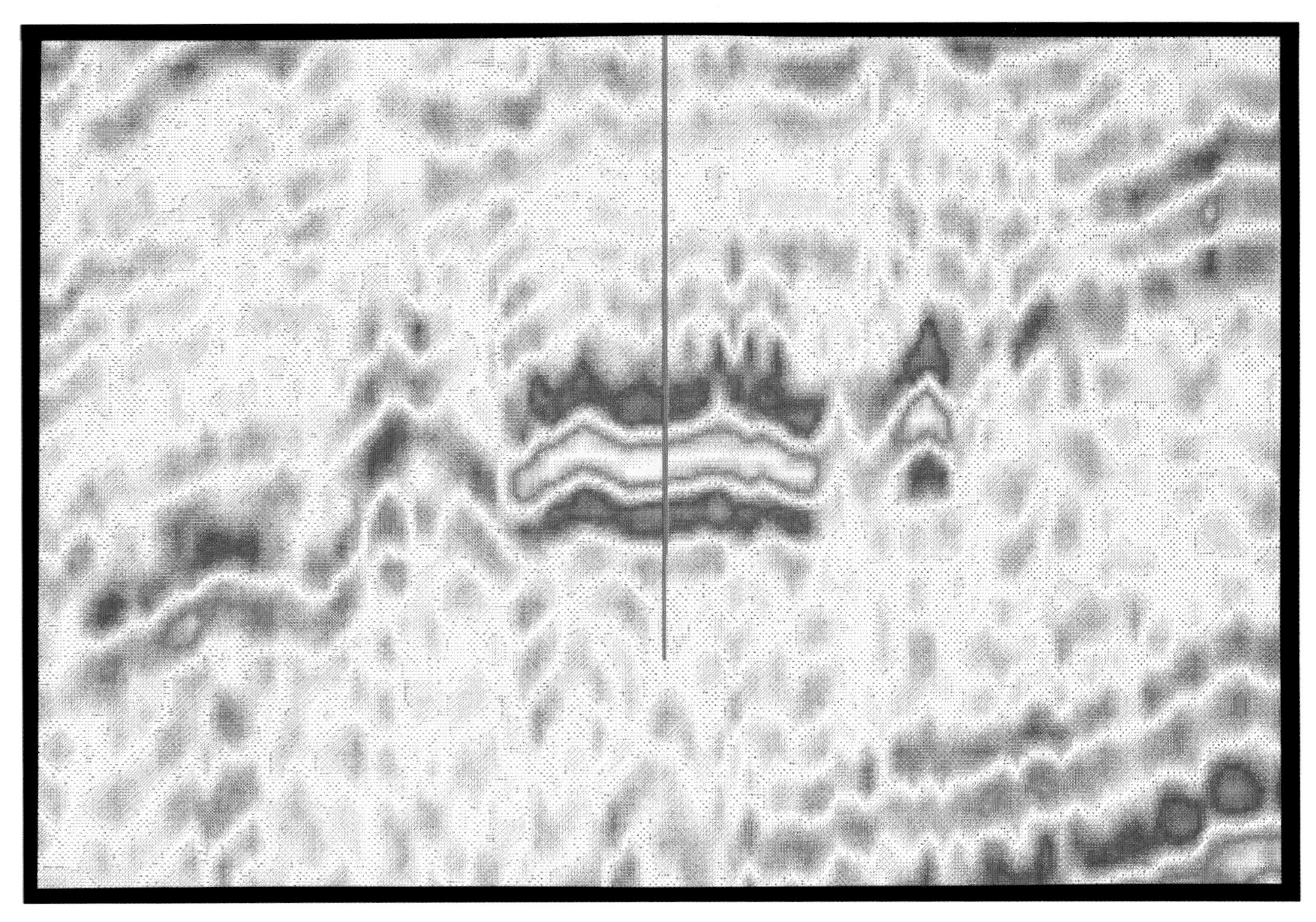

Fig. 2-27. Gulf of Mexico Miocene gas reservoir reflections showing a phase of approximately 90°. (Courtesy Conoco Inc. and Digicon Geophysical Corp.)

tude over a Gulf of Mexico reservoir, but presented with three different color schemes. In Figure 2-30 these data are represented in a green gradational scheme to accentuate the lineations due to faulting. The gradational colors accentuate these lineations by using the full dynamic range of color density and allow the eye to integrate all of the data quickly.

Figure 2-31 shows the same data displayed with a gradational color scheme using a wider range of hues. Now the relative strength of the amplitudes has much more impact on the eye; the advancing reds and yellows appear much more interesting than the cooler greens and blues. By using this scheme, the large anomaly near the top of the section draws considerable attention. A successful well was targeted and drilled, based on this display.

Yet another display of the same data (Figure 2-32) shows that a large area of high amplitude may be considered prospective. Here the low amplitude zones have been colored with fairly neutral grays. Further drilling potential can be considered on the basis of this display if amplitude strength is the key to developing this reservoir.

Thus one horizon slice was used for three different purposes by employing three different color schemes. The first drew attention to the faulting, the second to a particular anomaly, and the third to total drilling potential. Separate features of the data were enhanced differently by the different uses of color.

References

Backus, M. M., and R. L. Chen, 1975, Flat spot exploration: Geophysical Prospecting, v. 23, p. 533-577.

Balch, A. H., 1971, Color sonograms; a new dimension in seismic data interpretation:

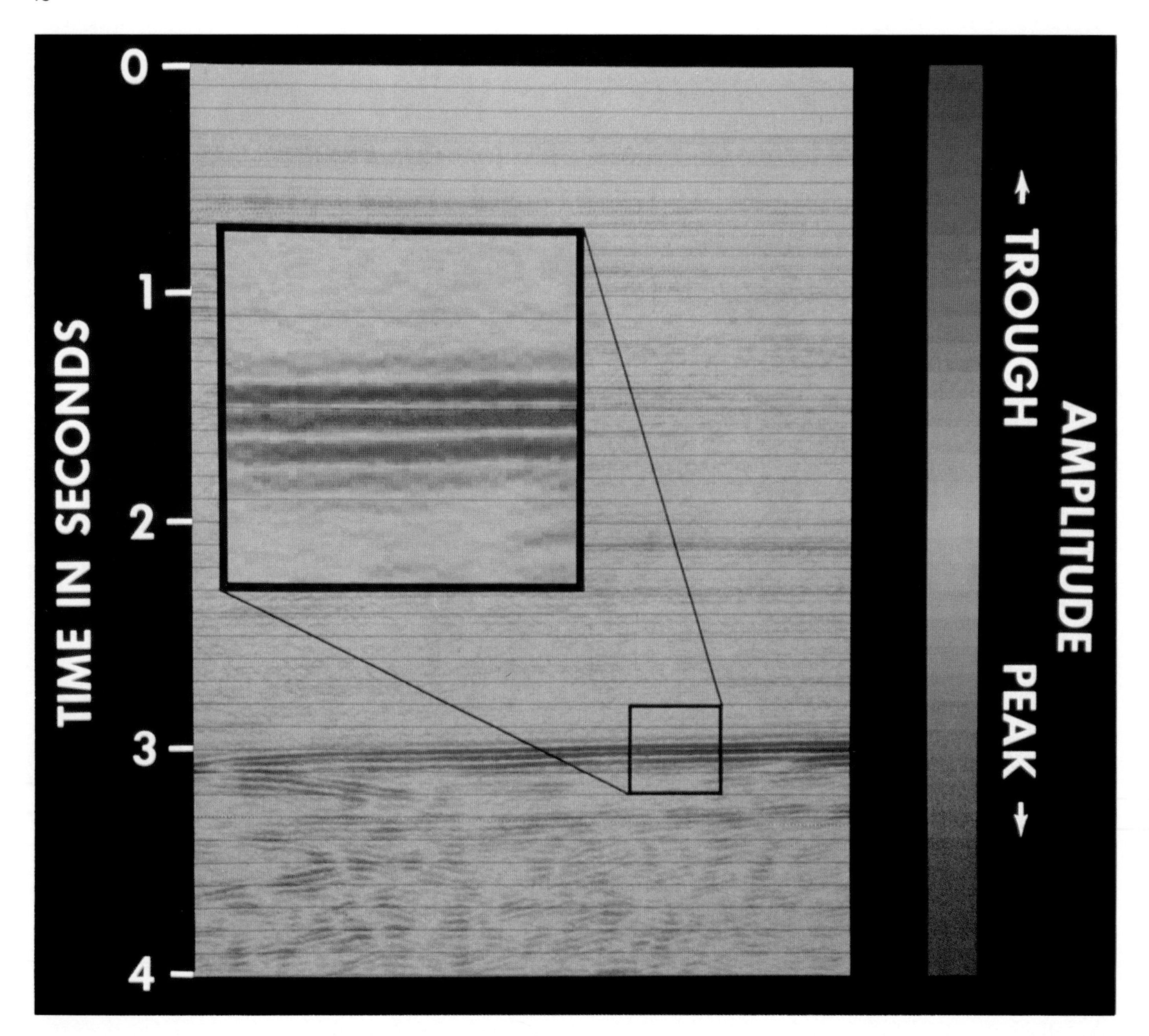

Fig. 2-28. Basement reflection displaying zero-phaseness. Note that the polarity of the data is the opposite of what is considered normal in this book, so the colors are reversed. (Courtesy Geophysical Service Inc.)

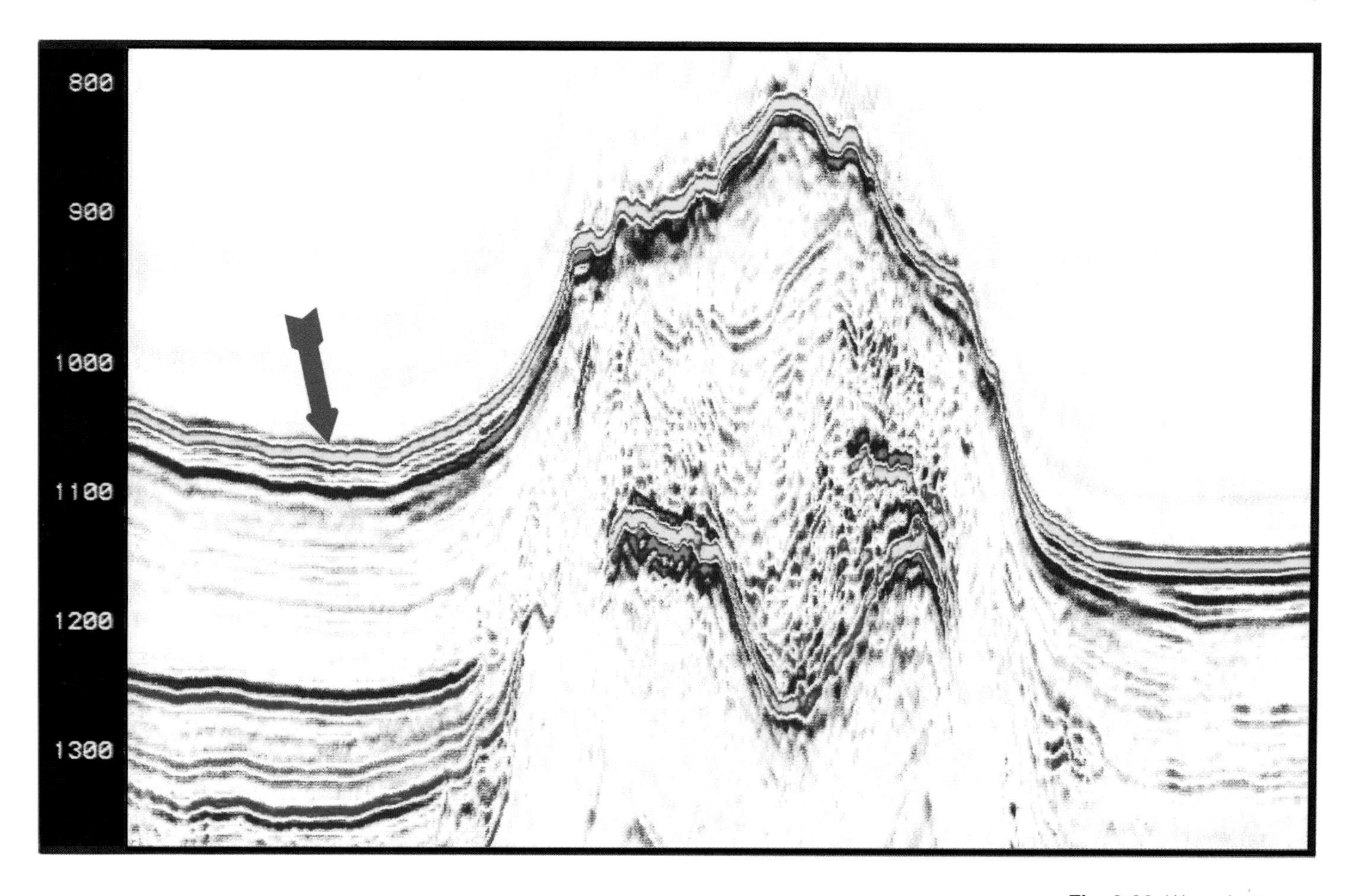

Fig. 2-29. Water bottom reflection in deep water displaying zero-phaseness. (Courtesy Conoco Inc.)

Geophysics, v. 36, p. 1074-1098.
Galbraith, R. M., and A. R. Brown, 1982, Field appraisal with three-dimensional seismic surveys offshore Trinidad: Geophysics, v. 47, p. 177-195.
Gerhardstein, A. C., and A. R. Brown, 1984, Interactive interpretation of seismic data: Geophysics, v. 49, p. 353-363.
Kallweit, R. S., and L. C. Wood, 1982, The limits of resolution of zero-phase wavelets: Geophysics, v. 47, p. 1035-1046.
Lindseth, R. O., 1979, Synthetic sonic logs—a process for stratigraphic interpretation: Geophysics, v. 44, p. 3-26.
Neidell, N. S., and J. H. Beard, 1985, Seismic visibility of stratigraphic objectives: Society of Petroleum Engineers Paper 14175.
Taner, M. T., and R. E. Sheriff, 1977, Application of amplitude, frequency and other attributes to stratigraphic and hydrocarbon determination, *in* C. E. Payton, ed., Seismic stratigraphy-applications to hydrocarbon exploration: AAPG Memoir 26, p. 301-327.
Wood, L. C., 1982, Imaging the subsurface, *in* K. C. Jain, and R. J. P. deFigueiredo, eds., Concepts and techniques in oil and gas exploration: Society of Exploration Geophysicists Special Publication, p. 45-90.

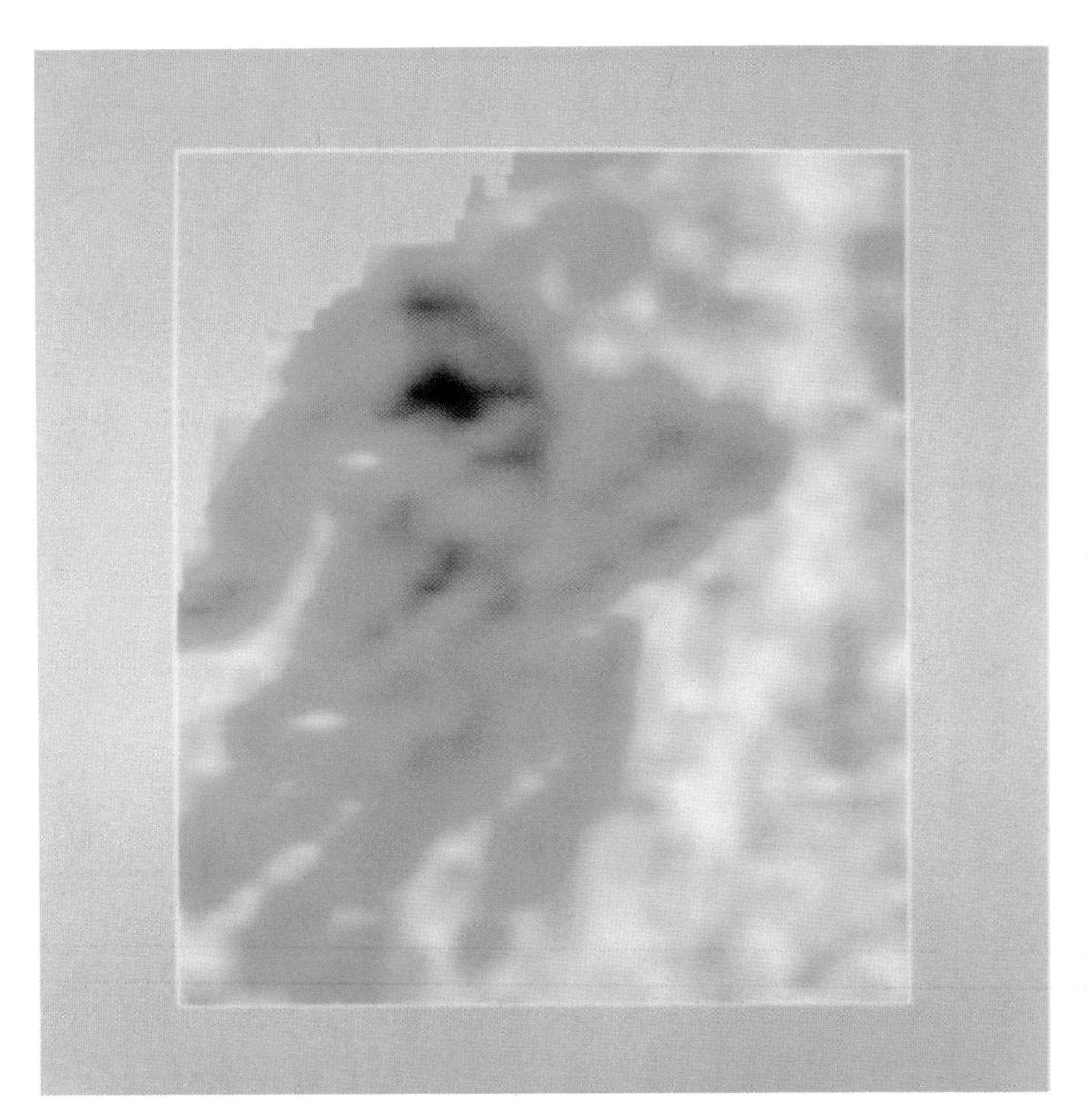

Fig. 2-30. Horizon slice displaying amplitude in gradational green to accentuate lineations due to faulting. (Courtesy Chevron U.S.A. Inc.)

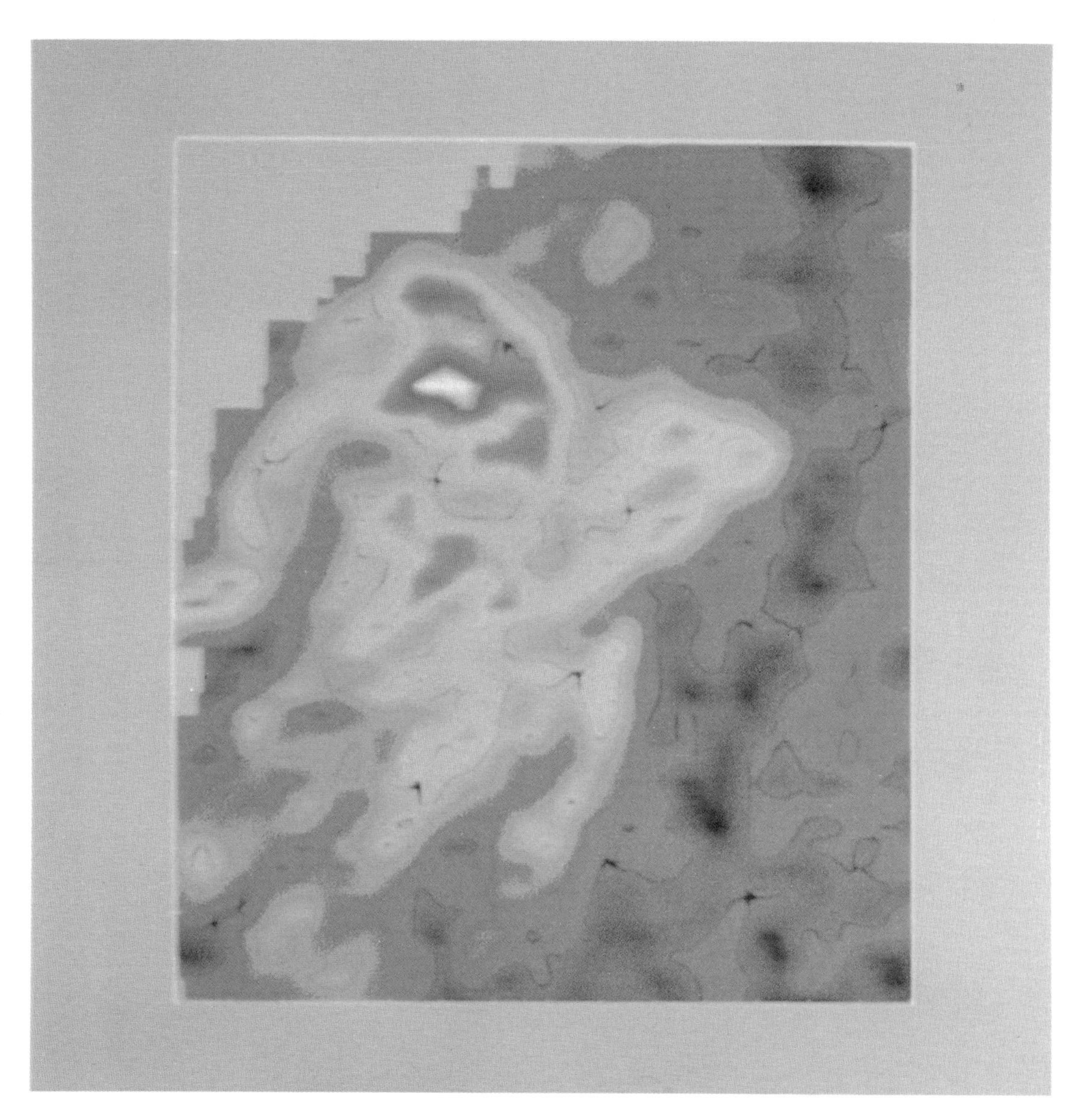

Fig. 2-31. Same horizon slice as in Figure 2-30 displayed in a wider range of hues to draw attention to the high amplitudes using advancing colors. (Courtesy Chevron U.S.A. Inc.)

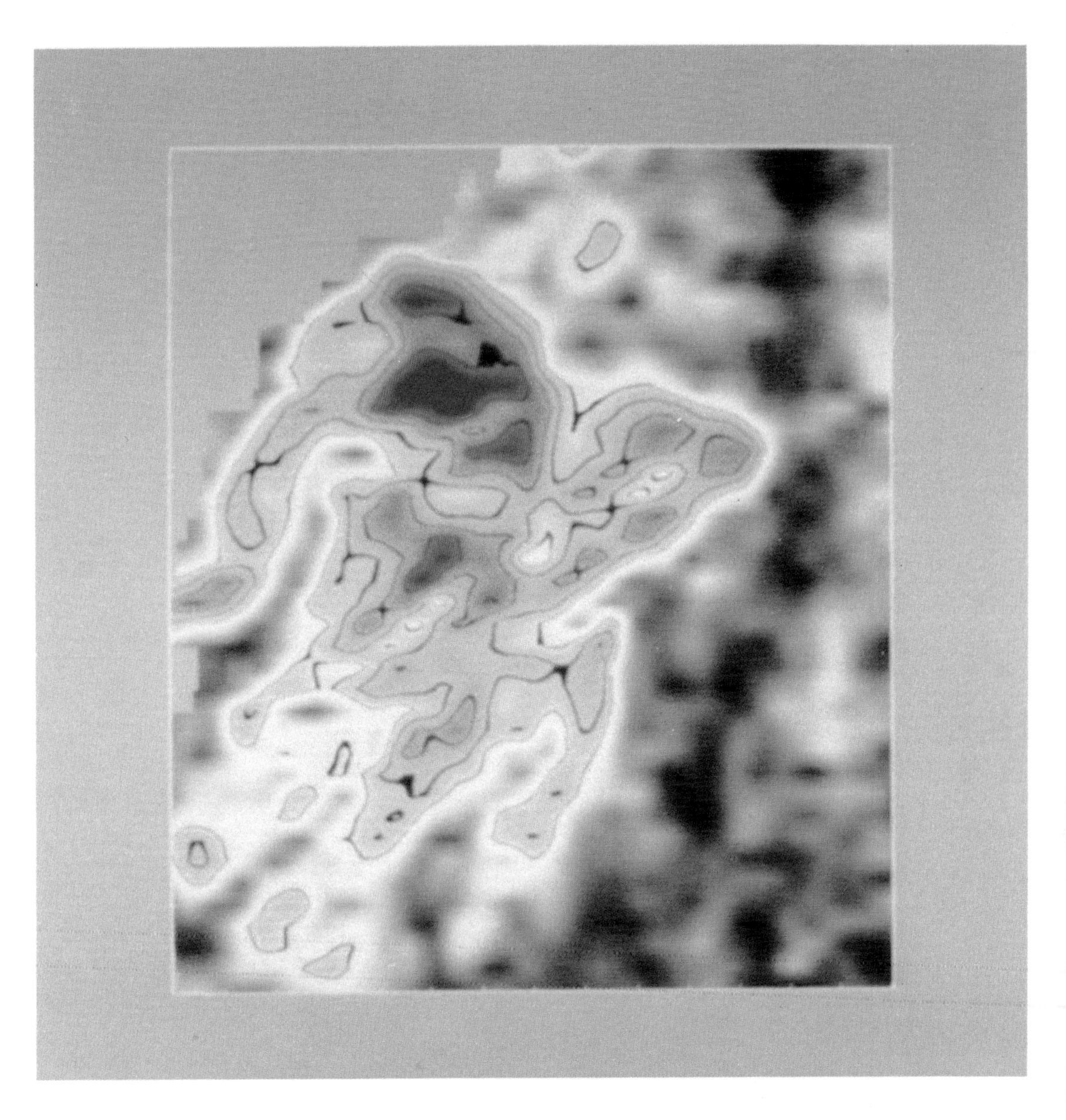

Fig. 2-32. Same horizon slice as in Figure 2-30 displayed in reds, yellows and grays to accentuate total drilling potential. (Courtesy Chevron U.S.A. Inc.)

STRUCTURAL INTERPRETATION

Direct Contouring and the Importance of the Strike Perspective

The 3-D seismic interpreter works with a volume of data. Normally this is done by studying some of each of the three orthogonal slices through the volume. This chapter explores the unique contribution of the horizontal section to structural interpretation. The interpreter of structure needs to be able to judge when to use horizontal sections and when to use vertical ones in the course of an overall interpretive project.

Figure 3-1 demonstrates the conceptual relationship between a volume of subsurface rock and a volume of seismic data. Consider the diagram first to represent subsurface rocks and the gray surface to be a bedding plane. The two visible vertical faces of the rectangular solid show the two dip components of the plane; the horizontal face shows the strike of the plane. Now consider the rectangular solid of Figure 3-1 to be the equivalent volume of seismic data. The gray plane is now a dipping reflection and its intersections with the three orthogonal faces of the solid show the two components of dip and the strike as before. Hence the attitude of a reflection on a horizontal section indicates directly the strike of the reflecting surface. This is the fundamental property of the horizontal section from which all its unique interpretive value derives.

Contours follow strike and indicate a particular level in time or depth. When an interpreter picks a reflection on a horizontal section, it is directly a contour on some horizon at the time (or depth) at which the horizontal section was sliced through the data volume.

Figure 3-2 shows three horizontal sections, four milliseconds apart. By following the semicircular black event (peak) from level to level and drawing contours at an appropriate interval, the structural contour map at the bottom of Figure 3-2 was generated. Note the similarity in shape between the sections and the map for the anticlinal structure and the strike east of the faults. In the central panel the peaks from 1352 ms are printed in black and the peaks from 1360 ms in blue/green. This clearly demonstrates the way in which the events have moved with depth.

Figures 3-3 and 3-5 provide one vertical section and several horizontal sections from which the relationship between the two perspectives can be appreciated. Line P (Figure 3-3) runs north-south through the middle of the prospect with south at the right. The time interval 2632-2656 ms shows some continuous reflections. Proceeding from south to north (right to left, Figure 3-3; bottom to top, Figure 3-5) the structure is first a broad closed anticline, then a shoulder, then a smaller anticline.

Figure 3-5 demonstrates a simple exercise in direct contouring from a suite of horizontal sections. The red event (trough) expanding in size from left to right has been progressively circumscribed in the lower part of the figure. The last frame is a raw contour map of this horizon. This first structural representation has been made quick-

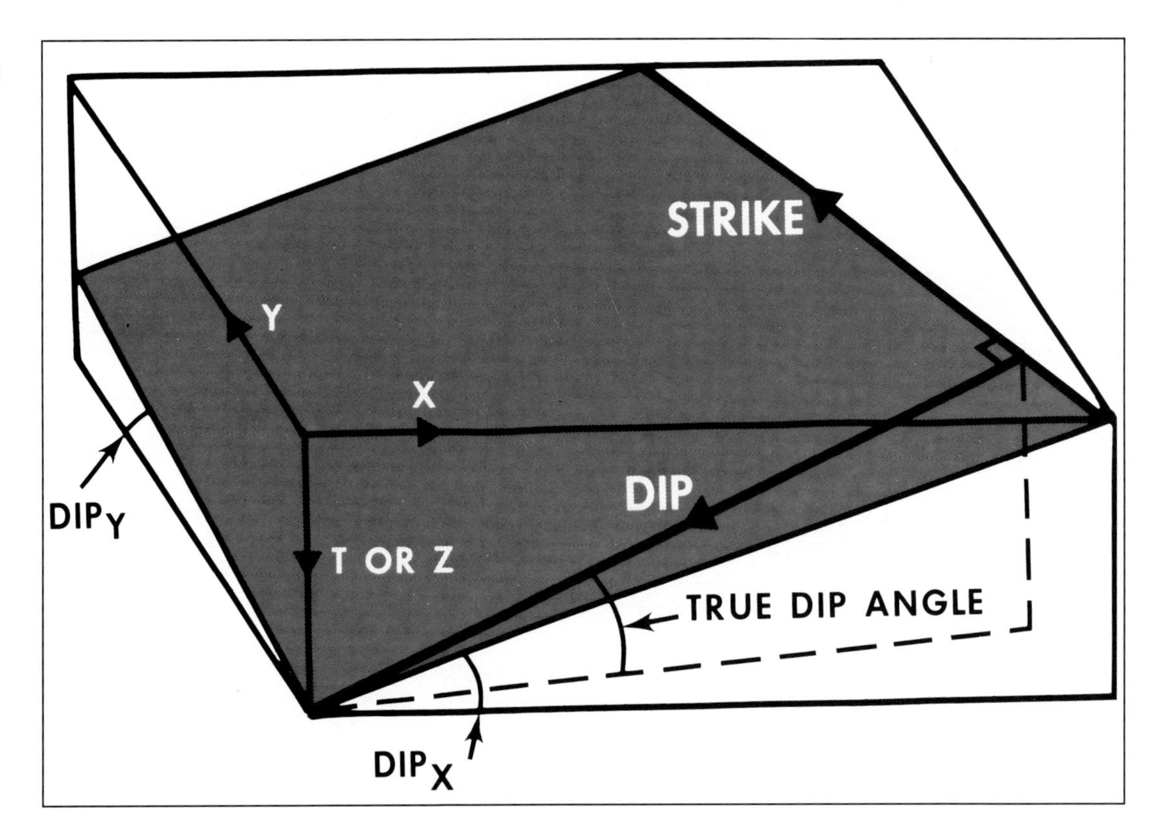

Fig. 3-1. Relation between dip and strike of a seismic reflector within a data volume.

ly and efficiently without the traditional intermediate tasks of timing, posting and contouring. When drawing structural contours from horizontal sections in this way, it is wise to visualize the three-dimensionality of the structure and to appreciate where on the seismic waveform the contour is being drawn (Figure 3-4). The latter problem applies particularly to the use of variable area displays as used, for example, in Figure 3-5. The contour is here drawn all the way around the red event only because the dip is down all the way around the structure; this is a consistent point on the seismic waveform, namely its upper edge (Figure 3-4).

Figure 3-6 shows 24 horizontal sections covering an area of about 5 sq mi (13 sq km). These can be used as a structural interpretation exercise. Obtain a small piece of transparent paper and register it over the rectangular area. Begin with the upper left frame and find the red event in its lower right corner. Mark this event by following its maximum amplitude and then mark its changed position from frame to frame until you reach 2160 ms. Your resultant contour map should show that the dip is generally northwest and that the strike swings about 40° toward the north over the structural range of the map. You will probably detect a fault toward the west of the area as well. If you study the arcuate events west of the fault, you will recognize a small anticline closing against the fault and a small syncline south of it. There is no way to establish the correlation across the fault.

An event on a horizontal section is generally broader than on a vertical section as dips are usually less than 45°. Figure 3-7 shows the effect of dip and frequency on the width of an event on a horizontal section. A gently dipping event is very broad and a steeply dipping event is much narrower. Increasing dip and increasing frequency both make horizontal section events narrower. The width of an event on a horizontal section is strictly half the spatial wavelength.

Because typical dips are much less than 45°, fewer horizontal sections than vertical ones are needed to study the full extent of a reflection within a given data volume. This gives horizontal sections greater efficiency than vertical sections in structural mapping. Combining this benefit with the fact that horizon tracks (picks) are directly contours, then the value of horizontal sections to structural interpretation is substantial.

Text continues on page 65

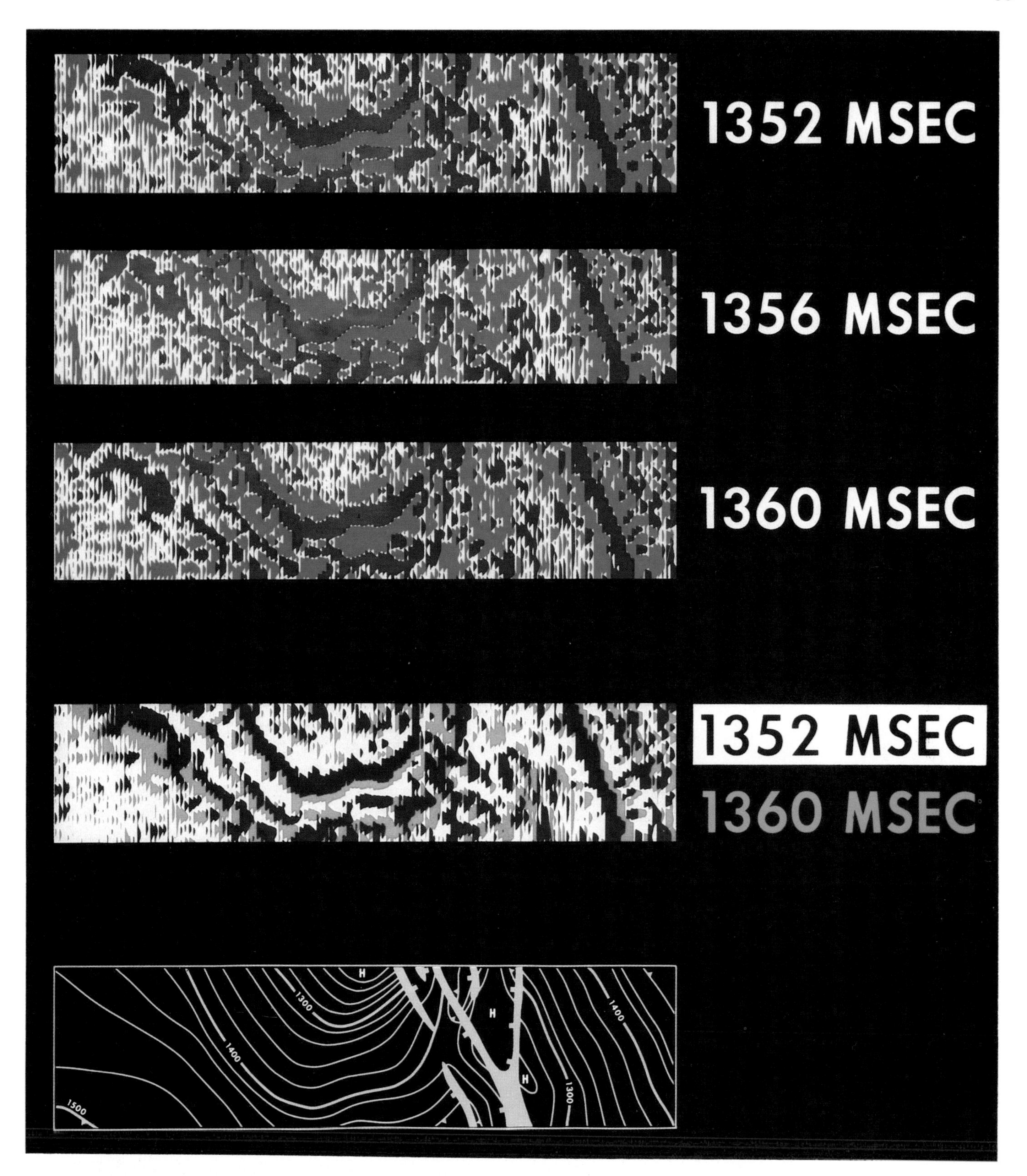

Fig. 3-2. Dual polarity horizontal sections from offshore Holland; two-level single polarity horizontal section, showing movement of events from 1352 ms to 1360 ms; interpreted contour map on horizon seen as strongest event on horizontal sections.

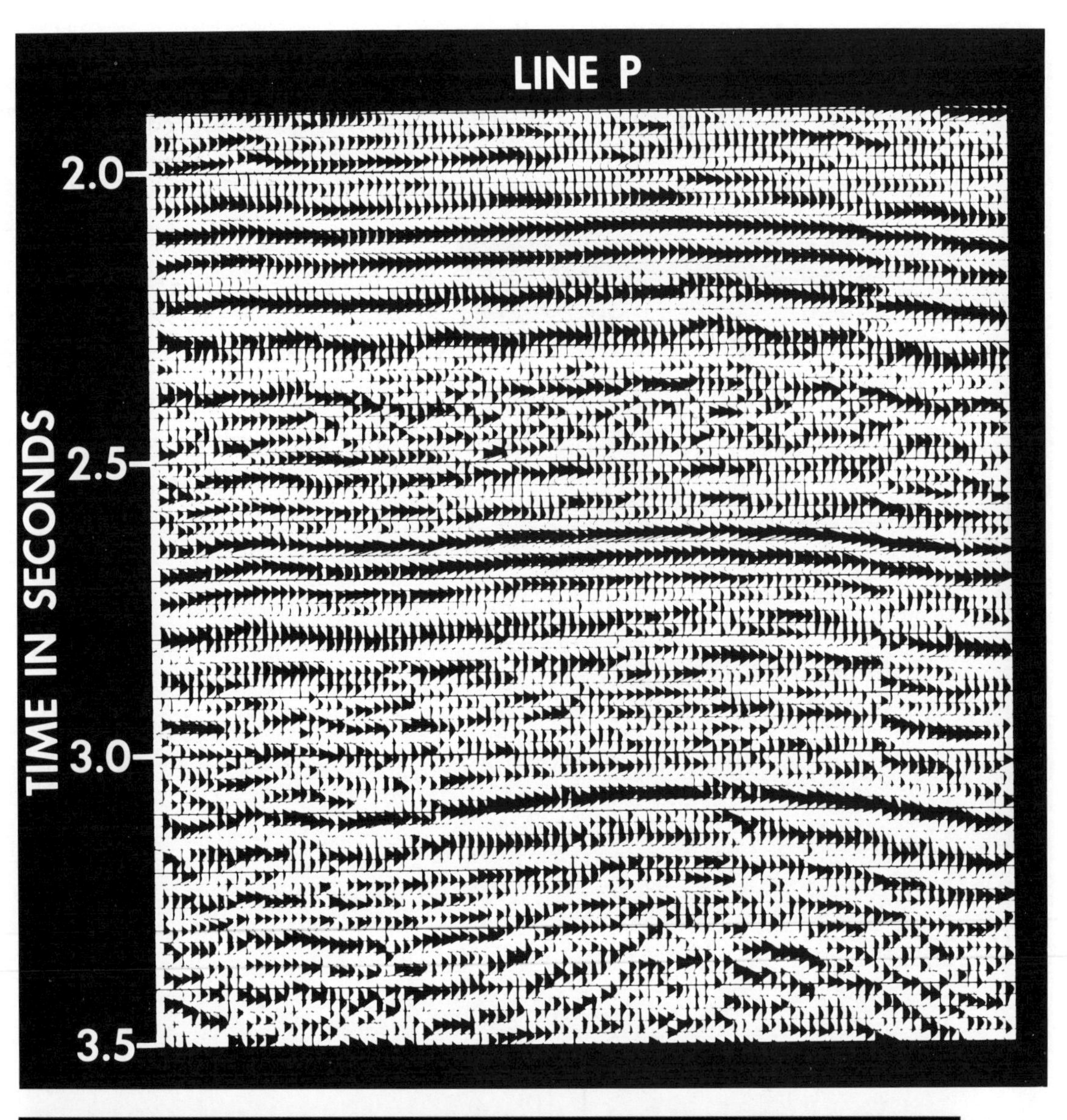

Fig. 3-3. North-south vertical section from Peru through same data volume sliced in Figure 3-4. (Courtesy Occidental Exploration and Production Company.)

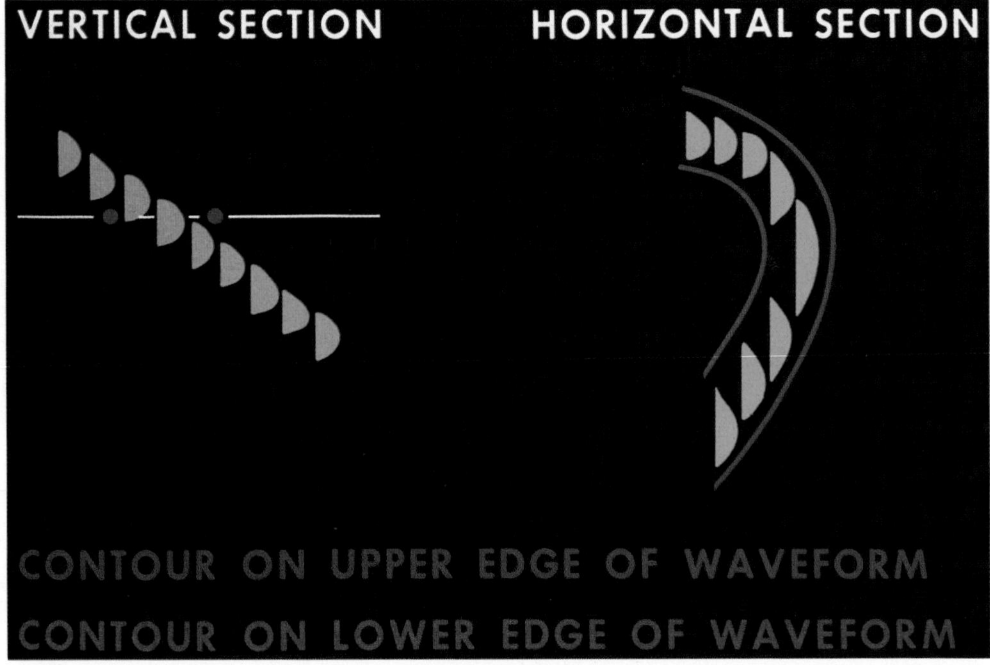

Fig. 3-4. Where on the waveform should one place the contour when working with variable area display?

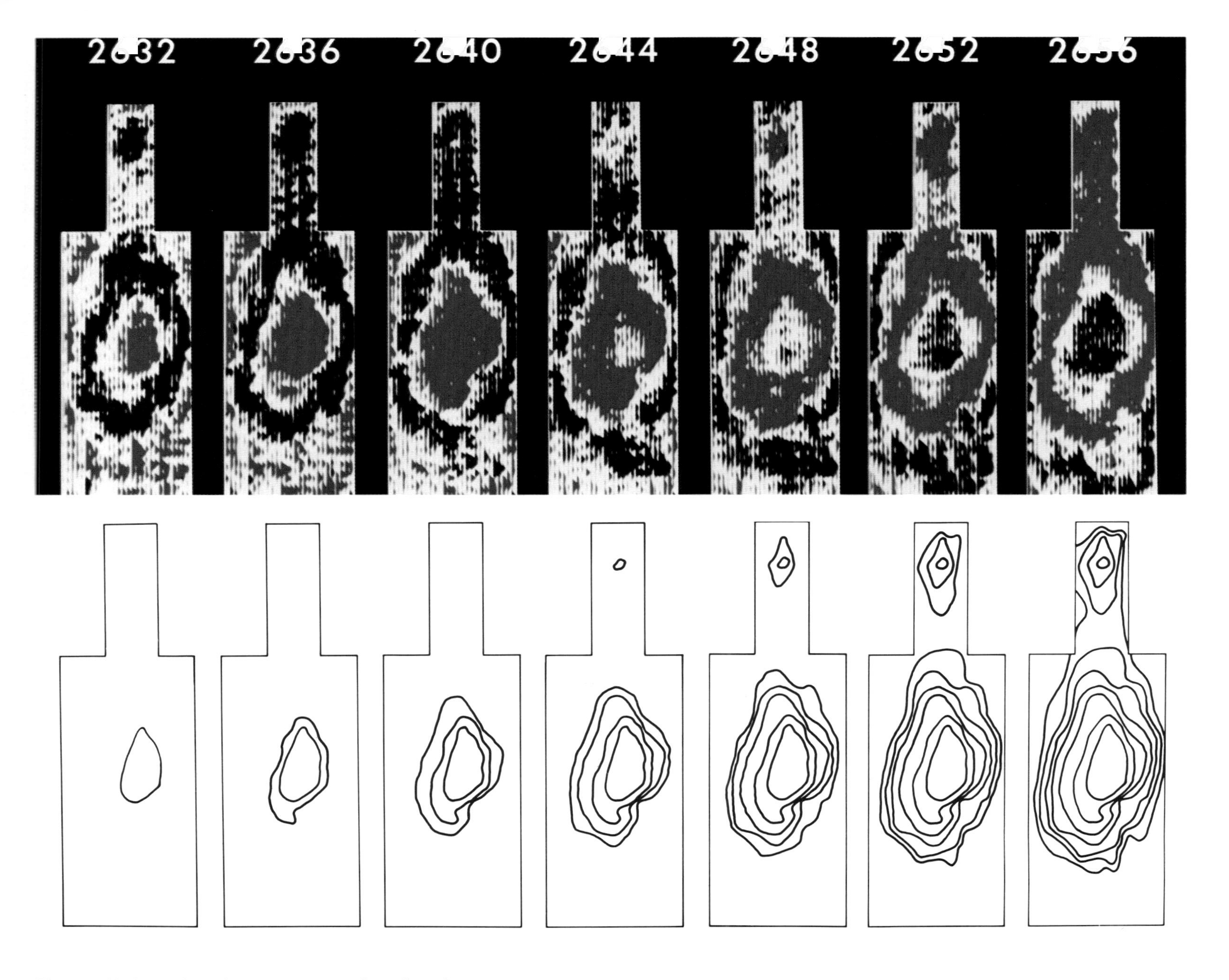

Fig. 3-5. Horizontal sections, 4 ms apart, from Peru (courtesy Occidental Exploration and Production Company) and raw interpreted contour map made by successively circumscribing the red event on each section.

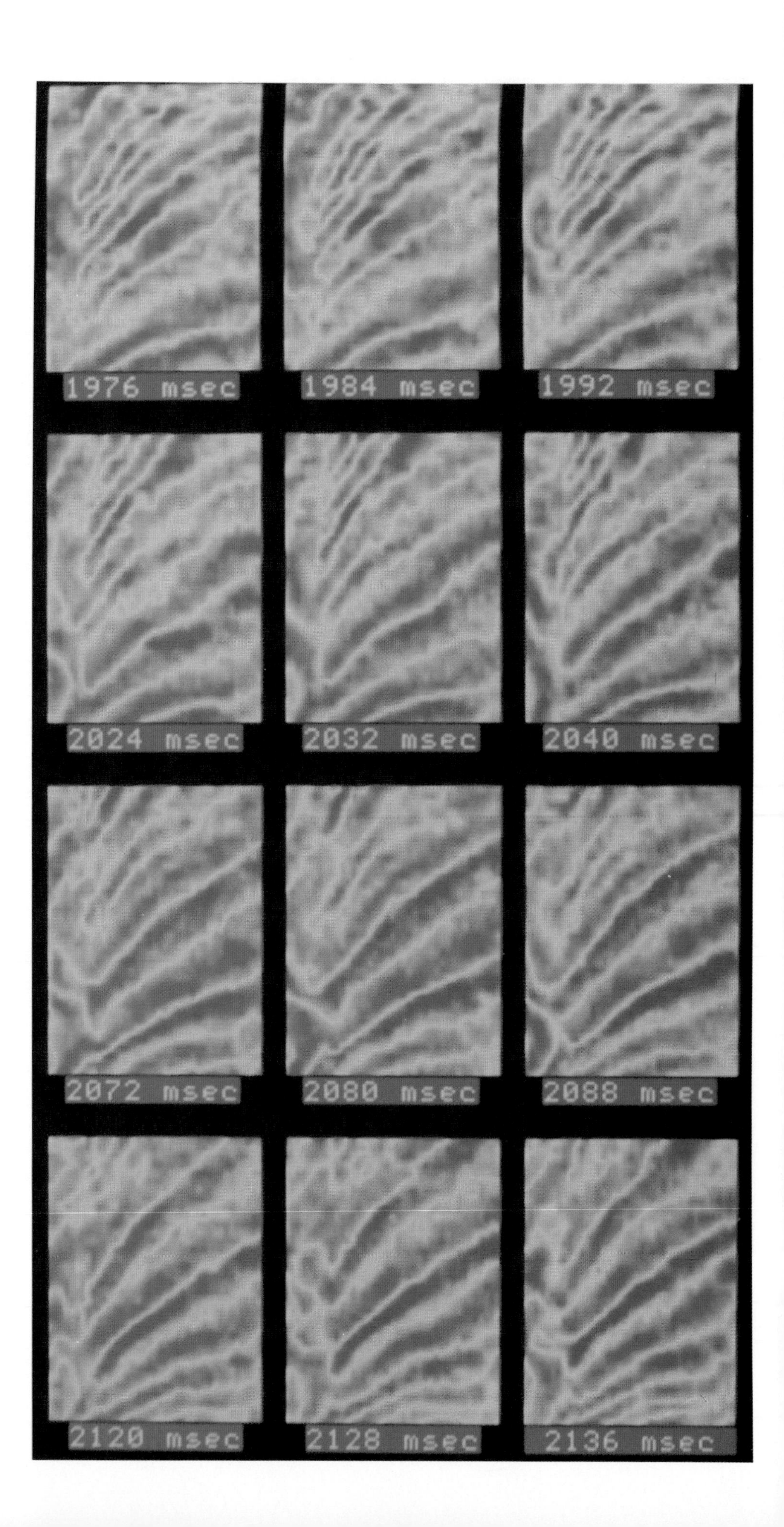

Fig. 3-6. Horizontal sections, 8 ms apart, from offshore Trinidad. (Courtesy Texaco Trinidad Inc.)

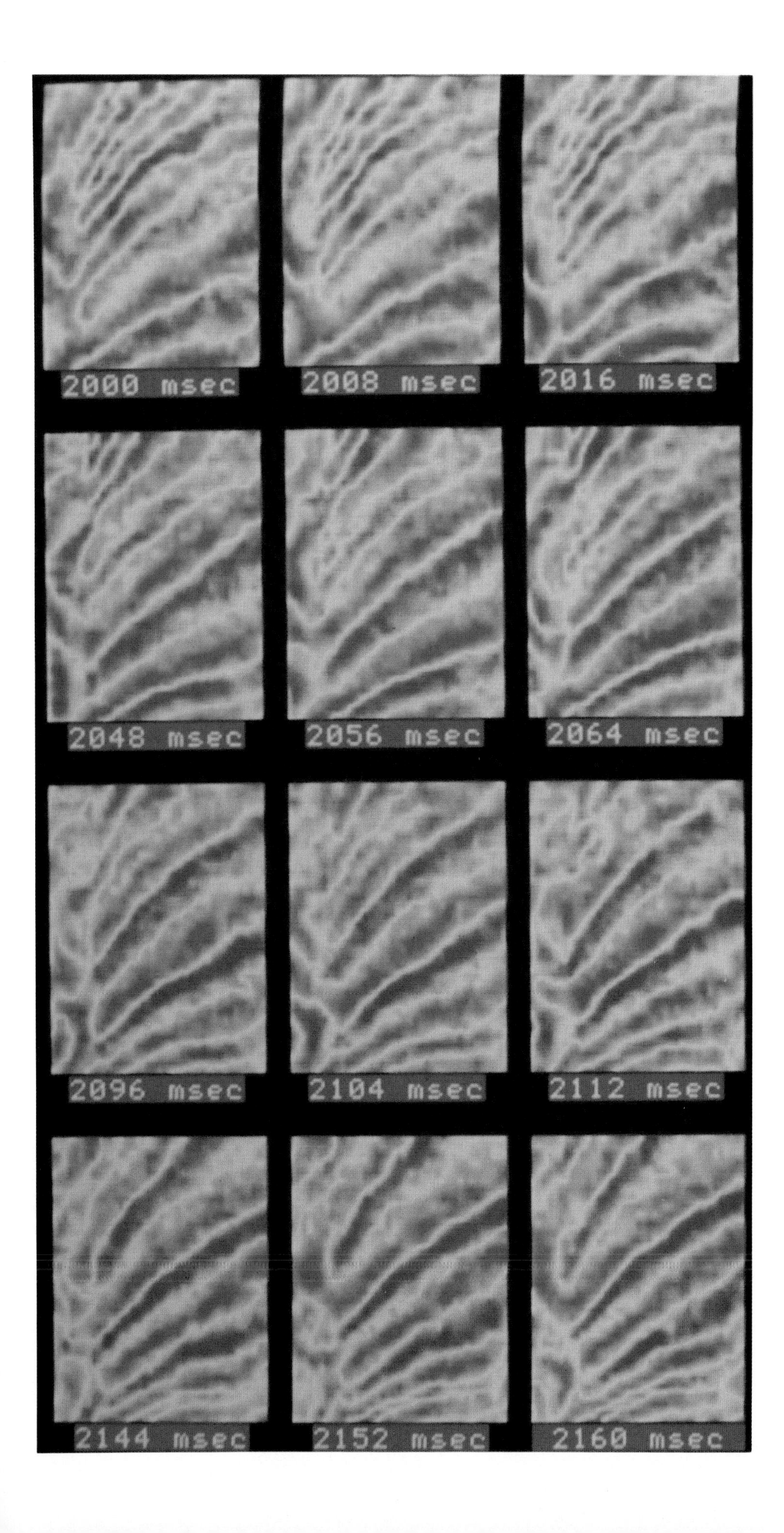
2000 msec
2008 msec
2016 msec
2048 msec
2056 msec
2064 msec
2096 msec
2104 msec
2112 msec
2144 msec
2152 msec
2160 msec

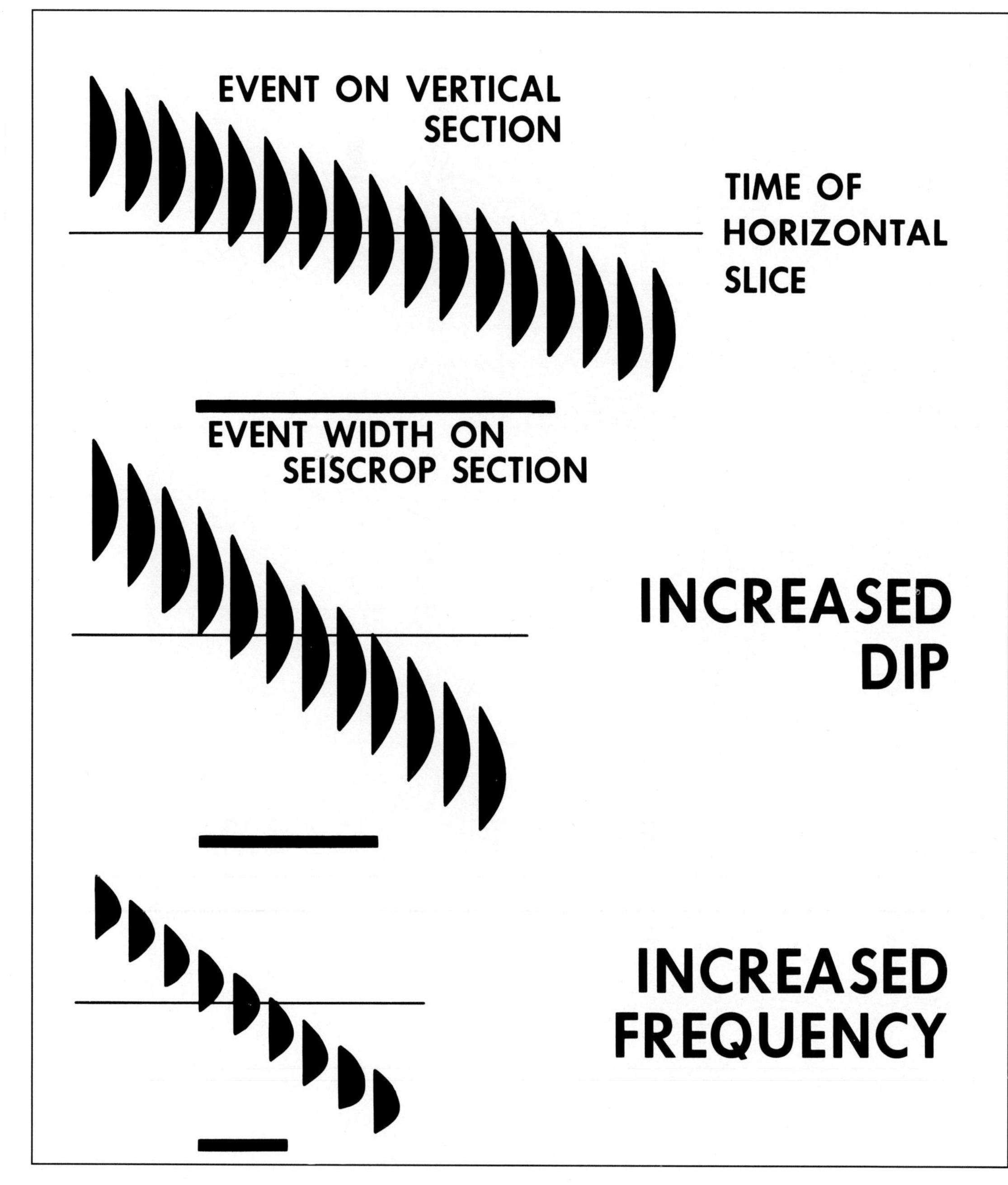

Fig. 3-7. The width of an event on a horizontal section decreases with increased dip and also with increased frequency.

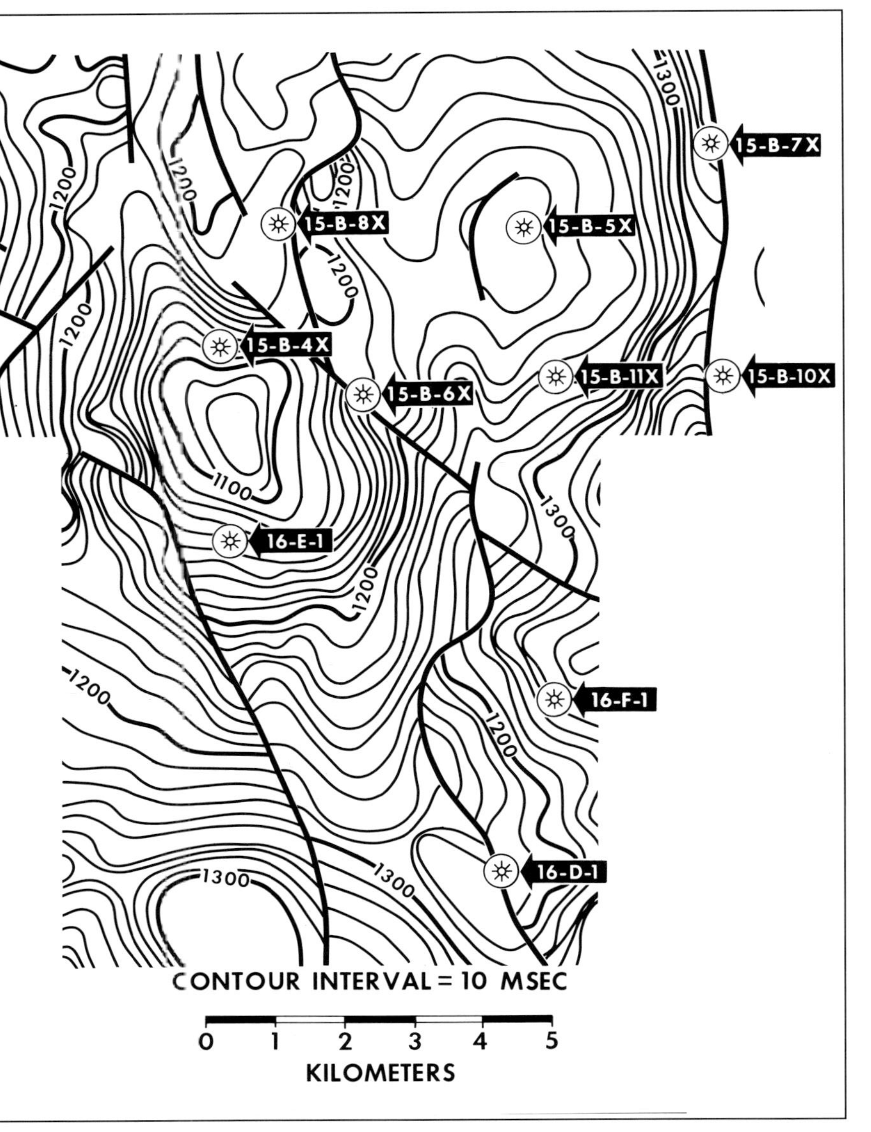

Fig. 3-8. Structural contour map derived from 2-D data from the Gulf of Thailand. (Courtesy Texas Pacific Oil Company Inc.)

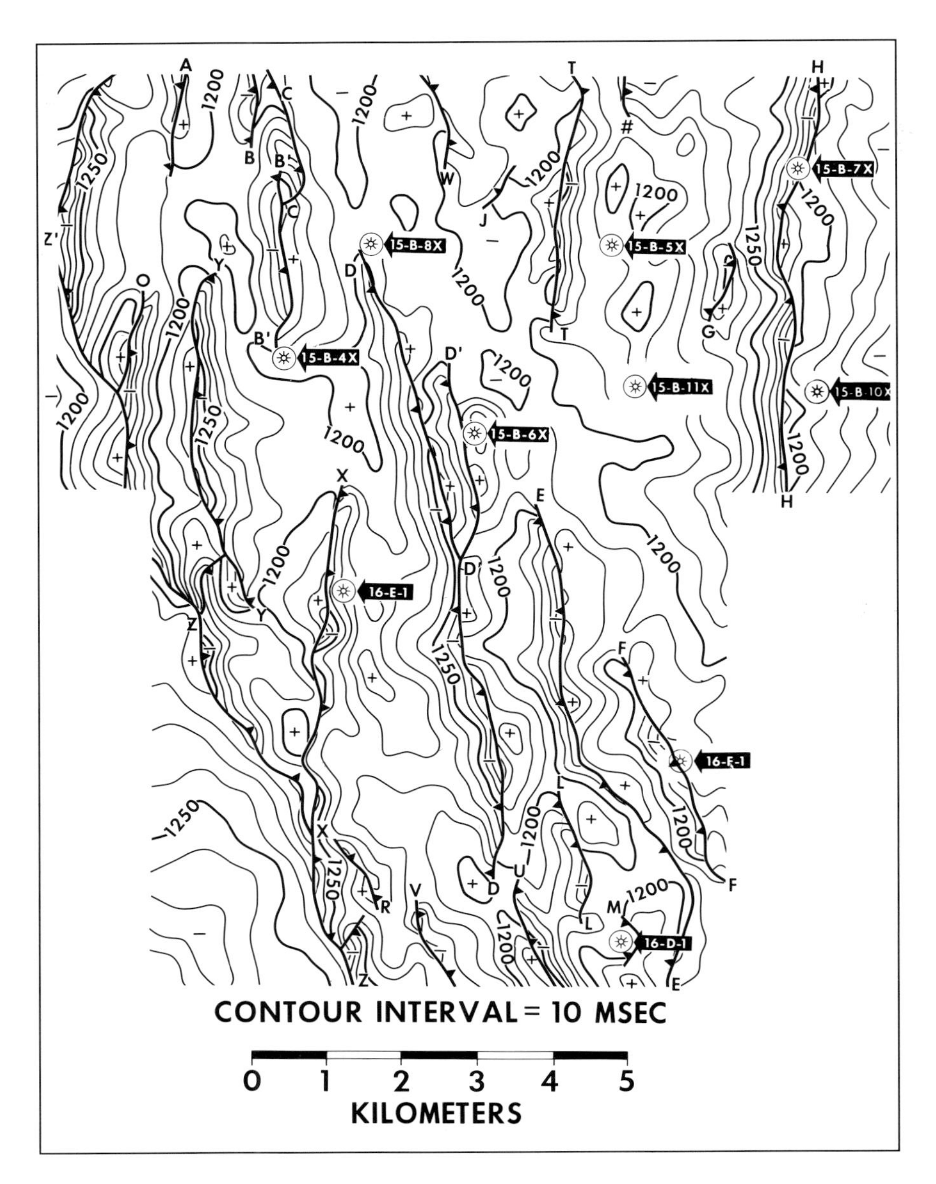

Fig. 3-9. Structural contour map derived from 3-D data from the Gulf of Thailand for the same horizon mapped in Figure 3-8. (Courtesy Texas Pacific Oil Company Inc.)

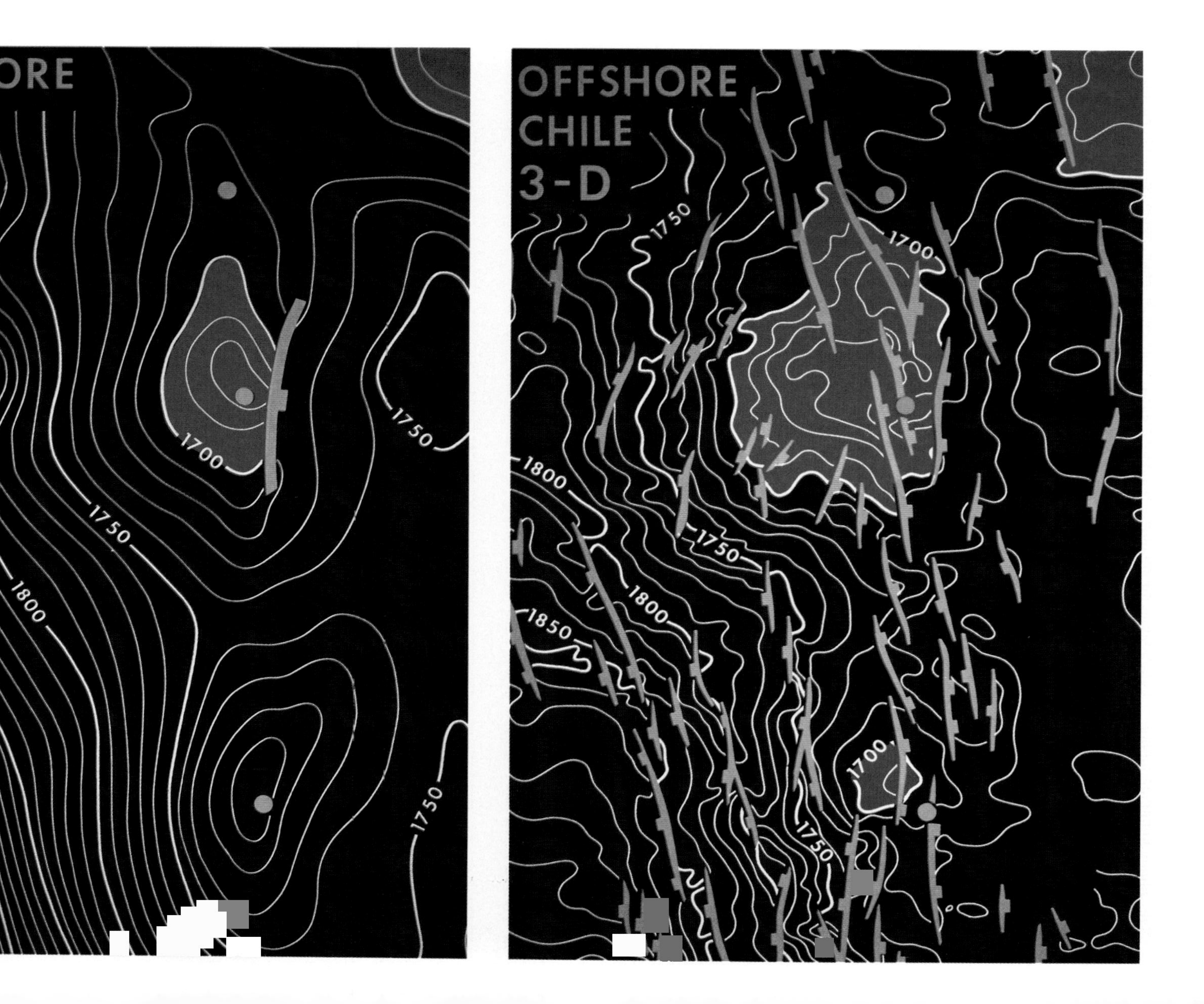

Fig. 3-10. Structural contour map derived from 2-D data from offshore Chile. (Courtesy ENAP).

Fig. 3-11. Structural contour map derived from 3-D data from offshore Chile for the same horizon mapped in Figure 3-10. (Courtesy ENAP).

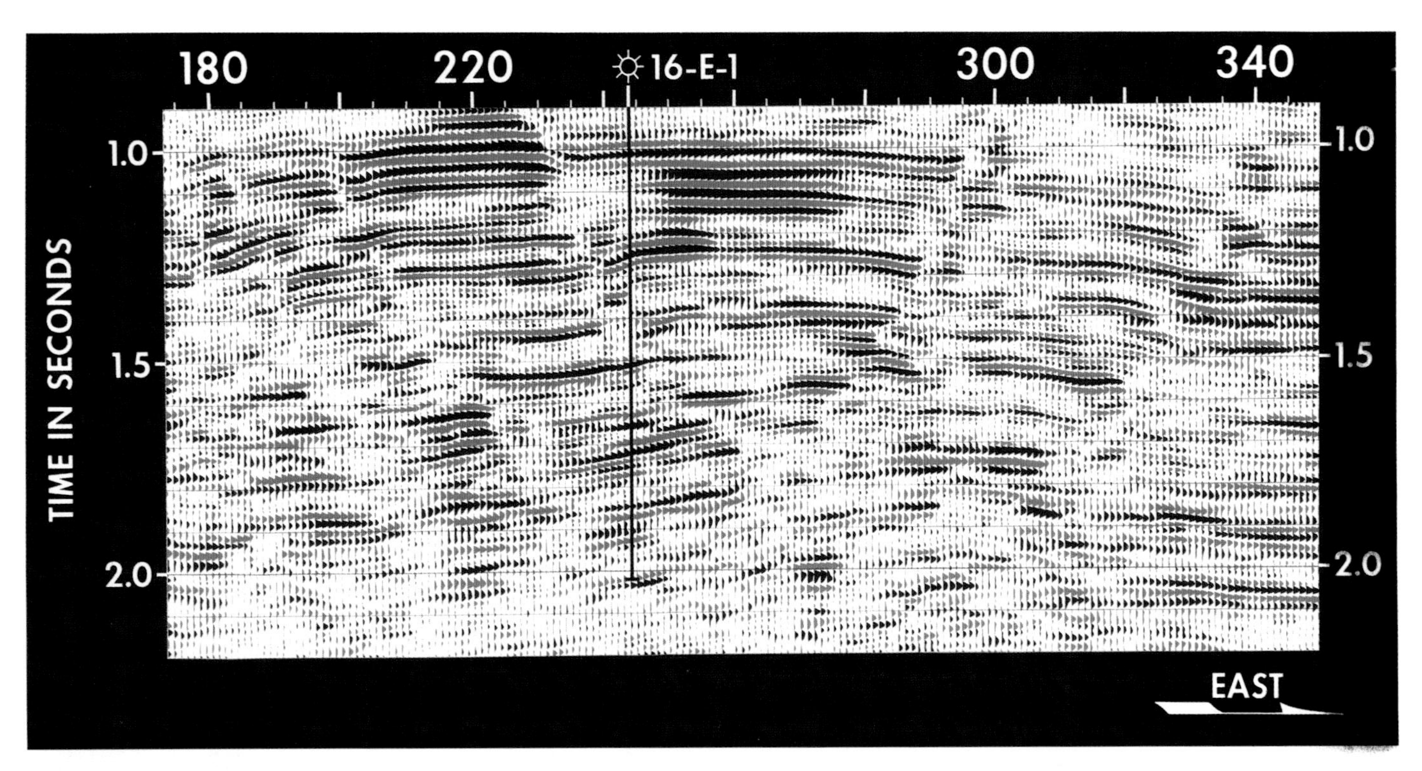

Fig. 3-12. Line 55 from Gulf of Thailand 3-D data. (Courtesy Texas Pacific Oil Company Inc.)

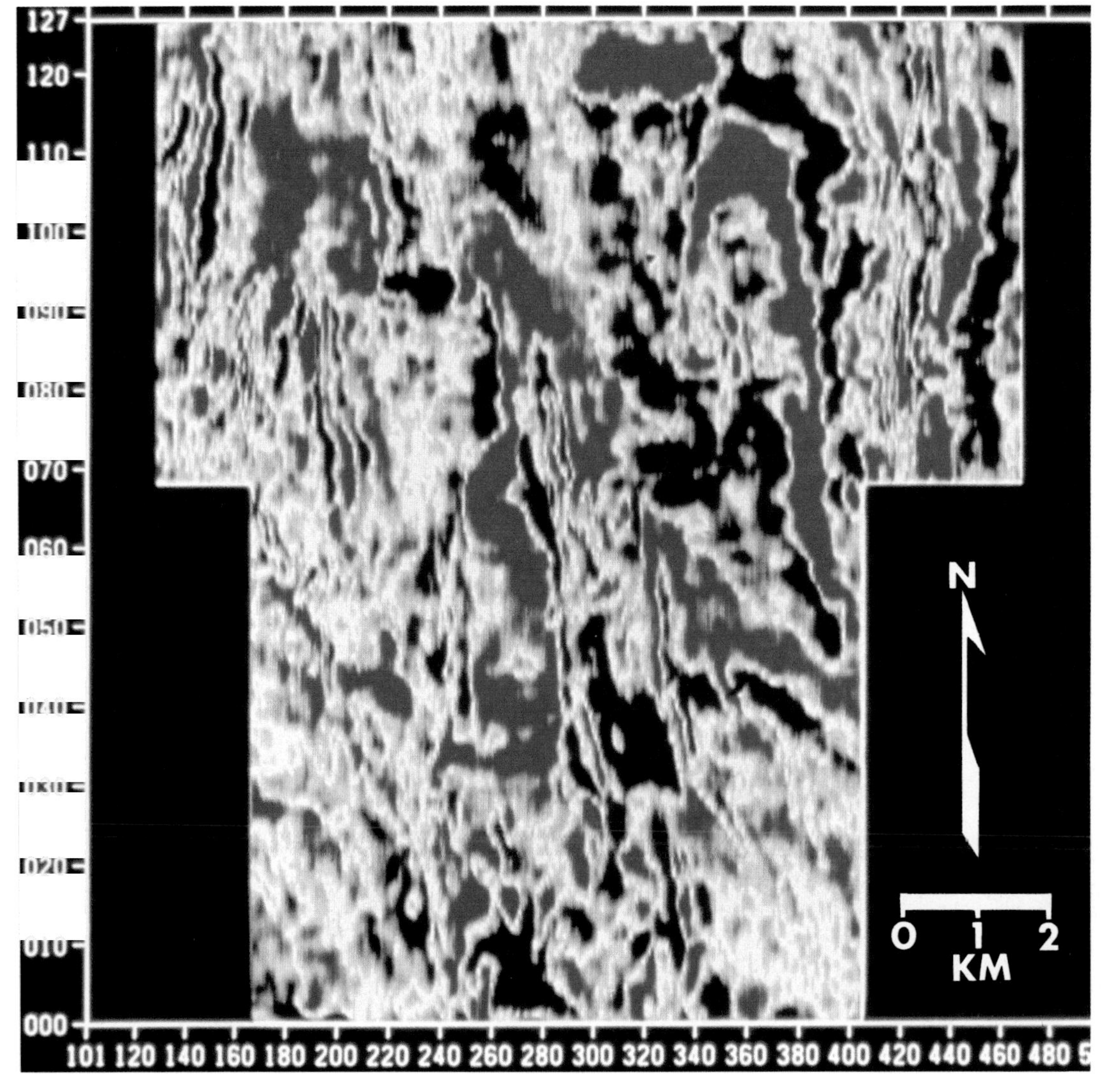

Fig. 3-13. Horizontal section at 1388 ms from Gulf of Thailand. (Courtesy Texas Pacific Oil Company.)

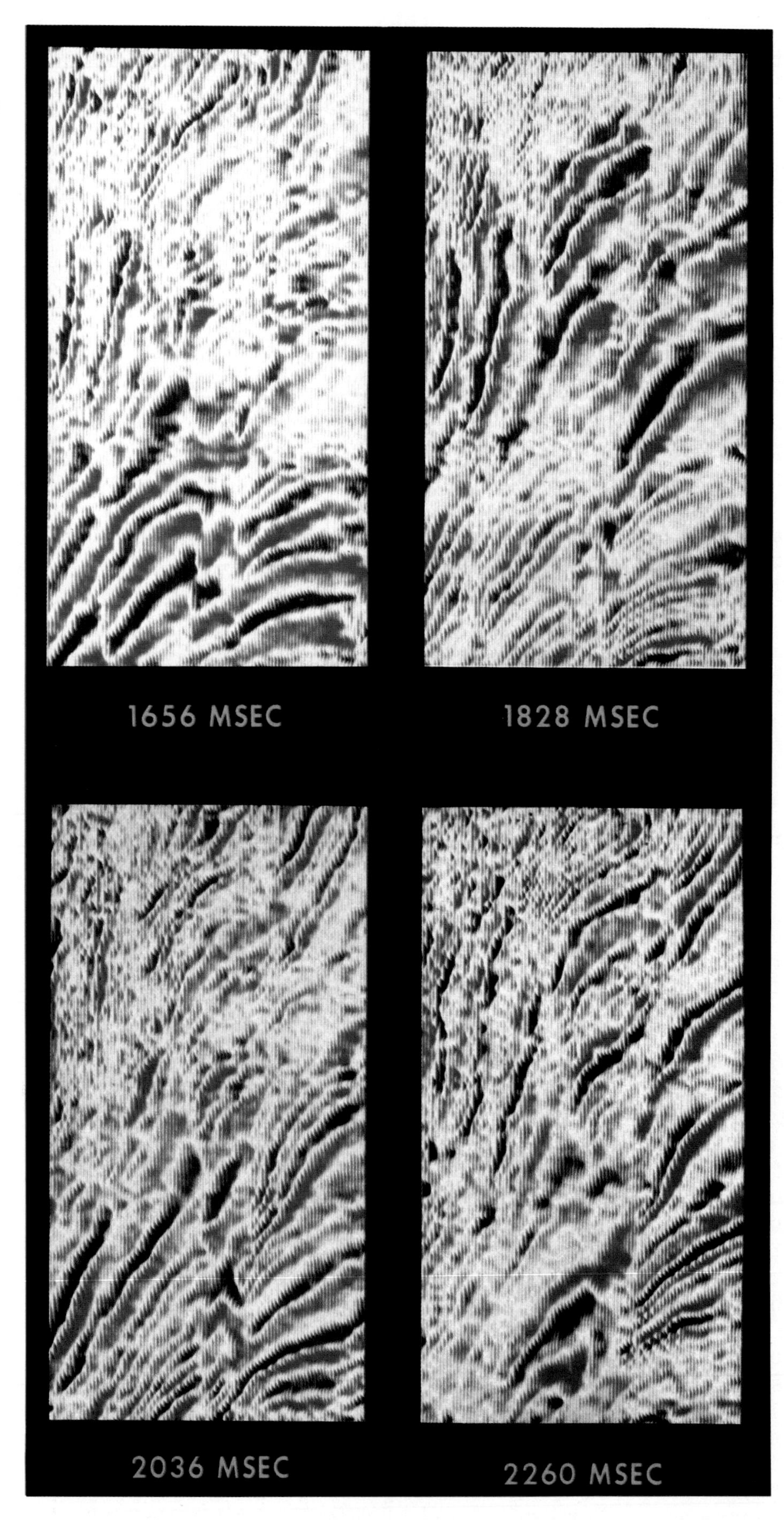

Fig. 3-14. Horizontal sections from offshore Trinidad. Event terminations indicate faulting. (Courtesy Texaco Trinidad Inc.)

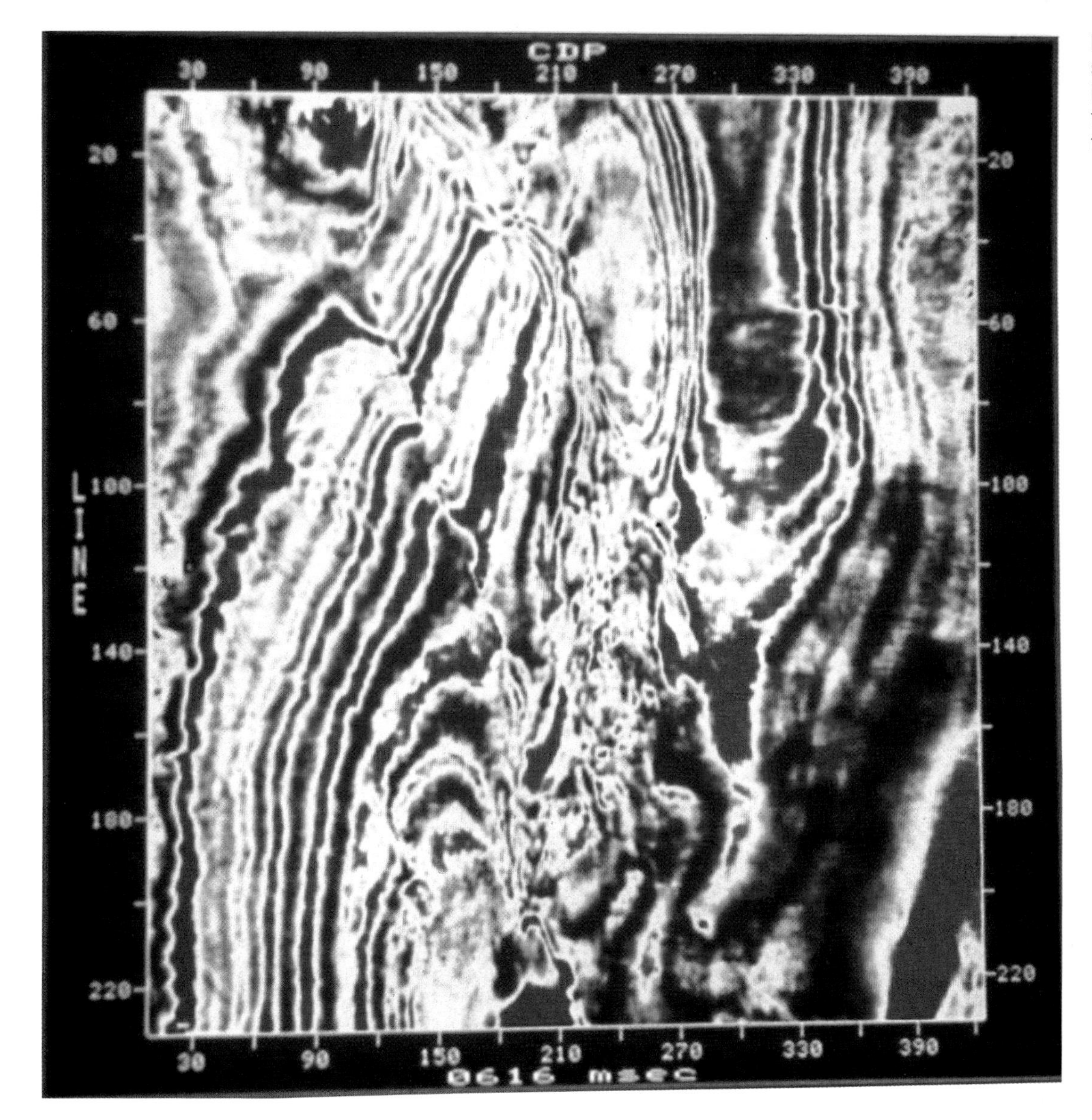

Fig. 3-15. Horizontal section from onshore Europe. Event terminations indicate faulting.

Fault Recognition and Mapping

When an interpreter works with 3-D data after having previously mapped from 2-D data over the same prospect, the most striking difference between maps is commonly the increased fault detail in the 3-D map. Figures 3-8 and 3-9 provide a typical comparison and also demonstrate increased detail in the shape of the structural contours. Comparison of Figures 3-10 and 3-11 also shows a considerable increase in the number of faults and in the structural detail. The three well locations indicated in blue appear structurally quite different on the 2-D and 3-D maps.

We expect to detect faults from alignments of event terminations. Figure 3-12 shows a vertical section from the 3-D data which provided the map of Figure 3-9. The event terminations clearly show several faults. The horizontal section of Figure 3-13 is from the same data volume and, in contrast, does not show clear event terminations. Figure 3-14 shows four horizontal sections from a different prospect but one in a similar tertiary clastic environment. Here event terminations clearly indicate the positions of three major faults on each of the four sections.

Why are event terminations visible at the faults in Figure 3-14 but not in Figure 3-13? The answer lies simply in the relationship between structural strike and fault strike. Any horizontal section alignment indicates the strike of the feature. If there is a significant angle between structural strike and fault strike, the events will terminate.

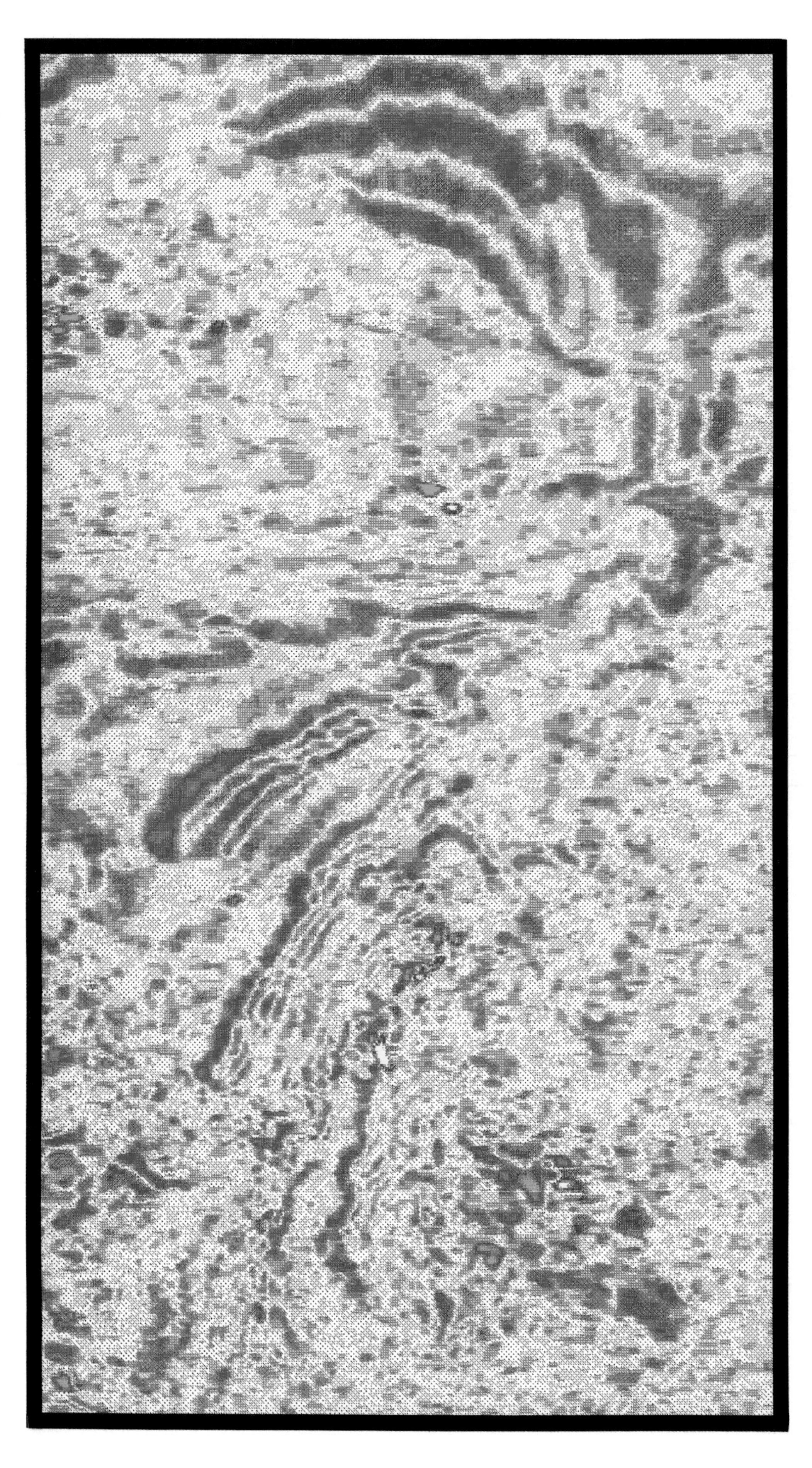

Fig. 3-16. Horizontal section at 1500 ms from Gulf of Mexico showing many clearly visible faults. At least 10 are identifiable. (Courtesy Conoco Inc. and Texaco U.S.A. Inc.)

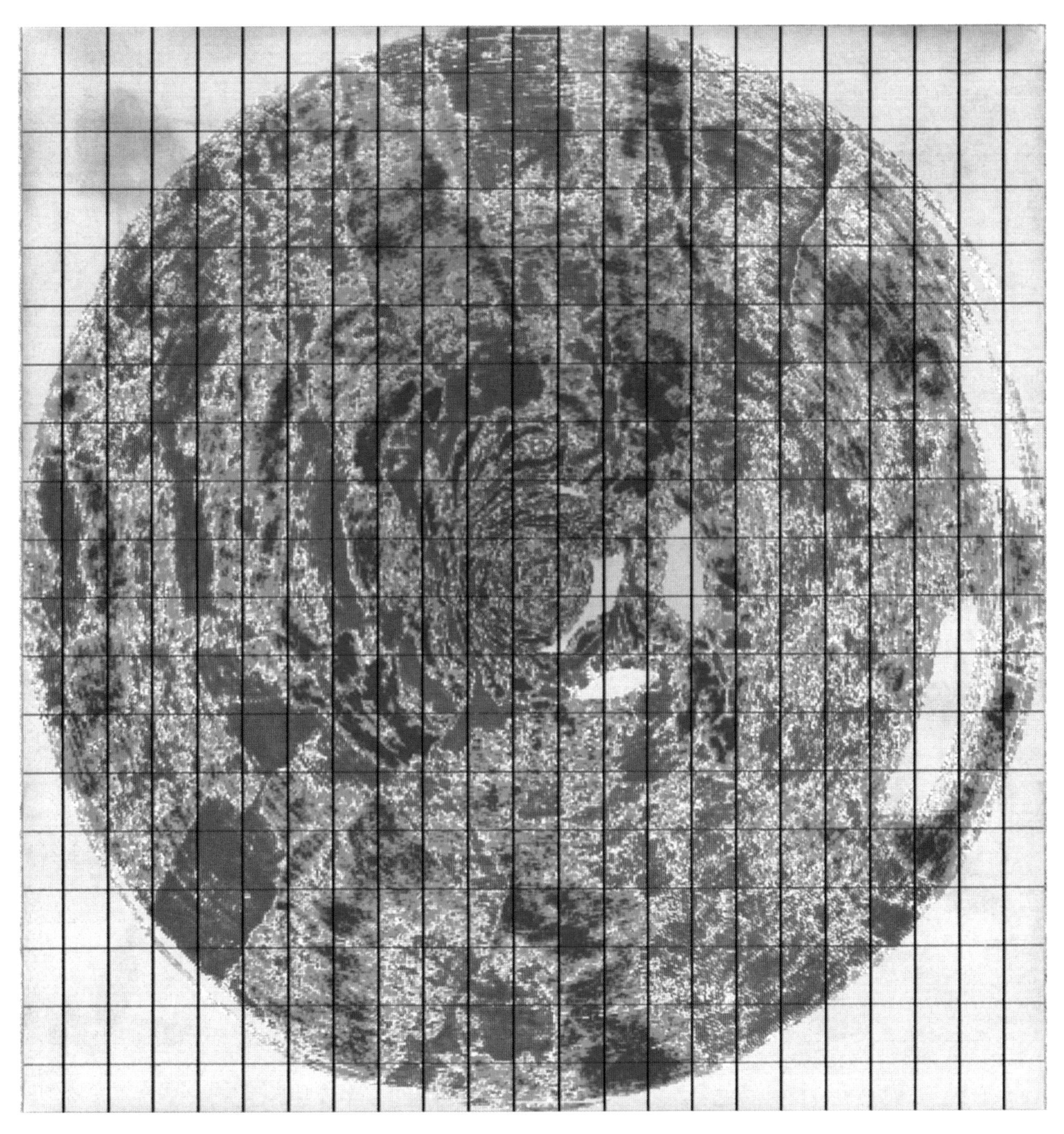

If structural strike and fault strike are parallel, or almost so, the events will not terminate but will parallel the faults. Comparison of Figures 3-13 and 3-9 demonstrates that situation. The difficulty of seeing faults on a time slice when they parallel structural strike is overcome by using the attribute coherence (see Chapter 8).

Because an alignment of event terminations on a horizontal section indicates the strike of a fault, the picking of a fault on a horizontal section provides a contour on the fault plane. Thus picking a fault on a succession of suitably spaced horizontal sections constitutes an easy approach to fault plane mapping. The faults evident in Figure 3-14 have been mapped in this way.

Fig. 3-17. Horizontal section at 1000 ms from Gulf of Mexico Concentric Circle Shoot showing many radial faults surrounding a salt dome. (Courtesy Tensor Geophysical Service Corporation.)

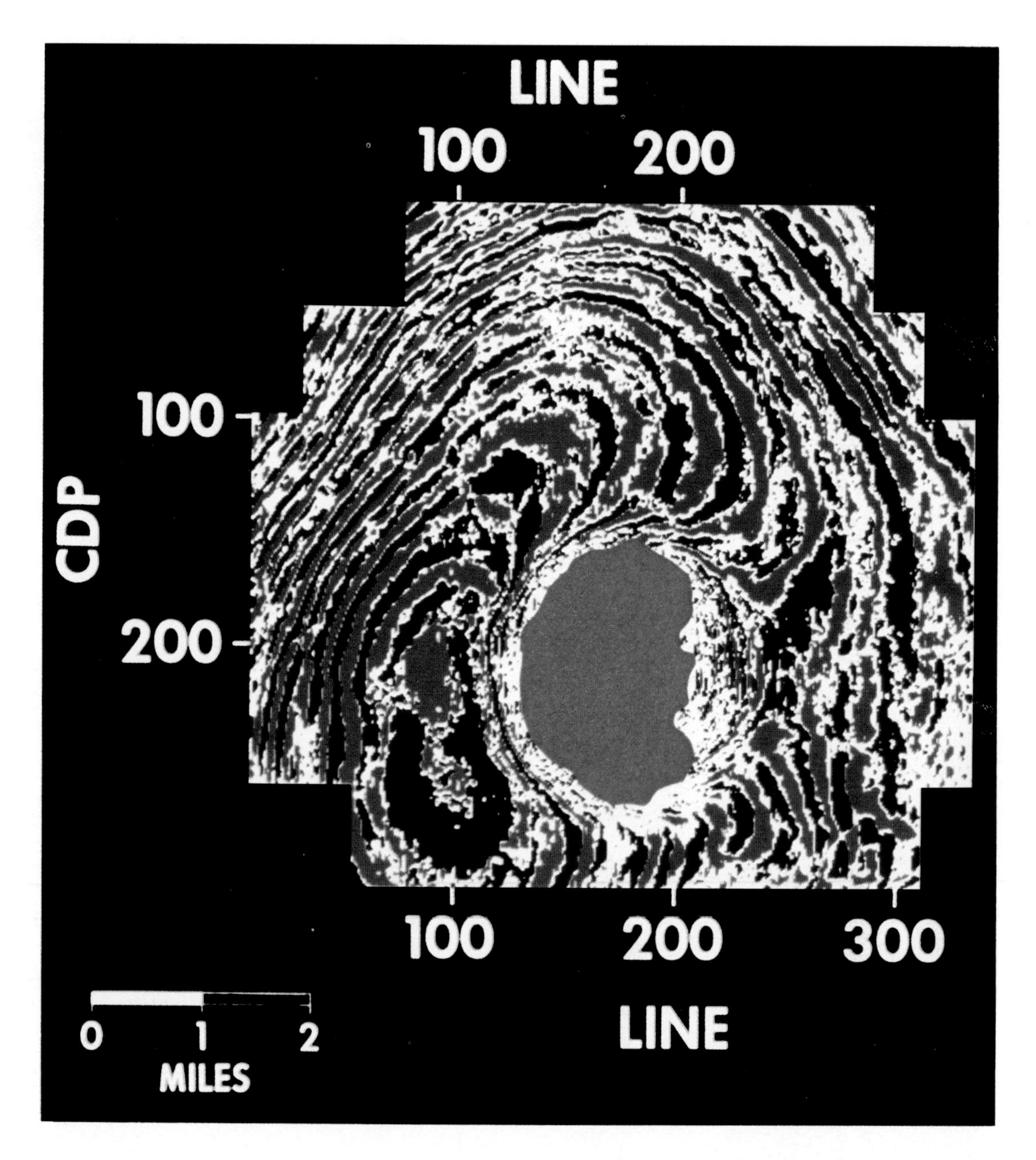

Fig. 3-18. Horizontal section at 3252 ms from Eugene Island area of Gulf of Mexico showing interpreted shape of salt plug. (Courtesy Hunt Oil Company.)

In the lower right corner of the horizontal section at 2260 ms (Figure 3-14) two fault blocks show events of quite different widths. This is the effect of dip which was explained by Figure 3-7. We also see a similar effect of dip in Figure 3-13 where the faults are mostly traced by narrow sinuous events striking approximately north-south.

Figure 3-15 shows a variety of structural features: prominent faults, more subtle faults, culminations and various character changes. It is very important that horizontal sections play their proper role in fault interpretation. In the early stages of structural interpretation of a prospect, the major faults will be identified on some widely-spaced vertical sections. The way in which these faults join up into a fault framework should then be established from horizontal sections. This is part of the overall recommended procedure of Figure 3-32. Lineations of event terminations will normally link the faults already recognized vertically. Figures 3-16 and 3-17 show clearly visible faulting that evidently could be used in this way.

Today's interactive workstations help in the coordinated use of vertical and horizontal sections by providing the capability of cross-posting. When a fault is picked on a vertical section, its intersection will appear on an intersecting horizontal section.

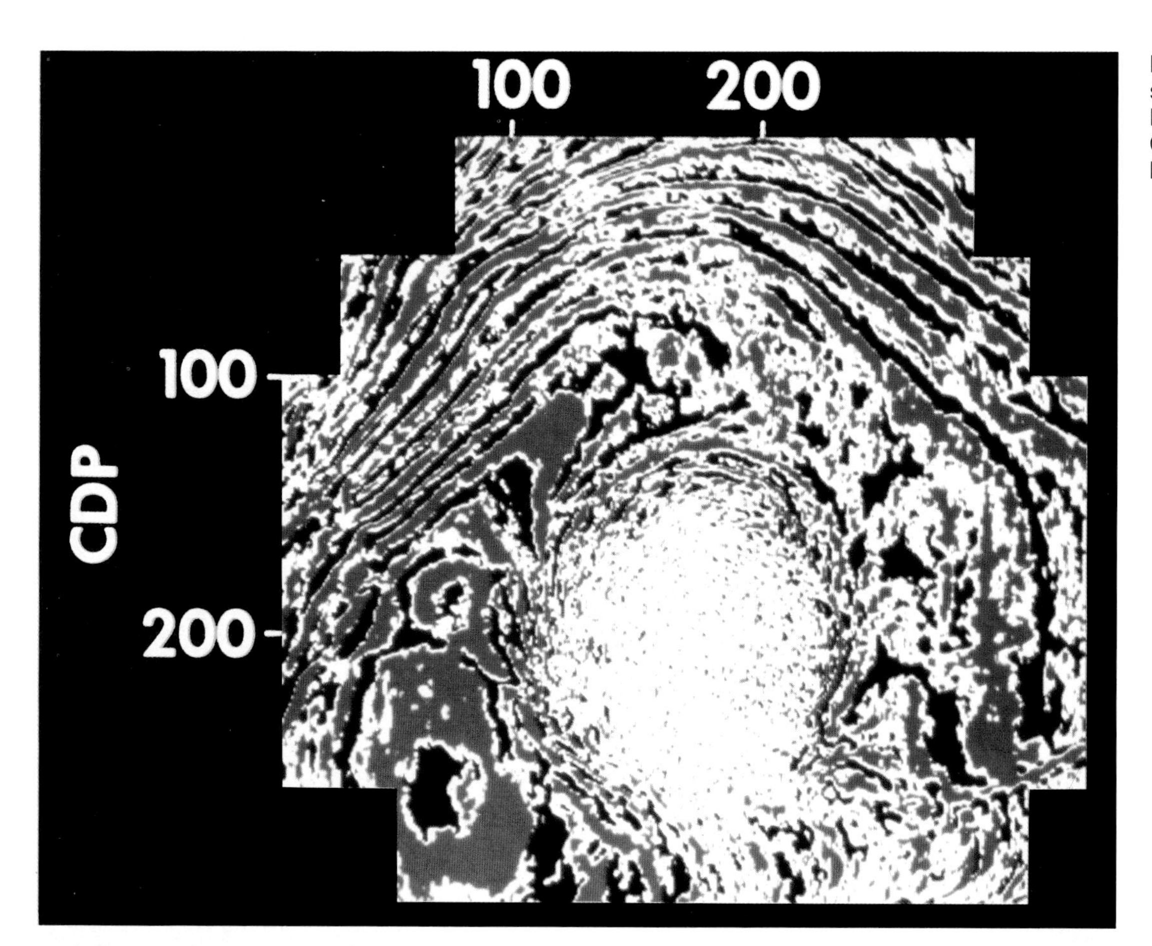

Fig. 3-19. Horizontal section at 3760 ms from Eugene Island area of Gulf of Mexico. (Courtesy Hunt Oil Company.)

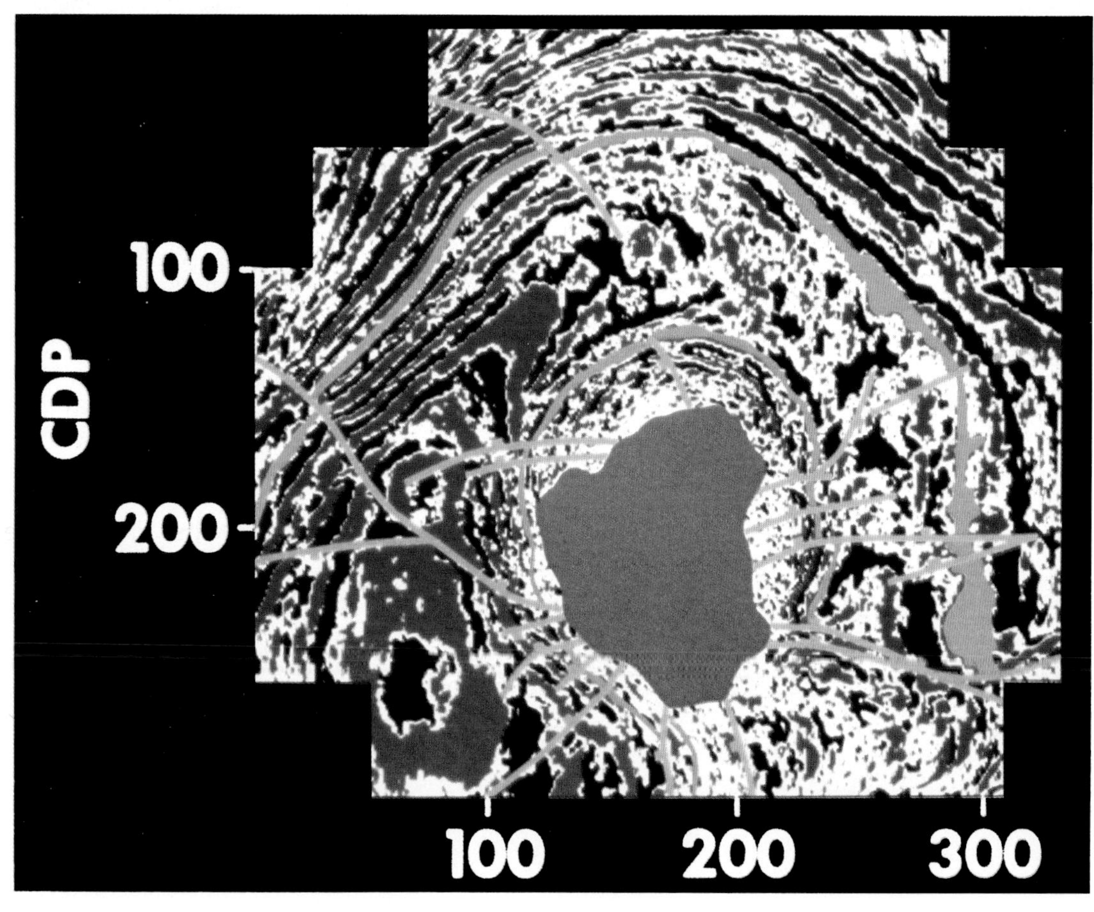

Fig. 3-20. Same horizontal section as Figure 3-19 with interpretation of faults and the green horizon. (Courtesy Hunt Oil Company.)

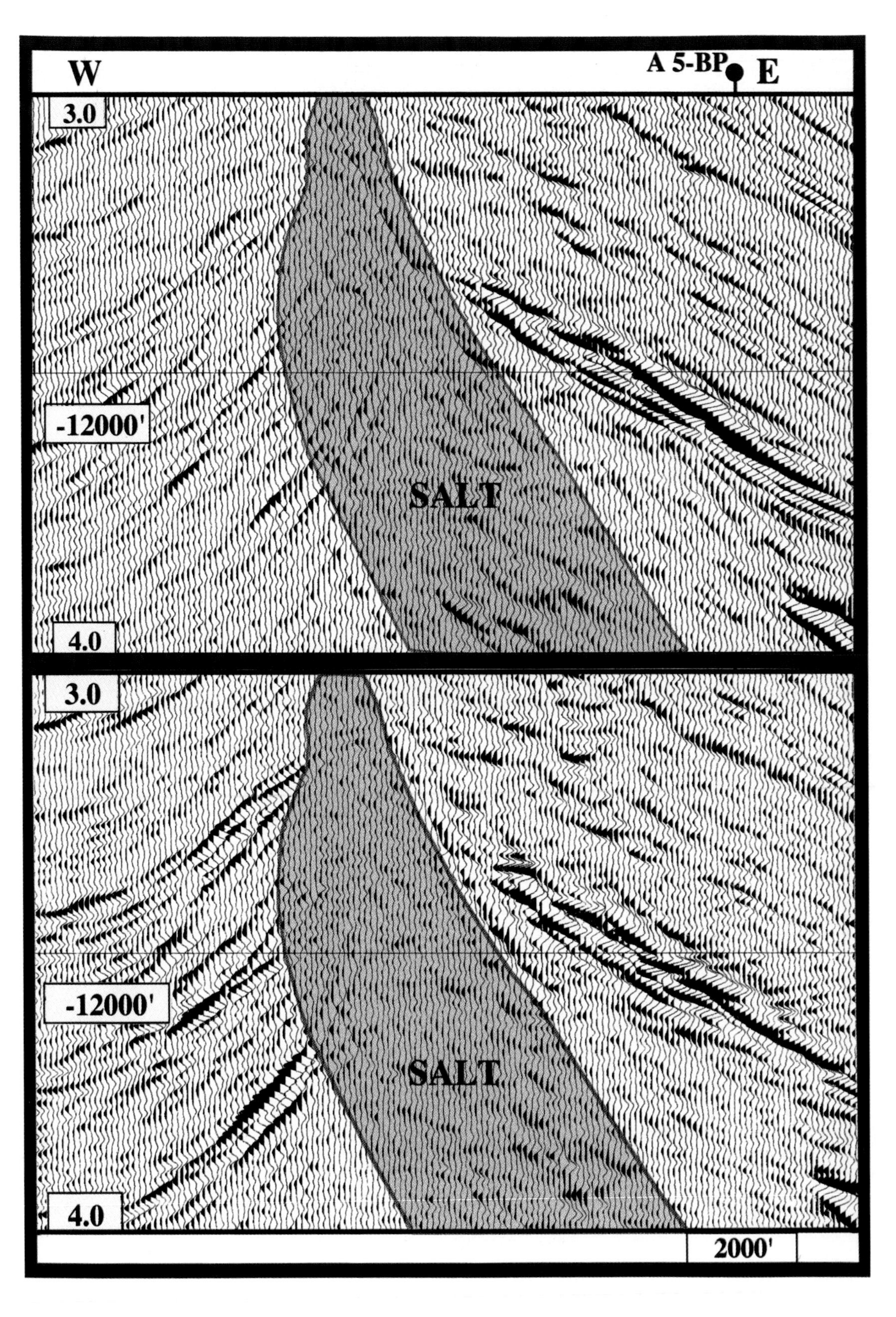

Fig. 3-21. Line 556 from the E-W survey at Bullwinkle (upper section). This line is extracted along the inline direction of this survey, hence the shooting direction is dip to the salt/sediment contact. Line 556 from the N-S survey at Bullwinkle (lower section). This line is extracted from the crossline direction of this survey. The shooting direction is perpendicular to the plane of the section and therefore strike to the salt/sediment contact. Note the improved sediment image along the western side of the overhung salt. This is attributed to less salt-related ray path distortion. (Courtesy Shell Oil Company.)

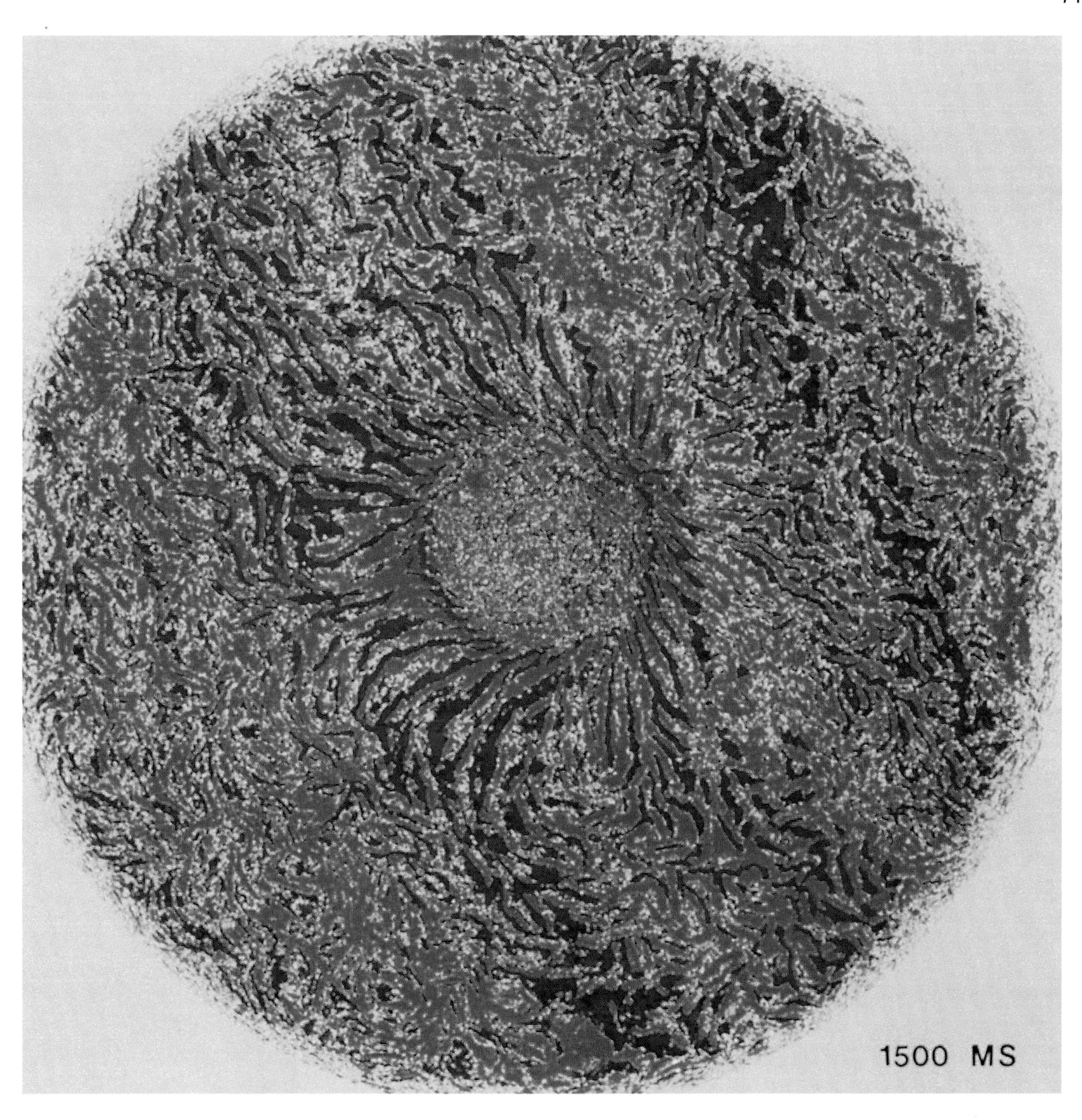

Fig. 3-22. Time slice 1500 ms from a circular 3-D survey in the North Sea. The shattered appearance results from the shattering of a thin limestone. (Courtesy Mobil North Sea Limited.)

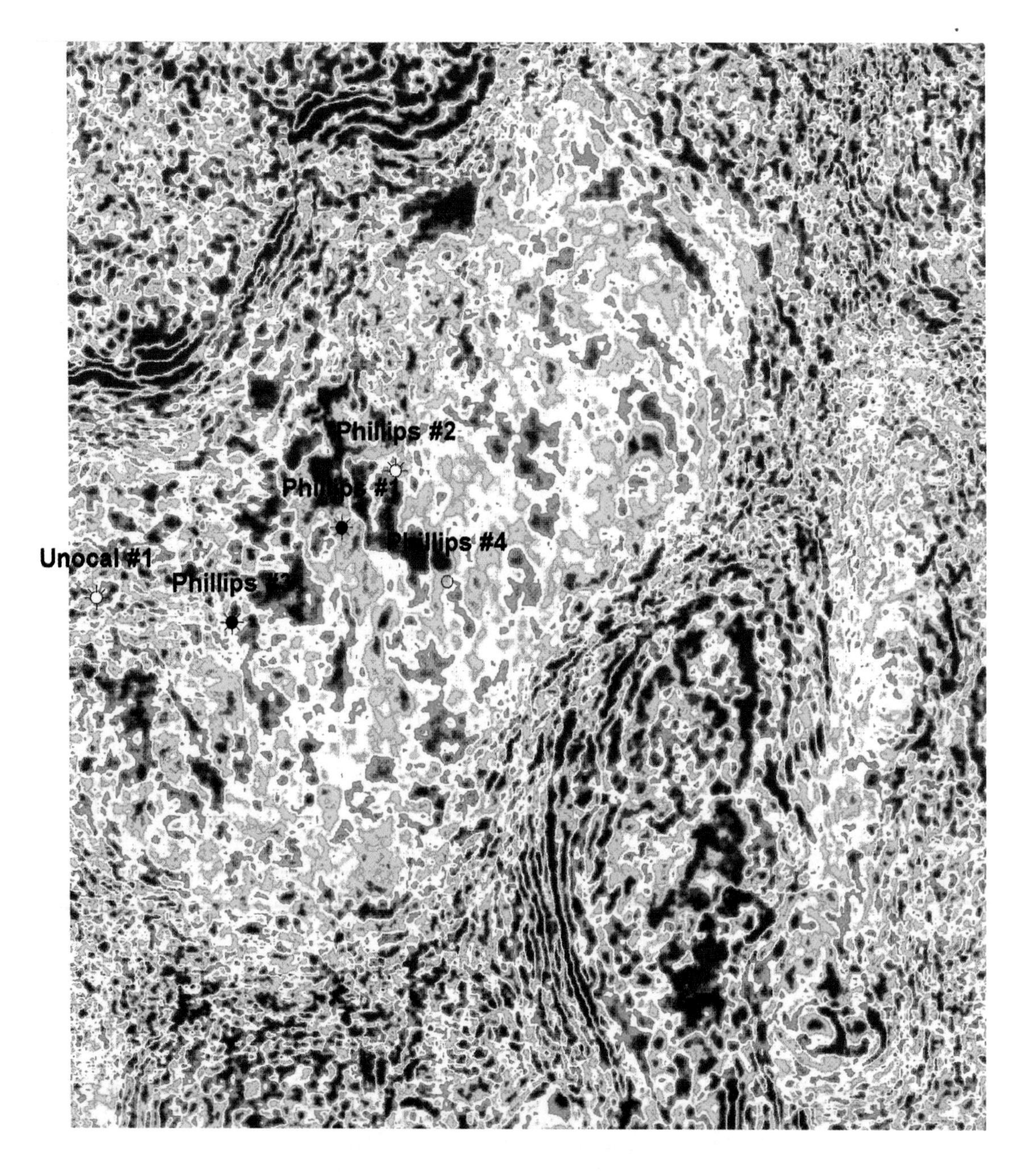

Fig. 3-23. Depth slice 4530 meters from 3-D prestack depth migrated data volume covering the Gulf of Mexico Mahogany prospect. (Courtesy Diamond Geophysical Service Corporation.)

When faults have been picked on several vertical and horizontal sections, the faults can be displayed as surfaces to check their geological validity.

Interpretation in the Vicinity of Salt

The horizontal section of Figure 3-18 shows a rim syncline surrounding a salt diapir. The narrow events around the salt indicate the steep dips near the intrusion. Figures 3-19 and 3-20 show a deeper horizontal section from the same volume without and with interpretation. The horizon of interest, marked in green on Figure 3-20, is intersected twice, once on either side of the rim syncline. The faulting at this level, marked in yellow, is complex but can be seen fairly well on this one horizontal section. From pre-existing 2-D data in the area only one of these faults had been identified (Blake, Jennings, Curtis, Phillipson, 1982).

Interpretation of seismic reflection terminations against salt is a very important matter because many hydrocarbon traps are found in this structural position. Numerous data collection and processing developments have been aimed at this problem

(French, 1990). For example, full one-pass 3-D migration is considered preferable to the more traditional two-pass approach.

Case History 11 in Chapter 9 discusses the importance of precise definition of the salt/sediment interface and shows success in doing so. Figure 3-21 also addresses this issue and demonstrates that, by collecting the data in a direction strike to the salt/sediment interface, the definition of reflections terminating at the salt is significantly improved.

Figure 3-22 shows time slice 1500 ms from a circular 3-D survey in the North Sea. In the center a salt diapir is visible. Collecting data in circles around a circular salt body means that the collection direction is consistently strike to the salt/sediment interface. The myriad of short arcuate features on Figure 3-22 show the effects of shattering of a thin limestone layer encased in shale.

Depth migration and pre-stack depth migration in 3-D have recently become economically feasible and have been used extensively for imaging under salt (Appendix A). It is the abrupt large velocity contrasts that make this more elaborate migration necessary. After such processing the whole data volume is in depth and thus horizontal sections become **depth slices**. Figure 3-23 shows a depth slice under the Mahogany salt sill in the Gulf of Mexico and the successful wells.

Composite Displays

The interpreter of 3-D data is not restricted to single slice displays. Because the work is done with a data volume, composite displays can be helpful in appreciating three-dimensionality and also in concentrating attention on the precise pieces of data that provide insight into the problem at hand.

Figure 3-24 is a composite of horizontal and vertical sections spliced together along their line of intersection. The vertical section shows that the circular structure is a syncline. The horizontal section pinpoints the position of its lowest point. The fault on the left of this structure can be followed across the horizontal section. Figure 3-25 provides a different view of the structure. The same horizontal section is here spliced to the portion of the vertical section above in the volume.

It is possible to make cube displays showing, simultaneously, three orthogonal slices through the volume (Figures 1-14 and 1-15, and 3-26). These can certainly aid in the appreciation of three-dimensionality but have limited application in the mainstream of the interpretation process, because two of the faces of any cube displayed on a monitor or piece of paper will always be distorted. An adaptation of the cube display concept is presented in Figure 3-27 and is known as the **chair**; it is really just the cube with a vertical section added above the horizontal section at the back. On Figure 3-27 the three-dimensional shape of a growth fault can be followed easily.

Figure 3-28 is a different kind of chair display. It is less dramatic than the chair of Figure 3-27 but probably more useful because none of the sections are distorted. The fault on the left can be followed clearly across the horizontal section indicating that the fault visible on line 75 (top, Figure 3-28) is the same as seen on line 110 (bottom, Figure 3-28). The other faults have a distinctly different azimuth and also traverse a shorter distance.

Figures 3-29 and 3-30 illustrate the study of a trio of normal faults. In Figure 3-29 one horizon has been tracked indicating the interpreted correlation across the faults. At the bottom of this figure a portion of the data from each of the four fault blocks is enlarged and again carries the interpreted track. Each block has been adjusted vertically to bring the track segments into continuity so that the correlation between these blocks of data can be assessed easily. Note how this display accentuates the apparent growth on the center fault of the three. In Figure 3-30 the composite horizontal and vertical section display permits the study of the horizontal extension of each of these three faults. A display customized to a problem usually helps significantly in the solution of that problem.

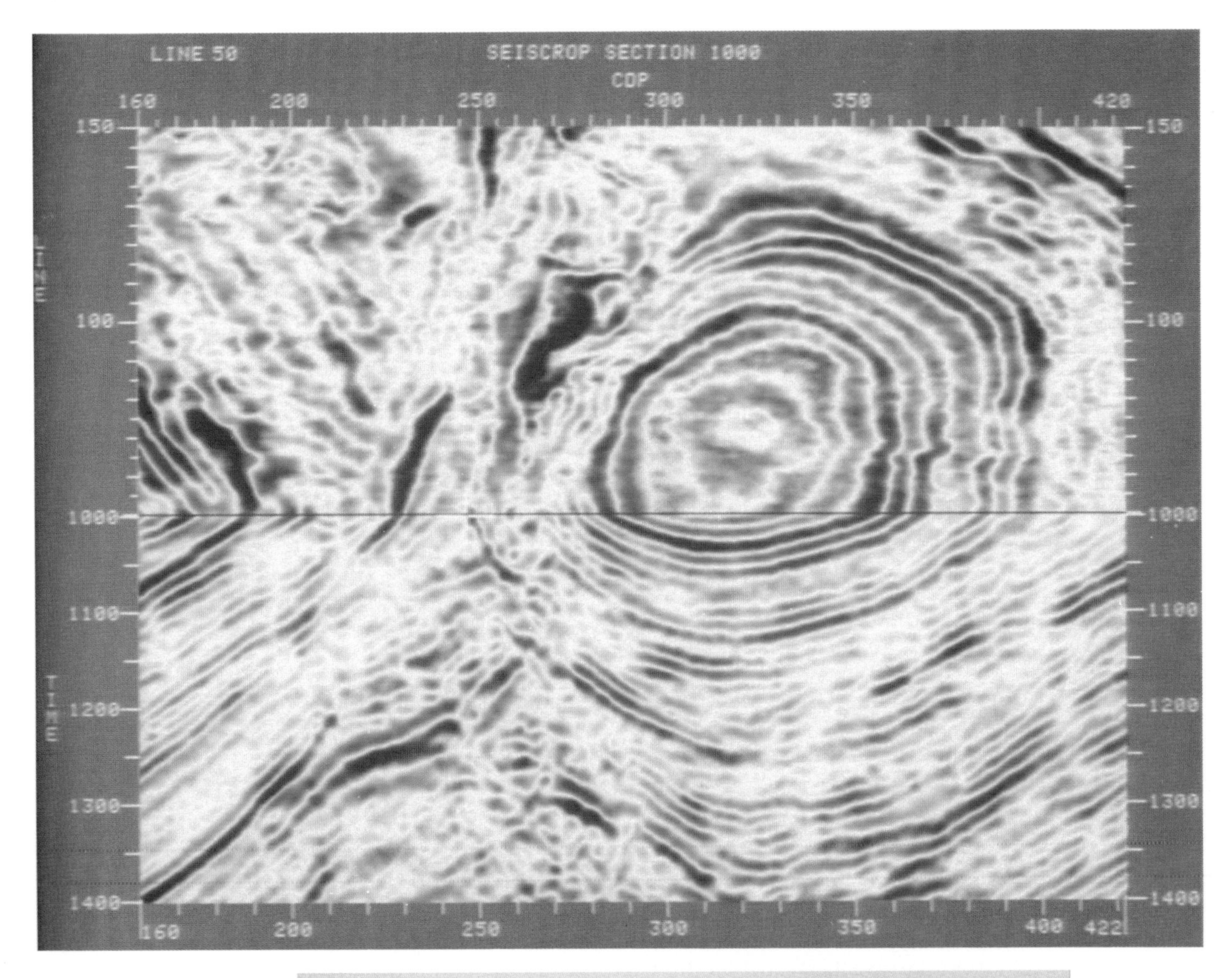

Fig. 3-24. Composite display of horizontal and vertical sections from onshore Europe. Vertical section segment lies beneath horizontal section.

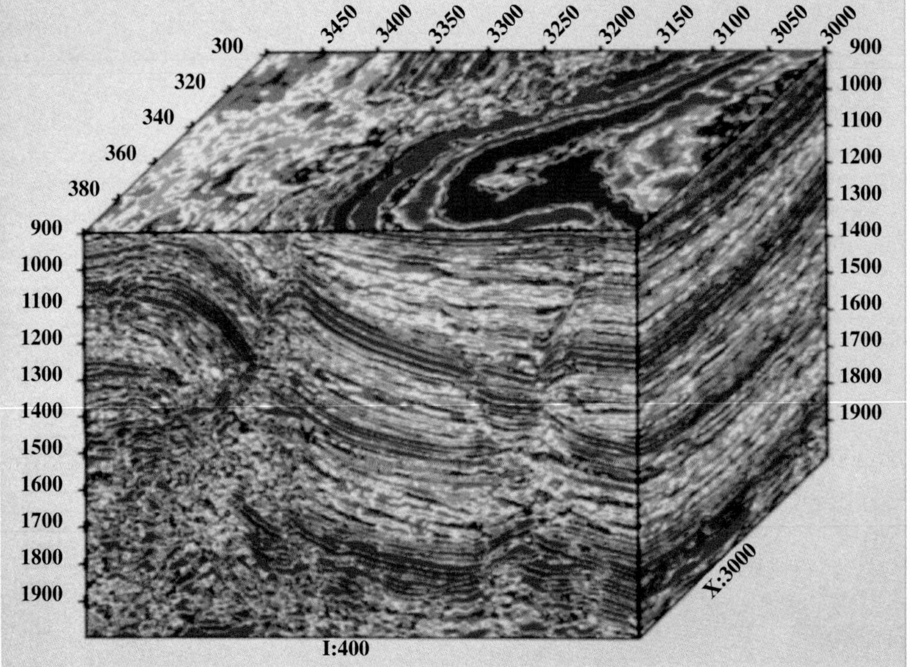

Fig. 3-26. Cube display made of two orthogonal vertical sections and one horizontal section. (Courtesy Western Atlas International.)

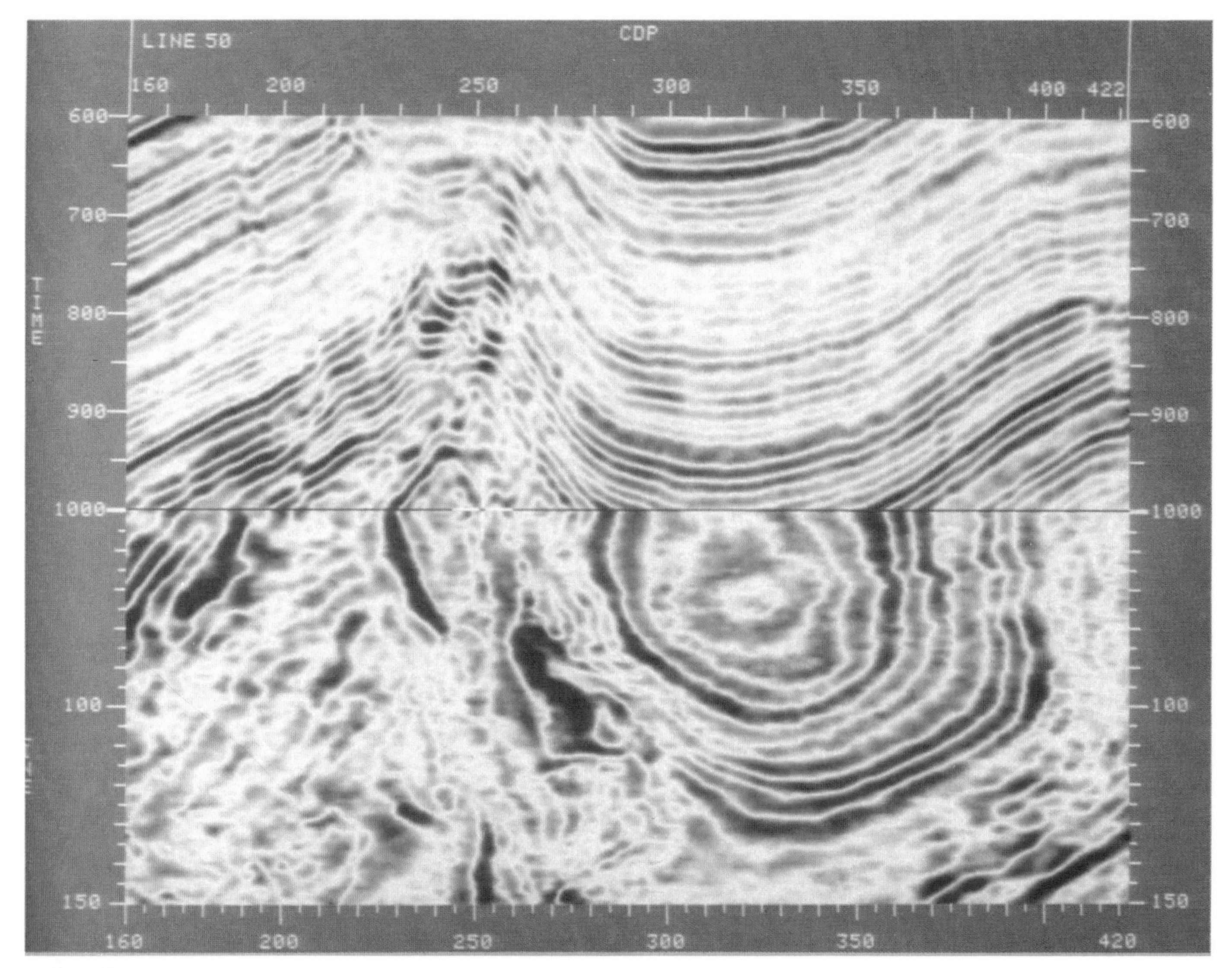

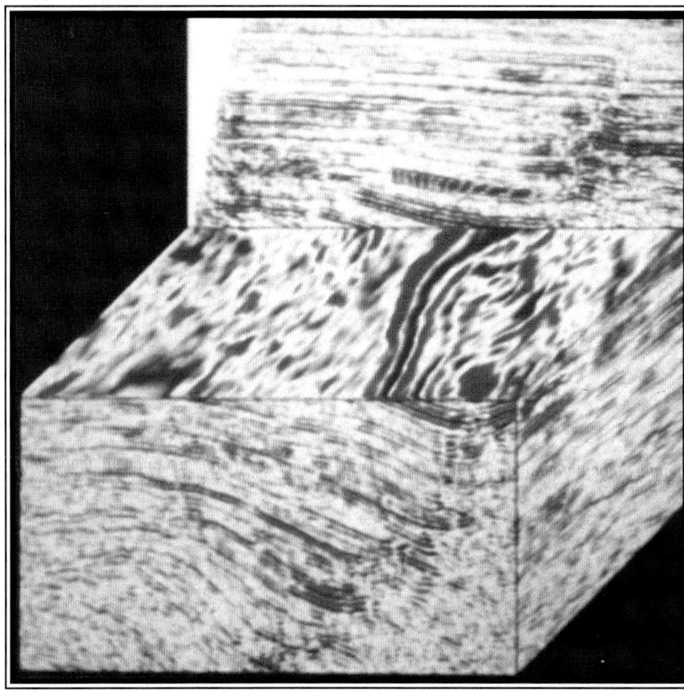

Fig. 3-25. Composite display of horizontal and vertical sections from onshore Europe. Vertical section segment lies above horizontal section.

Fig. 3-27. Chair display of Gulf of Mexico data made of two lines, one crossline and one horizontal section. (Courtesy Geophysical Service Inc.)

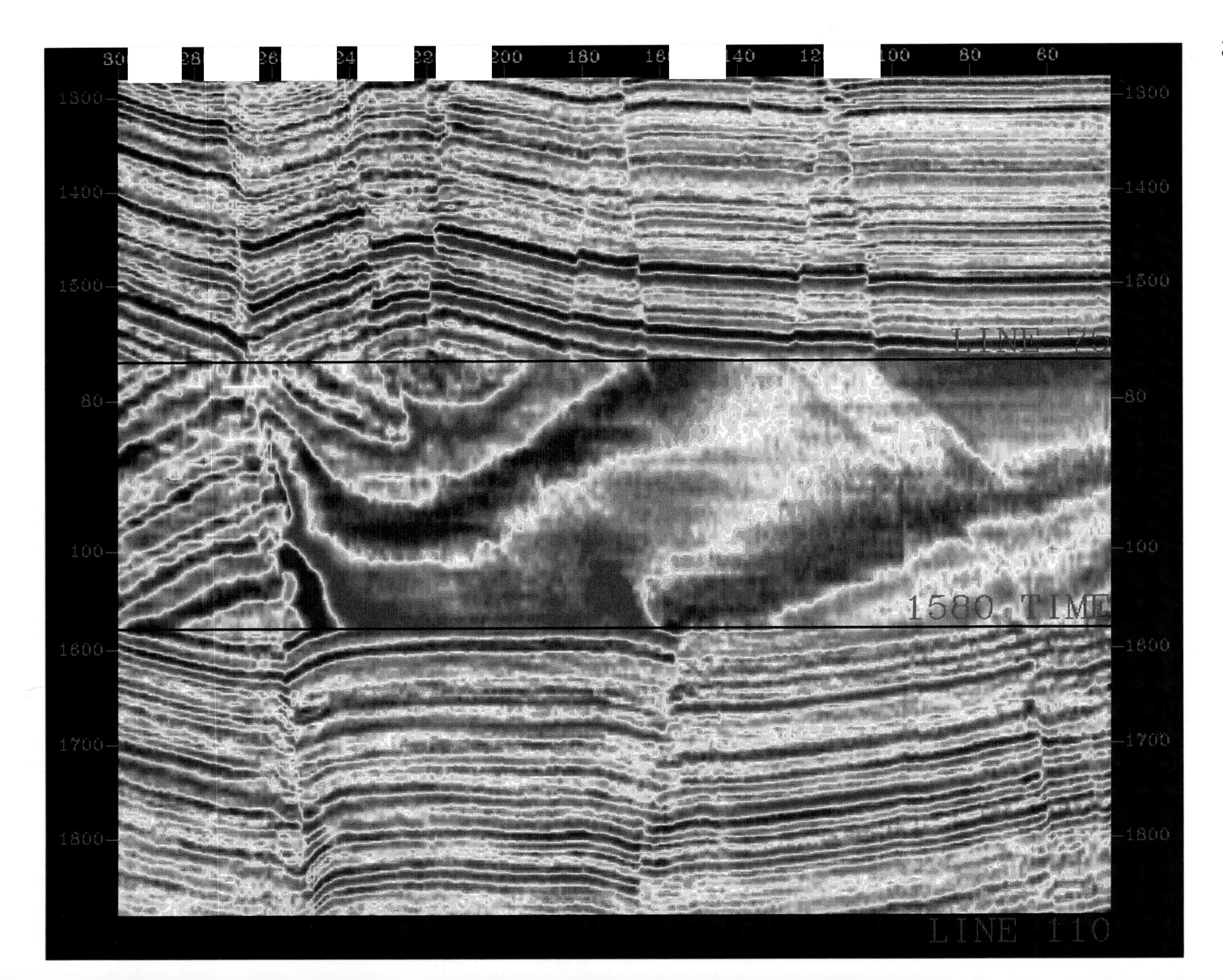
LINE 110
1580 TIME

Fig. 3-28. Chair display made of two vertical sections and one horizontal section. Compare this with Figure 3-27 and note that here all the three sections are undistorted. (Courtesy Landmark Graphics Corporation.)

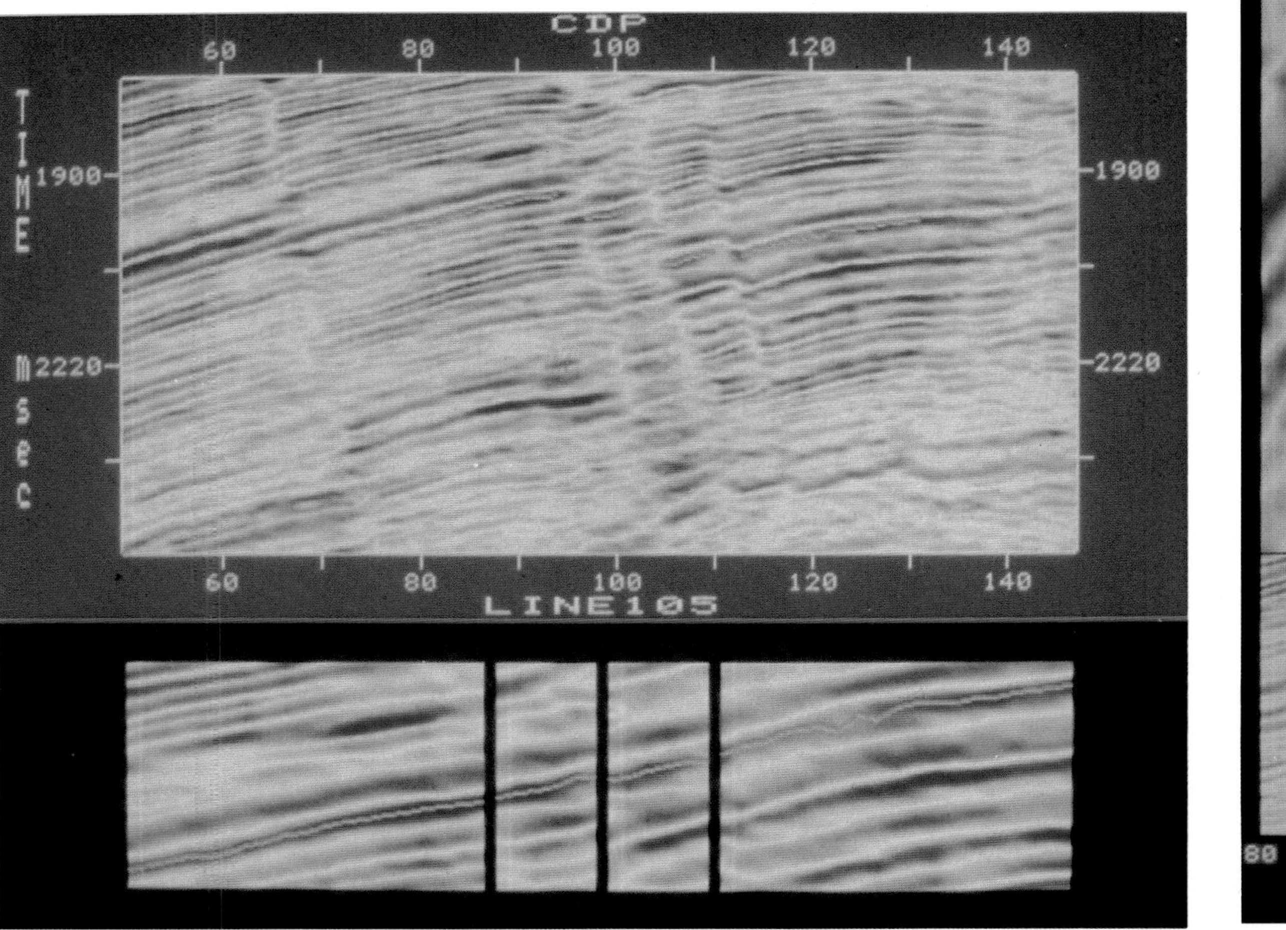

Fig. 3-29. Vertical section and magnified portions thereof des gned to study fault correlations offshore Trinidad. (Courtesy Texaco Trinidad Inc.)

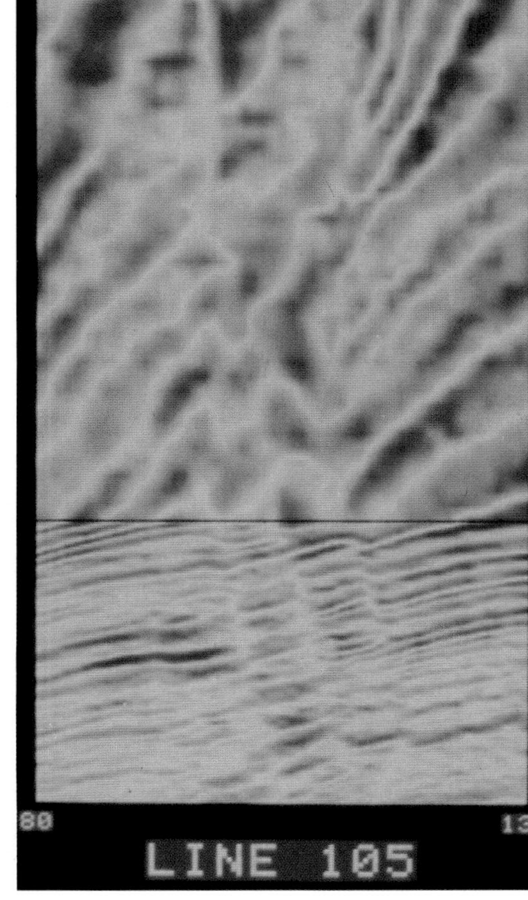

Fig. 3-30. Composite display of horizontal and vertical sections from offshore Trinidad showing horizontal extent of faults studied in Figure 3-28. (Courtesy Texaco Trinidad Inc.)

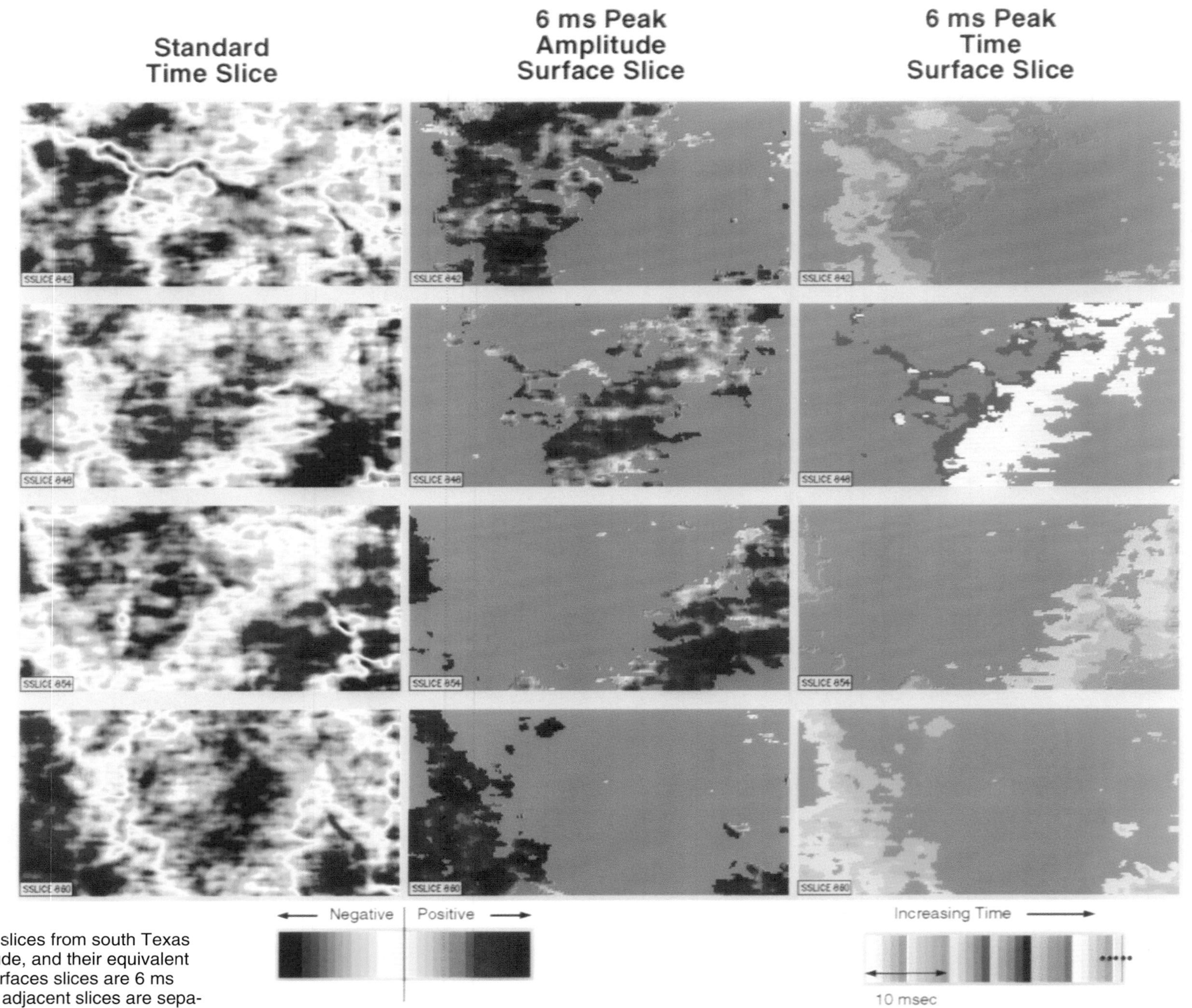

Fig. 3-31. Surface slices from south Texas in time and amplitude, and their equivalent time slices. The surfaces slices are 6 ms thick and vertically adjacent slices are separated by 6 ms. (Courtesy Exxon Production Research Company.)

RECOMMENDED PROCEDURE

1. Preview of data on composite displays and movies.
2. Horizon identification at wells.
   Assessment of data phase and polarity.
3. Recognition of major faults on widely-spaced vertical sections.
4. Fault framework by tying together with horizontal sections.
5. Initial horizon control using vertical and horizontal sections.
   Horizontal sections provide efficiency of coverage.
6. Automatic spatial tracking to complete horizon on every point.
   Autotracking provides precision picking.
   Get to this point as quickly as possible.
7. Scrutiny of intermediate horizon products for new features and for validation of tracking:
   Color-posted time structure (including lineations of untracked points)
   Color-posted horizon slice (for lineations and patterns in amplitude)
   High spatial frequency residual
   Dip magnitude and azimuth, difference, edge detection and illumination
8. Revision of horizons and faults, and rerun of autotracking.
9. Final time structure maps and horizon slices with chosen amounts of gridding or smoothing.
10. Isochron, isopach and depth maps.
11. Detailed stratigraphic and reservoir studies.

Fig. 3-32. Recommended interactive 3-D interpretation procedure.

Interpretation Procedures

The interpreter of 3-D data has a real opportunity to generate accurate subsurface structure maps but to do so a large amount of data must be studied. Furthermore *all* the data must be used to ensure proper information extraction. Thus the interactive workstation is an essential data management tool.

We will first discuss the interpretation procedure used with the Seiscrop Interpretation Table. Although this device is used rarely if at all today, it teaches an important component of 3-D structural interpretation—the use of the horizontal section. In the late 1970s and early 1980s, interpreters concentrated too much on horizontal sections through the use of the Seiscrop Interpretation Table. Today, most interpreters concentrate too much on vertical sections, because vertical sections can be manipulated and tracked easily on interactive workstations and because the interpreters' previous experience makes them prefer that perspective.

Initially the interpreter will pick faults and make a preliminary interpretation on a selected set of vertical sections in the line and crossline directions, for example on a one kilometer grid. This will provide the approximate extent of the first fault block in which mapping will begin. Normally the interpreter will identify the horizon to be followed at a well. Using the selected set of vertical sections the approximate fault locations are marked on the base map on the screen of the table. The event to be mapped is then identified on one horizontal section and followed up and down within the first fault block, drawing contours from the horizontal sections at the desired

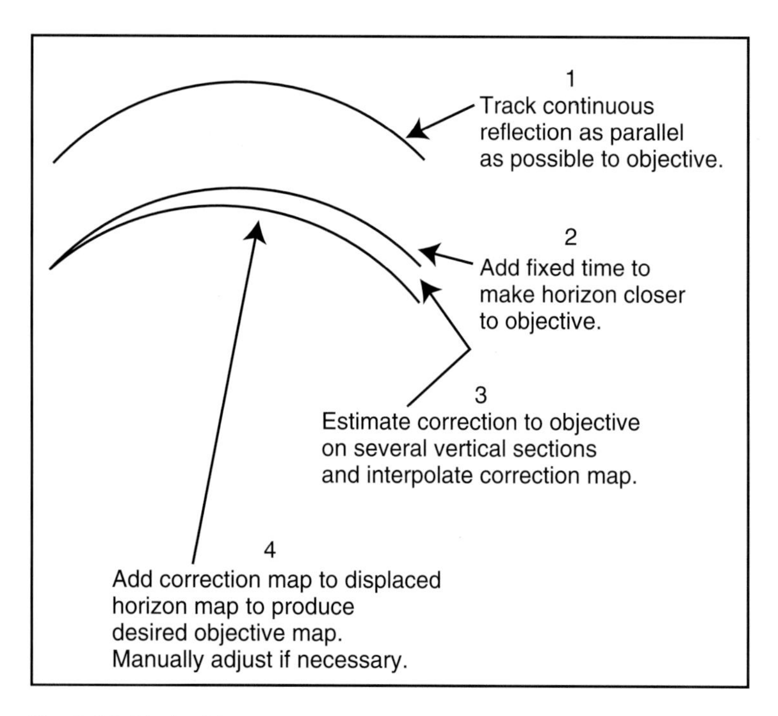

Fig. 3-33. Method for establishing a horizon through poor data or along an unconformity - the 3-D equivalent of a phantom horizon.

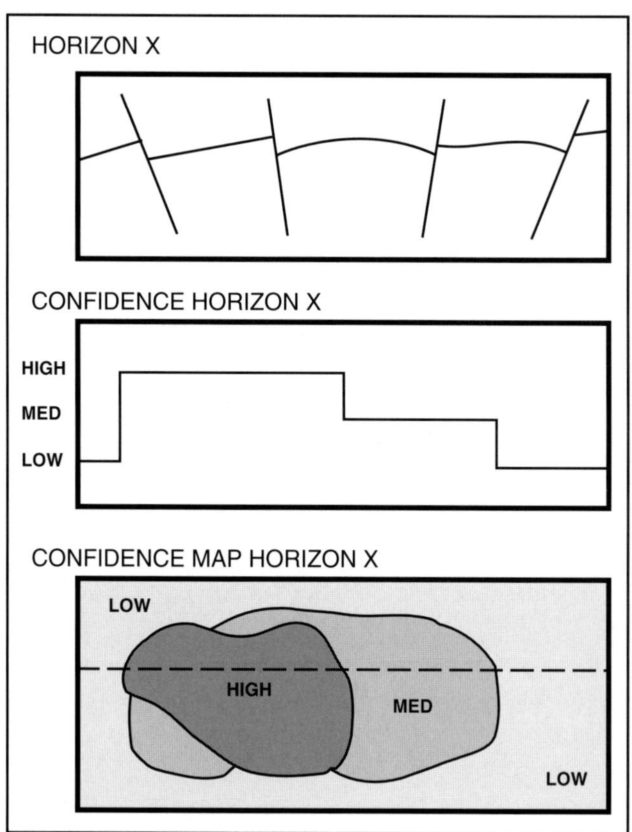

Fig. 3-34. Method for generating a confidence map to help qualify interpreted results.

interval. The faults surrounding the first fault block are marked in detail at the same time. Several iterations through the sections covering the structural relief of the horizon in this first fault block may be necessary before the interpreter is satisfied with the contours drawn. Selected vertical sections are revisited to establish the correlation into the next fault block and the procedure then repeats in that fault block. The interpreter thus works from fault block to fault block until the prospect is covered; alternatively the same horizon may be carried in two or more blocks at the same time. When the interpreter encounters a problem in understanding the data at a particular location, reference to vertical sections through that point in line, crossline, and other directions is made. Arbitrary lines may be specially extracted from the data volume for the purpose. Once the problem is resolved, the interpreter should be able to return to the horizontal sections to continue contouring.

This experience with time slices (horizontal sections) underlines the value of the strike perspective discussed earlier. Continuing experience in 3-D interpretation consultation has taught the author that proper use of time slices is one of the most difficult but important hurdles for everyone wishing to master the technology. Not only do time slices provide an essential view of the data which reveals various subtle features, they also add significant efficiency to the total project. Most interpretation procedures today involve use of automatic spatial tracking to ensure that all the data is used and that high precision is obtained. The autotracker needs control or seed points and horizontal sections should be used along with vertical sections for this purpose. Once some vertical sections have been examined to establish general structural relationships, spatial continuity should be followed on appropriate time slices. These directly drawn structural contours then give the autotracker plenty of seed points to tie it down. A successful way of building time slices into the interpretation procedure is to use composite displays such as Figures 3-24 and 3-25 or chair displays such as Figure 3-28. In this way the relationship between horizontal and vertical section is apparent.

Another way of incorporating the efficiency of the strike view into interpretation procedures is to use **surface slices** (Stark, 1996). A surface slice displays either time or amplitude of one chosen point on the seismic waveform, for example the crest of a peak, over a defined time interval (typically three samples). A surface slice can thus, in a sense, be regarded as a thick time slice. However, the surface slice displays one data phase only over that thickness, in contrast to the time slice which shows all seismic phases at one unique time. A surface slice thus has no frequency dependence and provides true dip magnitude. Figure 3-31 contains a sequence of time slices and the equivalent amplitude surface slices and time surface slices. The surface slices are 6 ms thick and contain only peaks, either their horizon amplitudes or their travel times. Slices on the same row are from the same time, and the rows are separated by 6 ms. Vertically adjacent surface slices are thus independent and it can be seen that the edges of the colored areas fit together. This is the procedure with surface slices - a kind of jigsaw puzzle of horizon segments.

Figure 3-32 charts a recommended procedure for 3-D interpretation using an interactive workstation. The interactive capabilities required to follow this procedure include:

(1) automatic and manual tracking of horizons on vertical and horizontal sections;
(2) automatic spatial horizon tracking and editing through a 3-D data volume;
(3) correlation of vertical sections with well data;
(4) extraction, storing and manipulation of seismic amplitudes;
(5) manipulation of maps;
(6) flexible use of color; and,
(7) extraction and use of seismic attributes.

This approach incorporates many of the notions from the previous procedures but utilizes the greatly extended capabilities. The procedure of Figure 3-32 also addresses several areas of stratigraphic and reservoir interpretation which will be discussed in later chapters.

Some of the important principles implicit in the procedure of Figure 3-32 are that you

- understand the phase of data *before* embarking on the mainstream interpretation,
- use horizontal sections to full advantage; benefit from the efficiency of strike,
- study only as many vertical and horizontal sections as is necessary to provide initial input control for automatic spatial tracking,
- use intermediate horizon products to full advantage for refining the interpretation,
- do not smooth any map or map-style product until degree of smoothing required can be judged intelligently, and
- engage in stratigraphic and reservoir studies in order to get the most out of the data.

Autotracking is central to the procedure of Figure 3-32. More discussion on its use appears later in this chapter and also in Chapter 8. Today's autotrackers are fairly robust and will handle poorer data than many people think. However, there are always horizons of interest that have continuity too poor for acceptable autotracker performance.

In 3-D interpretation we studiously avoid drawing old-fashioned phantom horizons. Figure 3-33 shows the preferable alternative. We track a horizon above the objective horizon and displace it down with a constant time shift. Then we estimate the mistie between this and the desired horizon at several locations and interpolate a

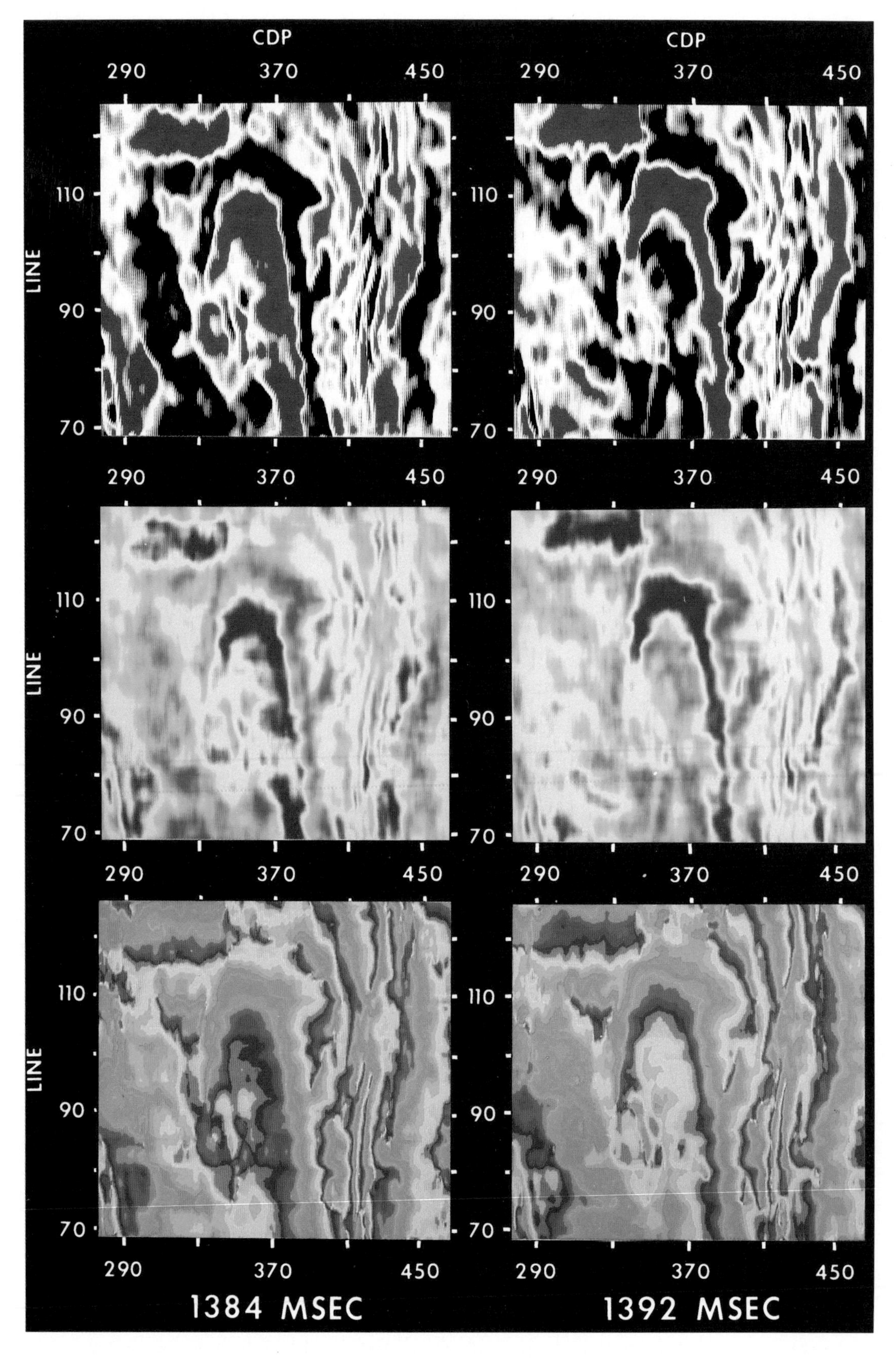

Fig. 3-35. Horizontal sections, 8 ms apart, from Gulf of Thailand displayed in dual polarity variable area (upper row), with seismic amplitude coded to color (middle row), and with instantaneous phase coded to color (lower row). (Courtesy Texas Pacific Oil Company Inc.)

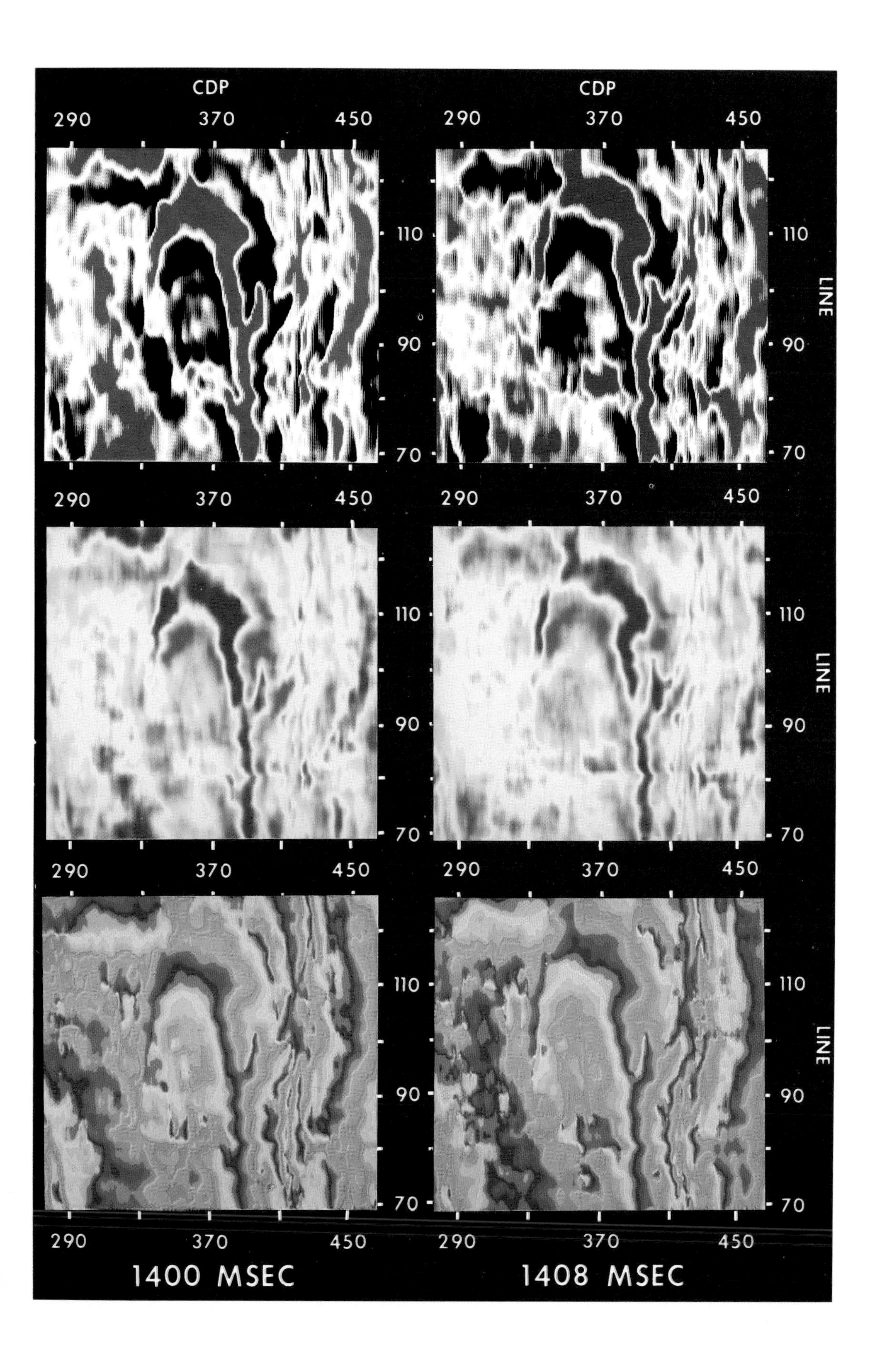
CDP
290
370
450
110
90
70
LINE
1400 MSEC
1408 MSEC

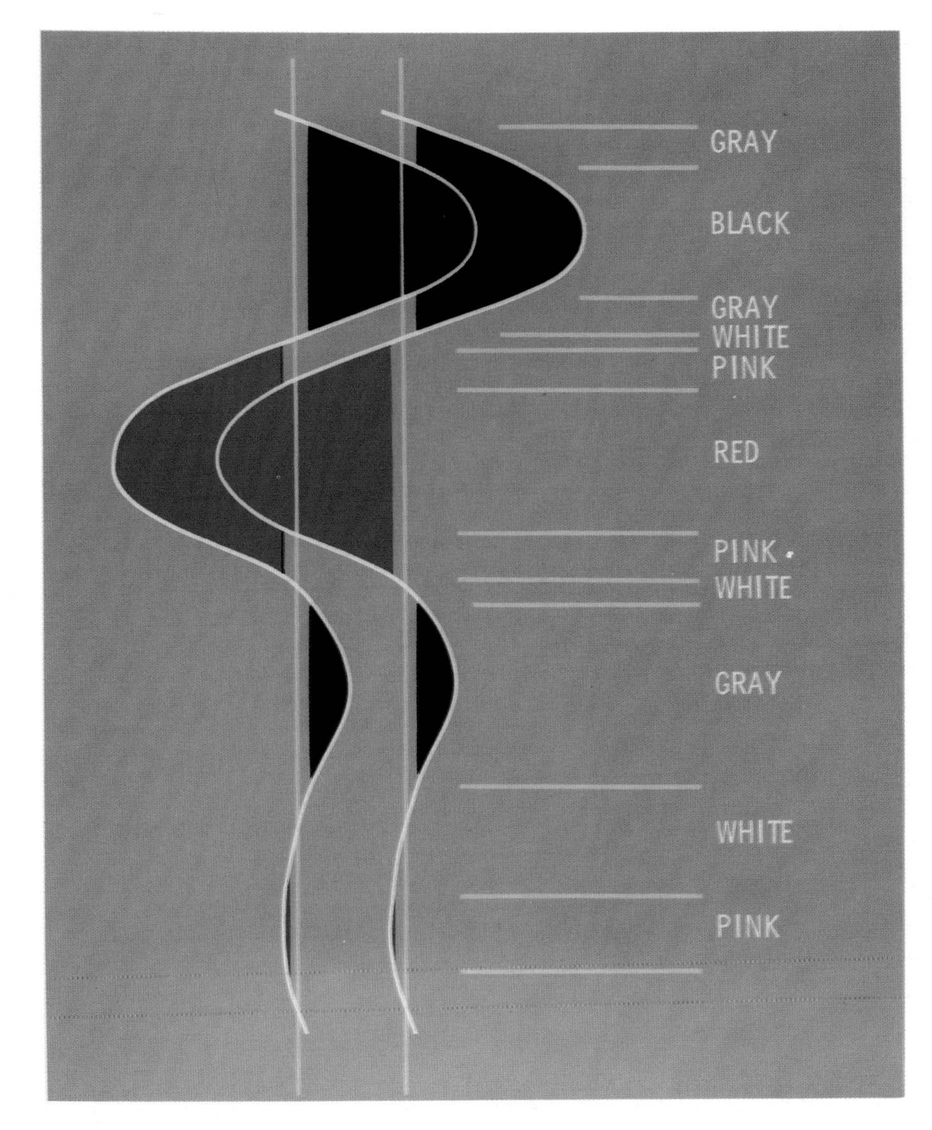

Fig. 3-36. Waveform definition using dual polarity variable area sections. The troughs are shown as excursions to the left; in practice sections are displayed with the troughs rectified and hence swinging to the right.

mistie, or correction, map. By adding this to the displaced hoizon map we should obtain something close to what is desired. We can then manually adjust as necessary, but the amount of manual work should be small compared with the traditional manual approach for the whole project.

Figure 3-34 illustrates a way of generating a confidence map to communicate to managers, engineers, clients, and others the confidence we have in the interpreted maps we produce. Qualitative assessment of confidence, here using three levels, is made occasionally based on data quality and whatever interpretation difficulties may occur. The resultant interpolated confidence map can be included with final maps for the benefit of those who later use them.

Advantages and Disadvantages of Different Displays

With increasingly successful amplitude preservation in seismic processing, interpreters are increasingly suffering from the limited optical dynamic range of conventional seismic displays. Too common are the variable area sections where some events of interest are heavily saturated and others have barely enough trace deflection to be visible. This applies to all displays, vertical and horizontal, made with variable area techniques. Horizontal sections, historically, were first made with variable area using one polarity only, normally peaks. This soon evolved into dual polarity variable area

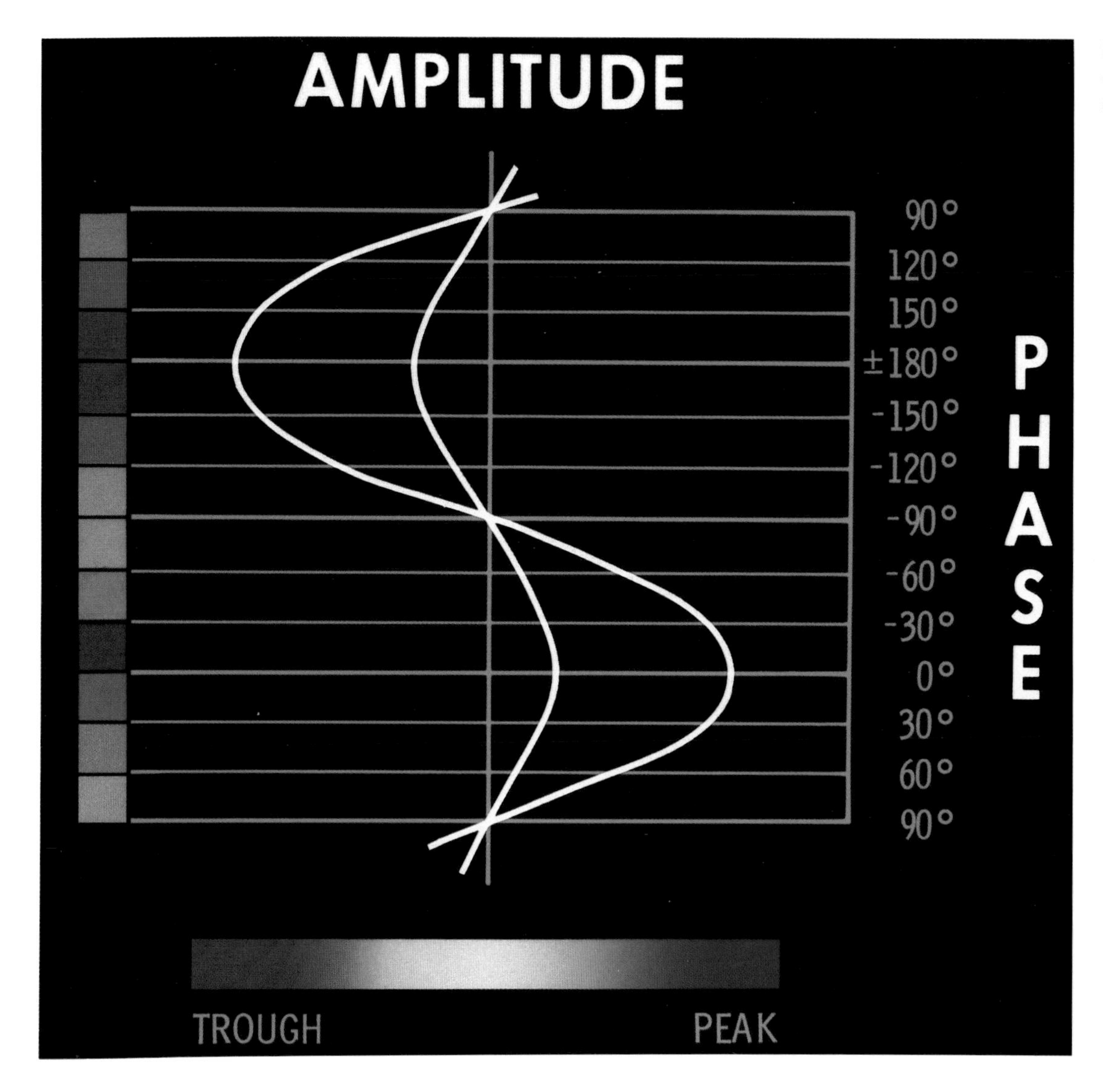

Fig. 3-37. Waveform definition using amplitude and phase color sections.

giving equal weight to peaks and troughs (see Chapter 2). This is exemplified by the upper row of sections in Figure 3-35 and explained in detail by the diagram of Figure 3-36.

Dual polarity variable area provides five clearly discernible amplitude levels. The highest amplitude peaks are saturated and appear as continuous black areas; the medium amplitude peaks do not coalesce and appear as discontinuous black areas which look gray; the lowest amplitudes are below the variable area bias level and appear white; the medium amplitude troughs appear pink; and the highest amplitude troughs are continuous red areas. In structural mapping the interpreter must pick a consistent point on the seismic waveform. Most commonly, a red event (trough) is followed and picked on the edge of the pink, that is close to, but not quite at, the zero crossing. This point is fairly consistent in phase as the amplitude of the event changes with position over the prospect. It is more consistent than the edge of the red, the other simple option with this mode of display.

If the detail in the seismic waveform provided by dual polarity variable area is inadequate, which is commonly the case today, then the increased dynamic range of full variable intensity color is required. The many ways of using color to interpretive advantage are discussed in Chapter 2. Gradational blue and red is a most useful application; this is illustrated in the middle row of sections in Figure 3-35 and explained in detail by the diagram of Figure 3-37. On such a display the interpreter can see the local amplitude maxima of a peak (or a trough) and draw a contour along

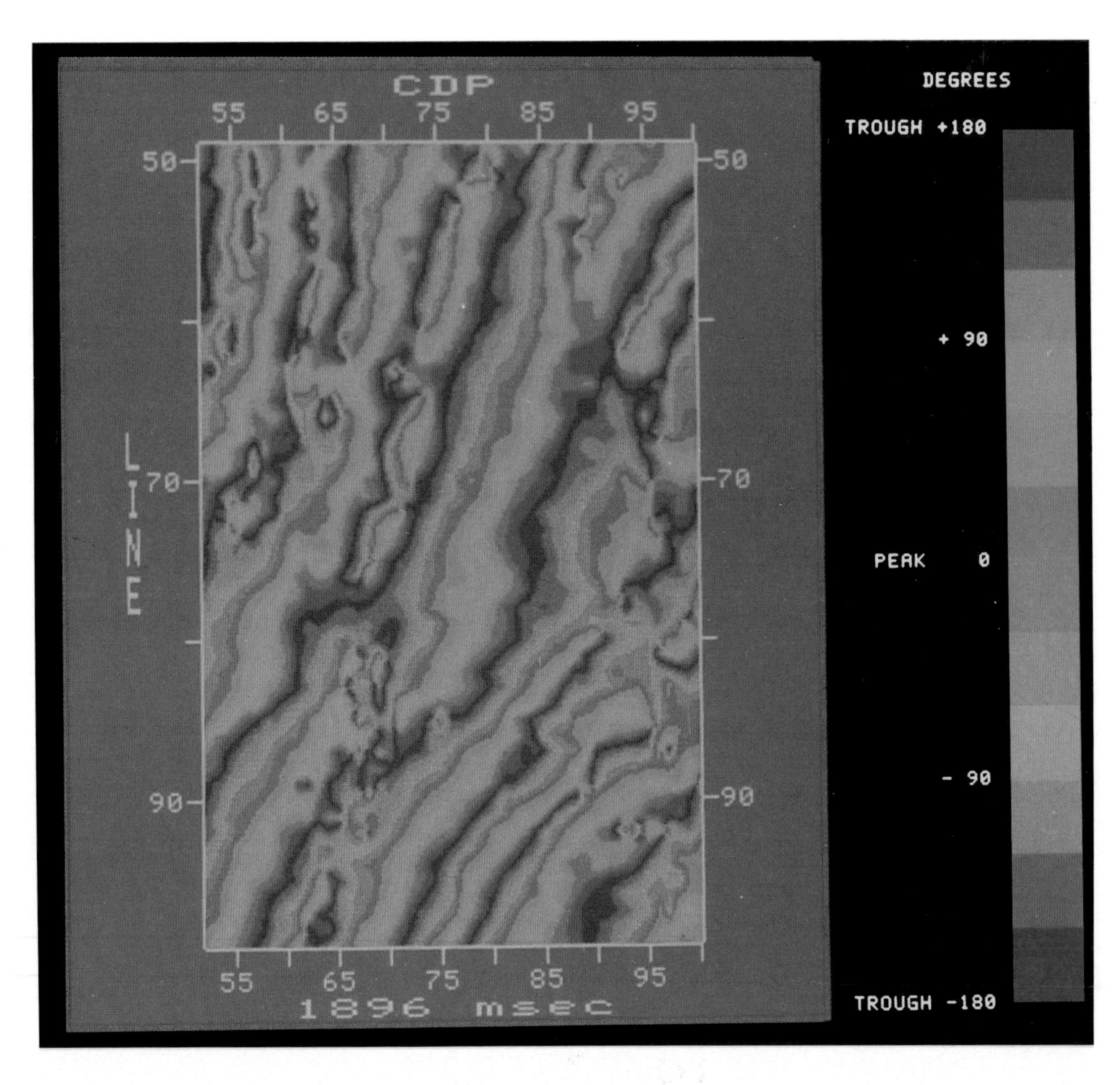

Fig. 3-38. Horizontal section at 1896 ms in instantaneous phase from offshore Trinidad. (Courtesy Texaco Trinidad Inc.)

the locus of those maxima, thus picking the crest of the seismic waveform.

A further option available to the structural interpreter is horizontal sections displayed in phase, using instantaneous phase derived from the complex trace (Taner, Koehler and Sheriff, 1979). This approach is illustrated by the lower row of sections in Figure 3-35 and explained in detail by the diagram of Figure 3-37. Phase indicates position on the seismic waveform without regard to amplitude, making a phase section like one with fast AGC (Automatic Gain Control), destroying amplitude variations and enhancing structural continuity. A phase section is displayed with color encoded to phase over a given range, for example 30°. Color boundaries occur at significant phase values such as 0° (a peak), 180° (a trough), +90° and –90° (zero crossings). By following a chosen color boundary on a horizontal section displayed in this way, the interpreter is drawing a contour for his horizon map picked at a specific phase point. Thus the interpreter can also, if necessary, compensate for any estimated amount of phase distortion in the seismic wavelet.

Figure 3-38 is a horizontal phase section from a different area; the structural continuity is clear. Figure 3-39 shows the same section in edited phase, a simple modification of the display colors. A few degrees of phase centered on 0° have been colored black; a few degrees of phase centered on 180° have been colored red; and all other phases have been colored white. This gives the appearance of an automatically picked section with all the peaks and troughs at that level indicated. The interpreter simply selects the one he wants. A combination of these phase and amplitude displays is provided by Figure 3-40, where edited phase highlights the positions of the maximum amplitudes of peaks and troughs.

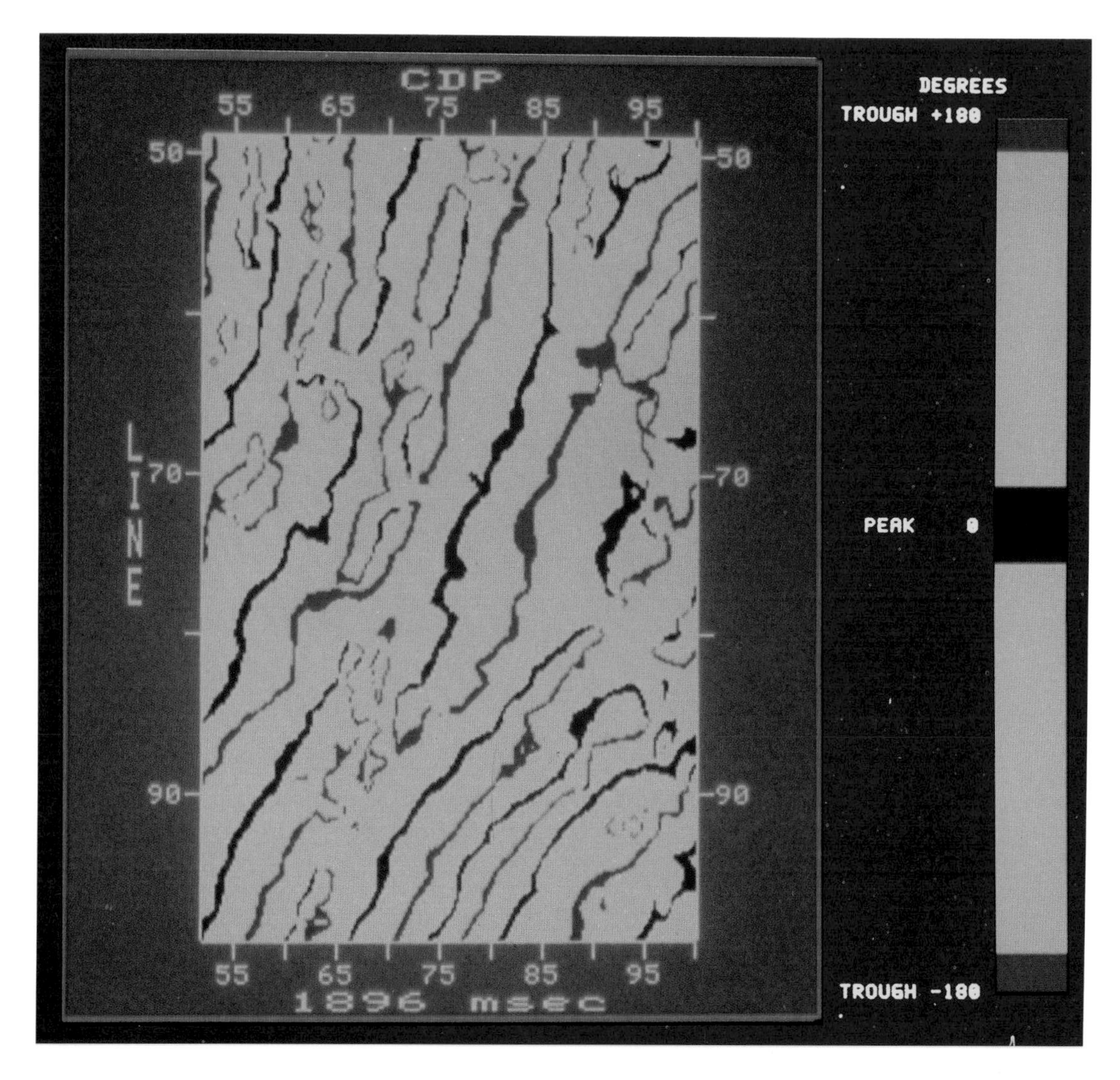

Fig. 3-39. Horizontal section at 1896 ms in edited phase from offshore Trinidad. (Courtesy Texaco Trinidad Inc.)

Horizontal Sections From Widely Spaced Data

Because of the interpretive benefits of horizontal sections, the question of their possible construction from widely-spaced data naturally arises—data such as a conventional 2-D grid. The interpreter would like to slice all the data values at a single time from the 2-D lines, lay each down at its appropriate horizontal position and fill the grid cell spaces by some data interpolation procedure. In general, 2-D data will not satisfy the 3-D sampling requirements between lines as discussed in Chapter 1. Aliasing will therefore occur across the 2-D grid cells during the data interpolation. This can usually be prevented by using a special type of interpolation, generally known as intelligent interpolation, but the resultant data still have the resolution only of the widely-spaced input data and thus are normally useful only for studying broad structural trends.

There exists an adaptation of 3-D surveying in which the parallel recording lines are more widely spaced than the sampling theory demands. Intelligent interpolation is then used during processing to fill the gaps with data at a suitably close spacing for 3-D migration and horizontal section generation. This type of horizontal section is useful for structural interpretation and will have no aliasing problems. However, the interpreter must recognize the spacing at which the original lines were collected and not expect to see structural detail with dimensions smaller than this. These widely-spaced 3-D data (Figure 3-41) are being used for regional studies of salt tectonics and for correlating around salt domes from one intersalt basin to another. They are opening up new styles of interpretation. Although this widely-spaced 3-D is little used today, it has led to very extensive coverage of properly-sampled 3-D in some areas. In the Gulf of Mexico, very large speculative 3-D surveys provide more 3-D data for

Fig. 3-40. Horizontal section in edited phase superimposed on amplitude. The edited phase in cyan follows the maximum amplitude of the peaks which are blue. The edited phase in yellow follows the maximum amplitude of the troughs which are red.

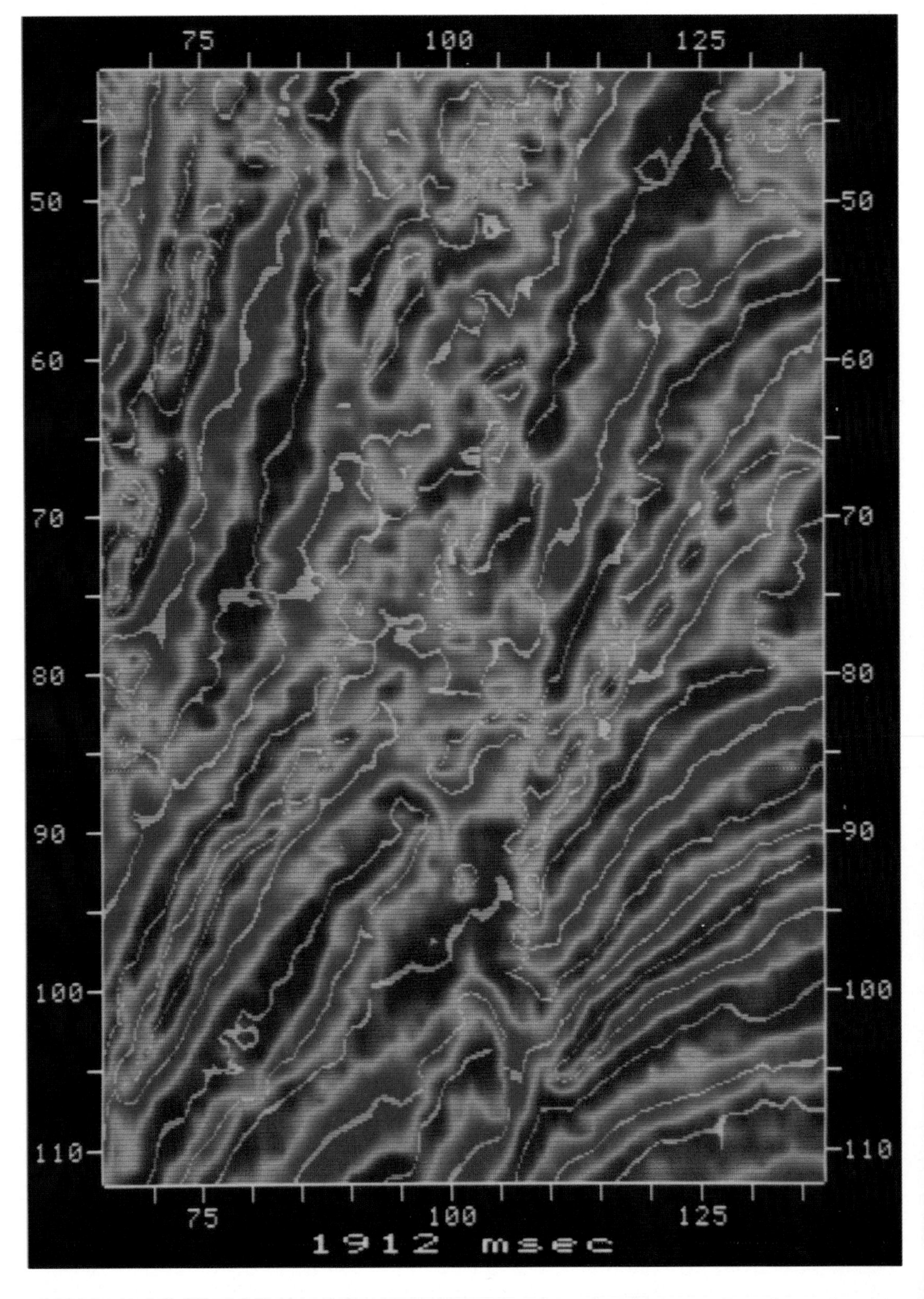

detailed prospect studies but also the broad view for visualizing regional trends as in Figure 3-41. On land, surveys are being spliced together to give the same kind of broad view over large areas, for example in Oman (Figure 8-13).

Subtle Structural Features

Figures 3-42 and 3-43 are horizontal sections from a data volume in which a subtle, small-throw fault became a significant part of the interpretation at the target level. On

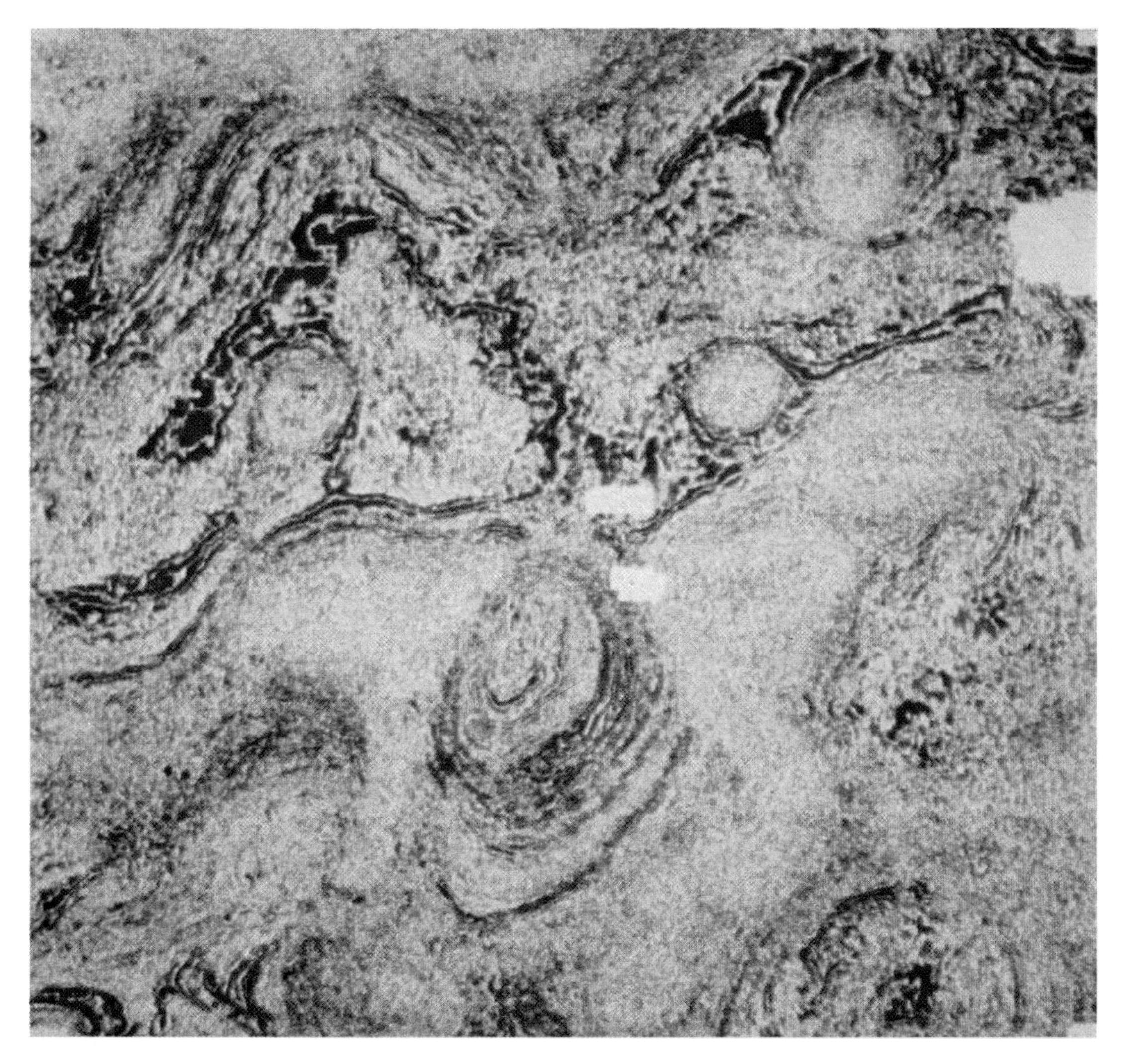

Fig. 3-41. Horizontal section from Gulf of Mexico widely-spaced 3-D data. This section covers an area of 1480 sq km (570 sq mi) and is used for regional studies of salt tectonics. (Courtesy Geophysical Service Inc.)

both figures the subtle fault is seen as a minor discontinuity in one peak (black) and one trough (red) between Lines 720 and 760 and Crosslines 40 and 55. Crossline 45 (Figure 3-44) shows this fault, in the middle of the section between 2.3 and 2.4 seconds, to be a really subtle feature. The interpreter working on these data first noticed the discontinuity on the horizontal sections and considered it a real geologic feature because it preserved its characgter over many contiguous slices. Hence, the interpretation of this nearly north-south fault was incorporated into the final structure map, as shown in Figure 3-45.

In an area of gentle dips, such as this one, the horizontal section events are broad and a discontinuity in them, namely in strike, may be more easily visible than a discontinuity in the events on the vertical sections. This demonstrates that horizontal sections have a role in the identification of subtle faults, just as they do in the recognition of subtle stratigraphic features, as discussed in Chapter 4. Figure 3-46 shows several straight lineations, principally through the black structural event, that are caused by subtle fautling and jointing. These are so subtle they would never be recognized on vertical sections. Here they are identified by the straight linear pattern that appears in the strike view.

Horizon and formation attributes are tools of major importance in the identification and study of subtle faults. These are discussed in Chapter 8.

Visualization and Autotracking

Visualization pervades all stages of 3-D interpretation starting with the volume of data (Chapter 1) and concluding with the extensive discussion on horizon attributes

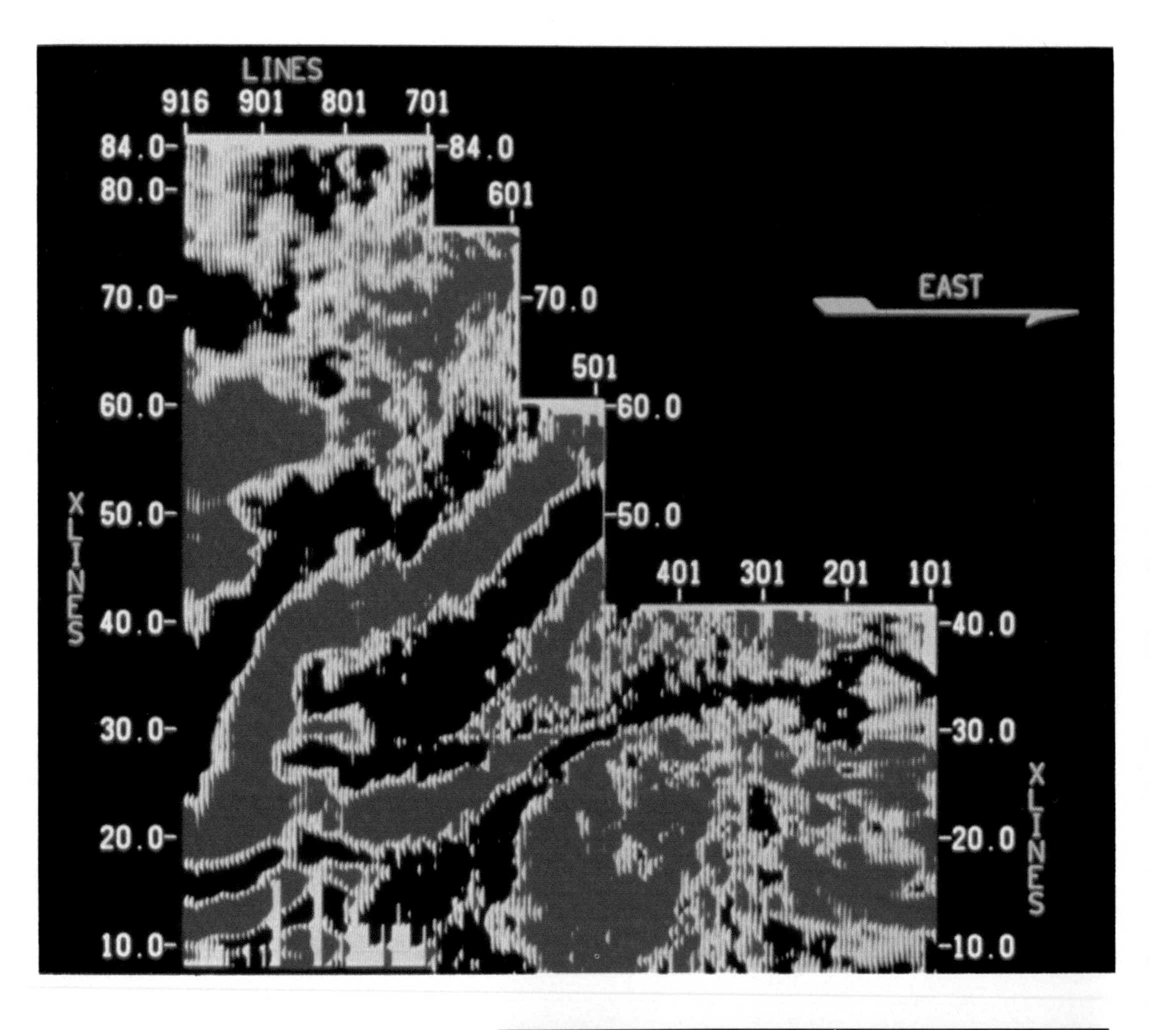

Fig. 3-42. Horizontal section at 2332 ms from south Louisiana marsh terrain. (Courtesy Texaco Inc.)

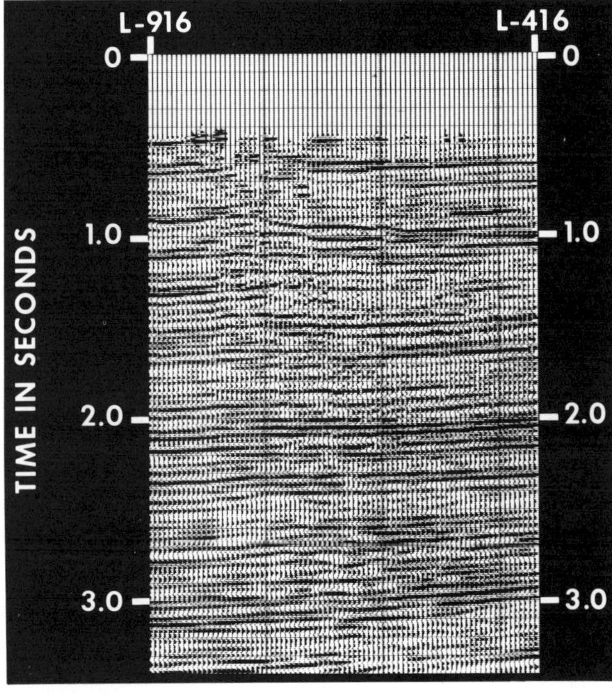

Fig. 3-44. Crossline 45 showing subtle fault identified on horizontal sections of Figures 3-42 and 3-43. (Courtesy Texaco Inc.)

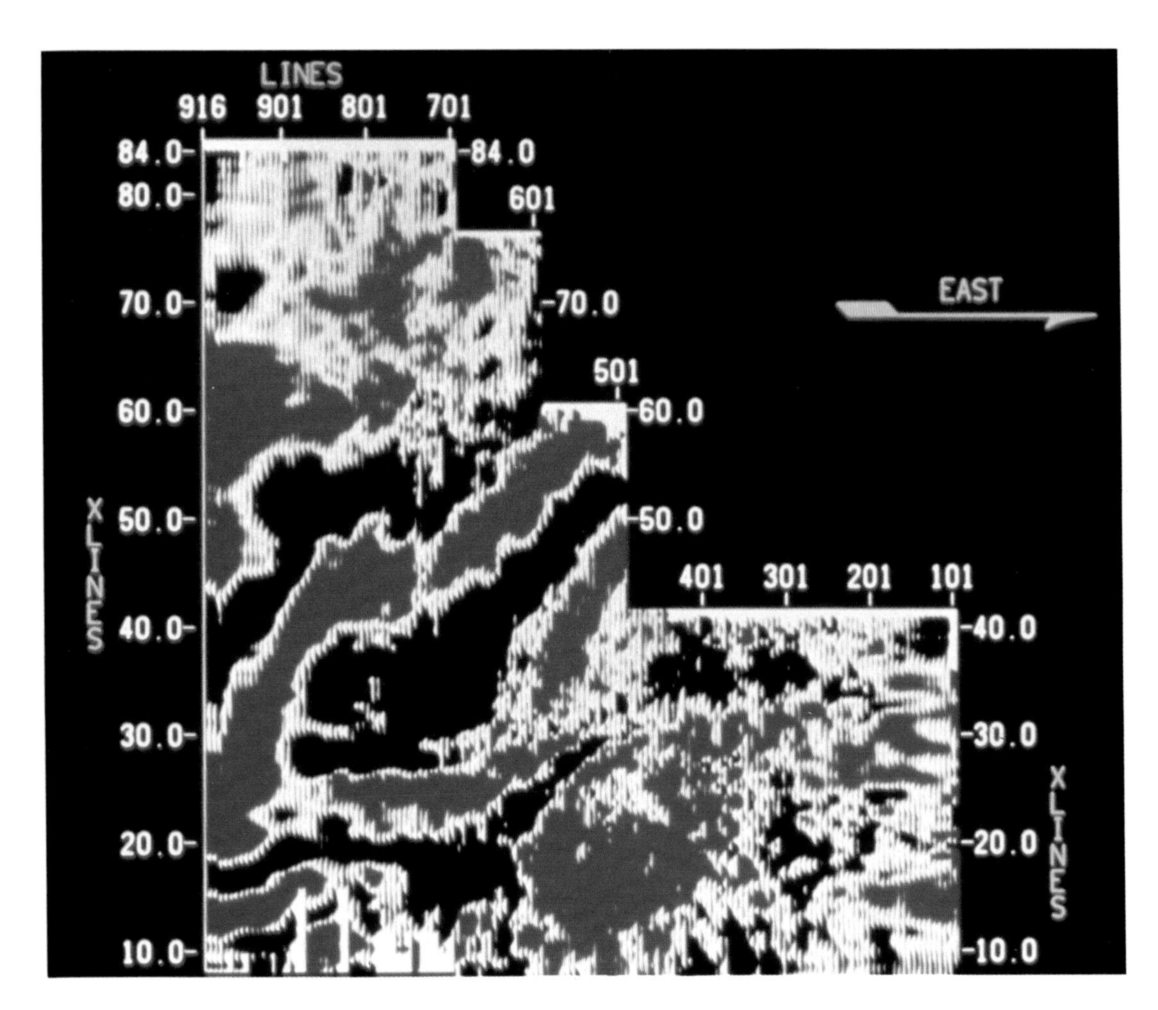

Fig. 3-43. Horizontal section at 2340 ms from south Louisiana marsh terrain. (Courtesy Texaco Inc.)

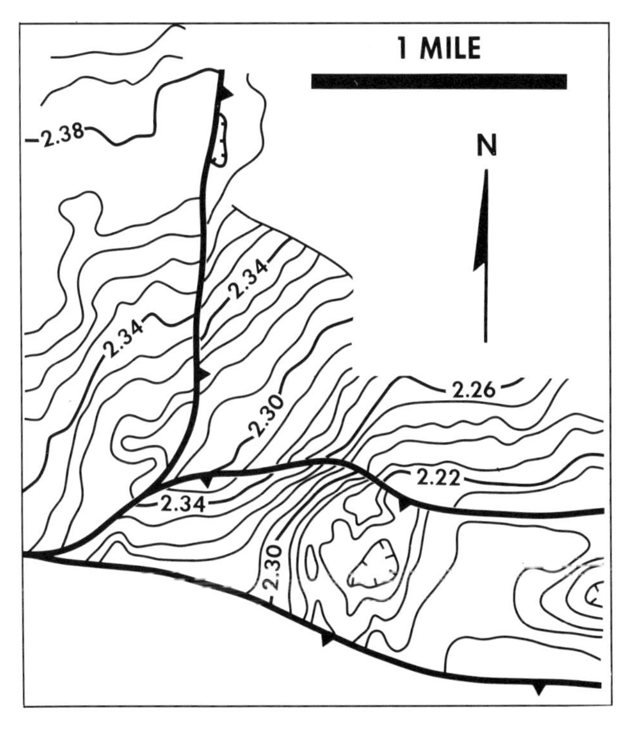

Fig. 3-45. Structural contour map showing subtle fault identified on horizontal sections of Figures 3-42 and 3-43. (Courtesy Texaco Inc.)

Fig. 3-46. Horizontal section at 646 ms from high resolution 3-D survey at Ekofisk field in the North Sea. Note lineations due to faulting and jointing. (Courtesy Phillips Petroleum Company Norway.)

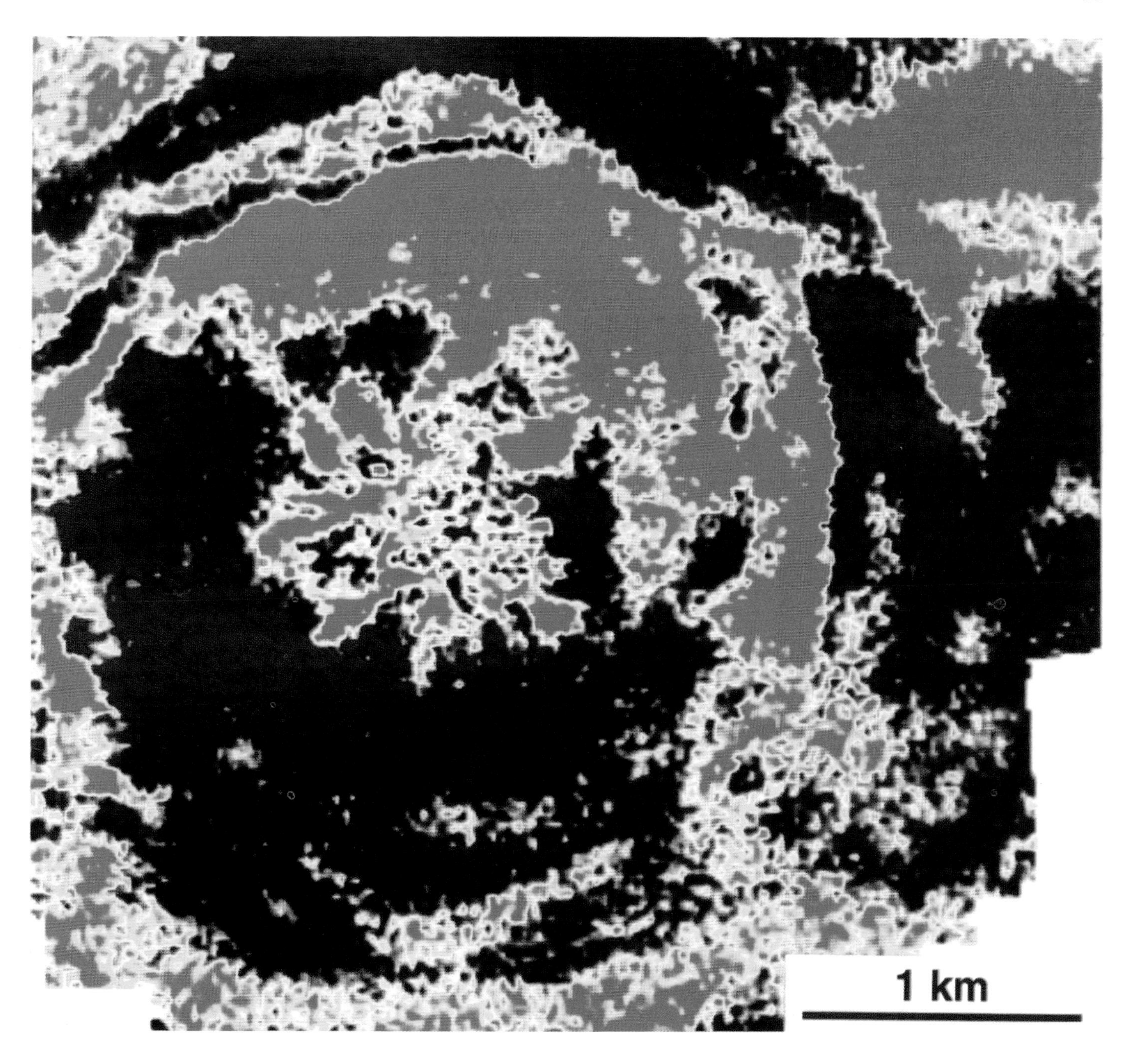

Fig. 3-47. Impact crater of Devonian age seen on time slice 776 ms from the USA mid-continent. (Courtesy Texaco Exploration and Production Inc.)

(Chapter 8). The time slice, or horizontal section, so central to this chapter on structure, is itself a major visualization aid and should be used as such. One vertical section and one horizontal section often provide an early visualization of the structure under study.

Figure 3-47 is a time slice at 776 ms showing a distinctly circular feature about 3 km in diameter and occurring at a depth of about 1000 m. It is a very striking view of an impact crater, or astrobleme, of Devonian age. The radial patterns in the center strongly suggest ejecta from the central uplift.

Once horizons and faults have been interpreted they can be visualized as surfaces (Figure 3-48). The relationship between the surfaces can then be studied to help in validation of the interpretation and in placement of the wells to intersect multiple objectives.

The horizon surfaces above will have been produced using automatic spatial track-

Fig. 3-48. Interpreted horizons and faults seen together as surfaces for validation and visualization. (Courtesy Landmark Graphics Corporation.)

Fig. 3-49. Time structure map generated by automatic spatial tracking operating in an extrapolatory manner from minimum seed points located in the northeast. (Courtesy Landmark Graphics Corporation.)

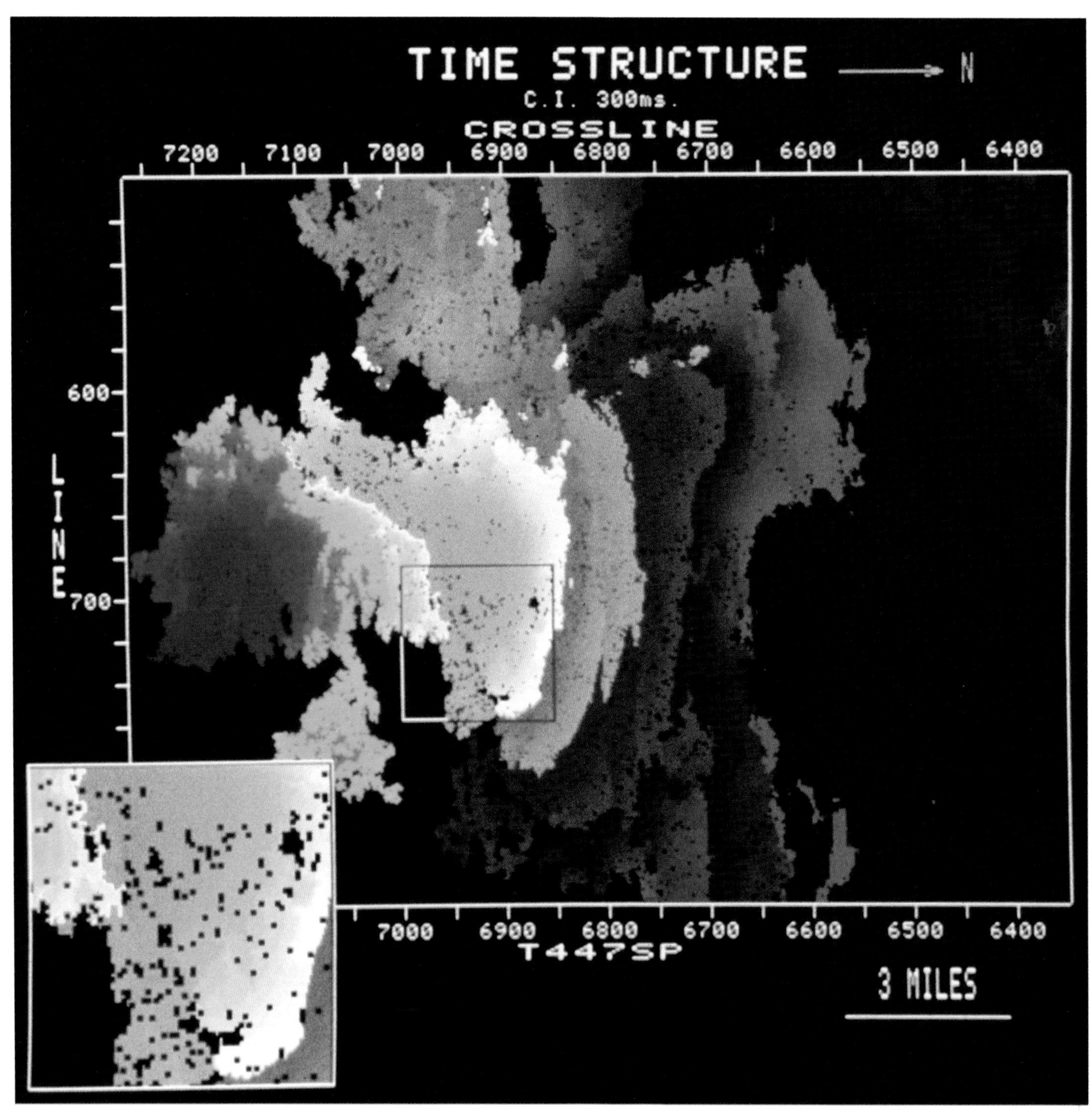

Fig. 3-50. Time structure map generated by automatic spatial tracking showing lineations of untracked points. (Courtesy Geophysical Service Inc.)

ing starting from seed, or control, points on vertical and horizontal sections. This procedure, discussed earlier, uses the tracker in a controlled interpolatory manner, that is it is operating between points where the interpretation has been prescribed. With good data the automatic spatial tracker can be used in an extrapolatory manner from minimum seed points. Figure 3-49 is an example where the tracker was seeded in the northeast and moved outwards independently to define several fault blocks and the faults between them. Notice along the faults, and in some other places, untracked points where the tracking criteria could not be satisfied.

Untracked points are also evident in Figure 3-50 and in several places they line up. These lineations of untracked points indicate places where the tracker had difficulty and may indicate subtle faults, sharp changes of dip, facies changes or other boundaries.

References

Blake, B. A., J. B. Jennings, M. P. Curtis, and R. M. Phillipson, 1982, Three-dimensional seismic data reveals the finer structural details of a piercement salt dome: Offshore Technology Conference Paper 4258, p. 403-406.

Denham, J. I., and H. R. Nelson, Jr., 1986, Map displays from an interactive interpretation: Geophysics, v. 51, p. 1999-2006.

French, W. S., 1990, Practical seismic imaging: The Leading Edge, v. 9, no. 8, p. 13-20.

Johnson, J. P., and M. R. Bone, 1980, Understanding field development history utilizing 3-D seismic: Offshore Technology Conference Paper 3849, p. 473-475.

Stark, T. J., 1996, Surface slice generation and interpretation—a review: The Leading Edge, v. 15, p. 818-819.

Taner, M. T., F. Koehler, and R. E. Sheriff, 1979, Complex seismic trace analysis: Geophysics, v. 44, p. 1041-1063.

STRATIGRAPHIC INTERPRETATION

Recognition of Characteristic Shape

Where a vertical seismic section intersects a stratigraphic feature the interpreter can normally find a small amplitude or character anomaly. The expression of a sand-filled channel or bar, for example, is therefore normally so subtle that it takes a considerable amount of interpretive skill to detect it. In contrast, a horizontal section reveals the spatial extent of an anomaly. The interpreter can thus observe characteristic shape and relate what he sees to geologic experience. A shape or pattern which is unrelated to structure may prove to be interpretable as a depositional, erosional, lithologic or other feature of significance. Klein (1985) and Broussard (1975), among others, have provided depositional models on which the interpreter can base his recognition of depositional features. The study of horizontal sections and horizon slices can provide a bird's-eye view of ancient stratigraphy, analogous to the view of modern stratigraphy obtained out of an airplane window.

Figure 4-1 shows five adjacent vertical seismic sections from a small 3-D survey in the Williston basin of North Dakota. Note that the reflections indicate largely flat-lying beds. At 1.8 seconds there is a very slight draping of reflections which is only just discernible. Figure 4-2 shows two single-polarity horizontal sections superimposed on each other. The data from both levels reveal the same almost circular shape. This is the outline of a carbonate buildup measuring approximately one kilometer in diameter.

Figures 4-3 and 4-4 are horizontal sections from a 3-D survey recorded in the Gippsland basin offshore southeastern Australia (Sanders and Steel, 1982). Many small circular features are strikingly evident. These appear as small depressions on the vertical sections which attract little attention. It is the characteristic circular shape when viewed horizontally that attracts the interpreter's eye. The circular features measure 200 to 500 m in diameter and are interpreted as sinkholes in a Miocene karst topography. The beds in which these features exist are dipping from upper left to lower right (east) in Figures 4-3 and 4-4. The width of the reflection is a function of seismic frequency and structural dip (see Chapter 3). The visibility of the sinkholes in the presence of this structure is because their diameters are each less than the reflection width.

Figure 4-5 shows a bifurcating channel close to a Gulf of Mexico salt dome. The salt dome's semi-circular expression results from the intersection of the horizontal section at 416 ms with the dipping structural reflections adjacent to the dome. Away from the salt dome the beds are close to flat-lying, so the horizontal section is sliced along the bedding plane. As a result, the channel is almost completely visible. In fact, the bedding is not exactly flat and some parts of the channel are more clearly seen on the adjacent section at 412 ms. Simple addition of these two horizontal sections improved the continuity of the channel (Figure 4-6). Adding together of horizontal sections is a useful approach to the enhancement of stratigraphic features if, *but only if*, the structural

Text continues on page 106.

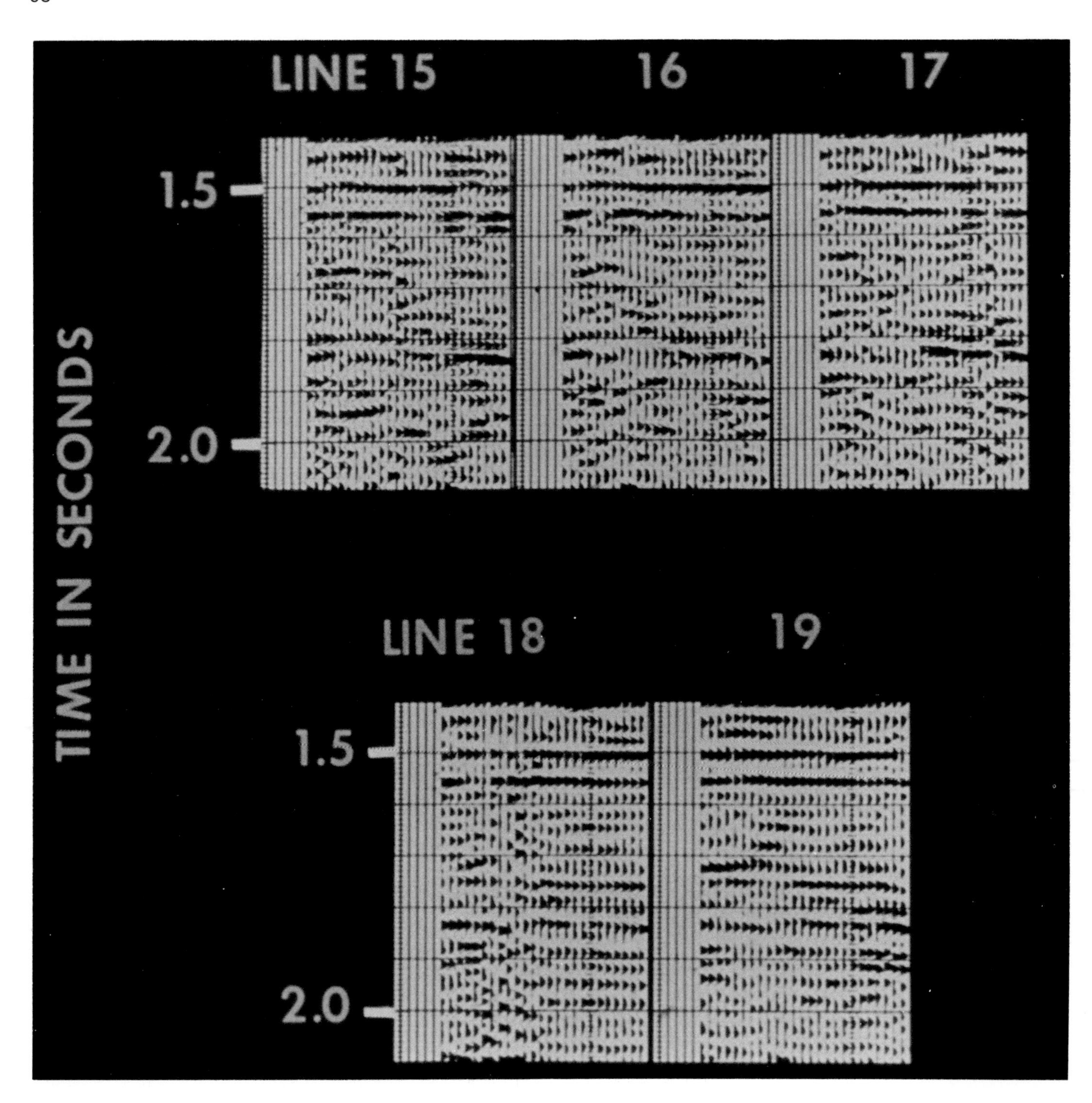

Fig. 4-1. Five adjacent vertical sections from 3-D survey in the Williston basin of North Dakota. (Courtesy Geophysical Service Inc.)

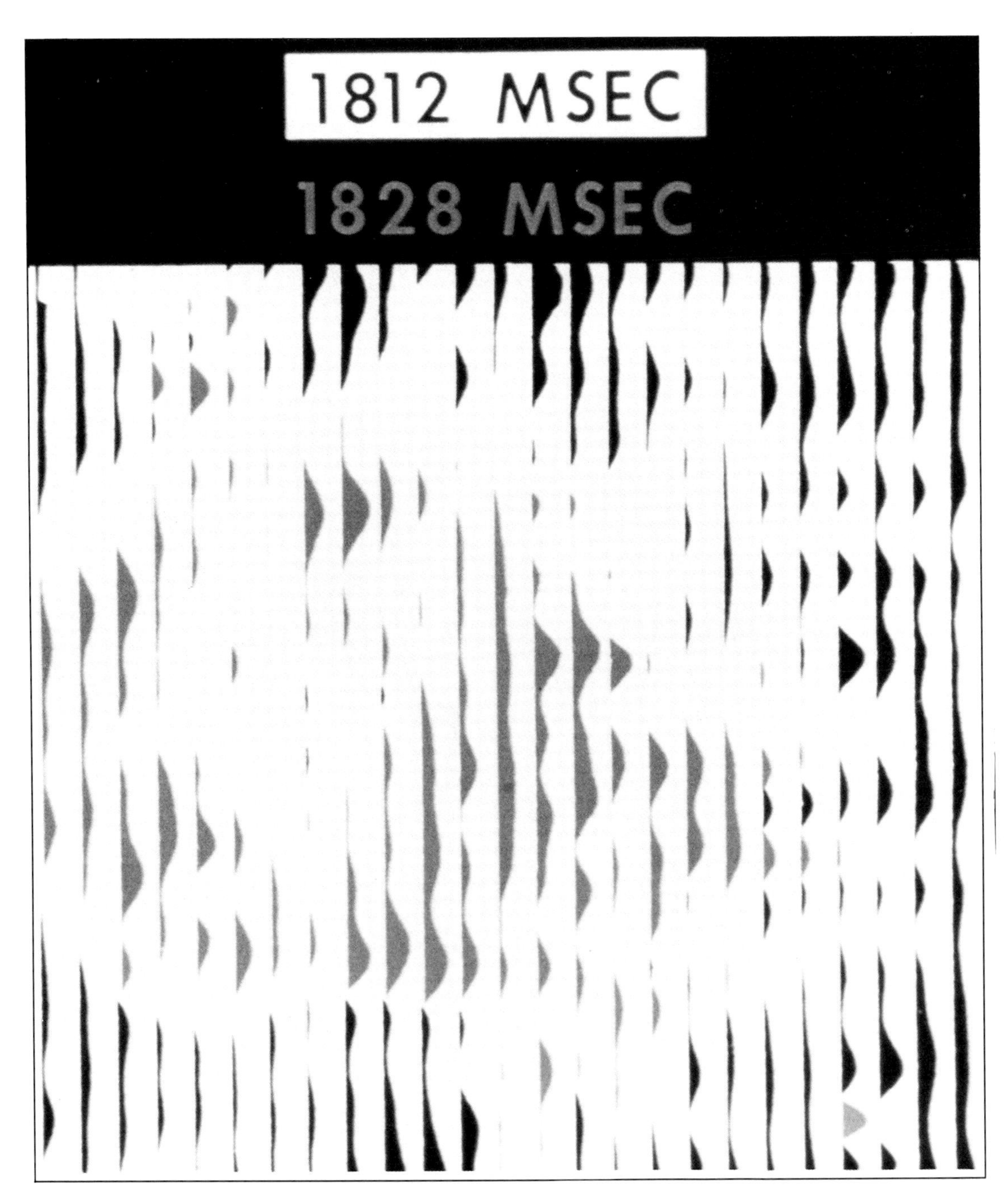

Fig. 4-2. Horizontal sections from 1812 and 1828 ms from North Dakota, each showing positive amplitudes only. The approximately circular outline between the black and the gray indicates the shape of a carbonate buildup. (Courtesy Geophysical Service Inc.)

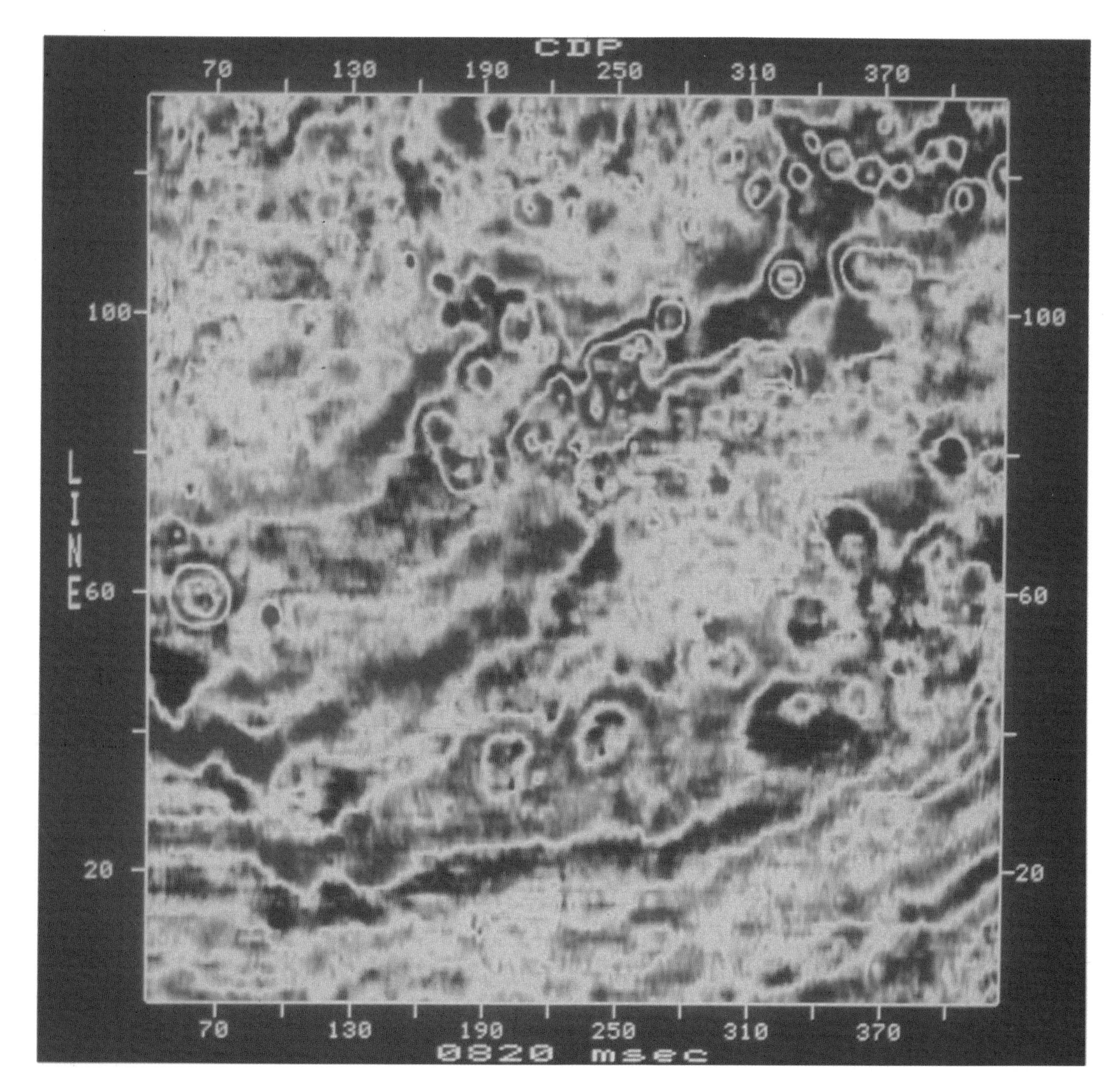

Fig. 4-3. Horizontal section at 820 ms from 3-D survey over Mackerel field in offshore Gippsland basin, southeastern Australia. Circular objects are interpreted as sinkholes in karst topography. (Courtesy Esso Australia Ltd.)

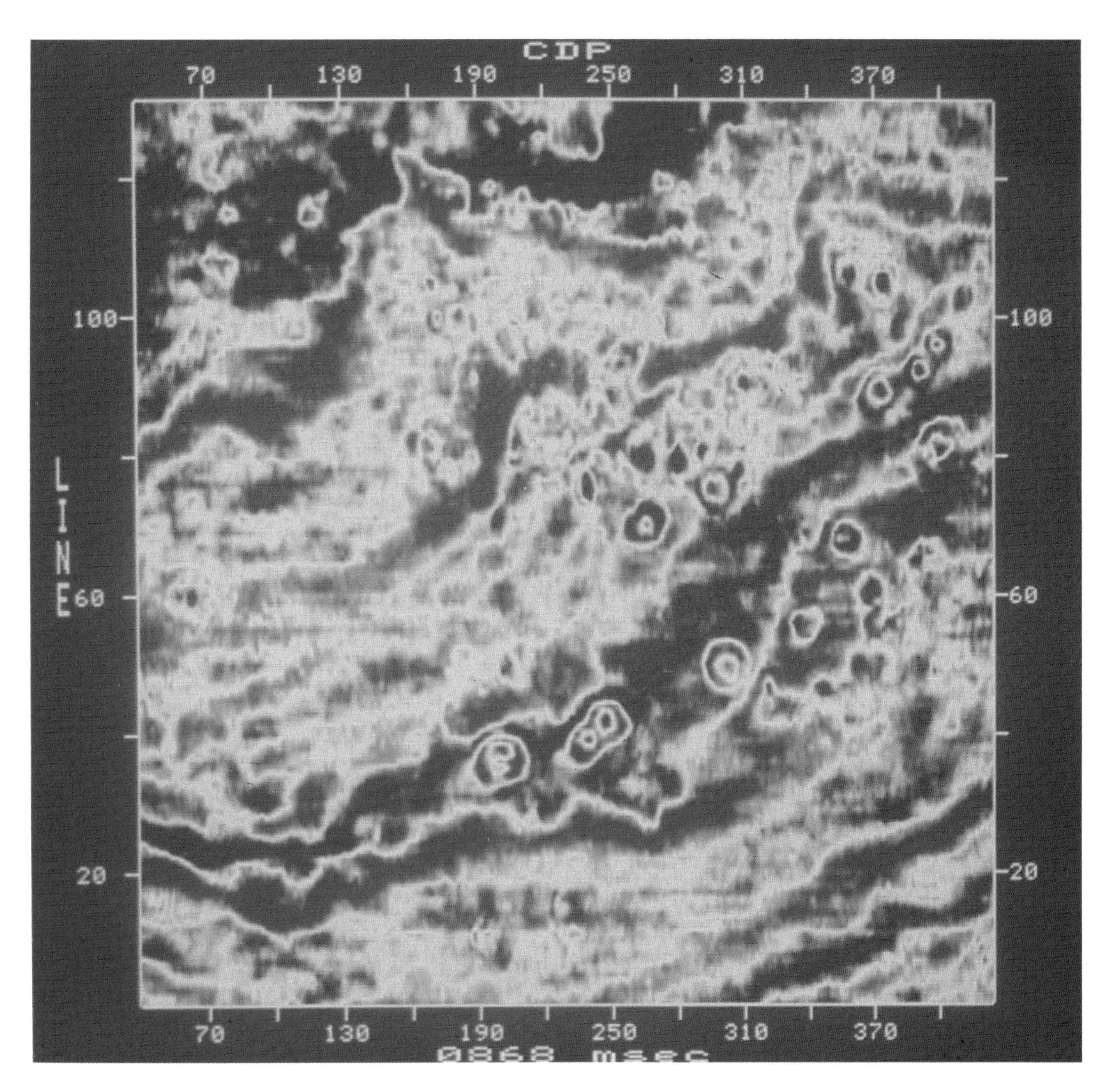

Fig. 4-4. Horizontal section at 868 ms from 3-D survey over Mackerel field in offshore Gippsland basin, southeastern Australia. Circular objects are interpreted as sinkholes in karst topography. (Courtesy Esso Australia Ltd.)

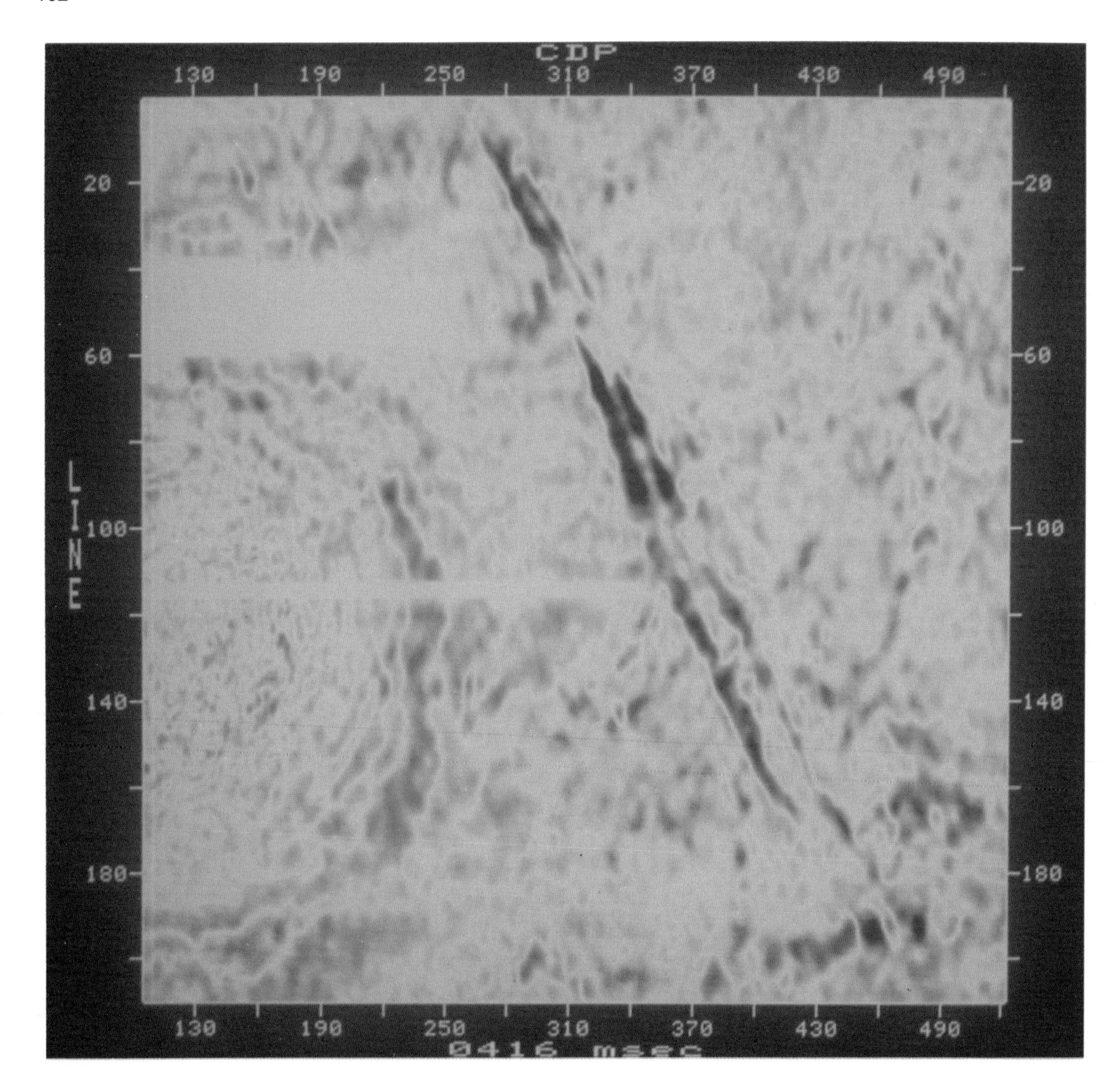

Fig. 4-5. Horizontal section at 416 ms from 3-D survey in the Gulf of Mexico. The bifurcating channel is seen close to the edge of a salt dome. (Courtesy Chevron U.S.A. Inc.)

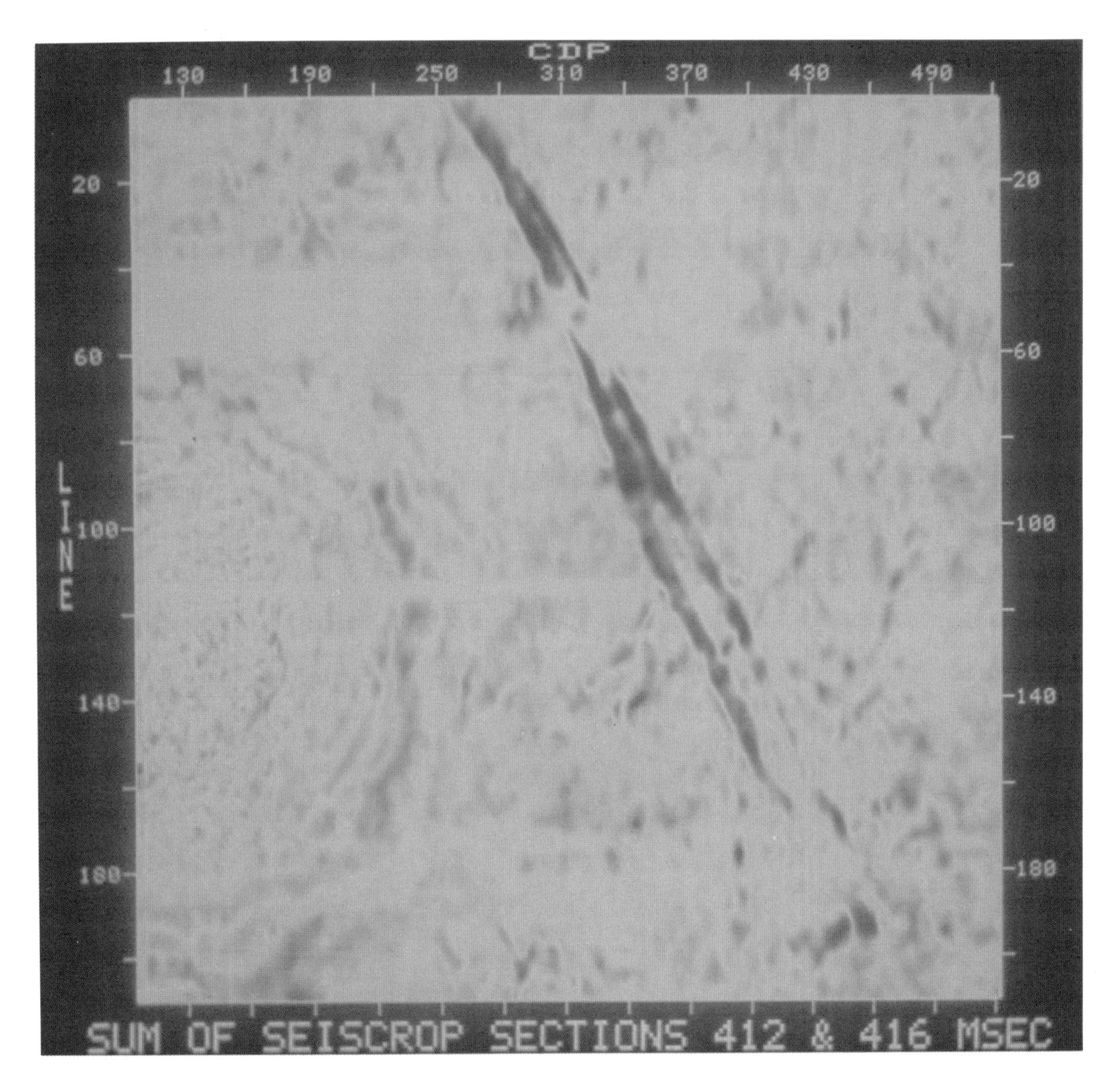

Fig. 4-6. Sum of horizontal sections at 412 and 416 ms from same survey as Figure 4-5 showing enhancement of the channel. (Courtesy Chevron U.S.A. Inc.)

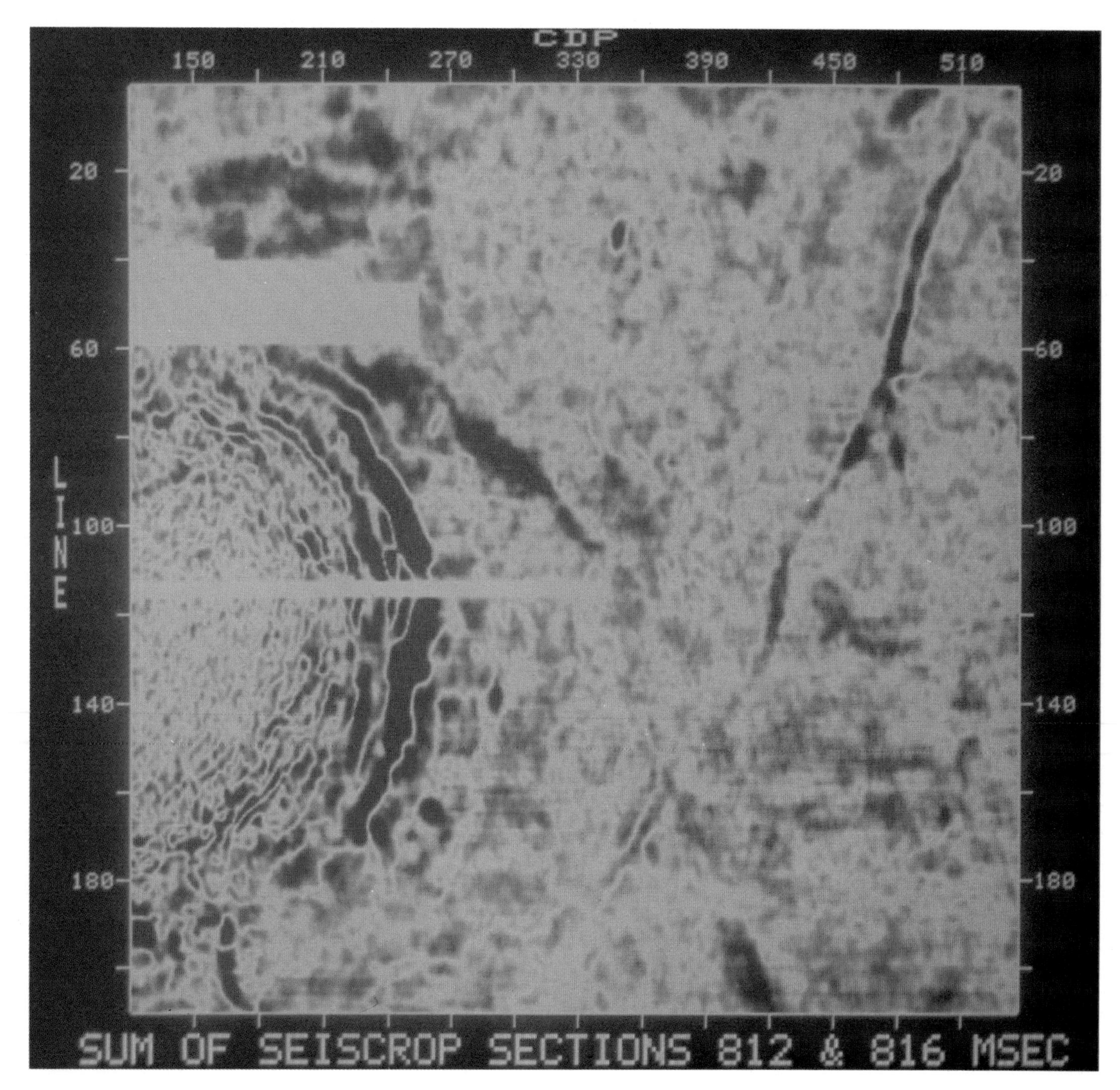

Fig. 4-7. Sum of horizontal sections at 812 and 816 ms from same survey as Figure 4-5 showing a branching channel. (Courtesy Chevron U.S.A. Inc.)

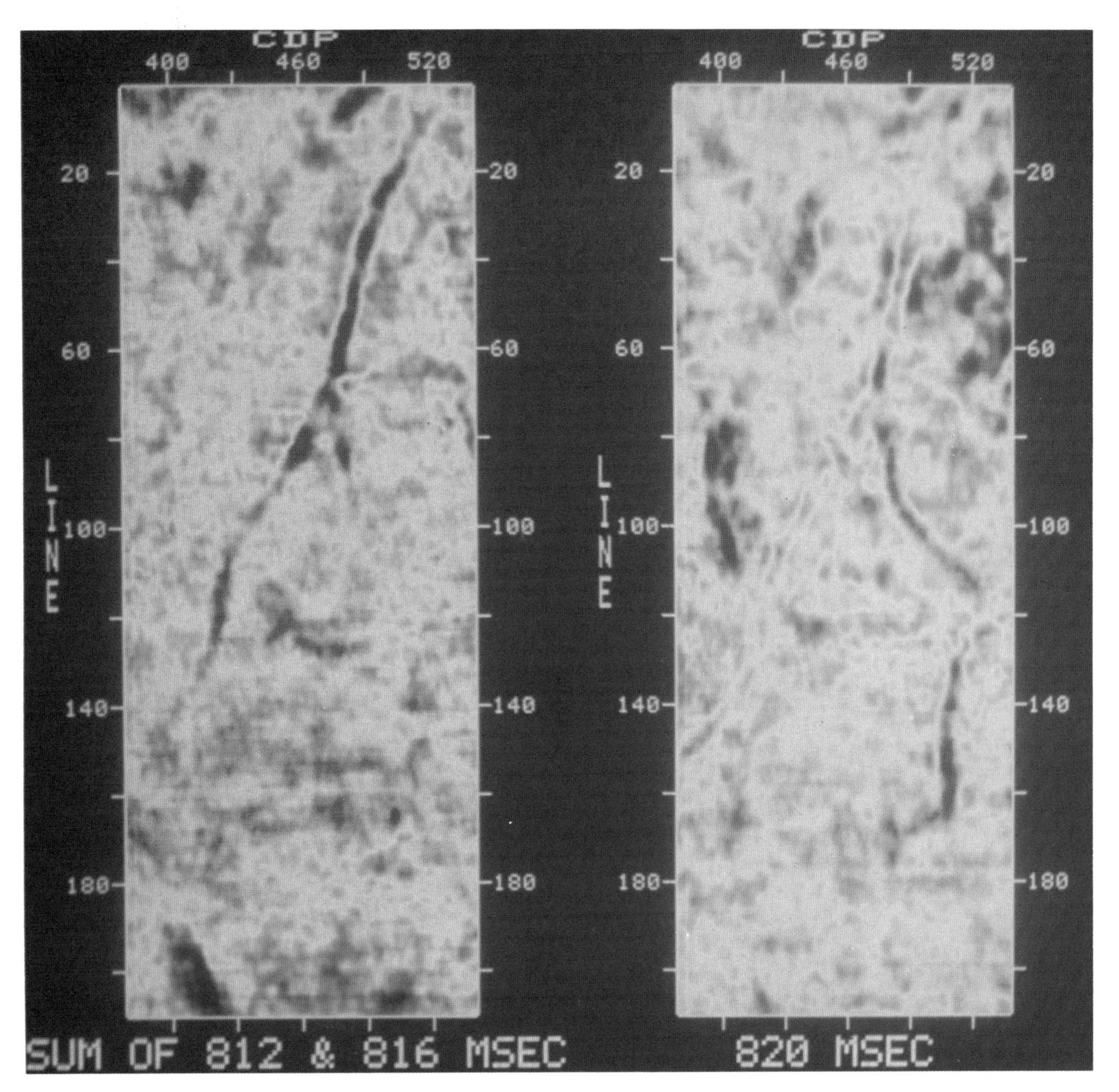

Fig. 4-8. Composite display of horizontal sections at 812 and 816 ms showing western branch of channel and at 820 ms showing eastern branch. (Courtesy Chevron U.S.A. Inc.)

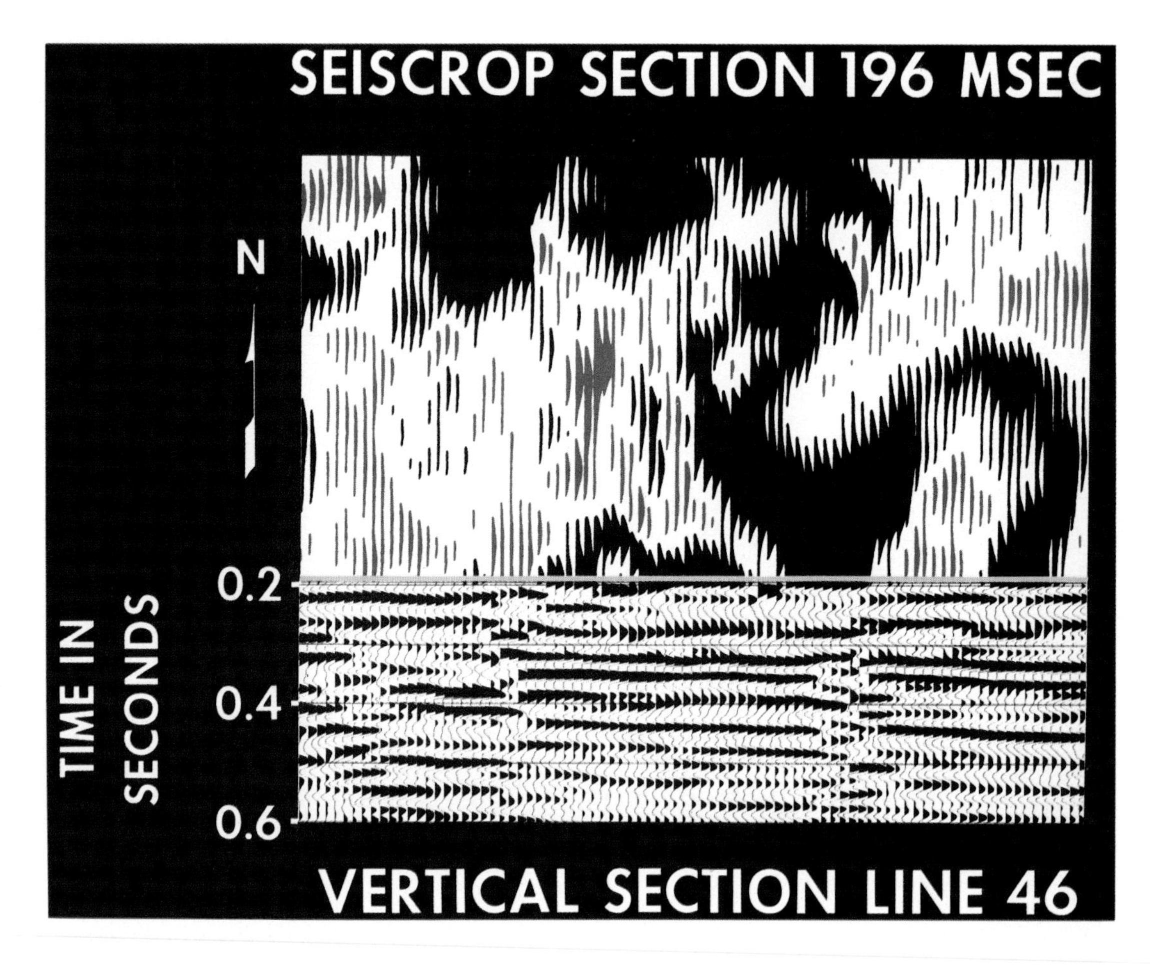

Fig. 4-9. Composite display of vertical and horizontal sections from Gulf of Thailand showing spatial continuity of vertical section event segments. (Courtesy Texas Pacific Oil Company Inc.)

variation across the feature is less than half a period of the appropriate seismic signal.

Figure 4-7 shows another channel deeper in the same data volume. Enhancement again resulted from adding together the horizontal sections from 812 and 816 ms. The channel branches at Line 70, CDP 470, but the eastern branch is not visible. Figure 4-8 shows just the portion of the survey area covering the channel system and includes the horizontal section at 820 ms. Here the eastern branch is clearly visible showing that it is structurally slightly deeper than the western branch. This indicates that the depositional surface containing this channel dips away from the salt dome, which dip was presumably induced by the movement of the salt. Thus, in order to view the entire channel system, several horizontal sections covering the structural range of this depositional surface are required.

Figures 4-9 through 4-16 show examples of depositional features observed on horizontal sections through flat-lying beds in the Gulf of Thailand. The vertical section in Figure 4-9 shows that the beds are flat-lying and that around 200 ms there are some abrupt character changes. The attached horizontal section shows that these reflection segments have spatial continuity. Figure 4-10, covering the whole prospect area, makes it clear that the continuity is part of a meandering channel system. Anyone who has flown over the Mississippi River will immediately relate Figure 4-10 to observations made from the airplane window.

In the Gulf of Thailand there is a regional unconformity in the mid-Miocene and above that unconformity the beds in this prospect area are largely flat-lying. Therefore, many horizontal sections above 900 ms directly reveal depositional features because the sections are parallel to bedding planes. Figure 4-11 is a schematic composite of the features observed. The interpretation of these in sequence indicated a delta prograding across the survey area from southwest to northeast during the mid-Miocene to Pleistocene.

Examples of the depositional features observed are presented in Figures 4-12

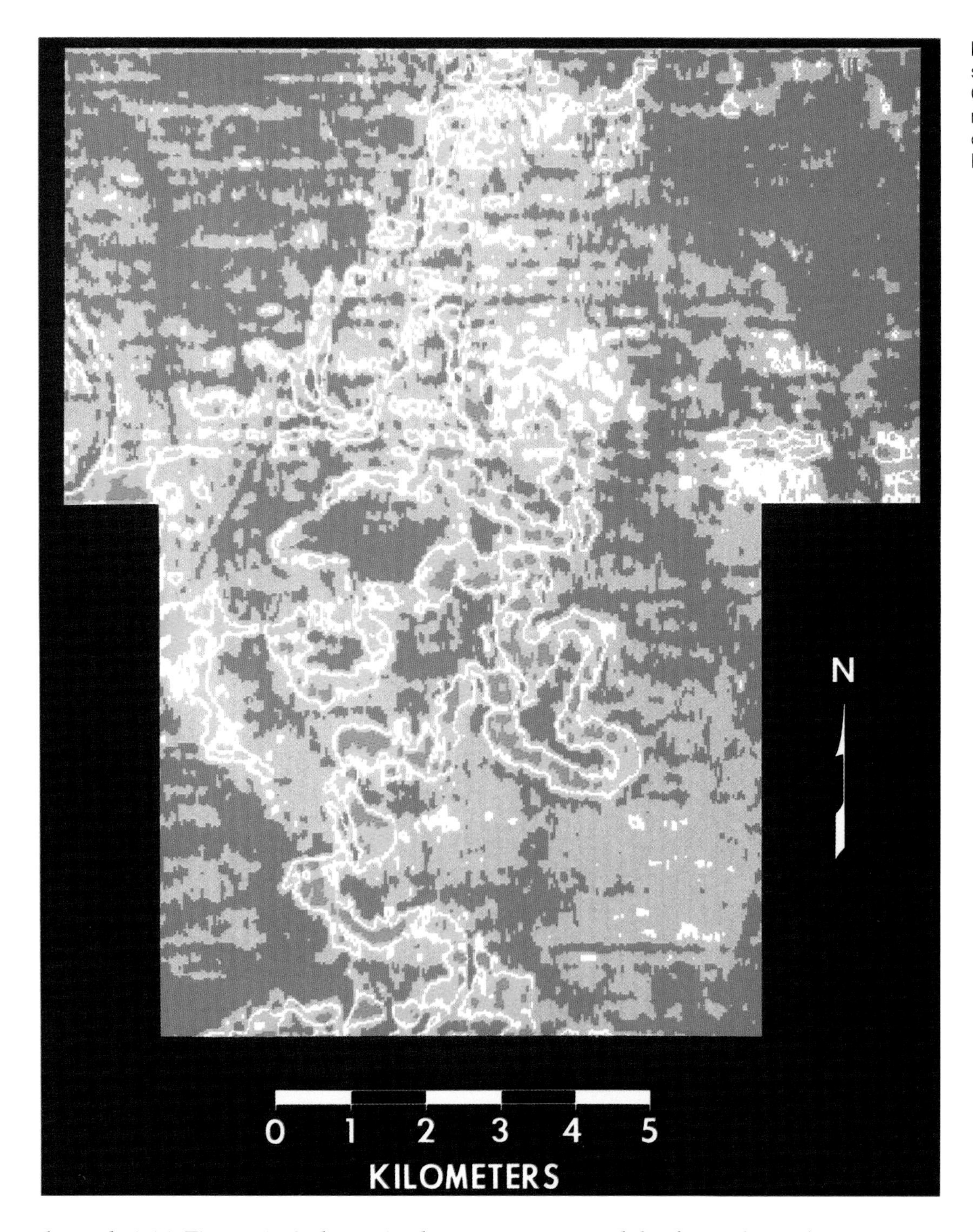

Fig. 4-10. Horizontal section at 196 ms from Gulf of Thailand showing meandering stream channel. (Courtesy Texas Pacific Oil Company Inc.)

through 4-16. Figure 4-12 shows in the upper center a delta front channel. Figure 4-13 shows a large offshore bar trending northwest-southeast, transverse to the direction of delta progradation. Figure 4-14 shows two smaller bars, center and lower left, with the same orientation. Figure 4-15 shows, in the upper right, a reworked bar; toward the bottom are straight linear features suggestive of distributary channels. Figure 4-16 shows many twisting channels, some of them very narrow.

Figure 4-17 shows part of the Mahakam delta in Indonesia. At this time (about 18,000 years ago) deposition was clearly occurring in this part of the delta. Patterns are very similar to those visible in the present Mahakam delta (Figure 4-18). However, in another part of the ancient delta (Figure 4-19) erosion was occurring as evidenced by the dendritic patterns of canyons.

Figure 4-20 shows a shallow horizontal section from a part of the Gulf of Thailand

Text continues on page 115.

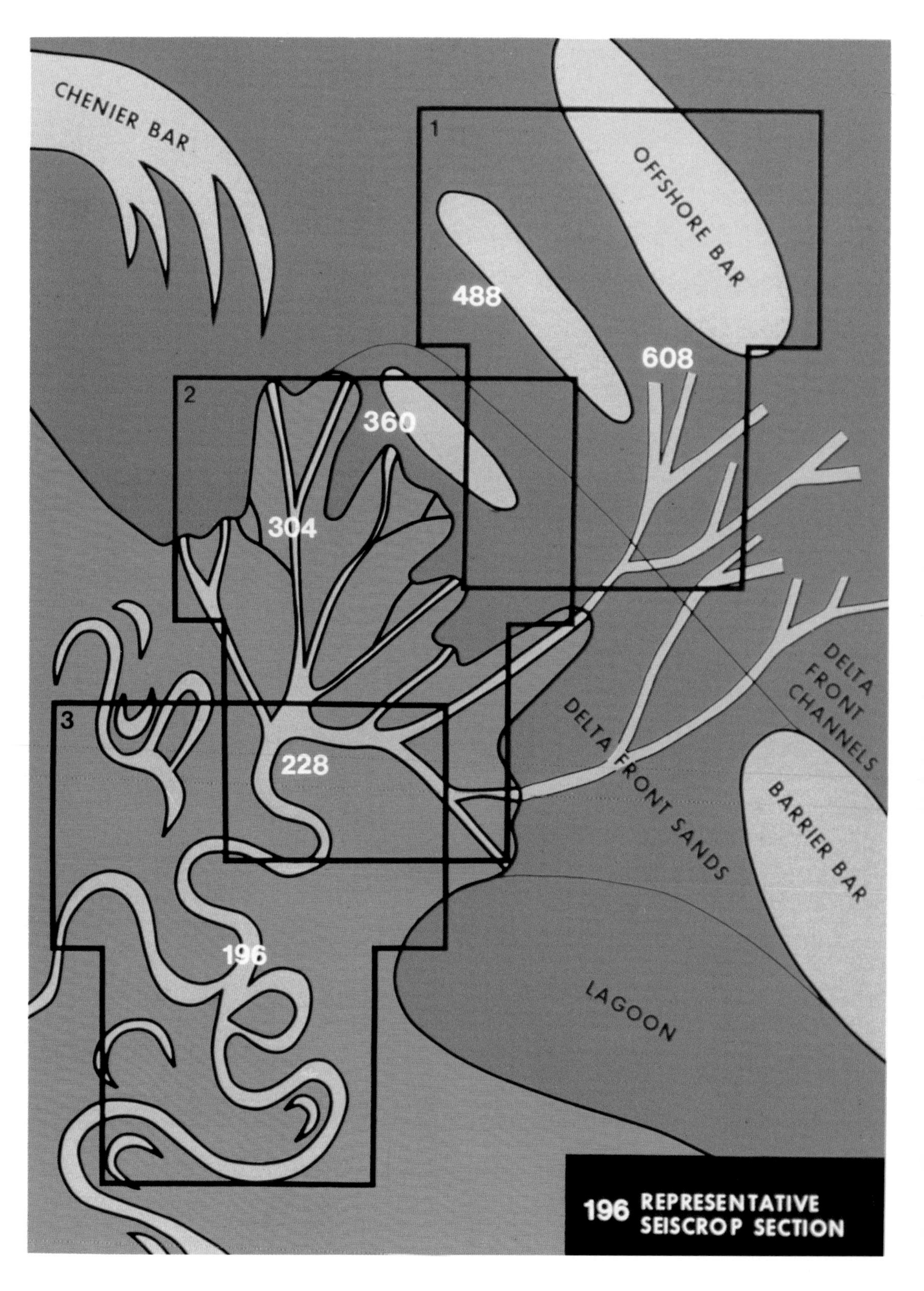

Fig. 4-11. Schematic diagram of delta prograding across the Gulf of Thailand 3-D survey area between mid-Miocene and Pleistocene.

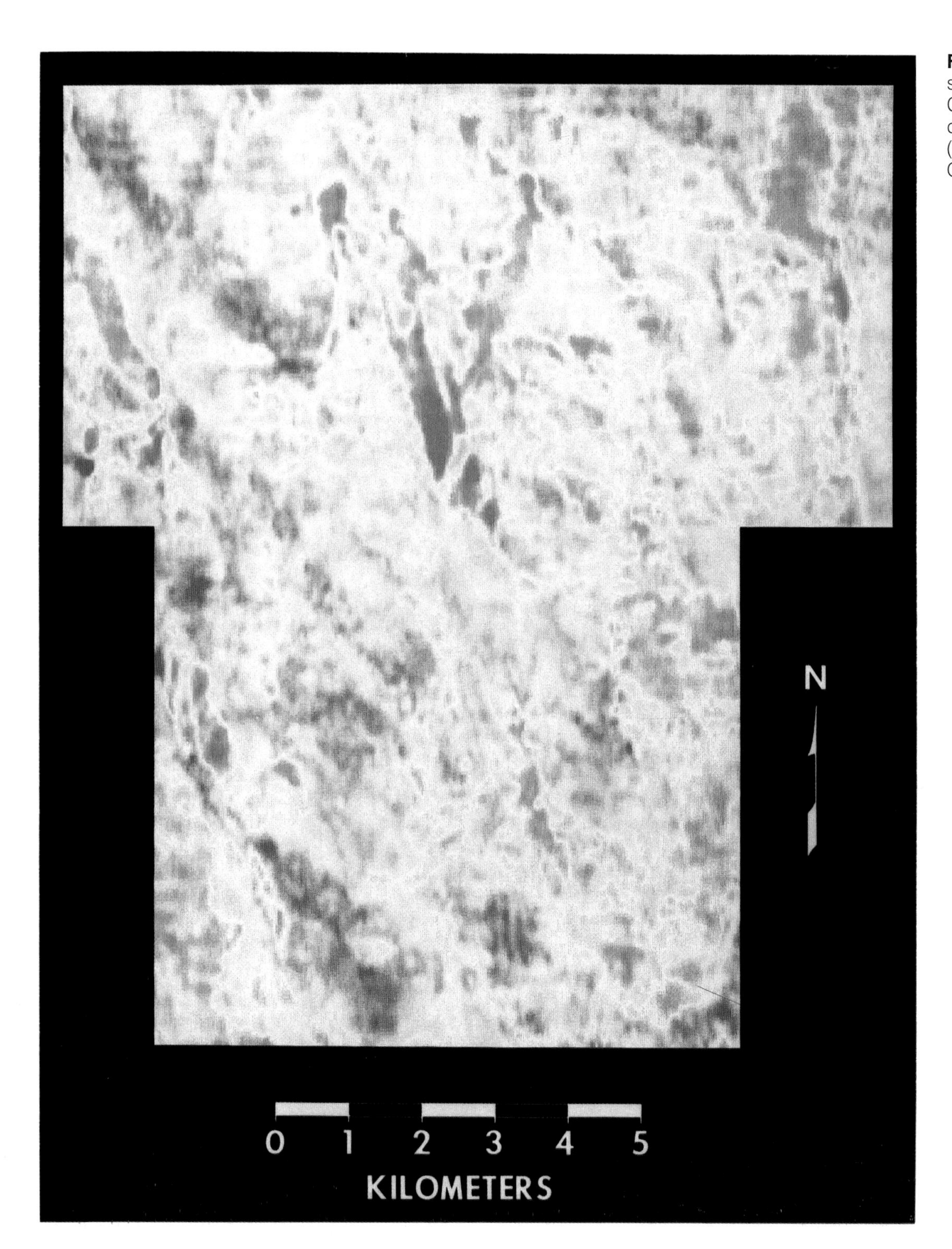

Fig. 4-12. Horizontal section at 608 ms from Gulf of Thailand showing delta front channel. (Courtesy Texas Pacific Oil Company Inc.)

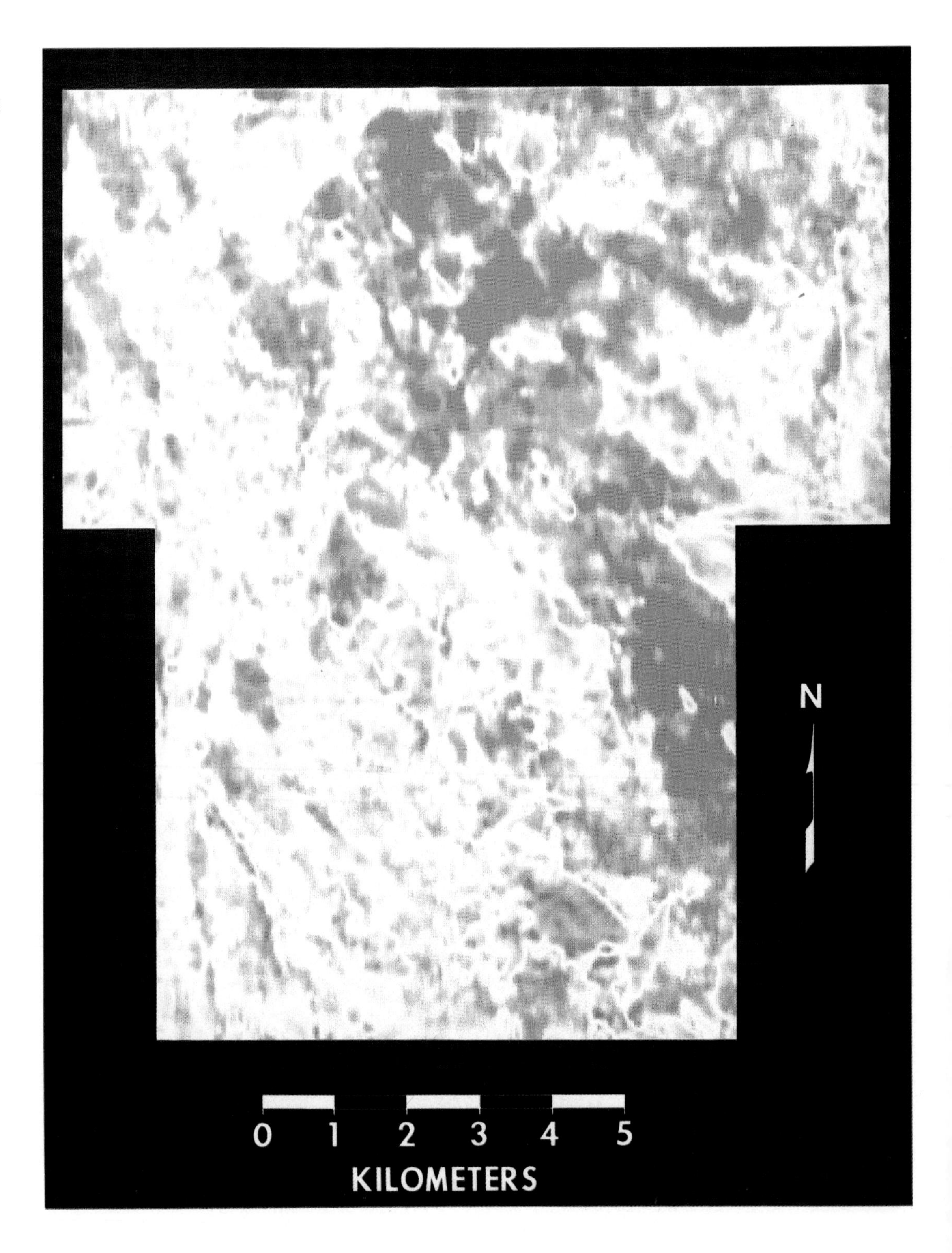

Fig. 4-13. Horizontal section at 488 ms from Gulf of Thailand showing large offshore sand bar. (Courtesy Texas Pacific Oil Company Inc.)

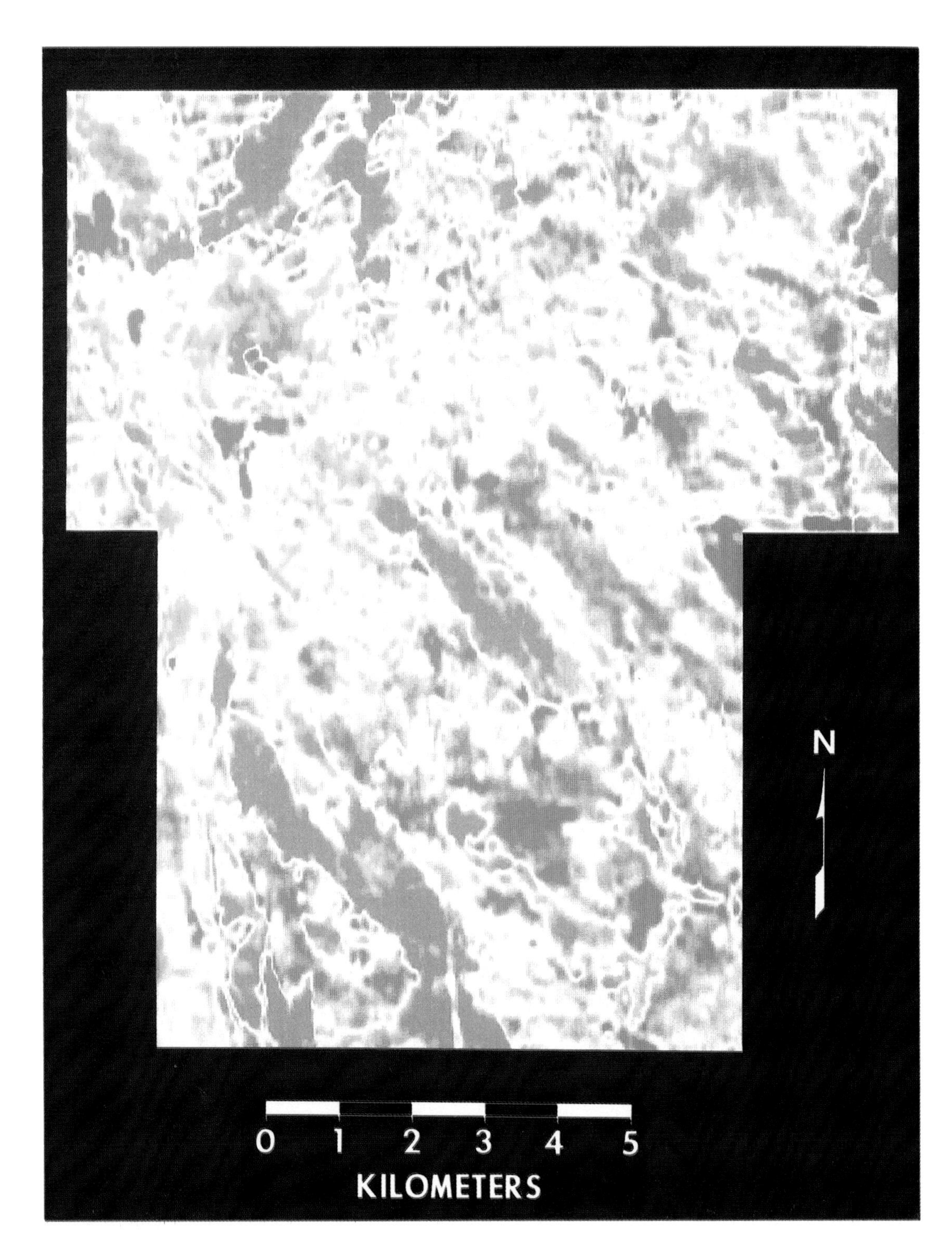

Fig. 4-14. Horizontal section at 360 ms from Gulf of Thailand showing small sand bars. (Courtesy Texas Pacific Oil Company Inc.)

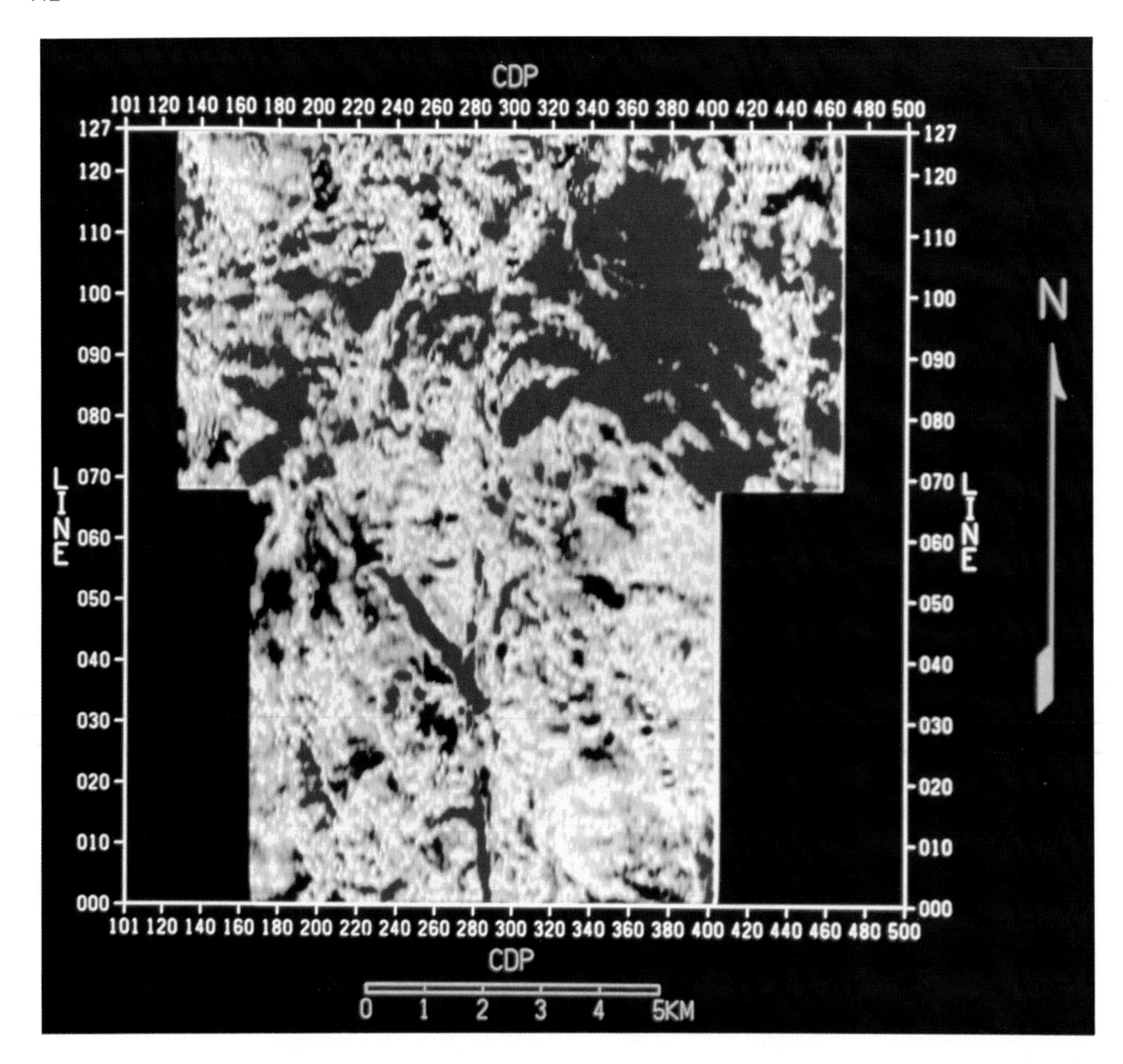

Fig. 4-15. Horizontal section at 304 ms from Gulf of Thailand showing a reworked bar and distributary channels. (Courtesy Texas Pacific Oil Company Inc.)

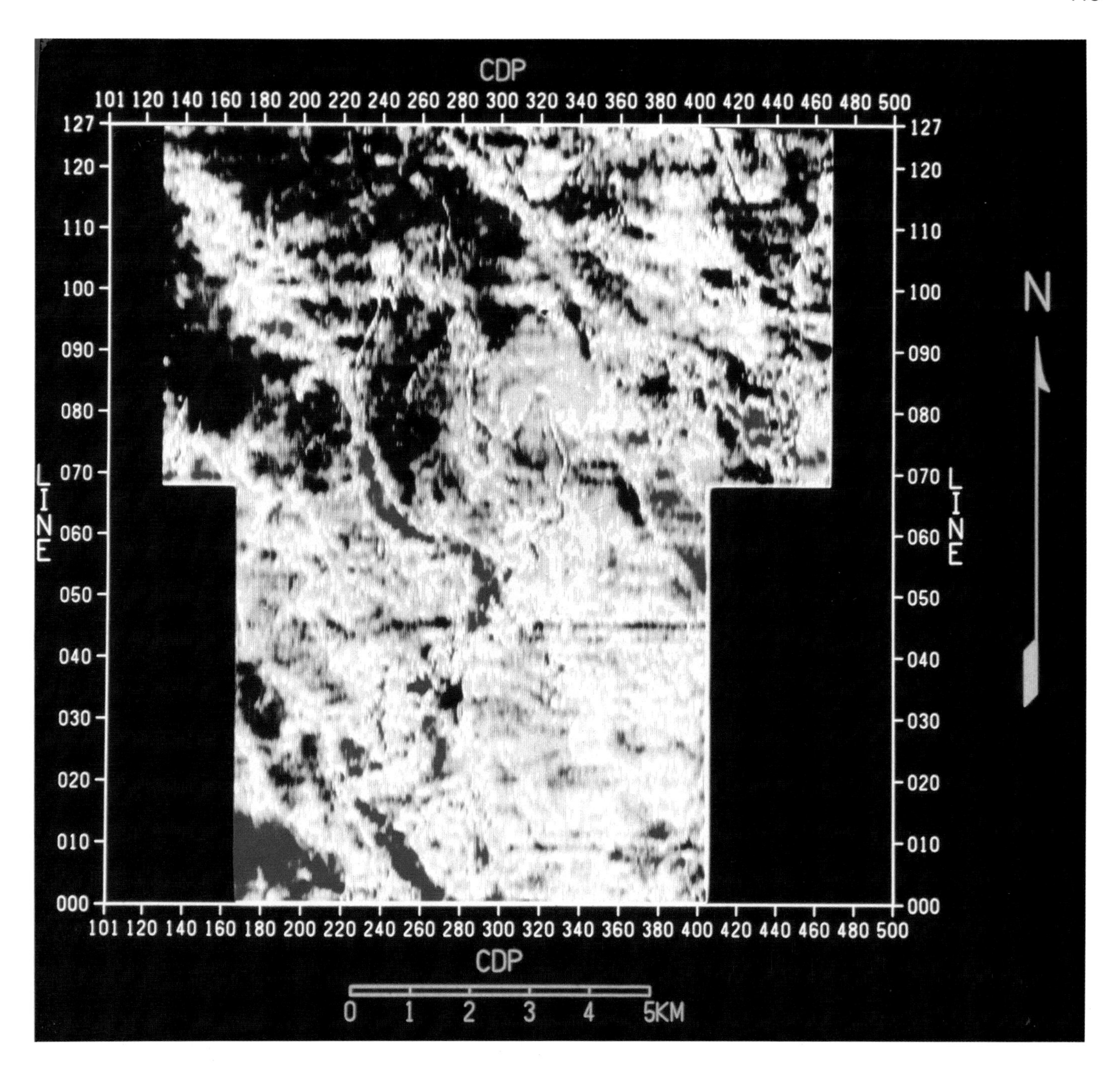

Fig. 4-16. Horizontal section at 228 ms from Gulf of Thailand showing several channels, large and small. (Courtesy Texas Pacific Oil Company Inc.)

Fig. 4-17. Horizontal section at 100 ms from Peciko 3-D survey recorded in the Mahakam delta offshore Kalimantan, Indonesia. The deltaic features seen here are about 18,000 years old. (Courtesy Total Indonesie.)

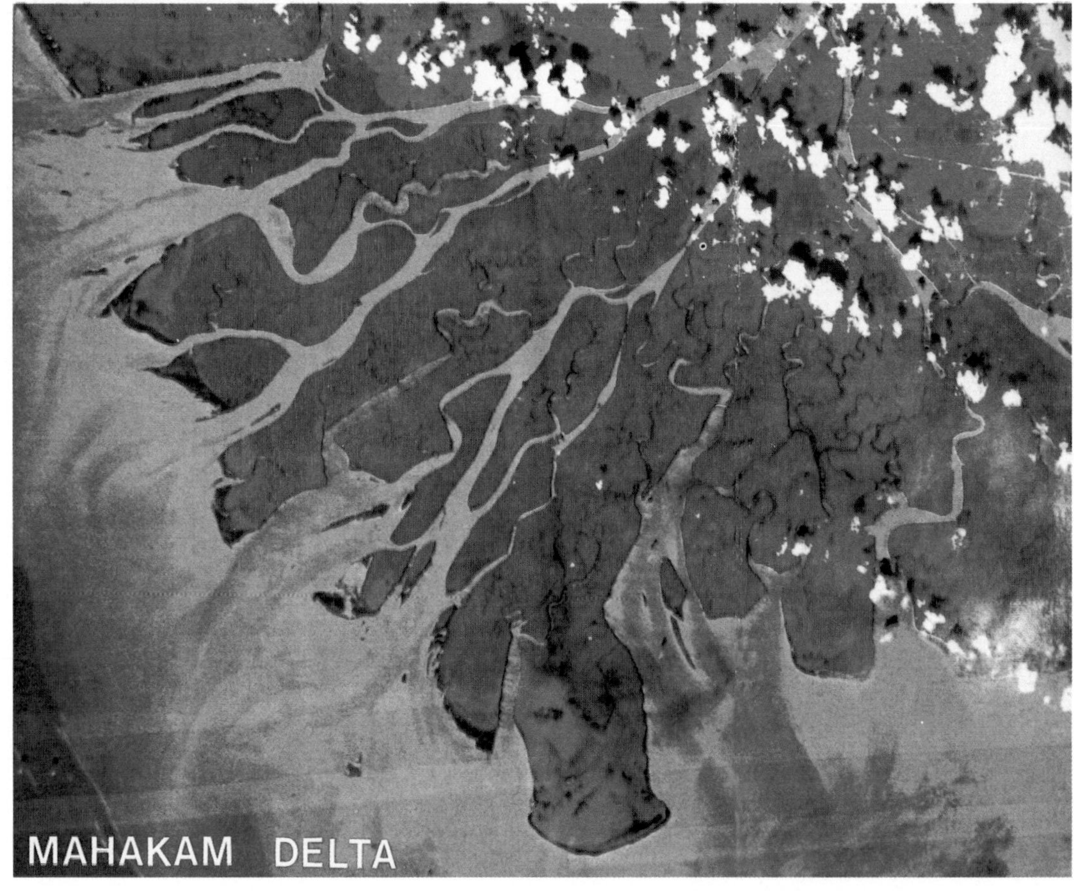

Fig. 4-18. Satellite photograph of part of present Mahakam delta for comparison with Figure 4-17. (Courtesy Total Indonesie.)

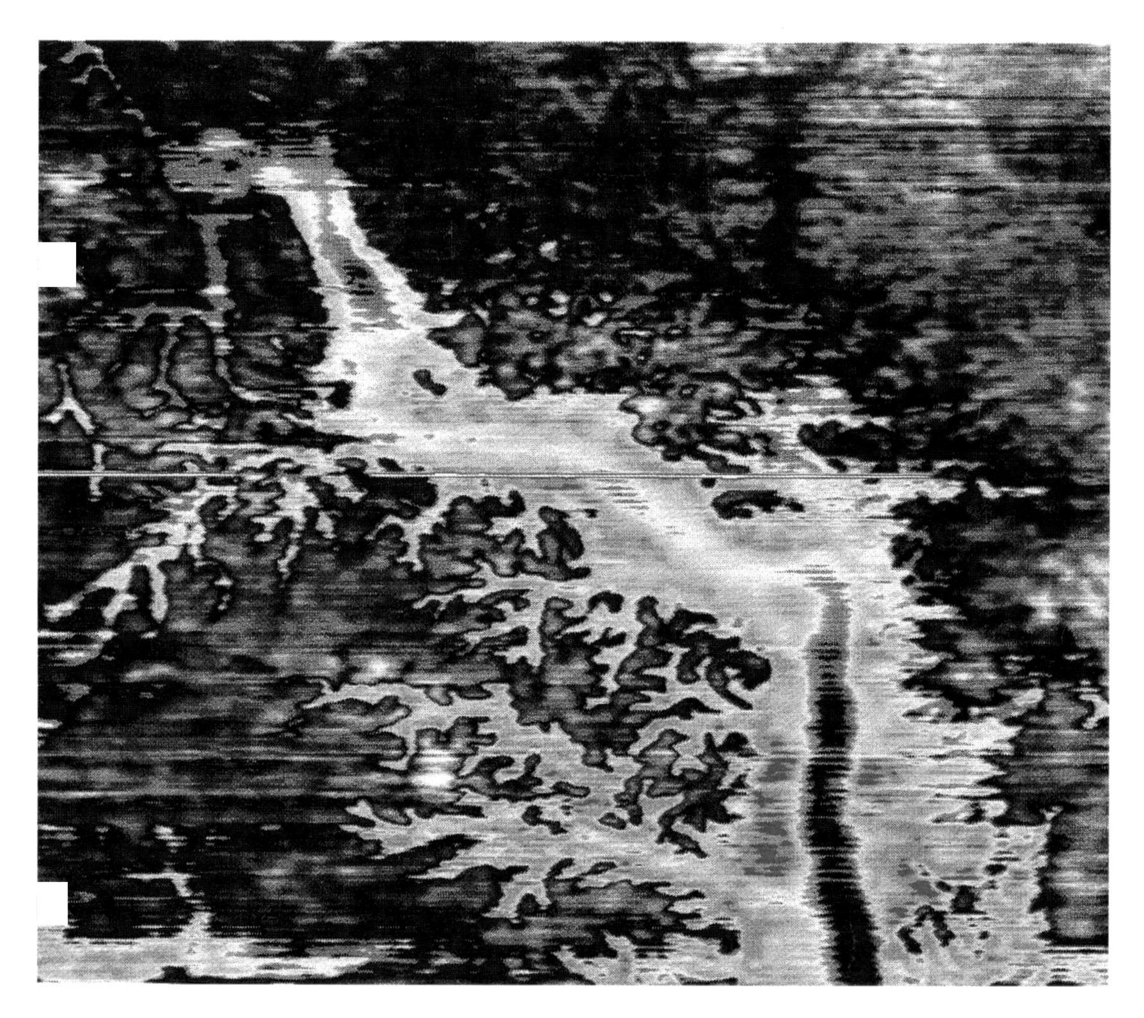

Fig. 4-19. Horizontal section at 104 ms from Nubi 3-D survey recorded in the Mahakam delta offshore Kalimantan, Indonesia. Note the dendritic patterns of incised canyons. (Courtesy Total Indonesie.)

different from that discussed above; it covers a much larger area than other sections in this chapter, which is evidenced by the collage of eight panels. There is a plethora of depositional features clearly visible. In the lowermost and uppermost parts of the figure, channels cross each other. This is evidence that a horizontal section observes a slab of the subsurface of finite thickness (around a quarter of a wavelength) during the deposition of which, in this area, conditions changed significantly. On the right center of Figure 4-20 a meandering channel is visible. Where this channel turns into the center of the figure, it passes point bars inside the meander loops and crevasse splays outside them.

Figure 4-21 is a horizontal section from the Gulf of Mexico showing another clearly visible channel. The channel is fairly difficult to observe on the companion vertical section of Figure 4-22. This demonstrates again the unique value of the strike perspective in recognizing characteristic stratigraphic patterns.

Figure 4-23 shows the interpretation of several Miocene deltaic fans. They are visible on one horizontal section because the structural dip is very gentle. Figure 4-24 shows one deltaic fan from deeper within the same area. The single gray scale used for display of these two examples was beneficial for the overall fan morphology because much of the stratigraphic patterns were in low amplitudes (refer to the discussion of color schemes in Chapter 2).

Reconstituting a Depositional Surface

In general, stratigraphic features, after being deposited on a flat-lying surface, will be bent and broken by later tectonic movements. Stratigraphy and structure then become confused and the interpretive task comes in separating them. The structure must be interpreted before stratigraphy can be appreciated.

Fig. 4-20. Shallow horizontal section from Gulf of Thailand showing channels, point bars and crevasse splays. (Courtesy Unocal Thailand Ltd.)

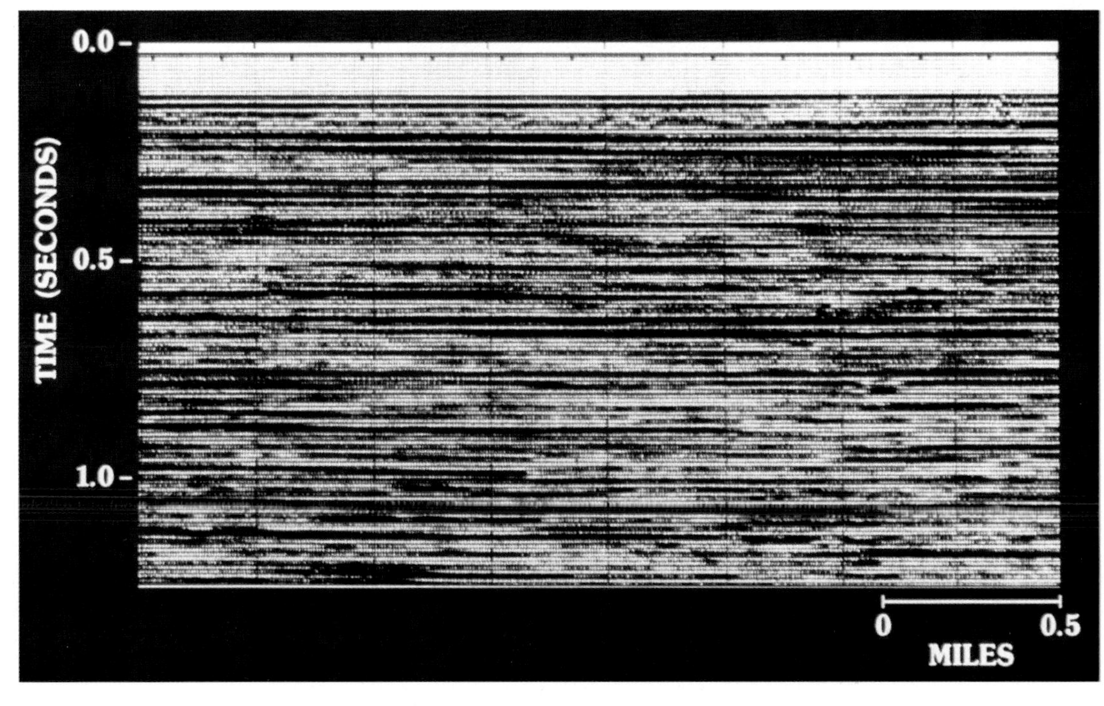

Fig. 4-21. Horizontal section from Matagorda Block 668, offshore Texas, showing prominent channel. It is a useful and interesting challenge to locate the channel intersection on the vertical section of Figure 4-22. (Courtesy ARCO Oil and Gas Company.)

Fig. 4-22. Vertical section from Matagorda Block 668, offshore Texas. (Courtesy ARCO Oil and Gas Company.)

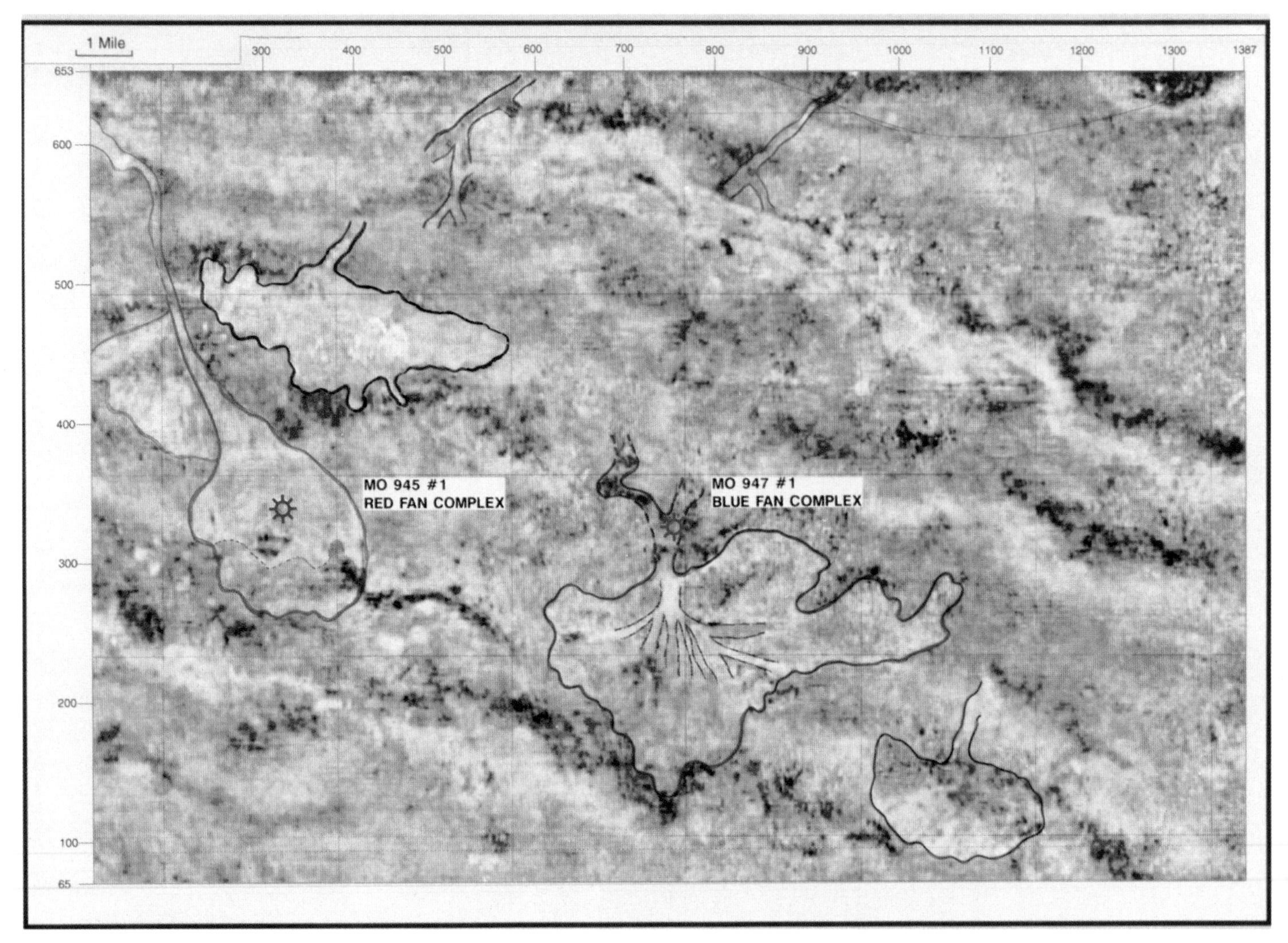

Fig. 4-23. Horizontal section at 936 ms from Mobile area, offshore Alabama, showing interpretation of numerous Miocene deltaic fans. (Courtesy Conoco Inc. and Digicon Geophysical Corp.)

Figure 4-25 illustrates schematically how a channel can be recognized and delineated in the presence of structure. In this example the interpreter has horizontal sections at 4 ms intervals from 1240 to 1260 ms. The selected event at 1240 ms for the horizon under study is traced to provide the contour as shown for 1240 ms. A high amplitude anomaly is recognized and marked at the position of the green blob. This procedure is repeated at 1244, 1248, 1252, 1256, and 1260 ms. At each of these levels the interpreter found an amplitude anomaly; together these arranged themselves into the curvilinear feature marked by the orange lines in Figure 4-25. This is manual amplitude mapping but the interactive workstation gives us several tools to do this in an efficient way.

Figure 4-26 shows a vertical section interpreted on three horizons. The Shallow Horizon, marked in blue, was selected on the basis of both structural and stratigraphic objectives. Figure 4-27 shows the structural contour map of the Shallow Horizon resulting from a full-scale structural interpretation of all the 3-D data. The desire then was to slice through the data volume along this structurally interpreted horizon in order to gather up all the seismic amplitudes associated with it. This is normally accomplished by the process of *amplitude extraction*, a menu-initiated search-and-gather operation on the interactive workstation. Alternatively, it is possible to flatten the data volume on the Shallow Horizon, as structurally interpreted in Figure 4-27, and then slice horizontally through the flattened volume at the level of the interpreted horizon.

The resultant section is known as a **horizon slice,** horizon amplitude map, or horizon Seiscrop section, where the critical word is **horizon**. This type of section, following one horizon, must be along bedding planes or it loses its value for stratigraphic

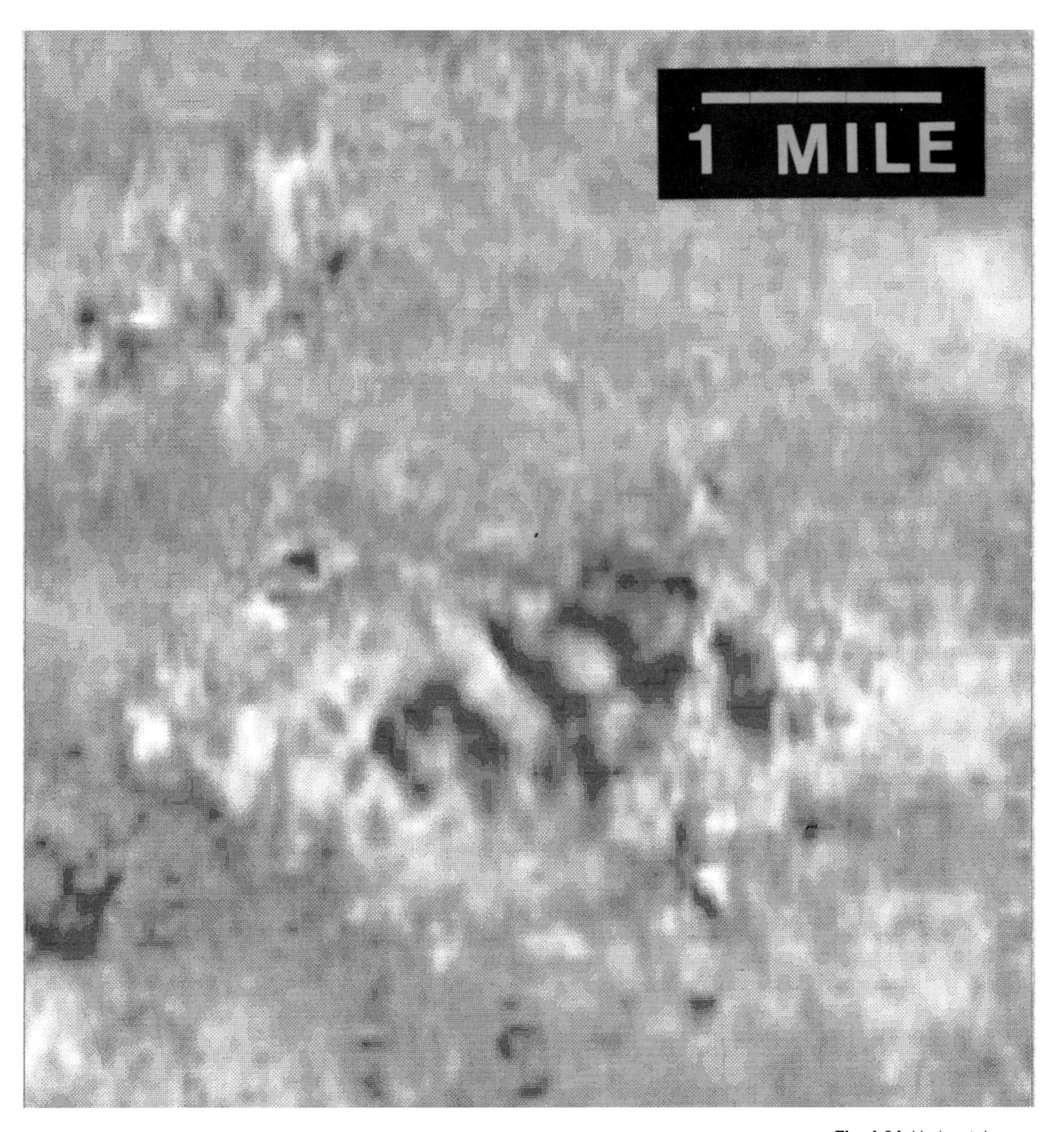

Fig. 4-24. Horizontal section at 1268 ms from Mobile area, offshore Alabama, showing one Miocene deltaic fan. Gas is being produced from one of the black channels. (Courtesy Conoco Inc. and Digicon Geophysical Corp.)

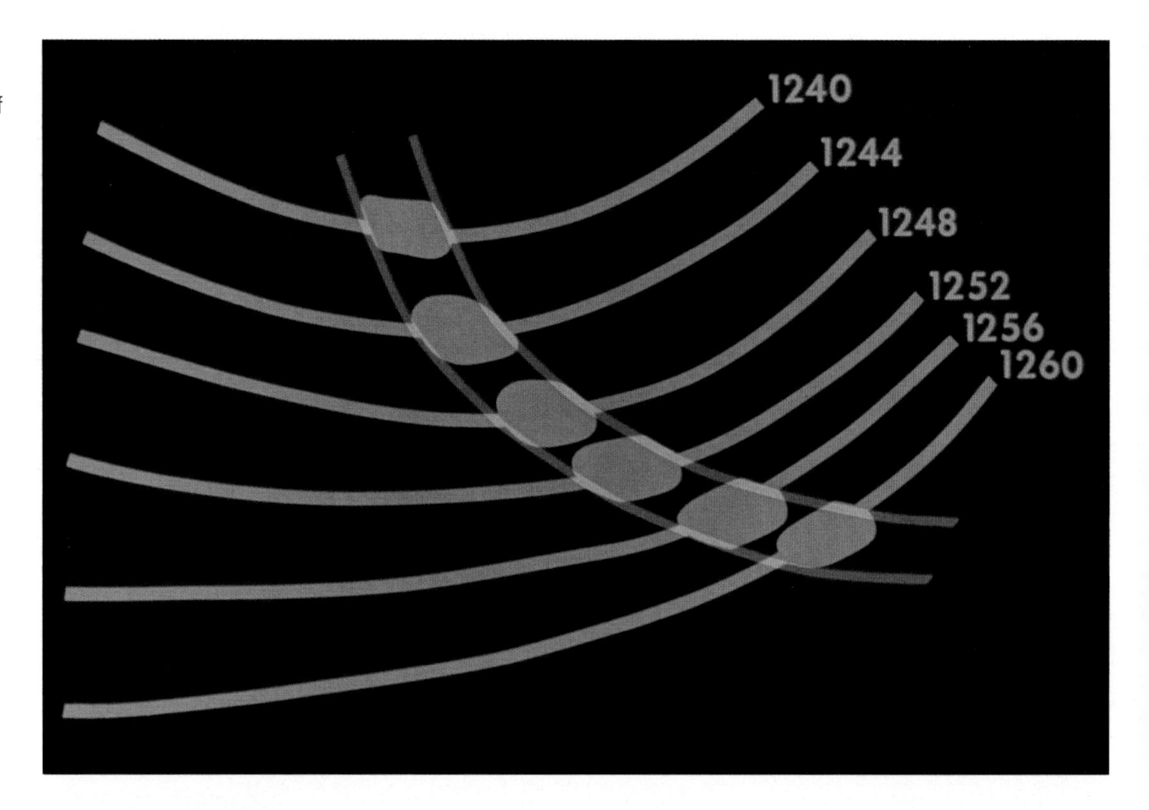

Fig. 4-25. How to follow an anomalous amplitude feature in the presence of structure on a sequence of horizontal sections.

interpretation. The importance of this approach was first stressed by Brown, Dahm, and Graebner (1981).

Figures 4-28 and 4-29 are horizon slices through adjacent conformable horizons both following the structural configuration of Figure 4-27. Both were sliced through peaks and hence all amplitudes are positive and show as varying intensities of blue; the darker blues indicate the higher amplitudes. The approximately north-south light-colored streaks are the faults; the width of a streak gives an indication of fault heave.

Figure 4-28 shows a broad high amplitude trending northwest-southeast toward the left of the section. This is interpreted as a sand bar. It is evident that this inferred bar has been dissected by several faults. The process of constructing the horizon slice has put the bar back together. Hence the construction of a horizon slice amounts to the reconstitution of a depositional surface.

Figure 4-29 shows more spatial consistency of the darker blues, indicating that this horizon follows a sheet sand. There is a curvilinear feature, somewhat the shape of a shepherd's crook, which runs northwest-southeast just to the west of well 5X. This is interpreted as an erosion channel in the sheet sand. The fact that this inferred channel is continuous across the fault just west of well 5X lends support that this horizon slice has correctly reconstituted the depositional surface into which the channel was cut.

Figure 4-30 indicates by two black arrows the two seismic horizons followed in the construction of the horizon slices of Figure 4-31. The high amplitude feature shaped somewhat like a hockey stick appears very similar on the two sections. It is invisible on other adjacent horizon slices (not shown). Hence the seismic signature of this inferred channel is trough-over-peak, which implies high velocity material, given the polarity convention implicit in these data. After inverting the whole data volume to seismic logs, a horizon slice through this velocity volume positioned between the horizon slices of Figure 4-31 generated the velocity horizon slice of Figure 4-32. The darker colors indicate the high velocity channel fill.

Methods of Making Horizon Slices

Automatic horizon tracking, now commonplace in interactive interpretation systems, has greatly facilitated the generation of horizon slices. When a horizon is tracked, the extreme amplitude as well as its time is stored in the digital database. Mapping of the times produces a structure map; mapping of the amplitudes produces a horizon slice. More commonly, only the time is stored as a result of horizon tracking and later the amplitudes are extracted from the data. In addition, it is possible to extract the amplitudes not coincident with the tracked horizon but parallel to it and shifted by a chosen number of milliseconds.

Figure 4-33 shows two lines from a Gulf of Mexico 3-D prospect, where a horizon is tracked one-and-a-half periods above a red blob considered to be of stratigraphic interest. The structural continuity is better for the horizon being tracked than for the blob, so the structure was defined at this level and the horizon slice made parallel to it through the blob at a fixed time increment deeper.

The resulting horizon slice is shown in Figure 4-34 and the interpreter can readily infer the existence of another channel. The black horizontal lines indicate the positions of the two vertical sections of Figure 4-33. The amplitude of the channel reflection is greater to the northeast; a discussion of this relative to implied gas content appears in Chapter 5.

A horizon slice is by definition a slice along a bedding plane, but the methods by which an interpreter may make such a slice are many and varied (Figure 4-35). If the slice is made at the tracking level, following automatic horizon tracking, the horizon slice is made up of truly crestal amplitudes and should thus be accurately along the bedding plane. However, if the structure is defined by tracking at one level where the continuity is clear but the slice is made parallel to that at another level, then the slicing and tracking levels must be sufficiently conformable for the horizon slice to adequately follow the bedding plane. This approach beneficially segregates the stratigraphic and structural components of an interpretation. Minor irregularities at the tracking level may not be paralleled at the slicing level, so spatial smoothing of the tracked times may be desirable before displacing the horizon down or up to the slicing level.

Slicing through a zone of poor reflection continuity (where tracking would have been impossible) that is parallel to a good reflection at the top or the base of the zone has in several cases yielded meaningful and interpretable stratigraphic patterns. This demonstrates that data that may appear poor and uninterpretable on vertical sections may in fact reveal significant stratigraphic information when viewed spatially over bedding plane surfaces. In the case of a poor continuity interval of nonuniform thickness it can be useful to track a reflection at the top and one at the base. Then the slice is made within that interval using a surface whose shape is based partly on the upper tracked surface and partly on the lower tracked surface, the proportions of each depending on where within the interval the slice is desired.

When a tracked horizon is displaced down by a constant time shift, the option exists to snap the displaced horizon to the exact crestal amplitude of the new reflection. Sometimes this will be the right course of action and sometimes it will be the wrong. Figure 4-36 illustrates diagrammatically a situation where it would be wrong. Good reflections exist at the sand/shale interface and the shale/limestone interface. However, at the latter patches of porosity form the exploration objective and also introduce character changes along the limestone reflection. The shale is of uniform thickness. The structure is followed on the sand/shale reflection, which is stratigraphically uncomplicated, and displaced down by a constant time shift to the top of the limestone. The amplitudes are then extracted without snapping to yield a horizon slice on which the porosity should show as low amplitude patches. Snapping would move the horizon down to the top of the unporous limestone and the low amplitude patches would be lost!

After amplitude has been extracted on the objective horizon corrections are sometimes required (Figure 4-35). Consider, for example, that a high amplitude bright spot on a deeper horizon slice is shadowed over part of its area by a shallower high amplitude anomaly. Some fraction of the amplitude extracted on the shallower horizon can be added to the amplitude on the deeper horizon to compensate for the shadowing effect. The fraction to use must be established empirically but the author has often used one-quarter. Lateral variations in amplitude caused by surface conditions can

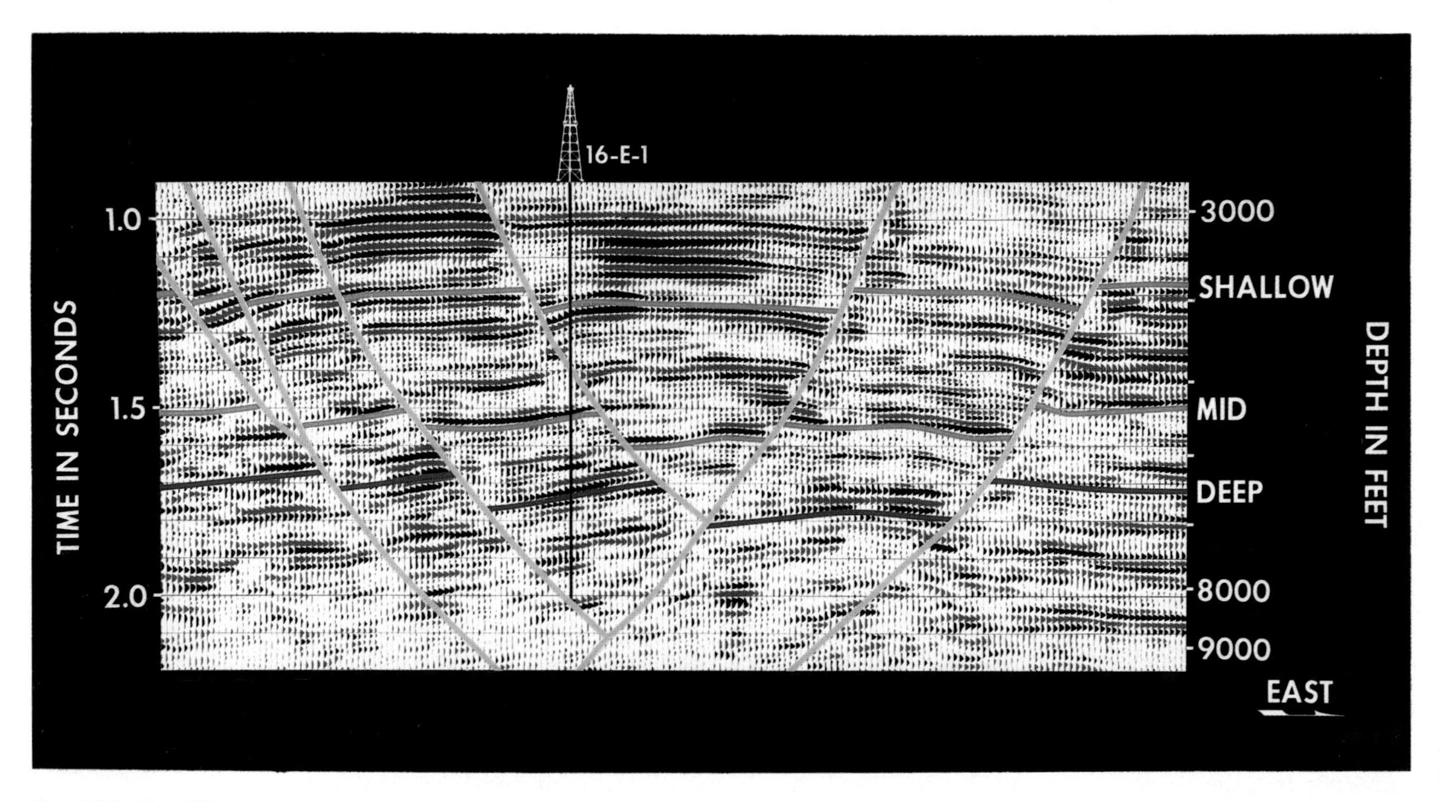

Fig. 4-26. Line 55 interpreted showing structure of Shallow Horizon. (Courtesy Texas Pacific Oil Company Inc.)

sometimes be removed successfully by normalization. The amplitude of a shallower horizon is assumed to be constant and then the amplitude ratio of the objective horizon to the shallow reference should remove the effect.

Figure 4-37 shows a sequence of faults affecting one horizon interpreted on a vertical section from a 3-D survey in the Gulf of Thailand. Figure 4-38 shows the time structure map resulting from the complete structural interpretation of the same horizon. The faults trending north-northwest to south-southeast divide the area into seven fault blocks. The corresponding horizon slice is shown in Figure 4-39. A meandering stream channel is evident and gas production from the channel has been established in two of the fault blocks.

The continuity of the channel confirms that the depositional surface has been correctly reconstituted. Clearly the value of such a horizon slice for stratigraphic purposes is critically dependent on the accuracy of the structural interpretation that was involved in its derivation. Here the stratigraphic and structural interpretation actually impacted each other iteratively. The first horizon slice generated for this level did not show the channel continuity of Figure 4-39 in one of the fault blocks. This suggested miscorrelation into that block. After re-examining the correlation and retracking the data in that block, the horizon slice shown as Figure 4-39 was obtained. The improved channel continuity indicated the relative correctness of the updated structural interpretation.

Figures 4-40 and 4-41 show the time structure map and horizon slice for one interpreted horizon in a Gulf of Mexico shallow water prospect. Two channels are evident, one of them intersected by a fault. The deeper channel lies between 2100 and 2200 ms which converts to depths around 2500 m (8,200 ft).

Figure 4-42 shows a Gulf of Mexico horizon slice with overlain structural contours. This is a particularly valuable form of display (compare Figure 5-20) because it permits interpretation of stratigraphic/reservoir patterns in their present-day structural context. Here the high amplitudes (reds and oranges) are caused by gas in several sand bodies. Note the sharp amplitude terminations toward the south, indicating the position of the gas-water contacts.

Figure 4-43 is an arbitrary line through three wells from a 3-D survey in southern

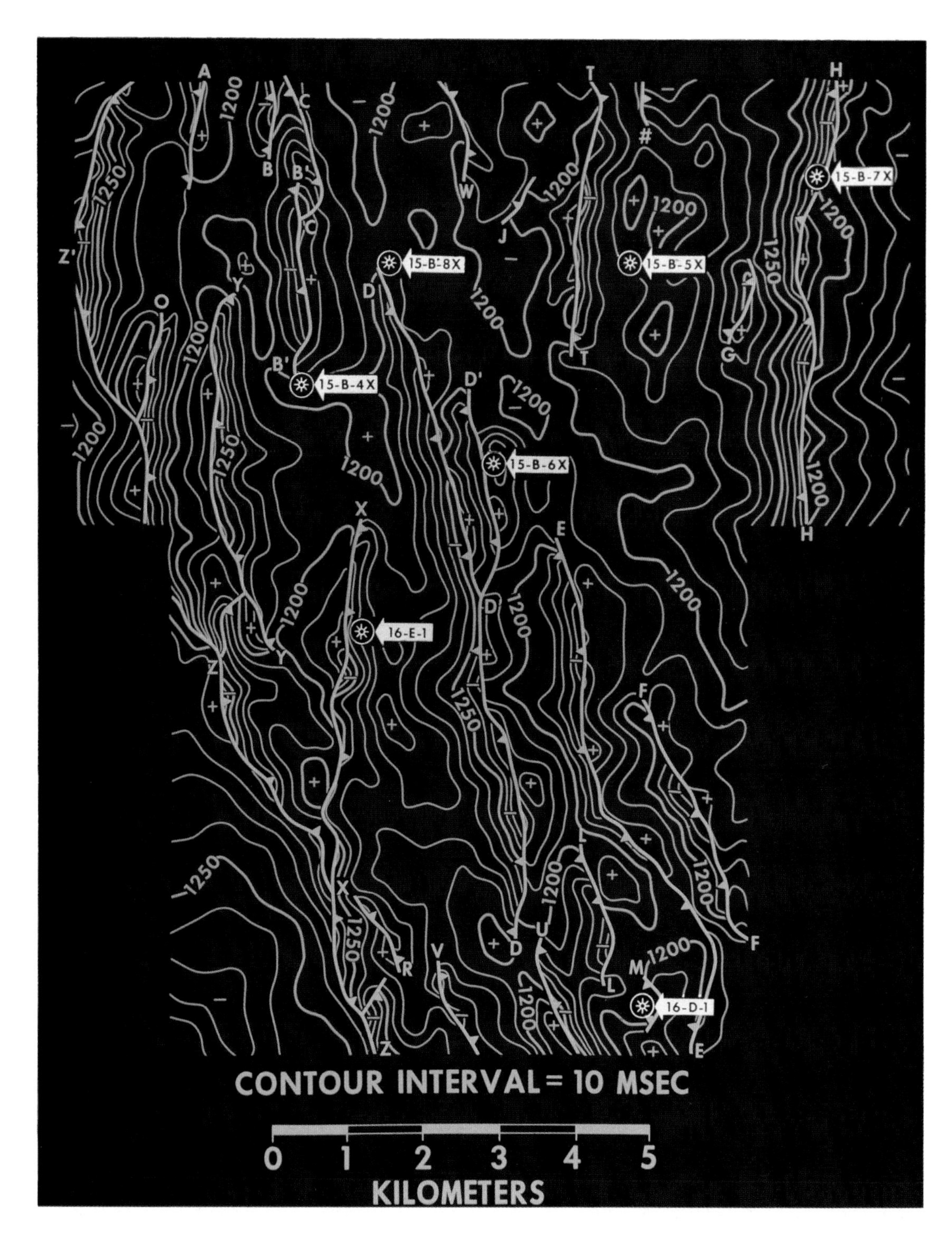

Fig. 4-27. Time structure map of Shallow Horizon. (Courtesy Texas Pacific Oil Company Inc.)

Canada. The structure was defined at the Base Bow Island reflection. A slice parallel to this through the Glauconite zone yielded the horizon slice of Figure 4-44. Here the stratigraphic patterns were not clearly apparent, but a further interpretation tied to well intersections yielded the superimposed stratigraphic descriptions.

Figure 4-45 is a mosaic of amplitudes from 3-D and 2-D data from Argentina and demonstrates the increased stratigraphic detail available from 3-D data (Gerster, 1995). Three 3-D surveys are shown; seven others exist in the immediate area or are under way.

Figure 4-46 is a horizon slice from offshore eastern Canada showing many channels and their levees, an old shore line and an abandoned lake (Enachescu, 1993).

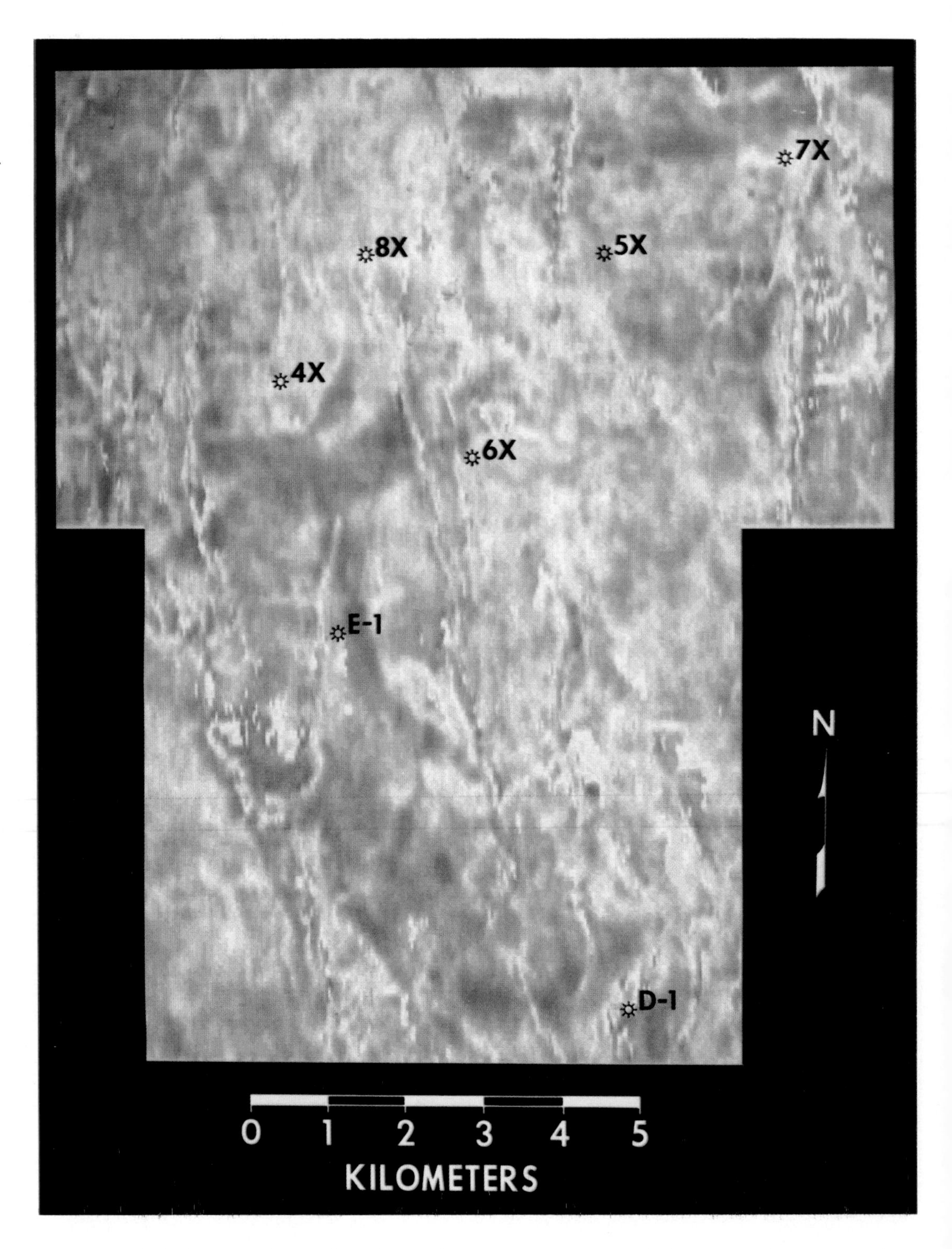

Fig. 4-28. Horizon slice 180 feet (60 m) below Shallow Horizon showing northwest- southeast-trending high amplitude interpreted as a sand bar. (Courtesy Texas Pacific Oil Company Inc.)

Figure 4-47 is a horizon slice from the Norwegian North Sea. The interesting fan-shaped feature is interpreted as a mass flow in the Danian (basal Tertiary) chalk.

The majority of the horizon slices presented in this chapter display seismic amplitude, and this also reflects the author's usage. However, it is possible to make horizon slices in other attributes. Figure 4-32, for example, displays inversion velocity. Sonneland (1990) defines a variety of attributes that he calls "classifiers," and discusses their applications to seismic interpretation. Today a plethora of amplitude-derived attributes are available and these are discussed in Chapter 8.

Unconformities

Figure 4-48 shows a horizon slice from yet another Gulf of Mexico prospect. The

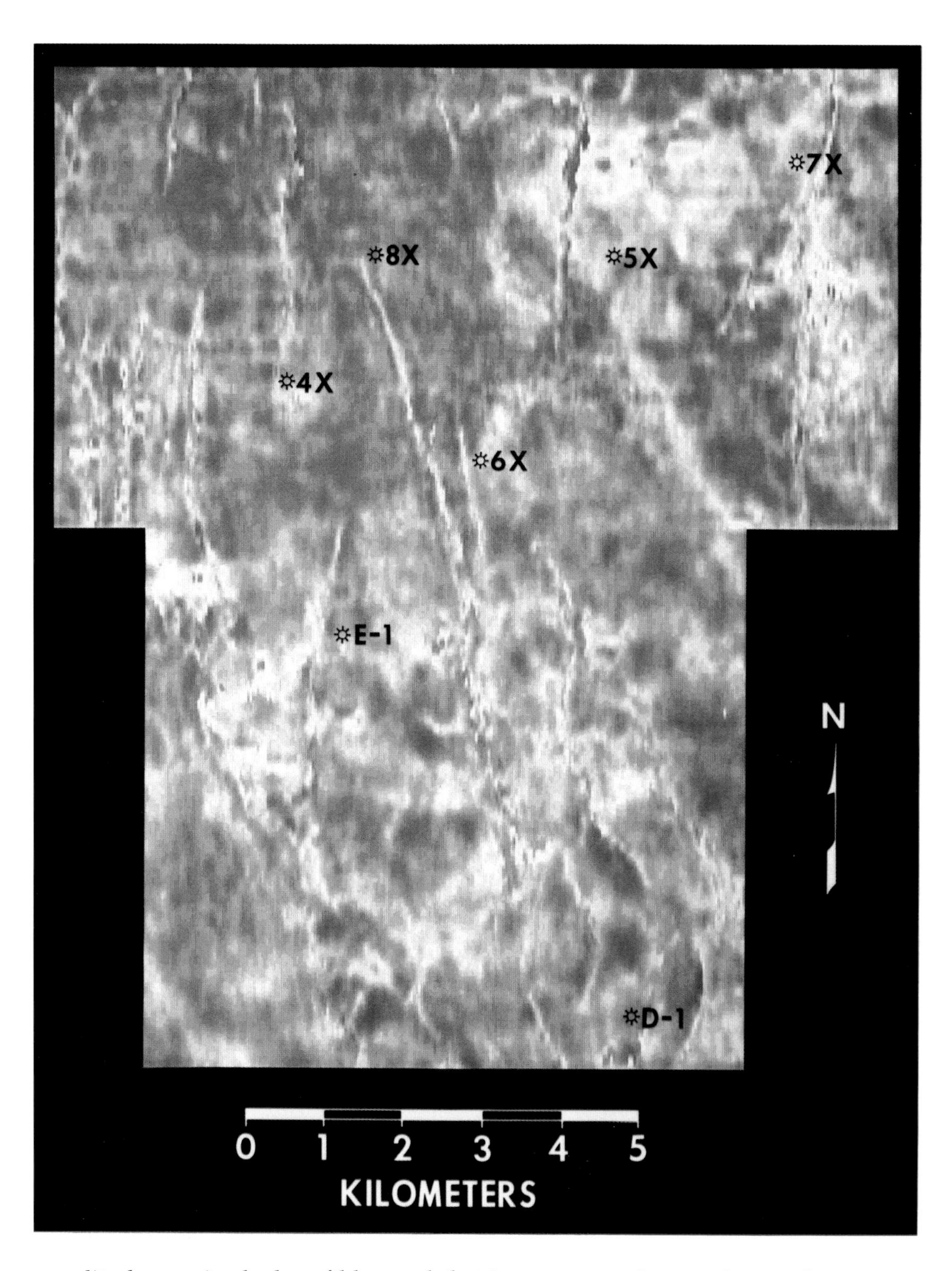

Fig. 4-29. Horizon slice through Shallow Horizon showing a partly eroded sheet sand. (Courtesy Texas Pacific Oil Company Inc.)

amplitudes are in shades of blue and the time structure is superimposed as contour lines with an interval of 100 ms. Several amplitude lineations are apparent. The ones running approximately east-west are faults as evidenced by the displacement of the contours.

The major lineation running north-northwest–south-southeast is apparently unrelated to the faulting. It is interpreted as the truncation of a sand dipping up from the east. It is probably a depositional edge but the erosional truncation of a sand at an unconformity would show in exactly the same manner. It is this lineation on the horizon slice which caught the interpreter's eye and thus begged for an explanation.

An excellent example of the variation in reflection character and amplitude across

Text continues on page 134.

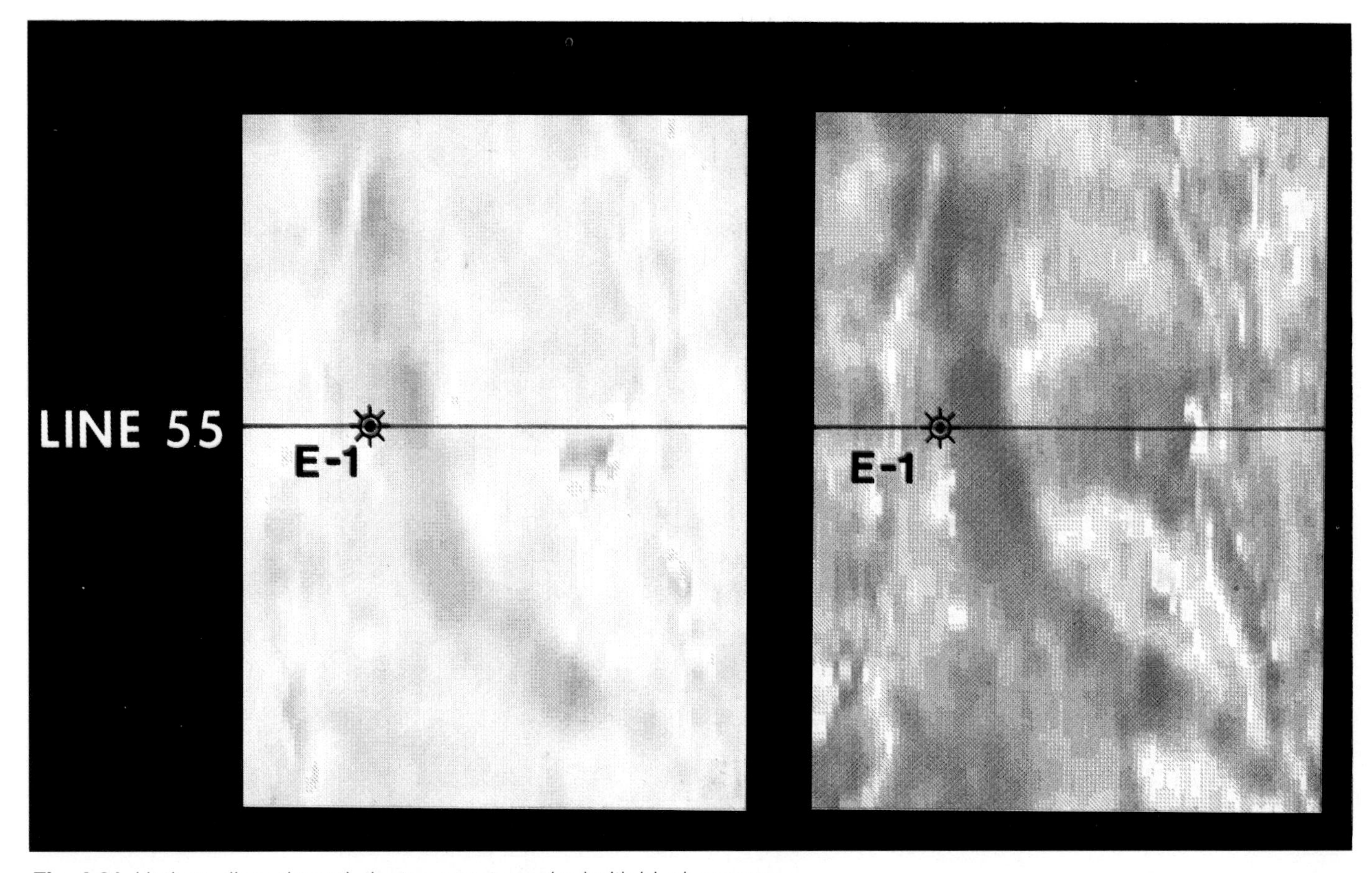

Fig. 4-31. Horizon slices through the two events marked with black arrows on Figure 4-30. The curvilinear features are interpreted as the reflections from the top and base of a channel. (Courtesy Texas Pacific Oil Company Inc.)

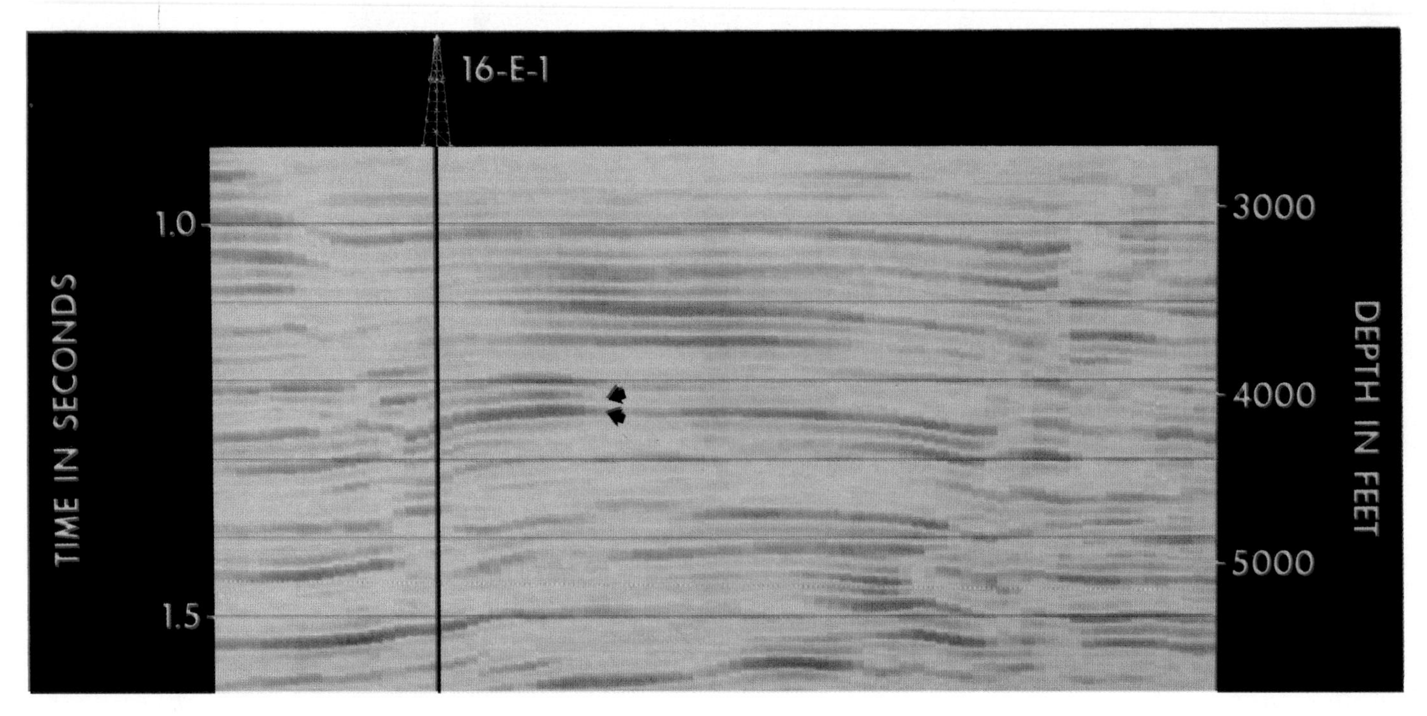

Fig. 4-30. A portion of Line 55 through the central graben of the 3-D prospect. (Courtesy Texas Pacific Oil Company Inc.)

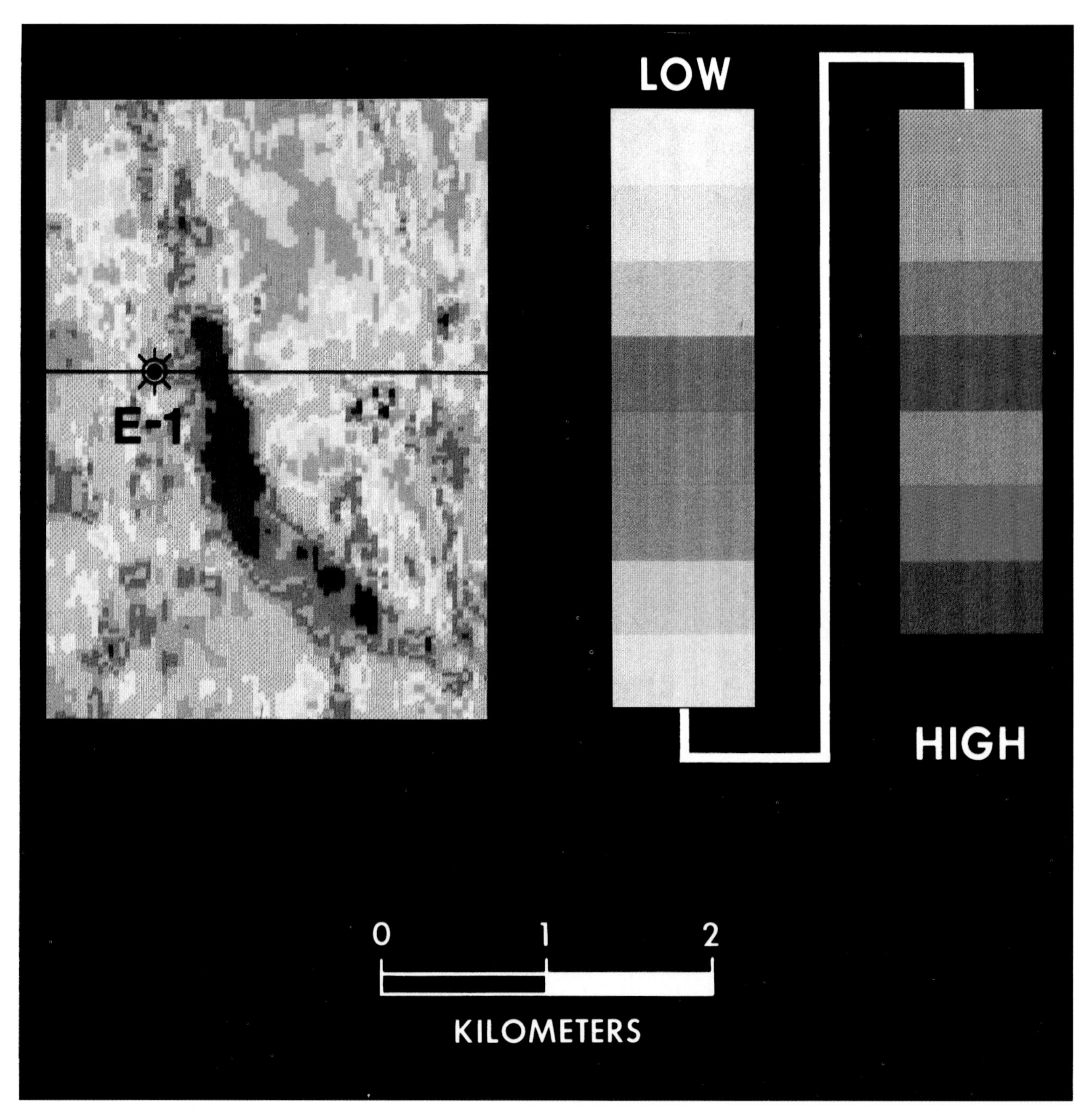

Fig. 4-32. Horizon slice in velocity positioned between the sections of Figure 4-31 and showing the extent of the high velocity channel fill. (Courtesy Texas Pacific Oil Company Inc.)

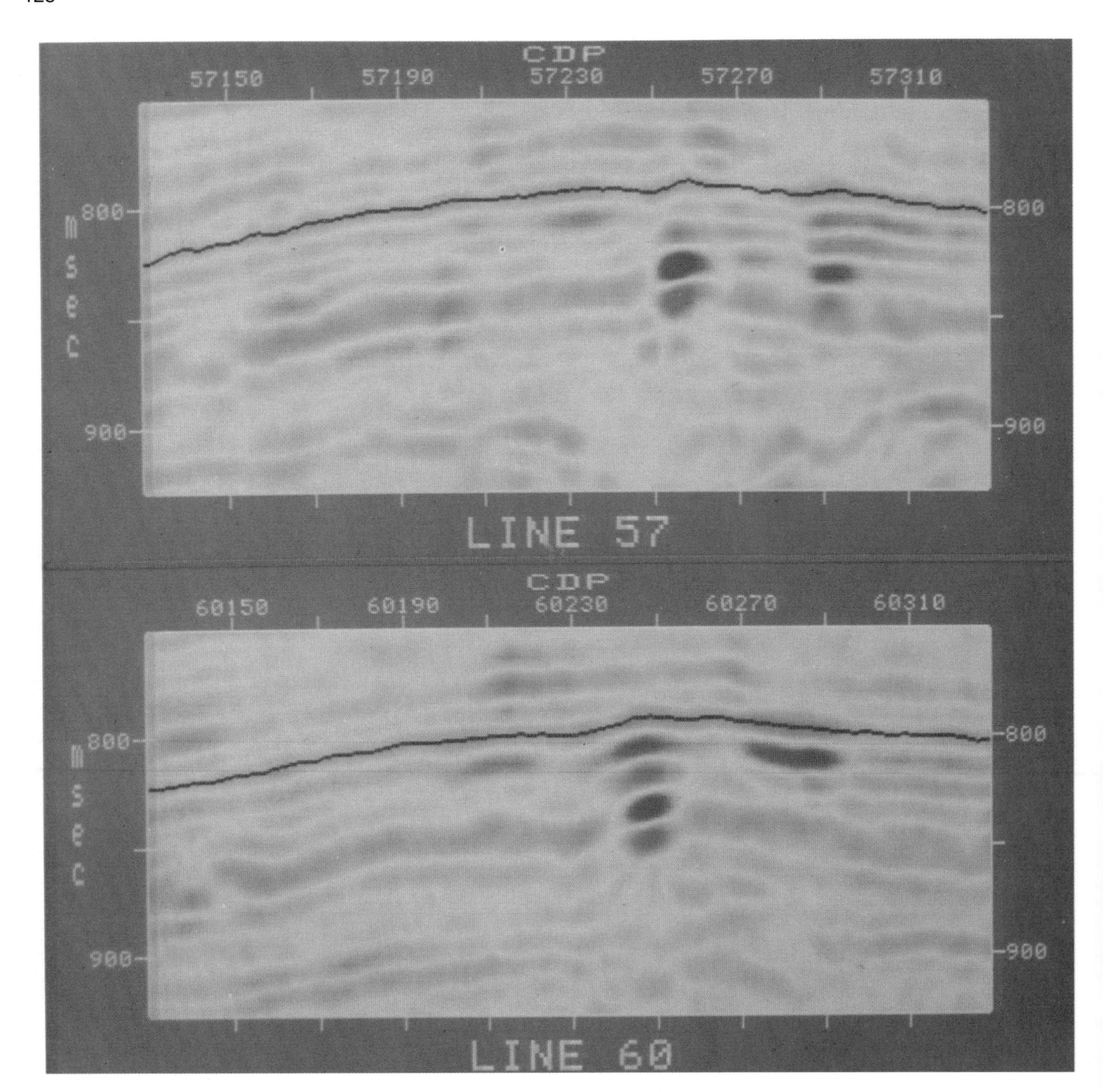

Fig. 4-33. Lines 57 and 60 from a 3-D survey in the Gulf of Mexico showing a tracked horizon above bright events indicating channel intersections. (Courtesy Chevron U.S.A. Inc.)

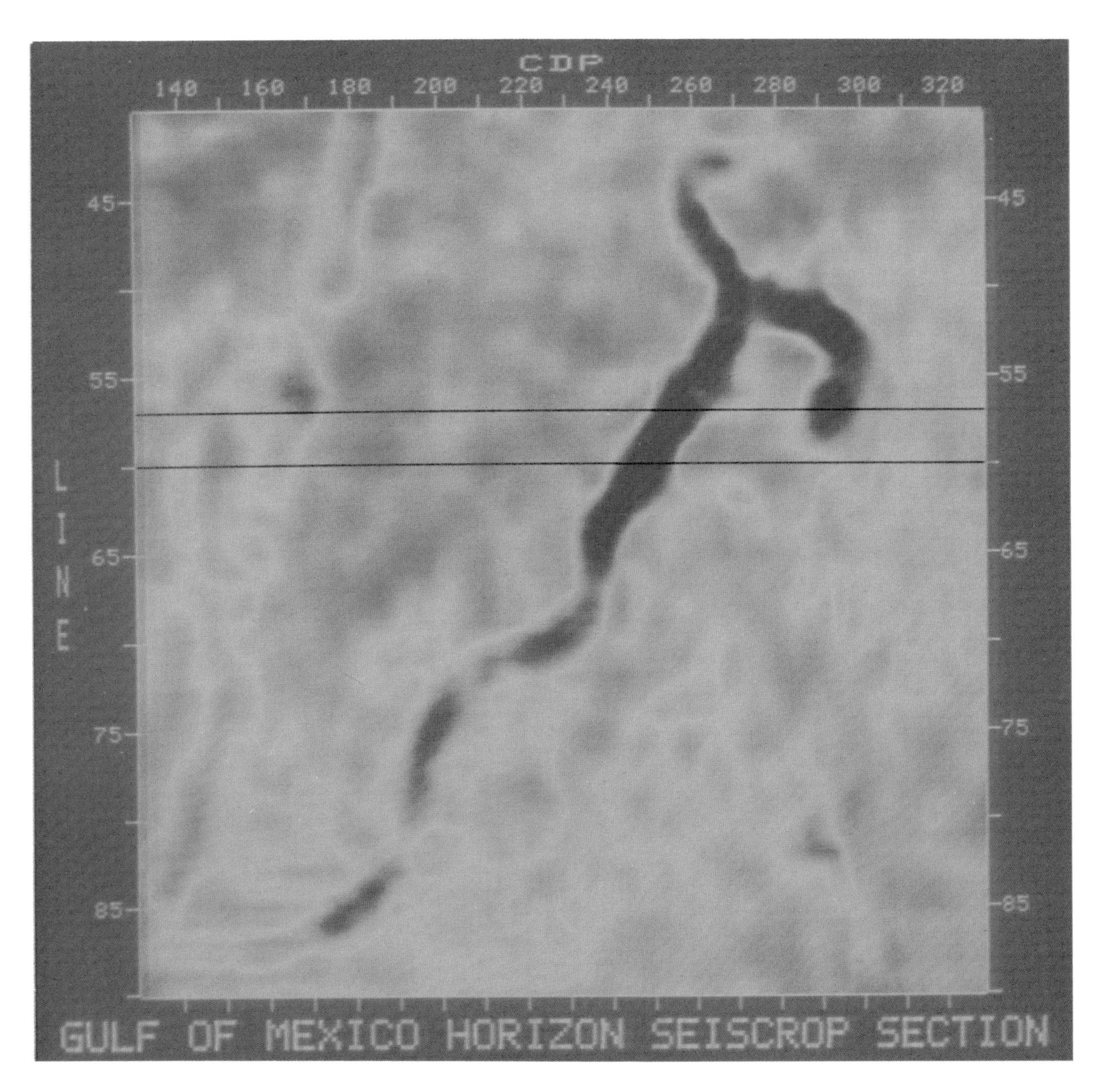

Fig. 4-34. Horizon slice showing channel intersected in Figure 4-33. (Courtesy Chevron U.S.A. Inc.)

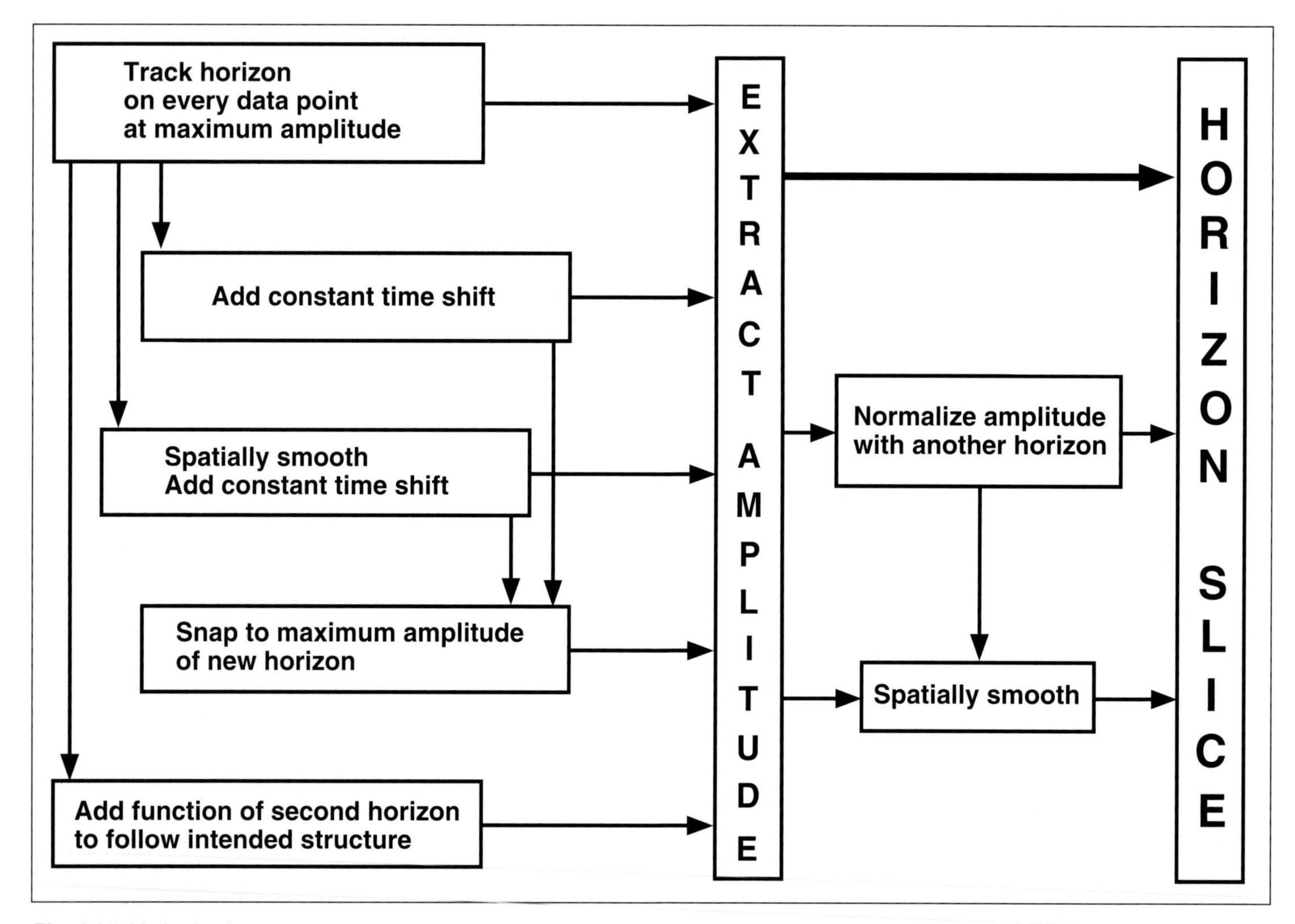

Fig. 4-35. Methods of making horizon slices. The tracking level and the slicing level need not be the same. Amplitude corrections may be necessary to compensate for shallower effects.

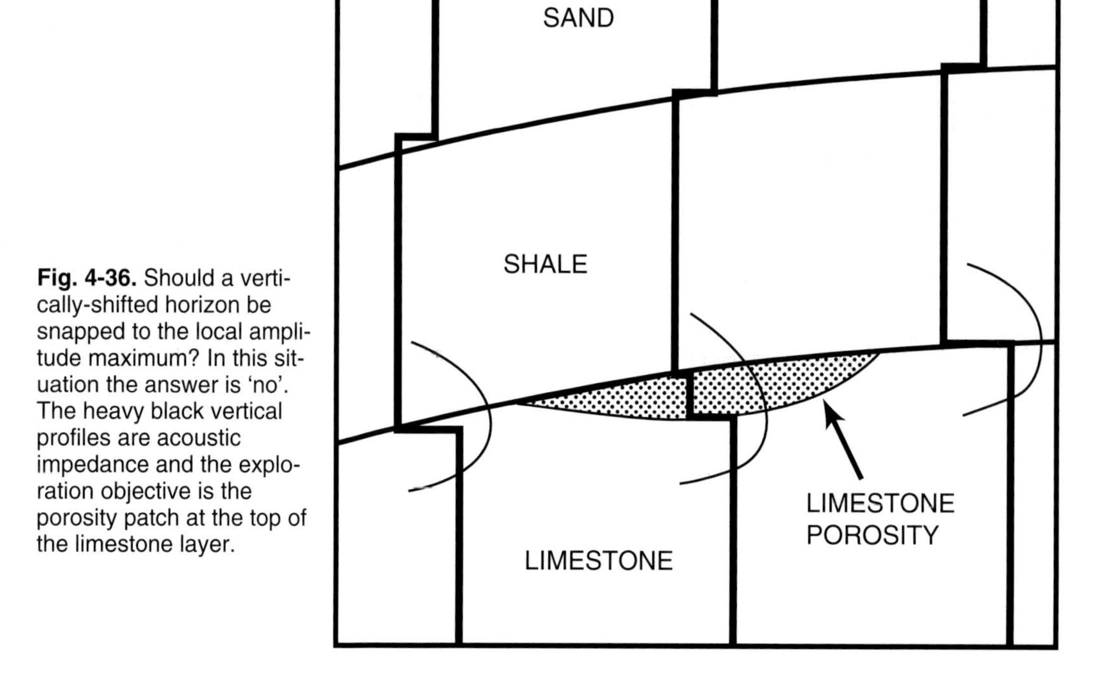

Fig. 4-36. Should a vertically-shifted horizon be snapped to the local amplitude maximum? In this situation the answer is 'no'. The heavy black vertical profiles are acoustic impedance and the exploration objective is the porosity patch at the top of the limestone layer.

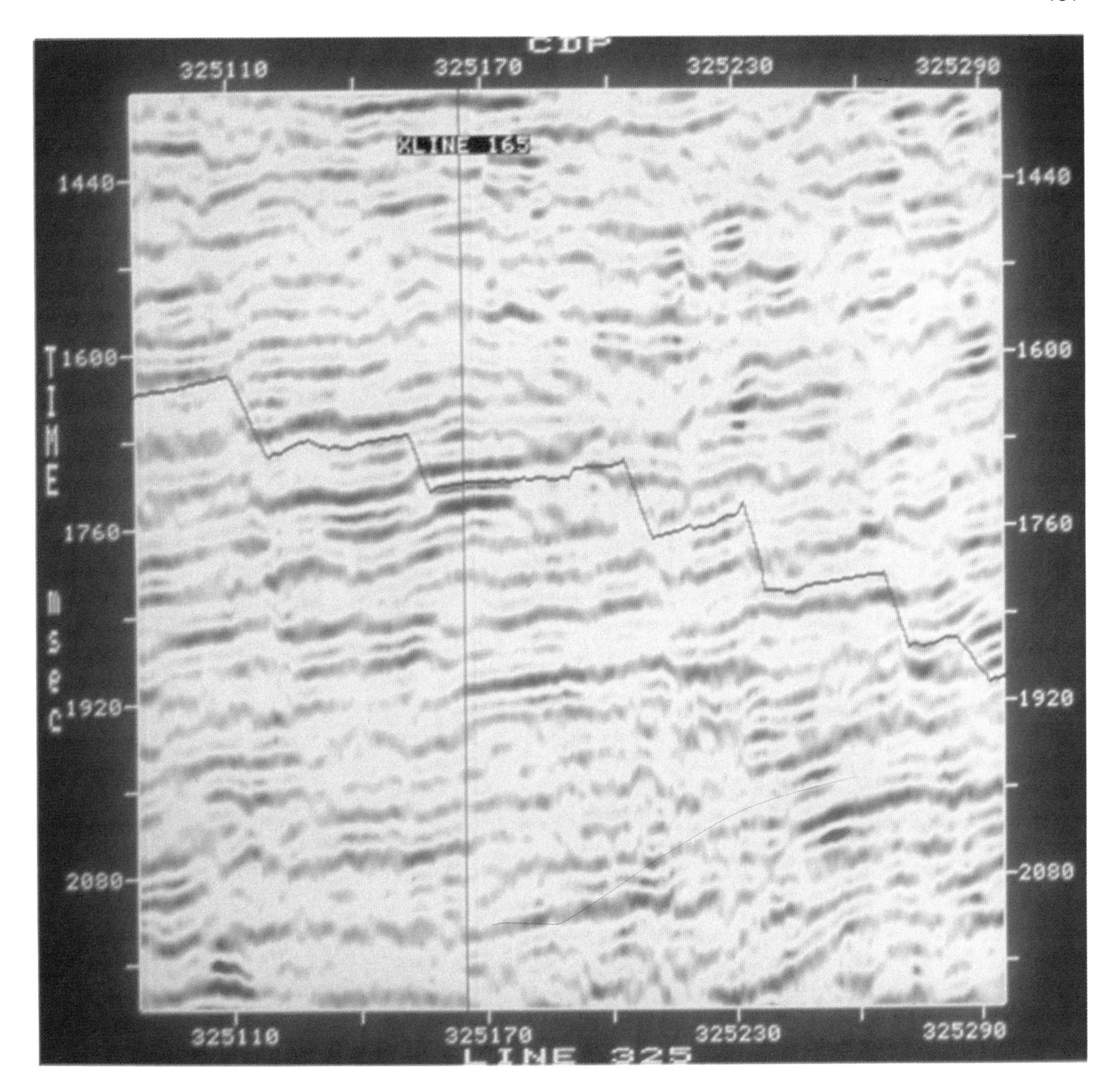

Fig. 4-37. Line 325 from 3-D survey in the Gulf of Thailand showing interpreted horizon through many fault blocks. (Courtesy Unocal Thailand Ltd.)

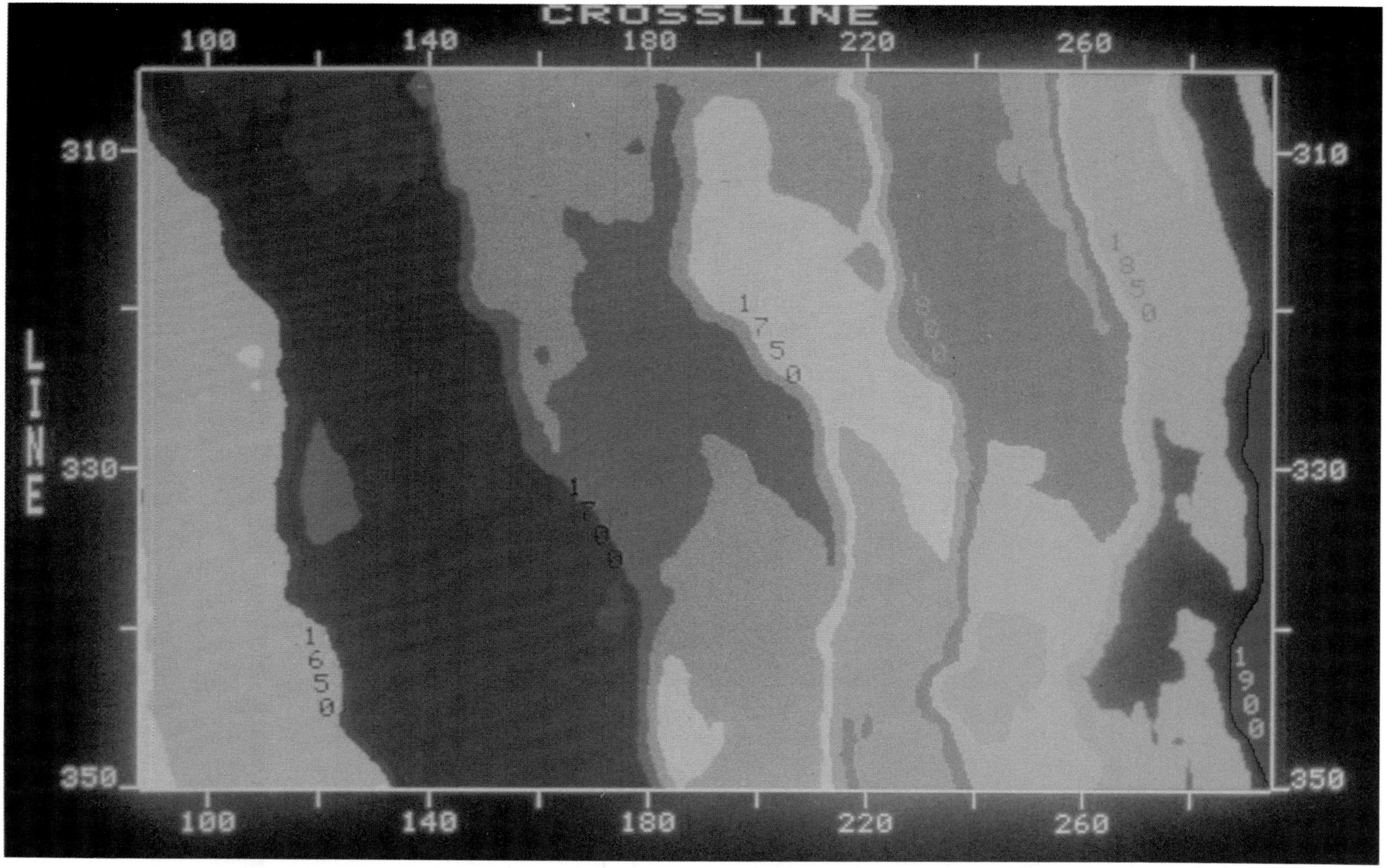

Fig. 4-38. Time structure map of horizon tracked in Figure 4-37. (Courtesy Unocal Thailand Ltd.)

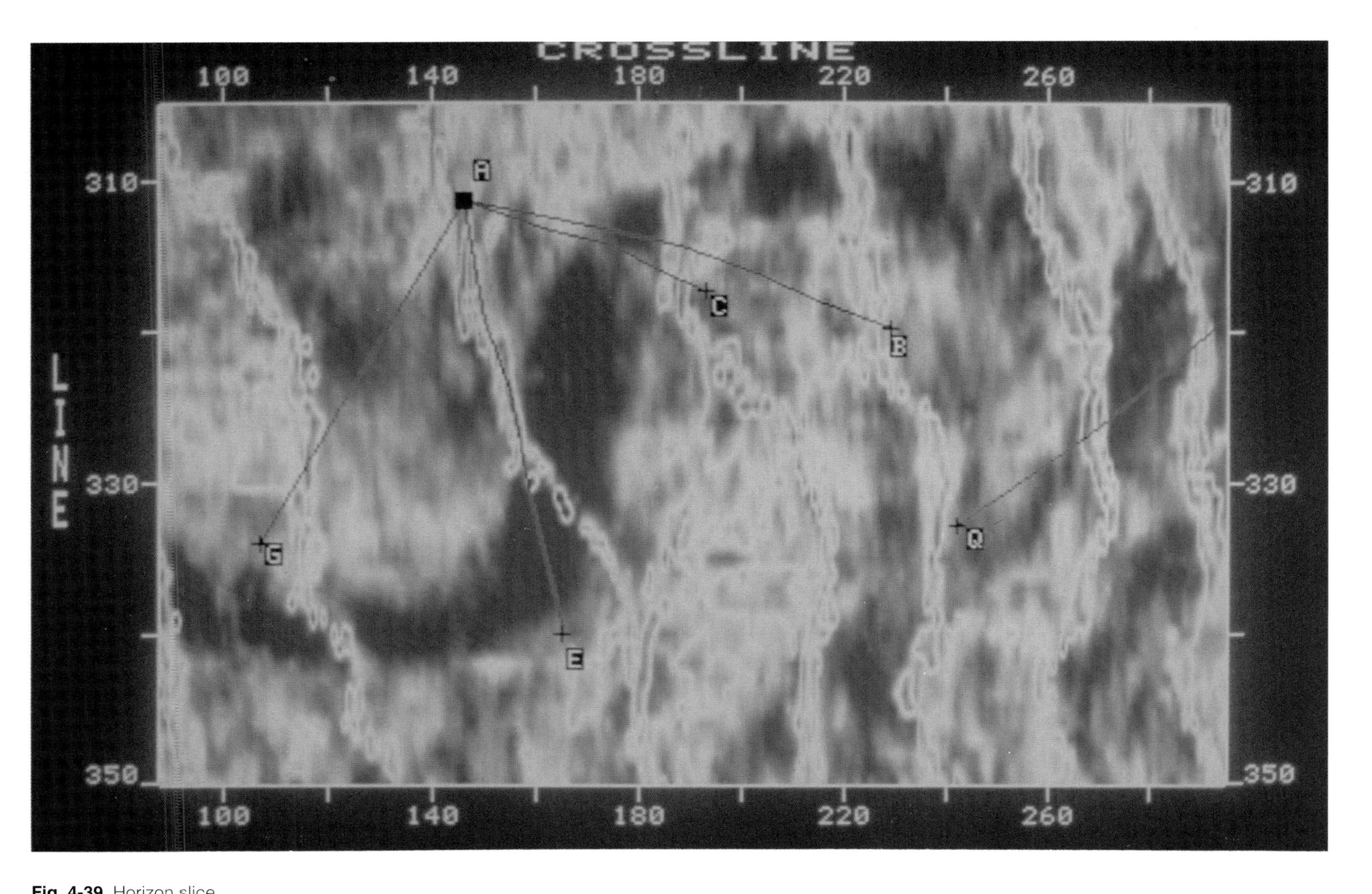

Fig. 4-39. Horizon slice showing spatial distribution of amplitude over the horizon mapped in Figure 4-38. Gas production has been established in the meandering channel. (Courtesy Unocal Thailand Ltd.)

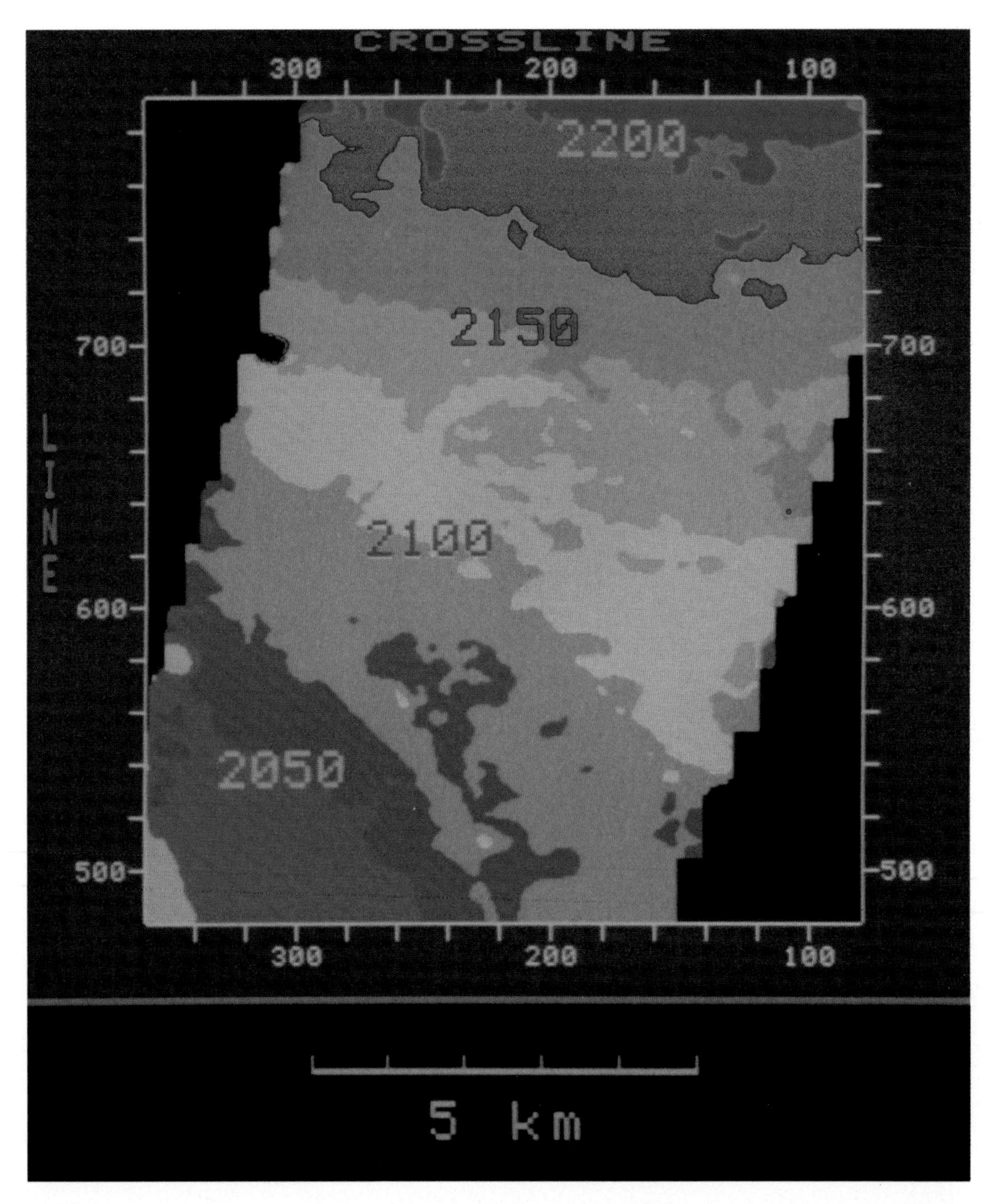

Fig. 4-40. Time structure map for a horizon interpreted from a 3-D survey offshore Louisiana. The large numbers are contour designations in milliseconds. (Courtesy Geophysical Service Inc.)

an angular unconformity comes from the Lisburne 3-D survey. The following description and Figures 4-49, 4-50, and 4-51 have been kindly provided by J. J. O'Brien of Standard Alaska Production Company, now BP Exploration (Alaska). The Lisburne Field is located on the North Slope of Alaska, adjacent to Prudhoe Bay Field and partly underlying the Prudhoe Bay gas cap. The Lisburne reservoir consists of a thin-bedded limestone/dolomite/shale sequence that was deposited in a subtidal to supratidal environment.

Prior to field startup in 1986, Standard Oil acquired a 3-D seismic survey over the eastern truncation zone of the field, where the Lisburne carbonate section is truncated by the Lower Cretaceous unconformity and the reservoir fluids are trapped by the overlying Cretaceous shale. Interpretation of the 3-D survey included tracking of key horizons on the migrated dataset using an interactive workstation and generation of time maps and horizon slices (amplitude maps) for these horizons. Horizon slices were found to possess significant information content. In particular, the amplitude of

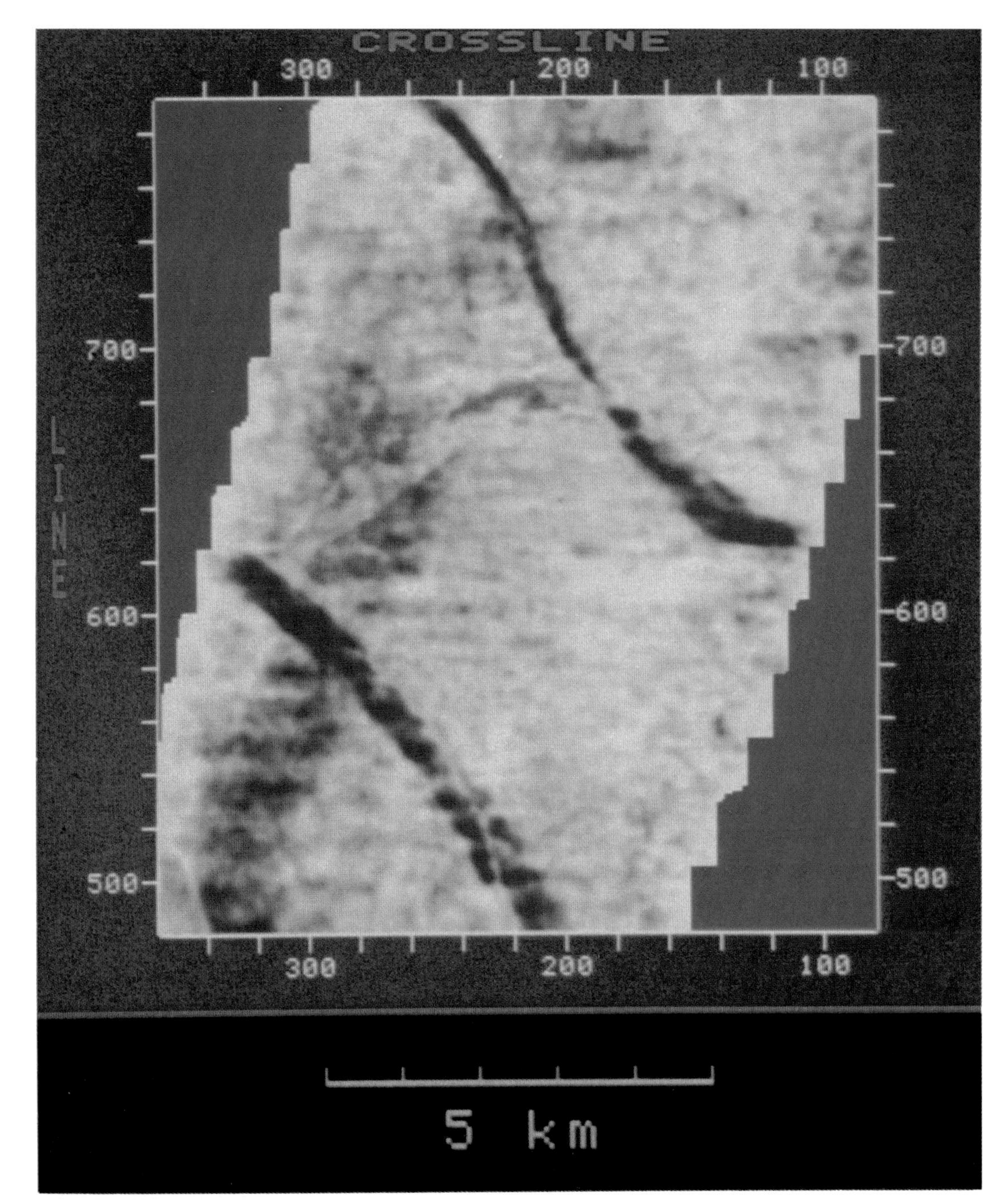

Fig. 4-41. Horizon slice showing spatial distribution of amplitude over the horizon mapped in Figure 4-40 and indicating two channels. (Courtesy Geophysical Service Inc.)

the reflection from the Lower Cretaceous unconformity (Figure 4-50) shows distinct lateral variations. Some of these are interpreted in Figure 4-51 and are referenced to a vertical section in Figure 4-49.

In the western portion of the survey the Lower Cretaceous unconformity truncates the clastic section overlying the Lisburne carbonates. In this area, the unconformity surface represents a clastic/clastic interface with a relatively low acoustic impedance contrast, resulting in a low amplitude seismic reflection. Moving eastward, the Lower Cretaceous unconformity truncates the Lisburne carbonate section, and its surface represents an interface between a thick, uniform shale and an underlying carbonate section. There is, in general, a higher contrast in acoustic impedance across the unconformity in this area and reflection amplitudes are therefore greater.

In the area where the Lower Cretaceous unconformity truncates the Lisburne section, variations in reflection amplitude are seen. A lower amplitude feature is observed trending from northwest to southeast, subparalleling the truncation; time

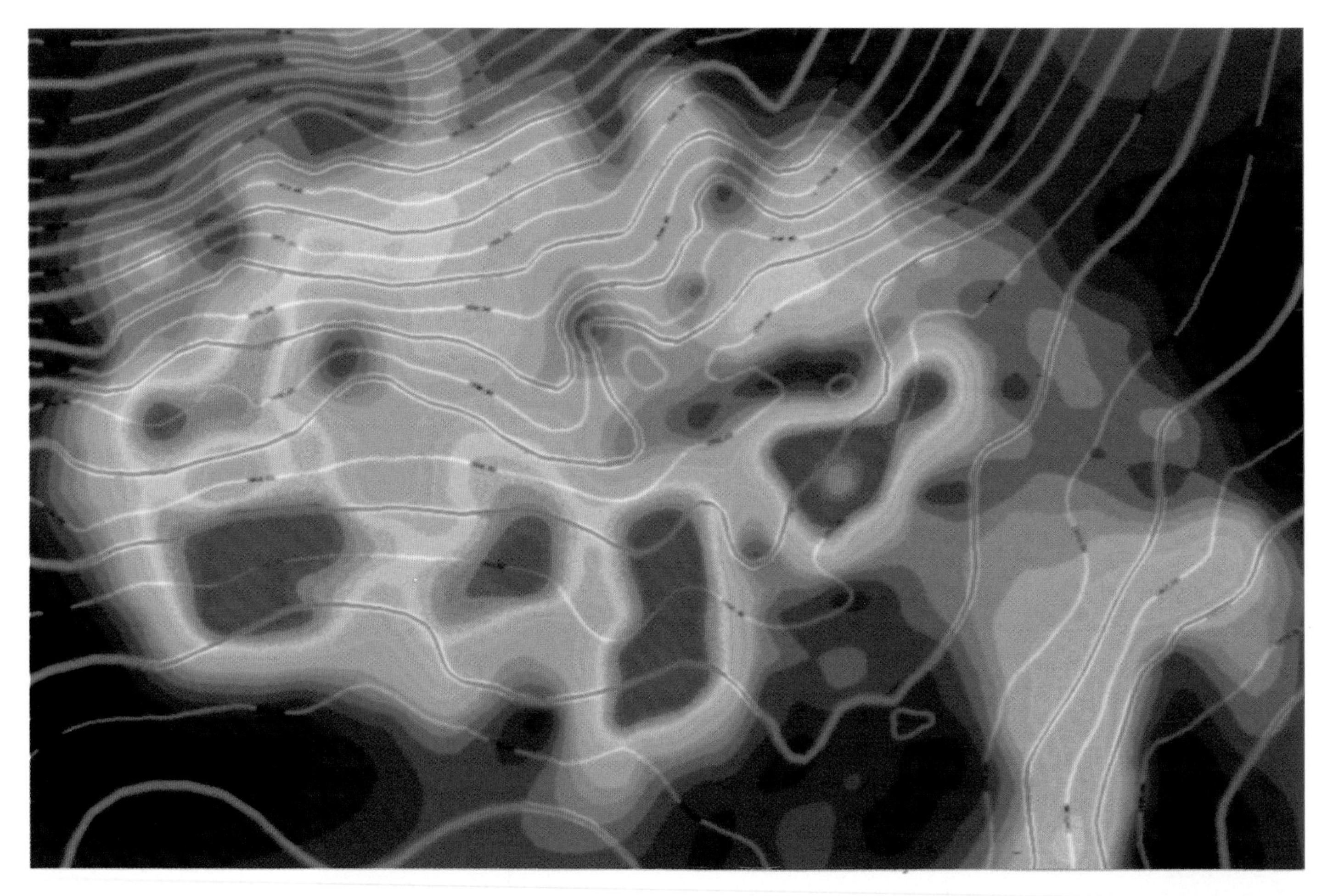

Fig. 4-42. Horizon slice from Gulf of Mexico with overlain structural contours showing high amplitudes caused by gas in several sand bodies. (Courtesy GeoQuest Systems Inc.)

mapping indicates that this trend coincides with the truncation of Zone 5 of the Lisburne reservoir. The L-7 well, which penetrated this amplitude trend, encountered a 29-m (95-ft)-thick section of Zone 5 with excellent porosity development immediately underlying the unconformity surface. Because the overlying formation is a marine shale that is uniformly low in acoustic impedance over the survey area, this low amplitude feature represents a lower acoustic impedance trend within the reservoir; the well data from L-7 suggest that it may be an enhanced porosity trend.

Farther east another low amplitude lineament is seen trending from northwest to southeast. This feature coincides with the truncation of the Green Shale, a low impedance unit that is 9-18 m (30-60 ft) thick. In addition, a number of narrower east-west–trending lineaments are observed that correspond to faults cutting the unconformity surface and that have throws of up to 23 m (70 ft).

References

Broussard, M. L., ed., 1975, Deltas: Houston Geological Society, 555 p.

Brown, A. R., C. G. Dahm, and R. J. Graebner, 1981, A stratigraphic case history using three-dimensional seismic data in the Gulf of Thailand: Geophysical Prospecting, v. 29, p. 327-349.

Enachescu, M. E., 1993, Amplitude interpretation of 3-D reflection data: The Leading Edge, v. 12, p. 678–685.

Gerster, R. L., 1995, The role played by 3-D and 2-D seismic data in the exploration and development of the largest oil field in Argentina: SEG Annual Meeting expanded abstract.

Klein, G. deV., 1985, Sandstone depositional models for exploration for fossil fuels, third edition: Boston, Massachusetts, International Human Resources Development Corporation, 209 p.

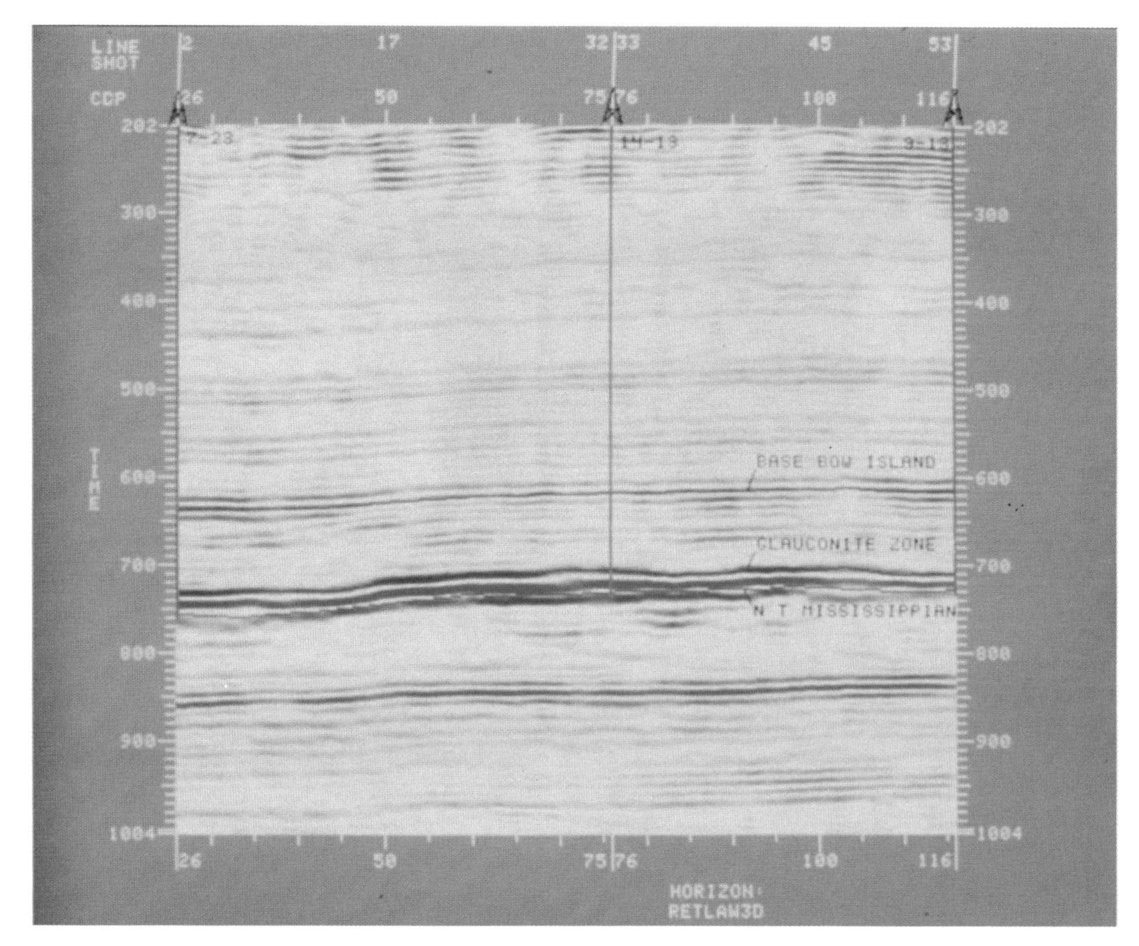

Fig. 4-43. Arbitrary line joining three wells in Retlaw prospect, southern Alberta, showing Base Bow Island reflection (the tracking level) and Glauconite Zone (the slicing level). (Courtesy Geophysical Service Inc.)

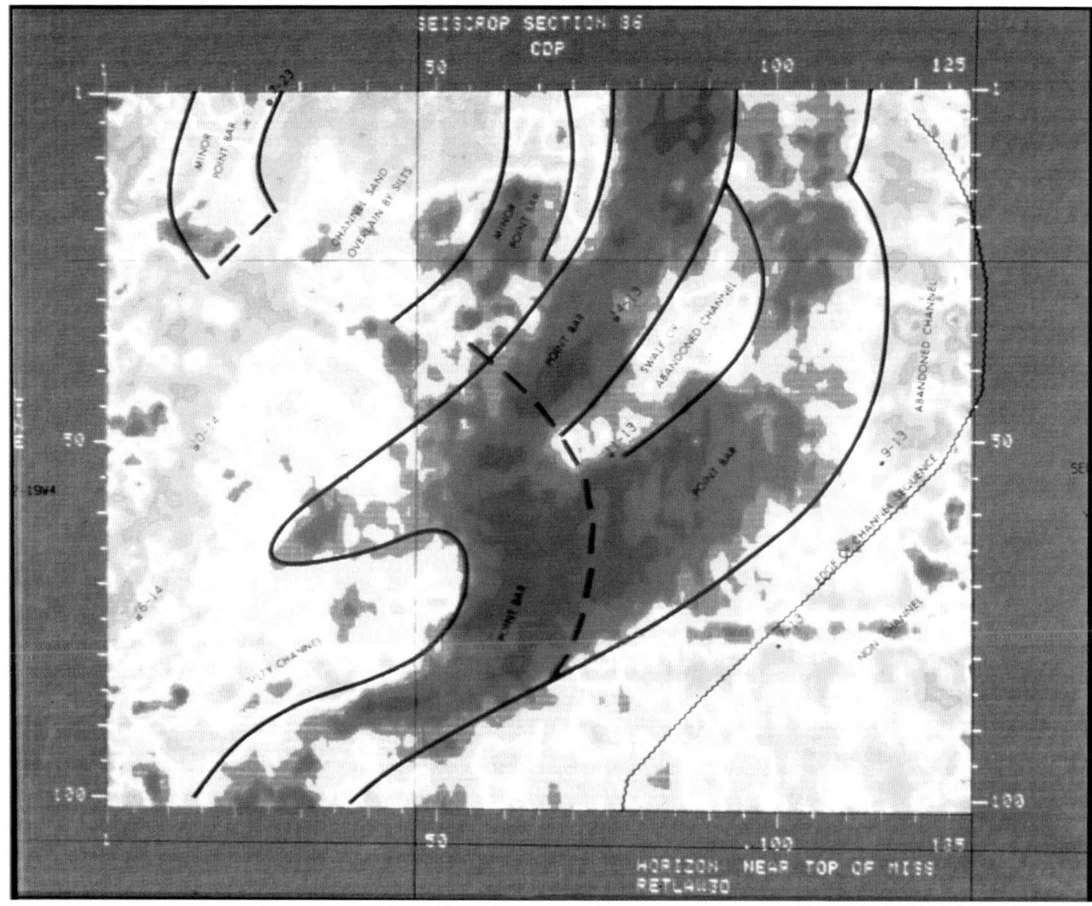

Fig. 4-44. Horizon slice through Glauconite Zone (near Top Mississippian) showing interpretation of depositional features. (Courtesy Geophysical Service Inc.)

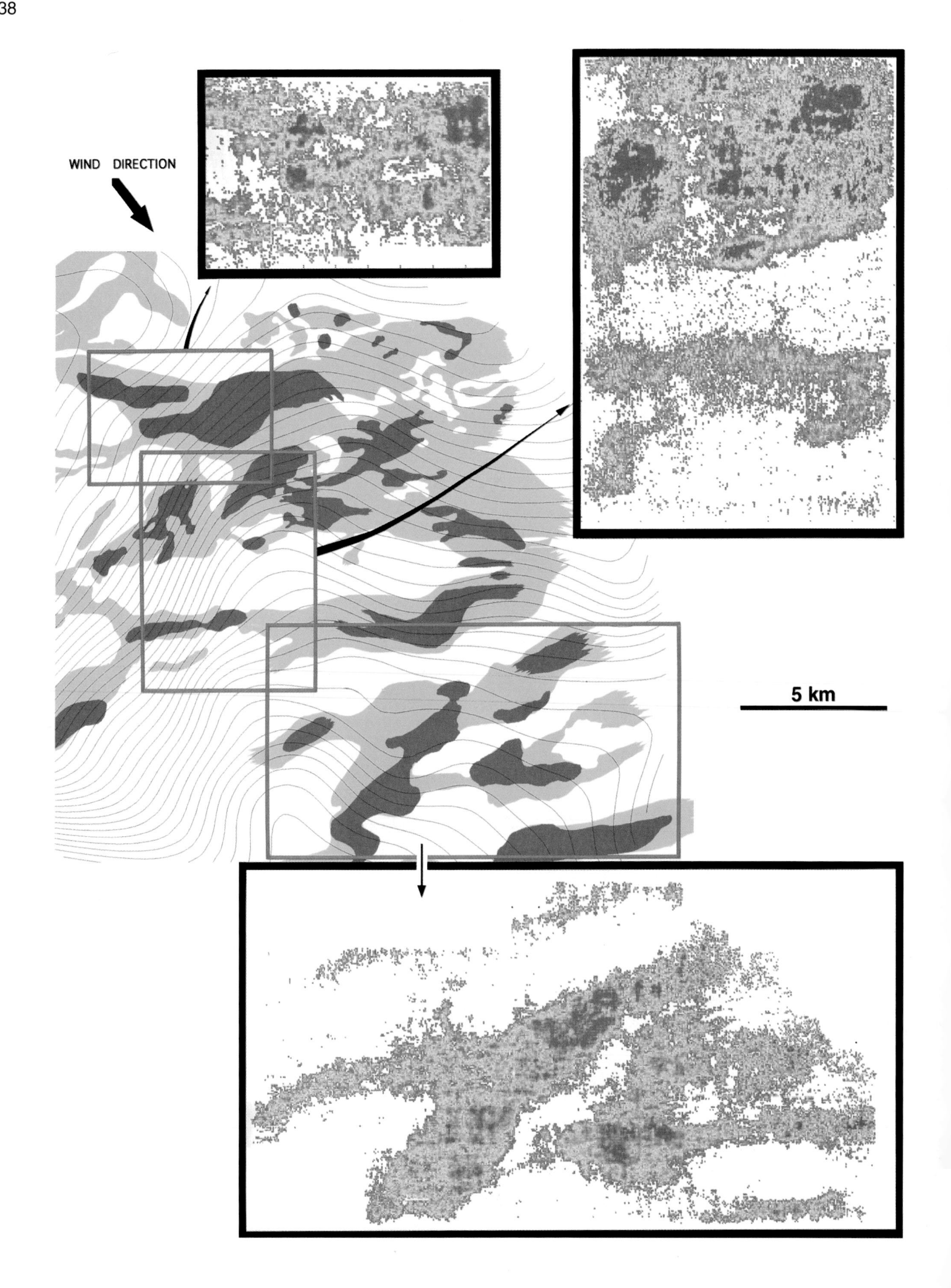
WIND DIRECTION
5 km

Fig. 4-46. Horizon slice from Jeanne d'Arc basin offshore eastern Canada showing a near-shore paleodrainage system. The arrow points along the ancient shore line. (Courtesy Husky Oil.)

Sanders, J. I., and G. Steel, 1982, Improved structural resolution from 3D surveys in Australia: Australian Petroleum Exploration Association (APEA) Journal, v. 22, p. 17-41.

Sonneland, L., and O. Barkved, 1990, Use of seismic attributes in reservoir characterization, *in* A.T. Buller et al., eds., North Sea oil and gas reservoirs II: Graham and Trotman, p. 125–128.

Fig. 4-45. Distribution of Avile sand in part of the Neuquen basin, Argentina. Note the elongated eolian dunes. Horizon slices from three 3-D surveys are compared with amplitude mapping from 2-D data. (Courtesy YPF S.A.)

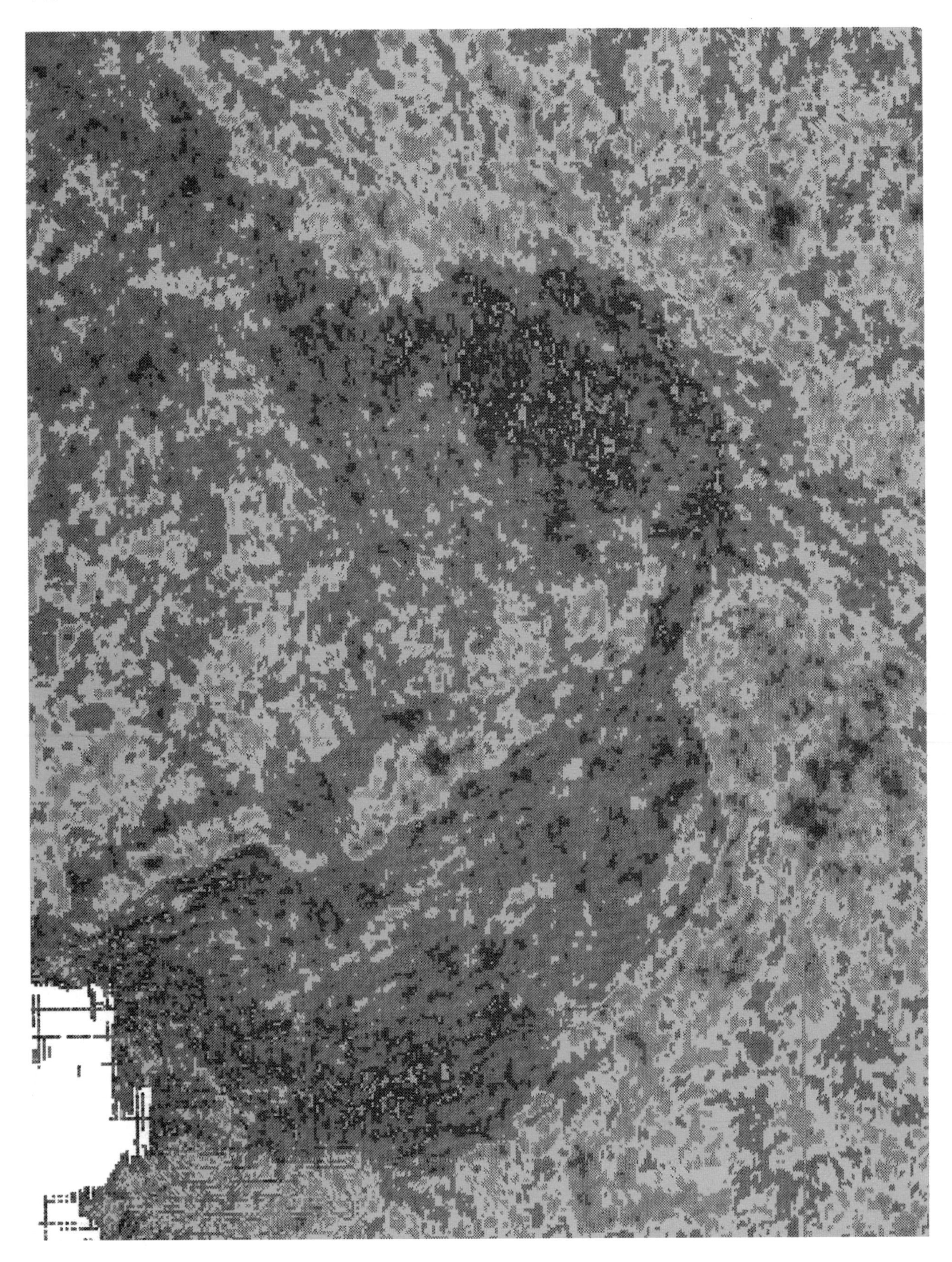

Fig. 4-47. Horizon slice from Central Graben area in Norwegian North Sea showing probable Danian intra-chalk mass flow. (Courtesy A/S Norske Shell.)

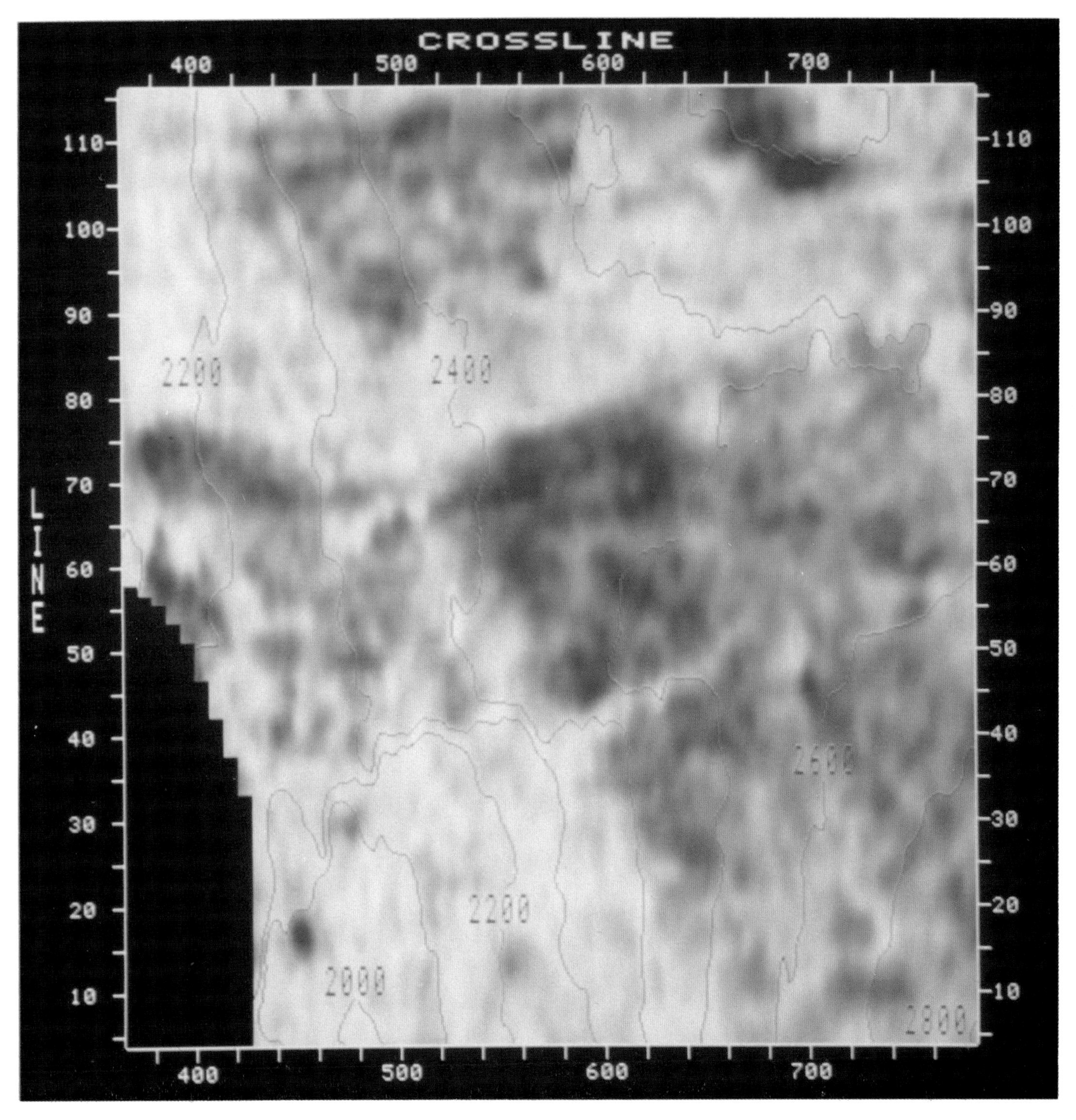

Fig. 4-48. Horizon slice and superimposed structural contours from a Gulf of Mexico 3-D survey. The ENE-WSW amplitude edges indicate faults. The NNW-SSE amplitude edge indicates a bed truncation. (Courtesy Chevron U.S.A. Inc.)

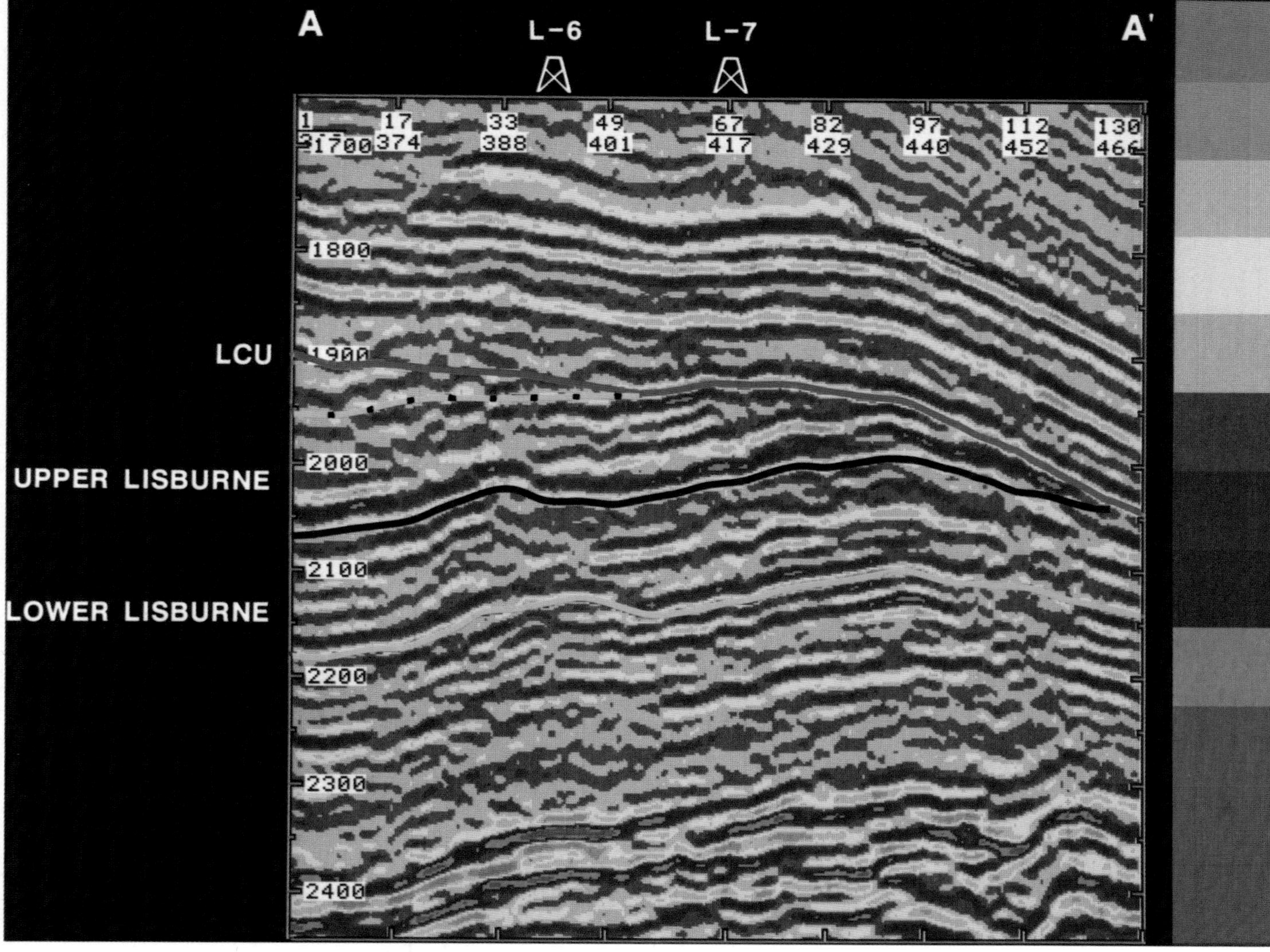

Fig. 4-49. Vertical section from Lisburne 3-D survey, North Slope of Alaska, showing Lower Cretaceous unconformity. (Courtesy Standard Alaska Production Company)

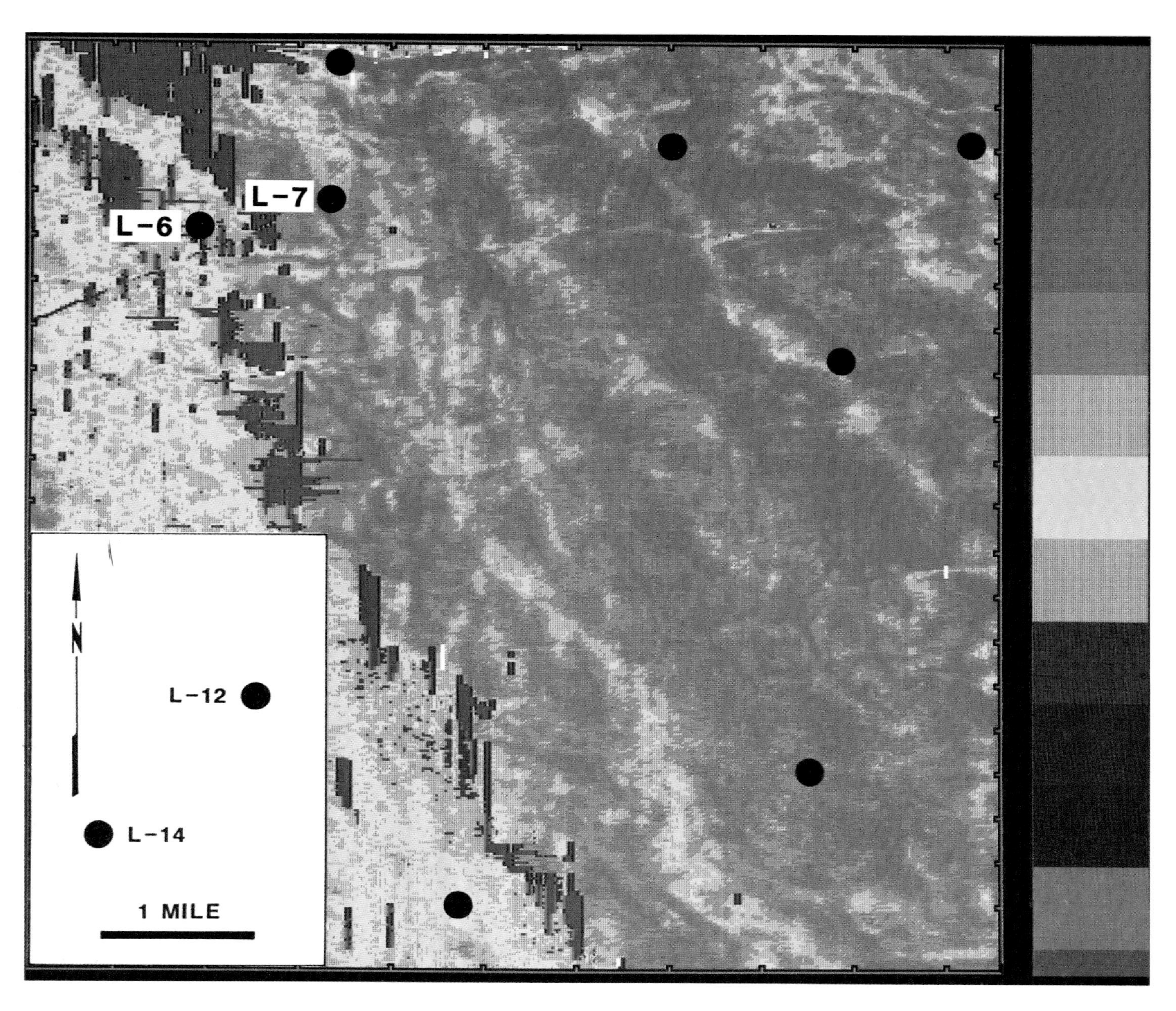

Fig. 4-50. Horizon slice (amplitude map) along Lower Cretaceous unconformity showing amplitude lineations trending from northwest to southeast. (Courtesy Standard Alaska Production Company.)

Fig. 4-51. Horizon slice (amplitude map) along Lower Cretaceous unconformity, showing interpretation of amplitude lineations in terms of truncations of reservoir units. (Courtesy Standard Alaska Production Company)

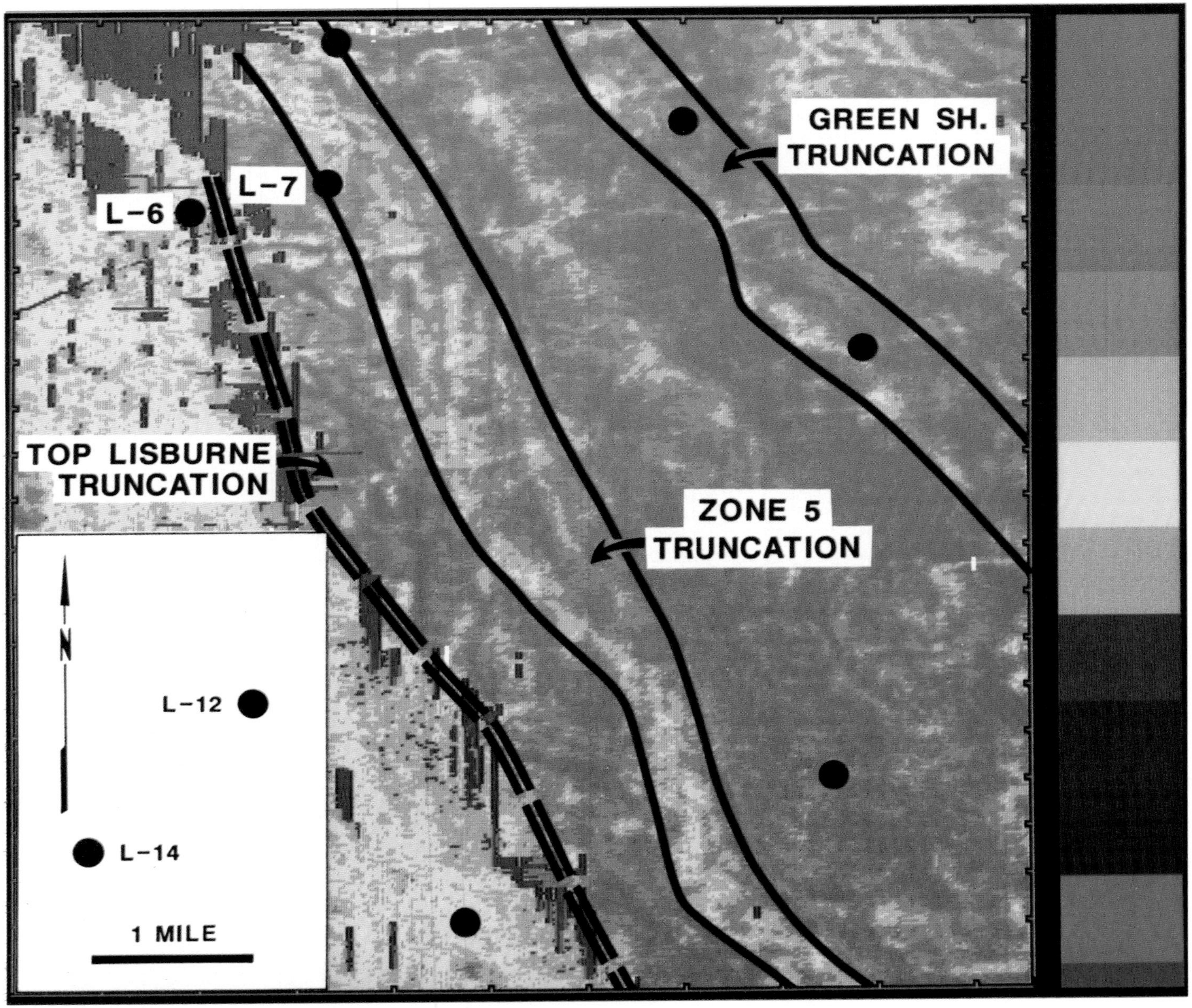

RESERVOIR IDENTIFICATION

Bright Spots As They Used To Be

Figure 5-1 shows a bright spot presented by Tegland (1973). This was one of the early examples studied and was observable because amplitude had been preserved in seismic processing. In earlier years, when records were normally made with automatic gain control, there was little opportunity for studying amplitudes. The bright spot of Figure 5-1 is actually a very good one for its era because it also shows a flat spot, presumably a fluid contact reflection. The flat spot terminates laterally at the same points as does the bright spot; we would consider this a simple form of bright spot validation, increasing the interpreter's confidence that the anomaly indicates the presence of hydrocarbons.

With the improvements in seismic processing over two decades, we can now consider polarity and phase as well as amplitude and spatial extent. Frequency, velocity, amplitude/offset and shear wave information can also help in the positive identification of hydrocarbon indicators. These are all subjects of this chapter and the direct observation of hydrocarbon fluids is now very widespread.

Most direct hydrocarbon indication relates to gas rather than oil reservoirs as the effect on acoustic properties of gas in the pore space is significantly greater than oil. Figure 5-2 (derived from Gardner, Gardner, and Gregory, 1974) summarizes the different effects of gas and oil and shows that the effect of either diminishes with depth.

Backus and Chen (1975) were very thorough in their discussion of the diagnostic benefits of flat spots, and Figure 5-3 shows a flat spot at 1.47 seconds that they discussed. Figure 5-4 is interpreted sufficiently to highlight the various hydrocarbon indicators on the section. The flat spot is easily identified by its flatness, and because it is unconformable with adjacent reflections. Hence it is a good indicator of the hydrocarbon/water contact. Nevertheless there is no high amplitude associated with it.

The reflection from the top of the reservoir (Figure 5-4) changes from a peak to a trough across the fluid contact and this again implies a significant change in acoustic properties between the gas sand above the hydrocarbon/water contact and the water sand beneath it. This phase change, or a polarity reversal, will be discussed in more detail in the next section.

The Character of Hydrocarbon Reflections

If the seismic data under interpretation have been processed to zero phase (see Chapter 2), then the detailed character of the bright spots, flat spots and other hydrocarbon indicators can be very diagnostic. Figure 5-5 shows diagrammatically the hydrocarbon indicators which may be associated with different relative acoustic impedances of gas sand, water sand and embedding medium. The polarity convention expressed diagrammatically in Figure 5-5 is the same as that explained in Chapter 2, namely that a decrease in acoustic impedance is expressed as a peak which is blue and an increase is expressed

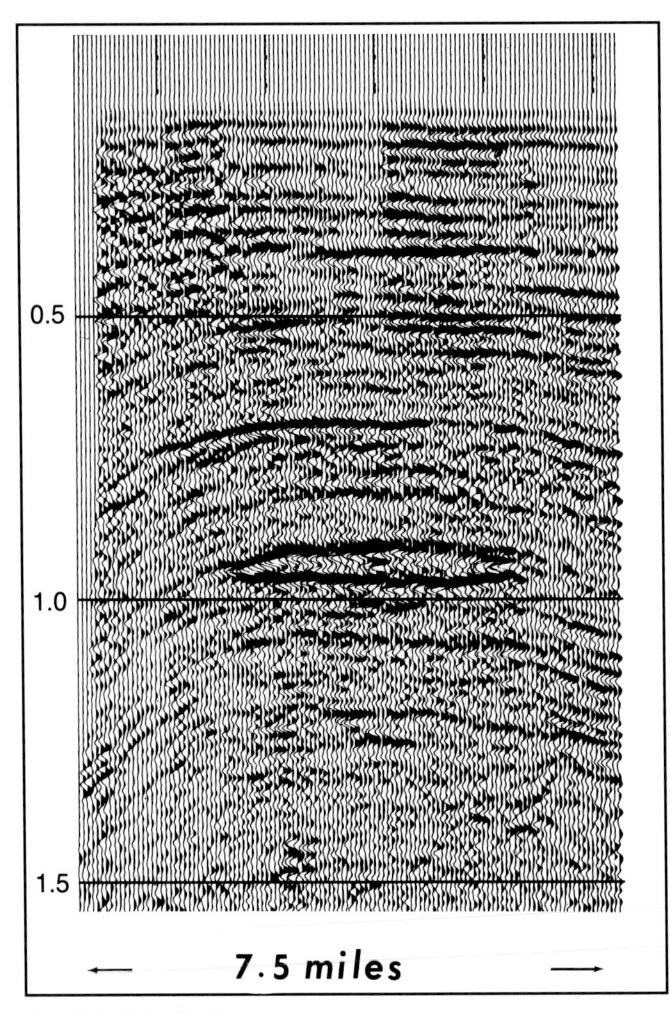

Fig. 5-1. A Gulf of Mexico bright spot and flat spot from the early 1970s.

Fig. 5-2. Percent velocity difference between sands saturated with different fluids (derived from Gardner, Gardner, and Gregory, 1974).

as a trough which is red. Peaks and troughs are symmetrical if they are the zero-phase expressions of single interfaces.

The top diagram of Figure 5-5 illustrates the most common situation: the water sand has an acoustic impedance lower than the embedding medium and the impedance of the gas sand is further reduced. For this situation the signature of the sand is peak-over-trough and, for the gas-filled portion, the amplitude is greater. This is the classical zero-phase bright spot with high amplitude for the top *and* base reflections. If the sand is thick enough for the top and base reflections to be separated, then a flat spot or fluid contact reflection should be visible between the gas sand and the water sand, that is at the point where brightening occurs. The flat spot reflection will be a trough because it must be an increase in acoustic impedance.

In the second diagram the situation is reversed; the water sand has a higher acoustic impedance than the embedding medium and hence has a signature of trough-over-peak. When gas replaces some of the water in the pores of the sand, the acoustic impedance is reduced, the contrast is reduced at the upper and lower boundaries, and the reservoir is seen as a dim spot. Again, if the sand is thick enough, a flat spot can be expected at the point where the dimming occurs and this again will be a trough.

In the third diagram the reduction in acoustic impedance of the sand, because of gas saturation, causes the acoustic impedance to change from a value higher than that of the embedding medium to one lower than that of the embedding medium. Hence the polarities of the reflections for the top and the base of the sand switch. The signature changes from trough-over-peak to peak-over-trough across the fluid contact. In order to observe such a phase change, or polarity reversal, in practice, the structural dip must be clearly determined from non-reservoir reflections just above and/or just below the sand under study. Again, if the sand is thick enough, a fluid contact reflection should be visible and it will be a trough.

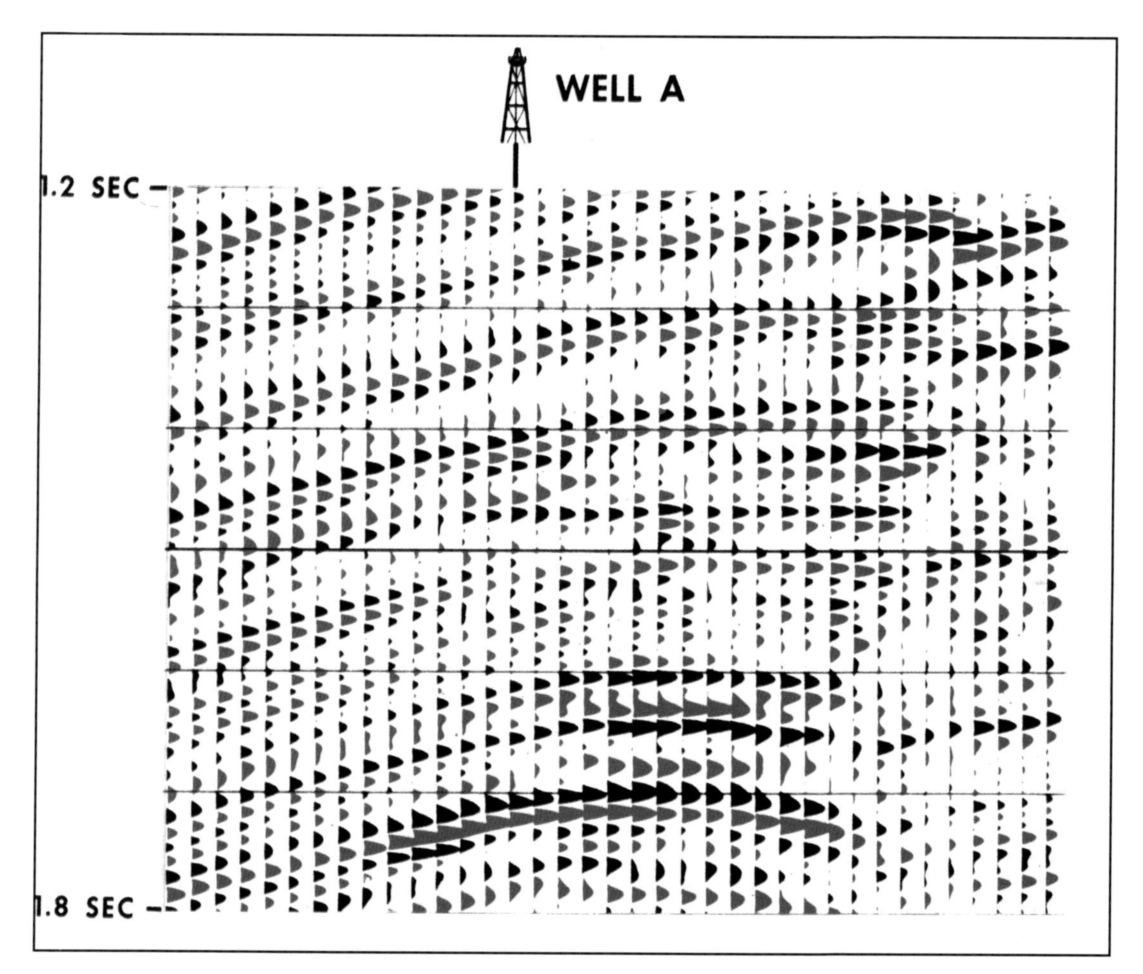

Fig. 5-3. Dual polarity section showing bright spot at 1.72 seconds and a flat spot at 1.47 seconds. (Courtesy Geophysical Service Inc.)

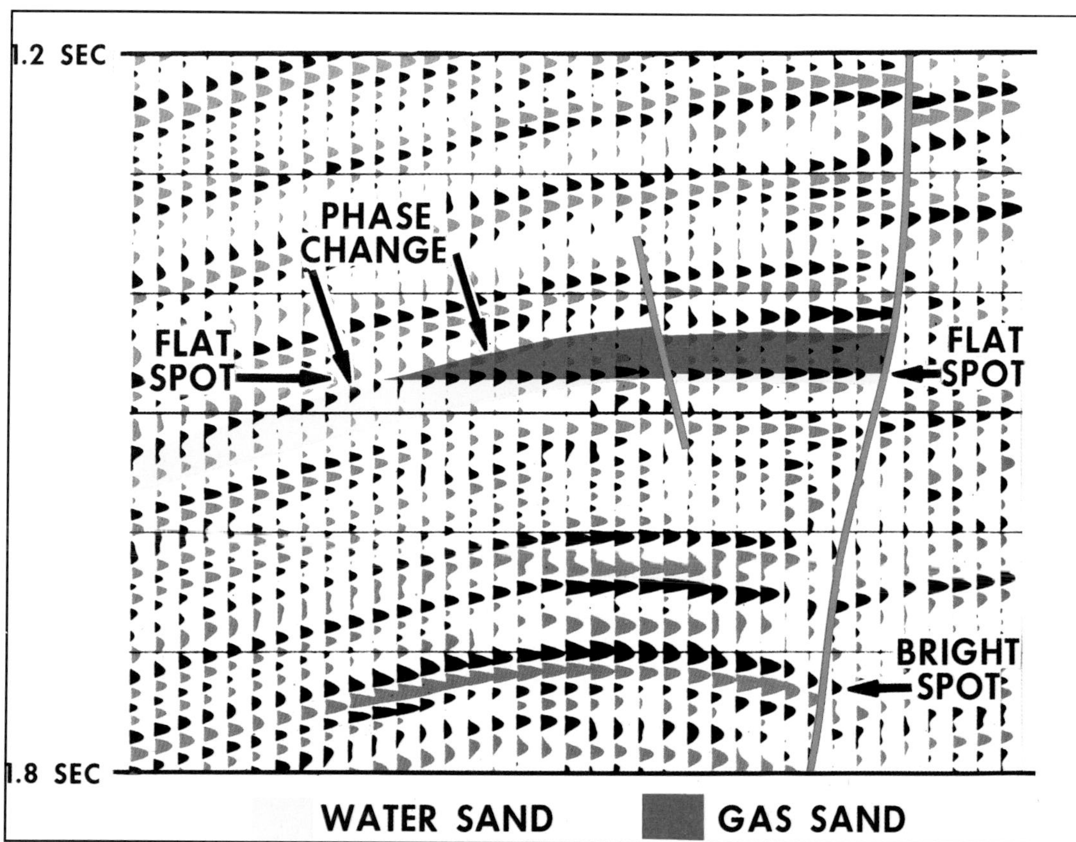

Fig. 5-4. Same section as Figure 5-3 showing the interpreted position of the gas reservoir and demonstrating a phase change between the reflections from the gas sand and the water sand. (Courtesy Geophysical Service Inc.)

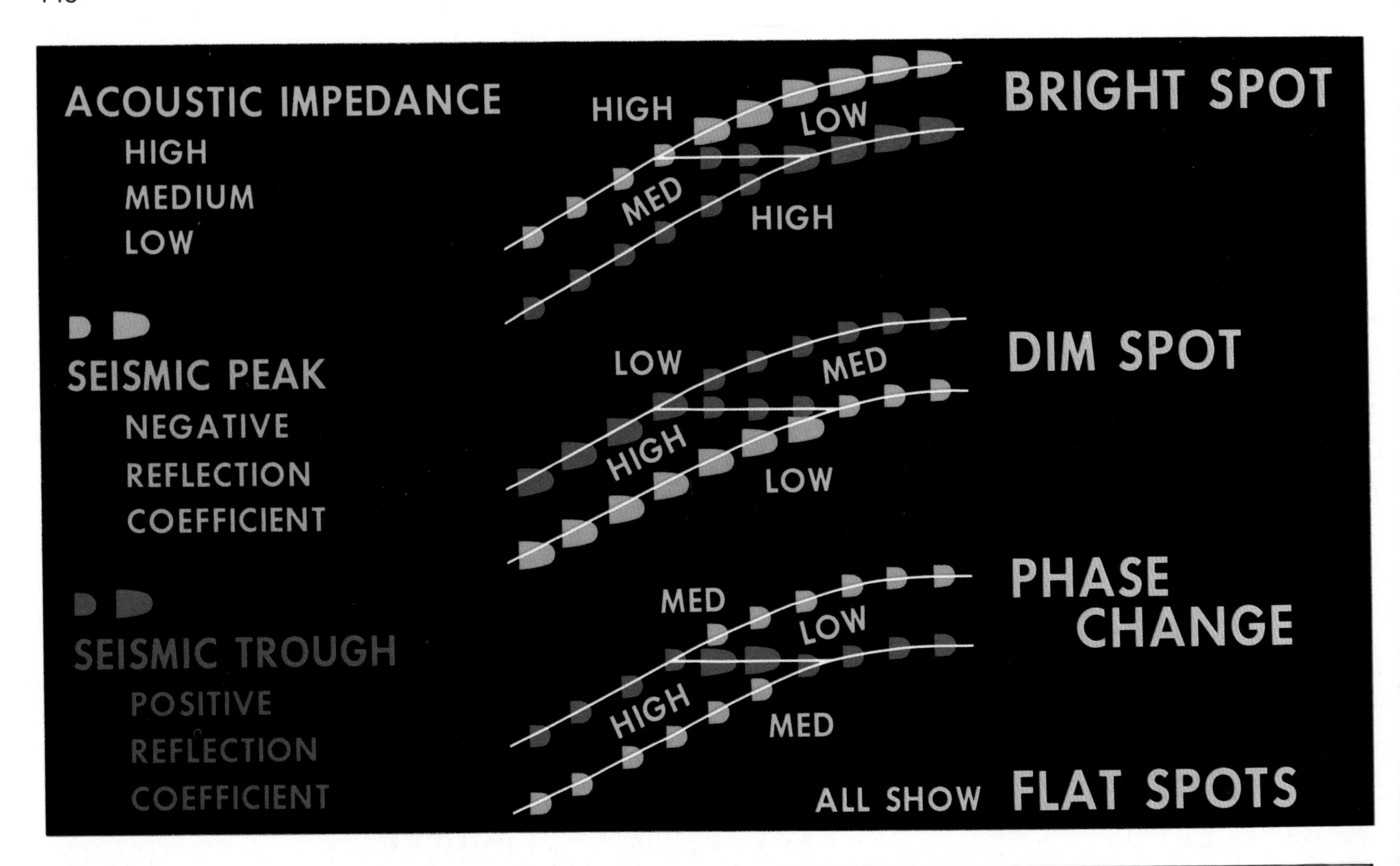

Fig. 5-5. Schematic diagram of the zero-phase response of reservoirs for different acoustic contrasts between the reservoir and the embedding medium.

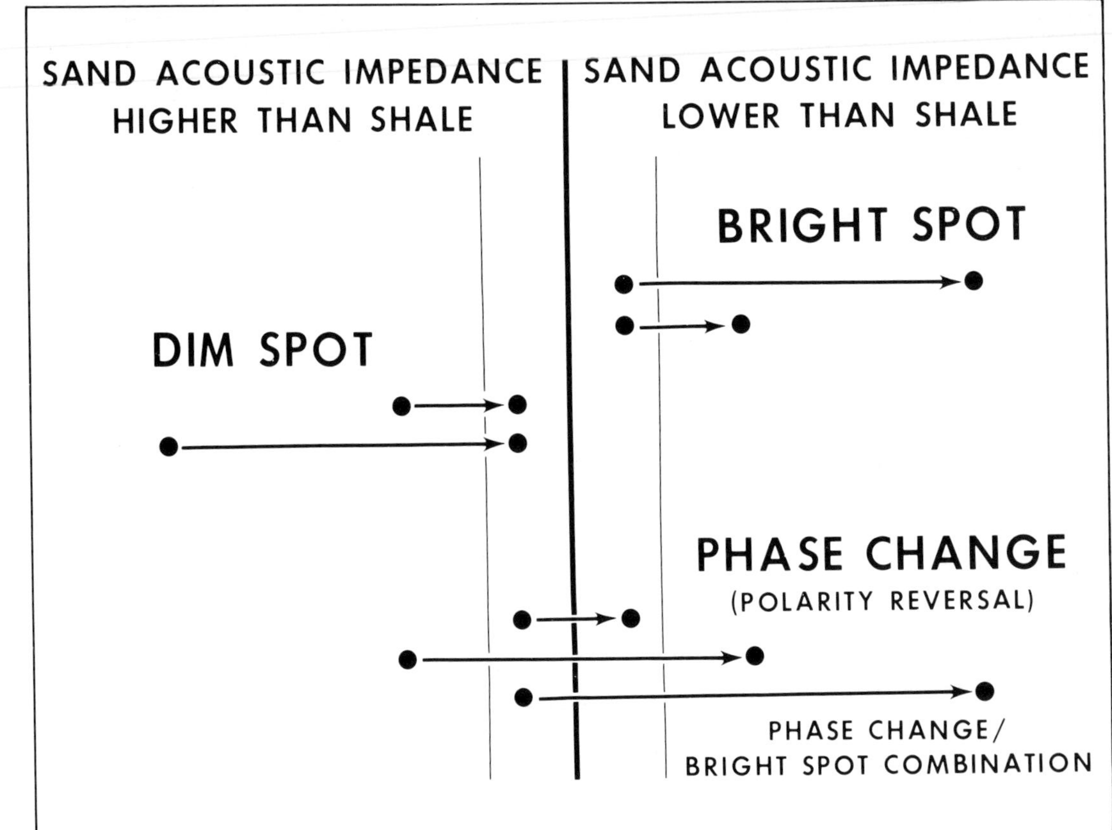

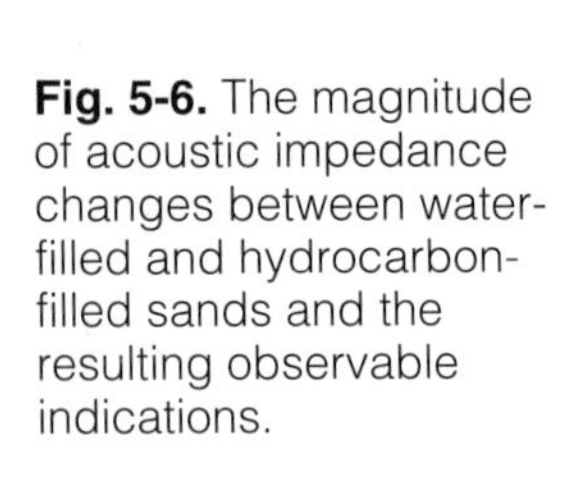

Fig. 5-6. The magnitude of acoustic impedance changes between water-filled and hydrocarbon-filled sands and the resulting observable indications.

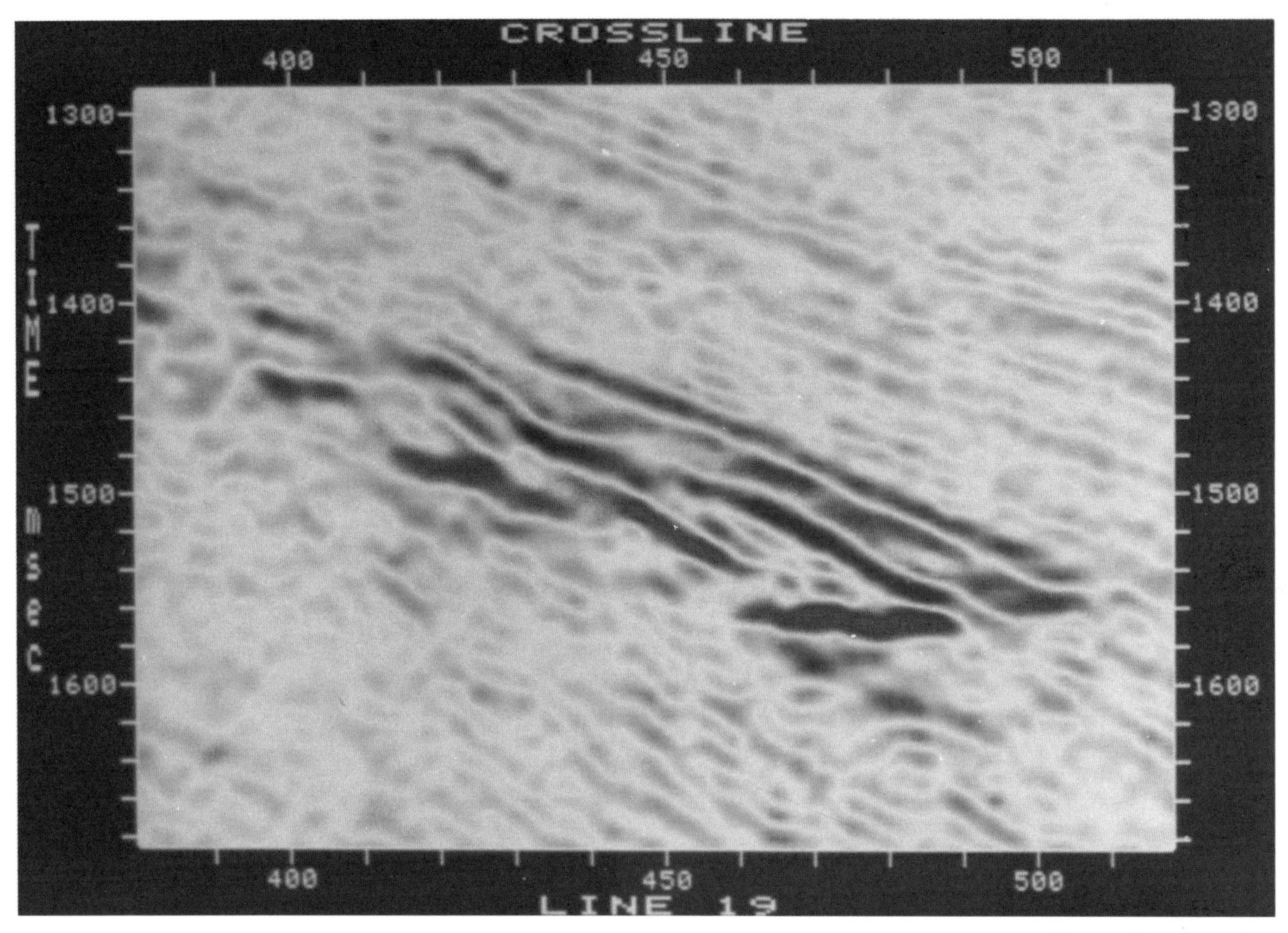

Fig. 5-7. Bright and flat reflections from a Gulf of Mexico gas reservoir known to be subdivided into upper and lower sand units. (Courtesy Chevron U.S.A. Inc.)

Figure 5-6 shows the magnitude of acoustic impedance changes between water sand and gas sand and hence the effect on seismic amplitude reflected from the interface between either of them and a uniform embedding medium. The diagram is drawn with Tertiary clastics in mind, but it also has generality. Tertiary sands and shales normally have rather similar acoustic properties and thus on the relative scale of Figure 5-6 lie between the narrow lines, that is, not far from the heavy line of acoustic impedance equality in the center. For a bright spot (without phase change) the water sand is located just right of the center line and the gas sand is much farther to the right. For a phase change, or polarity reversal, the movement from water sand to gas sand must be from left to right across the center line. In the last situation illustrated in Figure 5-6, the bright spot is exactly the same in amplitude and phase as the one illustrated at the top of the figure; the difference is that the last one labeled *phase change/bright spot combination*, came from a water sand with higher acoustic impedance than the embedding shale and was thus located on the left of the center line. Dim spots must start from a water sand significantly to the left on the relative scale of Figure 5-6, so that a visible movement still leaves them left of center. This is unusual in Tertiary clastics and thus is the reason we do not see many dim spots in that environment. In older rocks, however, much greater differences in acoustic impedance between sands and shales are normal, so that dim spots are more commonly observable.

Examples of Bright Spots, Flat Spots, Dim Spots and Phase Changes

Figure 5-7 shows a Gulf of Mexico bright spot known to be a gas reservoir. The reservoir reflections have very high amplitude and hence the interference from other nearby reflections, multiples or noise is small. The bright reflections show the zero-phase response of two reservoir sands, each peak-over-trough and located one on top of the other. The upper sand is fairly thin so there is only a hint of a flat spot reflection at the downdip limit of

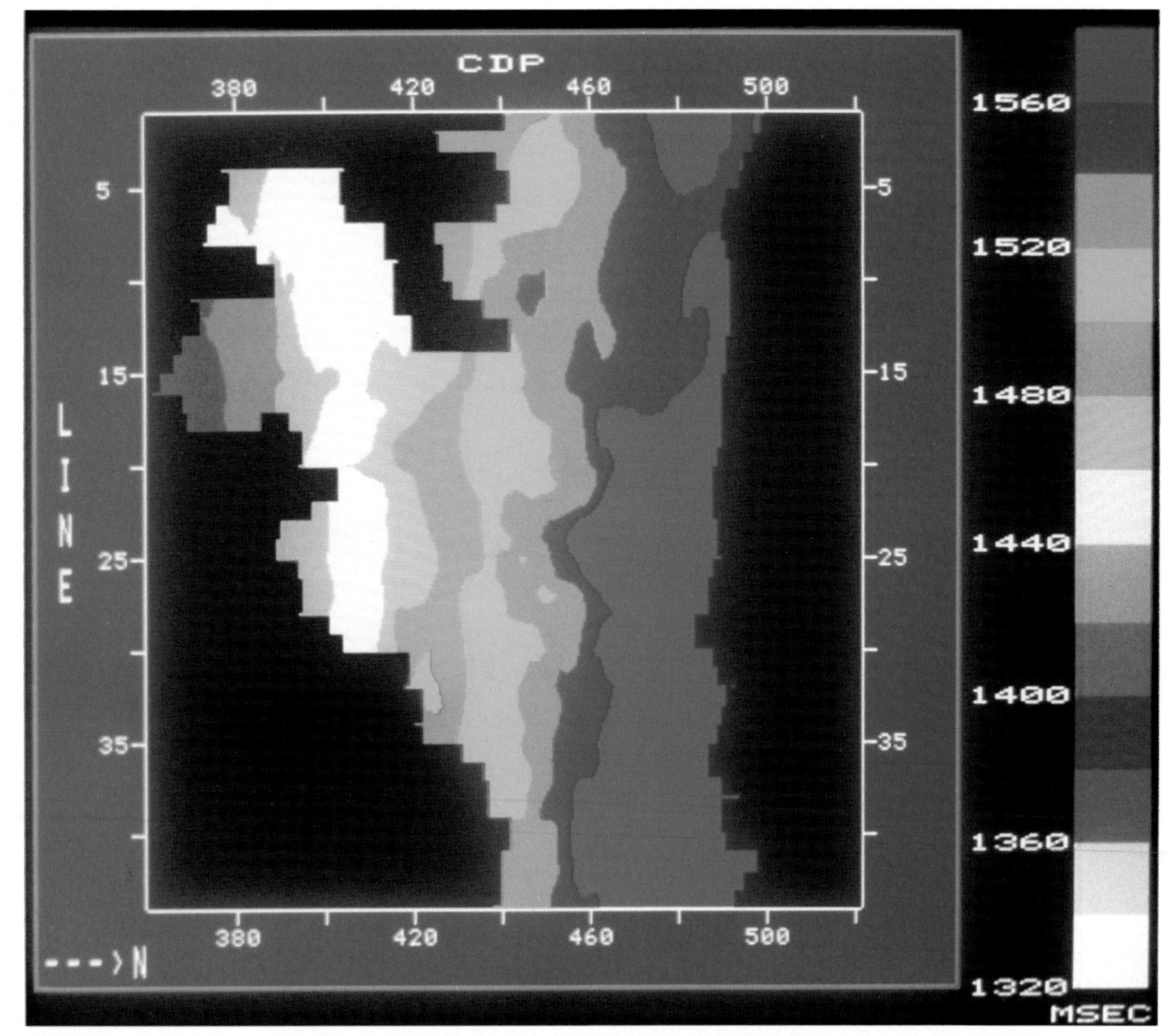

Fig. 5-8. Structure map of the base of the lower gas sand showing the areal extent of the flat spot seen in Figure 5-7. (Courtesy Chevron U.S.A. Inc.)

brightness. The lower sand is much thicker and the flat spot reflection is very clear.

Flat spot reflections are highly diagnostic indicators of gas but the interpreter should make several validity checks before drawing a conclusion. In Figure 5-7 the flat spot reflection is flat, bright and shows one symmetrical trough. It occurs at the downdip limit of the bright events and is unconformable with them. Figure 5-8 is a structure map of the base of the gas; it shows structural consistency for the flat spot reflection in the extent of the purple color.

Figure 5-9 illustrates other Gulf of Mexico bright spots and flat spots. These data are also zero phase, but the polarity is the reverse of that seen in Figures 5-7 and diagrammed in Figure 5-5. Hence, in Figure 5-9 flat spots are black and reflections from the top of gas reservoirs are red. Note particularly the prominent reservoir reflections between times of 1.5 and 1.6 on Line 49 and on Line 51 and on horizontal section 1.520 in the lower right against a salt dome.

Direct observation of hydrocarbon fluids through bright spots and flat spots is very well established in Tertiary clastic basins, of which the Gulf of Mexico is the most extensively studied. However, hydrocarbons are directly observable in many other parts of the world and in many older rocks. Figure 5-10 shows a fine flat spot and associated

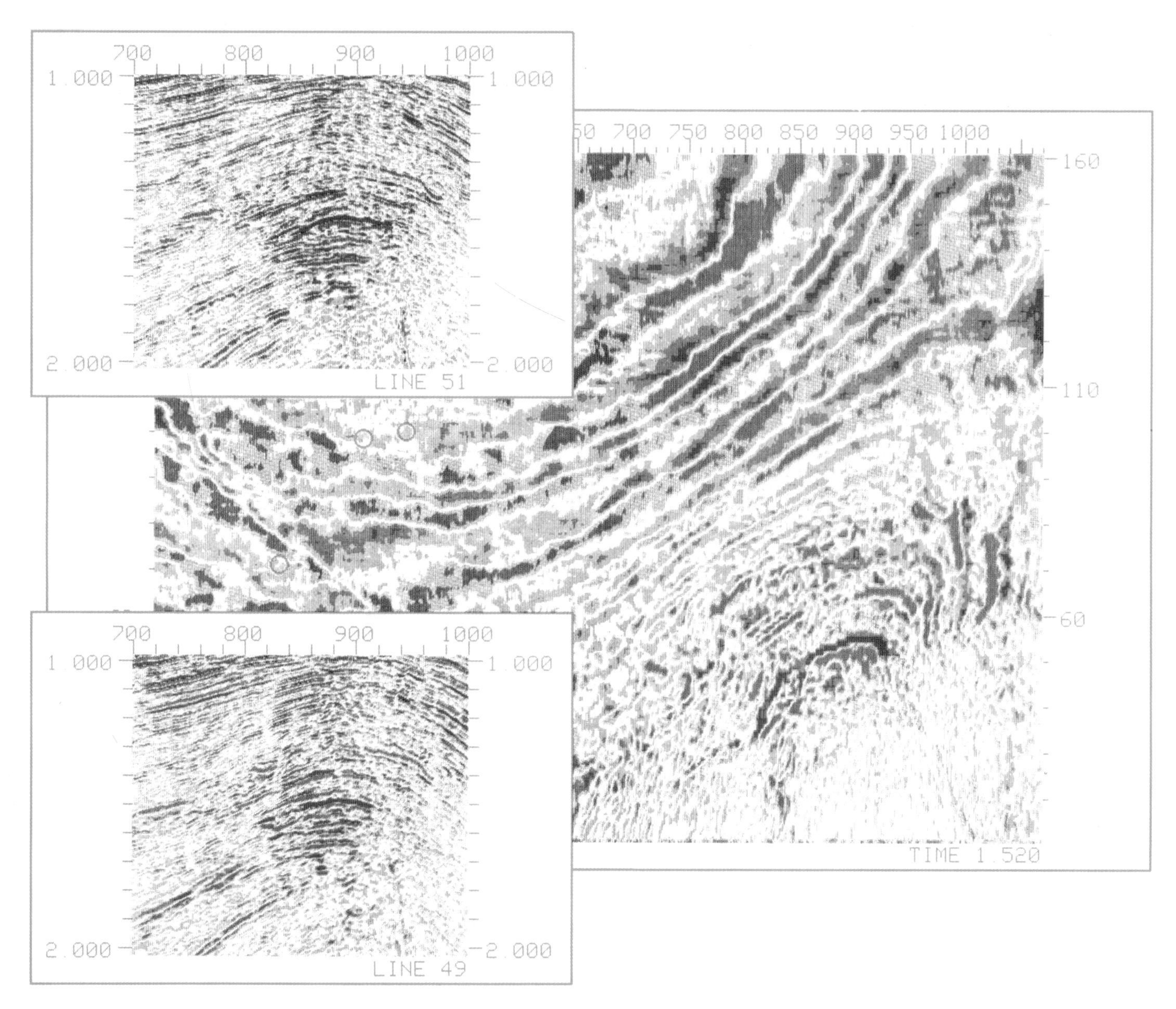

Fig. 5-9. Gulf of Mexico data indicating bright spots on vertical and horizontal sections. (Note that the polarity of the data is the opposite of what is considered normal in this book, so the colors are reversed.) (Courtesy Texaco U.S.A. Inc.)

bright spot in Jurassic rocks offshore Norway. Early observation of this flat spot on 2-D data (Birtles, 1986) contributed to the discovery of the Troll Field. Now with 3-D data, zero-phaseness and color the flat spot is outstandingly clearer. The flat spot is not very flat because of the tuning effects and velocity effects but it is clearly unconformable with structural reflections. Often unconformability is a more important diagnostic that flatness in the identification of fluid contact reflections.

Figure 5-11 shows a flat spot and associated bright spot in Permian rocks in The Netherlands. (The strongest reflections on the section are from an anhydrite layer; be careful to look just under these strong reflections in the center to find the flat spot and bright spot.) The change in amplitude of the black (top reservoir) reflection coincident with the flat spot is an important diagnostic observation. The author is aware of direct observation of hydrocarbon fluids in rocks up to Pennsylvanian or Carboniferous in age. As the quality of seismic data improves further, the age theshold should be pushed to even greater ages.

Figure 5-12 shows two examples of bright spots and flat spots indicating the reflections from the top and base of proven hydrocarbons in the North Sea. Line 182 shows continuity of the fluid contact across the reservoir, whereas line 137 shows it interrupted in the center where the reservoir is full to base. Figure 5-13 shows horizon slices from the top reservoir reflection and the fluid contact reflection. Similar patterns in amplitude

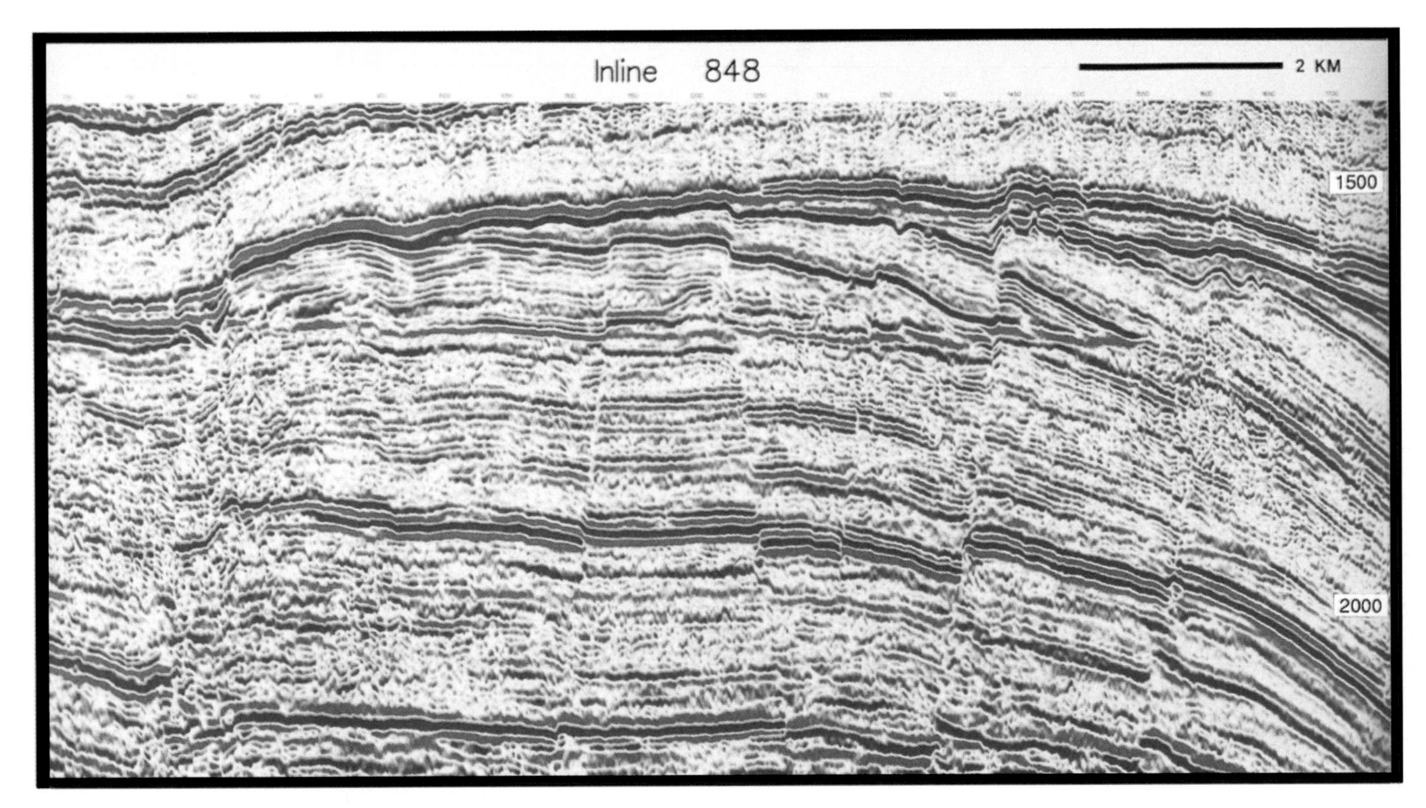

Fig. 5-10. Vertical section across Troll gas field offshore Norway. Note 7 km long flat spot at about 1700 ms. The reservoir is Jurassic in age with an average porosity of 28%. (Courtesy Norsk Hydro a.s.)

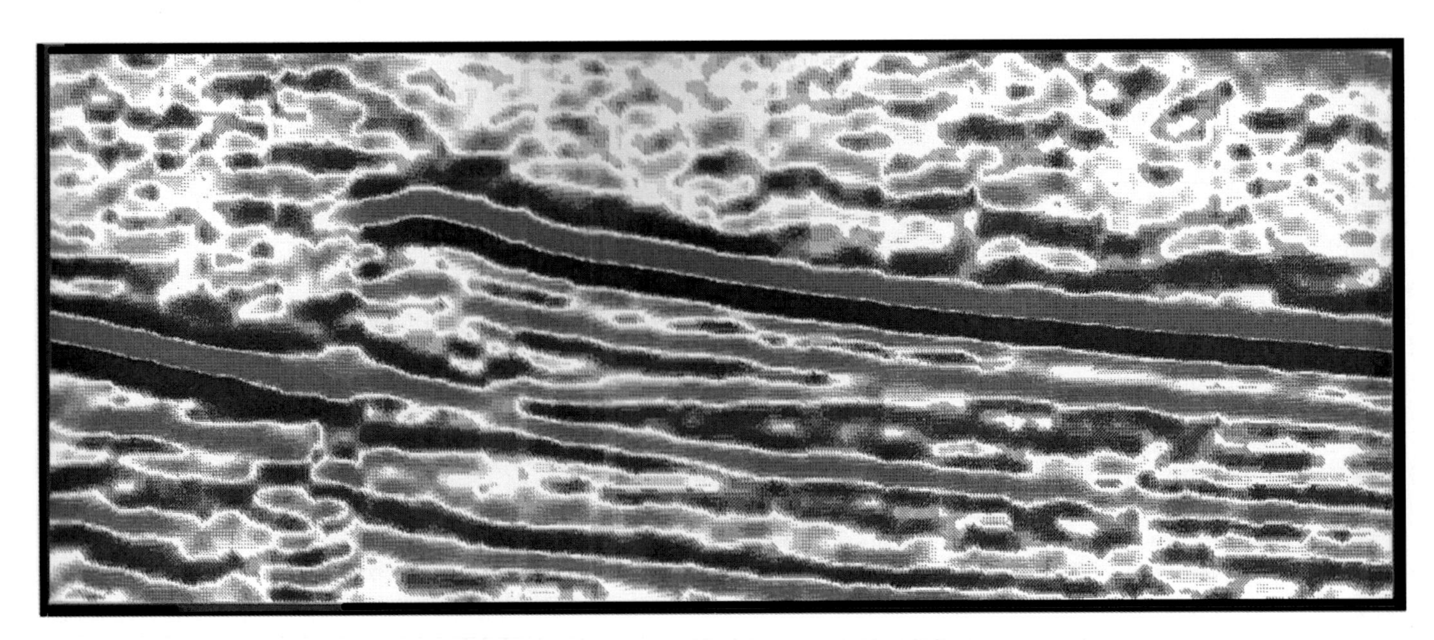

Fig. 5-11. Vertical section across a gas field in the Netherlands. The flat spot visible in the center is at a time of 1950 ms and a depth of 2850 m. The reservoir is within the Rotliegendes (lower Permian) sandstone and has an average porosity of 18%. (Courtesy Nederlandse Aardolie Maatschappij B. V.)

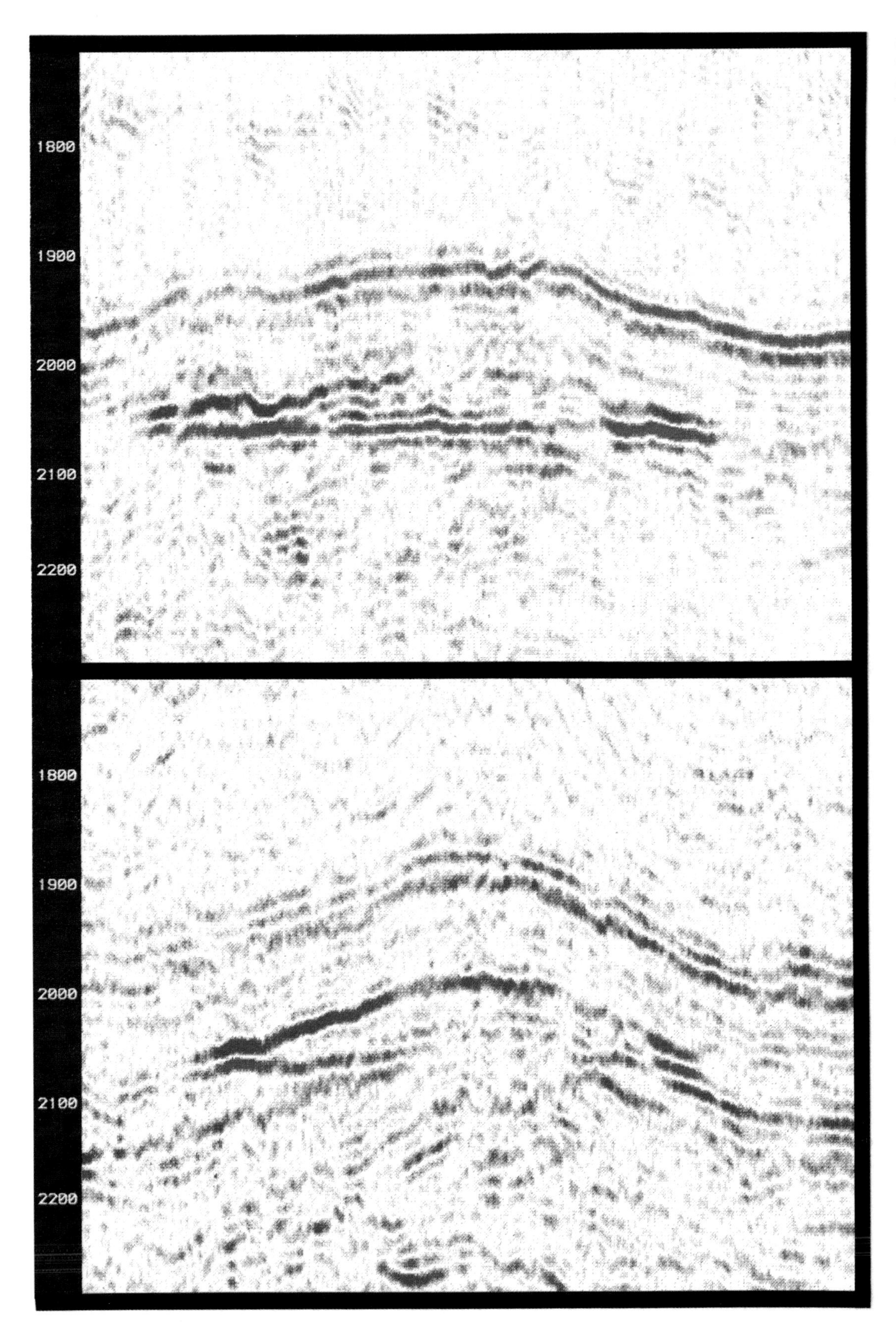

Fig. 5-12. Line 182 (upper) and line 137 (lower) over the Heimdal field in the Norwegian North Sea. The reflection from the top of the gas condensate reservoir is blue; the fluid contact reflection is red. (Courtesy Elf Aquitaine Norge a/s.)

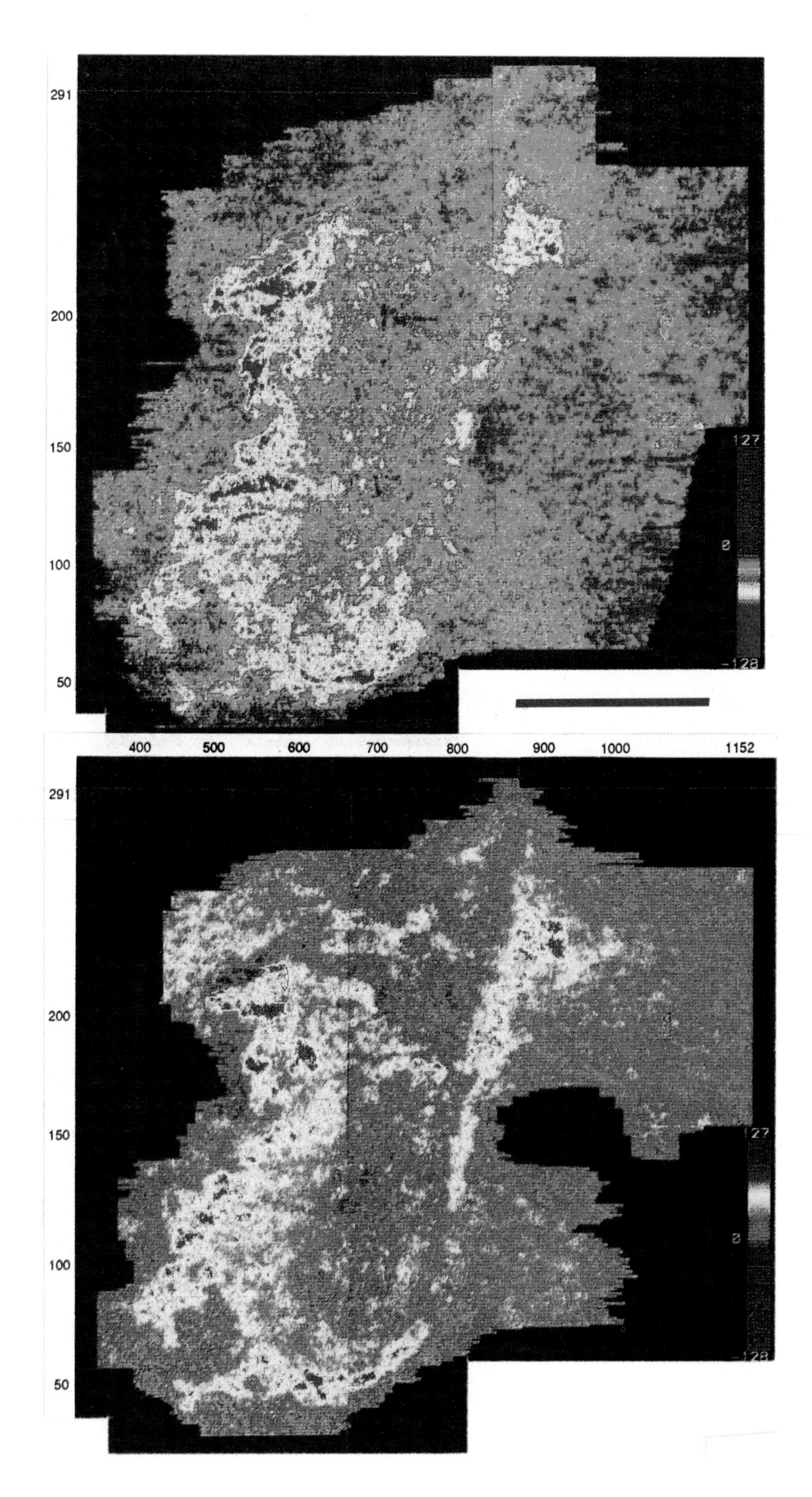

Fig. 5-13. Horizon slice through top reservoir reflection (upper) and through fluid contact reflection (lower) for the Heimdal field in the Norwegian North Sea. Red bar is 3 km. (Courtesy Elf Aquitaine Norge a/s.)

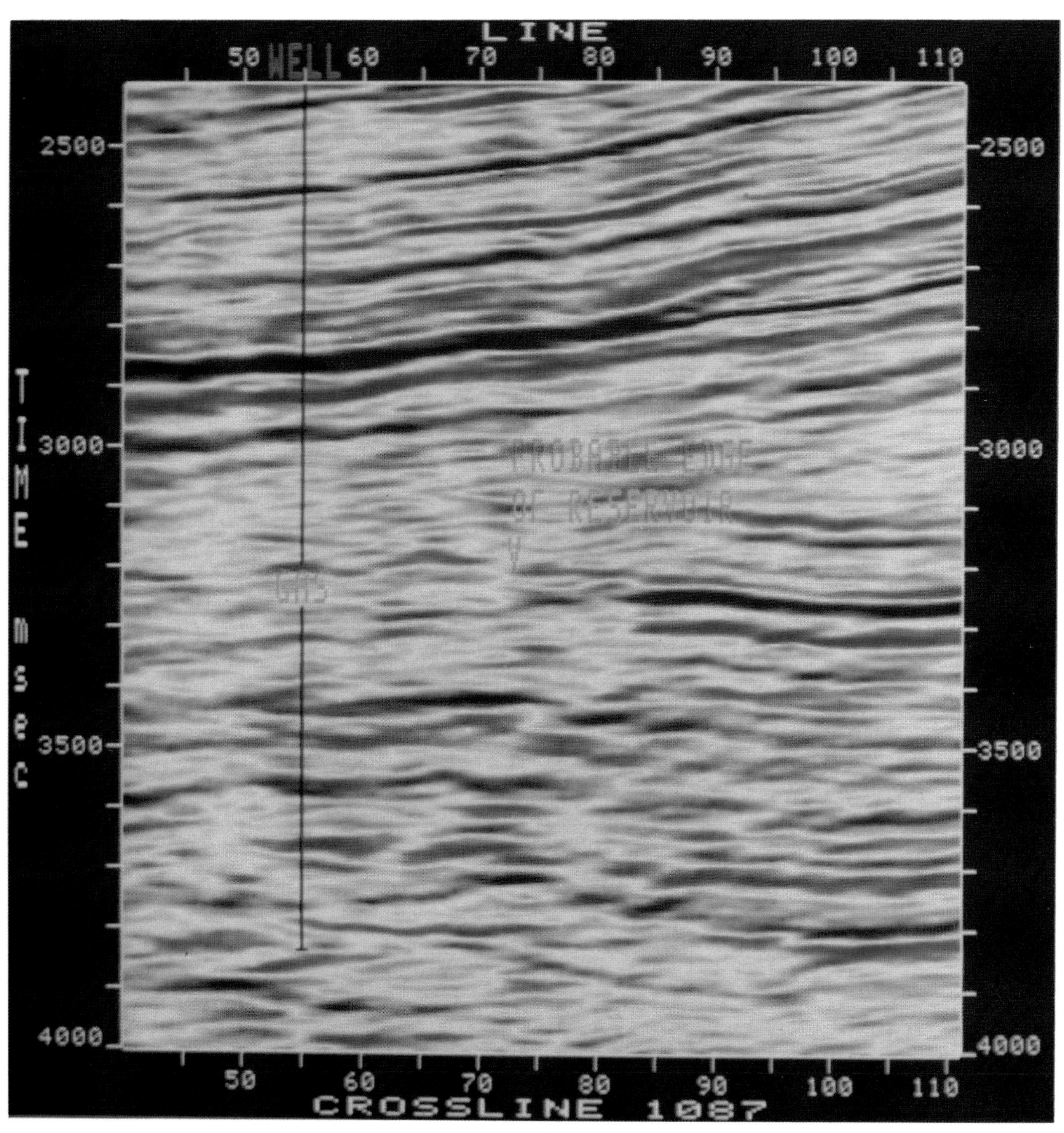

Fig. 5-14. A dim spot from a known gas reservoir offshore Trinidad. (Courtesy Texaco Trinidad Inc.)

confirm that these horizon slices indeed follow the top and bottom of the same interval.

Figure 5-14 is a practical example of a dim spot. The discovery well penetrates a gas column of about 400 ft (130 m) but the acoustic contrast of the gas sand with its embedding medium is small. Outside the reservoir the contrast between the sand and the embedding medium is much greater, as the amplitudes indicate. Here the reservoir is cemented, so the increase in acoustic impedance is caused partly by lack of gas and partly by the cementation. Figures 5-15 and 5-16 illustrate dim spots from the Northwest Shelf of Australia. In both cases the reservoir sands truncate at an unconformity overlain by soft marine shale of acoustic impedance similar to, but in fact even slightly lower than, the gas sands. Hence, the amplitude of the reflection from the unconformity dims

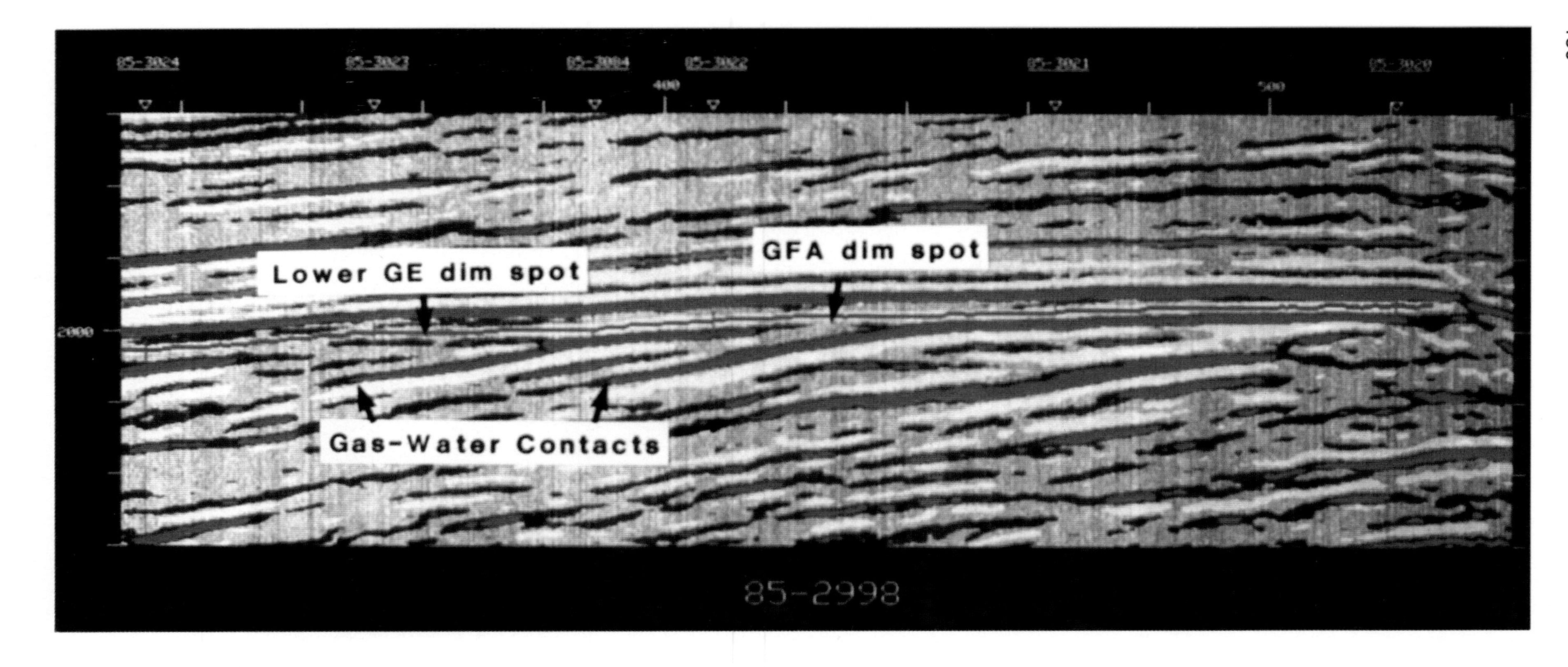

Fig. 5-15. Dim spots from Goodwyn gas field, Northwest Shelf, Australia, caused by gas sands truncating at an unconformity overlain by soft marine shale. (Courtesy Woodside Offshore Petroleum Pty., BP Development Australia Ltd., BHP Petroleum Pty. Ltd., Shell Development [Australia] Pty. Ltd., California Asiatic Oil Company, Japan Australia LNG [MIMI] Pty. Ltd., and Woodside Petroleum Ltd.)

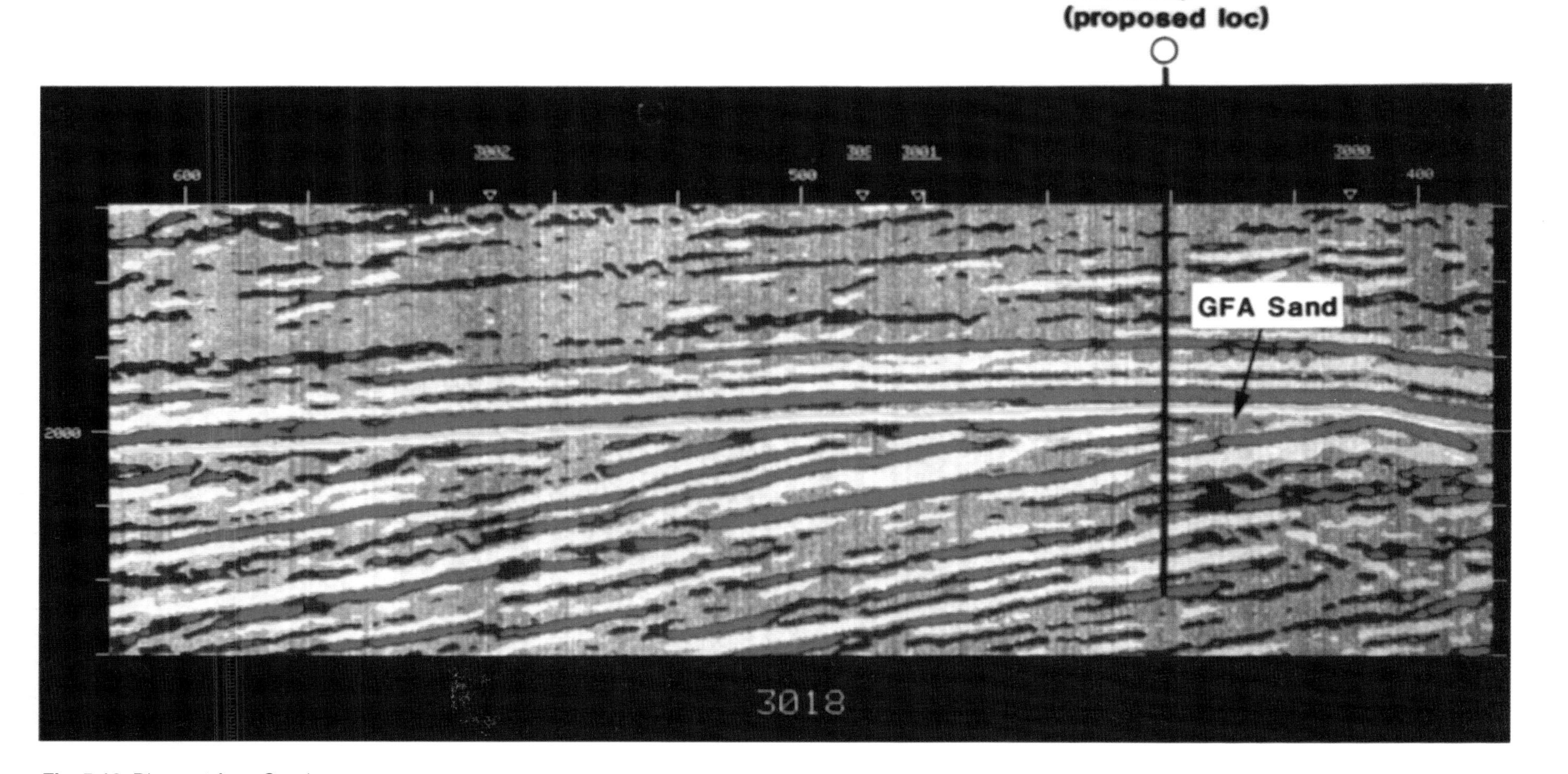

Fig. 5-16. Dim spot from Goodwyn gas field, Northwest Shelf, Australia, targeted as drilling location. (Courtesy Woodside Offshore Petroleum Pty., BP Development Australia Ltd., BHP Petroleum Pty. Ltd., Shell Development [Australia] Pty. Ltd., California Asiatic Oil Company, Japan Australia LNG [MIMI] Pty. Ltd., and Woodside Petroleum Ltd.)

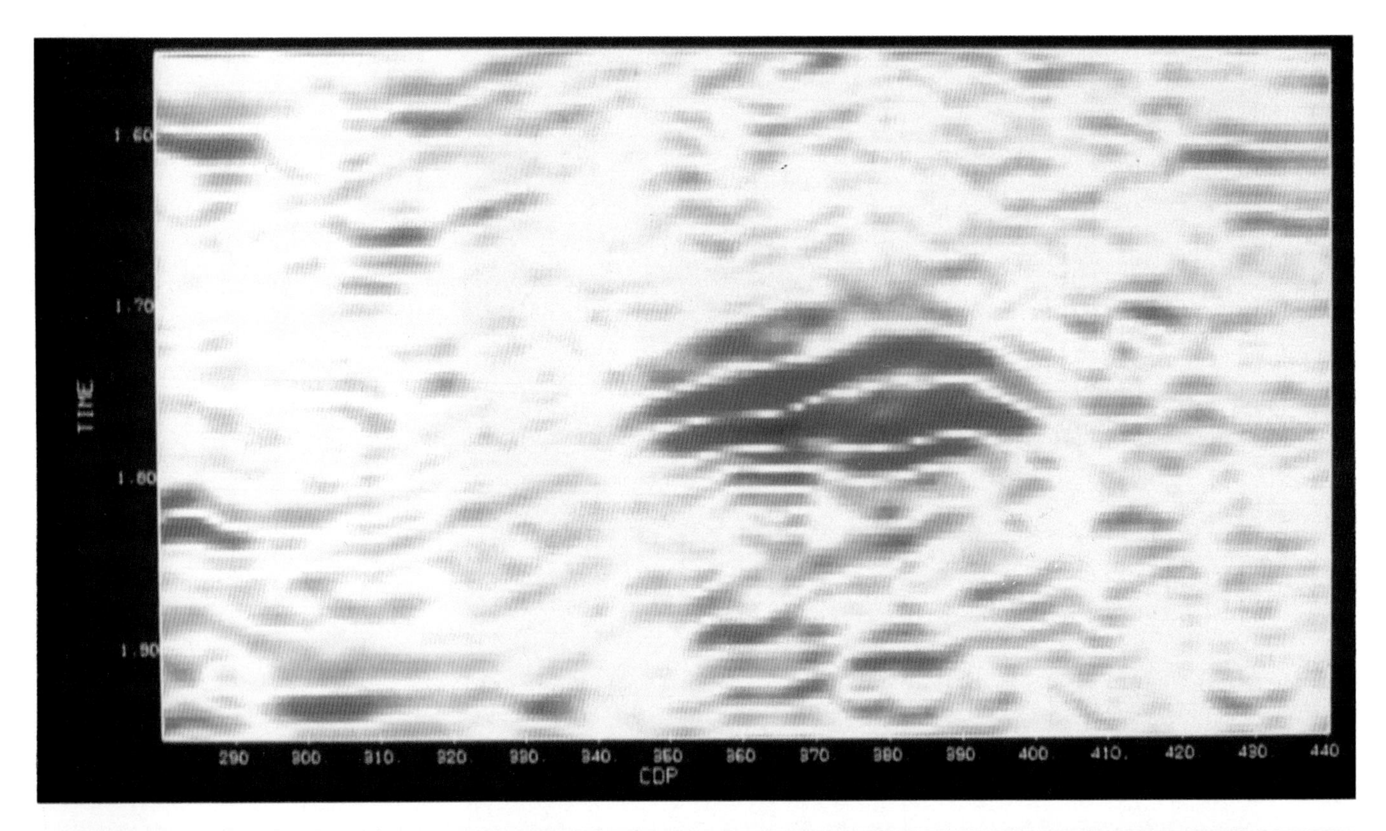

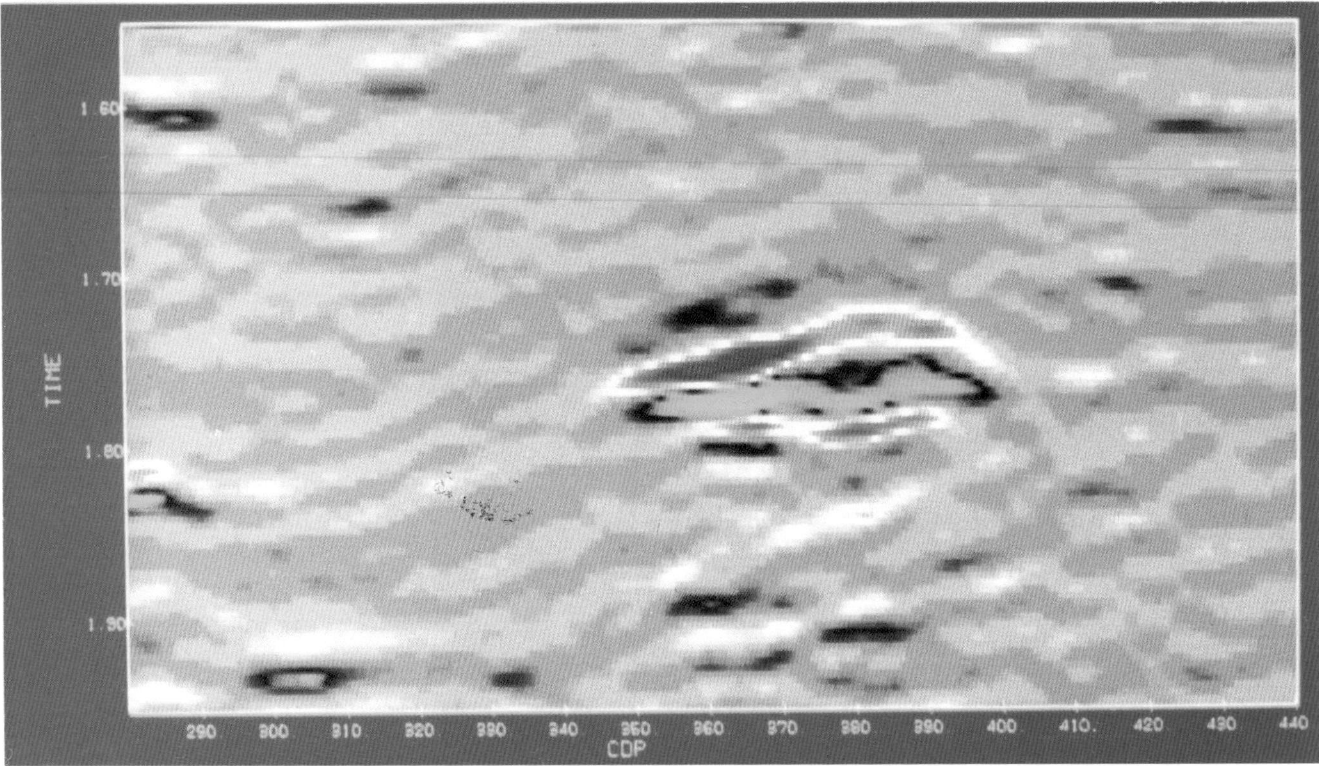

Fig. 5-17. (Top) Gulf of Mexico bright spot displayed in gradational blue and red with the gain set to maximize visual dynamic range and hence increase prominence of the amplitude anomaly. (Courtesy Chevron U.S.A. Inc.)

Fig. 5-18. (Bottom) Same bright spot as Figure 5-17 displayed in yellow, green and gray also in order to increase the prominence of the amplitude anomaly. (Courtesy Chevron U.S.A. Inc.)

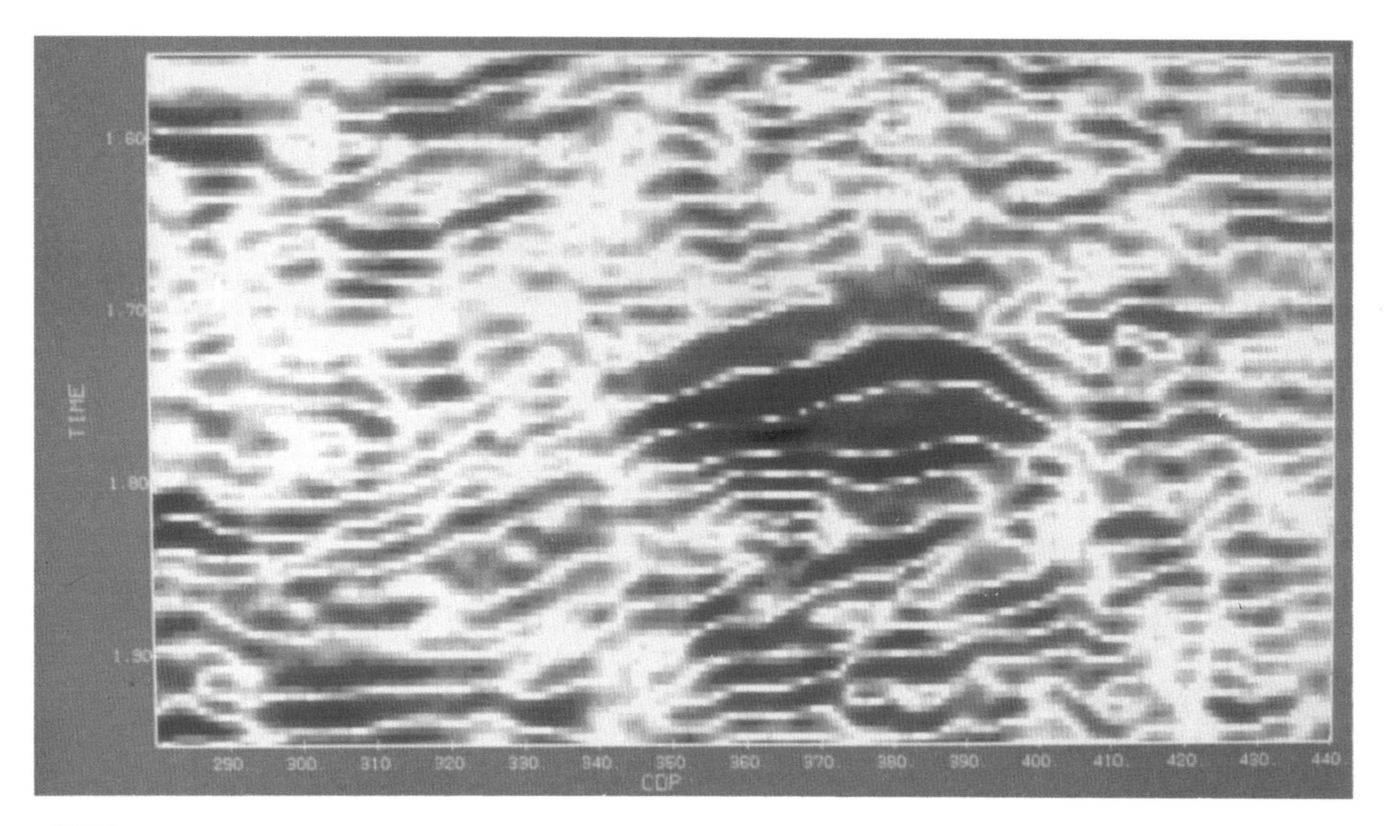

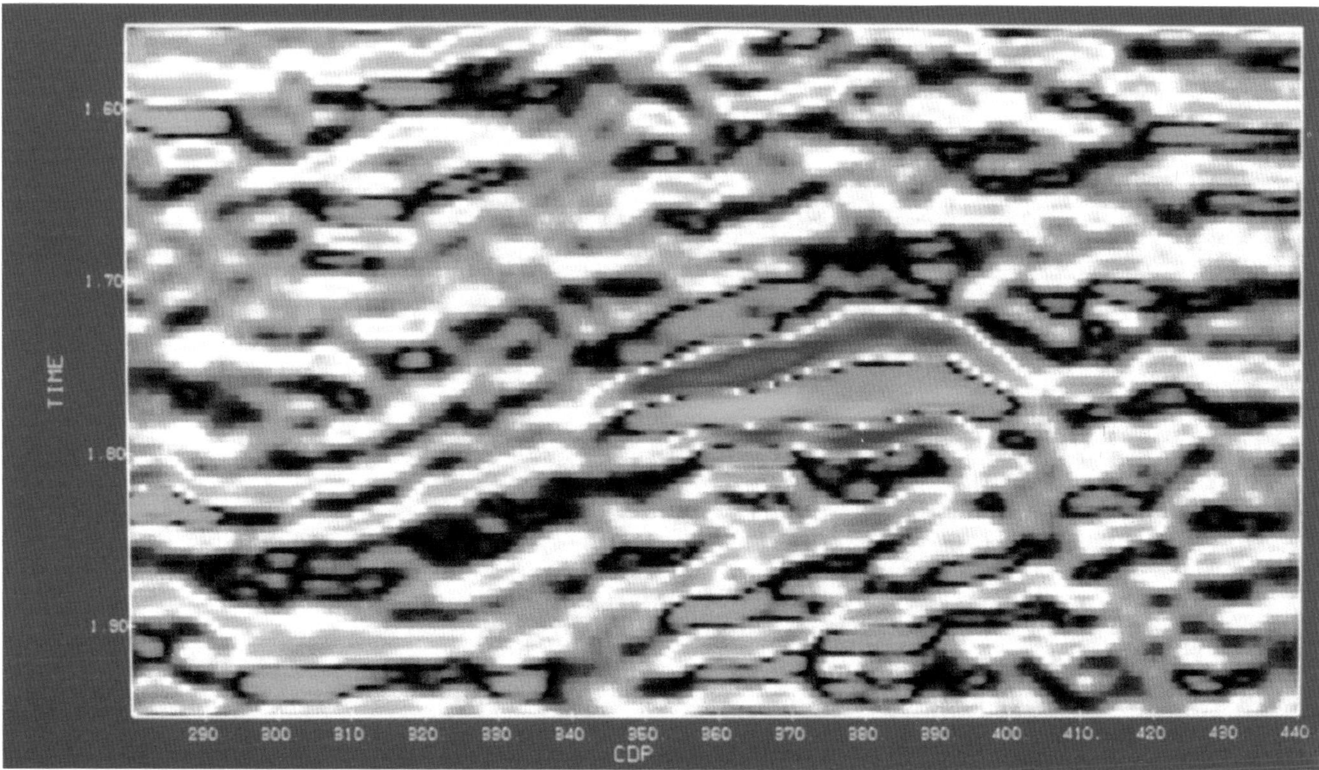

Fig. 5-19. (Top) Same bright spot and color scheme as Figure 5-17 but with the gain increased to study the continuity of reflections off the flank of the bright spot. Blue correlates with red and vice versa downdip indicating a phase change or polarity reversal at the edge of the bright spot. (Courtesy Chevron U.S.A. Inc.)

Fig. 5-20. (Bottom) Same bright spot as Figure 5-17, same color scheme as Figure 5-18 and same gain as Figure 5-19. The correlation of reflections downdip from the bright spot again indicates a phase change at the edge of the reservoir. (Courtesy Chevron U.S.A. Inc.)

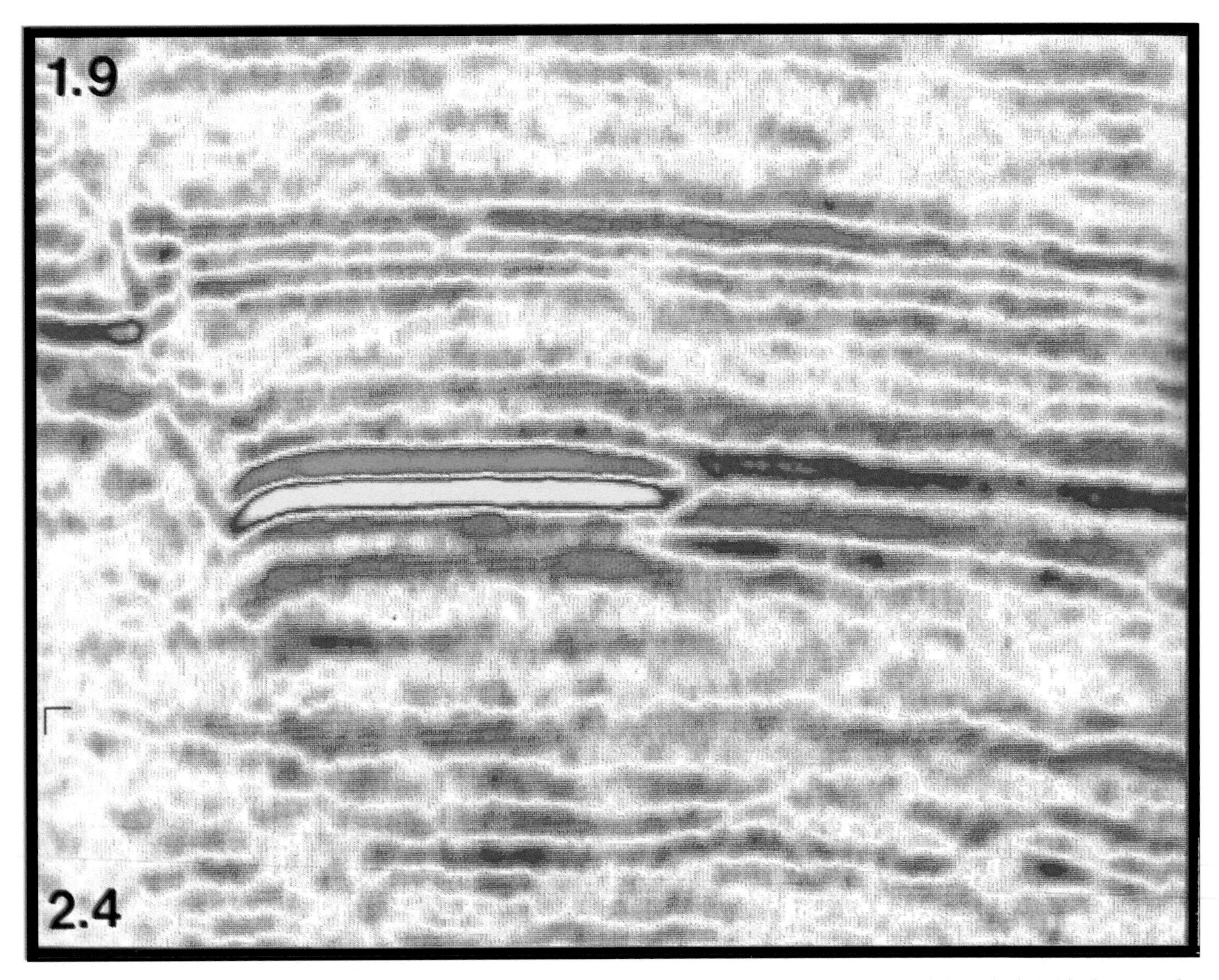

Fig. 5-21. Phase change (polarity reversal) caused by gas in unconsolidated sandstone of the Gulf of Mexico. (Courtesy Conoco Inc.)

as an indication of the truncating gas reservoirs. Tilbury and Smith (1988) discuss the geology and seismic modeling in support of this interpretation.

Figures 5-17, 5-18, 5-19 and 5-20 illustrate a phase change; all four figures are exactly the same piece of data displayed with different colors and gains. Figure 5-17 uses the standard blue and red gradational scheme and the amplitude anomaly is clearly visible. Its visibility is perhaps enhanced further by the yellow, green and gray color scheme of Figure 5-18. In order to check for a phase change, or polarity reversal, it is necessary to judge the structural continuity from the bright reflections to their non-bright equivalents downdip. There is a very great difference in amplitude between these, causing a great difference in color intensity. Figures 5-19 and 5-20 use the same colors respectively as Figures 5-17 and 5-18 but with a higher gain applied to the data. This makes it easier to judge the downdip continuity on the left of the bright spot and hence to observe that red correlates with blue (Figure 5-19) and green correlates with yellow (Figure 5-20). In this way a polarity reversal is established.

Figures 5-21 and 5-22 show striking phase changes (polarity reversals) respectively in the Gulf of Mexico and on the Northwest Shelf of Australia. The two data sets are opposite polarity from each other. Figure 5-21 shows the gas, which is low impedance, as blue-over-red; this is the polarity most commonly used in this book. Figure 5-22 shows the gas as red-over-blue. Figure 5-21 shows the gas contained by a fault on the west and the phase change on the east. Figure 5-22 shows the gas contained by a fault on the east and the phase change on the west. Some interpreters may feel inclined to identify both these

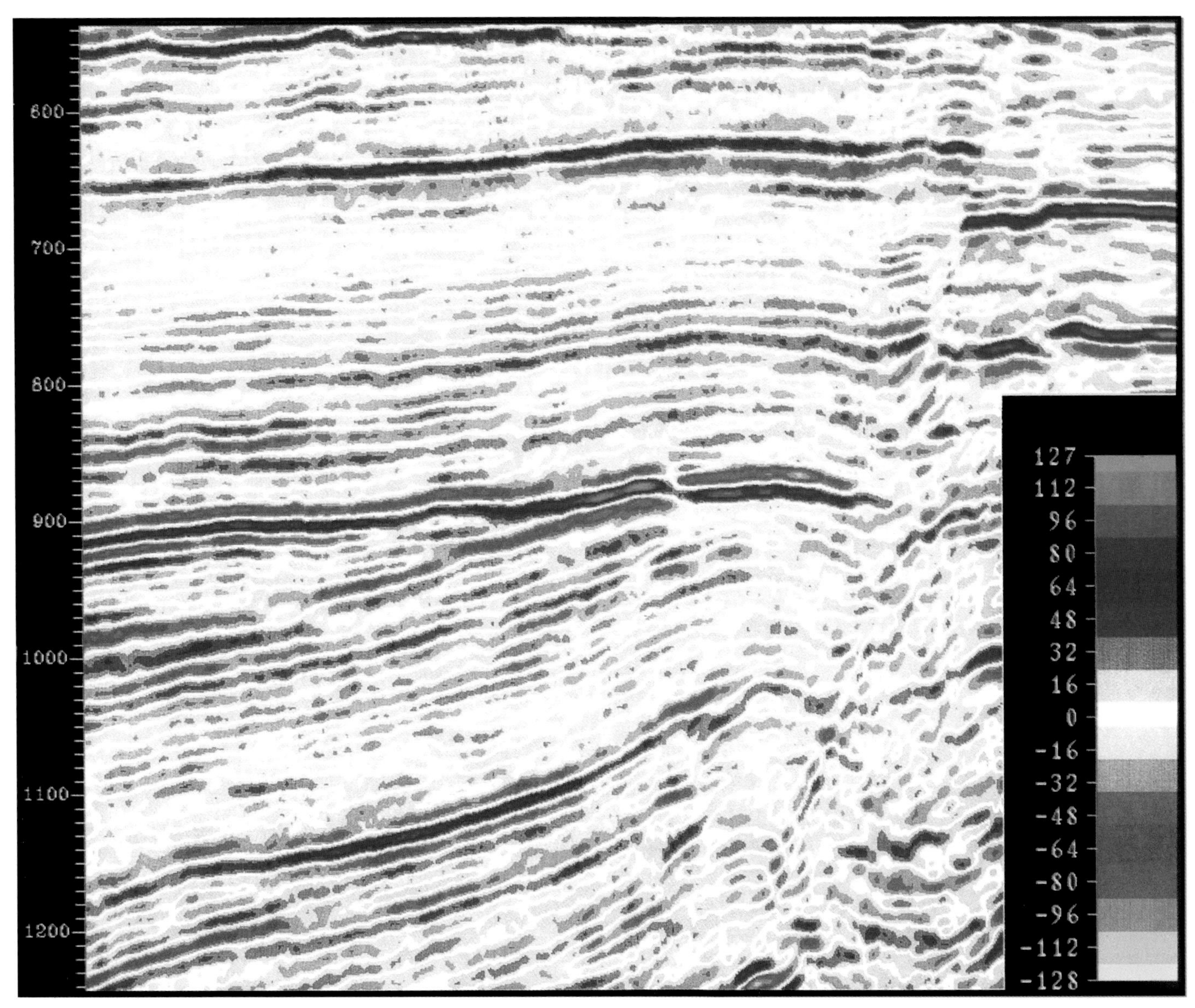

Fig. 5-22. Phase change (polarity reversal) caused by gas on the Northwest Shelf of Australia. (Courtesy West Australian Petroleum Pty. Ltd.)

phase changes as faults antithetic to the main faults. However, with experience the phase change explanation emerges as preferable. Structural/stratigraphic conflicts like this one are common and interpreters need to overcome their bias towards the structural solution.

Figure 5-23 shows a horizon slice indicating a channel. To the northeast the channel is bright, to the southwest it is not. The structural contours for this horizon have been superimposed and they demonstrate that the bright part of the channel is structurally above the dim part. This combination of structural and stratigraphic information helps validate gas content. Figure 5-24 is another way of graphically illustrating the same relationship; the representation of the channel in amplitude is superimposed on the structural configuration of the horizon surface.

Figure 5-25 demonstrates gas velocity sag on a flat spot reflection. The trough (red event) dipping west between 1560 and 1600 ms should presumably be flat in depth but is depressed in time by the increased travel through the low velocity, wedge-shaped gas sand. Flat spot dip caused in this way will always be in the opposite direction to structural dip. Figure 5-26 is another example of gas velocity sag. Here the high amplitudes are still in blue and red but the lower amplitudes are expressed in gradational gray tones. This provides the double benefit of highlighting the bright reflections and also helping establish fault definition by increasing the visibility of low amplitude event terminations. This section also demonstrates another phenomenon: there are bright events

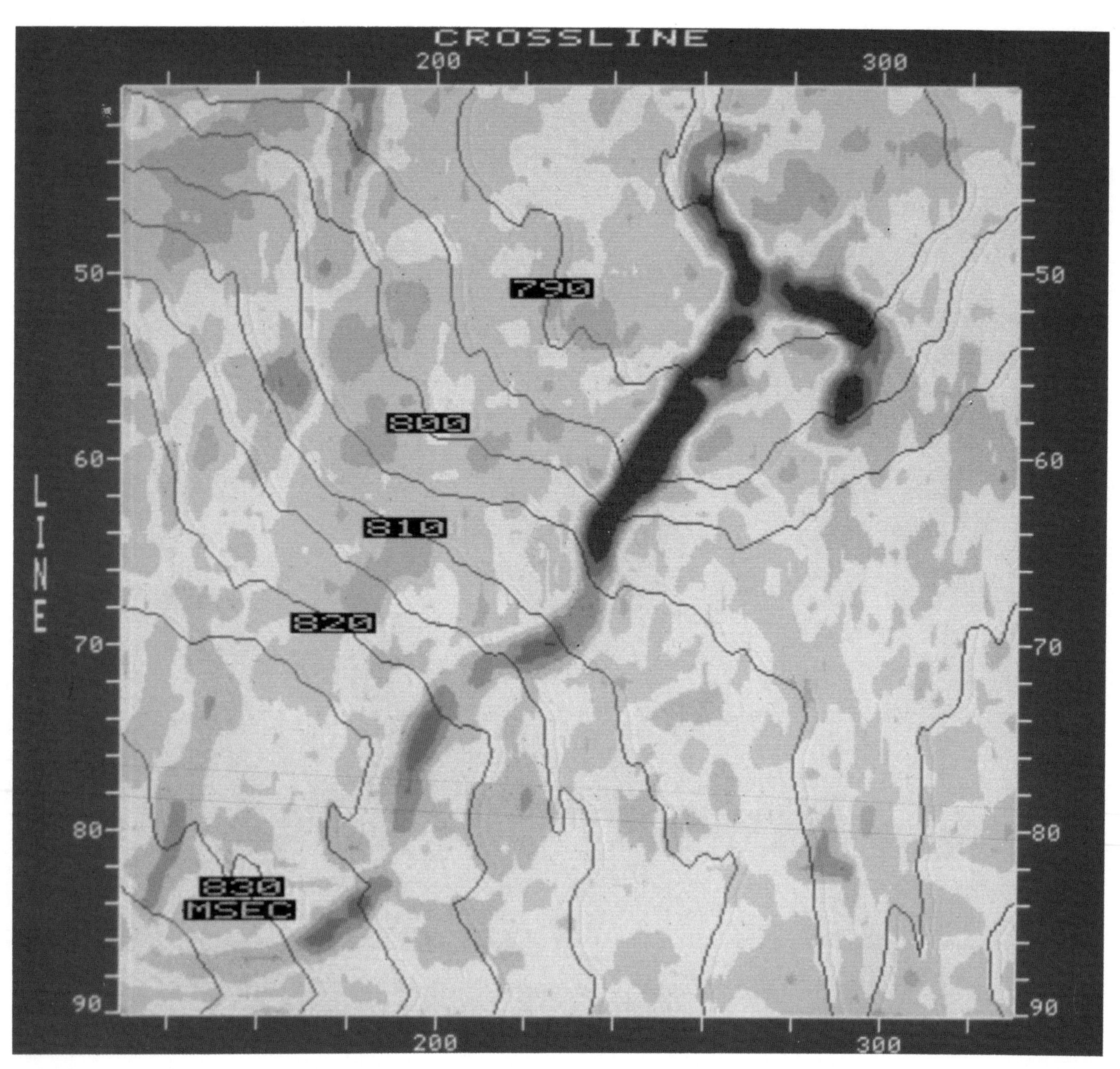

Fig. 5-23. Horizon slice showing Gulf of Mexico channel discussed in Chapter 4. The superimposed structural contours indicate that the bright part of the channel is shallower than the dim part. (Courtesy Chevron U.S.A. Inc.)

within the reservoir which have little expression outside. This illumination of internal layering is fairly common in clastic reservoirs and is discussed further in Chapter 7.

Figure 5-27 is a horizon slice showing a high-amplitude area surrounding known gas in well 7X; at this level well 10X is dry.

Figure 5-28 is a horizon slice showing a high-amplitude area known to be caused by oil. This is Bullwinkle field in Green Canyon, blocks 65 and 109, in the Gulf of Mexico. Oil is trapped in turbidite sands of Pleistocene age. O'Connell et al. (1990) have demonstrated that in this field, seismic amplitude is a robust indicator of oil. Clark (1992) and Greenlee et al. (1994) have both also demonstrated bright spots caused by oil.

Phase Problems and Multiple Contacts

All the discussion of the detailed character of hydrocarbon reflections so far in this chapter has been of zero-phase character. Unfortunately data phase is not always what it is supposed to be. Data processed to zero phase fairly often is close to 90° phase. Figure 2-24 shows a 90° phase flat spot and associated bright spot. It is the red and blue reflections

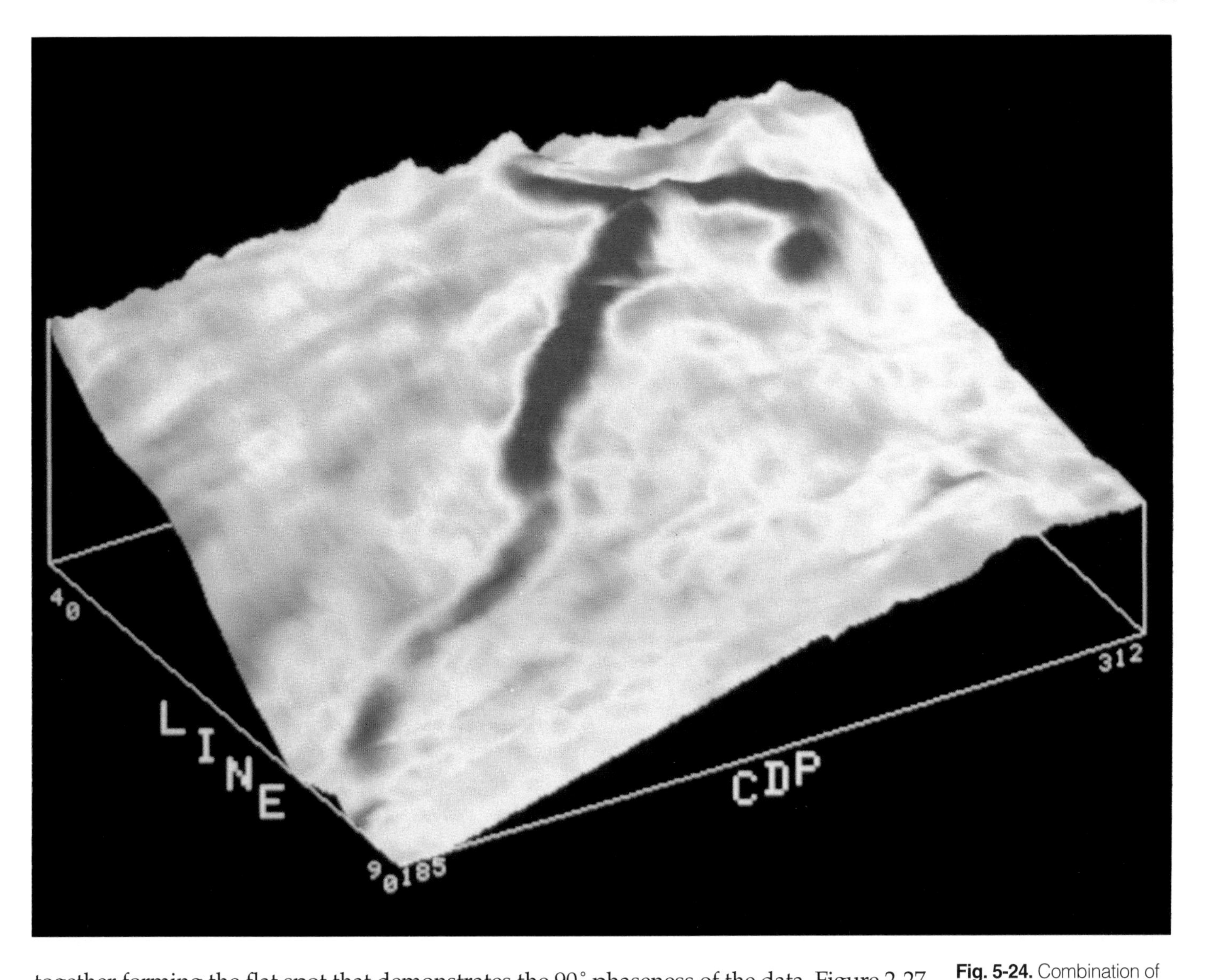

Fig. 5-24. Combination of the same horizon amplitude and structural information as Figure 5-23 using different colors and a three-dimensional perspective surface. (Courtesy Chevron U.S.A. Inc.)

together forming the flat spot that demonstrates the 90°-phaseness of the data. Figure 2-27 shows the 90° phase response of a thin gas reservoir. The three events for the two interfaces (top and base reservoir) indicate the 90°-phaseness as discussed in Chapter 2.

Figure 5-29 shows a flat spot with 90° phase character, yellow-over-red, at the green arrow. This is further confused, e.g. by comparison with Figure 2-24, by the fact that here the flat spot appears to be broken into four pieces. This is in fact caused by interference of strong internal reflections with the fluid contact reflection.

When we observe what appears to be two flat spots (e.g. Figures 2-24 and 5-29) the question arises as to whether we could be seeing two fluid contact reflections, for example gas-oil and oil-water. In fact a fluid contact is always an increase in acoustic impedance and thus two contacts in the same reservoir will always have the same character; so one red contact and one blue contact is impossible.

Figure 5-30 shows two flat spots at the green arrows indicating two contacts in an Indonesian gas reservoir. They are both red. In fact the upper contact is at the base of producible gas and the lower contact is at the base of trapped gas, perhaps a paleo gas-water contact.

Use of Frequency, Amplitude Variations With Offset and Shear Waves

Gas reservoirs attenuate high frequencies more than rocks without gas saturation. Following this principle, Taner, Koehler, and Sheriff (1979) have shown that low instantaneous frequency immediately below a suspected reservoir can be a good indicator of gas. The author has found this to be a rather unreliable indicator; several gas reservoirs studied

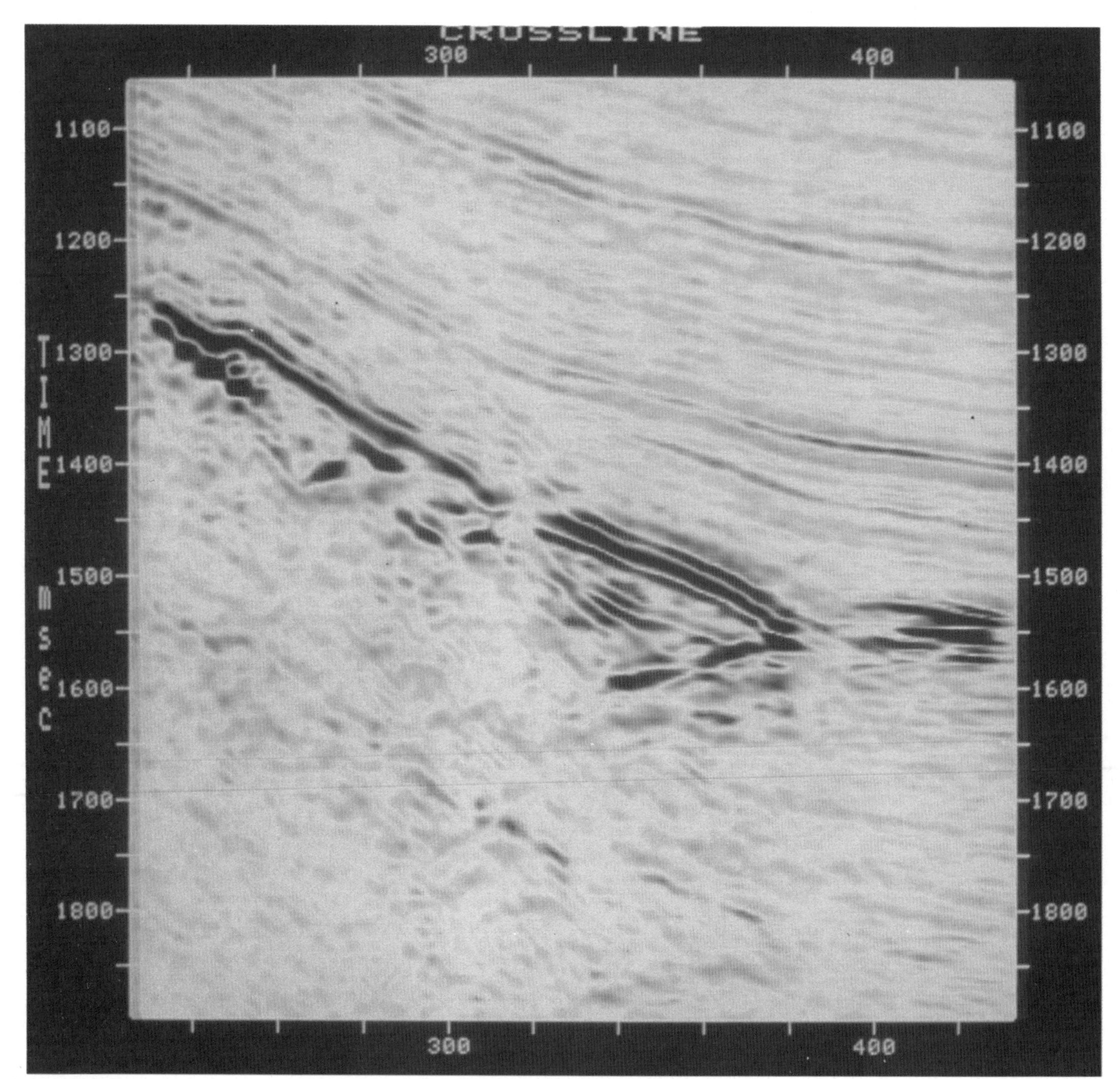

Fig. 5-25. Bright spot from a rather thick and complex gas sand. The red event dipping right-to-left is a flat spot displaying gas velocity sag. (Courtesy Chevron U.S.A. Inc.)

with good data have yielded ambiguous results in instantaneous frequency. Further discussion of instantaneous frequency and other frequency attributes follows in Chapter 8.

Interval velocity is reduced if a low velocity gas sand is included in the interval studied. For many years RMS velocities derived from normal moveout have been used to compute interval velocities, and for gross effects and trends this is valuable. However, the stability of interval velocities gets progressively worse for greater depths and also for thinner beds. This generally means that interval velocities are not sufficiently accurate to play a useful role in bright spot validation.

The variation of amplitude with recording offset has recently become a popular subject because of the possibility of extracting a significant amount of lithologic information from this kind of data. However, there are many difficulties both of a theoretical and practical nature (Backus and Goins, 1984). Among the practical issues, the data are prestack and hence have a lower signal-to-noise ratio, and, being multidimensional, there are many possible modes of display.

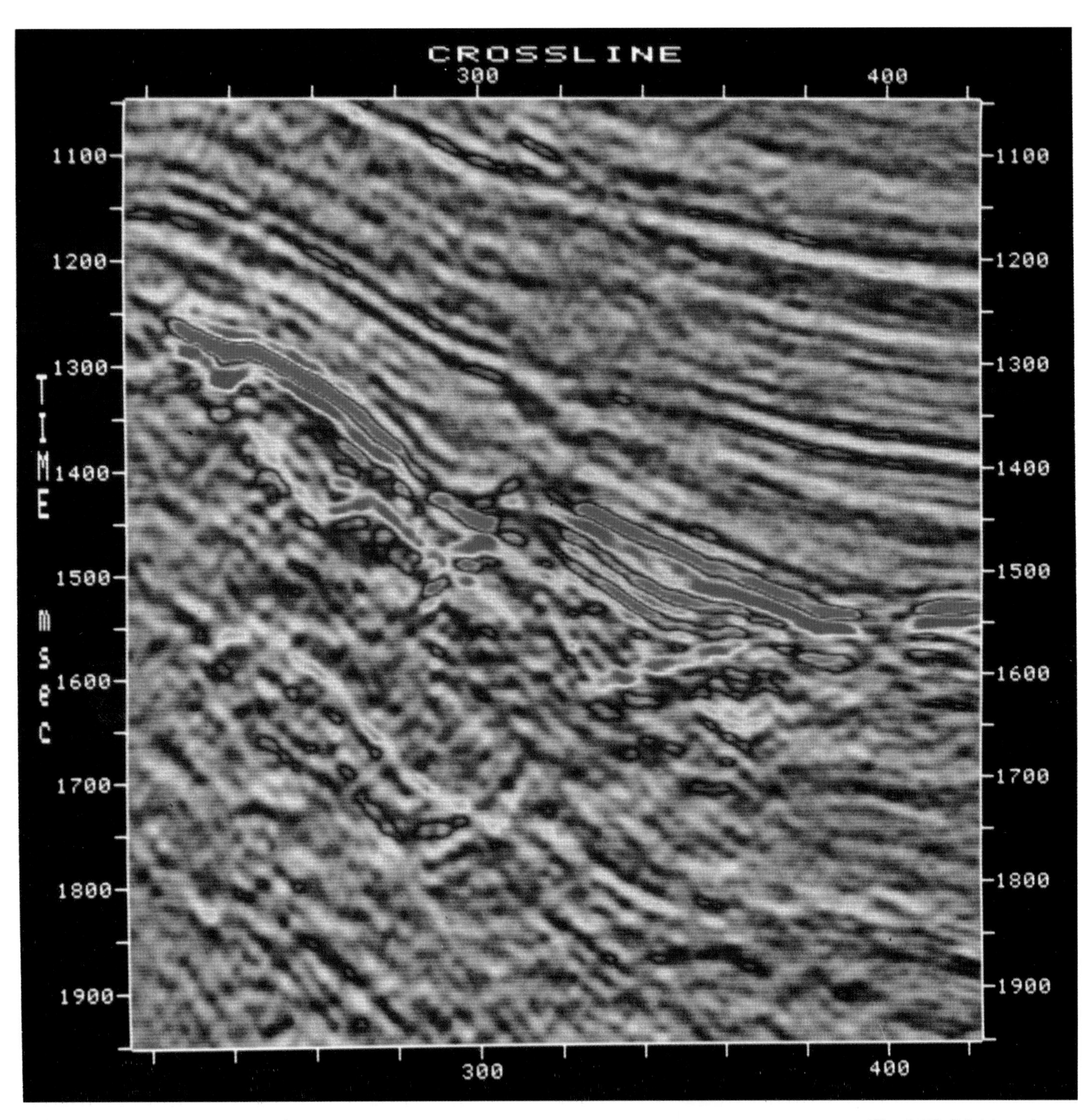

Fig. 5-26. Bright spot showing similar phenomena to Figure 5-25; the lower amplitudes are here displayed in gray tones. (Courtesy Chevron U.S.A. Inc.)

Ostrander (1984) demonstrated that in many practical cases gas sands show an increase of amplitude with offset and that this can be used as a means of identifying gas reservoirs. He studied the data in the form of common-depth-point gathers, normally stacking together common and adjacent offsets to improve signal-to-noise ratio. Common-depth-point gathers corrected for normal moveout but without any stacking are shown in Figure 5-31. The somewhat bright reflections at and below the black arrow are from a known Gulf of Mexico gas sand. Increase of amplitude with offset is just visible.

The application of the horizon slice concept has significantly increased the visibility of amplitude/offset effects for one horizon. Consider a volume of one line of prestack seismic data (Figure 5-32). The three dimensions are (1) CDP position along the line, (2) traveltime and (3) recording offset. The shape of one reflection without normal moveout cor-

Fig. 5-27. Horizon slice showing spatial extent of the high amplitude associated with gas. Well 7X is productive but well 10X is not productive at this level. (Courtesy Texas Pacific Oil Company Inc.)

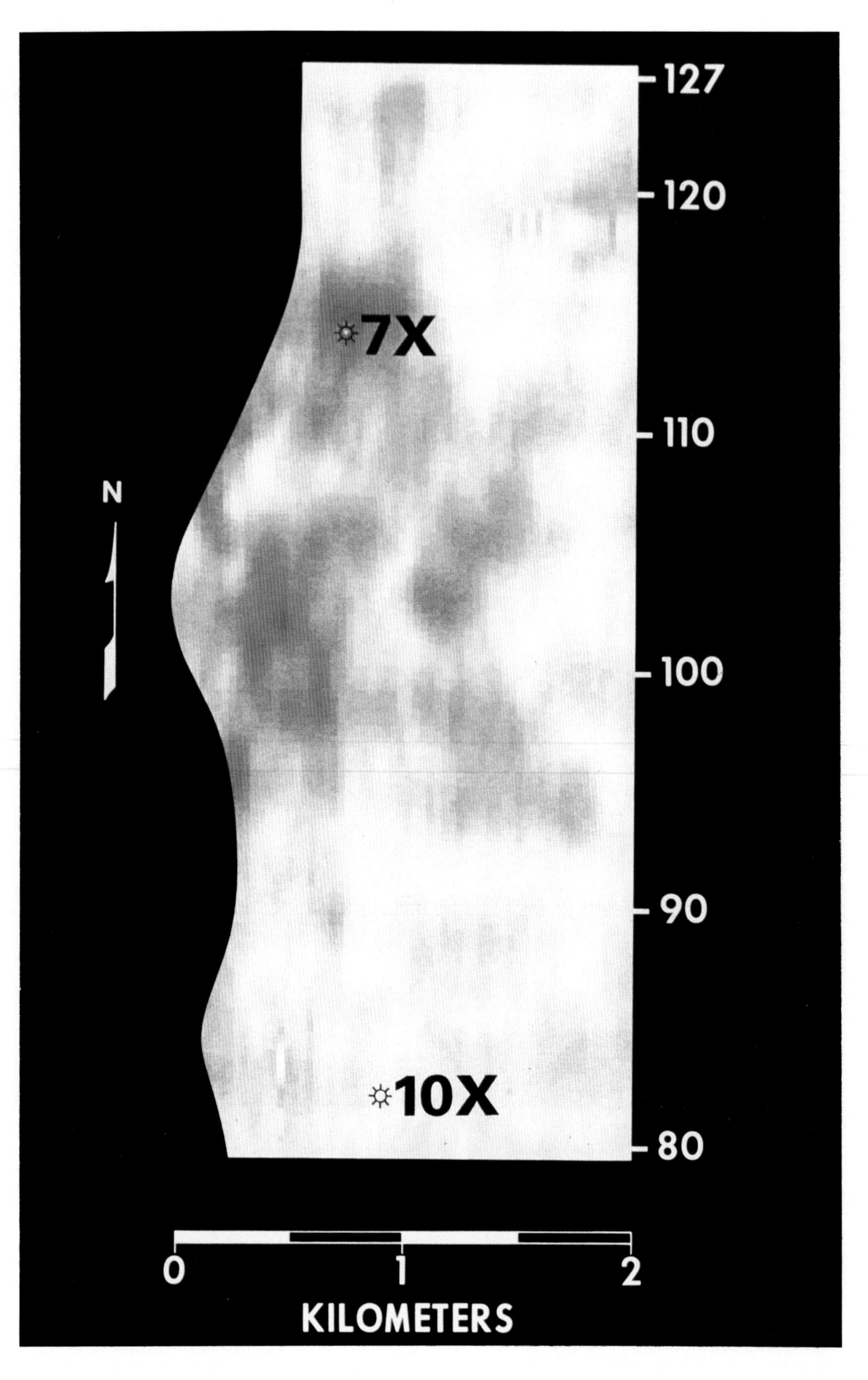

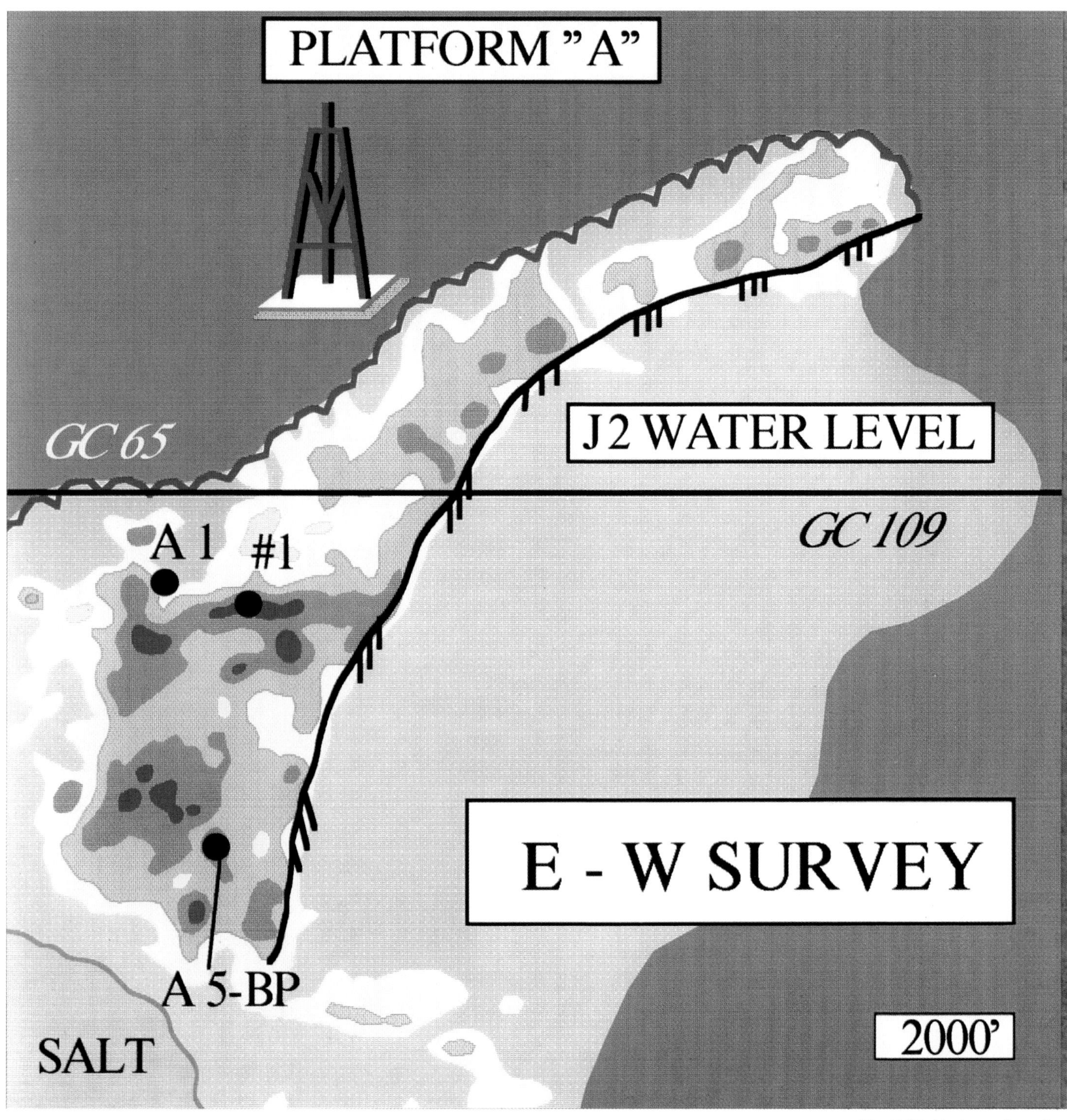

Fig. 5-28. Horizon slice of the main oil pay level at Bullwinkle field in the Gulf of Mexico. Note the sharp water level termination and updip amplitude dimming near the unconformity. Confirmation that the amplitude anomaly is caused by oil was provided by O'Connell et al. (1990). (Courtesy Shell Oil Company.)

rections is a cylindrical hyperbola as shown. By tracking this horizon and displaying the resultant amplitudes as if it were a horizon slice, a **horizon offset section** is obtained.

A horizon offset section prepared in this way is shown in Figure 5-33. The variation in amplitude with CDP position and with offset (approximately converted to incident angle) is shown for the trough immediately below the black arrow in Figure 5-31. The horizon offset section has been spatially smoothed, as an alternative to partial stacking, for increase of signal-to-noise ratio. The interpreter can observe, on this one section, the variation of amplitude with offset over many depth points for this horizon of interest. The amplitude increases with offset for most of the depth points and is hence consistent with gas content.

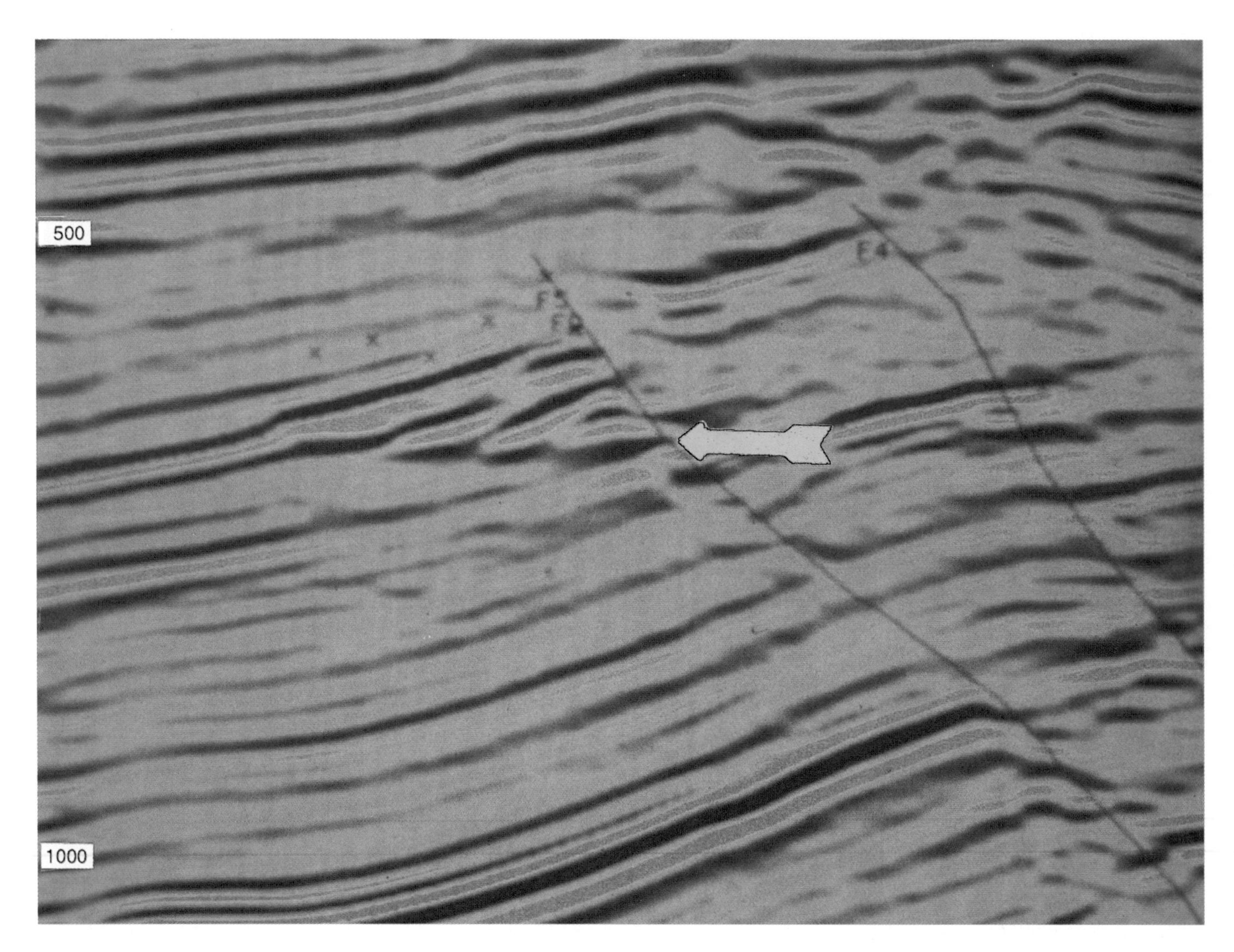

Fig. 5-29. Flat spot and bright spot visible in 90° phase data from offshore China. (Courtesy Bureau of Geophysical Prospecting, People's Republic of China.)

This method of validation requires that the gas sand and the embedding medium have very different Poisson's ratios. Because this is not always the case this method lacks certainty, even on theoretical grounds. On practical grounds poor signal-to-noise ratio is a common problem. Chiburis (1984) used amplitude ratio between the target horizon and a reference horizon and, because he used 3-D data, extensive smoothing was possible. In this way he delineated a gas cap from the amplitude/offset data and found reasonable agreement with engineering results.

Onstott, Backus, Wilson, and Phillips (1984) have used color in a novel way to study amplitude/offset information. They made vertical section substacks of near, mid and far traces and assigned to each of these one of the additive primary colors—red, green and blue. After combining these colored sections, the final section carried amplitude/offset information encoded in the color of every event.

The full and proper application of AVO to 3-D data involves both 3-D imaging and 3-D display. It is a complicated and computationally-demanding activity. The variation of amplitude with offset can be expressed as a single gradient factor and horizon slices in AVO gradient can be produced. Up to the present time the demonstrated value of these displays has been small so not very much has been done. Far offset–near offset amplitude difference is similar to AVO gradient. This, as a 3-D AVO attribute, is discussed and illustrated in Chapter 8.

Interpretation of shear wave amplitudes in conjunction with conventional compressional wave amplitudes can provide another method of bright spot validation. On land, S-wave data have generally been collected in a separate operation. S-waves are not trans-

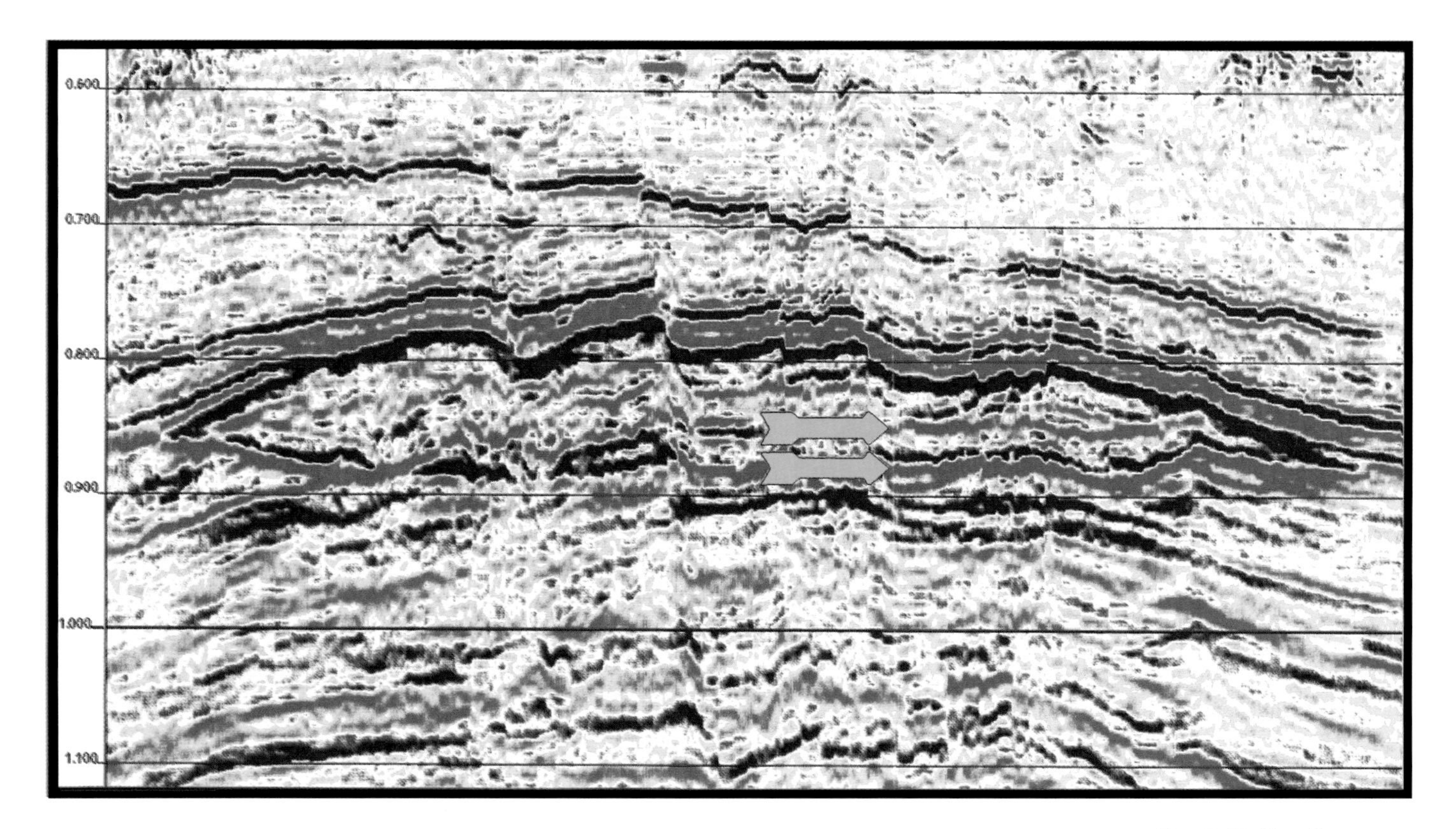

Fig. 5-30. Two flat spots in same gas reservoir offshore Indonesia indicating contact at the base of producible gas and at the base of trapped gas. (Courtesy Atlantic Richfield Bali North Inc.)

mitted through water so, at sea, it is necessary to use PSSP waves, mode converted at the water bottom. Tatham and Goolsbee (1984) have separated mode-converted S-wave data from P-wave data collected where the water bottom was hard.

Figure 5-34 summarizes the response of a water sand, a gas sand, a lignite bed, and a basalt bed to P- and S-wave energy; it should be studied in conjunction with Figure 5-5. Lignite has very low velocity and can be confused with a gas sand on the basis of P-wave response alone. Basalt, although high velocity, may also show a similar response if the polarity and phase of the data are not well understood.

The diagnostic comparison between P- and S-wave sections for a reflection from a gas sand is the presence of a P-wave bright spot and the absence of an amplitude anomaly for the correlative S-wave event. Figures 5-35 and 5-36 show comparative P-wave and S-wave sections reprinted from Robertson and Pritchett (1985). The 3rd Starkey reflection is bright on the P-section but not on the S-section, indicating gas.

In fact, the fundamental underlying principle is that compressional waves are sensitive to the type of pore fluid within rocks, whereas shear waves are only slightly affected. Hence the S-wave response of a reservoir sand will change little from below to above the gas/water contact, while the P-wave response normally changes greatly. Referring to Figure 5-5, it is clear that the P-wave dim spot would correlate on an S-section with a higher amplitude reflection. Where a phase change occurs across the gas/water contact on the P-section, the correlative P-wave and S-wave reflections from the gas sand will have opposite polarity. This is the situation interpreted by Ensley (1984).

Philosophy of Reflection Identification

Traditional approaches to reflection identification involve sliding a synthetic seismogram up and down on the real trace seeking a character match; we generally try to minimize the time mistie. With the all-too-common polarity and phase errors that may exist in our data, a more general and flexible approach will be more reliable. Let us not simply assume that the data phase and polarity are what they are supposed to be. The interpreter should attempt to assess them in, for example, the ways discussed in Chapter 2. If synthetic seismograms are used, then ones made with many different phases of wavelet should be compared to the real data. Color is very helpful in detecting detailed character

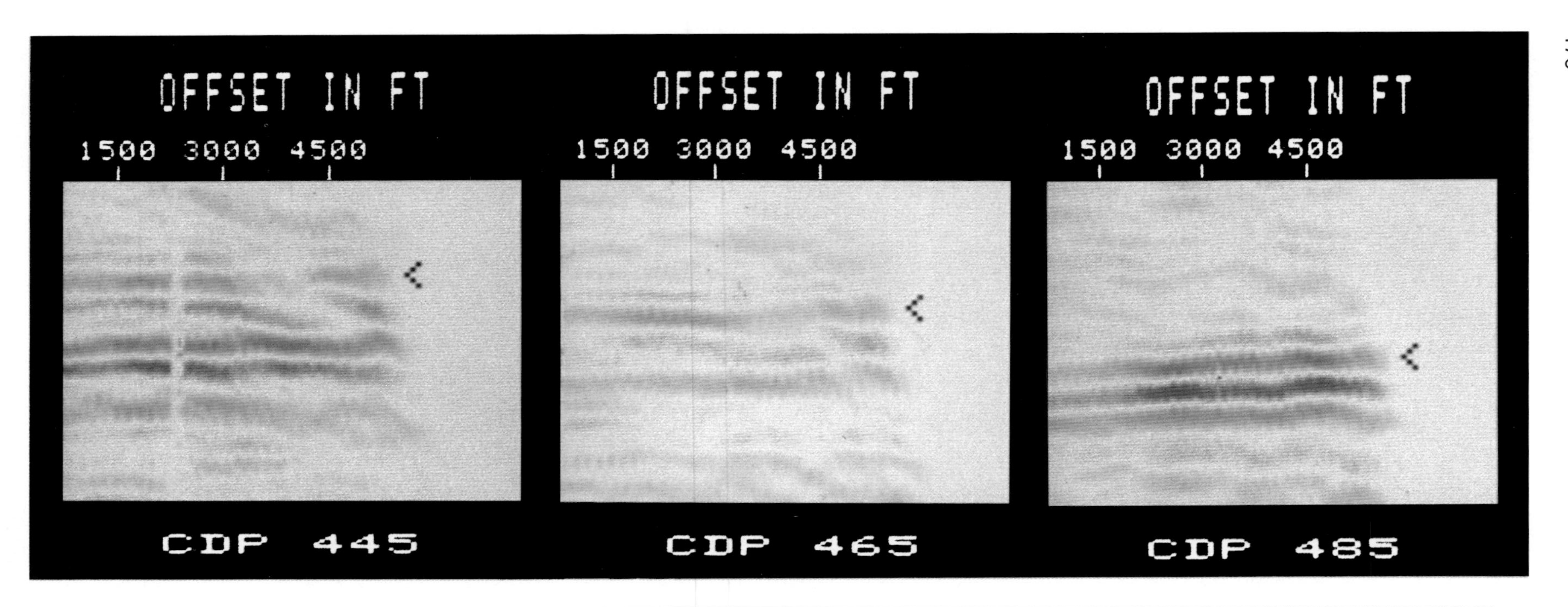

Fig. 5-31. Common-depth-point gathers corrected for moveout showing variation of amplitude with offset for several horizons.

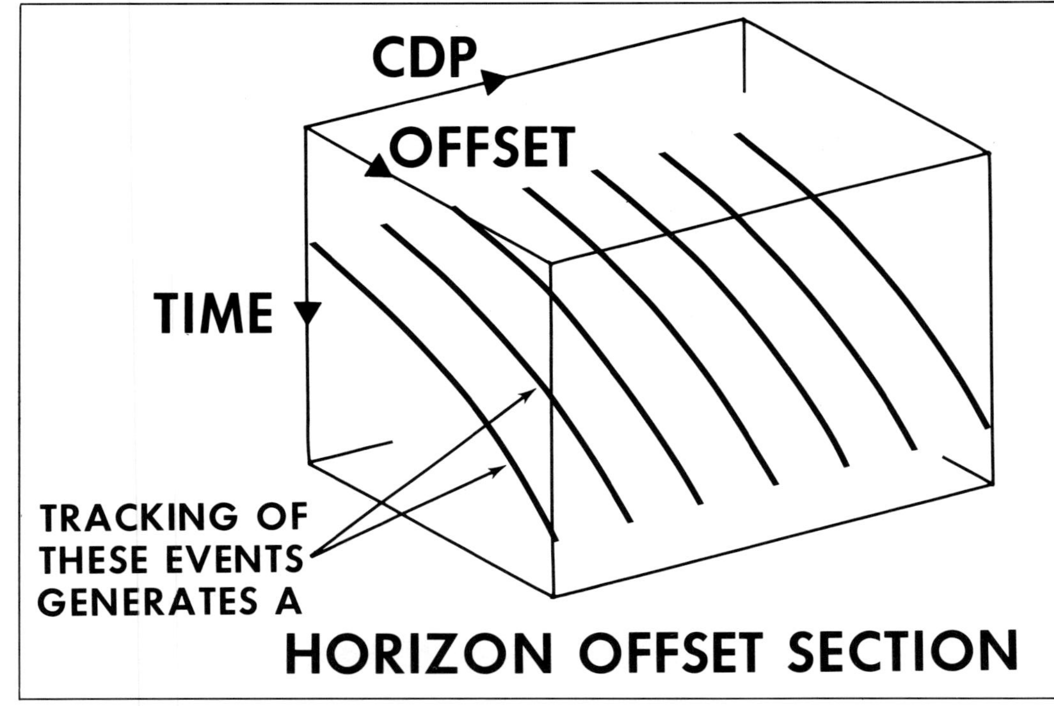

Fig. 5-32. The concept of a horizon offset section generated by tracking a horizon on a sequence of common-depth-point gathers.

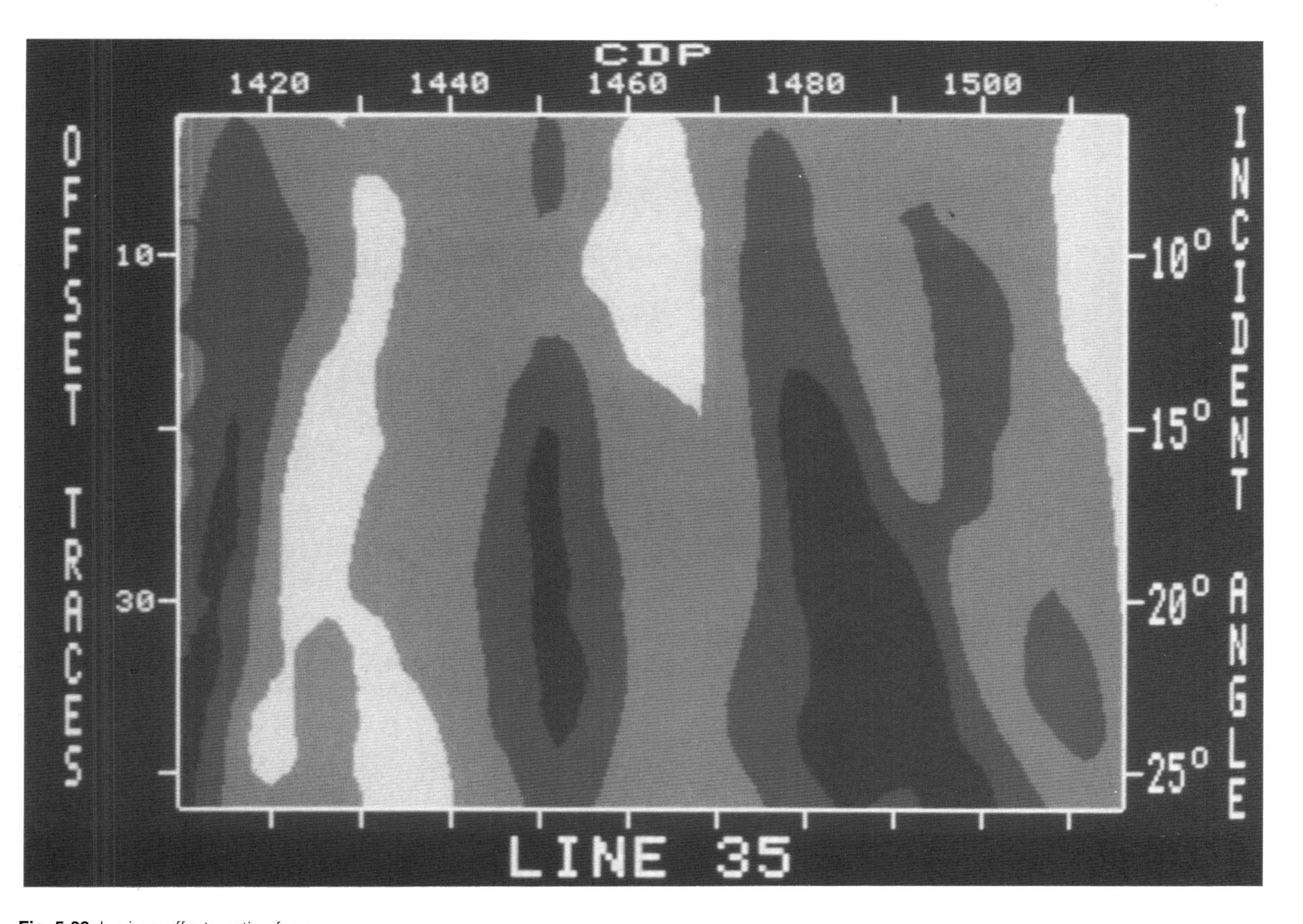

Fig. 5-33. Horizon offset section for a Gulf of Mexico bright spot showing increase of amplitude with offset for most depth points.

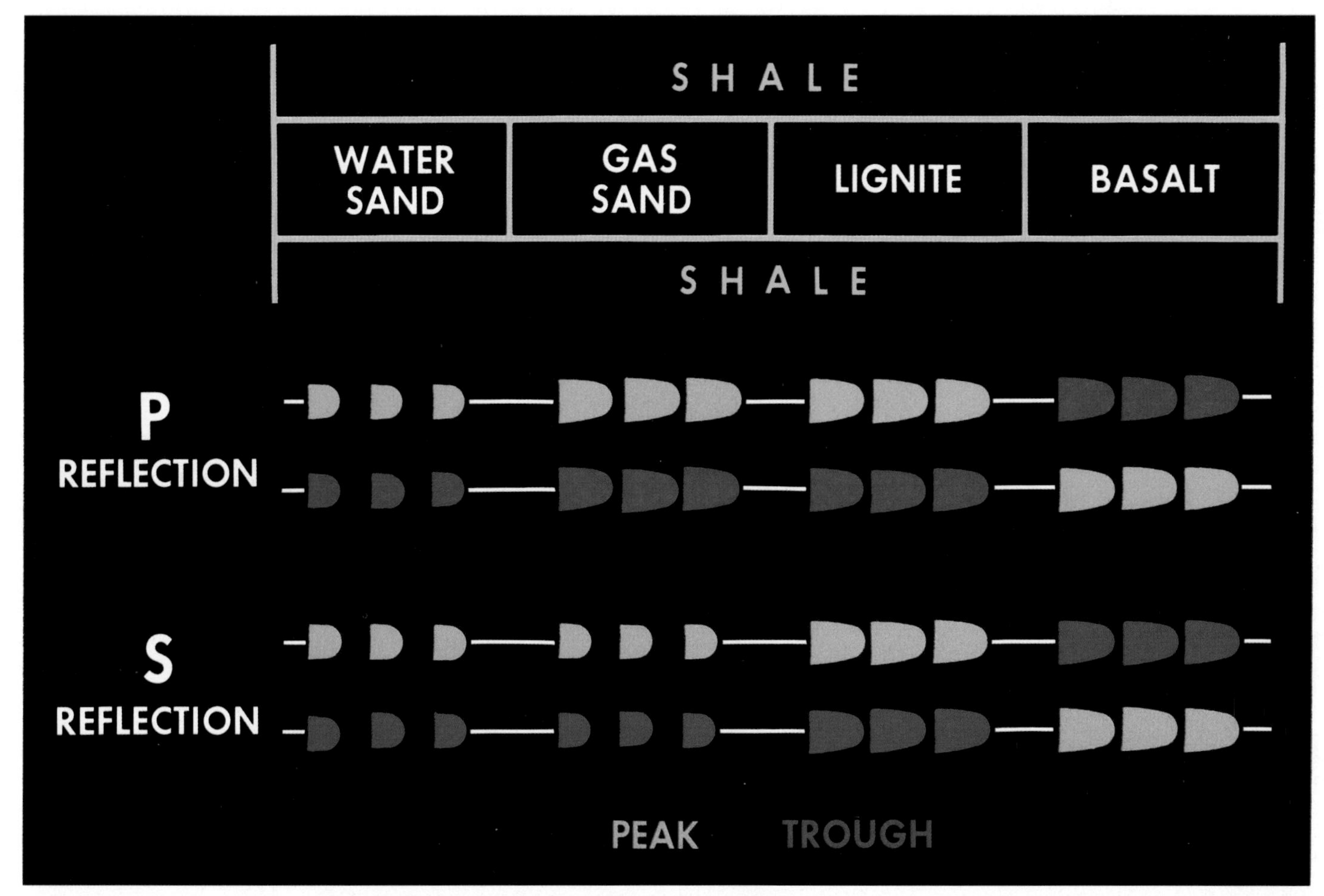

Fig. 5-34. Schematic diagram of the P-wave and S-wave zero-phase response for different beds encased in shale.

and recognizing phase and polarity errors.

The interpreter should be able to understand the complete character of the seismic data in the region of the geologic interface being tied. All the local reflections need to be understood; the more they cannot be understood, the more questions hang over the reflection identification. Clearly we must be confident of our reflection identification before any reservoir evaluation or characterization project, based on that reflection, can have any chance of success.

Let us consider an example: we find a seismic event of the right polarity (we think) at approximately the right time, so this becomes the top reservoir reflection. We know the approximate thickness of our reservoir so we look at the appropriate number of milliseconds deeper to find the reflection from the base. We find one of the right polarity at approximately the right time, so this becomes the base reservoir reflection. We know that we will need both top and base reflections for the study which is pending (Figure 5-13 and Chapter 7). All the wells show that the reservoir is a single, fairly massive unit so we would not expect any significant reflections from internal layers. However, on the seismic data, there is a reflection between top reservoir and base reservoir whose amplitude is greater than either top or base. Furthermore this "internal" reflection is widespread. The fact that we can then provide no satisfactory explanation for this reflection throws into question all the previous reflection identification!

Some general recommendations for seismic reflection identification then include:

- Don't necessarily assume the data polarity and phase are what they are supposed to be.

- If you are going to use synthetic seismograms, use a suite of seismograms of various phases and of both polarities.

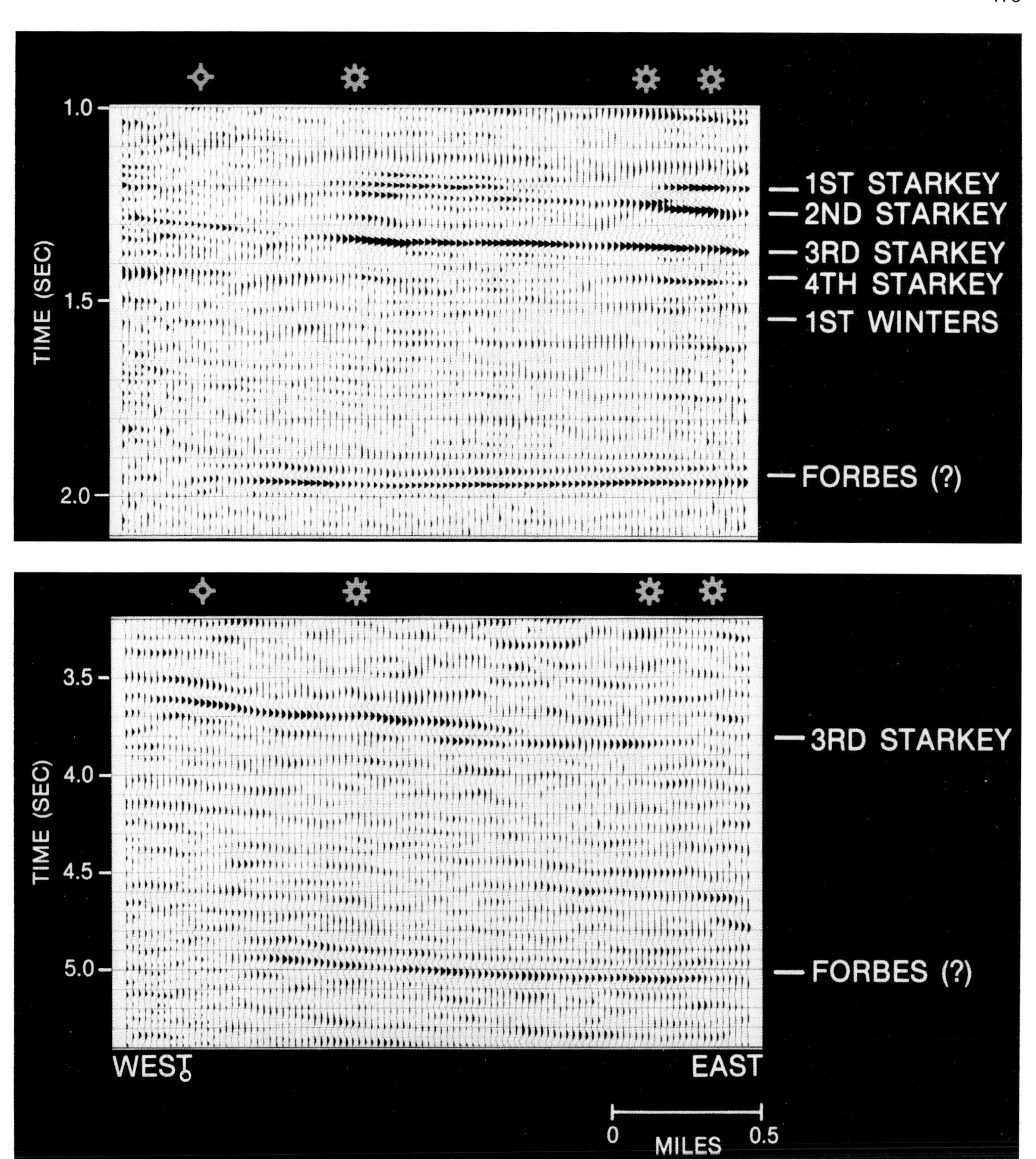

Fig. 5-35. (Top) P-wave section over California bright spot (from Robertson and Pritchett, 1985).

Fig. 5-36. (Bottom) S-wave section from same location as Figure 5-35 showing no amplitude anomaly for correlative event (from Robertson and Pritchett, 1985).

- If you have the ability to make comparisons in color, do so because much more detailed character will be visible and you will more readily recognize polarity and phase problems.
- Consider all the local character, that is, seek to explain **all** the reflections in the neighborhood of the objective.

Questions an Interpreter Should Ask in an Attempt to Validate Hydrocarbon Indicators

(1) Is the reflection from the suspected reservoir anomalous in amplitude, probably bright?
(2) Is the amplitude anomaly structurally consistent?
(3) If bright, is there one reflection from the top of the reservoir and one from the base?
(4) Do the amplitudes of the top and base reflections vary in unison, dimming at the same point at the limit of the reservoir?
(5) Are the data zero phase?
(6) Is a flat spot visible?
(7) Is the flat spot flat or dipping consistently with gas velocity sag or tuning?
(8) Is the flat spot unconformable with the structure but consistent with it?
(9) Does the flat spot have the correct zero-phase character?
(10) Is the flat spot located at the downdip limit of brightness (or dimness)?
(11) Is a phase change (polarity reversal) visible?
(12) Is the phase change structurally consistent and at the same level as the flat spot?
(13) Do bright spot, dim spot, or phase change show the appropriate zero-phase character?
(14) Is there a low-frequency shadow below the suspected reservoir?
(15) If the reservoir is thick, are there significant reflections inside?
(16) Do statistical crossplotting techniques indicate a flat spot?
(17) Is there an anomaly in moveout-derived interval velocity?
(18) Is a study of amplitude versus offset on the unstacked data likely to yield further validation evidence?
(19) Are any shear wave data available for further validation evidence?

Every hydrocarbon indicator is potentially a reservoir, but any one indication can be spurious. Confident identification of a hydrocarbon necessarily involves the accumulation of evidence. The more questions on the above list to which you can answer "Yes," the greater should be your confidence. Every negative answer needs to be satisfactorily explained or the identification falls into serious question.

Figure 5-37 contains several suspected hydrocarbon reservoirs. Try asking the above questions for these data. You should find affirmative answers for questions 1 through 14 for many separate reservoirs. How many did you find? While scrutinizing these data, it is useful to bear in mind simple reservoir models, such as those portrayed in Figure 5-5. An effective way to interpret the reservoirs from Figure 5-37 is to draw an overlay for the reservoir reflections. On the assumption of zero-phaseness, interfaces should be drawn along crestal amplitudes, and it is important to mark the top and bottom of each suspected hydrocarbon reservoir *and* the top and bottom of the correlative aquifer in each case.

For comparison with Figure 5-37, Figure 5-38 presents the same data at a normal display aspect ratio and in variable area/wiggle trace. This demonstrates not only the value of color for identifying the reservoir reflections in Figure 5-37 but also the value of an expanded vertical scale.

Figure 5-39 presents the same section as Figure 5-37, but with the results of a well inserted. This demonstrates that there are seven stacked gas reservoirs at this offshore California location.

The Occurrence of Hydrocarbon Indicators

The nature of hydrocarbon indication—that is, whether the phenomenon is bright spot, phase change, or dim spot—depends on the relative acoustic impedances of hydrocarbon sand, water sand, and shale (Figure 5-5). Each of these acoustic

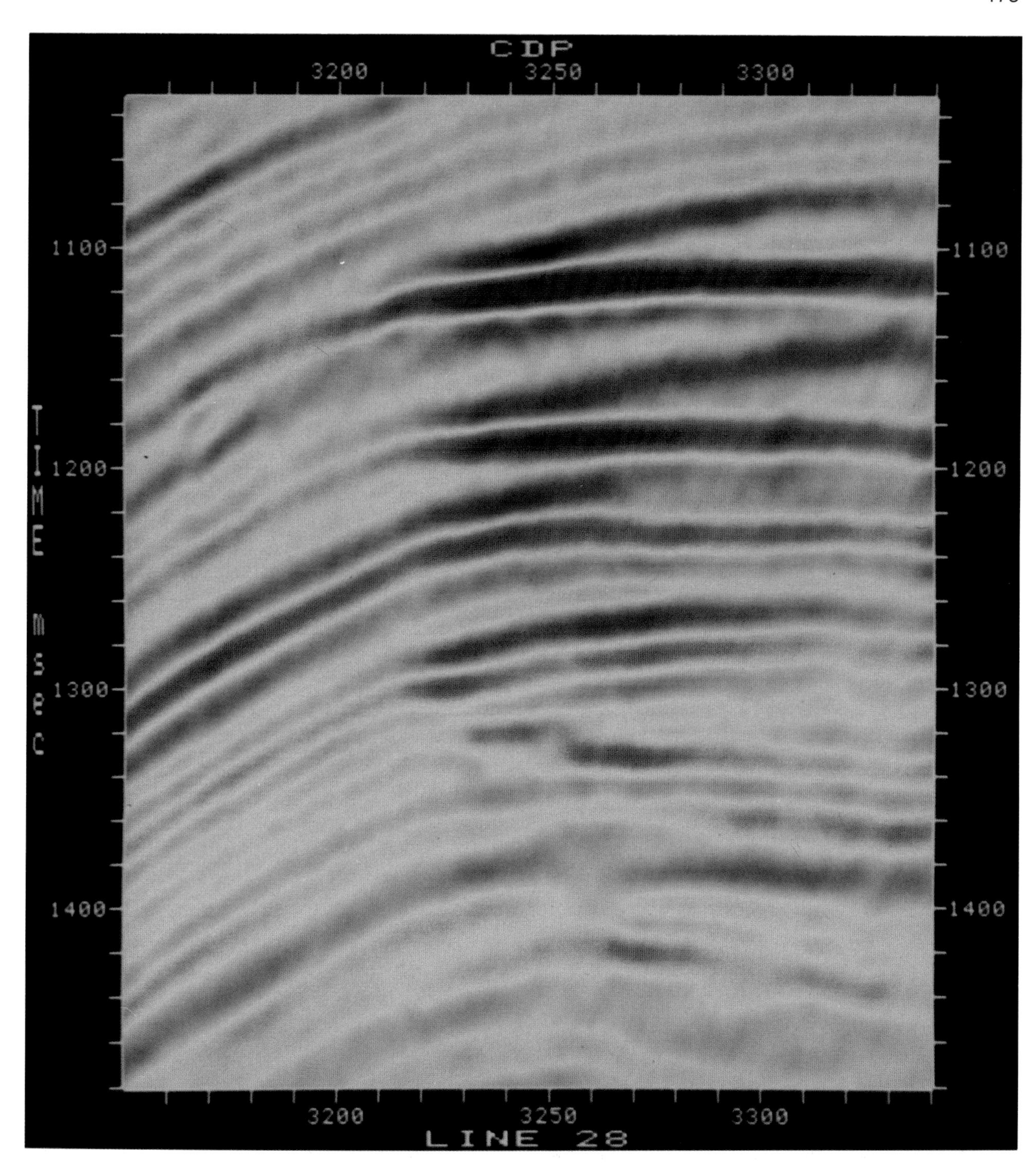

Fig. 5-37. Hydrocarbon indicators offshore California. It is intended that the reader interrogates this section with the "Questions an interpreter should ask in an attempt to validate hydrocarbon indicators."

impedances increases with depth (Figure 5-40) and they also each increase with rock age. It is difficult to be quantitative because they are also dependent on lithology, porosity, and local environment. Figure 5-40 is thus plotted for the qualitative product of depth and age. The effect of compaction on the shale causes its acoustic impedance to increase less rapidly than that of the sand. Below the point where the shale acoustic impedance crosses that of the water sand, phase changes must occur. Below the point where the shale acoustic impedance crosses that of the hydrocarbon sand, dim spots must occur. Of course, all the phenomena are reducing in visibility with depth and age, and somewhere there is a cut-off below which no hydrocarbon observations will be possible. However, Figure 5-40 provides a likely pattern of occurrence. Once one direct observation has been made in a given area it is probable that deeper and older observa-

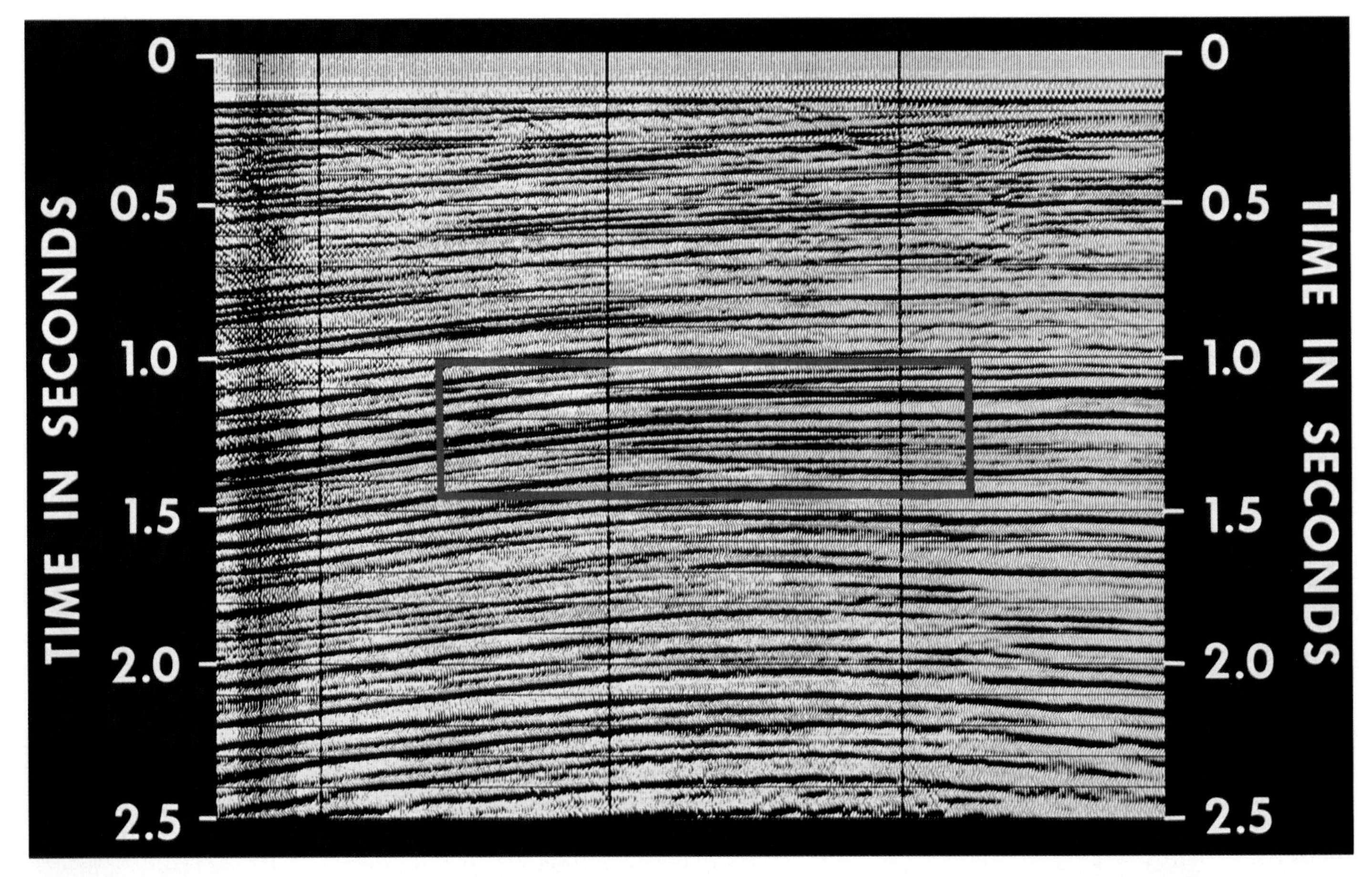

Fig. 5-38. Vertical section from offshore California in variable area/wiggle trace and normal aspect ratio. Data within red rectangle is the same as that presented in color in Figure 5-37 demonstrating that in the latter the vertical scale is exaggerated.

tions will fit the trend here illustrated.

Figure 5-41 is an attempt to separate the effects of depth and age. The depth is the depth of maximum burial, and old rocks are unlikely to have been at a shallow depth for all their geologic history. Nevertheless, Figure 5-41 indicates that bright spots will occur at great depths for very young rocks. It also indicates that hydrocarbon phenomena will occur in older rocks that are reasonably shallow. The author is aware of Gulf of Mexico proven hydrocarbon bright spots at 5500 m in the Pleistocene. Furthermore, hydrocarbon indicators have been observed in Permian rocks as previously discussed (Figure 5-11).

References

Backus, M. M., and R. L. Chen, 1975, Flat spot exploration: Geophysical Prospecting, v. 23, p. 533-577.

Backus, M. M., and N. Goins, 1984, Change in reflectivity with offset, Research Workshop report: Geophysics, v. 49, p. 838-839.

Birtles, R., 1986, The seismic flat spot and the discovery and delineation of the Troll Field, *in* Habitat of hydrocarbons on the Norwegian continental shelf, A.M. Spencer et al. (eds.): Graham and Trotman, p. 207-215.

Chiburis, E. F., 1984, Analysis of amplitude versus offset to detect gas/oil contacts in the Arabian Gulf: Proceedings, SEG 54th Annual Meeting, p. 669-670.

Clark, V. A., 1992, The effect of oil under in situ conditions on the seismic properties of rocks: Geophysics, v. 57, p. 894–901.

Ensley, R. A., 1984, Comparison of P- and S-wave seismic data: a new method for detecting gas reservoirs: Geophysics, v. 49, p. 1420-1431.

Gardner, G. H. F., L. W. Gardner, and A. R. Gregory, 1974, Formation velocity and density—the diagnostic basics for stratigraphic traps: Geophysics, v. 39, p. 770-780.

Greenlee, S.M., G. M. Gaskins, and M. G. Johnson, 1994, 3-D seismic benefits from exploration through development—an Exxon perspective: The Leading Edge, v. 13, p. 730-734.

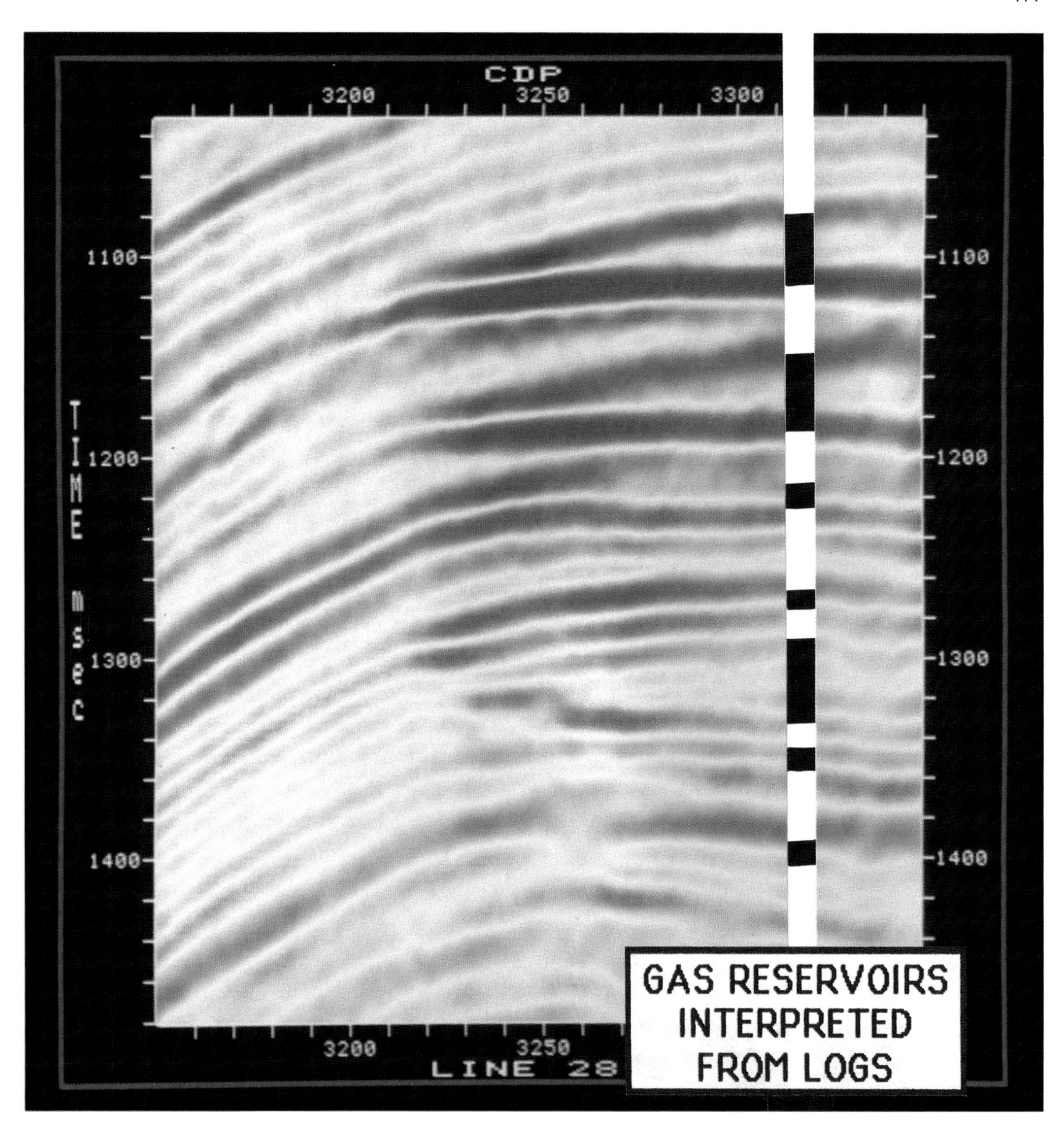

Fig. 5-39. Same section as Figure 5-37 with results of well inserted.

O'Connell, J. K., M. Kohli, and S. W. Amos, 1990, Bullwinkle: a unique 3-D experiment: Proceedings, SEG 60th Annual Meeting, p. 756-757.

Onstott, G. E. , M. M. Backus, C. R. Wilson, and J. D. Phillips, 1984, Color display of offset dependent reflectivity in seismic data: Proceedings, SEG 54th Annual Meeting, p. 674-675.

Ostrander, W. J., 1984, Plane-wave reflection coefficients for gas sands at non-normal angles of incidence: Geophysics, v. 49, p. 1637-1648.

Robertson, J. D., and W. C. Pritchett, 1985, Direct hydrocarbon detection using comparative P-wave and S-wave seismic sections: Geophysics, v. 50, p. 383-393.

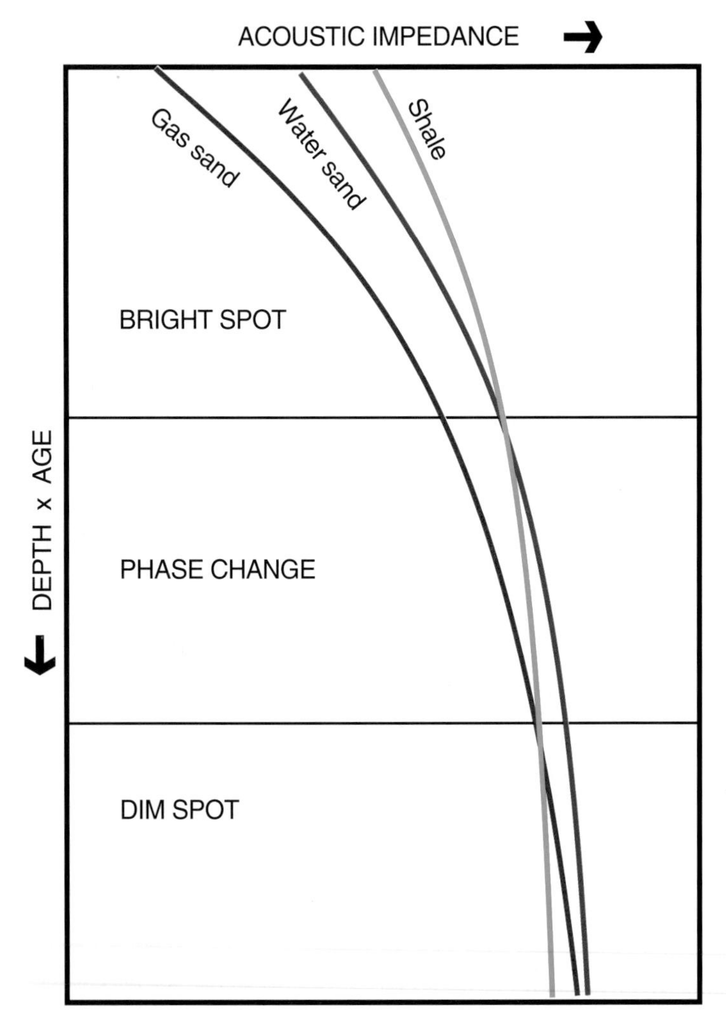

Fig. 5-40. Gas sand, water sand, and shale acoustic impedances all increase with depth and age but at different rates. The crossover points define whether the hydrocarbon indication is bright spot, phase change, or dim spot.

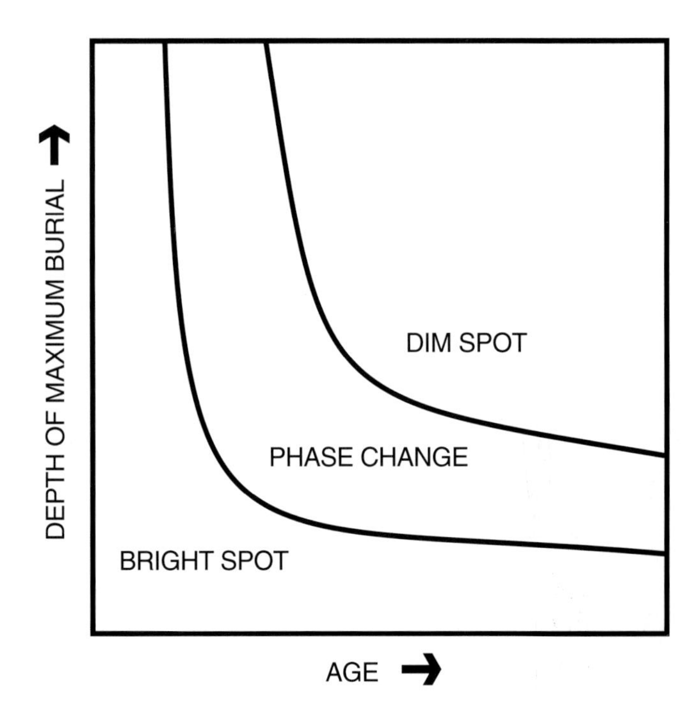

Fig. 5-41. Qualitative assessment of bright spot, phase change, and dim spot regions in terms of depth and age.

Taner, M. T., F. Koehler, and R. E. Sheriff, 1979, Complex seismic trace analysis: Geophysics, v. 44, p. 1041-1063.

Tatham, R. H., and D. V. Goolsbee, 1984, Separation of S-wave and P-wave reflections offshore western Florida: Geophysics, v. 49, p. 493-508.

Tegland, E. R., 1973, Utilization of computer-derived seismic parameters in direct hydrocarbon exploration and development, *in* Lithology and direct detection of hydrocarbons using geophysical methods: Dallas Geophysical and Geological Societies symposium.

Tilbury, L. A., and P. M. Smith, 1988, Seismic reflection amplitude (dim spot) study over the Goodwyn Gas Field, Northwest Shelf, Australia: Proceedings, ASEG/SEG Conference (Adelaide), p. 179-183.

TUNING PHENOMENA IN RESERVOIRS

Effect of Tuning on Stratigraphic Interpretation

Widess (1973) demonstrated the interaction of closely-spaced reflections. In his classic paper, "How thin is a thin bed?," he discussed the effect of bed thickness on seismic signature. For a bed thickness of the order of a seismic wavelength or greater there is little or no interference between the wavelets from the top and the bottom of the bed and each is recorded without modification. For thinner beds these wavelets interfere both constructively and destructively. Considering wavelets of opposite polarity, the amplitude of the composite wavelet reaches a maximum for a bed thickness of one-quarter wavelength (one-half period) and this is known as the tuning thickness. For beds thinner than this the shape of the composite wavelet stays the same but its amplitude decreases. Clearly, the bed thicknesses at which these phenomena occur depend on the shape of the wavelet in the data and hence on its frequency content.

These tuning phenomena are of considerable importance to the stratigraphic interpreter. They must be recognized as effects of bed geometry as opposed to variations in the acoustic properties of the medium. Figure 6-1 shows a sedimentary pod. As the reflections from the top and the base come together (within the black square) the amplitude abruptly increases; this is interpreted as tuning between the top and base reflections.

Convergence of reservoir reflections around the periphery of reservoirs is commonplace. Figure 6-2 illustrates the tuning phenomena in amplitude and time that occur between a top reservoir reflection and a fluid contact reflection near the downdip reservoir limit. At the tuning thickness the amplitude maximizes (for a given acoustic contrast), and for the parts of the reservoir thinner than this the amplitude will decrease. The tuning thickness is also the closest possible approach of the two seismic wavelets, so that, as the reservoir thins, the seismic reflections no longer will coincide with the reservoir interfaces. For zero-phase data this divergence will be disposed symmetrically between the top and fluid contact reflections, as shown in Figure 6-2.

The limit of seismic visibility indicated on Figure 6-2 is considered in more detail in Figure 6-3. For reservoirs with a higher acoustic contrast with the embedding medium, thinner parts of the reservoir will be visible, the exact thickness depending on the noise level in the data and the nature of the wavelet. Considering a common situation in Tertiary clastic reservoirs where the top and fluid contact reflections are equal in amplitude and opposite in phase, the actual downdip limit is invisible but can be found by extrapolating to zero the amplitude gradient observed between the tuning thickness and the limit of visibility.

Figure 6-4 shows some bright spots which are reflections from the top and base of gas sands of variable thickness. The base of the gas sands (the bright red events) are fluid contacts at most of the downdip limits. Hence the top and base reflections in many places constitute thinning wedges. Close inspection of Figure 6-4 reveals several local amplitude maxima close to the downdip limits of brightness. At these points the apparent dip also changes. The interfering wavelets are unable to approach each other more closely than a half period. Therefore, the composite bed signature for each of these thin beds assumes a dip attitude which is the mean of the

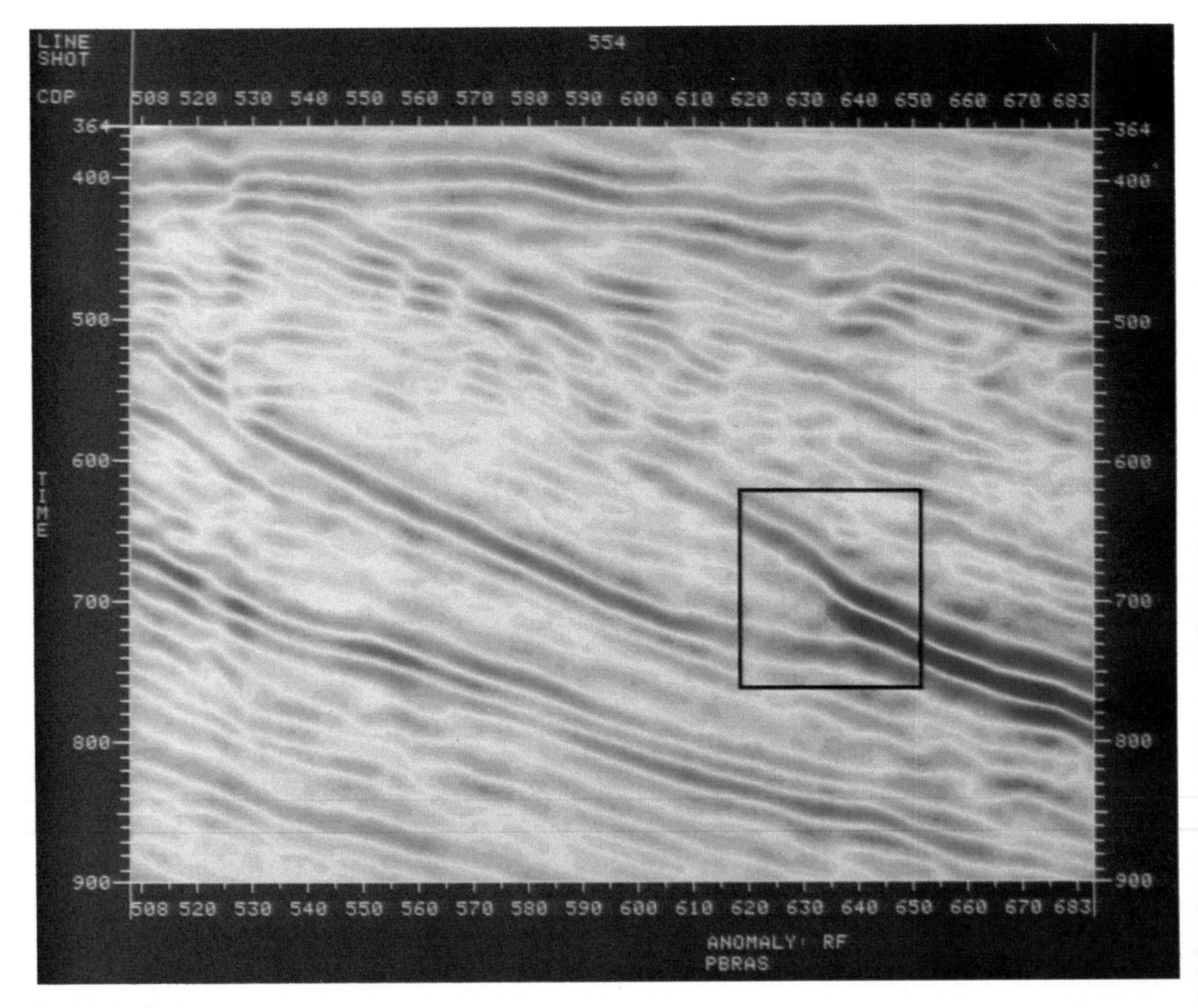

Fig. 6-1. Amplitude increases as reflections converge because of tuning. (Courtesy Petroleo Brasileiro.)

real dips of the top and the base of the bed. Because the base gas sand is flat at the downdip pinchout, it is easy to see the dip of the composite wavelet turning to assume this intermediate value.

Tuning amplitudes are easily recognized on horizon slices. Figure 6-5 is a horizon slice over the reflection from the top of a Gulf of Mexico gas reservoir. Dip is to the north, which is to the right in this figure. Horizon tracking was stopped at the limit of visibility, which is the edge of the various shades of blue along the north. Close to that edge and parallel to it is a lineation of locally higher amplitudes, visible as locally darker blues and indicated by a red arrow. This is the tuning thickness trend along which the reservoir thickness is equal to one-half of the seismic period, namely about 15 m (49 ft).

Tuning effects are not always a nuisance; in fact, they can be used to increase the visibility of thin beds. Amplitude tuning occurs for a layer thickness of one-half period of the dominant seismic energy, as already discussed. Frequency tuning, on the other hand, occurs for layer thicknesses of one-quarter period or less. Robertson and Nogami (1984) used instantaneous frequency sections to study thin, porous sandstone lenses based on this phenomenon.

Deterministic Tuning Curves

Tuning phenomena are usually described by graphs such as those of Figure 6-6. In this simple form the principles of tuning are well understood and widely published (for example, see Neidell and Poggiagliolmi, 1977). Figure 6-6 shows that measured thickness, indicated by the

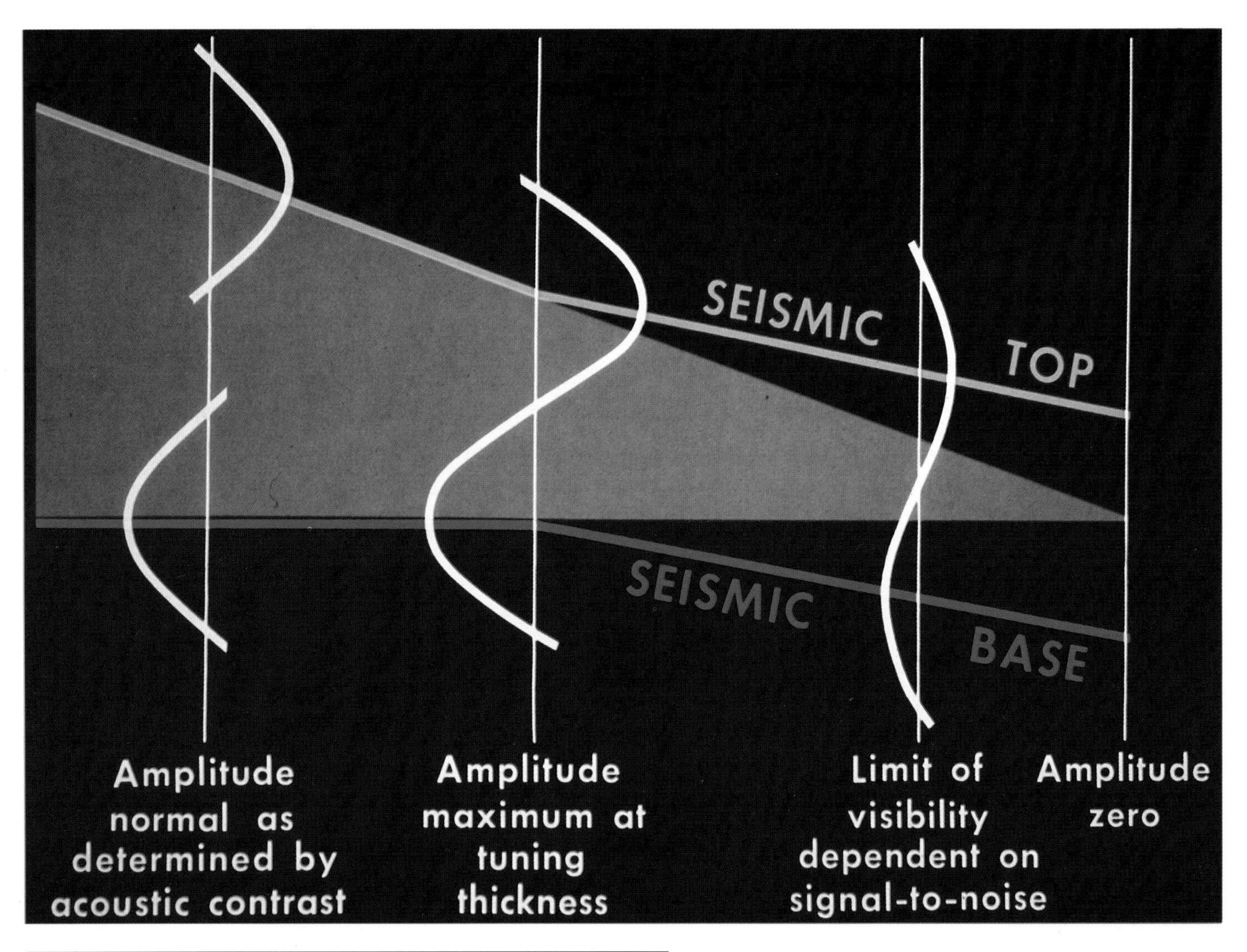

Fig. 6-2. Tuning effects in both amplitude and time applicable to zero-phase wavelets for a thinning wedge such as occurs between reservoir top and fluid contact reflections near a downdip reservoir limit.

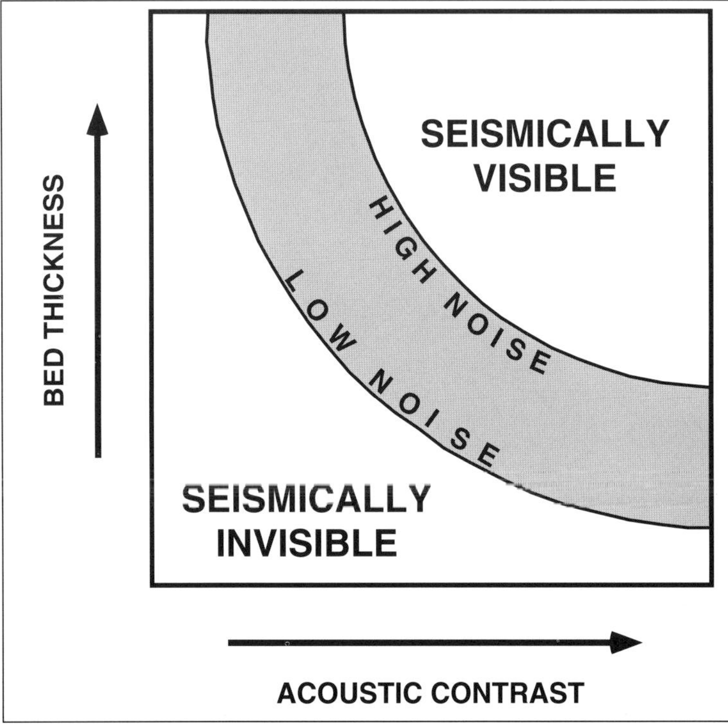

Fig. 6-3. Limit of seismic visibility depends on acoustic contrast of reservoir interfaces, noise level and wavelet shape. (After Meckel and Nath, 1977.)

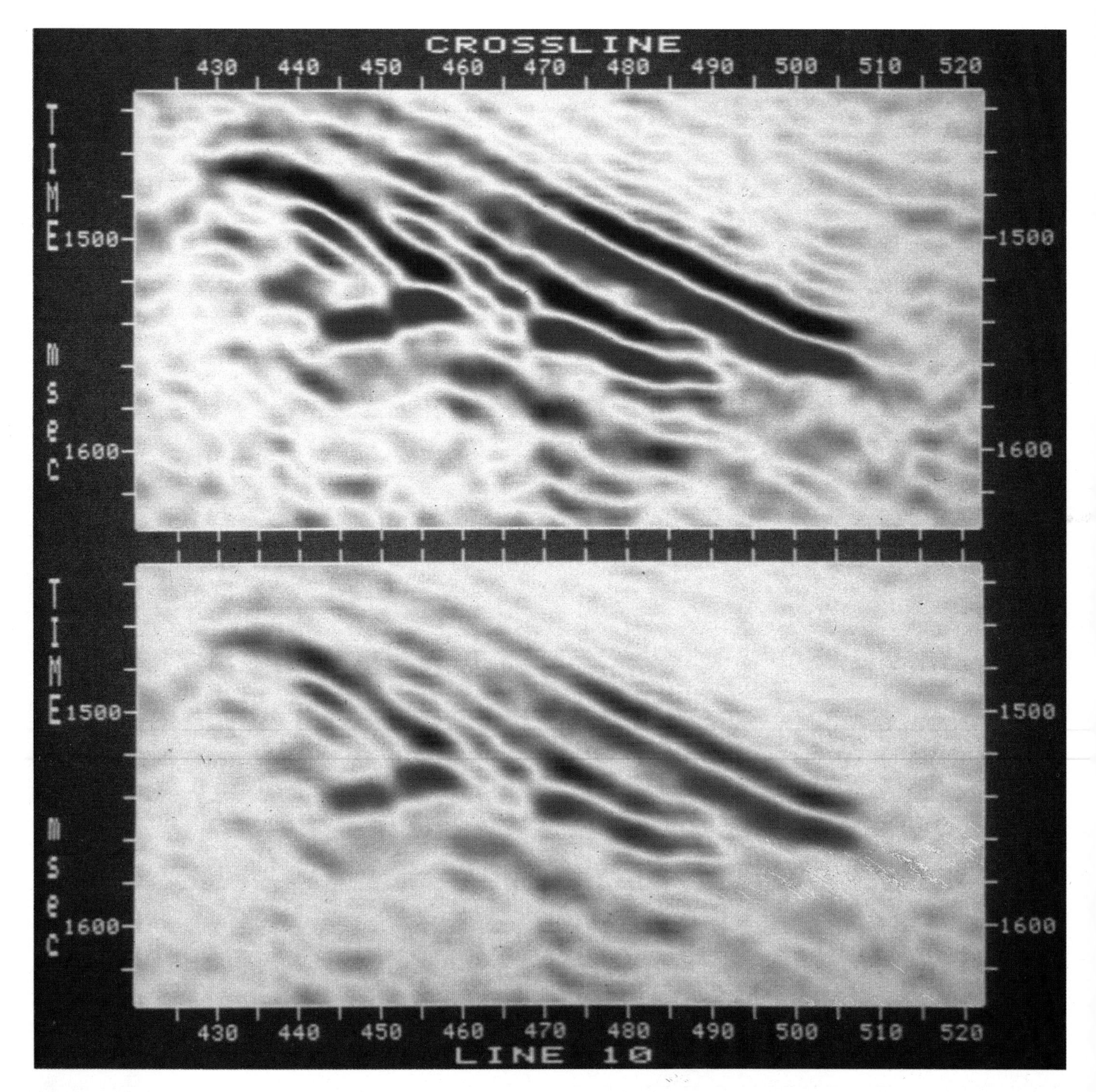

Fig. 6-4. High amplitude reflections from gas sands of variable thickness showing tuning effects in amplitude and time as events converge. The two panels are the same data with different gains to aid the observation of some of the subtle amplitude effects on the printed page. (Courtesy Chevron U.S.A. Inc.)

time separation of the reflections from the top and base of a bed, is only an acceptable measure of the true thickness of the bed for thicknesses above the tuning thickness. Also at tuning thickness the amplitude of the reflections reaches a maximum due to constructive interference between the reflected energy from the top and bottom of the bed.

The upper diagram of Figure 6-7 shows how the wavelets from the top and the base of a sand bed must be aligned to produce the principal tuning amplitude maximum; here it is assumed that the reflection coefficients are equal in magnitude and opposite in polarity. It is apparent that the shape of the tuning curve is dependent on the shape of the side lobes of the wavelet. Constructive interference occurs when the central peak of the wavelet from the base of the sand is aligned with the *first* negative side lobe of the wavelet from the top of the sand.

The lower diagram of Figure 6-7 shows how a second tuning maximum is caused. In this

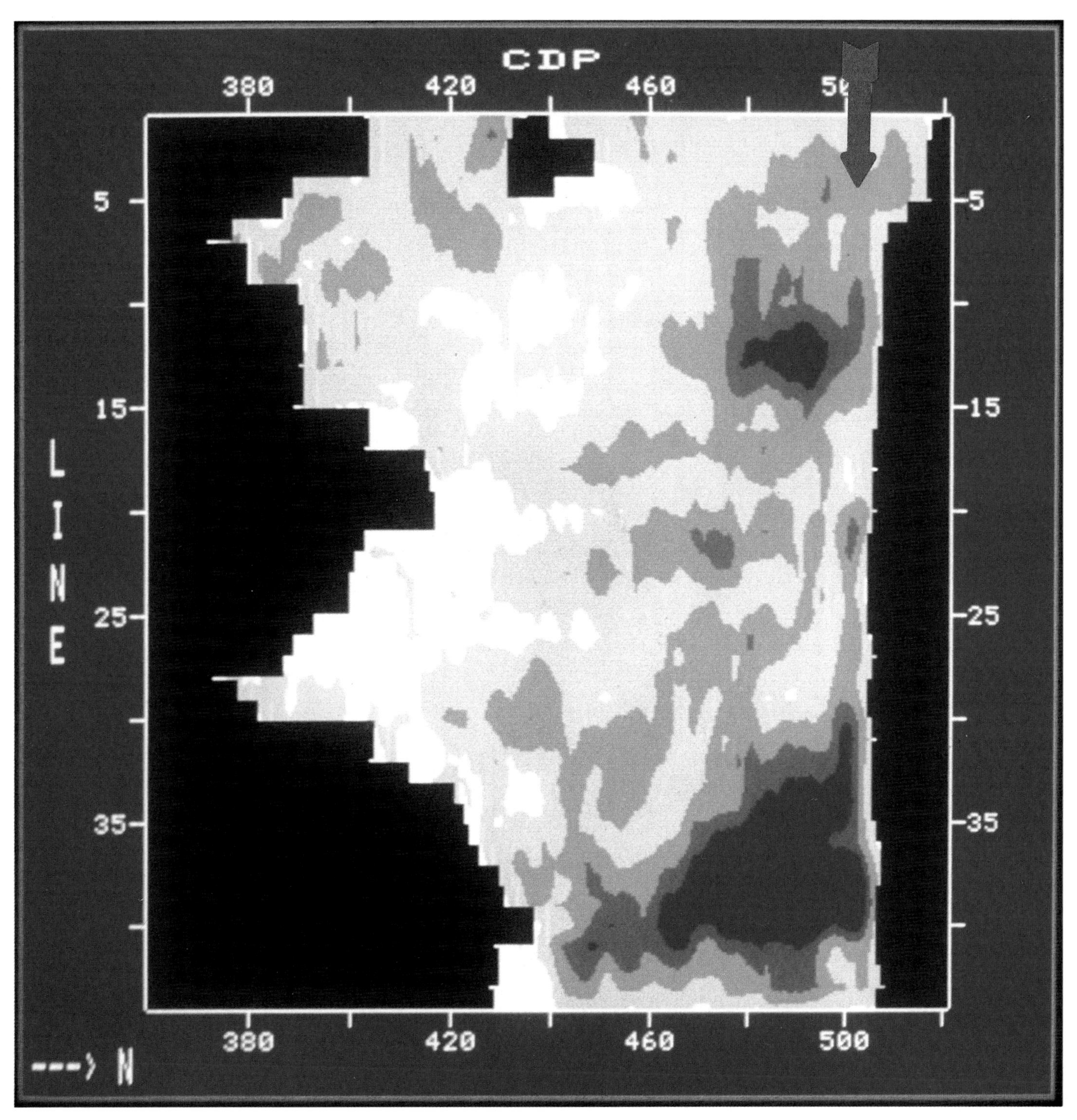

case the central peak of the wavelet from the base of the sand is aligned with the *second* negative side lobe of the wavelet from the top. Hence multiple wavelet side lobes generate multiple maxima in the tuning curve. Kallweit and Wood (1982) studied the resolving power of zero-phase wavelets and reported multiple maxima in their tuning curves (Figure 6-8).

Figure 6-9 illustrates deterministic tuning curves derived from four different wavelets. At the top of the page the Ricker wavelet has no side lobes beyond the first and consequently the tuning curve determined from it has only one maximum. This is the classical type of tuning curve, similar to that illustrated in Figure 6-6 and to that published by Meckel and Nath (1977).

The second wavelet in Figure 6-9 is a zero-phase wavelet derived from four corner frequencies defining a band-pass filter. It has, as can be seen, the same width of central peak as the Rick-

Fig. 6-5. Horizon slice over reservoir top where structural dip is to the north (right). Close to the downdip limit running top-to-bottom, an amplitude lineation in locally darker blues (in line with the red arrow) indicates the trend along which the reservoir thickness is equal to the tuning thickness. (Courtesy Chevron U.S.A. Inc.)

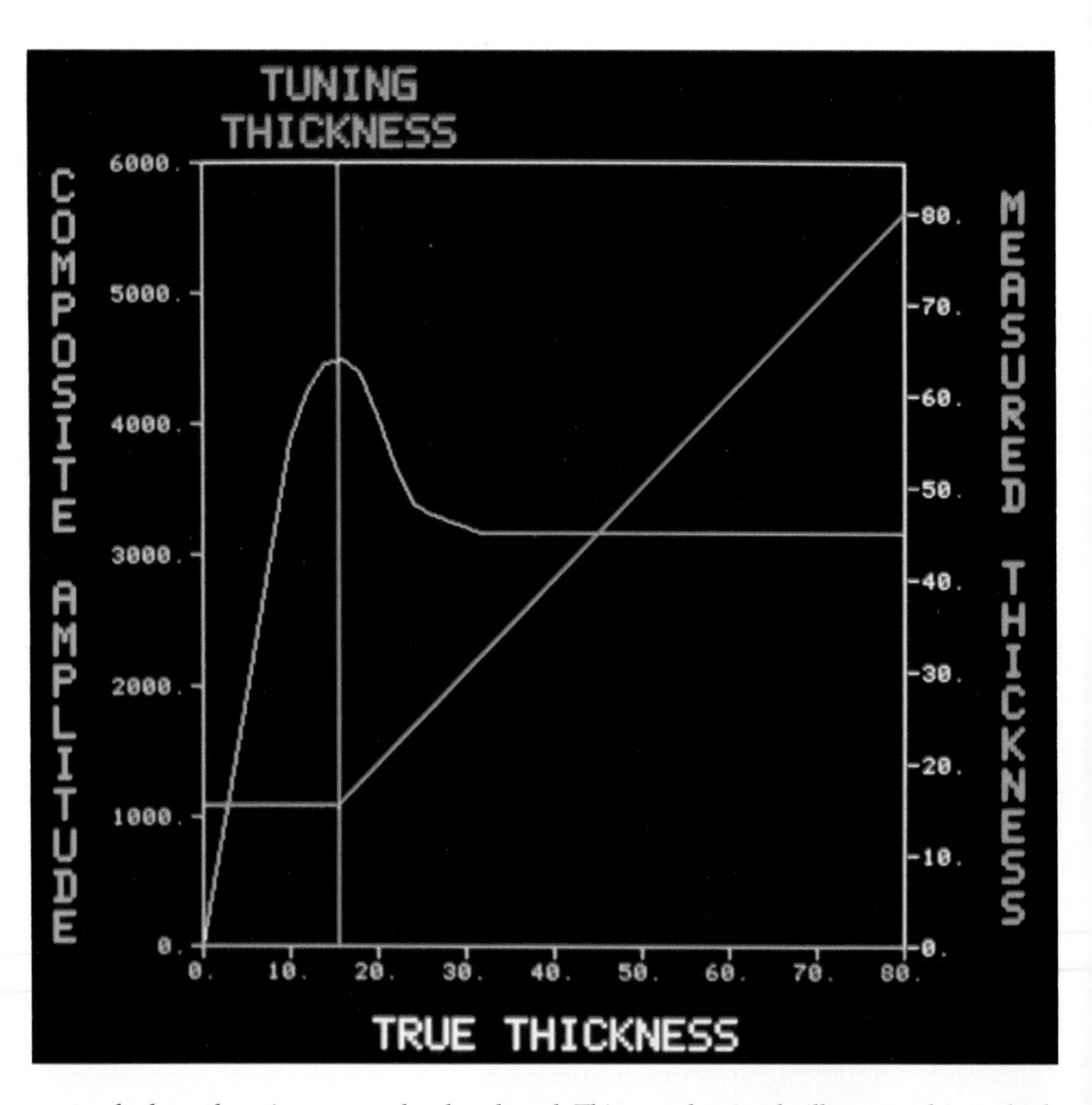

Fig. 6-6. Basic concepts of tuning for thin beds.

er wavelet but otherwise was randomly selected. This wavelet simply illustrates that multiple side lobes in the wavelet generate multiple maxima in the deterministic tuning curve. In fact it is interesting to note the similarity in shape between the tuning curve and half of the wavelet upside-down.

The third wavelet is again zero phase. Its simpler shape generated only two maxima in the tuning curve.

The fourth wavelet in Figure 6-9 was extracted from zero-phase data by a cross-correlation technique between the processed seismic trace and the synthetic seismogram at a well. The wavelet is seen to be almost, but not quite, zero phase. The deterministic tuning curve derived from the extracted wavelet shows some complexity but principally two maxima.

An amplitude spectrum was generated from this tuning curve. By interpreting this spectrum in terms of four corner frequencies it was possible to compute an ideal zero-phase equivalent wavelet and its tuning curve. For the extracted wavelet at the bottom of Figure 6-9 the ideal zero-phase equivalent wavelet is shown directly above as the third wavelet on the page.

In a practical situation the interpreter may be striving for a tuning curve applicable to the zone of interest over some broad area of a prospect. Inevitably, the interpreter will wonder whether a deviation from zero-phaseness such as that shown by the extracted wavelet at the bottom of Figure 6-9 is applicable to the whole area. He may reasonably consider the ideal zero-phase equivalent wavelet and its tuning curve to be more universal.

Statistical Tuning Curves

Tracked horizon data in time, amplitude and other attributes are normally mapped before drawing conclusions from it. It is also possible to crossplot one attribute against another from

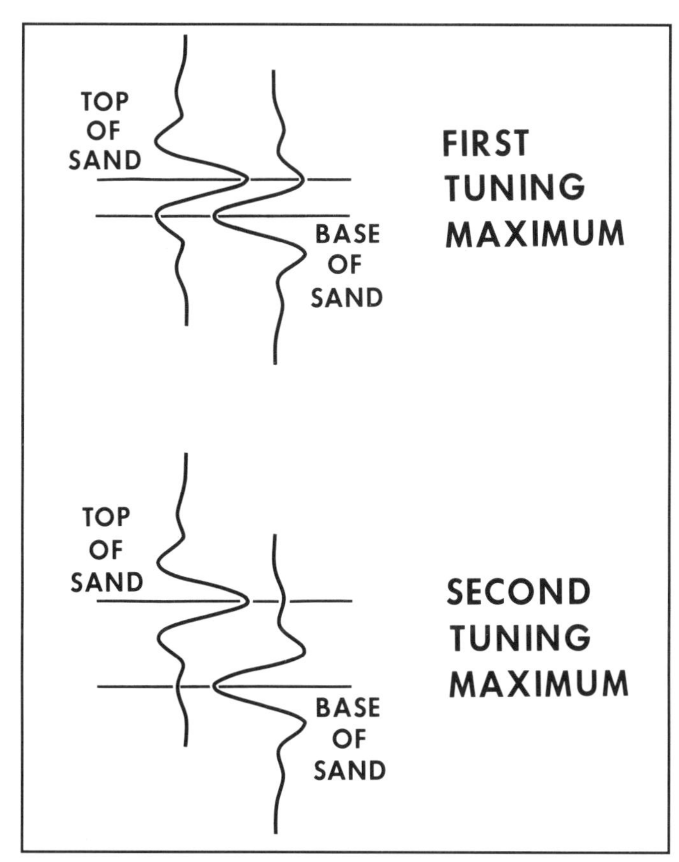

Fig. 6-7. Constructive interference of zero-phase wavelets to produce tuning maxima.

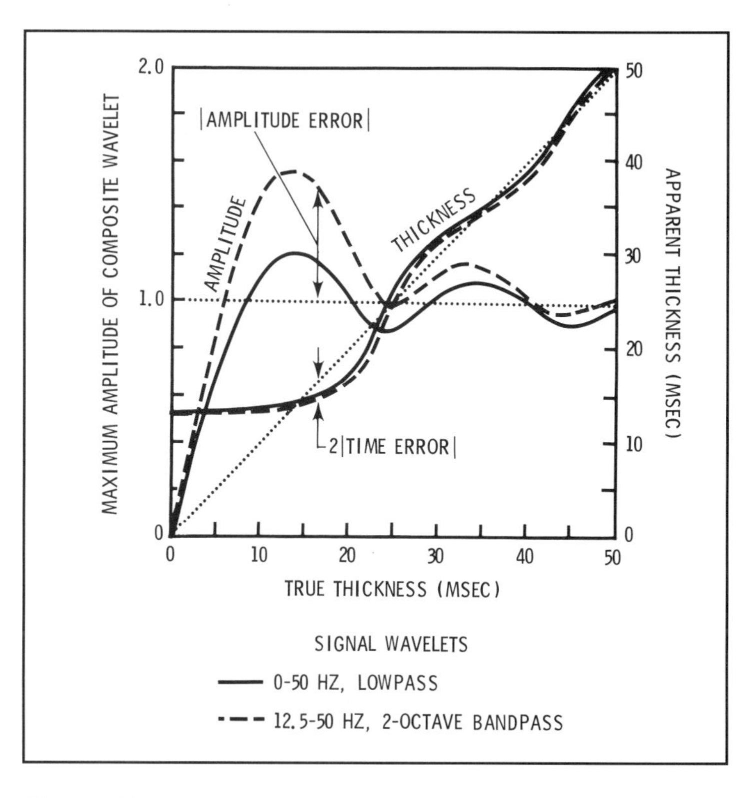

Fig. 6-8. Tuning curves for two zero-phase wavelets showing multiple amplitude maxima (after Kallweit and Wood, 1982).

the same subsurface position. Crossplotting operates within a user-specified subsurface area. With this capability statistical analysis of horizon data can be an important part of interactive interpretation.

In studying the detailed character of bright spots and the tuning phenomena therein, it may be desirable to make the simplifying assumption that lateral variations in amplitude are due to lithologic changes in the reservoir or to tuning effects, and *not* due to changes in the acoustic properties of the embedding media. Figure 6-10 shows a crossplot of the top sand amplitudes against the base sand amplitudes for a particular reservoir. The general proportionality between the two as indicated by the extension of the points along the diagonal yellow line indicates that, to a first approximation, the lateral changes in amplitude do result from lateral changes within the reservoir rather than in the encasing material.

In pursuing the quantitative study of reservoirs (Chapter 7), the absolute value summation of top and base reservoir amplitudes accentuates the properties of the reservoir, lithologic or geometric, relative to those of the encasing material. This absolute value summation is referred to as composite amplitude. Figure 6-11 shows a crossplot of composite amplitude against gross isochron (that is, measured thickness). These are the parameters for studying tuning (Figure 6-6). This crossplot incorporates many thousands of points, so it would be a daunting task to plot manually.

The principal maximum in composite amplitude (Figure 6-11) occurs at 16 ms, the tuning thickness. In addition there is a second maximum evident at about 35 ms. The meaning of these two maxima in terms of wavelet interaction was explained schematically in Figure 6-7. The first interpretation of a statistical tuning curve from this crossplot is then the envelope of the plotted points (Figure 6-12). This is based on the assumption that the highest amplitude points all indicate the maximum acoustic response of the interval under study and therefore that the variable shape of this envelope with isochron indicates geometric effects alone.

The horizontal blue line to the right of Figure 6-12 is the baseline and indicates the maximum untuned amplitude. The ratio of the amplitude of the tuning maximum to this baseline value is

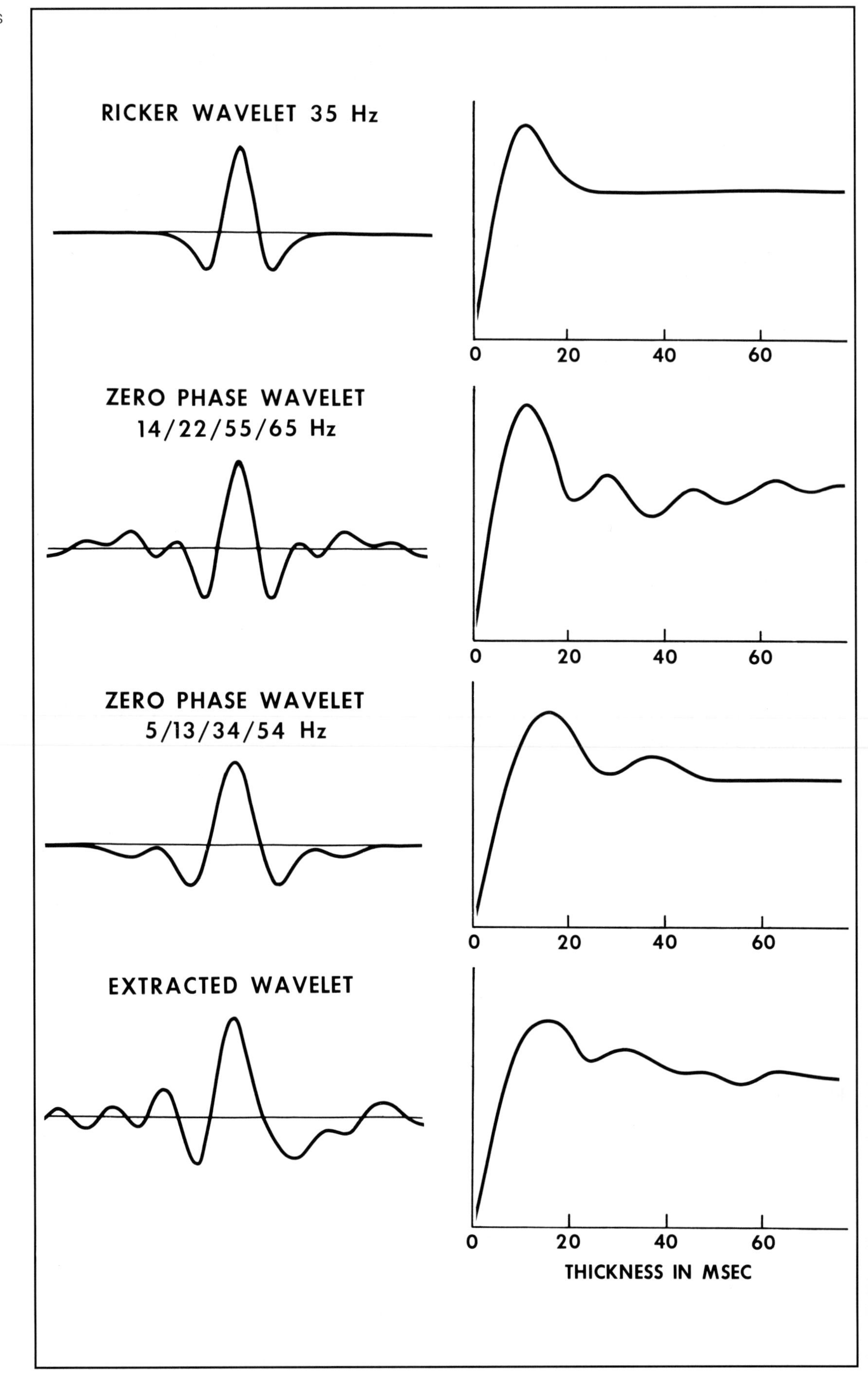

Fig. 6-9. Various wavelets and their corresponding deterministic tuning curves.

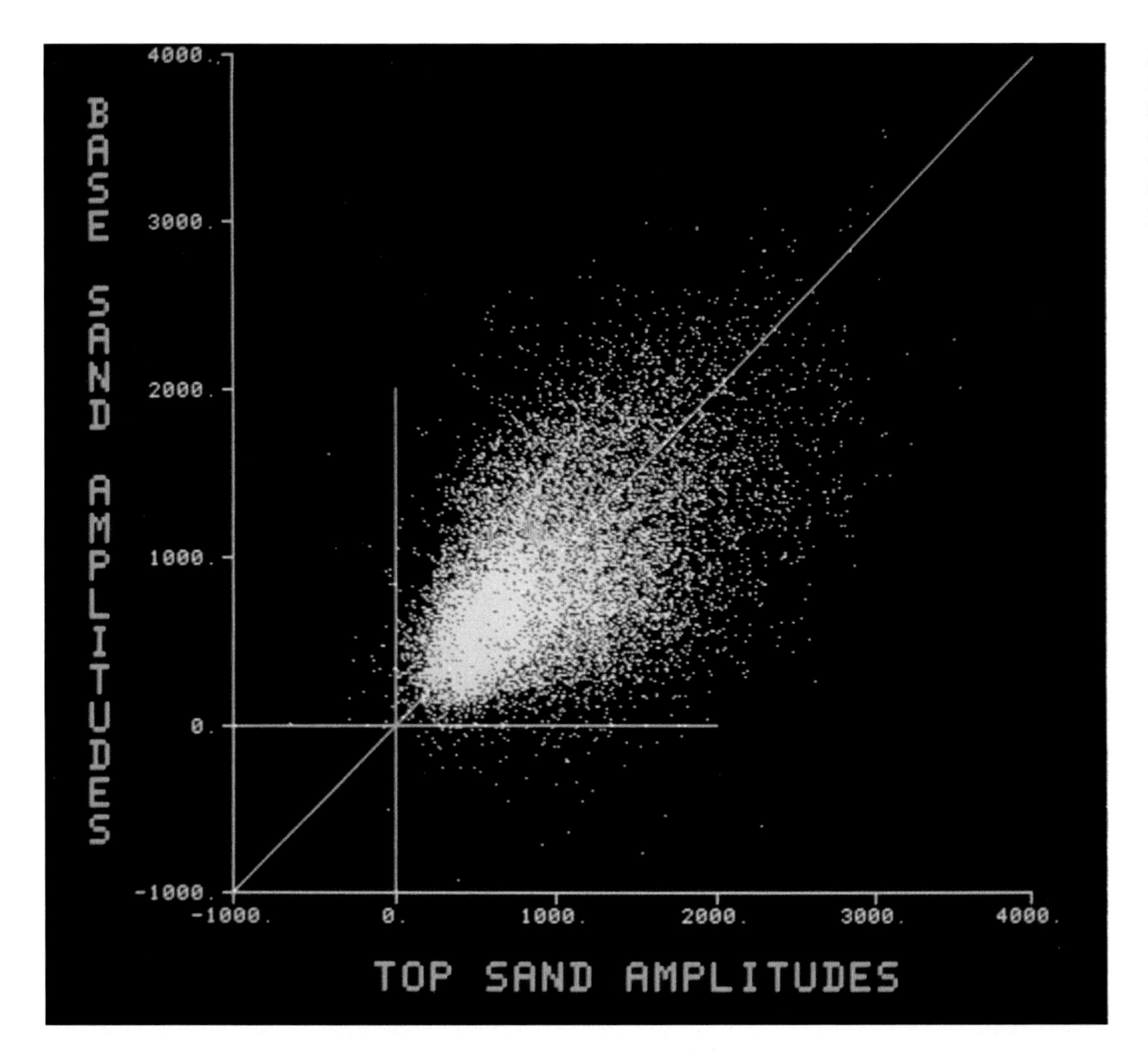

Fig. 6-10. Interactive crossplot of base sand amplitudes against top sand amplitudes demonstrating approximate proportionality.

controlled by the side lobe levels of the interfering wavelets. On this basis the tuning maxima as drawn in Figure 6-12 are too high. Considering the very large number of points plotted for isochrons in the 10-40 ms range, it is reasonable that some amplitudes are spuriously high because of constructive interference of the already-tuned reflections with nonreservoir interfaces, multiples or noise. In Figure 6-13, 99th percentile points computed over 2 ms isochron gates are plotted as blue asterisks. They fall at more reasonable levels relative to the untuned baseline.

Hence the existence of two maxima in the tuning curve was indicated by the raw crossplot, but a statistical analysis of the points guided by the knowledge of the deterministic tuning curve was required to establish the shape of the final curve. This final interpreted curve is shown in yellow in Figure 6-13.

Figure 6-14 shows the deterministic tuning curve points and the final interpreted curve superimposed on the same crossplot. Deterministic tuning curves have arbitrary vertical scales. Hence it was necessary to interpretively judge the factor by which the deterministic points must be scaled so that they could be plotted on the same composite amplitude axis as the crossplot points. This was done by matching the deterministic points to the crossplot envelope at the greater thicknesses where little or no tuning effect exists, and was confirmed by plotting the model response at a control well. The yellow curve in Figures 6-13 and 6-14 is the same. It is repeated to demonstrate how the final interpreted tuning curve for the area under study tied both the statistical and deterministic inputs.

Understanding the Magnitude of Tuning Effects

Interpreters sometimes try to explain as reservoir tuning effects amplitude phenomena which really cannot be explained that way. Understanding wavelet shape is the key to understanding what reasonably can or cannot be interpreted as a tuning effect.

Figure 6-15 shows four zero-phase wavelets of different bandwidths. The lowest frequency

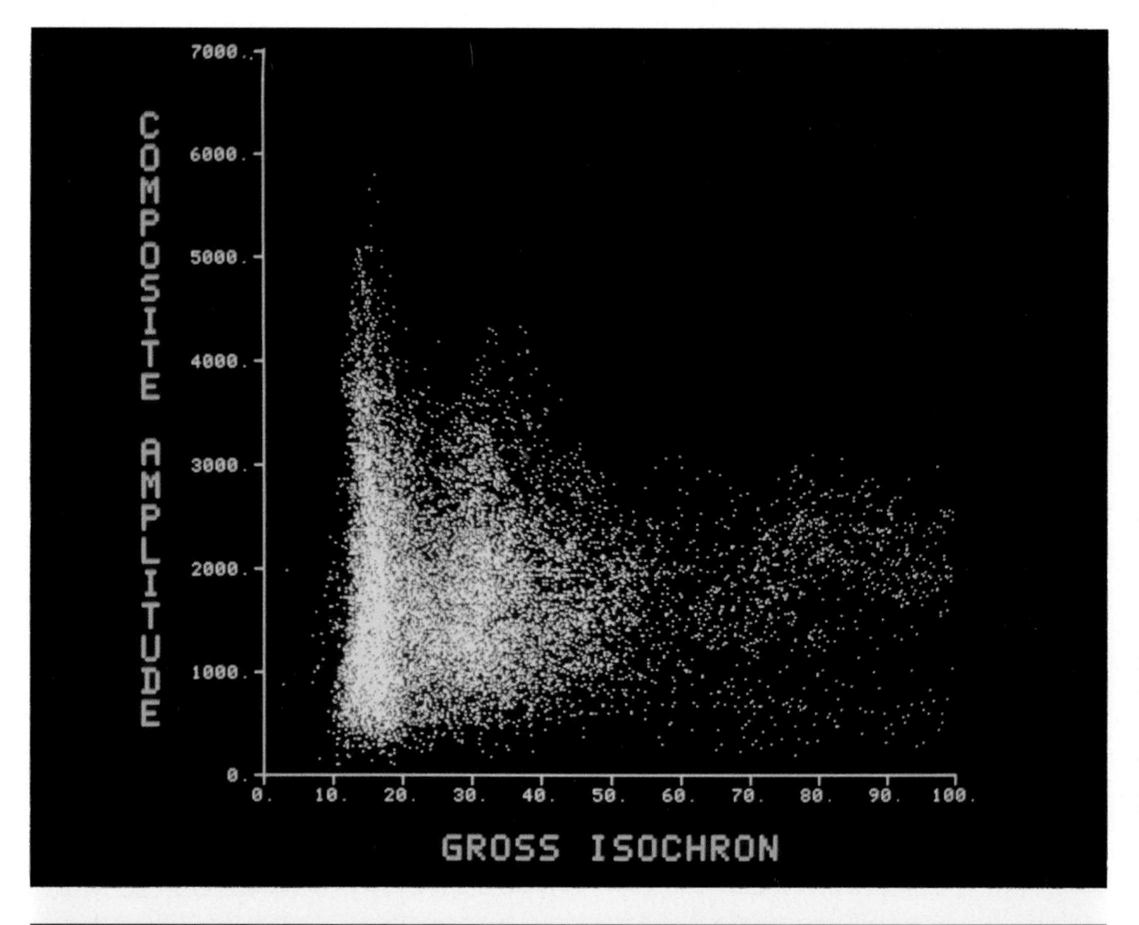

Fig. 6-11. Interactive crossplot of composite amplitude against gross thickness of a reservoir interval for all the interpreted data points in a prospect.

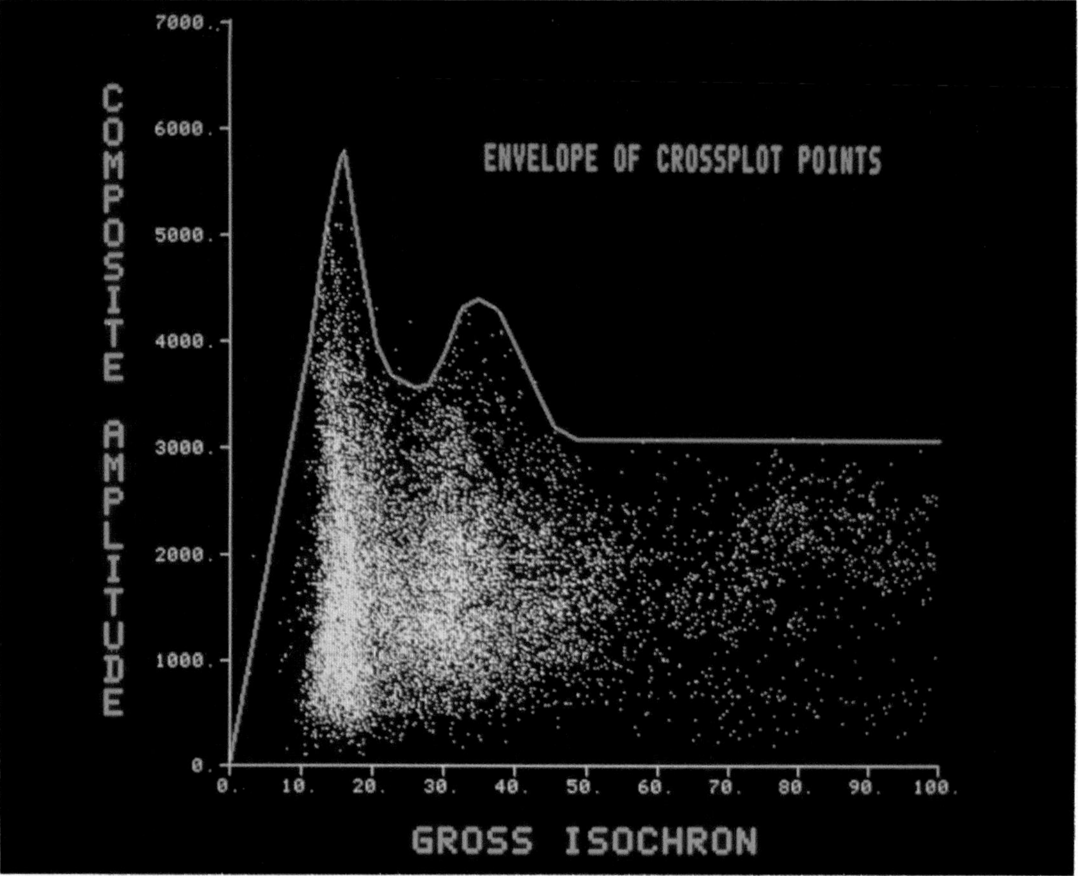

Fig. 6-12. Same crossplot as Figure 6-11 with upper envelope drawn as a first interpretation of a statistical tuning curve.

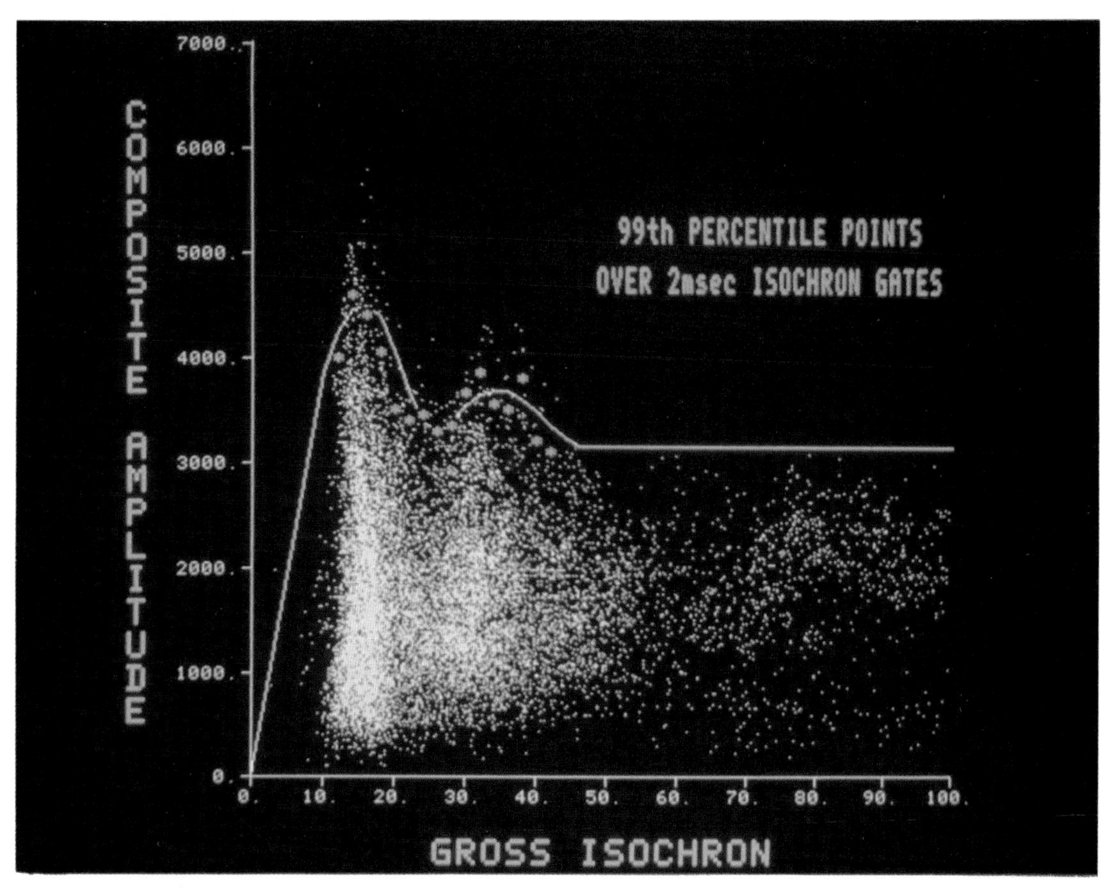

Fig. 6-13. Same crossplot as Figure 6-11 analyzed to yield 99th percentile points which show a more realistic peak-to-baseline ratio as required by deterministic studies.

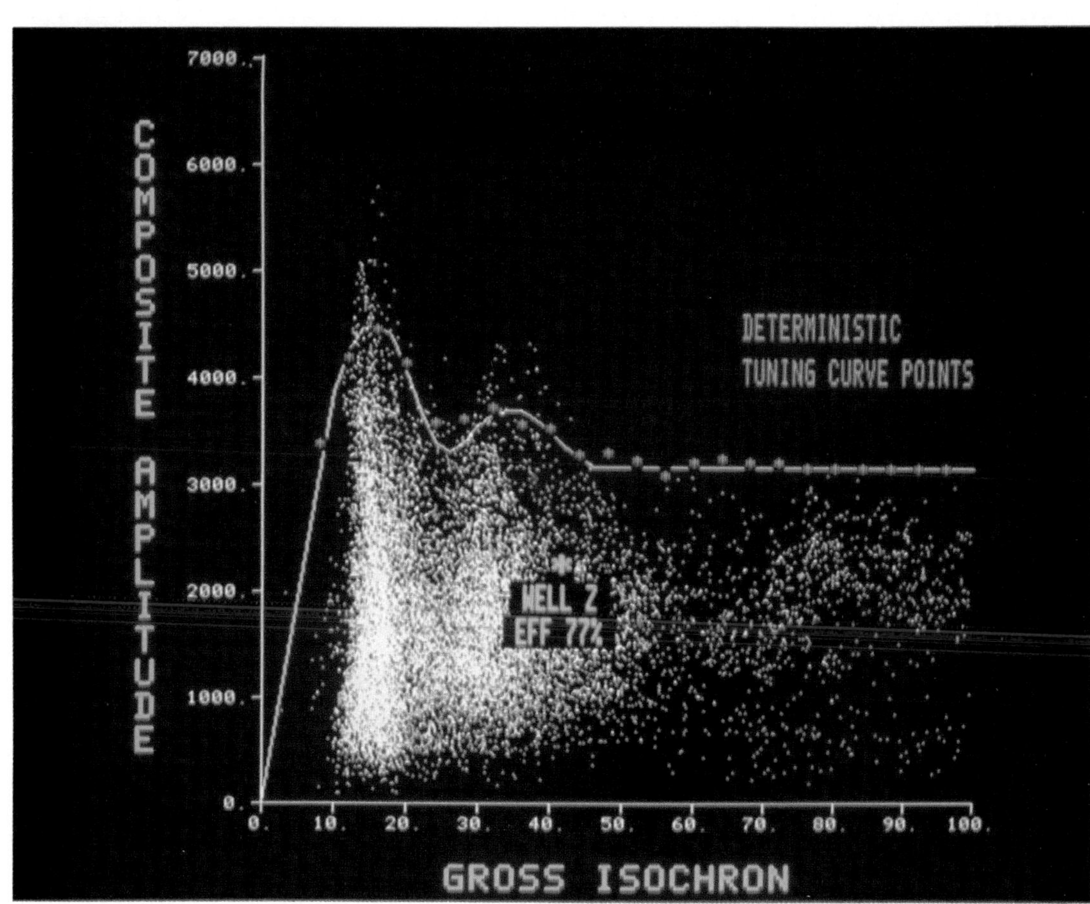

Fig. 6-14. Same crossplot as Figure 6-11 with deterministic tuning curve points computed from an extracted wavelet. The yellow final interpreted tuning curve is the same as in Figure 6-13 to provide comparison between statistical and deterministic tuning points.

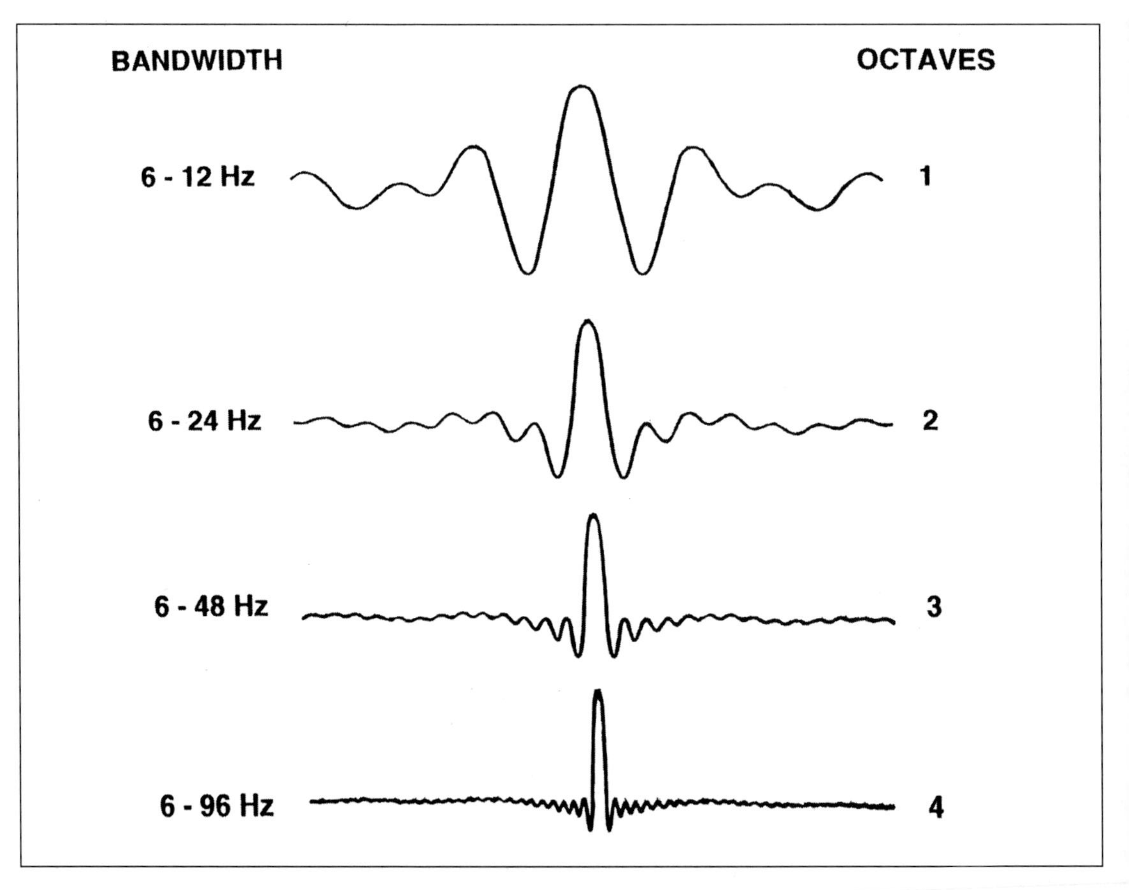

Fig. 6-15. Four zero-phase wavelets of different bandwidths demonstrating the effect of bandwidth on side lobe size and hence the magnitude of tuning effects.

in the band is held the same while the highest frequency is progressively increased by a full octave from one wavelet to the next. The center lobe of the wavelet narrows considerably but what we are concerned with here is the size of the first side lobe. Notice that the first side lobe is about the same size as the center lobe for the one octave wavelet; the first side lobe is about half the center lobe for two octaves; it is about a third of the center lobe for three octaves; and it is about a quarter for four octaves. Thus we can establish a generalized rule-of-thumb, that *the first side lobe as a fraction of the center lobe of a zero-phase wavelet is the reciprocal of the octave bandwidth.* Furthermore a typical bandwidth for good but not exceptional seismic data today is two-and-one-half octaves; this gives a side lobe level of 40% relative to the wavelet's center lobe.

I call this the "40% rule" and it is a very approximate but very useful rule-of-thumb. Figure 6-7 demonstrates that the size of the wavelet side lobe controls the magnitude of the tuning effects. Thus the 40% rule means that 40% is a common increase in amplitude at tuning for zero-phase wavelets reflected from the top and base of a reservoir layer. Increases in amplitude much greater than this must be explained some other way. A lateral amplitude increase of 40% is not dramatic even when using the highest dynamic range color scheme. It follows that dramatic amplitude changes which we observe in our data are unlikely to be simple tuning effects.

References

Kallweit, R. S., and L. C. Wood, 1982, The limits of resolution of zero-phase wavelets: Geophysics, v. 47, p. 1035-1046.

Meckel, L. D., Jr., and A. K. Nath, 1977, Geologic considerations for stratigraphic modeling and interpretation, *in* C. E. Payton, ed., Seismic stratigraphy—applications to hydrocarbon exploration: AAPG Memoir 26, p. 417-438.

Neidell, N. S., and E. Poggiagliolmi, 1977, Stratigraphic modeling and interpretation—geophysical principles and techniques, *in* C. E. Payton, ed., Seismic Stratigraphy—applications to hydrocarbon exploration: AAPG Memoir 26, p. 389-416.

Robertson, J. D., and H. H. Nogami, 1984, Complex seismic trace analysis of thin beds: Geophysics, v. 49, p. 344-352.

Widess, M. B., 1973, How thin is a thin bed?: Geophysics, v. 38, p. 1176-1180.

RESERVOIR EVALUATION

Reservoir Properties Deducible from Seismic Data

Reservoir evaluation, reservoir characterization or reservoir property mapping is an important use of 3-D seismic data. As presented here it is an extension of reservoir reflection identification (Chapter 5), horizon slicing techniques (Chapter 4) and tuning phenomena (Chapter 6). Further issues on reservoir evaluation follow in Chapter 8.

Figure 7-1 gives a summary list of reservoir properties needed for effective development and production of hydrocarbon reserves. The basic information in seismic data is only time, amplitude and perhaps frequency, so there are many fewer measurements than there are unknowns. Seismic attributes (Chapter 8) may be helpful but they are not new additional information. Seismic traveltime addresses structure and structural boundaries. Amplitude and frequency are left to address the other properties. Amplitude is well understood and is used widely; frequency is less well understood and tends to be noisy.

The principal reservoir properties that affect seismic amplitude can be divided into two groups:

GROUP A	GROUP B
nature of fluid	porosity
gross lithology	net pay thickness or net-to-gross ratio
pressure	lithologic detail
temperature	hydrocarbon saturation

The properties in Group A are those which, to a first approximation, affect the reservoir as a whole. The difference between gas and oil was discussed in Chapter 5. The gross lithology of a reservoir rock generally does not change much within one

RESERVOIR INFORMATION NEEDED	WILL 3-D SEISMIC PROVIDE ?
Structure	Generally yes
Reservoir boundaries	Generally yes
Fluid content & contacts	Yes for younger rocks
Porosity	Generally yes
Hydrocarbon saturation	Difficult but possible for younger rocks
Net pay or net sand	Generally yes
Permeability	No
Connectivity / flow units	Sometimes
Aquifer / drive mechanisms	Partially yes
Heterogeneities	all with resolution limitations

Fig. 7-1. Reservoir information needed for effective exploitation and the generalized ability of 3-D seismic data to help in providing it.

Fig. 7-2. Explanation of bright spot and dim spot regimes. In bright spot regime high amplitude is desirable; in dim spot regime low amplitude is desirable.

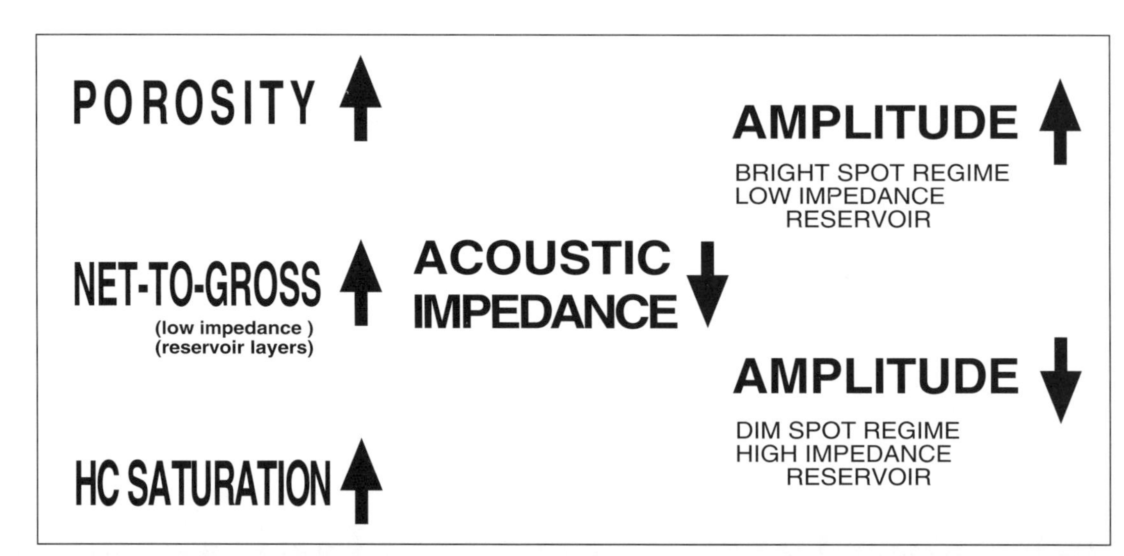

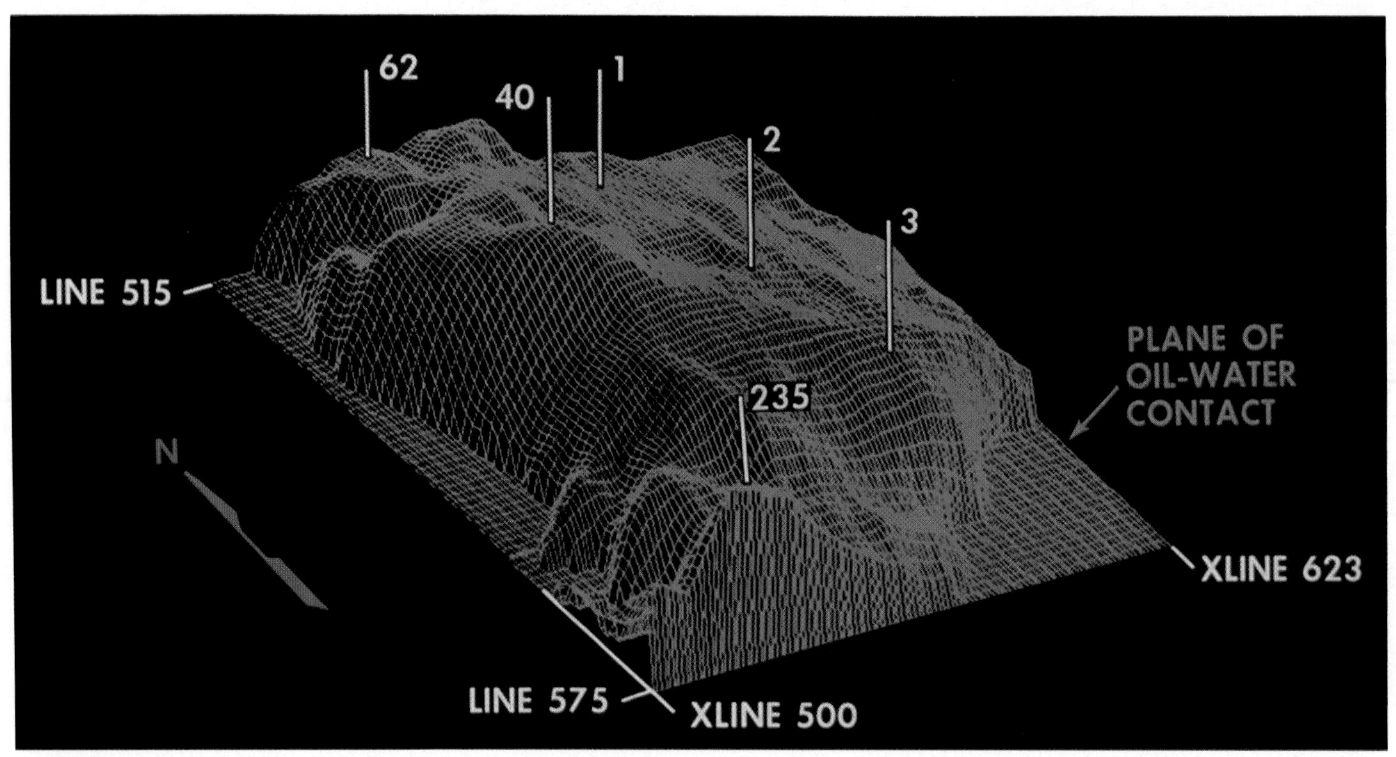

Fig. 7-3. Structural shape of the Macae calcarenite reservoir, Pampo oil field, offshore Brazil. (Courtesy Petroleo Brasileiro.)

reservoir; other related properties such as age, compaction and depth also will remain fairly constant. Anomalous reservoir pressure can affect seismic amplitude considerably but again this will generally affect the whole reservoir rather than only a part of it.

The properties in Group B are the ones which can vary laterally over short distances and therefore significantly affect the reserve estimates of a reservoir penetrated by a small number of wells. A major objective of development and production geophysics is to map these spatially-varying reservoir properties so that wells and platforms can be located optimally and reserve estimates can be made with greater precision.

Lateral changes in amplitude of reservoir reflections can be caused by changes in any one or more of these Group B properties, so there is an inherent ambiguity. Today's interpretive approach to reservoir evaluation thus requires that simplifying assumptions be made. Conveniently, the amplitude of a seismic bright spot is higher where hydrocarbon saturation is higher (although non-linearly; Domenico, 1974), where porosity is higher, and where net pay thickness is greater (with some complications due to tuning). It normally follows, therefore, that the brighter the bright

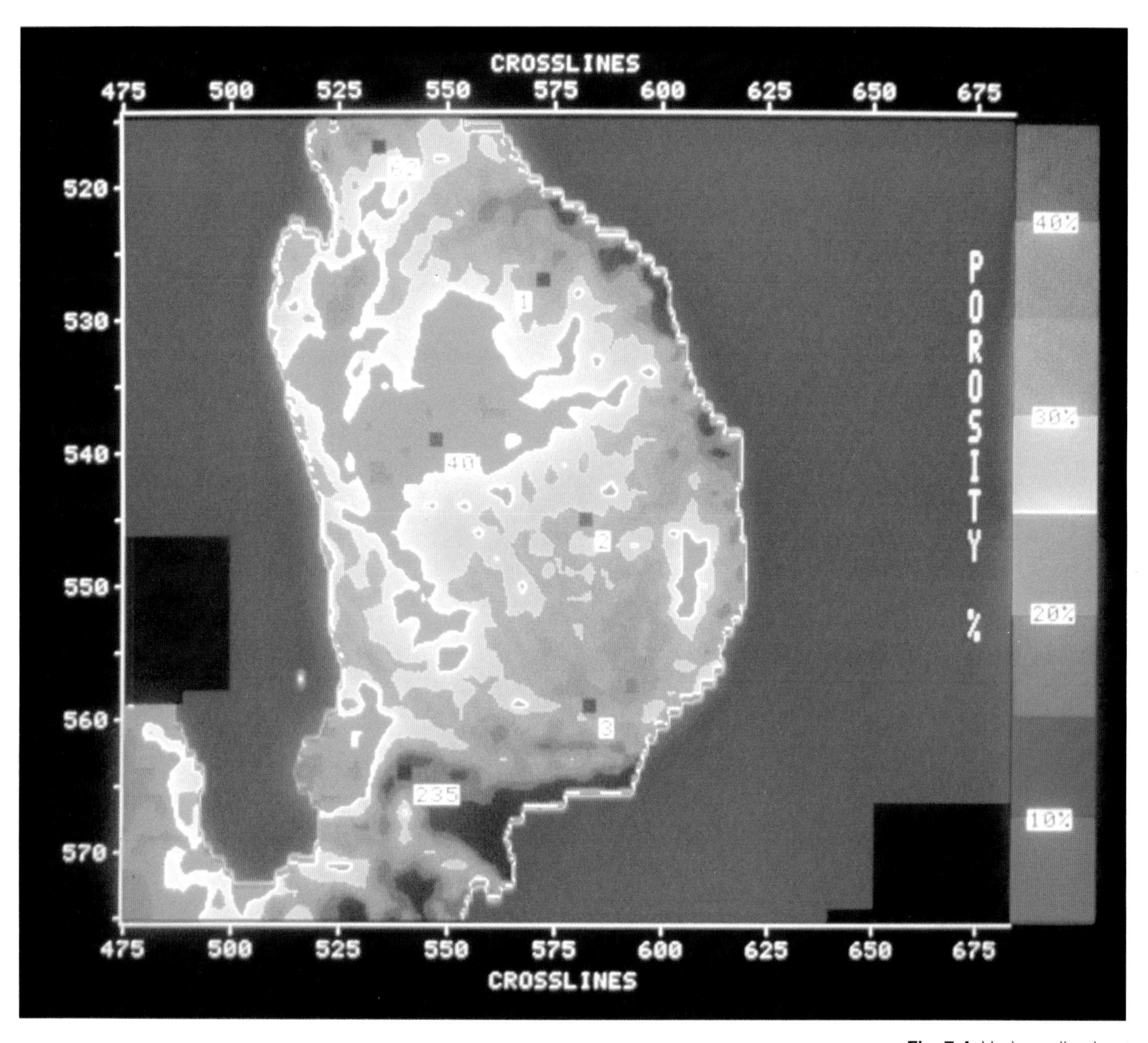

Fig. 7-4. Horizon slice in porosity through the Macae calcarenite reservoir. (Courtesy Petroleo Brasileiro.)

spot, the better the prospect. This relationship is generalized in Figure 7-2. Increasing porosity, net-to-gross and saturation all make the reservoir acoustic impedance lower. In the bright spot regime the amplitude of the reservoir reflections is then higher. This includes hydrocarbon bright spots but is not limited to them—any already low impedance reservoir will get lower and thus generate more amplitude. In the dim spot regime the reverse occurs. Here the reservoir is higher acoustic impedance than the embedding medium, so reducing impedance reduces contrast and thus amplitude. Clearly the most important seismic property of a reservoir is whether it is bright spot or dim spot regime. In the former high amplitude is good and the distribution of amplitude on the horizon slice guides development drilling (e.g. Figures 2-30 to 2-32). The reservoirs discussed in this chapter are in the bright spot regime. Case history 12 (Chapter 9) discusses a dim spot regime reservoir.

Porosity Using Inversion

Figure 7-3 shows the structural configuration of the Macae calcarenite reservoir in the Pampo oil field offshore Brazil (Curtis, Martinez, Possato, Saito, 1983). Amplitude

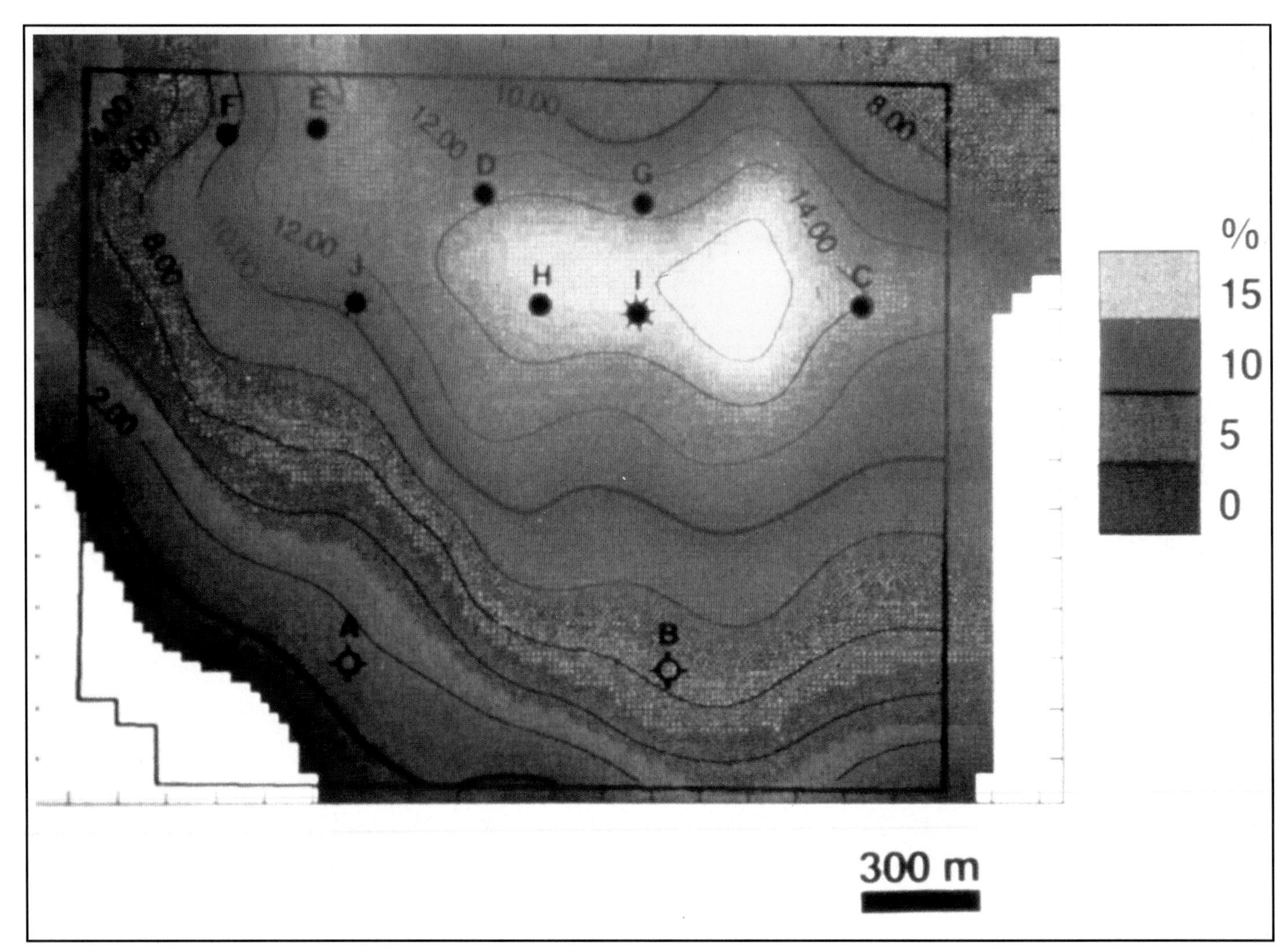

Fig. 7-5. Porosity map of Glauconite Formation, Alberta, Canada, derived from 3-D seismic data using two-dimensional, interpretively constrained inversion, Wyllie's equation in velocity and density, and correction for shale content. (Courtesy Western Atlas International.)

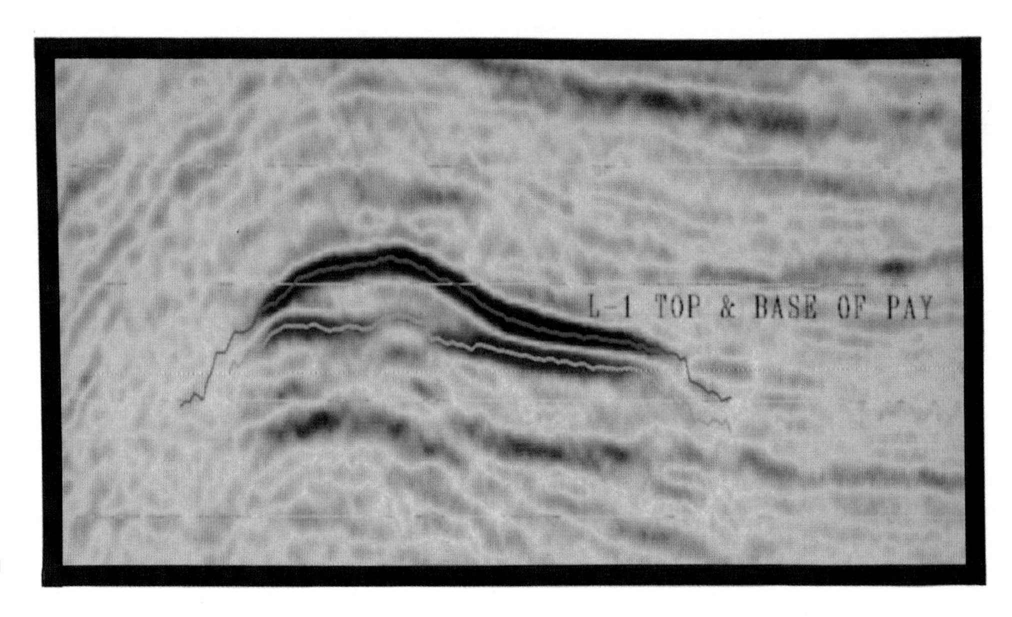

Fig. 7-6. Vertical section through Gulf of Mexico gas reservoir showing high amplitude reflections at the top and base of the gas. The automatic horizon track on the top reservoir reflection provided the amplitudes displayed in Figure 7-7. (Courtesy CNG Producing Company.)

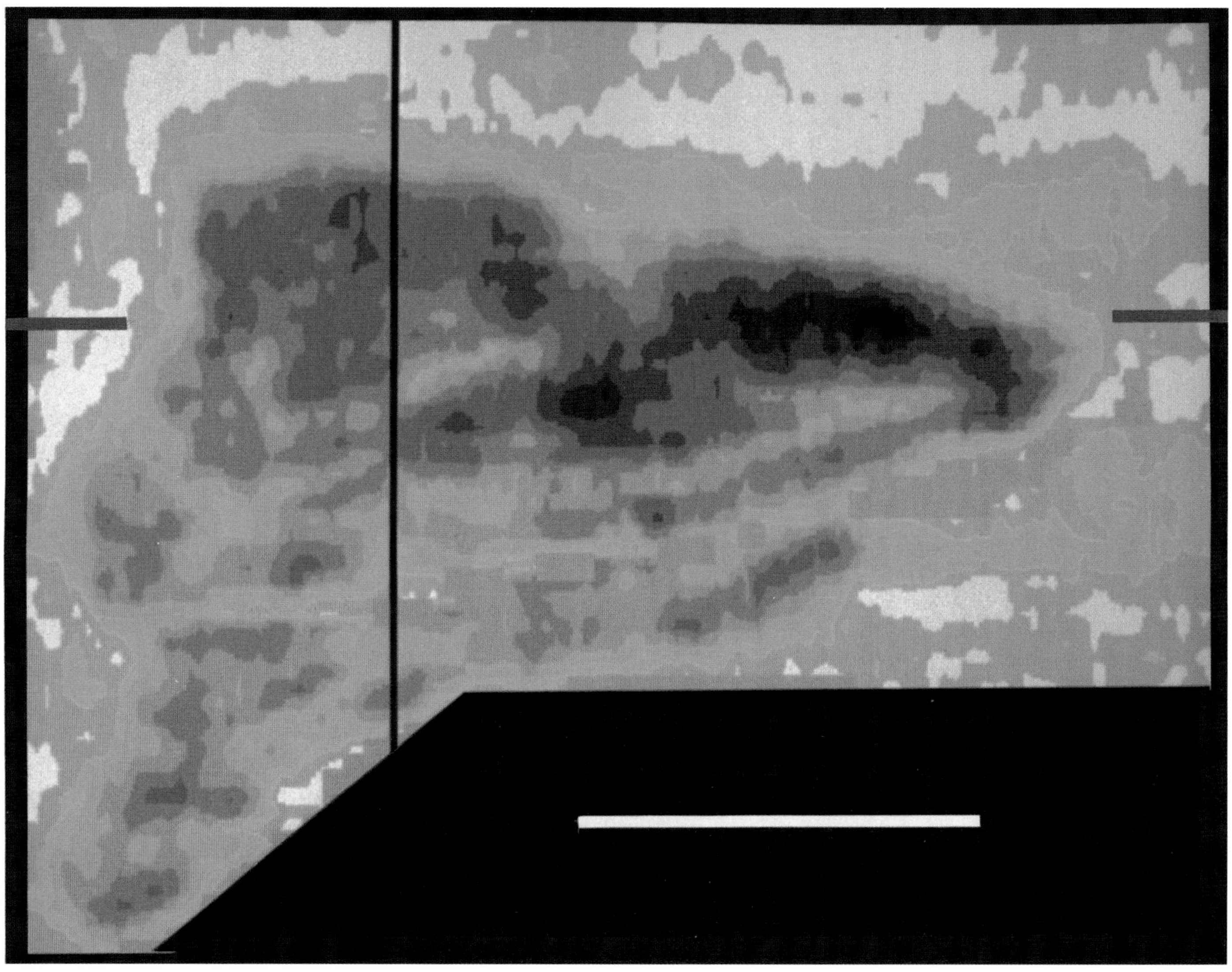

Fig. 7-7. Horizon slice from Gulf of Mexico Pleistocene gas reservoir showing patterns in amplitude interpretable as the effects of hydrocarbon, faulting, porosity, and tuning. Yellow bar is one km. Red lines indicate the position of vertical section in Figure 7-6. Black line is block boundary. (Courtesy CNG Producing Company.)

variations of the calcarenite event were considered to result primarily from porosity changes within the reservoir. The 3-D data volume was processed through recursive seismic inversion. The low-frequency interval velocity field originated from a 3-D inverse normal-incidence ray tracing procedure. The resultant velocities in the reservoir were then converted to apparent porosity using Wyllie's equation (Wyllie, Gregory, and Gardner, 1958). Figure 7-4 shows a horizon slice through the Macae calcarenite displaying apparent porosity variations within the reservoir. A decrease in porosity toward the reservoir core is evident and is confirmed by well data.

Some kind of seismic inversion is often considered useful when attempting to be quantitative about reservoir properties. The conversion from interface information (normal seismic amplitudes) to interval information (inverted seismic amplitudes) brings the seismic data into a more geologic form, one more readily correlated to well logs and to reservoir properties.

In the project leading to the porosity map of Figure 7-5, 3-D seismic data were inverted by a two-dimensional inversion process where the bandwidth was effectively broadened by interpretive constraint. The resultant data were in acoustic impedance and well data were available in velocity and density. Wyllie's equations in velocity and density were thus combined together. Both of these versions consider the addition of components of a mixture of sand grains, shale and hydrocarbon in porosity; the densities add up directly and the velocities add up as their inverse, namely transit time. The shale volume corrections were obtained from the well control. Fol-

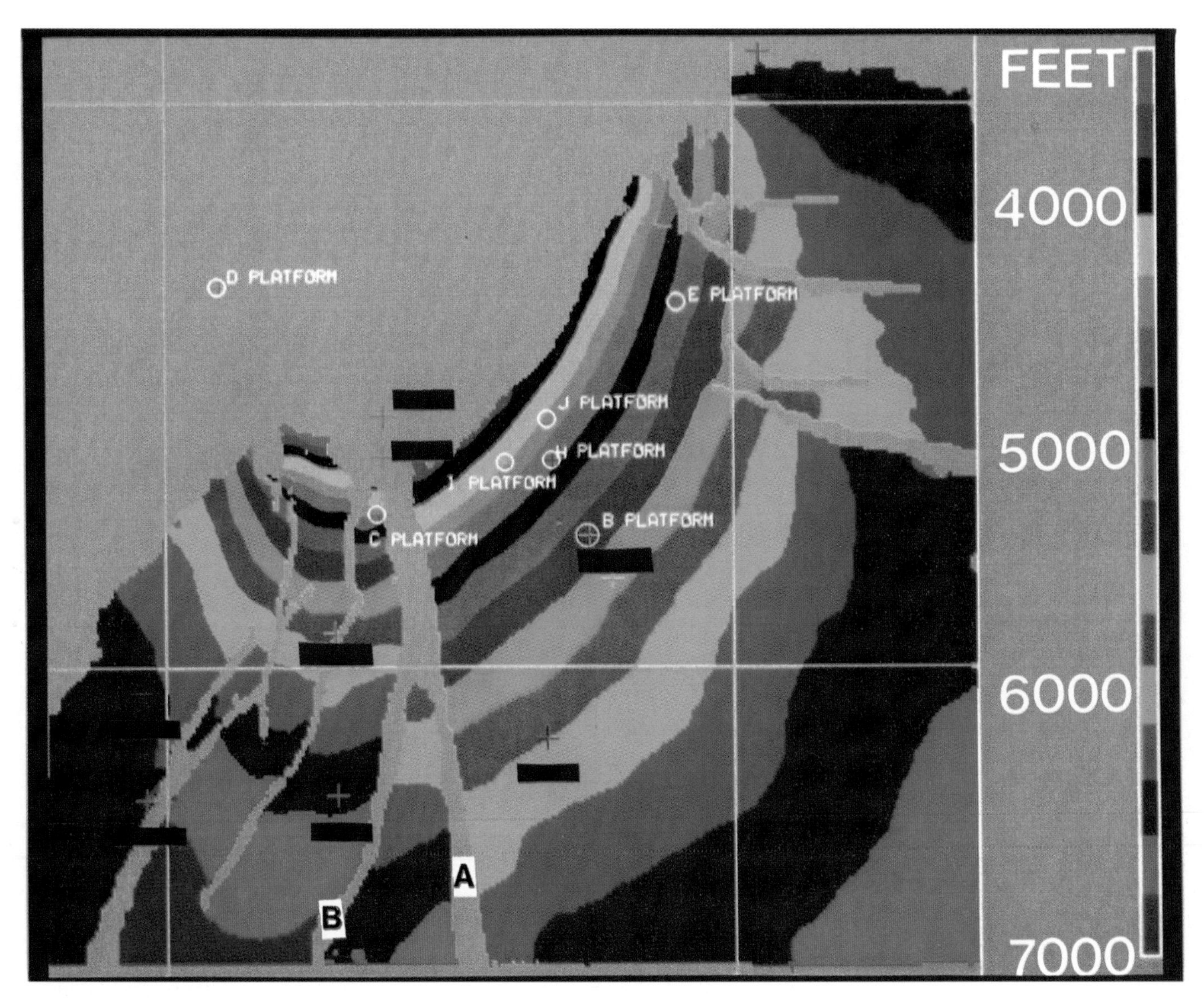

Fig. 7-8. Structure map of productive Gulf of Mexico sand showing setting in salt dome prospect. (Courtesy Chevron U.S.A. Inc.)

lowing this procedure, the reservoir porosity map of Figure 7-5 was obtained. It was used to predict successfully the porosity at two later wells.

Further porosity mapping projects are described in Case Histories 6 and 12 (Chapter 9).

Horizon Slices Over Reservoir Interfaces

A horizon slice showing the spatial distribution of seismic amplitude over a reservoir is an enormously valuable aid for studying that reservoir. The value comes not only from spatial continuity but also from the faithful amplitude reconstruction resulting from 3-D migration. For a low acoustic impedance reservoir, high amplitude is good because it may be caused by higher net-to-gross ratio, higher porosity, or higher hydrocarbon saturation.

Figure 7-6 shows high-amplitude reflections from a Pleistocene gas reservoir. The red reflection is from the reservoir top and the blue reflection is from the fluid contact. The horizon slice in Figure 7-7 follows the maximum amplitude of the top reflection. The overall high-amplitude triangular shape indicates the extent of the gas. Along the northern dip-controlled boundary, the amplitude gradient can be extrapolated to find the actual downdip reservoir limit. Some internal roughly northeasterly lineations indicate faulting. A high-amplitude trend running N60°E and crossing from the high amplitude gas area into the low amplitude surrounding area in the upper right corner is interpreted as a depositional trend of higher poros-

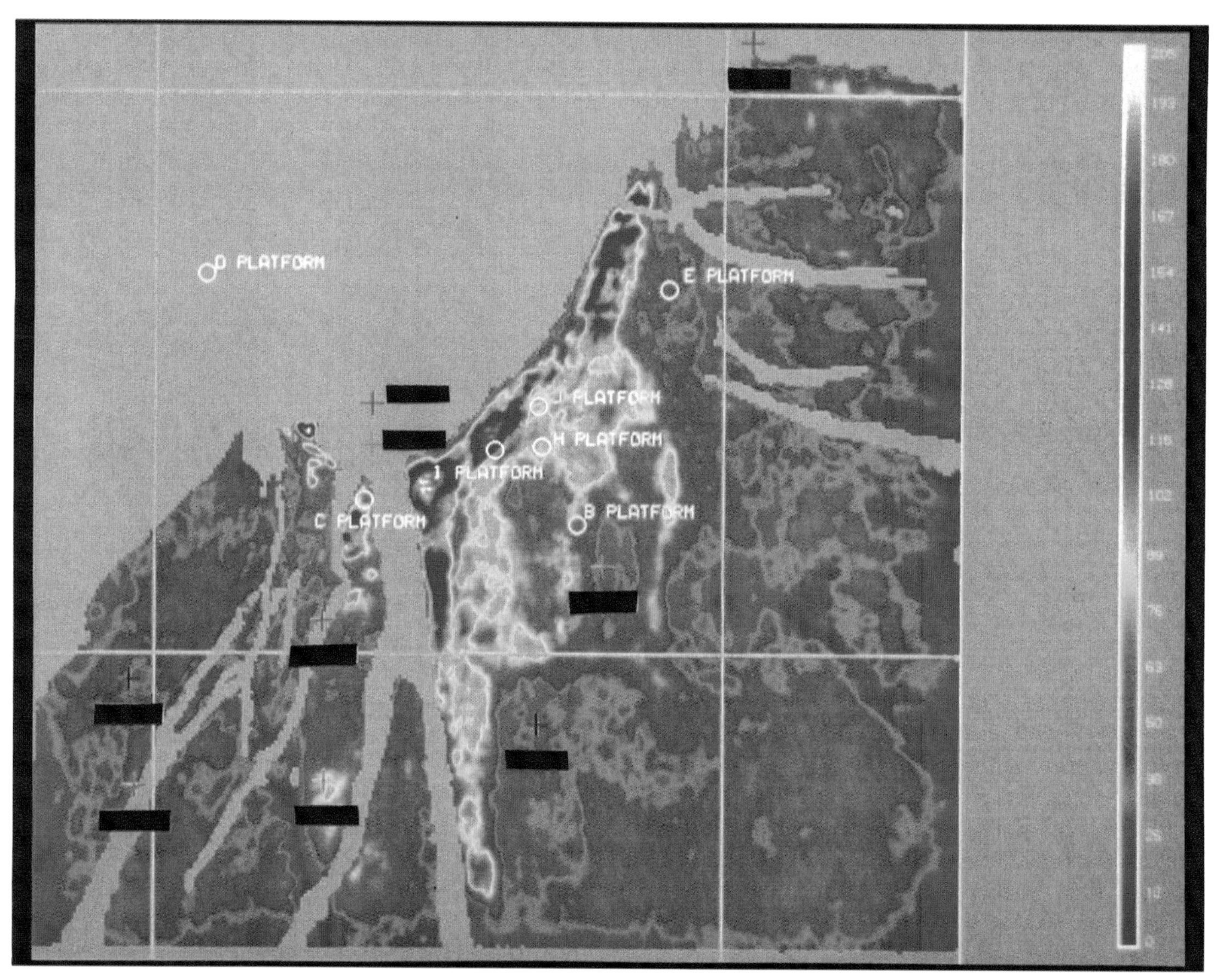

Fig. 7-9. Horizon slice corresponding to structure map of Figure 7-8. The red colors close to the salt indicate gas. The yellows indicate a porosity overprint. The high amplitudes along the upthrown side of fault A are interpreted as hydrocarbons in migration after flowing up the fault. (Courtesy Chevron U.S.A. Inc.)

ity. The high-amplitude trend running N80°W parallel to and just south of the northern reservoir boundary is caused by tuning. Separation of different effects in this way on the basis of interlocking patterns should be a normal part of horizon slice interpretation.

Figures 7-8 and 7-9 show a structure map and horizon slice from another Gulf of Mexico field. Again, different effects can be interpreted based on their different amplitude patterns. Salt is to the northwest and gas is trapped against the salt, as indicated in red. Variable porosity in the sand is probably responsible for the irregular yellow patterns downdip. The high amplitudes, principally in red, following the upthrown side of fault A and spanning a structural range of 2,000 feet, can only be interpreted as gas actively migrating toward the trap. This same observation and interpretation has been made on serveral other horizon slices in Tertiary clastic basins.

Net Pay Thickness

In an area of Pleistocene sediments offshore Louisiana the wells indicated that each reservoir sand interval was composed of several thin productive lobes and that the position of these lobes within the sands and their thickness varied laterally over a short distance (Brown, Wright, Burkart, Abriel, 1984). The top and base of the gross sand intervals generate the seismic reflections and the nonproductive segments within them are caused by the sands becoming tight and shaly. The aggregate thickness of the productive lobes is what matters economically. Therefore, the overall objective is

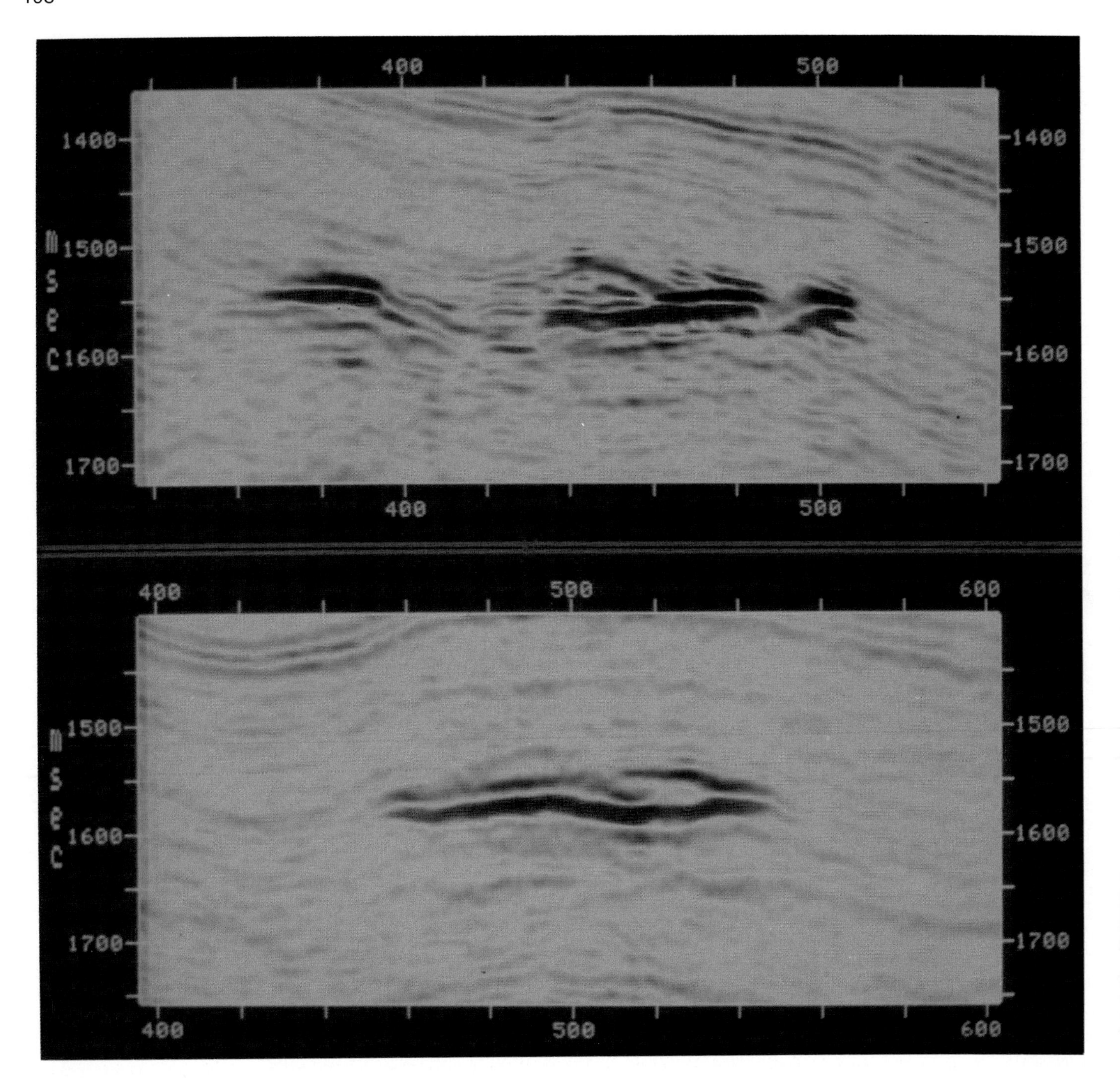

Fig. 7-10. Bright zero-phase reflections from the top and the base of one reservoir sand. (Courtesy Chevron U.S.A. Inc.)

to use amplitude measurements, coupled with time thickness measurements, to determine the spatial distribution of net producible gas sand from the seismic data.

The use of seismic amplitude to measure the proportion of sand within a sand/shale interval was demonstrated by Meckel and Nath (1977) for beds less than tuning thickness. Here the principle has been extended to thicker beds on the assumption that the individual lobes of producible gas sand are each below tuning thickness, that producible gas sand is a material of uniform acoustic properties, and that the internal layering is reasonably uniform.

Figure 7-10 shows bright reflections from one reservoir sand. The single peak-over-trough signature indicates zero-phaseness (see Chapter 2). Figure 7-11 shows two examples of two reservoir sands. Automatic tracking on an interactive interpretation system was used to track the bright reflection at the top and at the base of each reservoir interval. The tracker followed the maximum amplitude in the waveform while

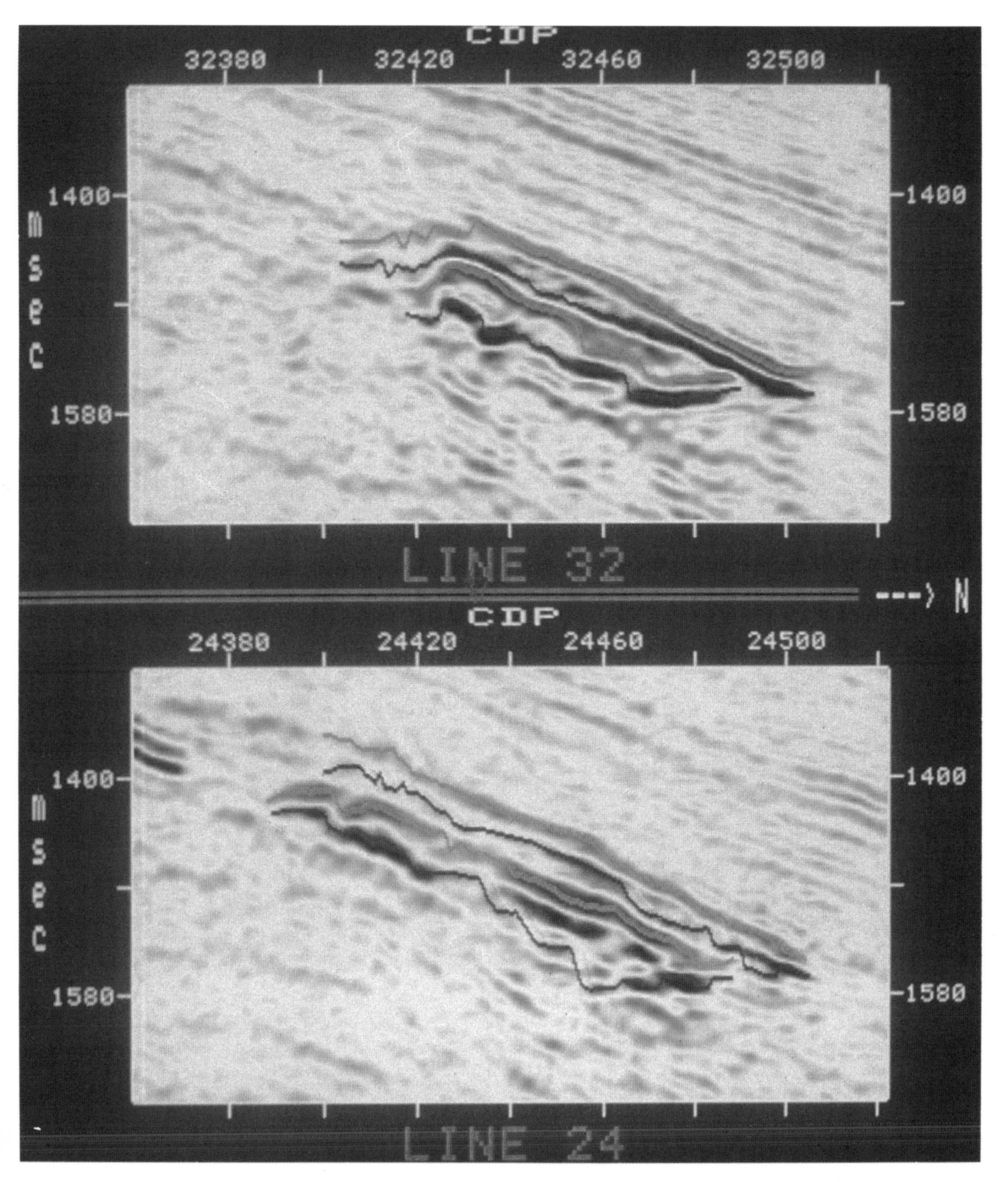

Fig. 7-11. Bright zero-phase reflections from the top and the base of two reservoir sands showing automatic tracks. (Courtesy Chevron U.S.A. Inc.)

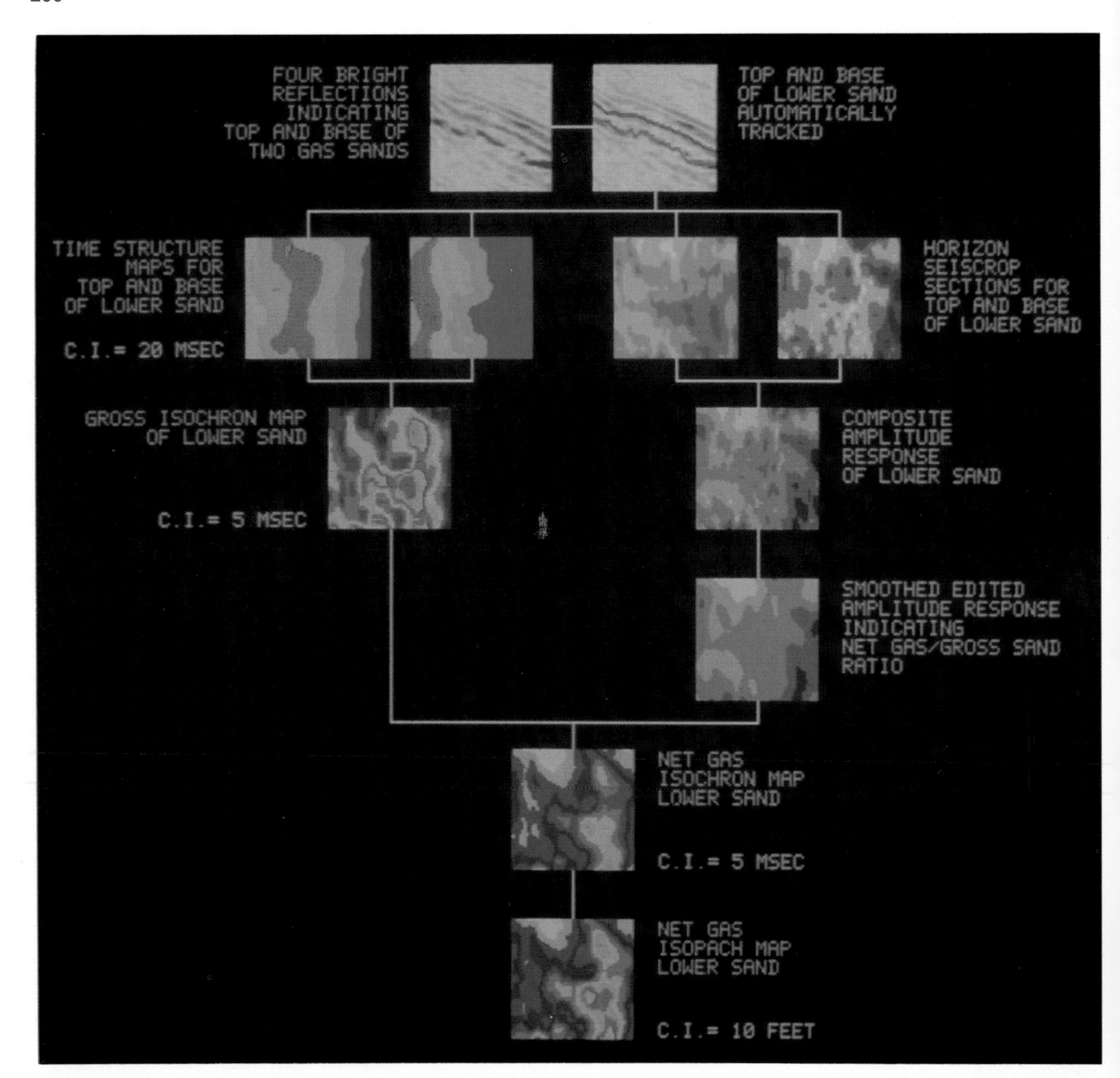

Fig. 7-12. Interpretive sequence used and intermediate products generated in the course of deriving net producible gas sand maps. C.I. = Contour Interval. Time structure maps show dip down to the right; the purple area is the flat spot at the base of the Lower Sand. The isochron and isopach maps have greens and blues indicating the thicker zones. All the four amplitude products have darker colors indicating the higher amplitudes. (Courtesy Chevron U.S.A Inc.)

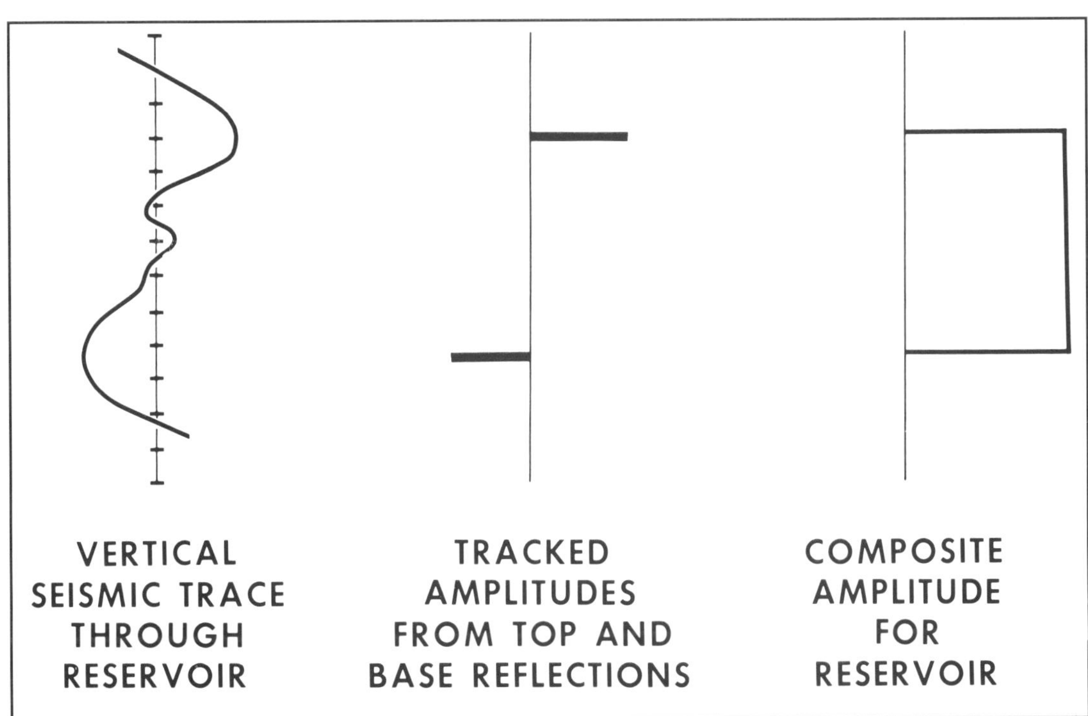

Fig. 7-13. The principle of tracking reservoir reflections and summing the absolute values of their maximum amplitudes to yield composite amplitude for the reservoir—a type of interpretively constrained seismic inversion.

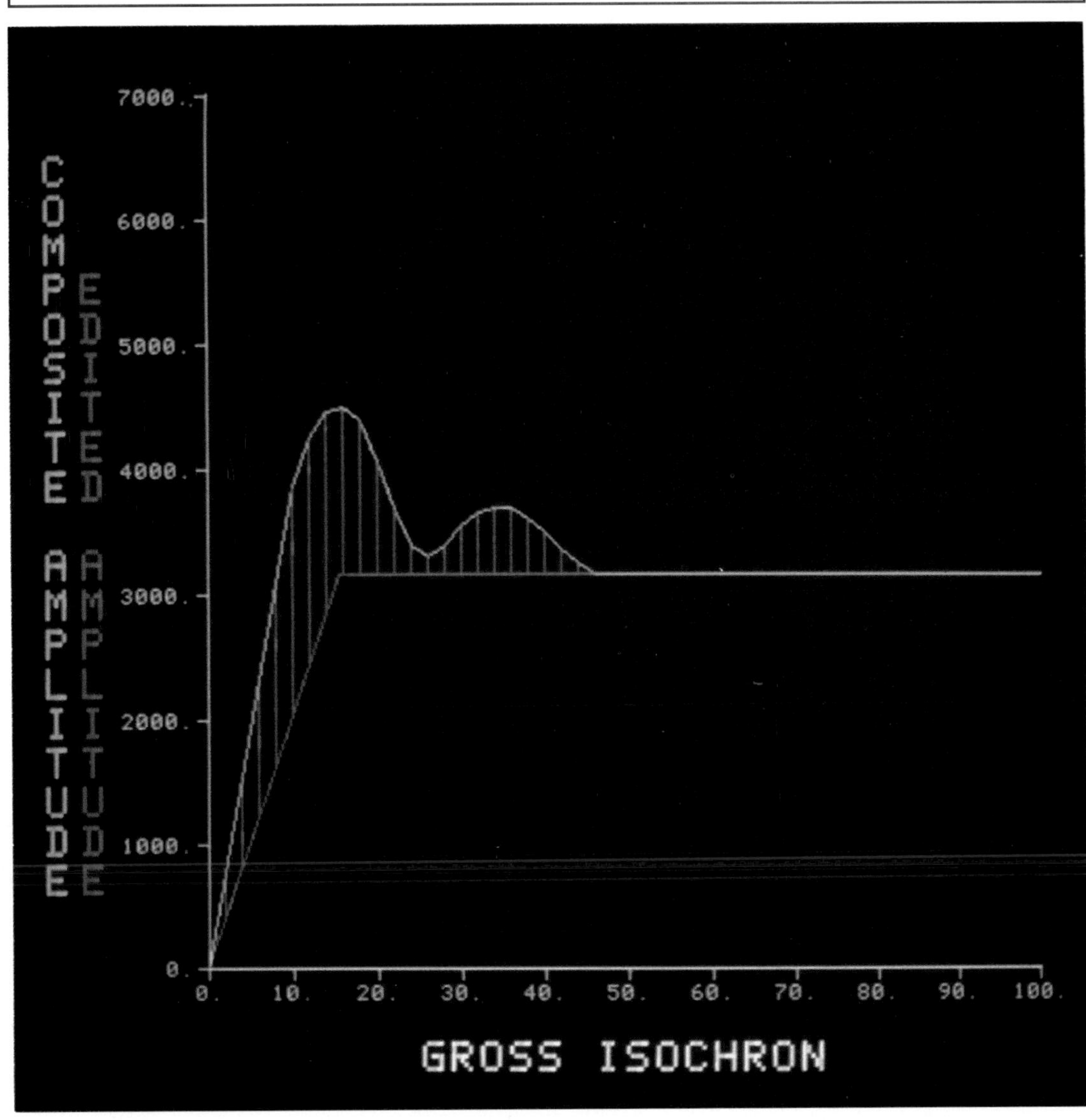

Fig. 7-14. The editing of tuning effects. The yellow is converted to the orange by a multiplier as a function of gross isochron.

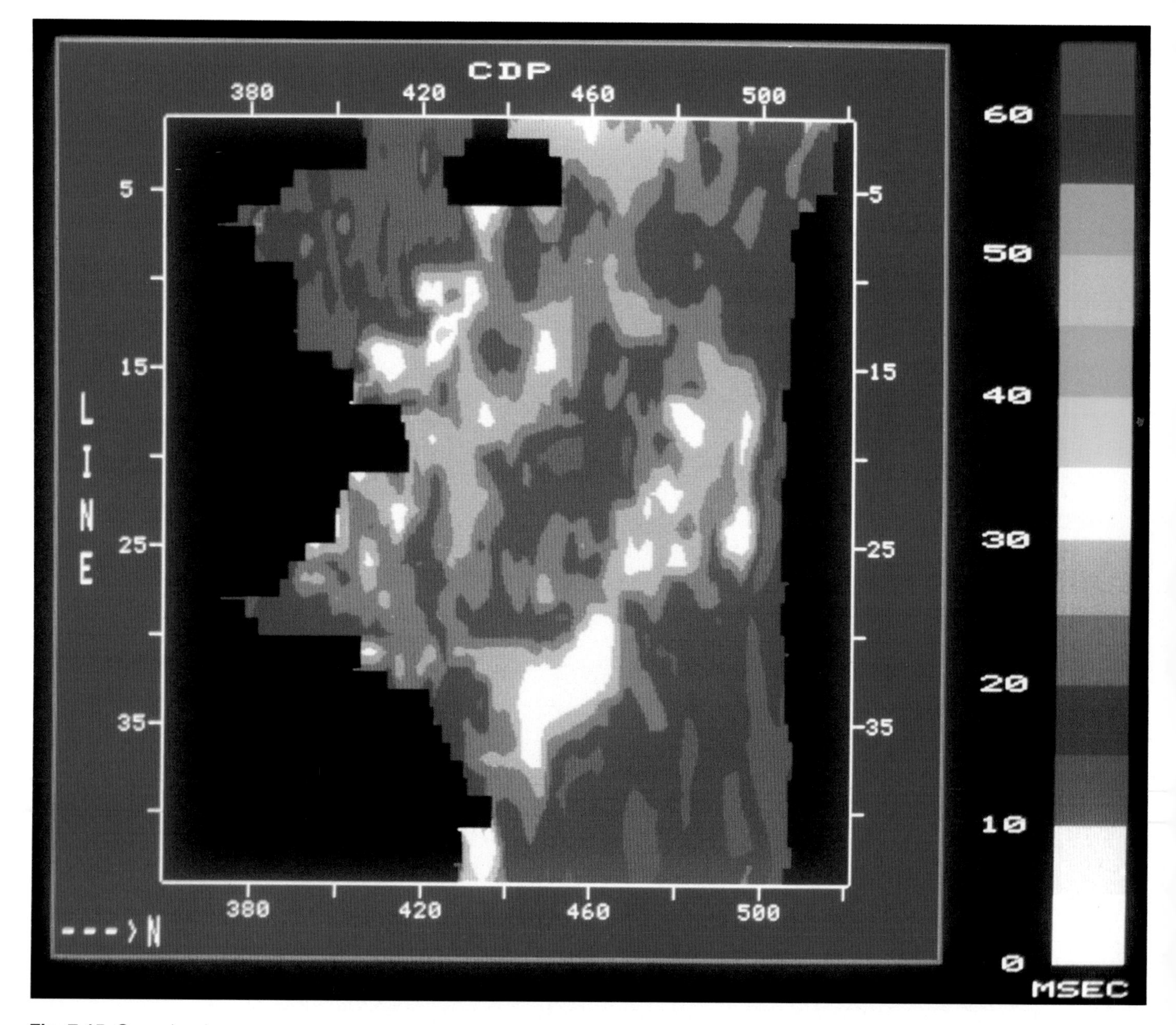

Fig. 7-15. Gross isochron map of Upper Sand showing two thickness trends. Area measures 2 × 2 km. (Courtesy Chevron U.S.A. Inc.)

the interactive system stored the time and the amplitude of that pick in the digital database. Given that the data were zero phase, the time of the maximum amplitude is the correct time for the reservoir interface.

Figure 7-12 shows the interactive interpretive sequence which was then applied to the times and amplitudes provided by the horizon tracking. For any one sand, the horizon times provided structure maps for the top and the base reflections. Subtraction of these time maps yielded the gross isochron (or time thickness) map for the sand. The horizon amplitudes provided horizon slices for the top and base sand reflections. These were then added together in absolute value to yield the composite amplitude response of the sand. This amplitude addition is, in effect, an interpretively constrained inversion; it provides an amplitude indicative of the properties of the interval between the top and base reservoir reflections as interpreted. The principle is diagrammed in Figure 7-13.

Tuning effects remained as a distortion in this composite amplitude response and had to be removed. The key is to understand the shape of the tuning curve in detail; this can be obtained deterministically from an extracted wavelet or statistically from a

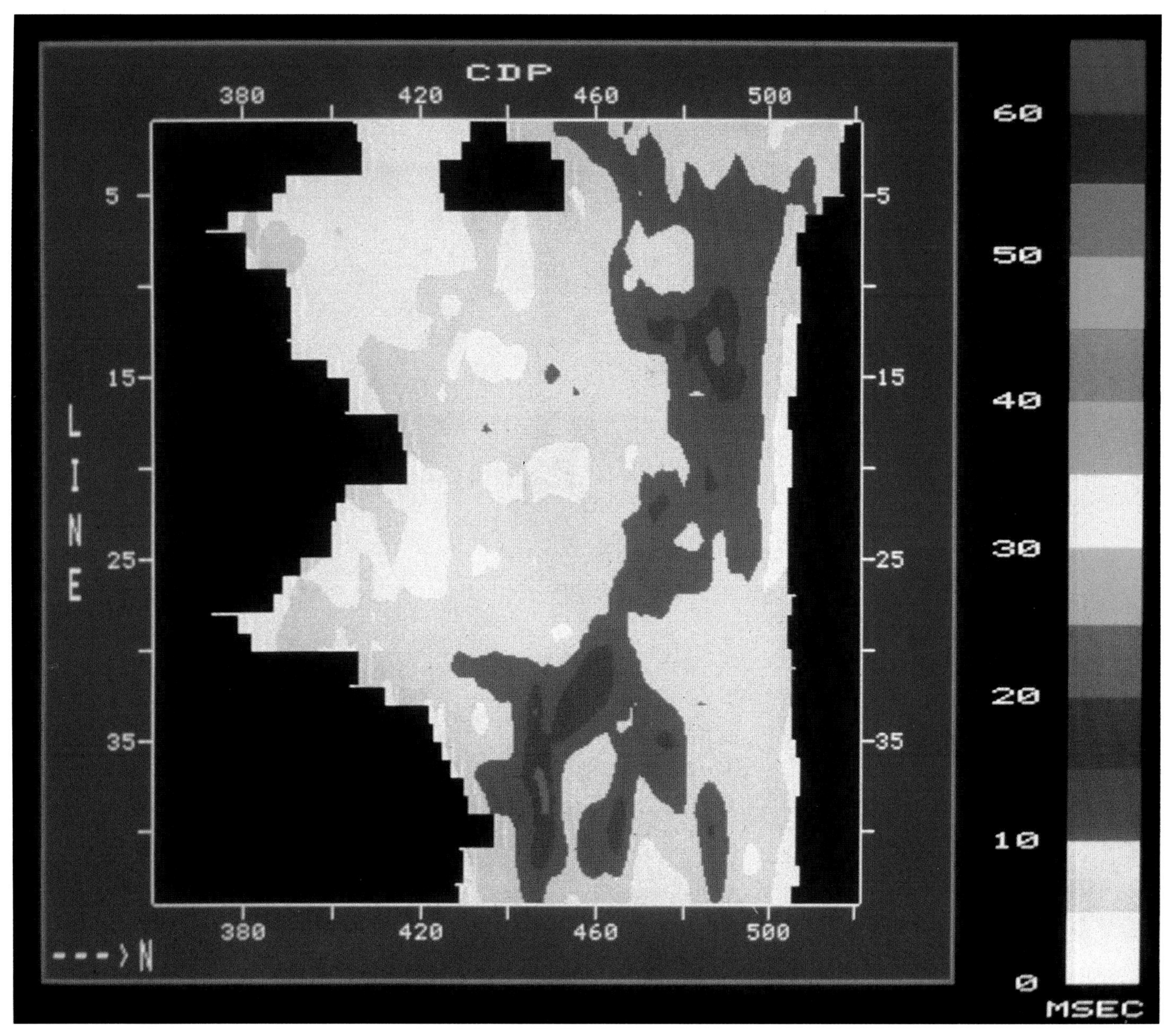

Fig. 7-16. Net gas isochron map of Upper Sand showing one thickness trend. (Courtesy Chevron U.S.A. Inc.)

crossplot (see Chapter 6). In this offshore Louisiana example both methods were used to yield the yellow tuning curve of Figure 7-14. Editing was then required to change the response from that shown in yellow to that shown in orange, so that the amplitude as a function of gross reservoir thickness alone is constant above tuning thickness and linearly decaying to zero for decreasing thickness below tuning.

In order to conclude the interpretive sequence of Figure 7-12, the composite amplitude response was edited according to Figure 7-14, and scaled with a single factor derived from one well to yield a map of net gas/gross sand ratio. Combining this by multiplication with the gross isochron map, a net gas isochron map was obtained. A constant gas sand interval velocity was then sufficient to convert this net gas isochron map to a net gas isopach map. In combining the gross isochron map with the net gas/gross sand ratio (derived by editing with the function in Figure 7-14), it should be remembered that there are no gross isochrons less than tuning thickness because of the tuning phenomenon itself (Figure 6-2). For actual gas sand thicknesses less than tuning, all the net gas sand information is encoded in the amplitude.

Figures 7-15 and 7-16 show comparable gross and net isochron maps, making clear

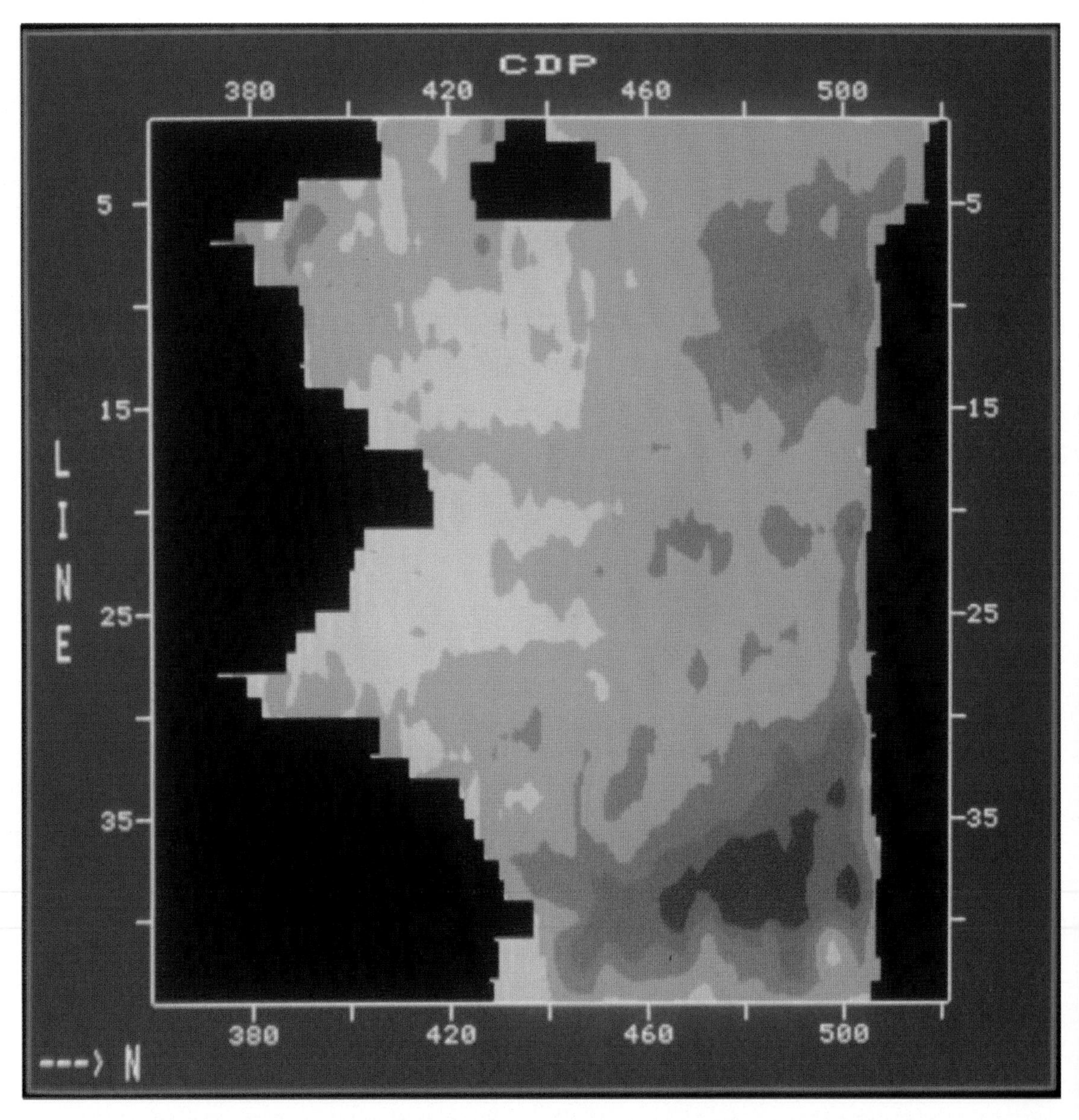

Fig. 7-17. Composite amplitude response (composite horizon slice) of Upper Sand showing high amplitudes in the north and east of the 2 × 2 km area. (Courtesy Chevron U.S.A. Inc.)

the contribution of the net gas/gross sand ratio derived from the composite seismic amplitude response (Figure 7-17). Note the two northwest-southeast thickness trends on the gross isochron map and then note that only one of them has survived in the net isochron map. This is caused directly by the higher amplitudes to the north and east as seen in Figure 7-17. Net gas sand maps derived in this way have been shown to tie well data acceptably. In practice, relative values are more accurate than absolute values because of the difficulty of determining the scale factor connecting edited amplitude to net gas/gross sand ratio.

When there is more than one mappable reservoir interval associated with the reservoir under study, each interval is treated separately and added together at net gas isopach stage. Figure 7-18 shows total net producible gas sand in color superimposed on the structural configuration of the top of the reservoir.

Integration of net gas isopach maps yields the volume of the reservoir. By integrating over chosen sub-areas, the reservoir volume over different lease blocks or areas of special interest can be readily determined.

Case Histories 7 and 11 (Chapter 9) report on the use of net gas sand mapping in other prospects. Different approaches to mapping net pay have been presented by Woock and Kin (1987) and by McCarthy (1984).

In the above method porosity was constant and therefore was not discussed. In

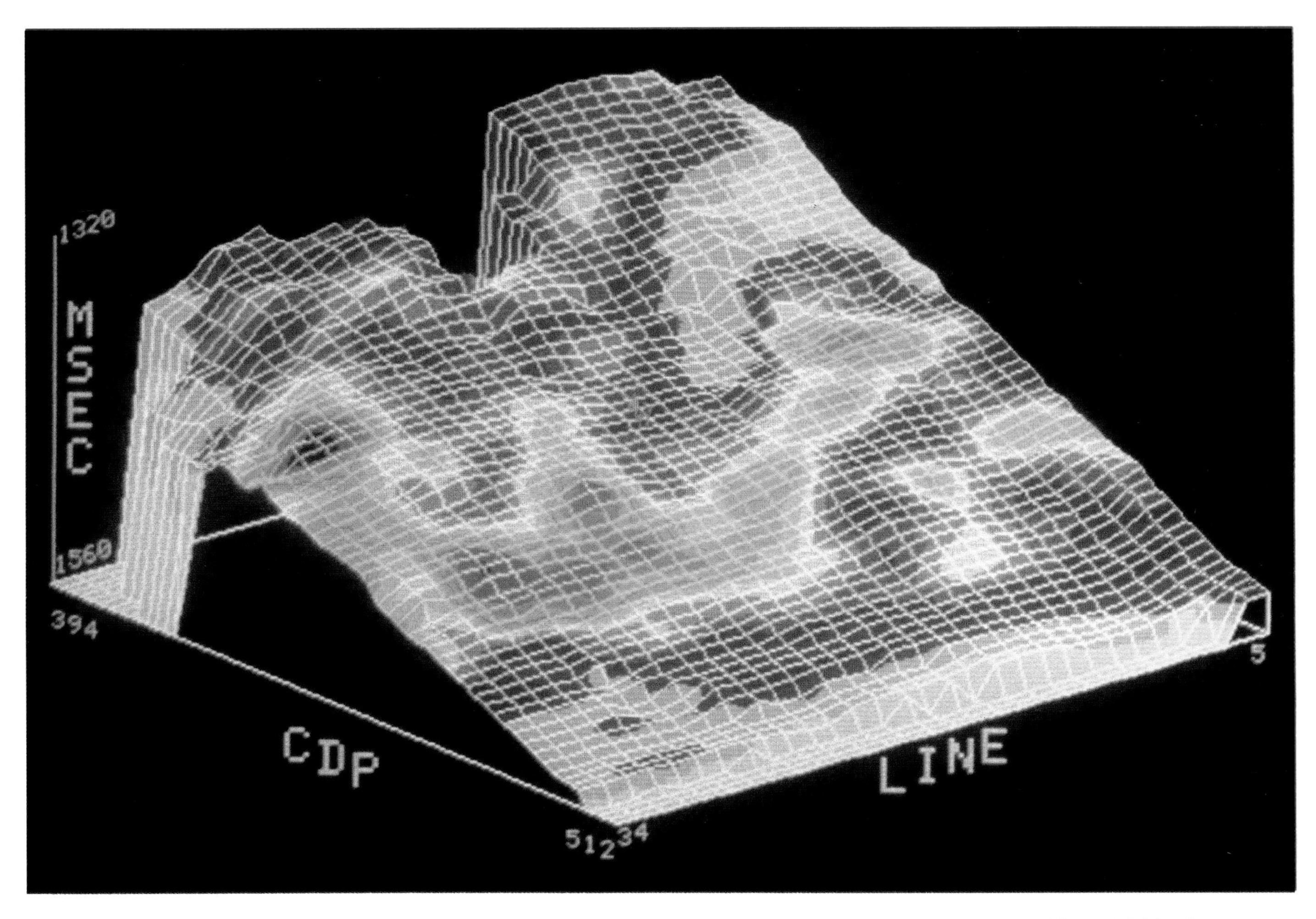

Fig. 7-18. Total net gas sand isopach map superimposed on the structure of the top of the reservoir. The greens and blues indicate the thicker net gas zones. (Courtesy Chevron U.S.A. Inc.)

general for the bright spot regime, porosity multiplies with net pay thickness to give porosity-thickness or pore volume. So the general statement of the method becomes: pore volume = amplitude x isochron. This is illustrated further in Figure 7-19. Composite amplitude with tuning effects edited out is normally the best amplitude to use. However, single horizon amplitude or windowed amplitude, with or without detuning, are all alternatives. Many variations have been tried and successes reported. For example, Neff (1990) starts with a modeled response at a well and thus attempts to handle vertical non-uniformity in the reservoir. Matteucci (1996) tried many seismic attributes and concluded that the composite amplitude-isochron product is "the attribute most strongly related to hydrocarbon pore volume."

Well Calibration

Seismic amplitude or other attributes extracted over one reservoir reflection from good quality data has high relative accuracy. However, it has no absolute accuracy and calibration with well data is necessary. There are many approaches and these are summarized in Figure 7-20. The author has had greatest success with the deterministic method using a single space-invariant scale factor. This factor can be determined from one well, where the amplitude is highest and the tie most reliable, from two wells, or from a crossplot using many wells. The single factor retains the relativities in the seismic data, its greatest strength.

If other properties of the reservoir and adjacent layers affect amplitude significantly and vary spatially, a space-variant scale factor, determined from the wells, will be necessary to remove them. A good discussion of secondary factors affecting amplitude and their removal appears in Case History 12. In projects like this one the relativities in the seismic amplitude are being changed by the different well ties. In gener-

Fig. 7-19. Reflection amplitude, preferably composite of reservoir top and base, times isochron thickness is normally the best measure of reservoir pore volume.

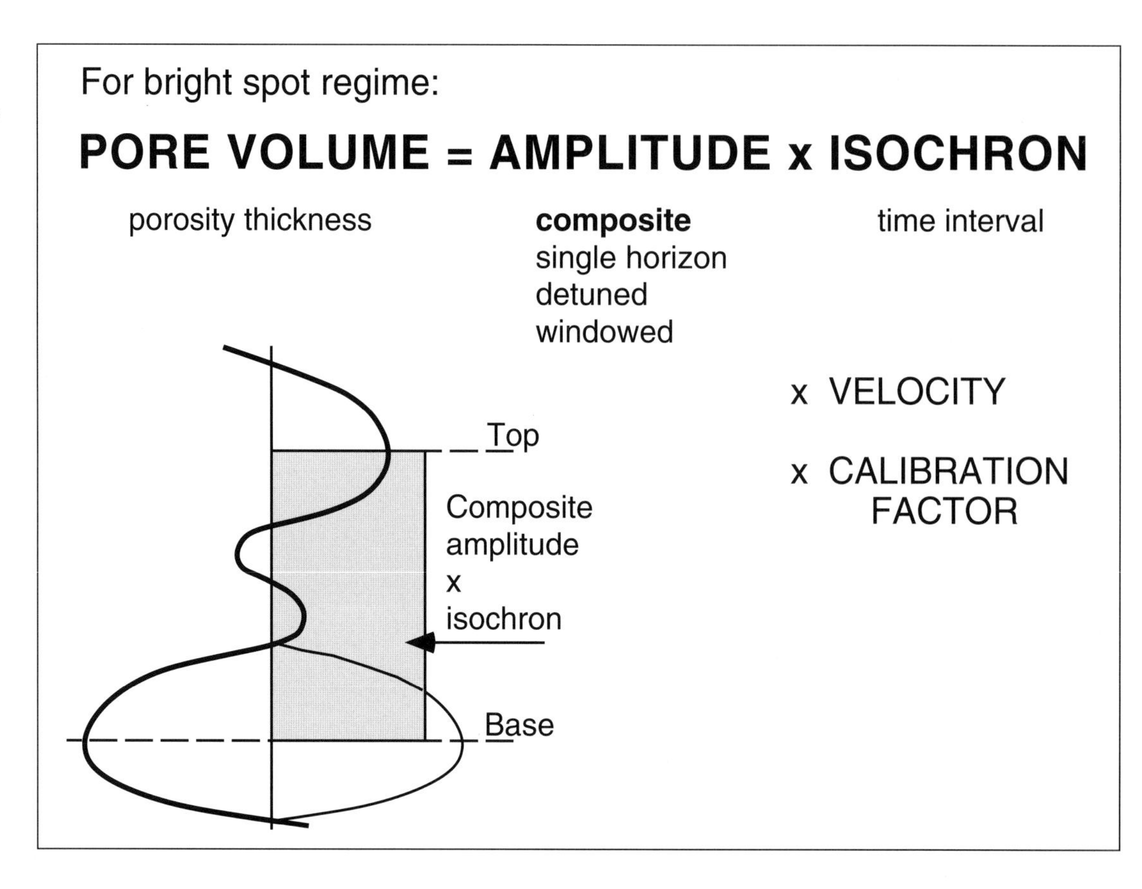

al the uncertainties in the position of each well and its measured property need to be considered, and also the uncertainties in the seismic measurement. The tie in the desired reservoir property between the seismic value and the well value needs only to be within each other's uncertainty and thus distortion of seismic relativities or flexing can be minimized.

In geostatistical approaches to well calibration (Figure 7-20) several reservoir property maps may be desired and several seismic attributes may be considered as a source of information. Crossplots establish relationships and variograms guide the spatial interpolation or cokriging. However, good statistical correlations are not sufficient for reliable extraction of multiple reservoir property maps; support from petrophysics and modeling is also required. Matteucci (1996) summed up this situation nicely: "It is very simple and easy to find relationships. It is much harder to estimate if they are statistically significant, robust, and geologically meaningful."

Statistical Use of Tracked Horizon Data

Once the time and amplitude of the zero-phase reflections from the top and base of a reservoir are stored in a readily-accessible digital database, statistical studies of the horizon data are straightforward. The value of interactive crossplotting for the statistical analysis of tuning phenomena was explained in Chapter 6.

Figure 7-21 shows an interactive crossplot of gross isochron against top sand time; that is, vertical thickness (in time) against structural position. The general effect is triangular with several clearly visible lineations. Interpretation of these lineations is a statistical assessment of the many thousands of data points included in this crossplot.

Figures 7-22 and 7-23 are crossplots of sub-areas of the prospect, each accompanied by an exemplary data segment. Figure 7-22 makes it clear that the lineation along the bottom of the crossplot illustrates that thicknesses can only be measured down to just below tuning.

Sub-area C (Figure 7-22) includes many lines of data illustrating good flat spot reflections, one of which is illustrated in the data insert. The orange straight line

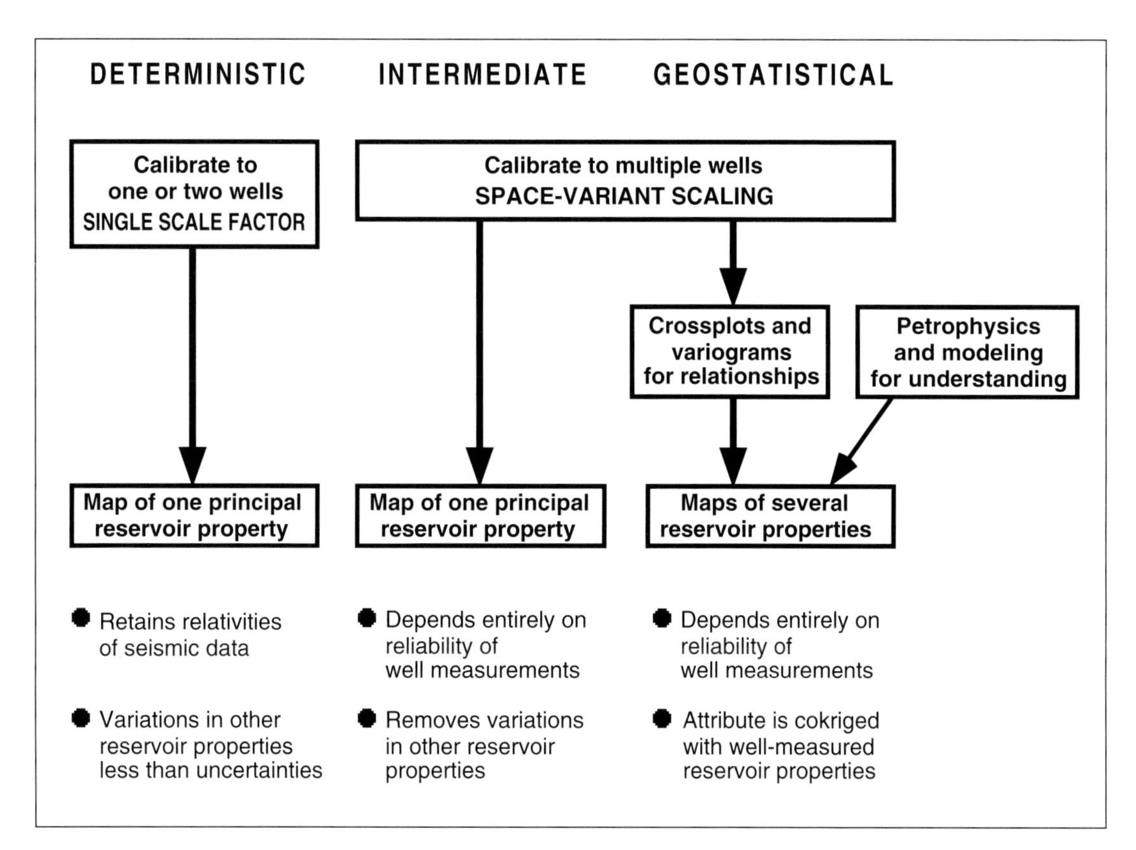

Fig. 7-20. Methods of calibration of seismic amplitude or other attribute to reservoir properties at wells. Tie should minimize seismic flexing to fit well value within uncertainties.

labelled FLAT SPOT is a line of the form, y = c – x, that is, one representing equal increments of gross isochron and top sand time. Because top sand time plus gross isochron equals base sand time and because base sand time is approximately constant for a flat spot reflection, it is indeed expected that a flat spot would plot along such a diagonal line of unit gradient.

Sub-area B (Figure 7-23) is an area of thicker sands. An orange line represents the line of equal increments of gross isochron and top sand time as before. Most of the points fall to the right of this line, indicating varying degrees of gas velocity sag. Extreme sag is illustrated in the data insert and marked on the crossplot by a pink line.

An interpreted version of the total area crossplot of Figure 7-21 is shown in Figure 7-24. This interpretation incorporates the observations from several sub-areas including the two discussed here. The convergence of many of the lineations in the lower right corner and also the great concentration of crossplot points at the same location suggest a common gas/water contact over a large part of the reservoir. This in turn suggests that much of the reservoir is in communication, at least in terms of flow rates effective over geologic time. The concentration of points in a swath along the lower part of the triangular pattern suggests that many of the sands were preferentially deposited with thicknesses of 30 ms (approximately 25 m) or less.

There is a suggestion of a lineation with the correct slope to be another flat spot reflection intersecting the top sand time axis at 1,440 ms. It is evident on both Figures 7-24 and 7-22. This observation can be considered to be the statistical identification of another possible fluid contact, which has obvious implications for segmentation of the reservoir.

Further Observations of Reservoir Detail

Figure 7-25 illustrates several aspects of a thick Gulf of Mexico gas reservoir. The upper right panel shows a vertical section indicating a wedge-shaped gas zone. The lower bright red reflection is from the fluid contact dipping to the left because of gas velocity sag. The structural dip is in the opposite direction. Layering within the

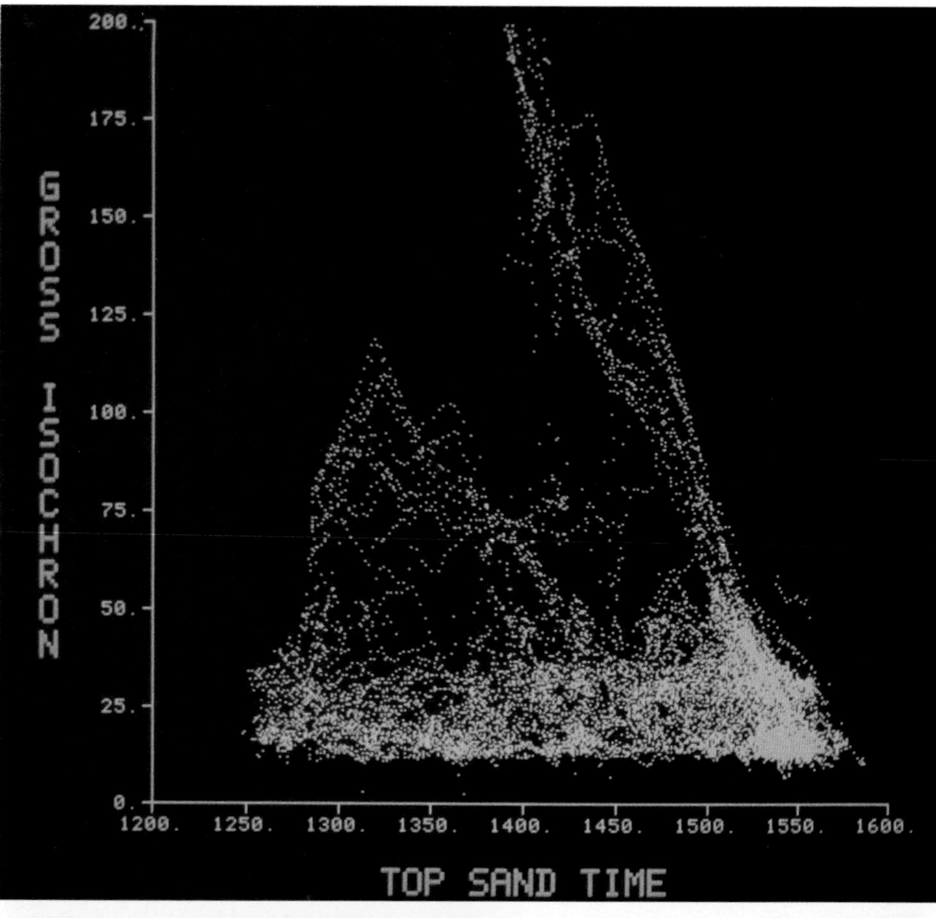

Fig. 7-21. Interactive crossplot of gross isochron against top sand time.

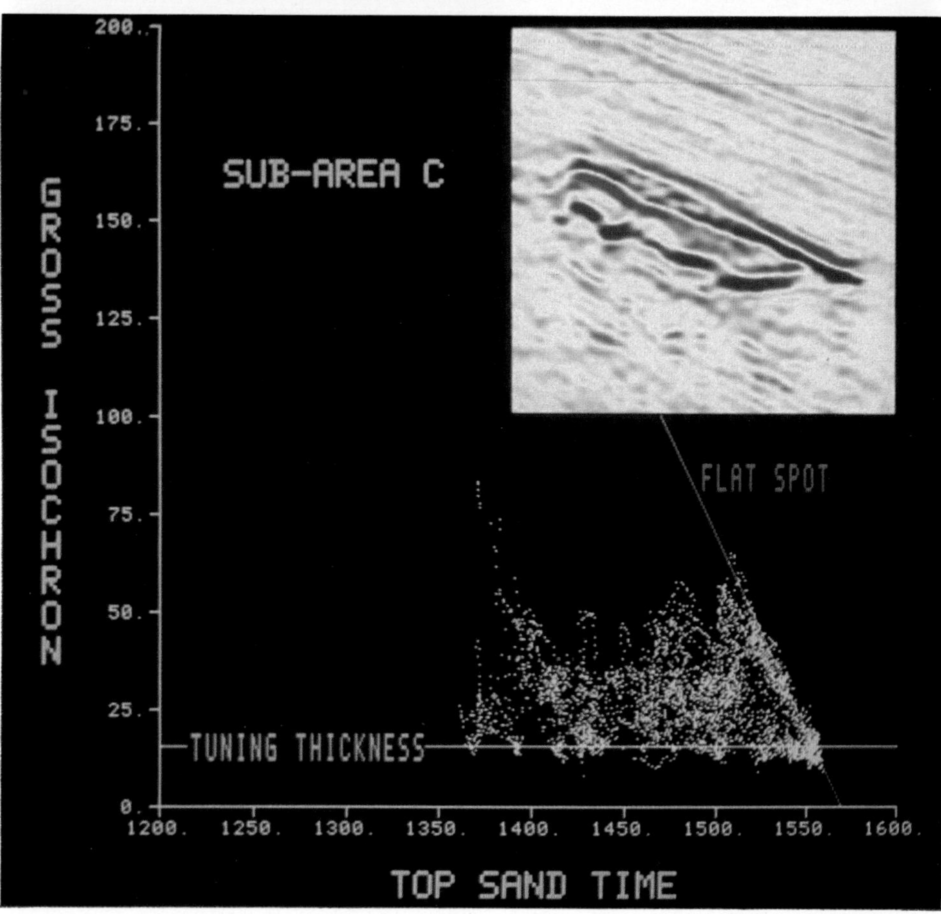

Fig. 7-22. Interactive crossplot of gross isochron against top sand time for sub-area C with exemplary data insert.

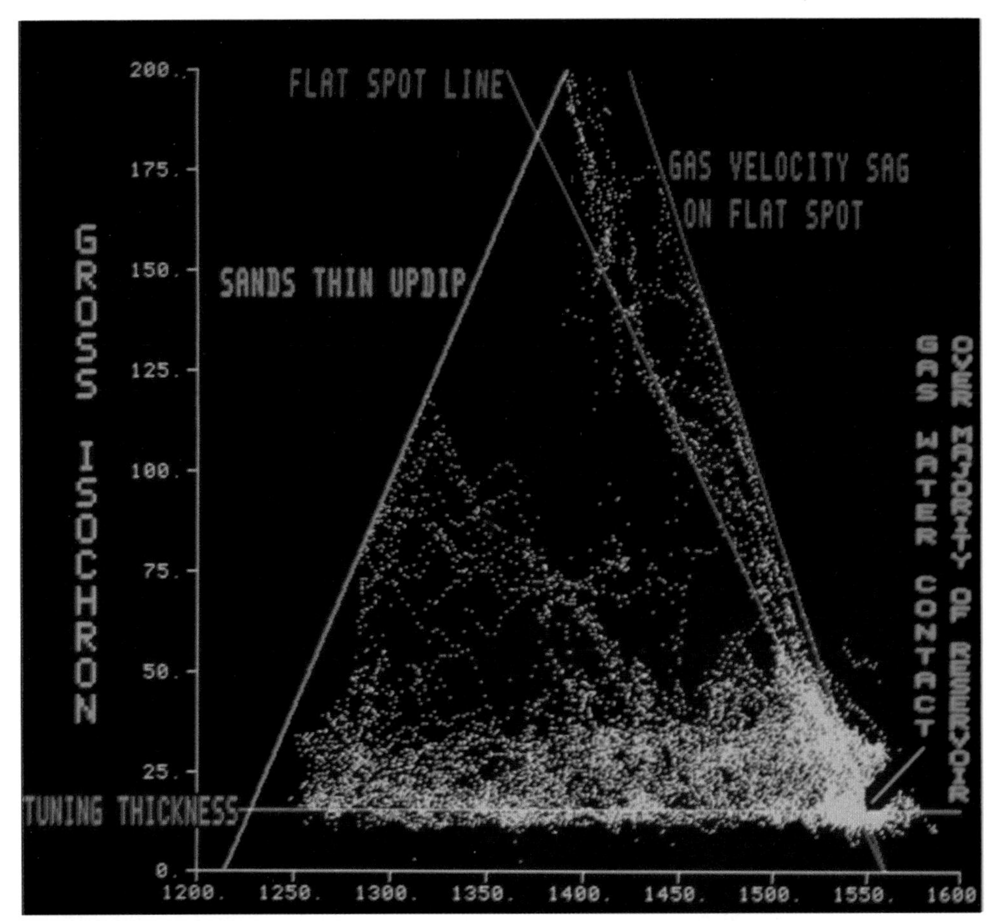

Fig. 7-24. Interactive crossplot of Figure 7-21 showing interpretation of lineations.

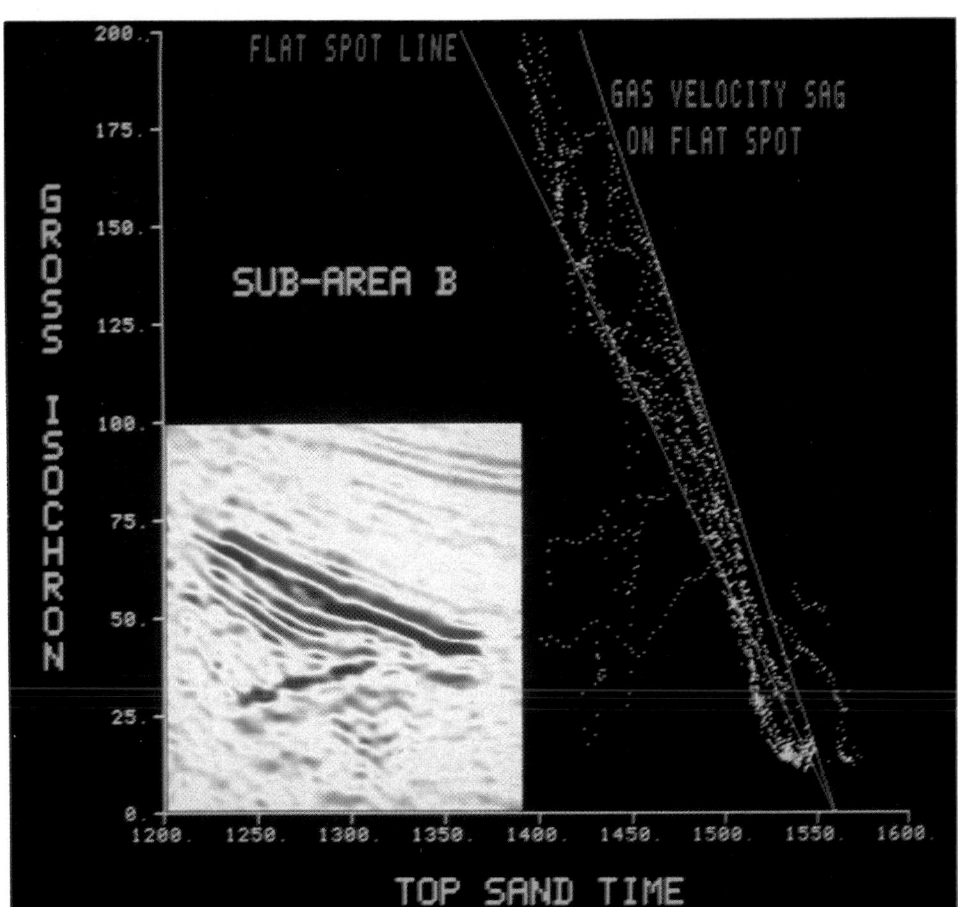

Fig. 7-23. Interactive crossplot of gross isochron against top sand time for sub-area B with exemplary data insert.

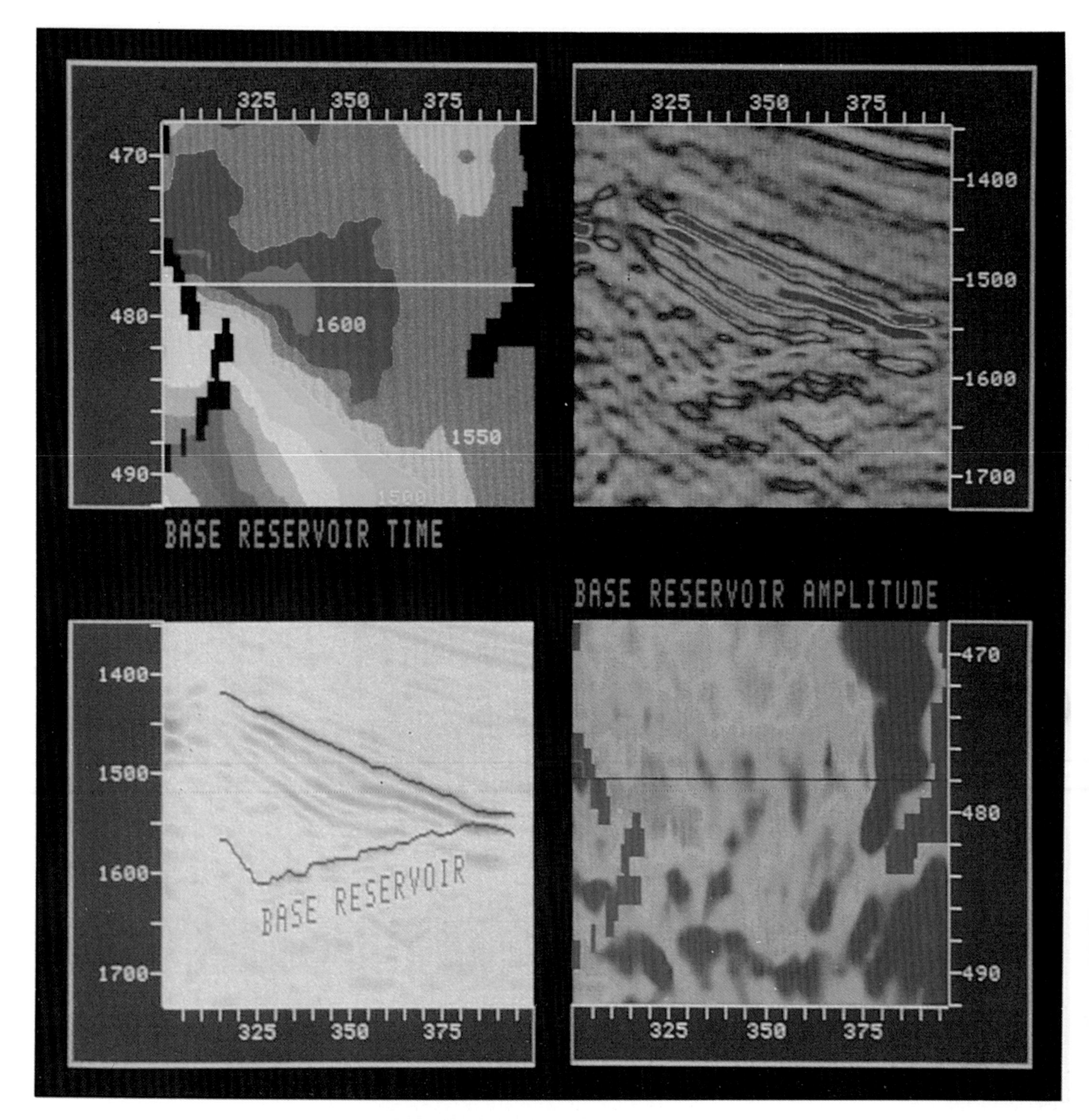

Fig. 7-25. (Upper Right) Vertical section through complex reservoir sand showing amplitude variation along fluid contact reflection.
(Upper Left) Time structure map on base reservoir reflection showing region of large gas velocity sag.
(Lower Left) Vertical section showing automatic tracks on top and base of gas.
(Lower Right) Horizon slice for base gas reflection indicating internal reservoir layering by patterns in amplitude.
(Courtesy Chevron U.S.A. Inc.)

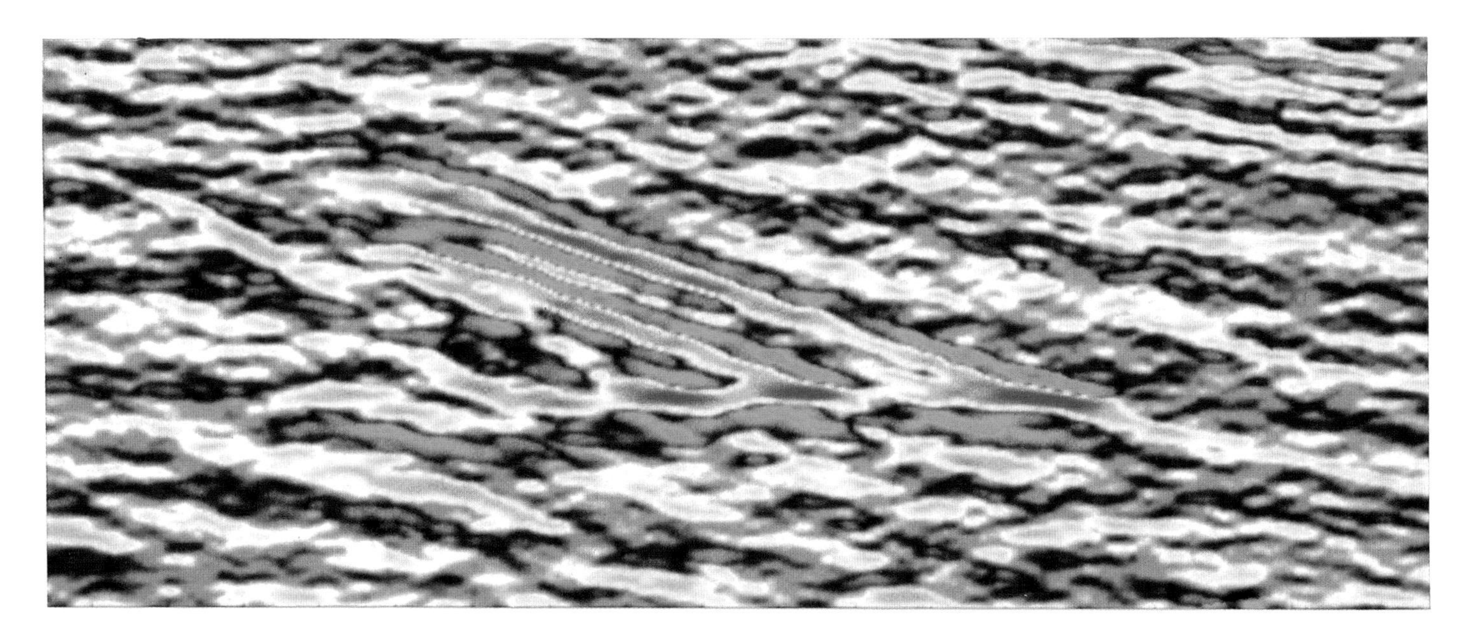

Fig. 7-26. Vertical section from Gulf of Mexico showing thick gas reservoir filled with strong internal reflections. (Courtesy Chevron U.S.A. Inc.)

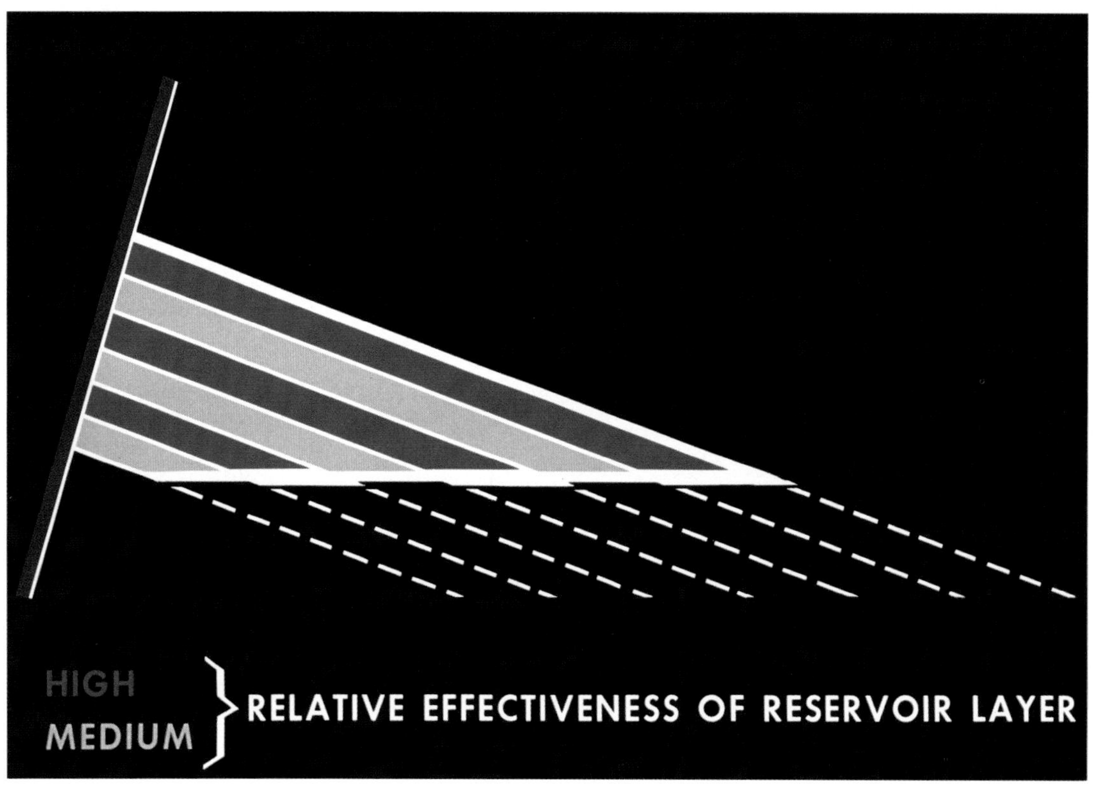

Fig. 7-27. Macro-layers of high and medium effectiveness (net-to-gross ratio) caused by micro-layers of gas sand interbedded with shale.

reservoir thus crosses the fluid contact and, because of different properties from layer to layer, causes varying amplitude along the fluid contact reflection.

Automatic tracks on the top and base of the gas reservoir are shown in the lower left panel of Figure 7-25. The time structure map for the base of the reservoir in the upper left panel illustrates the zone of major gas velocity sag by the area of dark blue. The horizon slice showing the spatial variation in amplitude over the base of the gas is seen in the lower right panel. Within the zone of major gas sag, approximately north-south high amplitude streaks illustrate the areas where layers of superior reservoir quality intersect the fluid contact reflection.

Figure 7-26 is a vertical section through another thick Gulf of Mexico gas reservoir. Note how the reservoir is filled with strong reflections that have little correlative amplitude outside the gas zone. This fairly common situation is explained by the dia-

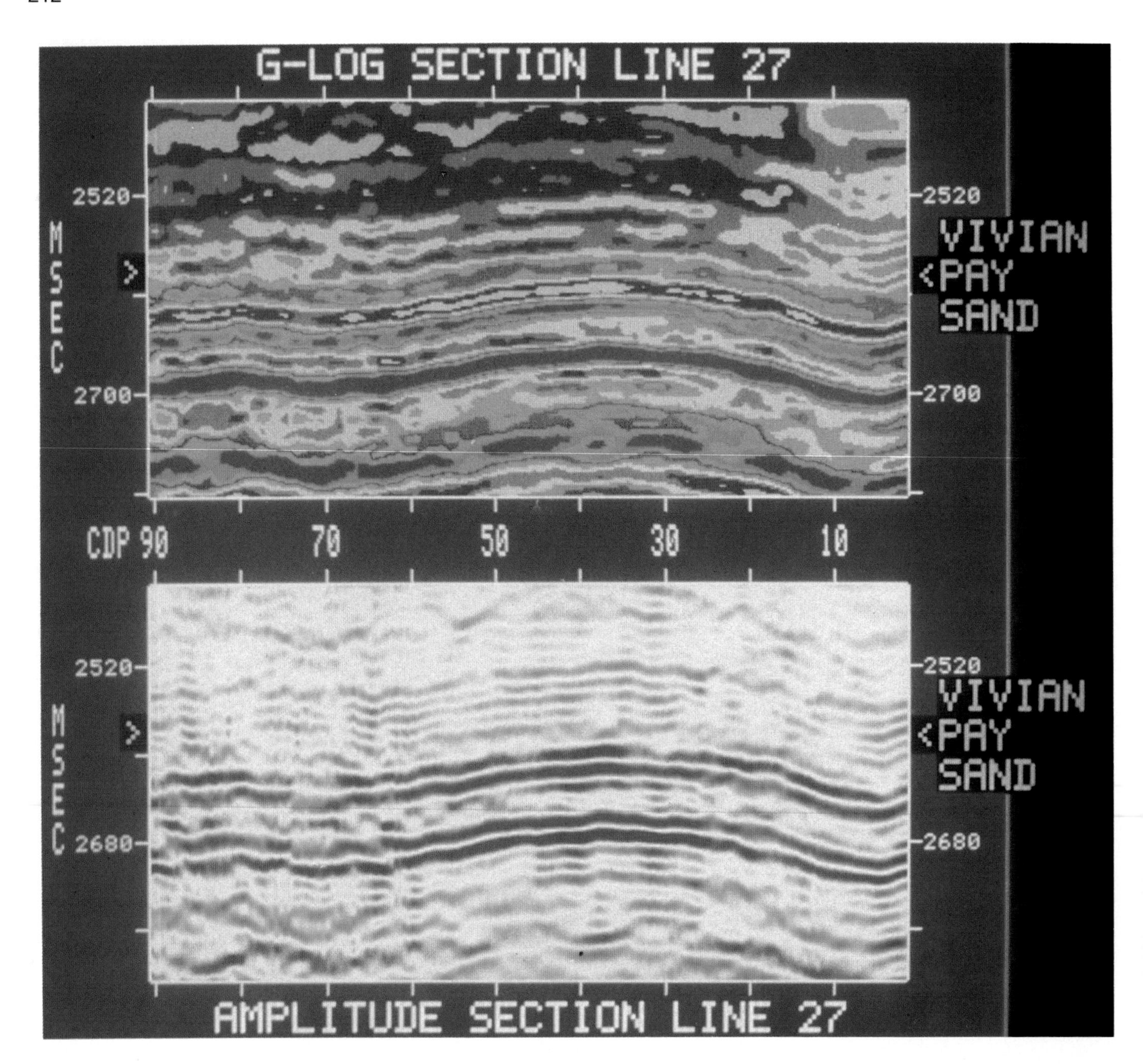

Fig. 7-28. Inversion velocity section and color amplitude section from 3-D survey in Peru indicating Vivian Pay Sand. Velocity legend is in Figure 7-29. (Courtesy Occidental Exploration and Production Company.)

gram of Figure 7-27. The seismic response indicates macro-layering, with individual layers about one-quarter of a wavelength in thickness. The true geologic layering is of layers much thinner than this. Gas in these very thin sand layers generates the seismically visible response of the thicker macro-layers, given that there is some vertical variability of net-to-gross ratio approximating these macro-layers. Note also that the fluid contact reflection appears stepped. This is caused by interference or tuning of the internal reflections with the flat spot. This observation is similar to that seen in Figure 5-29.

Figures 7-28 and 7-29 were derived from 3-D data from Peru. The Vivian pay sand is a proven hydrocarbon reservoir but the reflections from it, as illustrated in the lower panel of Figure 7-28, are unspectacular. The data were considered to be zero phase and thus appropriate for seismic inversion. The velocity section after inversion is shown in the upper panel of Figure 7-28. The lower velocity in the Vivian sand on top of the structure can be clearly seen. This is a case where seismic inversion has

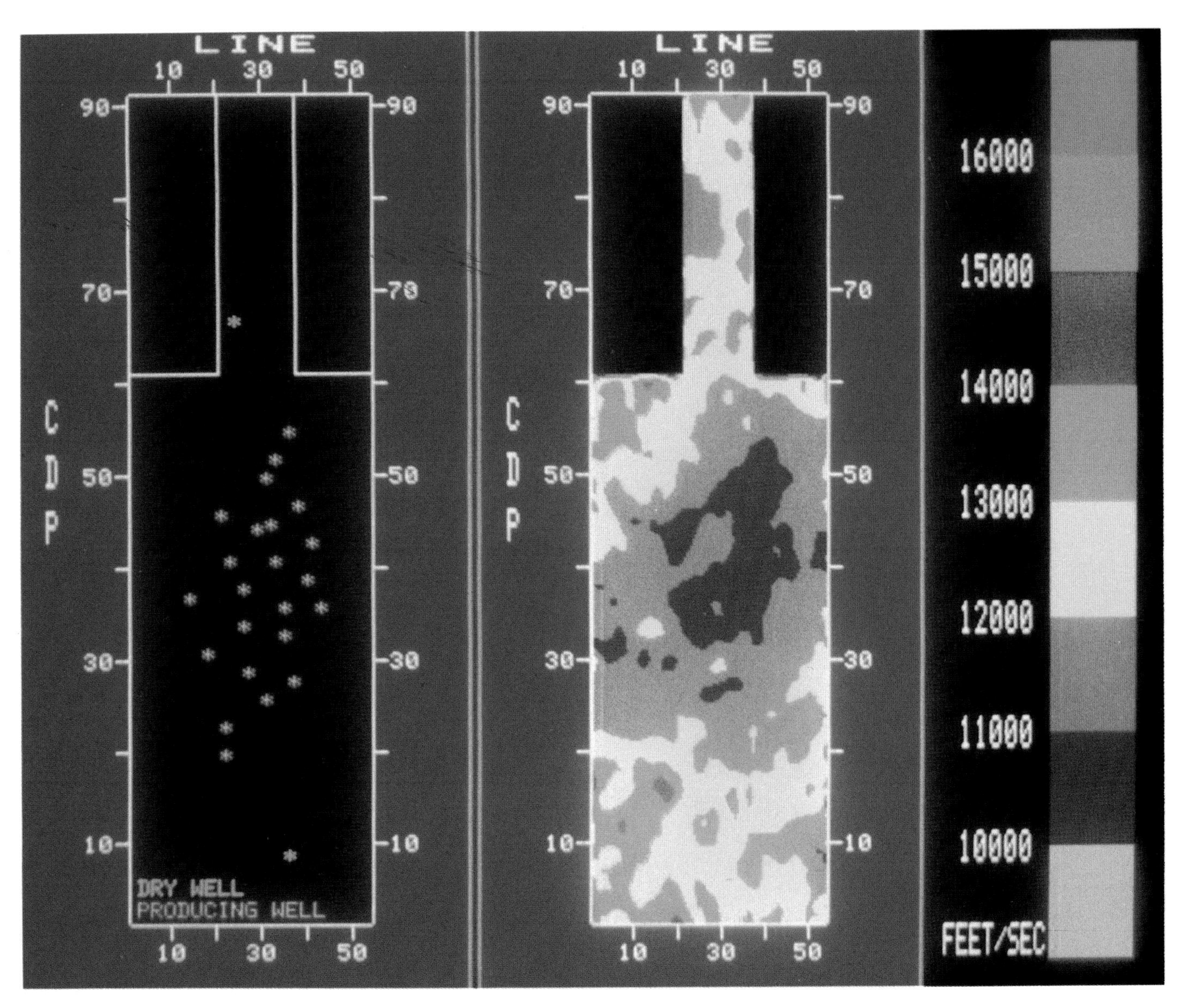

Fig. 7-29. Horizon slice in velocity through Vivian Pay Sand showing low velocity zone enclosing the area of producing wells. (Courtesy Occidental Exploration and Production Company.)

increased the visibility of a reservoir feature, because of the integration of the effects on the top and base reflections.

The whole 3-D volume was then inverted and the velocity low, identified as the Vivian pay sand, was sliced in a horizon-consistent manner. The resulting horizon slice in velocity, in the right panel of Figure 7-29, shows an oval-shaped area of low velocity. This field has now been substantially developed and the left panel of Figure 7-29 shows that the producing wells all lie within the low velocity zone.

Reservoir Surveillance

Multiple 3-D surveys recorded at different times during the production of a reservoir can be used to monitor its production performance. Case History 4 (Chapter 9) reports on the monitoring of a fire-flood enhanced oil recovery operation in north Texas. The geophysical objective was to observe the progressive expansion of an artificial bright spot caused by the combustion gases. Direct observation of the bright spot and also attenuation studies of a deeper reflection demonstrated anisotropic expansion of the burn front and also streaming in the upper part of the reservoir.

In northern Alberta, Canada, four 3-D surveys have been recorded on the same prospect to monitor the effects of heat resulting from the injection of steam. The

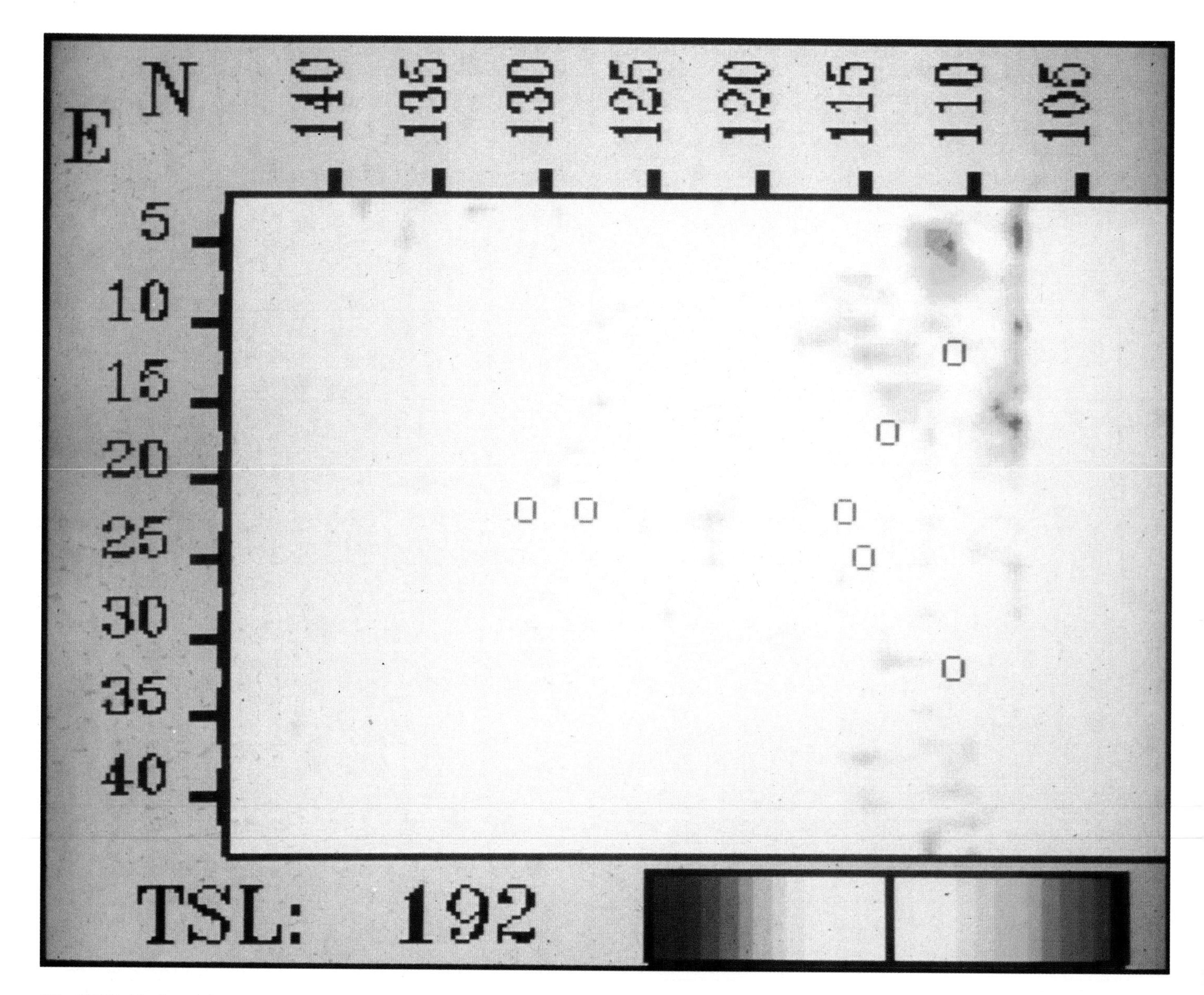

Fig. 7-30. Horizontal difference section at 192 ms through GLISP reservoir in northern Alberta, Canada, where the objective was to monitor steam injected into a tar sand reservoir. The colors show *differences* between the same section from two separate 3-D surveys recorded several months apart. The colors thus indicate the propagation of heat through the tar sand. (Courtesy Amoco Canada Petroleum Company Limited and N. E. Pullin.)

engineering objective is to increase the mobility of heavy tar, thereby making it producible by normal means. An increase in the temperature of the tar by 100°C decreases its velocity by 50%. This has been observed as push-down of a deeper reflector and also as increased amplitudes in the tar sand section. One way of observing these increased amplitudes is to generate the same horizontal section from two of the time-separated surveys and subtract one from the other. One such horizontal difference section, or time slice difference section, is shown in Figure 7-30; the yellows and reds are areas of *difference* between the two surveys and thus are direct observations of heat.

The Oseberg Field in the Norwegian North Sea was subject to gas flooding for reservoir pressure maintenance and Johnstad et al. (1995) have reported on seismic monitoring of production over a 30 month period. The lateral movement of the gas-oil contact was observed on horizon slice *difference* sections sliced from the 1992/1989 difference volume.

Fault Slicing

A **fault slice** is a slice through a 3-D data volume parallel to the interpreted position of a fault plane of interest. Fault slices have applications for mapping structure

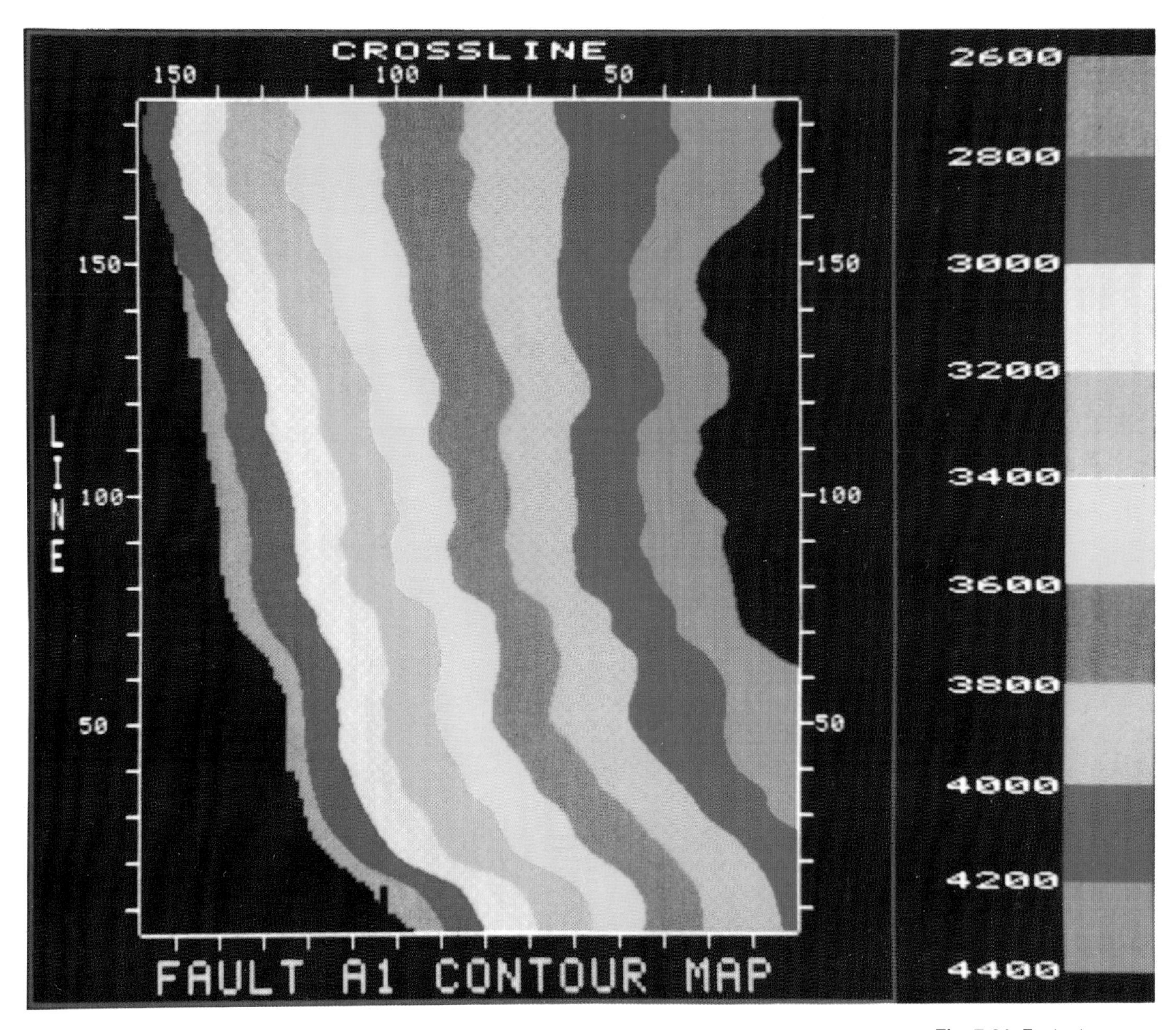

Fig. 7-31. Fault plane contour map, used as the reference surface for fault slicing. (Courtesy Texaco U.S.A. Inc.)

very close to a fault and fault throw, for identifying splinter faults, and for studying fault sealing or leaking.

First of all, the fault under study must be mapped (Figure 7-31). This serves as the reference surface parallel to which all the fault slices in the upthrown and downthrown fault blocks are generated. One of these, eight data points from the fault on upthrown side, is shown in Figure 7-32. This slice and others parallel to it proved useful for observing steep dips that led to the mapping of growth structure in the upthrown block. This benefit resulted from the uniform proximity of the fault slices to the parent growth fault. Horizon tracks on Figure 7-32, and those judged correlative on the fault slice eight data points from the fault on the downthrown side, were subtracted from each other to yield a map of throw across the fault plane, as shown in Figure 7-33.

Secondary, or splinter, faults are generated by movement on a major growth fault and may extend only a short distance from it. A fault slice, remaining uniformly close to the growth fault, slices this zone and intersects splinter faults branching off the parent. Figure 7-34 shows the interpretation of thirty splinters on one fault slice, each being supported by at least six event terminations. Figure 7-35 shows the mapping of one of

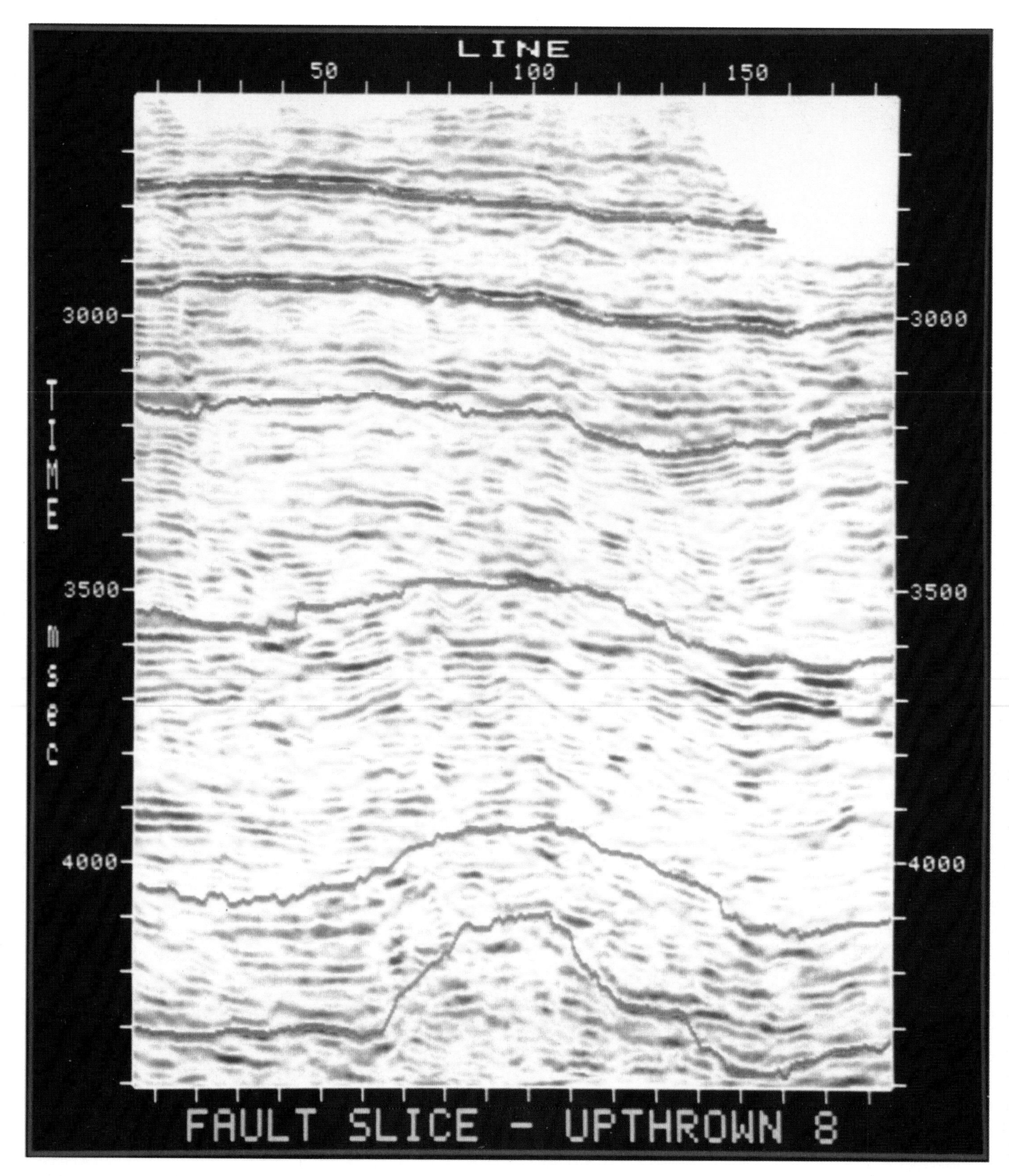

Fig. 7-32. Fault slice through Gulf of Mexico data volume parallel to the major growth fault mapped in Figure 7-31. Horizon tracks show deep structure caused by salt movement. (Courtesy Texaco U.S.A. Inc.)

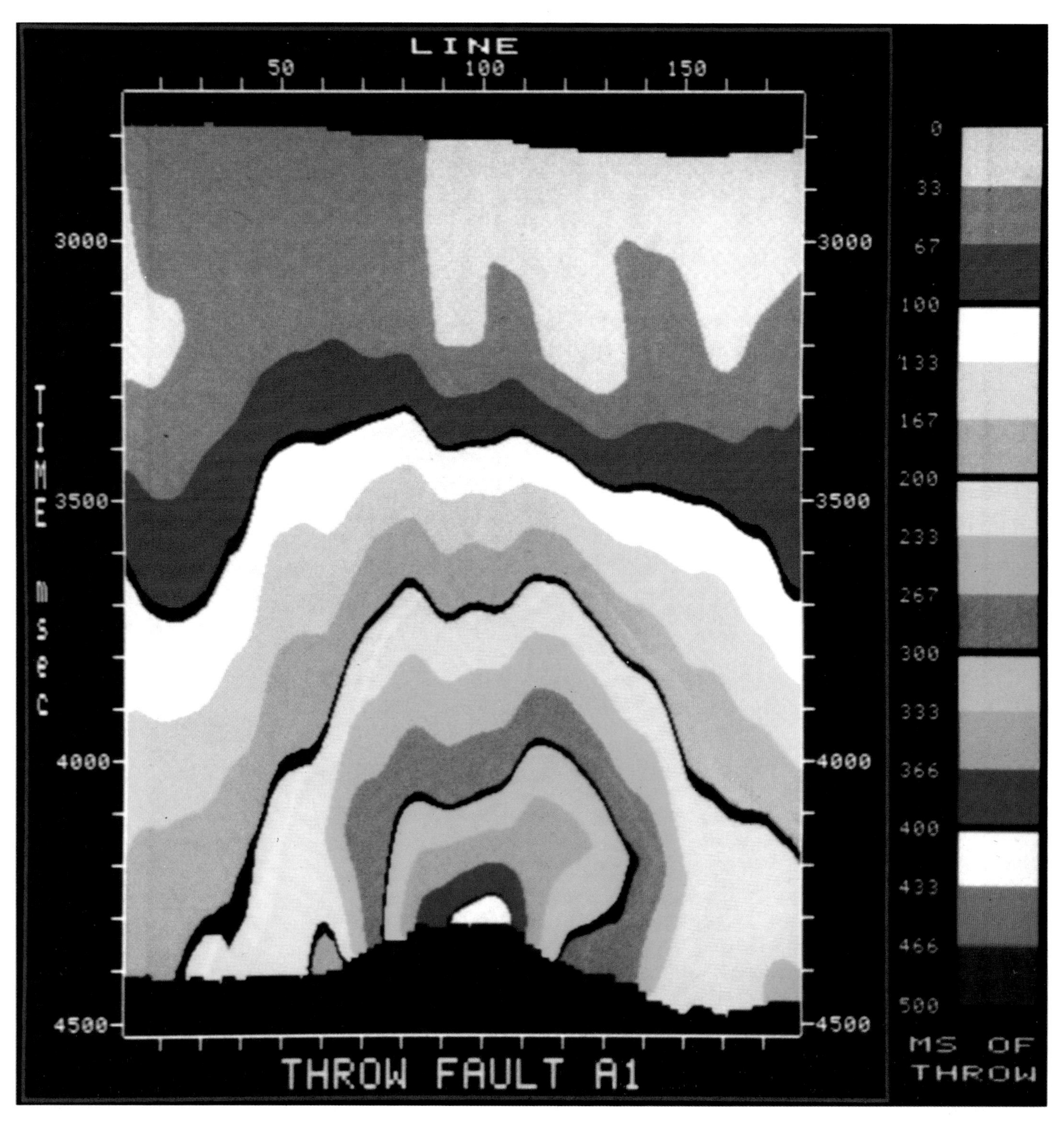

Fig. 7-33. Map of throw across the growth fault deduced from the horizon tracks in the upthrown fault block (Figure 7-32) and the correlative ones from a fault slice in the downthrown fault block. (Courtesy Texaco U.S.A. Inc.)

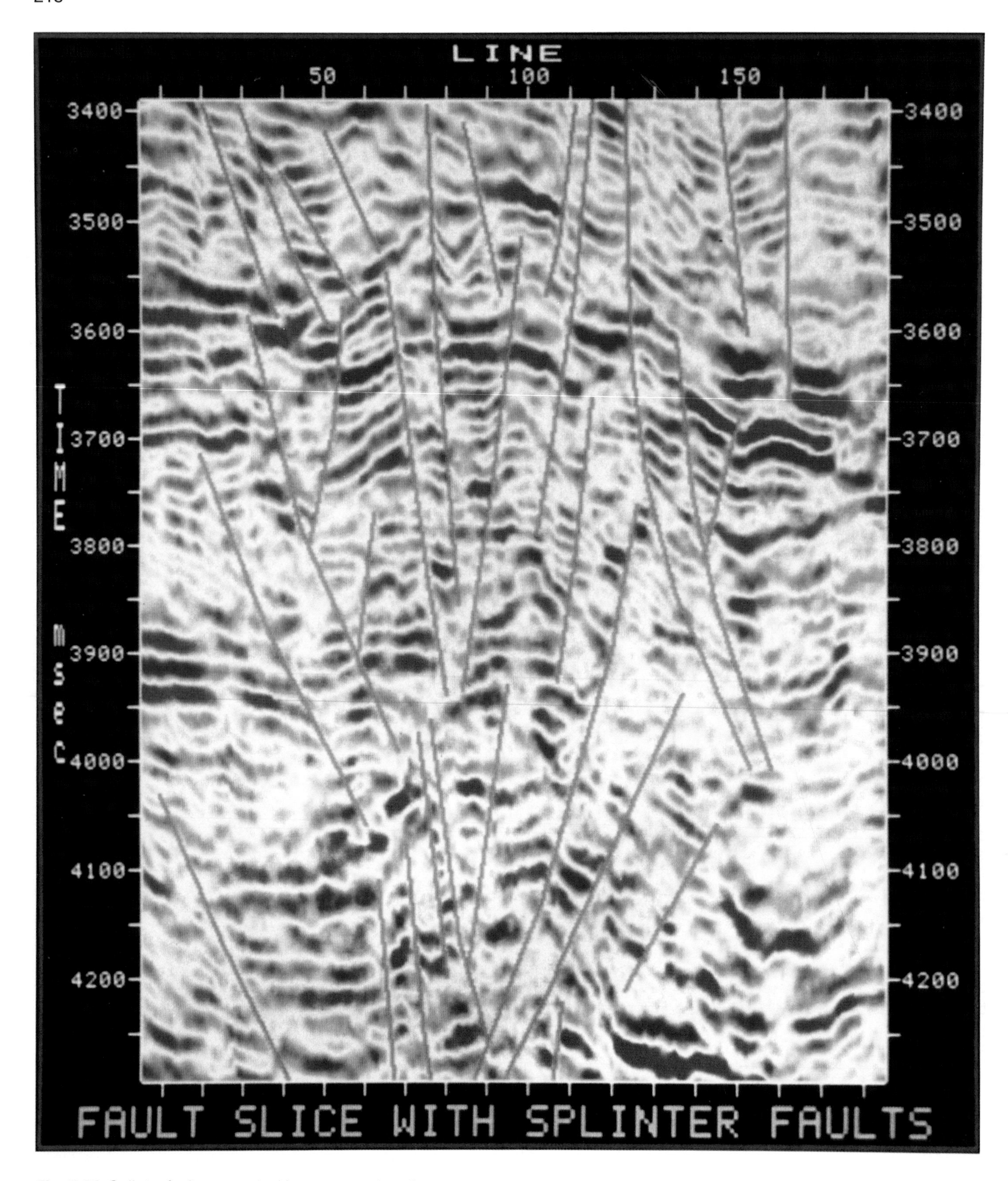

Fig. 7-34. Splinter faults generated by movement on the growth fault interpreted on one fault slice in the upthrown block. (Courtesy Texaco U.S.A. Inc.)

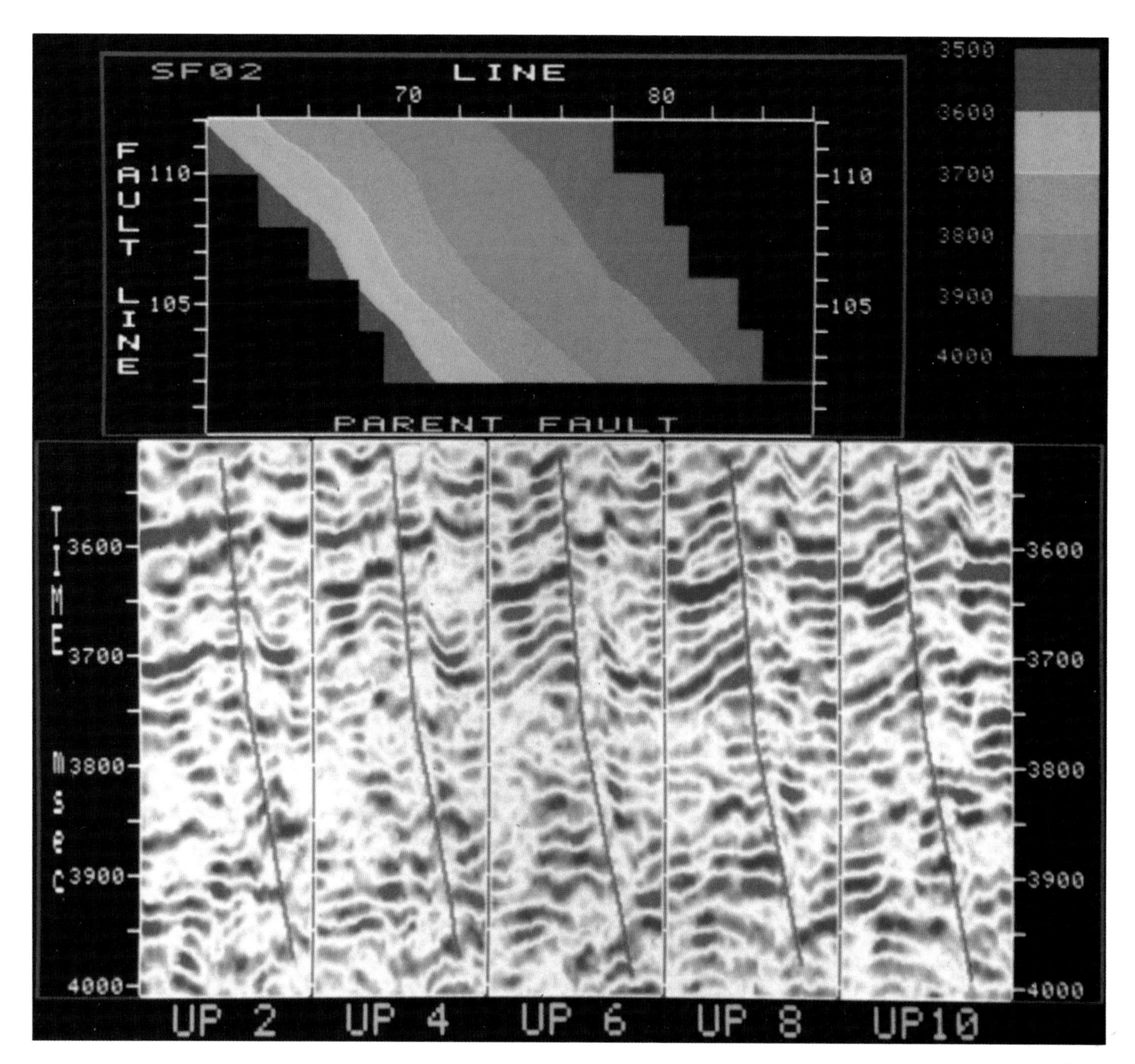

Fig. 7-35. Fault slice segments in the upthrown block used to map one splinter fault. The resultant map above is in coordinates relative to the growth fault, so that the contours show the relative strike of the splinter and its parent. (Courtesy Texaco U.S.A. Inc.)

them on a range of five fault slices covering the splintered zone. The resultant map in coordinates relative to the parent growth fault shows, by the attitude of the contours, the relative strike or azimuth of the splinter fault and its parent. A more thorough treatment of the method and benefits of fault slicing is provided by Brown, Edwards and Howard (1987).

An application of fault slicing to fault sealing and leaking has been presented by Bouvier et al. (1989, reference in Chapter 3). Figure 7-36 is a fault slice from Nigeria. It is unrolled, meaning that it is presented in terms of distance along the actual fault plane, compared with Figure 7-32, which has vertical time as its axis. The fault slice of Figure 7-36 is in depth, not time, and it is inverted to display acoustic impedance. Three well logs are superimposed and the thin black lines indicate the interpretation of sands and shales. This interpretation is presented on Figure 7-37 with the sands in white and the shales in black. Hydrocarbon fluids, proven and probable, are in various colors. Shale interpreted from a fault slice in the juxtaposed fault block is superimposed in brown.

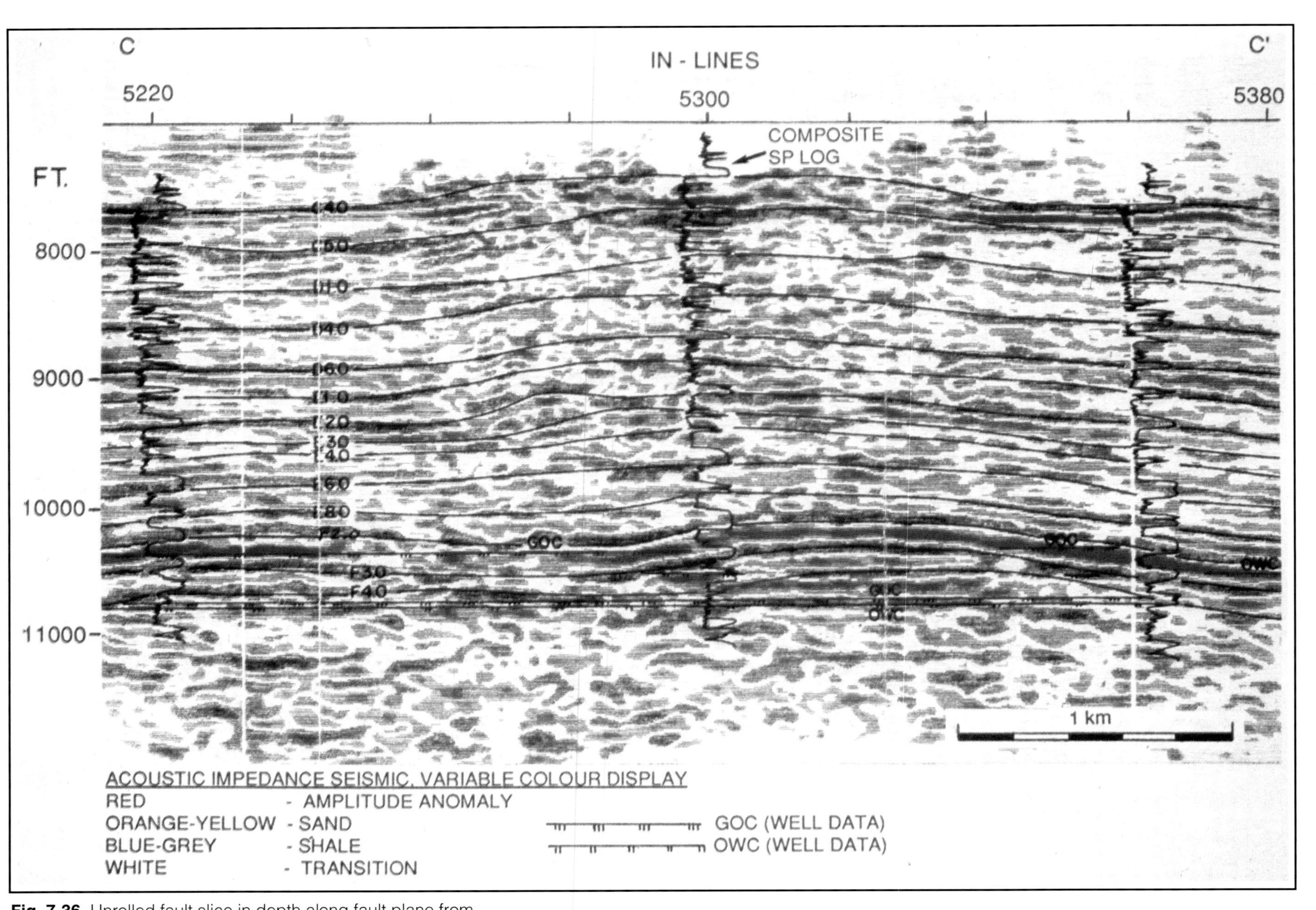

Fig. 7-36. Unrolled fault slice in depth along fault plane from Nigeria. Interpretation of sands and shales tied to three wells was for the study of fault sealing (Bouvier et al., 1989, reference in Chapter 3). (Courtesy Koninklijke/Shell.)

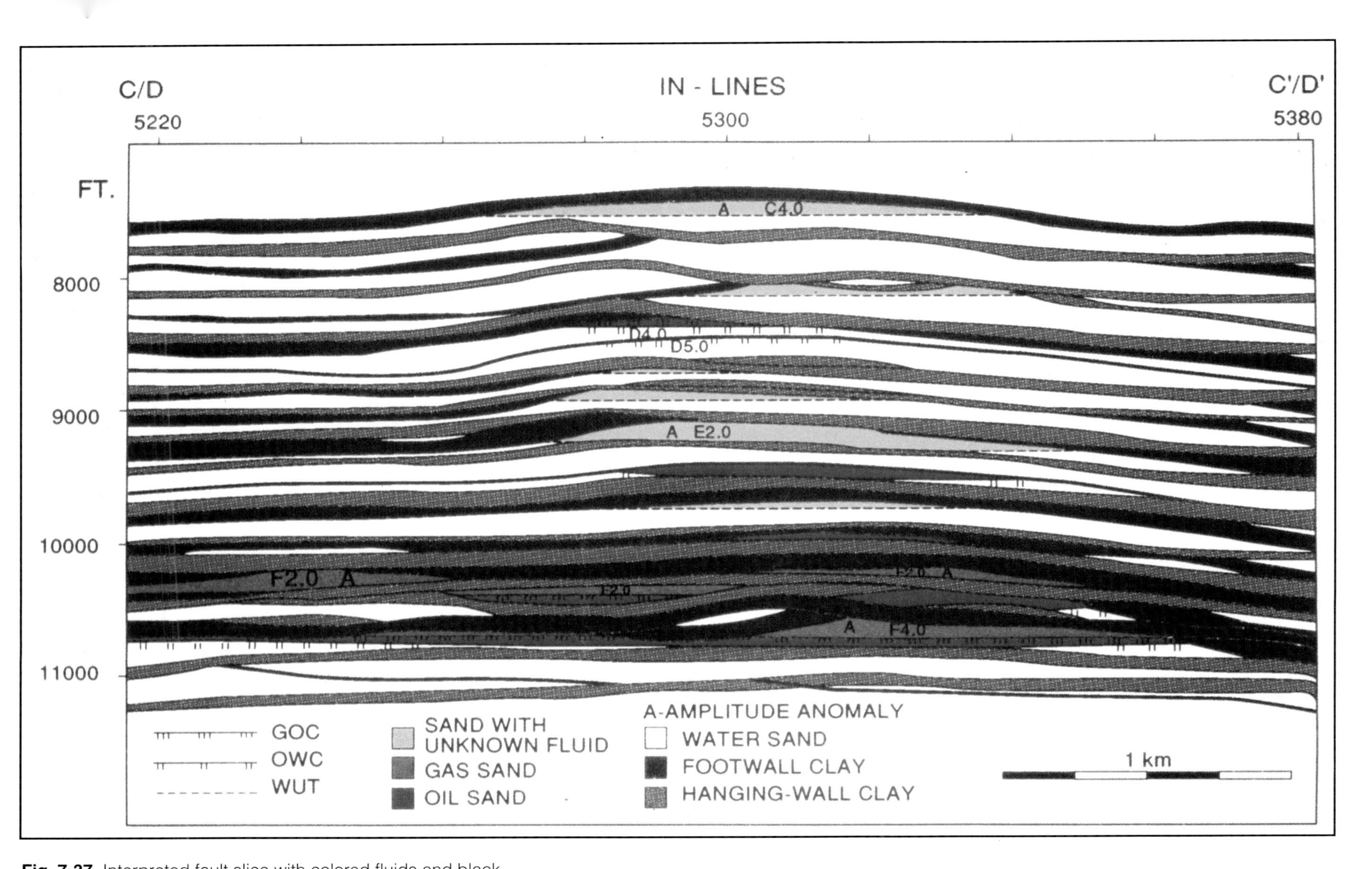

Fig. 7-37. Interpreted fault slice with colored fluids and black shale in one fault block. Superimposed in brown are the shale layers from the fault slice in the juxtaposed block. The fluids still visible potentially leak but may be sealed by clay smear (Bouvier et al., 1989, reference in Chapter 3). (Courtesy Koninklijke/Shell.)

The fluids still visible potentially will leak across the fault. However, shale may be smeared along the fault plane to seal these fluids also. Clay smear potential is a function of fault throw and shale-to-sand ratio. In this study, the fluids visible in red and green in the lower part of Figure 7-37 are in a region of high calculated clay smear potential (Bouvier, 1989, reference in Chapter 3).

References

Brown, A. R., G. S. Edwards, and R. E. Howard, 1987, Fault slicing—a new approach to the interpretation of fault detail: Geophysics, v. 52, p. 1319-1327.

Brown, A. R., R. M. Wright, K. D. Burkart and W. L. Abriel, 1984, Interactive seismic mapping of net producible gas sand in the Gulf of Mexico: Geophysics, v. 49, p. 686-714.

Curtis, M. P., R. D. Martinez, S. Possato and M. Saito, 1983, 3-dimensional seismic attributes contribute to the stratigraphic interpretation of the Pampo oil field, Brazil: Proceedings, SEG 53rd Annual Meeting, p. 478-481.

Domenico, S. N., 1974, Effect of water saturation on seismic reflectivity of sand reservoirs encased in shale: Geophysics, v. 39, p. 759-769.

Johnstad, S. E., R. H. Seymour, and P. J. Smith, 1995, Seismic reservoir monitoring over the Oseberg Field during the period 1989–1992: First Break, v. 13, p. 169–183.

McCarthy, C. J., 1984, Seismic prediction of pore fluid and gas thickness: Proceedings, SEG 54th Annual Meeting, p. 326-328.

Matteucci, G., 1996, Seismic attribute analysis and calibration—a case study from the Gulf of Mexico: Geophysical Society of Houston Technical Workshop, April 17.

Meckel, L. D., Jr., and A. K. Nath, 1977, Geologic considerations for stratigraphic modeling and interpretation, *in* C. E. Payton, ed., Seismic stratigraphy-applications to hydrocarbon exploration: AAPG Memoir 26, p. 417-438.

Neff, D. B., 1990, Estimated pay mapping using three-dimensional seismic data and incremental pay thickness modeling: Geophysics, v. 55, p. 567–575.

Woock, R. D. and A. R. Kin, 1987, Predictive isopach mapping of gas sands from seismic impedance: modeled and empirical cases from Ship Shoal Block 134 field: AAPG Bulletin, v. 71, p. 1143-1151.

Wyllie, M. R. J., A. R. Gregory and G. H. F. Gardner, 1958, An experimental investigation of factors affecting elastic wave velocities in porous media: Geophysics, v. 23, p. 459-493.

HORIZON AND FORMATION ATTRIBUTES

Classification of Attributes

The last few years (since the publication of the Third Edition of this book) have seen an enormous increase in the number of attributes available and their use. Attributes are used to pursue studies in structure, in stratigraphy and in reservoir properties. Therefore this chapter can be viewed as an extension of Chapters 3, 4 and 7.

An attribute is necessarily a derivative of a basic seismic measurement. All the horizon and formation attributes available (Figure 8-1) are not independent of each other but simply different ways of presenting and studying a limited amount of basic information. That basic information is **time**, **amplitude**, **frequency** and **attenuation** and these form the basis of our attribute classification.

As a broad generalization time-derived attributes provide structural information, amplitude-derived attributes provide stratigraphic and reservoir information. Frequency-derived attributes are not yet well understood but there is widespread optimism that they will provide additional useful stratigraphic and reservoir information. Attenuation is not used today but there is a possibility that in the future it will yield information on permeability. Most attributes are derived from the normal stacked and migrated data volume but variations of basic measurements as a function of angle of incidence (and hence source to receiver offset) provides a further source of information. The principal examples of these pre-stack attributes are AVO.

Post-stack attributes can be extracted along one **horizon** or summed over a **window** (Figure 8-1). The latter provides the concept of a formation attribute. In some cases the window is a constant flat time interval so that the display is effectively a thick time slice, sometimes termed a stat (statistical) slice. The window may be of a constant time interval but hung from one structurally-interpreted horizon so that the window properly follows a reservoir interval. The window may also be the interval between two structurally-interpreted horizons, for example the top and the base reservoir reflections. Within the window values may be summed to produce a **gross** attribute measurement, only some of the information may be extracted to provide a **selection** attribute, or the variation of the attribute within the window may be measured to assess a **distribution**. Many examples of the uses of these different types of attributes are provided in this chapter, but not every attribute in Figure 8-1 is illustrated. Some of those listed have yet to be understood.

Attributes are normally calculated and extracted from the data volume following automatic spatial tracking or snapping. These workstation capabilities help the interpreter use all of the data and provide horizon time values at the crest of a

Text continues on page 232

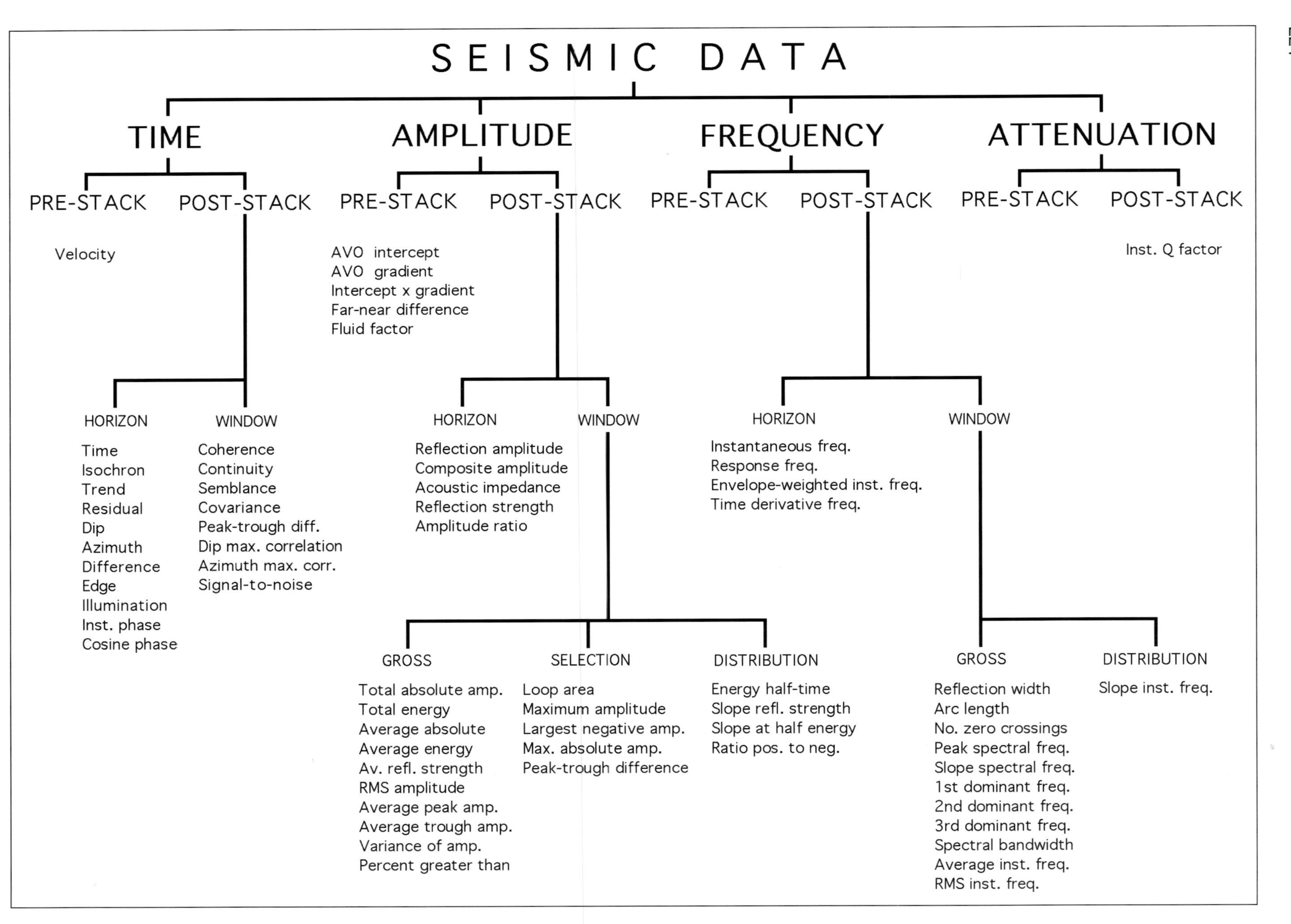

Fig.8-1. Seismic attributes derived from or related to the basic seismic information of time, amplitude, frequency and attenuation. Window can be constant time interval, constant interval hung from one horizon or interval between two horizons.

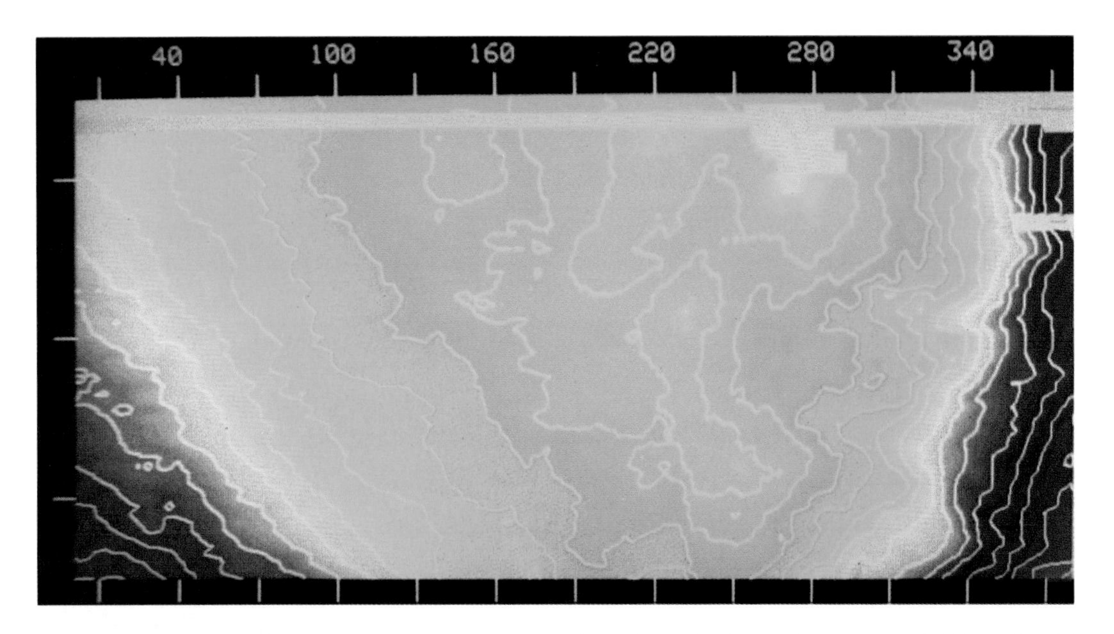

Fig. 8-2. Raw time structure map from Gippsland Basin, offshore SE Australia with value from every live trace. Time range is 1100ms to 1400ms. (Courtesy Landmark Graphics Corporation and BHP Petroleum Pty. Ltd.)

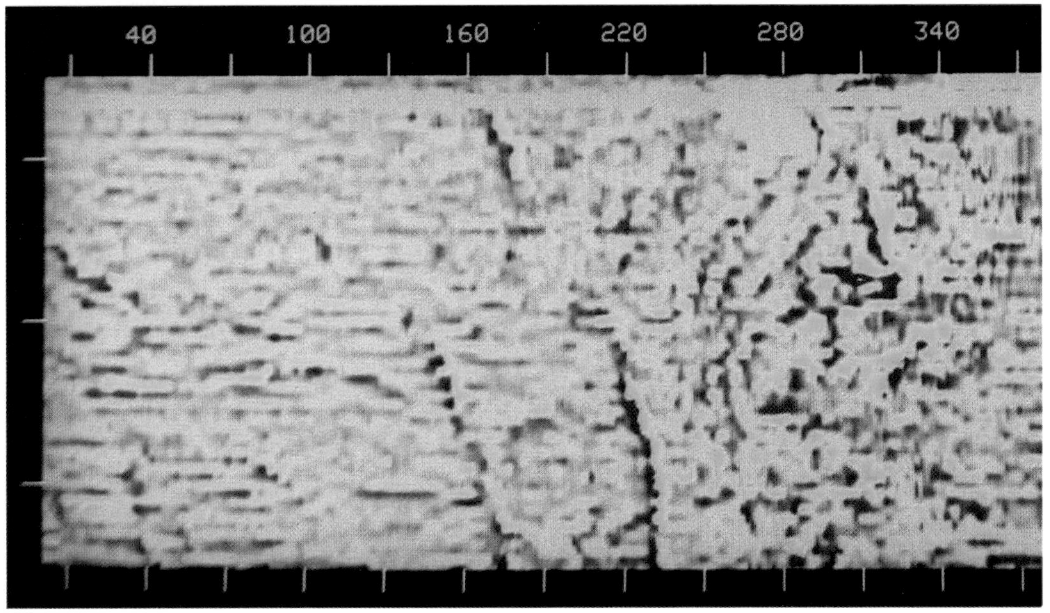

Fig.8-3. Residual map from Gippsland Basin created by subtracting raw and spatially-filtered horizon times. The NNW-SSE trends around crosslines 160 and 220 are faults; blue is upthrown side and red is downthrown side. The E-W linear trends are caused by data collection irregularities. (Courtesy Landmark Graphics Corporation and BHP Petroleum Pty. Ltd.)

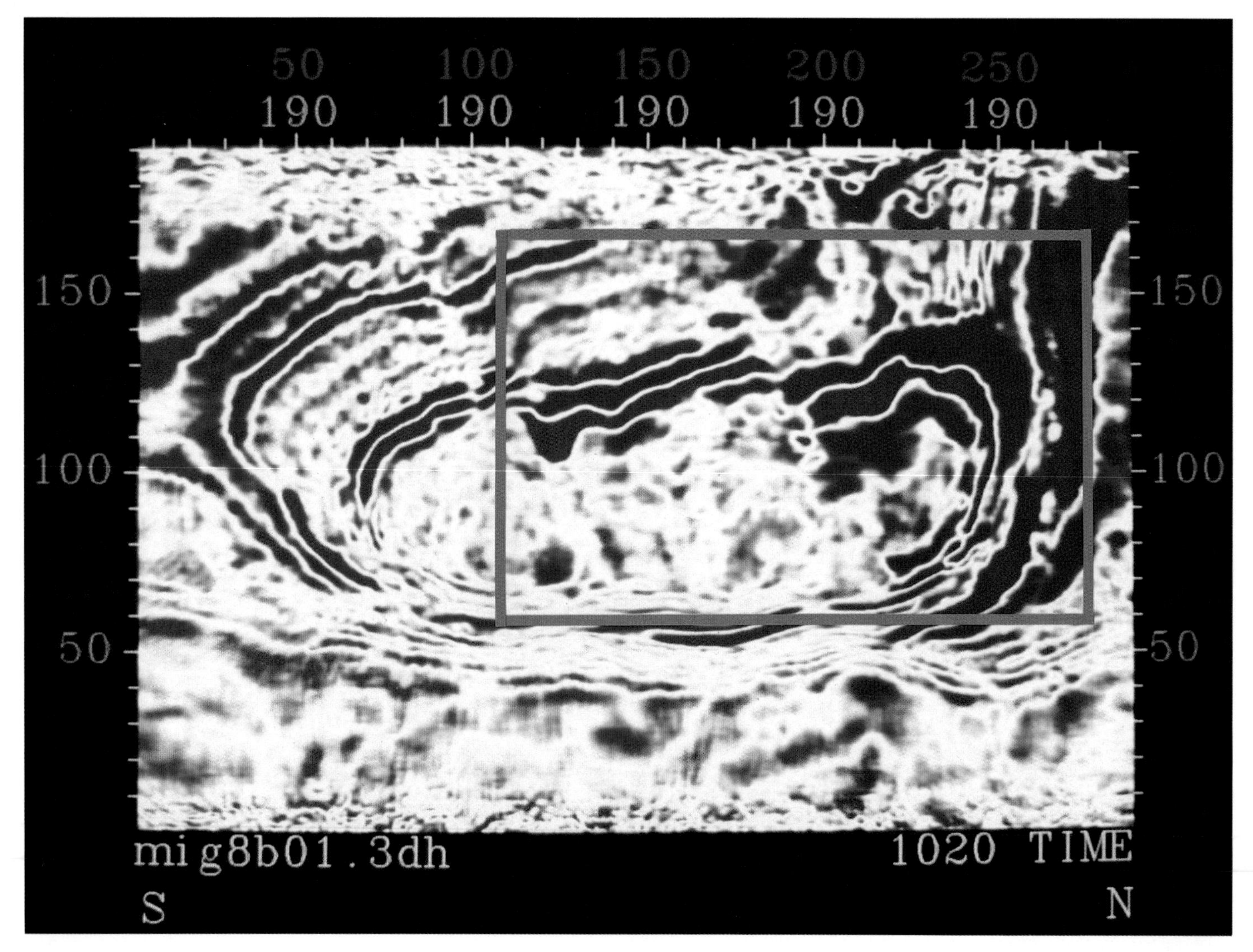

Fig. 8-4. Time slice at 1020ms from Dollarhide field in west Texas showing in green the outline of the area mapped in Figures 8-5 and 8-6. (Courtesy Unocal North American Oil & Gas Division.)

Fig. 8-5. (Opposite Top) Time structure map on Devonian reflection using raw times from automatic horizon tracking. (Courtesy Unocal North American Oil & Gas Division.)

Fig. 8-6. (Opposite Bottom) Residual map created by subtracting the raw and spatially-filtered time maps. Note particularly the lineations which fit existing fault patterns. (Courtesy Unocal North American Oil & Gas Division.)

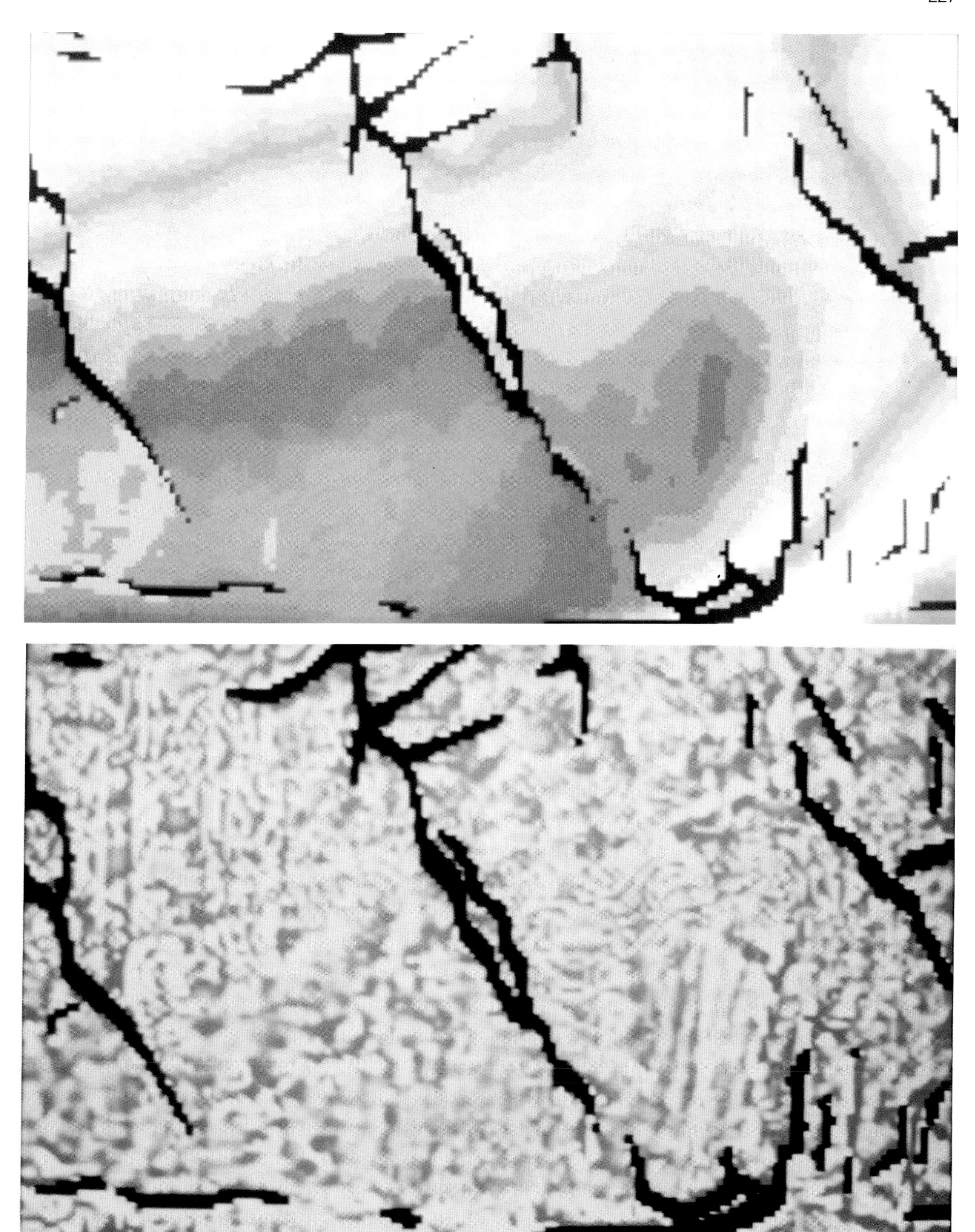

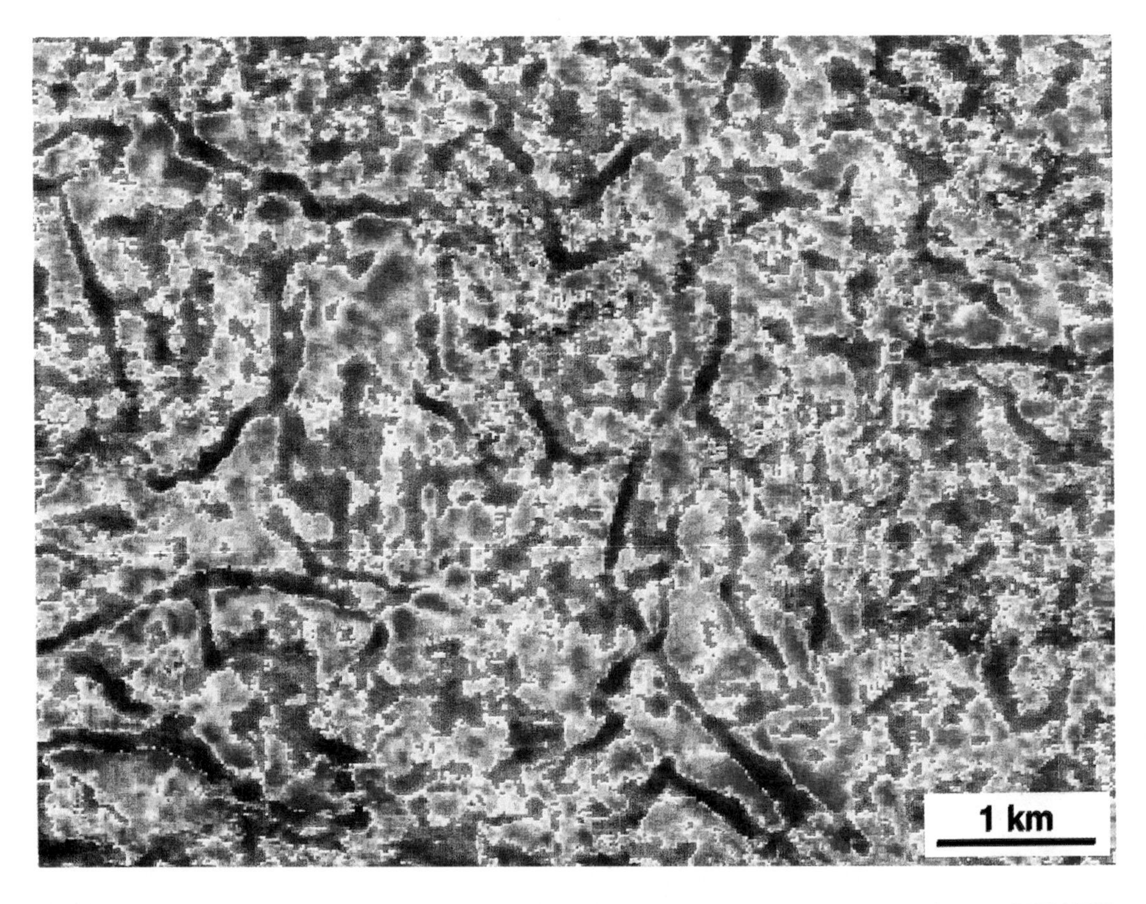

Fig. 8-7. Residual map from Eromanga Basin, South Australia. The linear blue features indicate grabens in the early Cretaceous C horizon occurring in this area around 1100ms. (Courtesy Santos Ltd.)

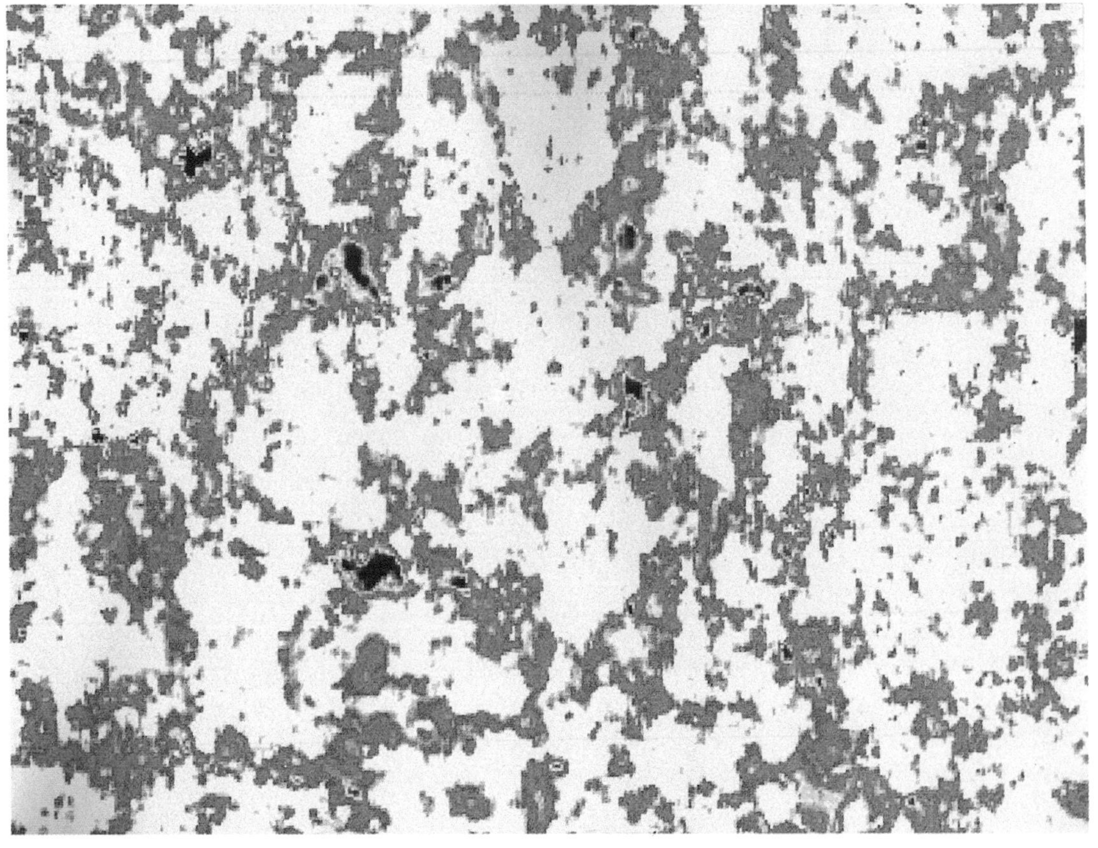

Fig. 8-8. Residual map on the DN horizon from Eromanga Basin, South Australia. The red and yellow patches indicate high spatial frequency drape over underlying bodies of calcified sandstone. (Courtesy Santos Ltd.)

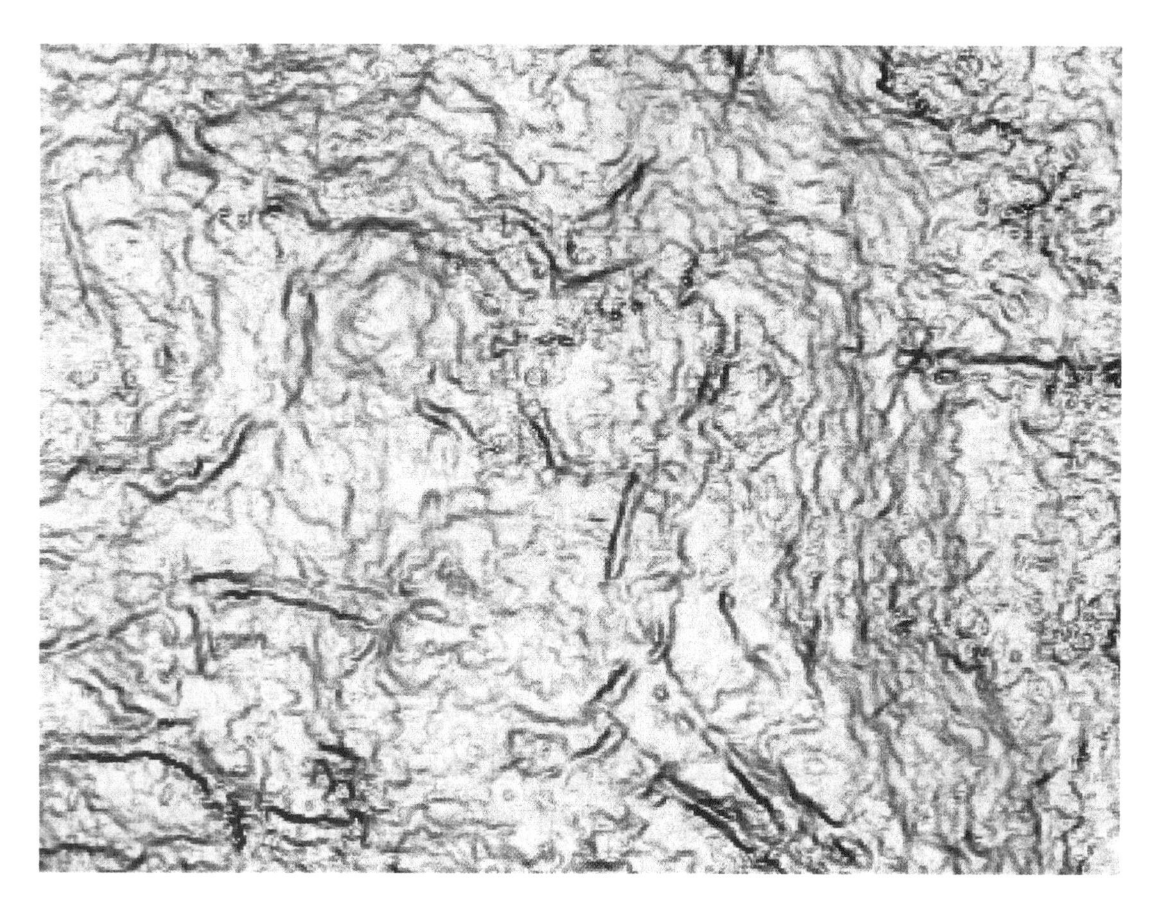

Fig. 8-9. Dip map on the C horizon from Eromanga Basin, South Australia. Note the double lineations indicating conjugate pairs of faults which correspond to the grabens of Figure 8-7. (Courtesy Santos Ltd.)

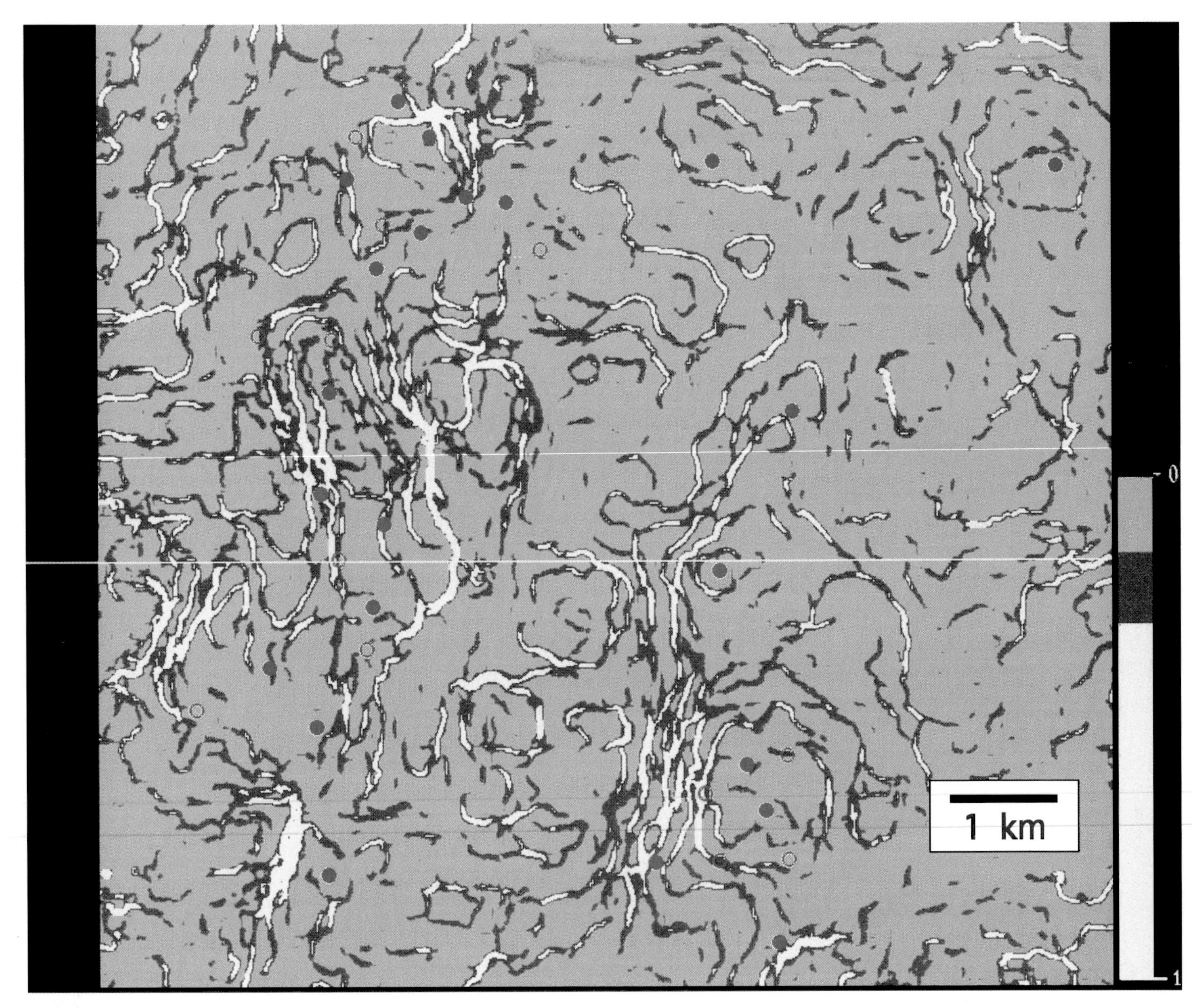

Fig. 8-10. Dip map of top reservoir reflection from Oman indicating large number of arcuate faults.(Courtesy Petroleum Development Oman LLC.)

Fig. 8-11. (Opposite Top) Same dip map as Figure 8-10 with manually-drawn red polygons around the significant dip anomalies. These polygons were then used in the final mapping of top reservoir. (Courtesy Petroleum Development Oman LLC.)

Fig. 8-12. (Opposite Bottom) Vertical section along yellow line in Figures 8-10 and 8-11. The blue horizon is top reservoir. A few sections like this one were used to validate faulting identified on the dip map. (Courtesy Petroleum Development Oman LLC.)

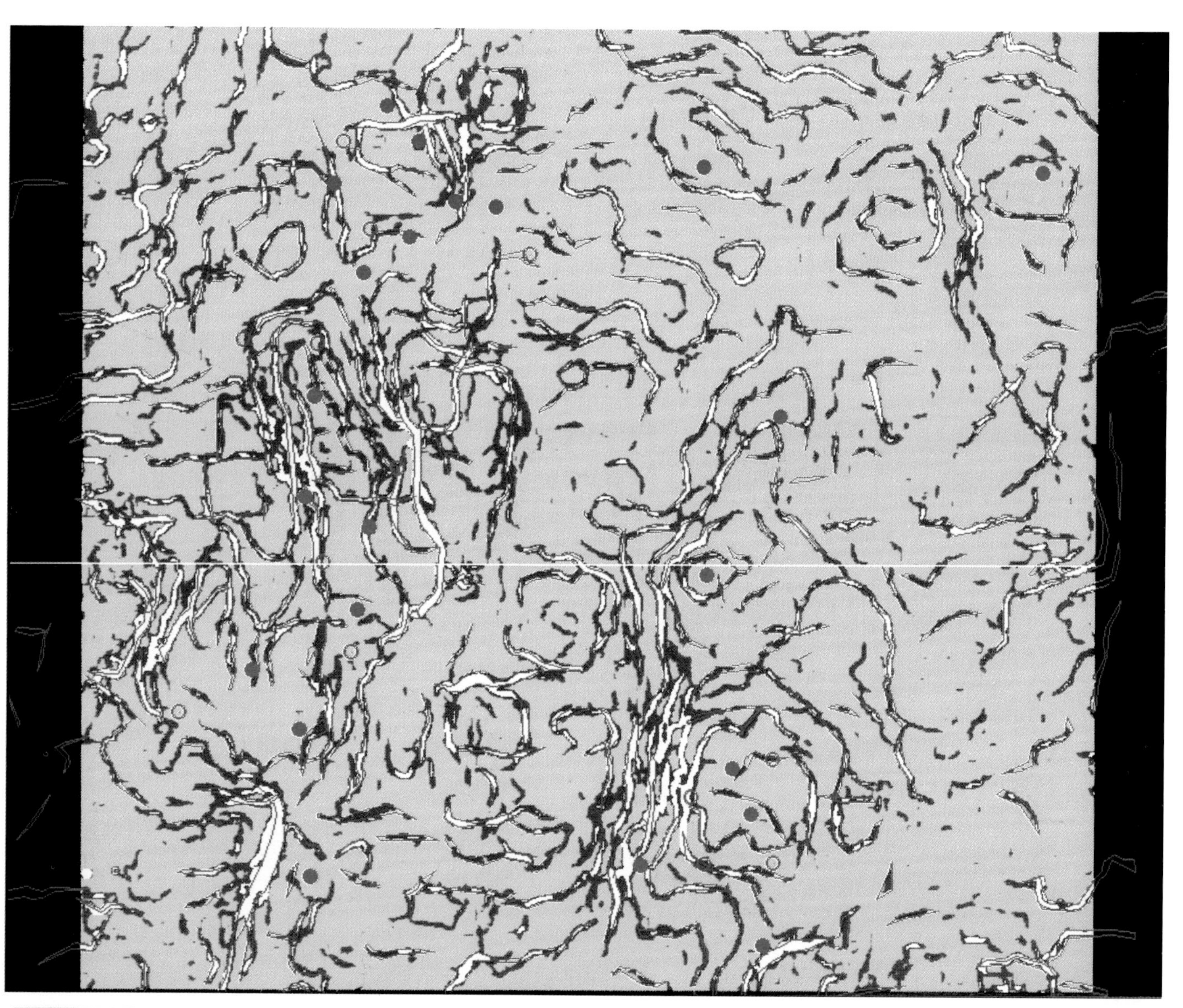

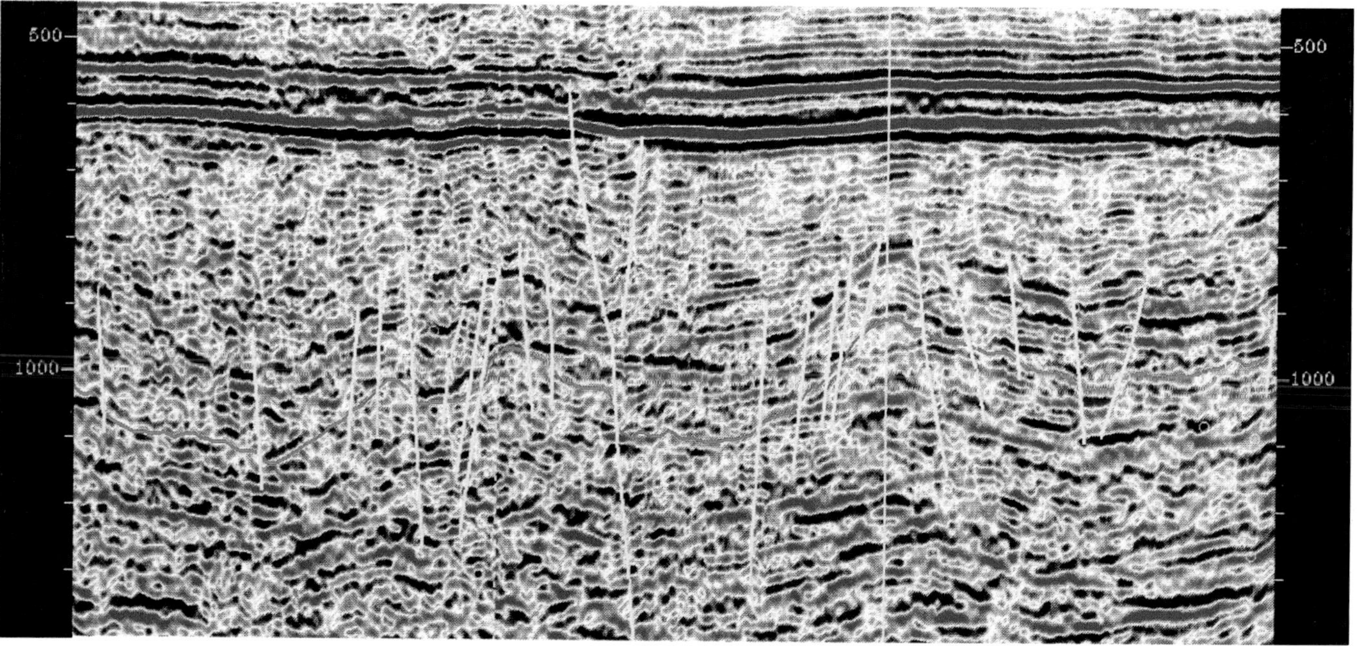
500
500
1000
1000

reflection with a precision of about a quarter of a millisecond. The horizon attribute extracted at the crestal time carries a similar precision. Proper use of machine precision is a key part of seismic attribute analysis as the interpretation objective is normally a rather detailed one. The quotation at the front of this book, "Believe everything to be geology until proved otherwise", applies here. It is wise to put all detailed information into the interpretation in the early stages and consider later what is geology and what is noise. Attribute displays can be very helpful in this important decision.

Time-derived Horizon Attributes

Residual, or high spatial frequency residual map, is the arithmetic difference between a high-precision automatically-tracked time map and its spatially-smoothed equivalent. Figure 8-2 illustrates such a time map from Australia and Figure 8-3 is the corresponding residual map (Denham and Nelson, 1986). There are two types of linear patterns on this residual. The NNW-SSE lineations are faults, mostly too subtle to be seen as contour disturbances on the time map. The exactly-parallel east-west lineations are in the data collection direction and are clearly caused by data collection irregularities; this is a form of noise and is therefore unwanted.

Figure 8-4 shows the area under study in the next example; Figure 8-5 is the auto-tracked high-precision time map; Figure 8-6 is the residual. The black features are gaps in the interpreted horizon which were recognized as faults in the mainstream of the interpretation using vertical and horizontal sections. A linear pattern in the residual parallel to the existing faults and towards the east end of the area is a rather clear indication of another fault or similar geologic feature. In contrast, the north-south patterns towards the west end are highly regular and parallel and do not fit the existing fault patterns; they must be identified as data collection lineations or noise.

Figure 8-7 is a residual map from a land 3-D survey in South Australia; it is the time difference between a snapped and a smoothed version of this early Cretaceous horizon. The dark blue linear features are grabens. Blue means that the snapped horizon is below the smoothed horizon thus establishing that these features are structurally low.

Figure 8-8 is a similarly-produced residual map on another deeper horizon from the same survey. The red and yellow patches indicate high-spatial-frequency drape over underlying bodies of calcified sandstone. The calcite varies rapidly both horizontally and vertically in this interval and therefore constitutes a serious problem because of its effect on depth-conversion velocities for deeper horizons. This residual map is helping to address the problem.

Dip, or dip magnitude, is another time-derived horizon attribute addressing issues of structural detail (Dalley et al., 1989). On the high-precision automatically-tracked time surface one time value is considered in relation to its immediate neighbors to form a local plane. The true dip of that local plane is the attribute dip; the direction of that dip is the azimuth, or dip azimuth, which will be discussed later.

Figure 8-9 shows the dip map for the same horizon as used in Figure 8-7. The double lineations indicate conjugate pairs of faults which correspond to the grabens of Figure 8-7 (Oldham and Gibbins, 1995).

Figure 8-10 shows the dip map for a top reservoir reflection from Oman. Many arcuate features are evident. The color scheme was set so that all that was clearly noise was suppressed into the gray-green; everything else was considered as a potential real fault. The red polygons in Figure 8-11 were drawn around the majority of the dip anomalies and were then used directly in the mapping process. In drawing the red polygons and in deciding which dip anomalies were real, a few vertical sections like Figure 8-12 were studied. The faults validated with a red polygon along the east-west yellow line in Figure 8-11 are shown in yellow in Figure 8-12.

Text continues on page 238

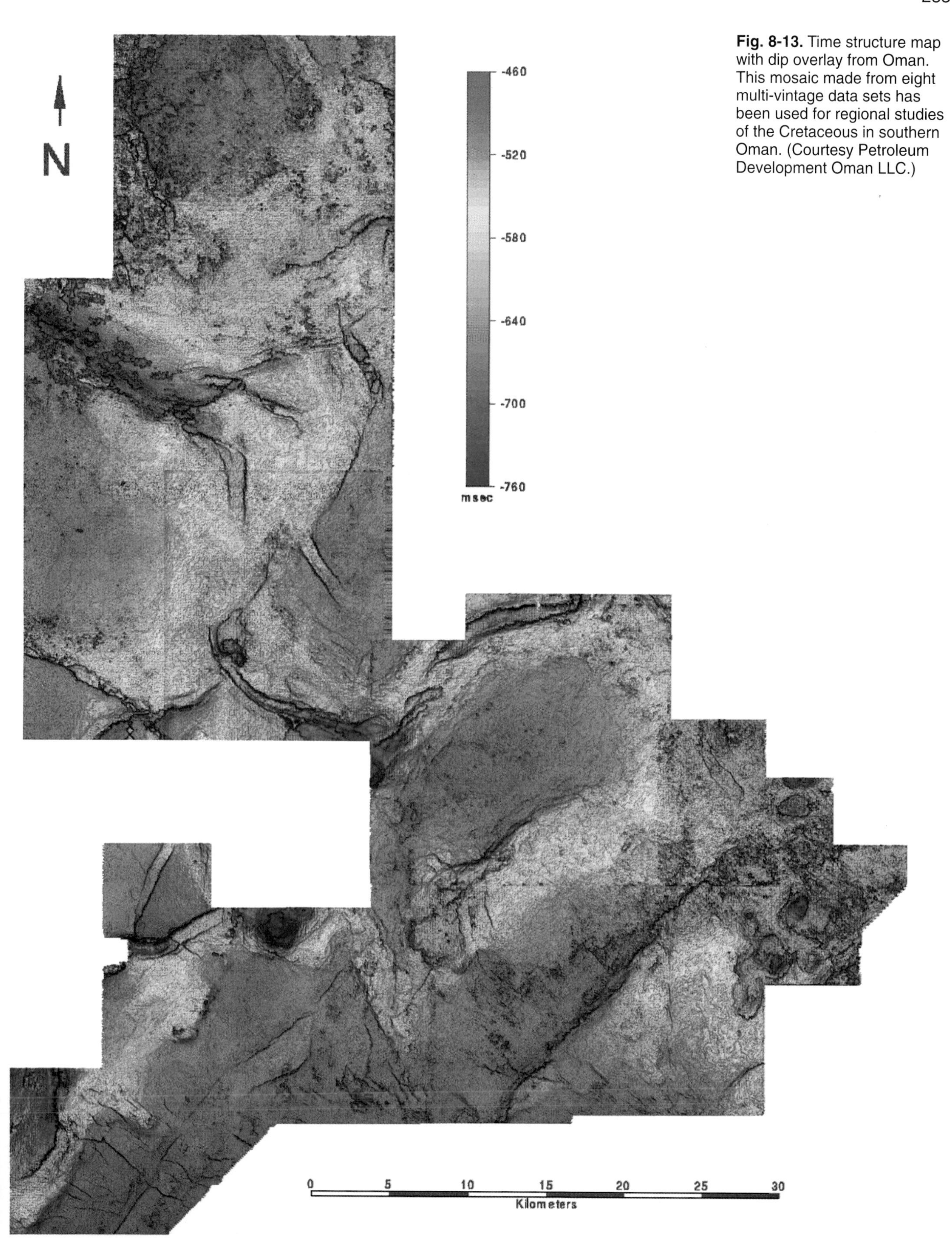

Fig. 8-13. Time structure map with dip overlay from Oman. This mosaic made from eight multi-vintage data sets has been used for regional studies of the Cretaceous in southern Oman. (Courtesy Petroleum Development Oman LLC.)

Fig. 8-14. Dip map used as quality control of automatic tracker performance. The very straight, high-dip pseudo-faults are tracking busts. (Courtesy Nederlandse Aardolie Maatschappij B.V.)

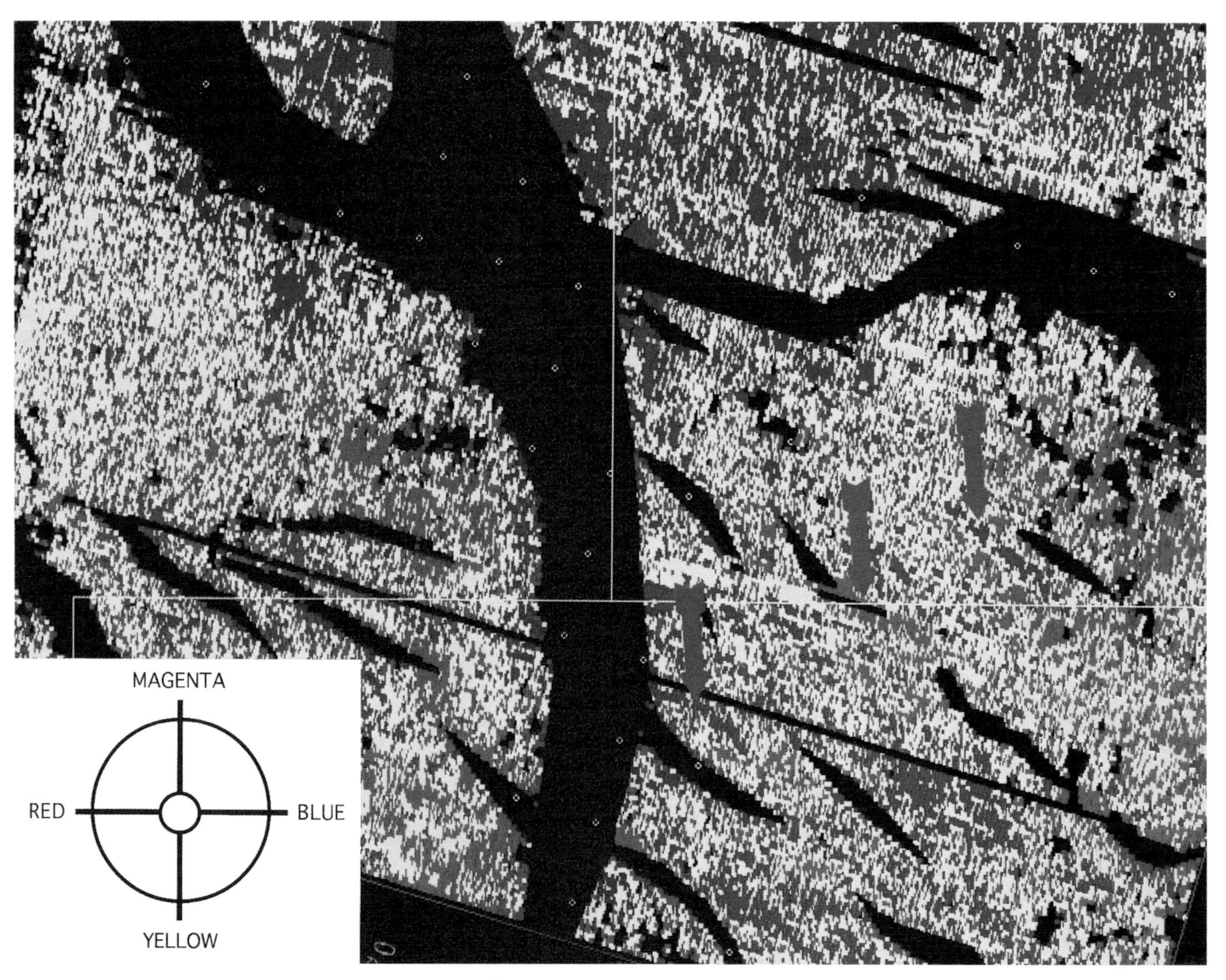

Fig. 8-15. Azimuth map from Lake Maracaibo, Venezuela. Red arrows indicate anomalies which conform well to existing fault patterns and are thus probable additional faulting. (Courtesy Maraven S.A.)

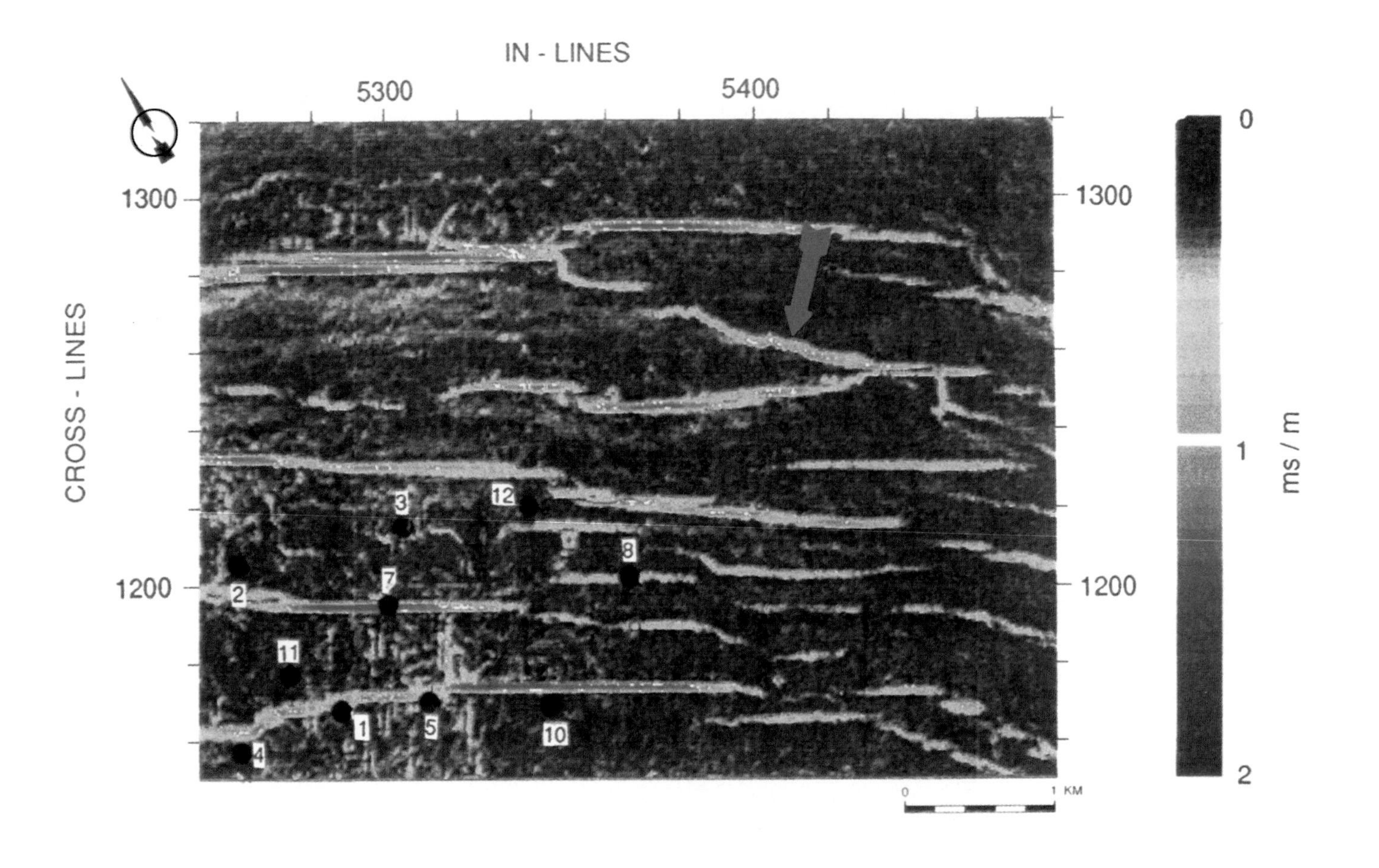
IN - LINES
5300
5400
CROSS - LINES
1300
1200
1300
1200
0
1
2
ms / m
0
1 KM
3
12
8
2
7
11
1
5
10
4

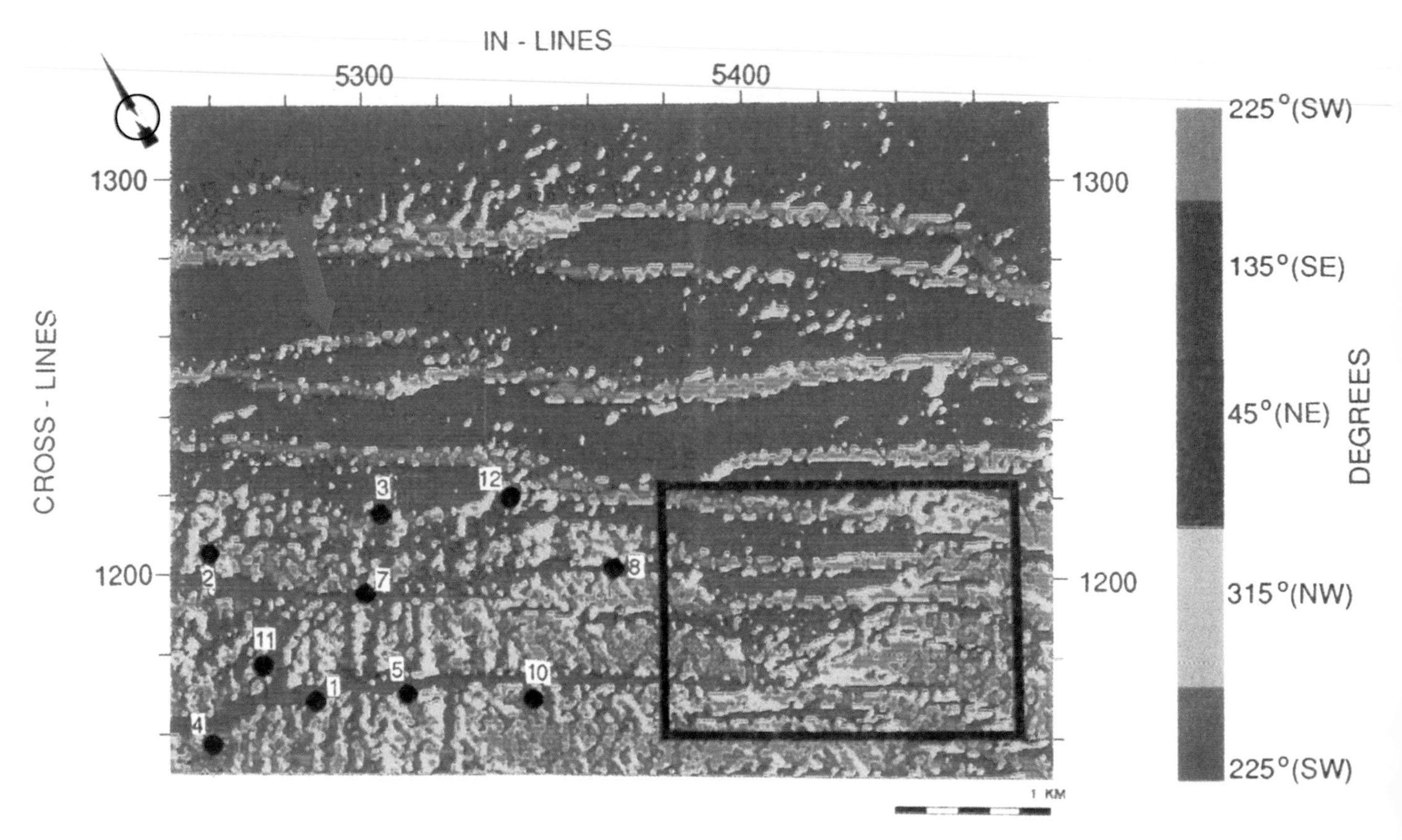
IN - LINES
5300
5400
CROSS - LINES
1300
1200
1300
1200
225°(SW)
135°(SE)
45°(NE)
315°(NW)
225°(SW)
DEGREES
1 KM
3
12
2
8
7
11
5
10
1
4

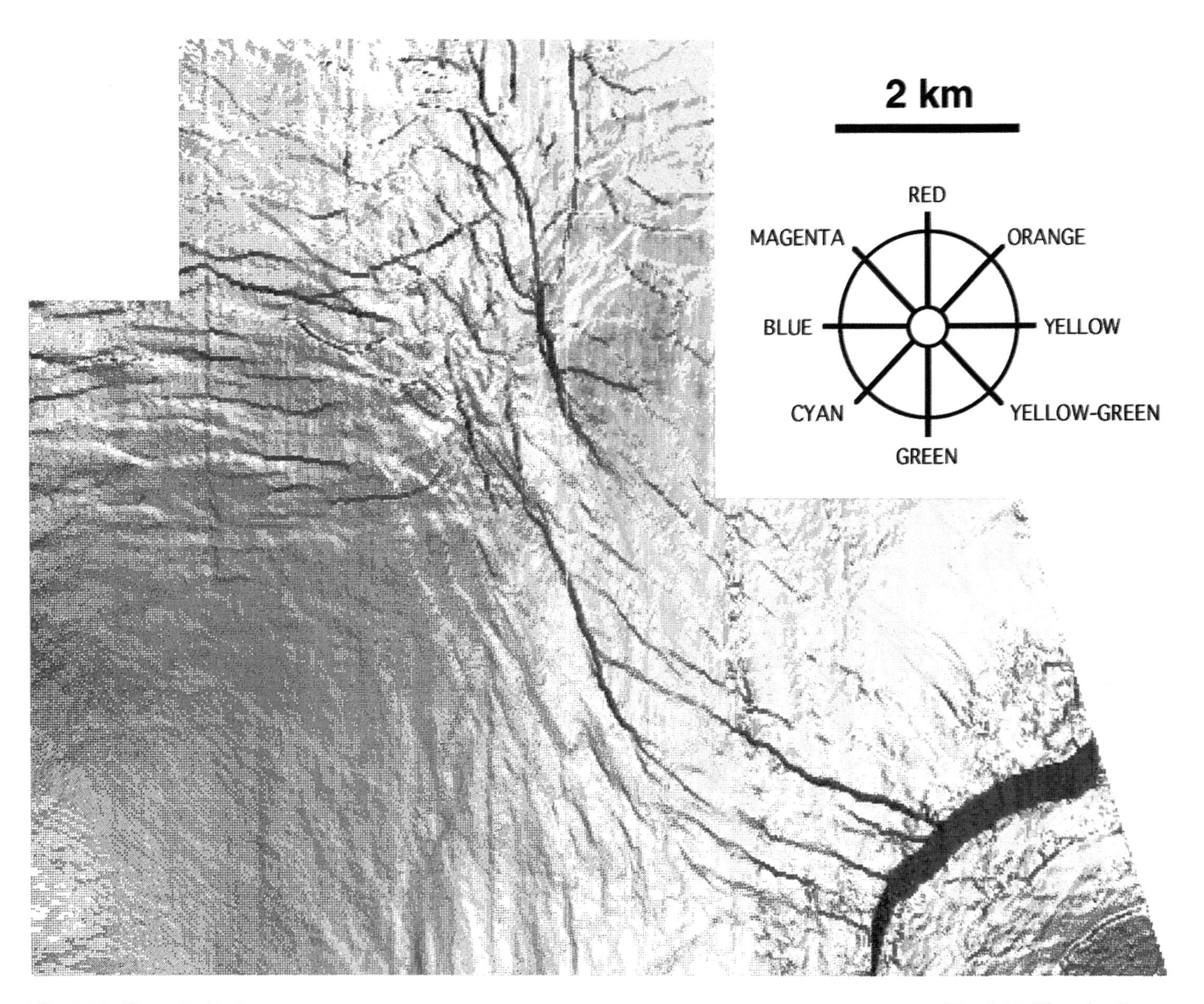

Fig. 8-16. (Opposite Top) Dip map from Nun River field, Nigeria, showing many faults (from Bouvier et al, 1989). Arrow indicates one fault visible on dip map but not on azimuth map in Figure 8-17. (Courtesy Koninklijke / Shell.)

Fig. 8-17. (Opposite Bottom) Azimuth map from Nun River field, Nigeria, showing generally the same faults as in Figure 8-16 (from Bouvier et al, 1989). Arrow indicates one fault visible in azimuth but not in dip. (Courtesy

Fig. 8-18. Dip-azimuth combination map from offshore The Netherlands. Azimuth is coded to hue according to legend; dip is coded to color density. The horizon is just above a salt dome located in the SW corner. (Courtesy Nederlandse Aardolie Maatschappij B.V.)

Figure 8-13 shows a mosaic of eight 3-D surveys from southern Oman covering a rather large area. The color represents time structure and the shaded overlay is dip indicating principally the faults. This combination display has been very useful in regional studies in the area.

Figure 8-14 shows the dip map of a horizon which was autotracked with little vertical and horizontal section control. Although the tracking was successful over much of the area, the tracker jumped onto the wrong event in several places. These tracking busts show as very straight high-dip pseudo-faults primarily in green and blue. This demonstrates that time-derived attribute displays such as dip can be used to quality control the performance of an autotracker as well as to establish further structural detail.

Azimuth, or dip azimuth, is used in a similar way to dip. Figure 8-15 shows an example of an azimuth map from Lake Maracaibo, Venezuela. The data is clearly noisy but nonetheless several meaningful patterns can be seen. The red arrows indicate anomalies which conform with existing fault patterns and are thus probable additional faulting. Two of the indicated faults are magenta and blue and thus dip to the northeast; the third is red and yellow and thus dips to the southwest.

Figures 8-16 and 8-17 show a pair of dip and azimuth displays from Nigeria. Both show many faults which are generally the same. However, red arrows indicate two faults, each of which is visible on only one of the displays. For a subtle fault the dip of the fault can accidentally concur with the dip of the surrounding horizon surface; likewise separately for the azimuth. It is thus wise to look at more than one type of attribute display in order to reduce the chance of missing something significant.

A **dip-azimuth** combination map (Figure 8-18) combines the separate dip and azimuth attributes onto the one display and is thus an alternative approach to the fault visibility problem discussed above. In Figure 8-18 the dip is coded to color density and the azimuth is coded to hue according to the circular legend. Many faults associated with salt movement are visible.

Edge, or edge detection, is a spatial operator usually occupying 9 points which operates like a spatial smoothing filter but has the opposite effect; spatial differences or edges are accentuated regardless of orientation. Figure 8-19 is an edge map on the top Tosca reflection in the Neuquen Basin of Argentina. In the west there is a swarm of north-south faults and several 7 km long arcuate faults. These arcuate faults look impressive on the edge map but have a barely-visible displacement on vertical sections. Similar arcuate patterns on the edge maps from deeper horizons indicate that these faults are in fact conical in shape pointing downwards with their apexes on an igneous intrusion.

Figure 8-20 is an edge map from southern Alberta in Canada here being used to delineate a reef.

Illumination, lighting, shaded relief or sun shading is a display technique already well known from topographic and other kinds of mapping. Portions of the autotracked time surface pointing towards the source are highlighted, those pointing away from the source are in shadow. Figure 8-21 has an illumination direction from the northwest. The fault running north-south down the west side is highlighted. Just east of this is a sequence of en echelon faults in shadow. Just southeast of center is a clearly-visible graben.

Figures 8-22 and 8-23 show the same time surface with different illumination directions as shown by the red arrows. Faults highlighted on one are in shadow on the other and vice versa. This is marine data collected in the northwest-southeast direction. With the illumination direction perpendicular to this the data collection lineations are visible (Figure 8-23); with the illumination direction parallel to the data collection the lineations are not visible (Figure 8-22).

Coherence and Continuity

Coherence, continuity, semblance and covariance are all rather similar. They aim to convert a volume of continuity (the normal reflections) into a volume of disconti-

Text continues on page 249

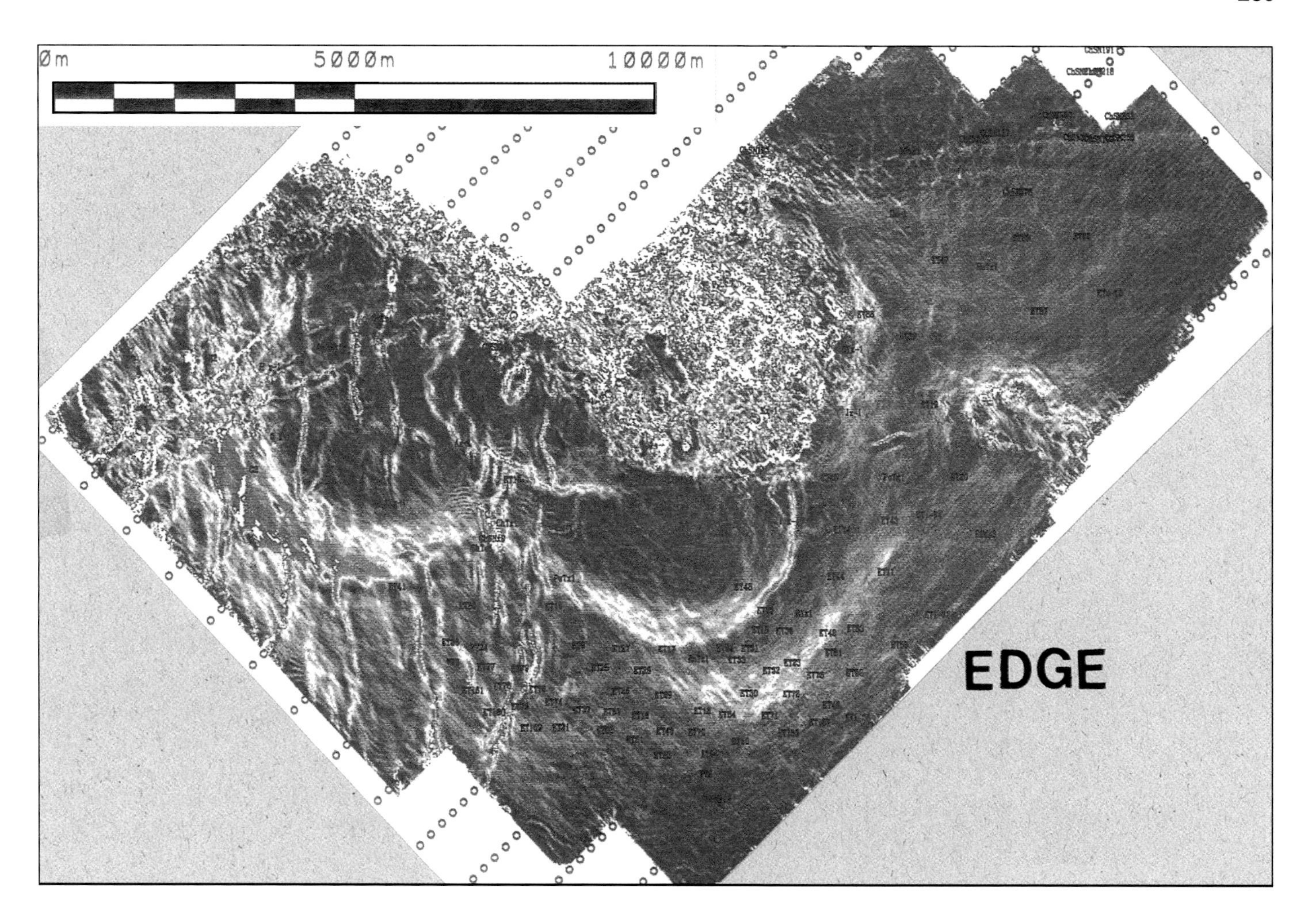

Fig. 8-19. Edge map from Neuquen Basin, Argentina. Arcuate faults about 7 km long have barely-visible displacements on vertical sections. (Courtesy Petrolera Argentina San Jorge S.A.)

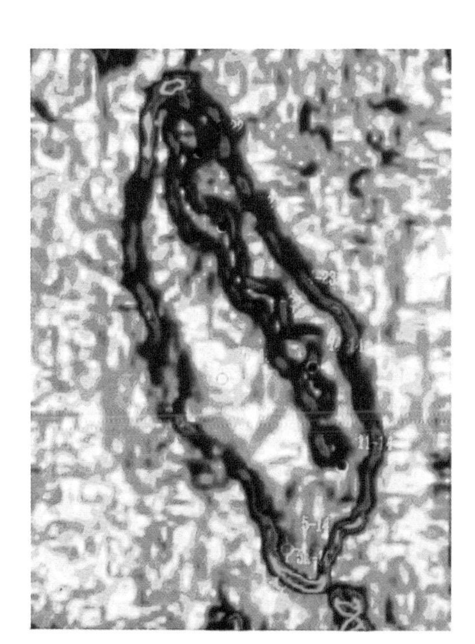

Fig. 8-20. Edge map from southern Alberta, Canada, outlining reef. (Courtesy Landmark Graphics Corporation.)

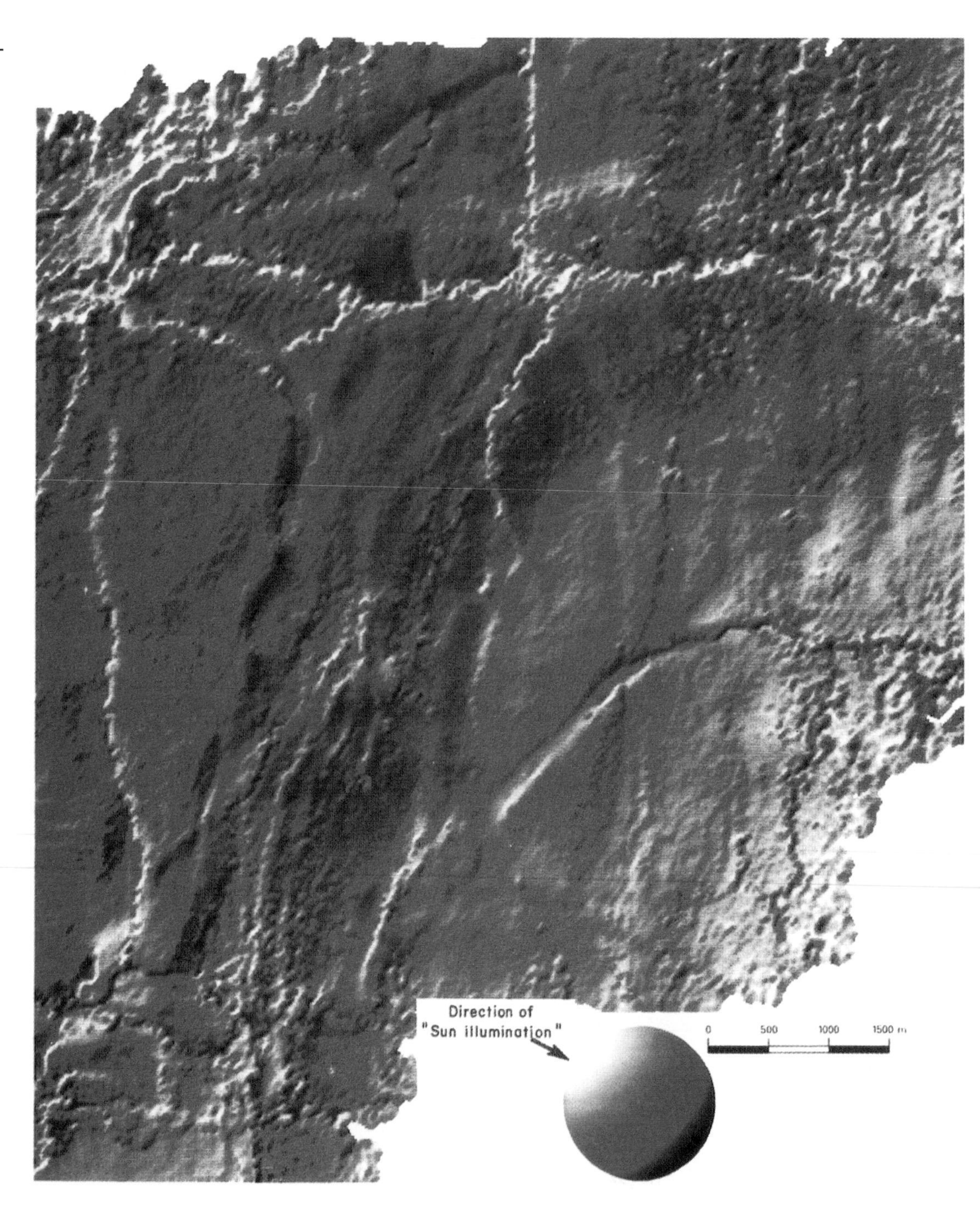

Fig. 8-21. Illumination display or shaded relief map on Base Zechstein refection from Annerveen gas field, The Netherlands (from Hoetz and Watters, 1992). The direction of illumination is from the northwest. Note the graben just SE of center. (Courtesy Nederlandse Aardolie Maatschappij B.V.)

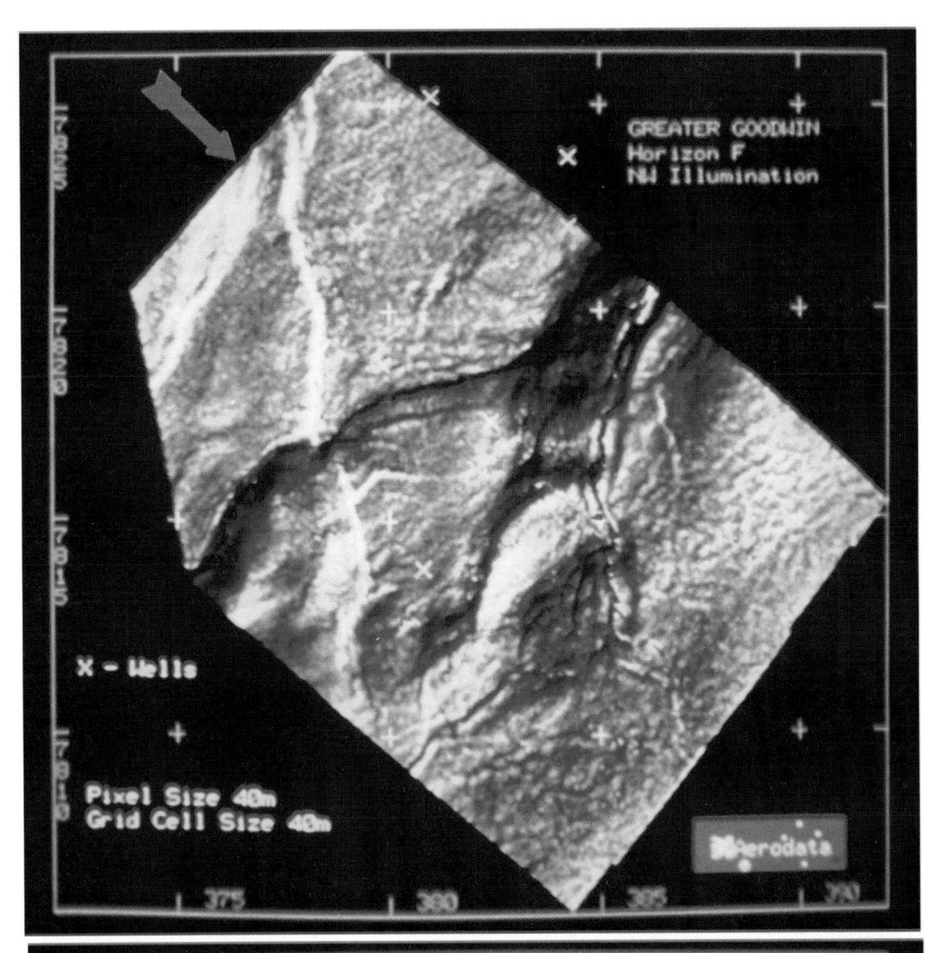

Fig. 8-22. Illumination display of horizon from Goodwyn field, Northwest Shelf, Australia. The direction of illumination is from the northwest. (Courtesy Woodside Offshore Petroleum Pty. Ltd.)

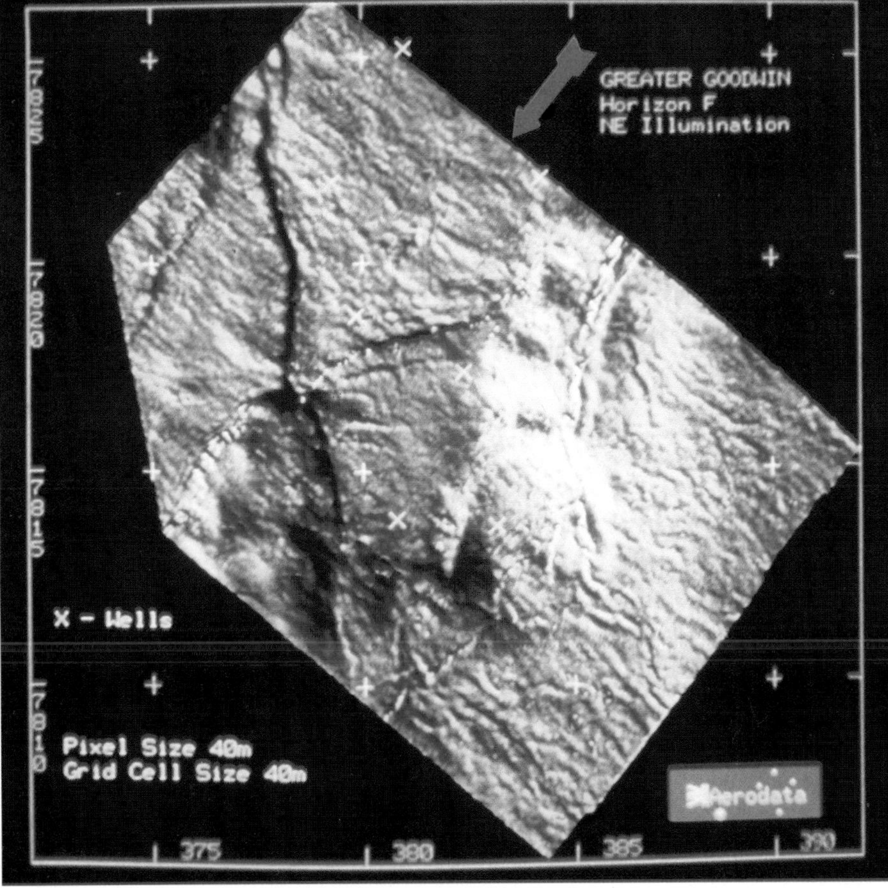

Fig. 8-23. Illumination display of horizon from Goodwyn field, Northwest Shelf, Australia. The direction of illumination is from the northeast for comparison with Figure 8-22. Note how the data collection lineations are now visible. (Courtesy Woodside Offshore Petroleum Pty. Ltd.)

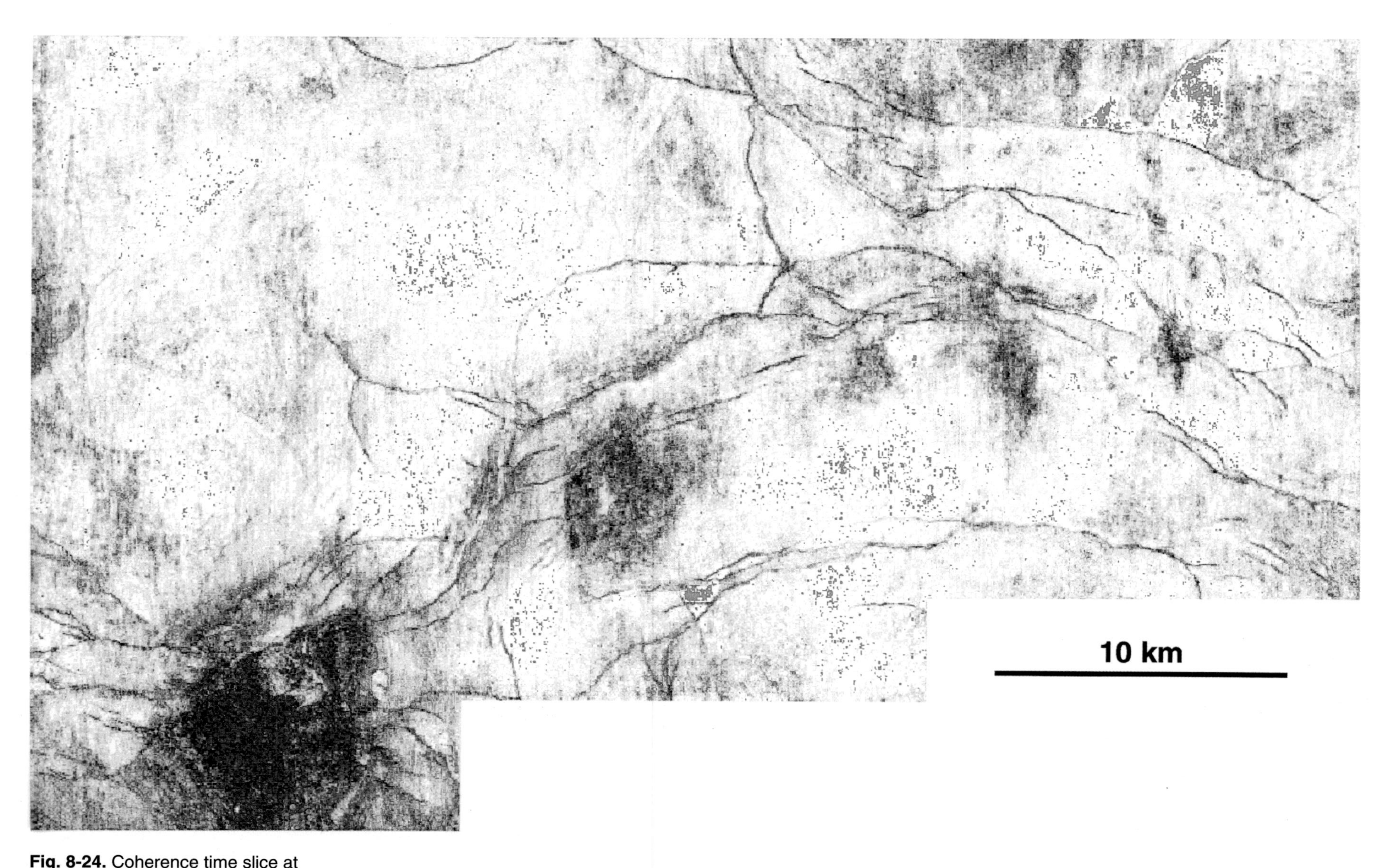

Fig. 8-24. Coherence time slice at 1525ms from Gulf of Mexico. Area covers 60 blocks. Note, by comparison with Figure 8-25, how faults are equally visible when parallel to or perpendicular to structural strike. (Courtesy Amoco Corporation.)

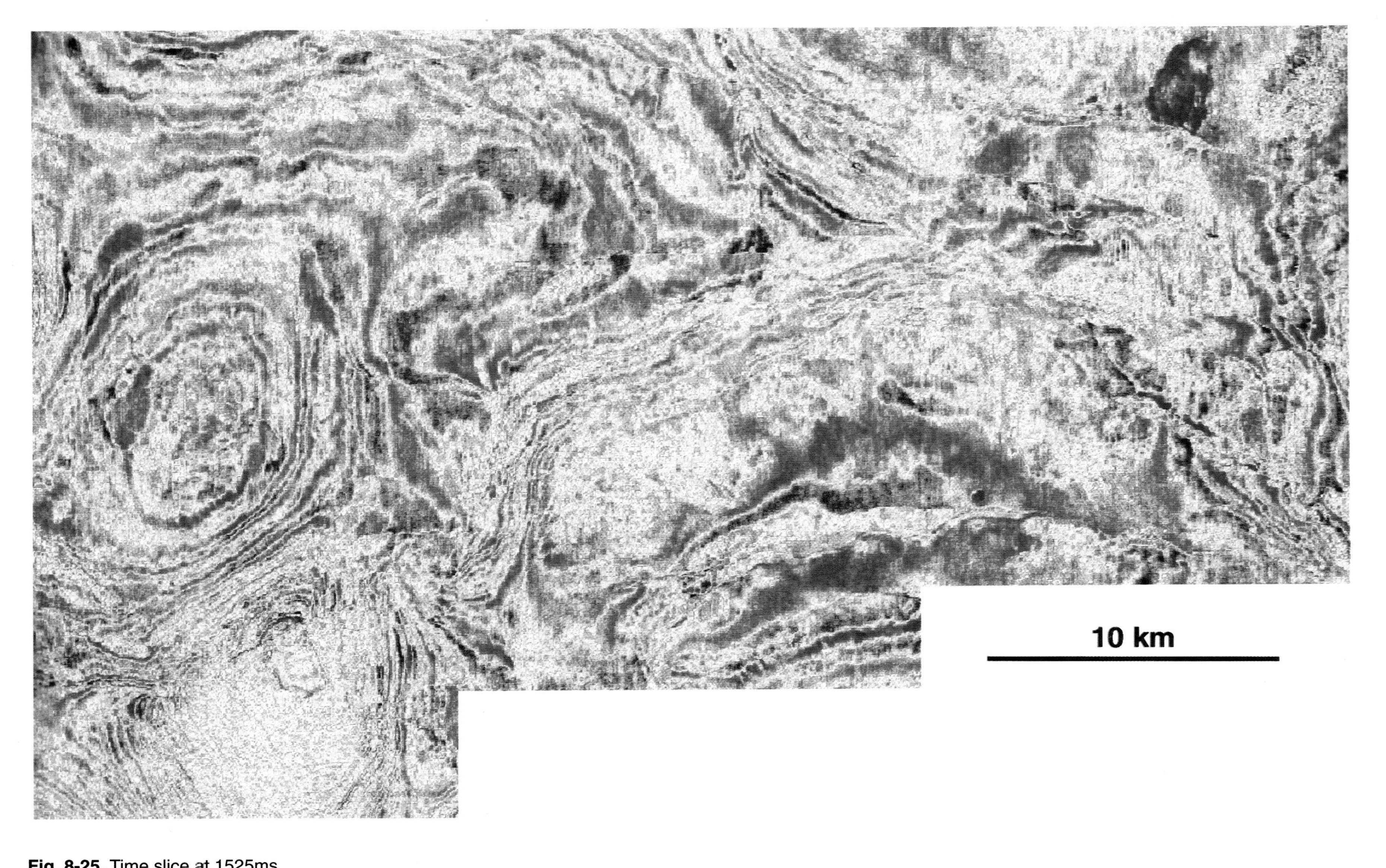

Fig. 8-25. Time slice at 1525ms from Gulf of Mexico for comparison with Figure 8-24. (Courtesy Amoco Corporation.)

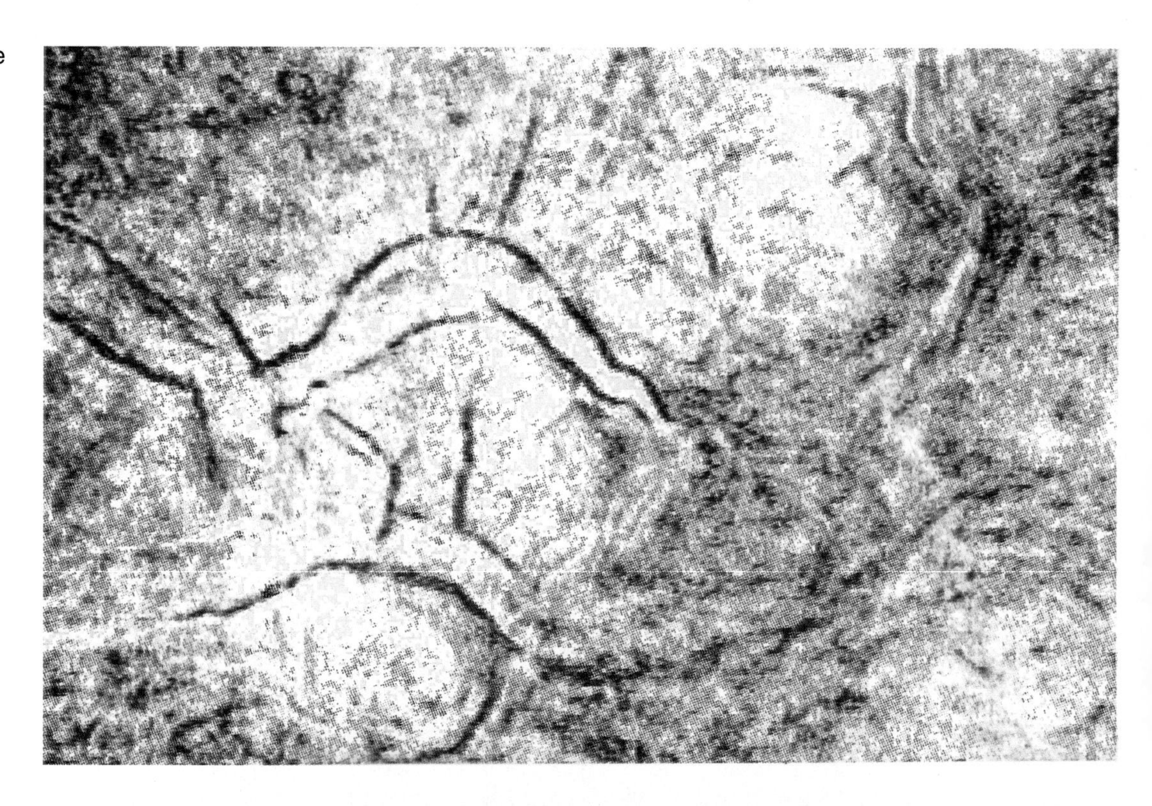

Fig. 8-26. Coherence time slice at 1640ms from North Sea (from Bahorich et al, 1995). (Courtesy Amoco Corporation.)

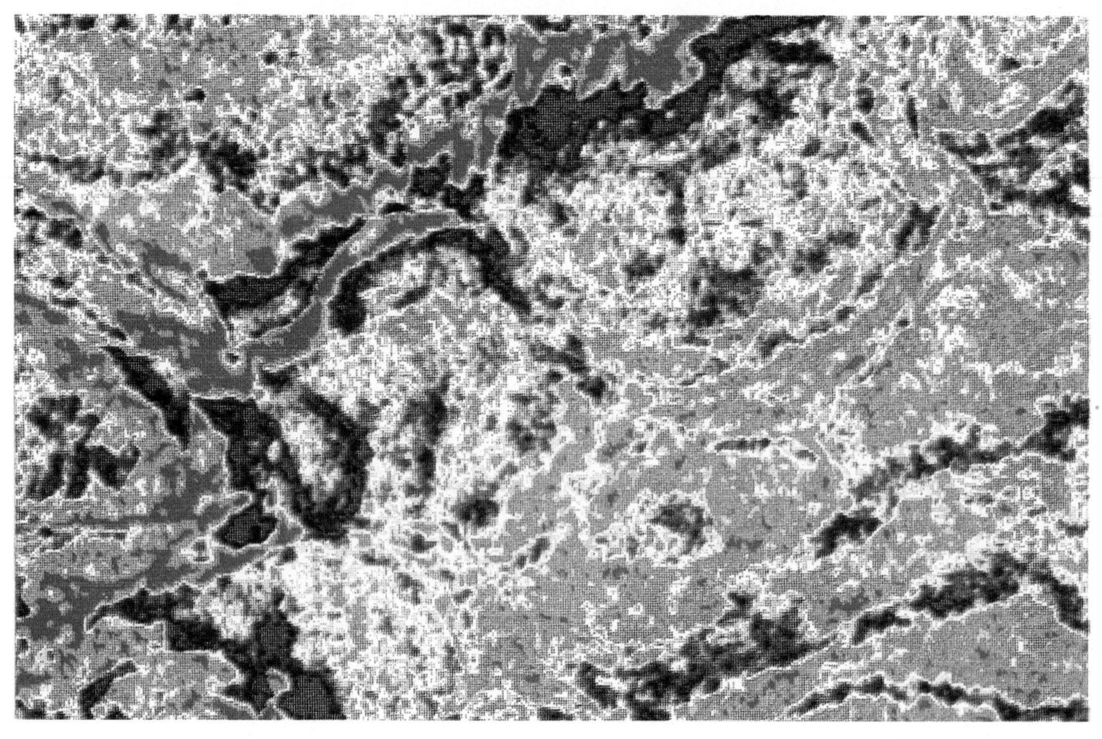

Fig. 8-27. Time slice at 1640ms from North Sea for comparison with Figure 8-26 (from Bahorich et al, 1995). (Courtesy Amoco Corporation.)

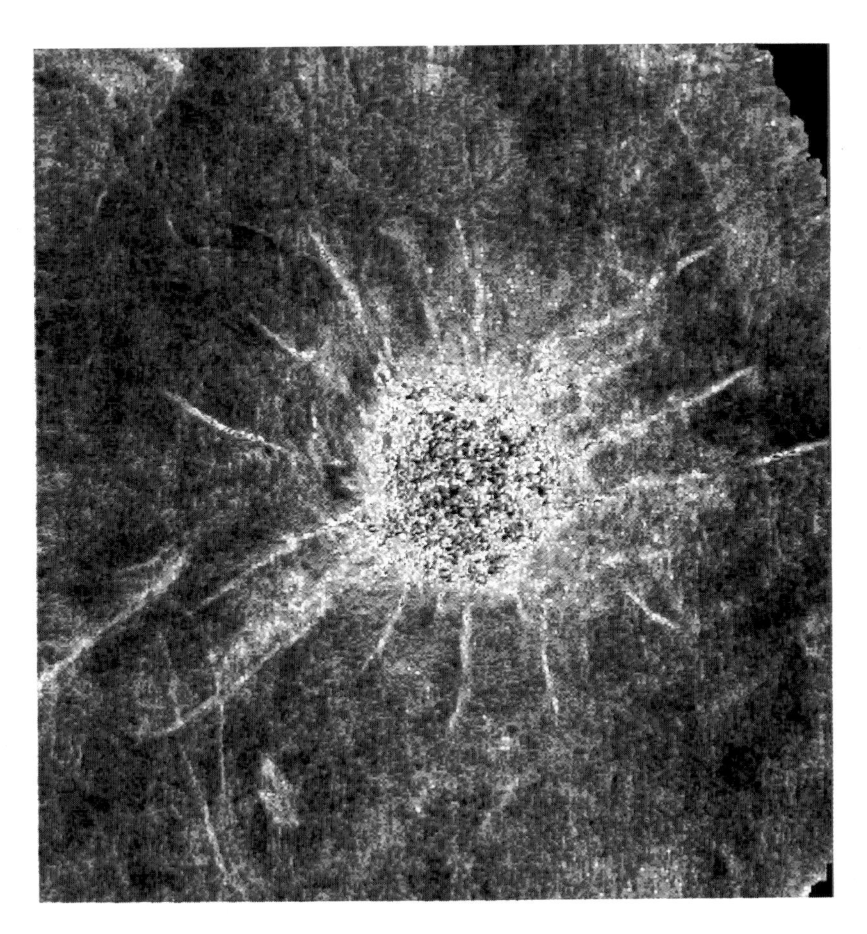

Fig. 8-28. Continuity time slice at about 1500ms from Gulf of Mexico showing many radial faults. (Courtesy CAEX Services and Landmark Advance Products Group.)

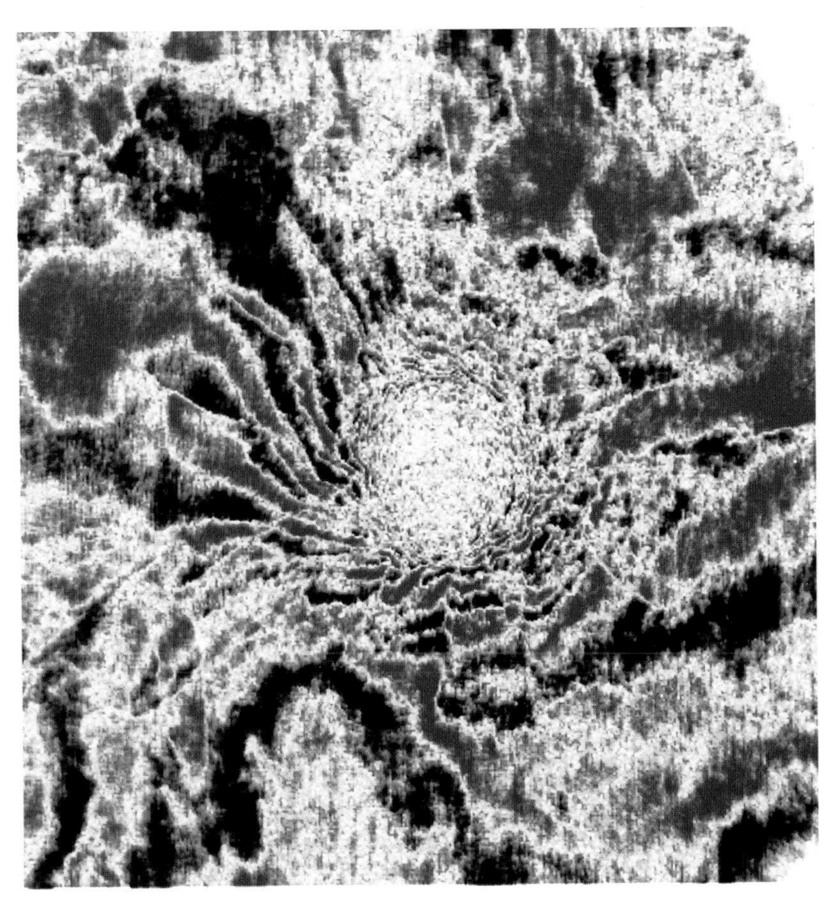

Fig. 8-29. Time slice at about 1500ms from Gulf of Mexico for comparison with Figure 8-28. (Courtesy CAEX Services and Landmark Advance Products Group.)

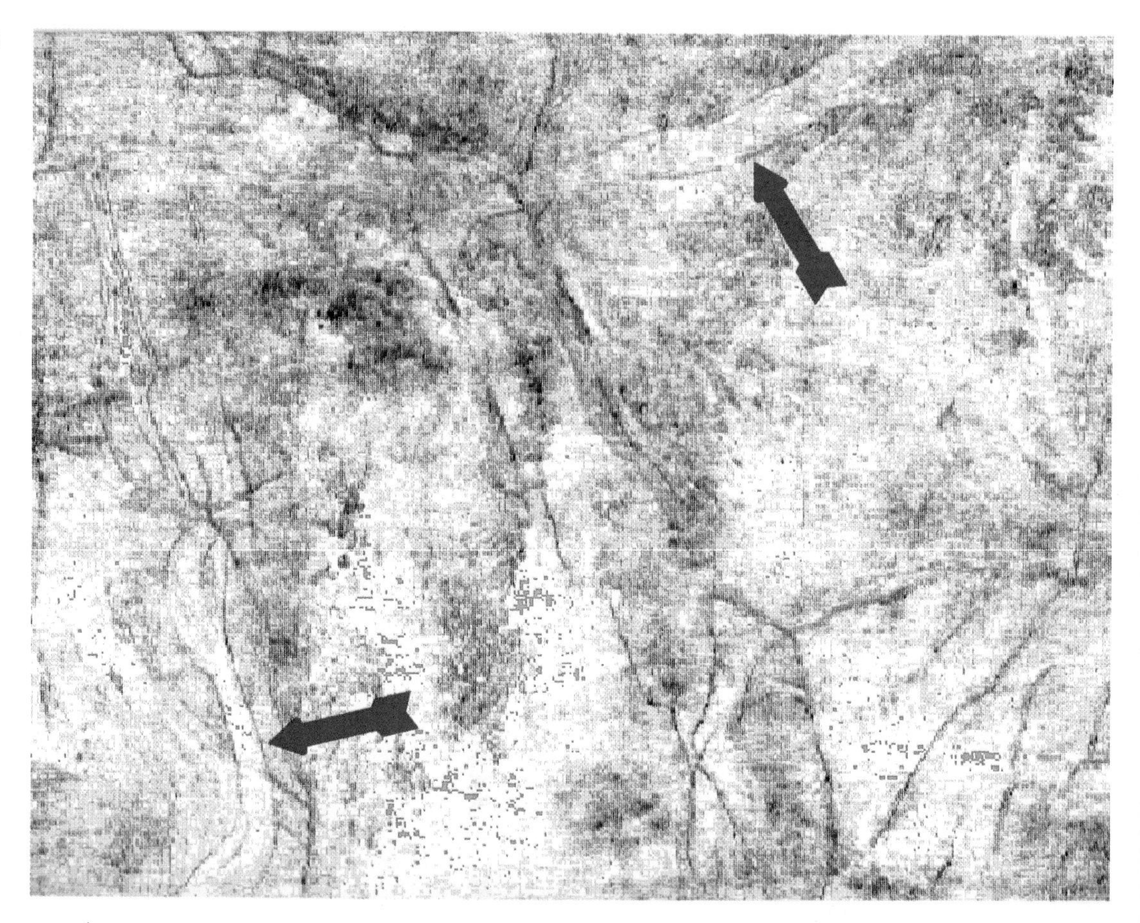

Fig. 8-30. Coherence time slice at 1250ms from Gulf of Mexico showing channels and faults (after Bahorich and Farmer, 1995). Note the channels indicated by the red arrows, how the channels are cut by faults, and also the point bars where one of the channels changes direction. (Courtesy Amoco Corporation. Data courtesy Geco-Prakla.)

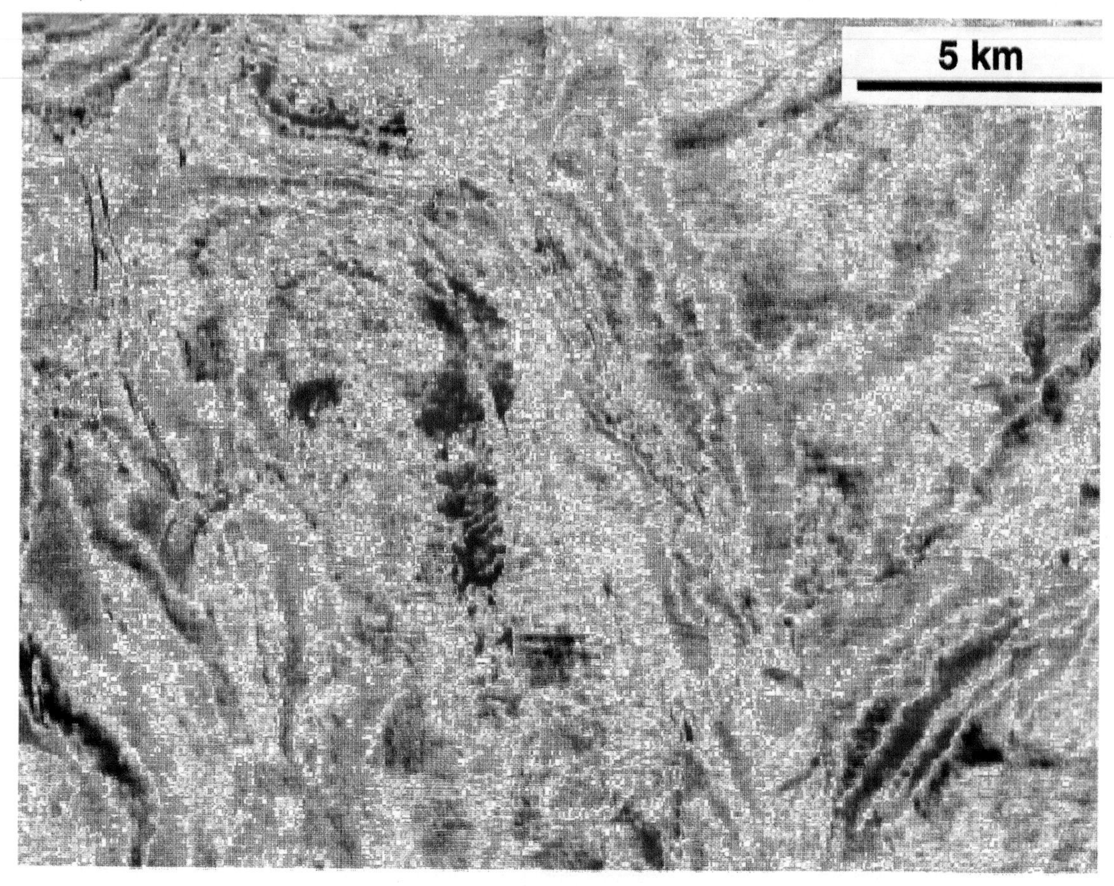

Fig. 8-31. Time slice at 1250ms from Gulf of Mexico for comparison with Figure 8-30 (after Bahorich and Farmer, 1995). (Courtesy Amoco Corporation. Data courtesy Geco-Prakla.)

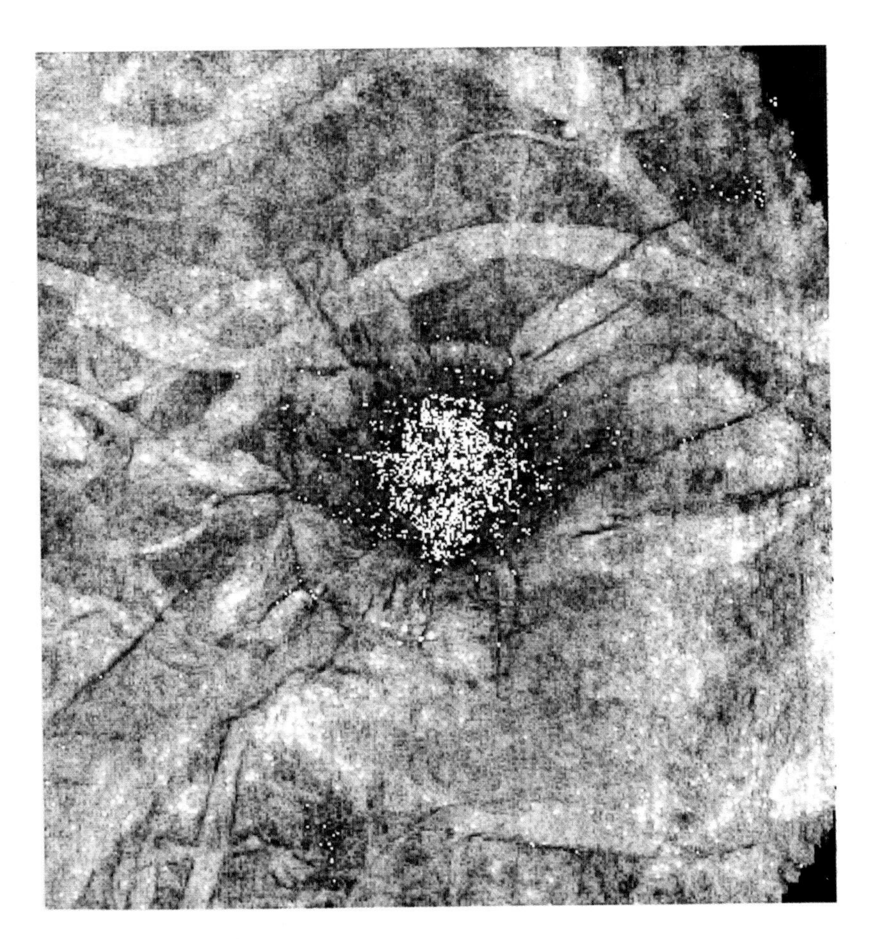

Fig. 8-32. Continuity time slice at about 1000ms from Gulf of Mexico showing channels and faults. (Courtesy CAEX Services and Landmark Advance Products Group.)

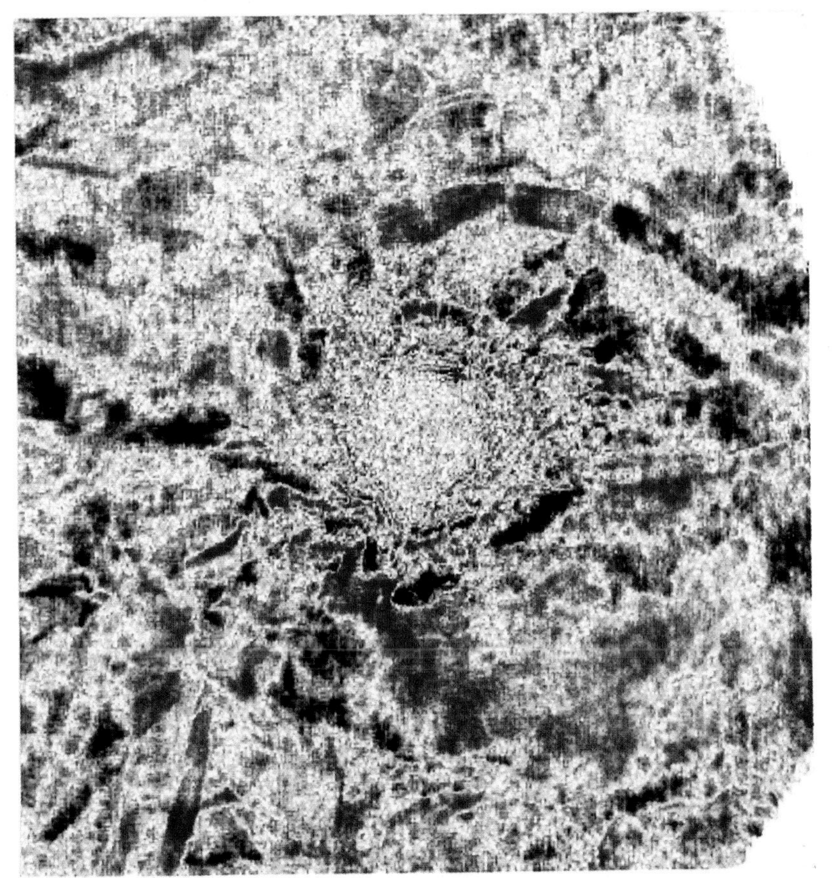

Fig. 8-33. Time slice at about 1000ms from Gulf of Mexico for comparison with Figure 8-32. (Courtesy CAEX Services and Landmark Advance Products Group.)

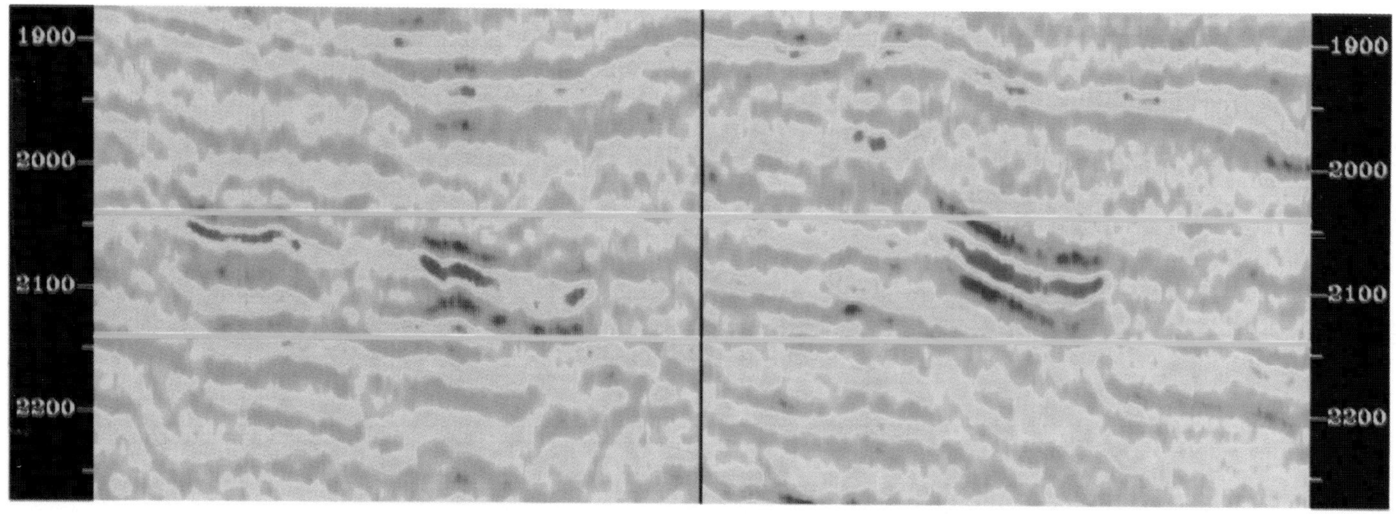

Fig. 8-34. Vertical seismic section from North Sea showing 100ms time window enclosing high-amplitude channel reflections. (Courtesy Landmark UK, Phillips UK and Western Geophysical Corporation.)

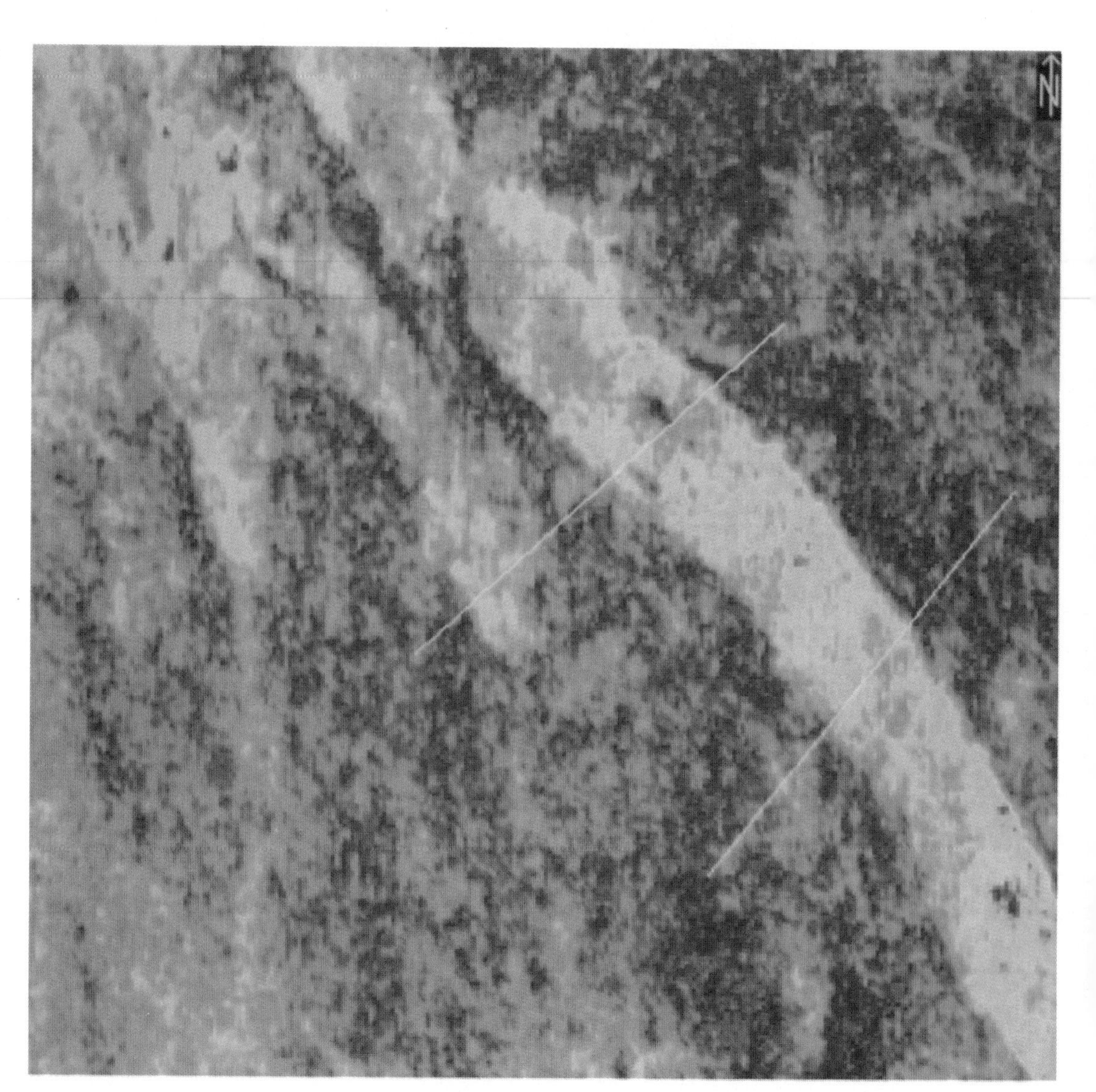

Fig. 8-35. Stat slice from North Sea displaying average absolute amplitude computed over time window indicated in Figure 8-34. Note definition of channel. (Courtesy Landmark UK, Phillips UK and Western Geophysical Corporation.)

nuity (the faults and other boundaries). These attributes operate within a time window and use a variety of mathematical approaches similar to correlation. Because the attributes are derived direct from the processed data they are free of interpretive bias, in contrast to the horizon attributes discussed in the last section which require an interpreted horizon as their input. By producing a volume where discontinuities have been given an apparent continuity, the autotracking of faults now becomes a possibility.

Figure 8-24 is a **coherence** time slice covering a large area of the Gulf of Mexico. Many faults are outstandingly visible. Figure 8-25 provides the regular time slice for comparison. Some of the faults are visible here but, as is normal with time slices, the fault visibility depends greatly on relative strike of fault and structure. On the coherence time slice of Figure 8-24 it is clear that the faults are equally visible regardless of their orientation relative to structural strike. This value of a coherence time slice is further exemplified by comparison of Figures 8-26 and 8-27 from the North Sea. Here the faults are very curved but nevertheless are equally visible along their entire length.

Figure 8-28 is a **continuity** time slice from the Gulf of Mexico showing radial faults around a salt dome. Continuity involves a straightforward multiple cross-correlation calculation which also yields related attributes such as the dip and the azimuth of the maximum correlation. Comparison of Figure 8-28 and 8-29 shows the improved fault visibility on the continuity time slice.

Figures 8-30 through 8-33 show examples of coherence and continuity time slices for delineating stratigraphic features and also the normal time slices for comparison. Some of the channels on Figures 8-30 and 8-32 are outstandingly clear. Furthermore the channel indicated by the lower left arrow in Figure 8-30 shows point bars where the channel changes direction.

Post-stack Amplitude Attributes

Reflection amplitude measured at the crest of an identified reflection is by far the most widely-used amplitude attribute. Reflection amplitude extracted over one horizon produces a display normally called a horizon slice. Horizon slices and their value in stratigraphic interpretation are addressed at length in Chapter 4. **Composite amplitude** is the absolute value summation of the amplitudes of reflections identified at the top and base of a reservoir, or other, interval. Its use is discussed in Chapter 7. Composite amplitude is the author's favorite amplitude-derived attribute for reservoir studies. **Acoustic impedance** derived from amplitude by seismic inversion is another way of combining information from reservoir top and base (with thickness limitations) and some examples of this are also shown in Chapter 7.

Amplitude attributes calculated in some way over a time window are many and varied (Figure 8-1). Most of them address the gross properties of the interval studied, some extract a selection of the amplitudes within the window, and some attempt to establish vertical internal distribution within the interval.

Figure 8-34 shows a vertical section from the North Sea showing some high-amplitude reflections from a channel. The horizontal blue lines indicate a constant time window of 100 ms enclosing the channel reflections. Figure 8-35 displays areally the **average absolute amplitude** calculated over this time window. This **stat slice** shows clearly the extent of the channel and its boundaries are sharply delineated. This approach has many applications when the objective has high amplitude and is rather discontinuous. For example, in tertiary clastic basins, the search for hydrocarbons depends very largely on a search for high amplitudes. In a reconnaisance mode absolute amplitude is summed over a sucession of intervals to provide the areal distribution of bright spots. This is accomplished much more quickly than tracking individual high amplitude reflections. However, the success of these stat slices depends critically on the amplitude of the objective reflections being dominant within the time window used.

Text continues on page 253

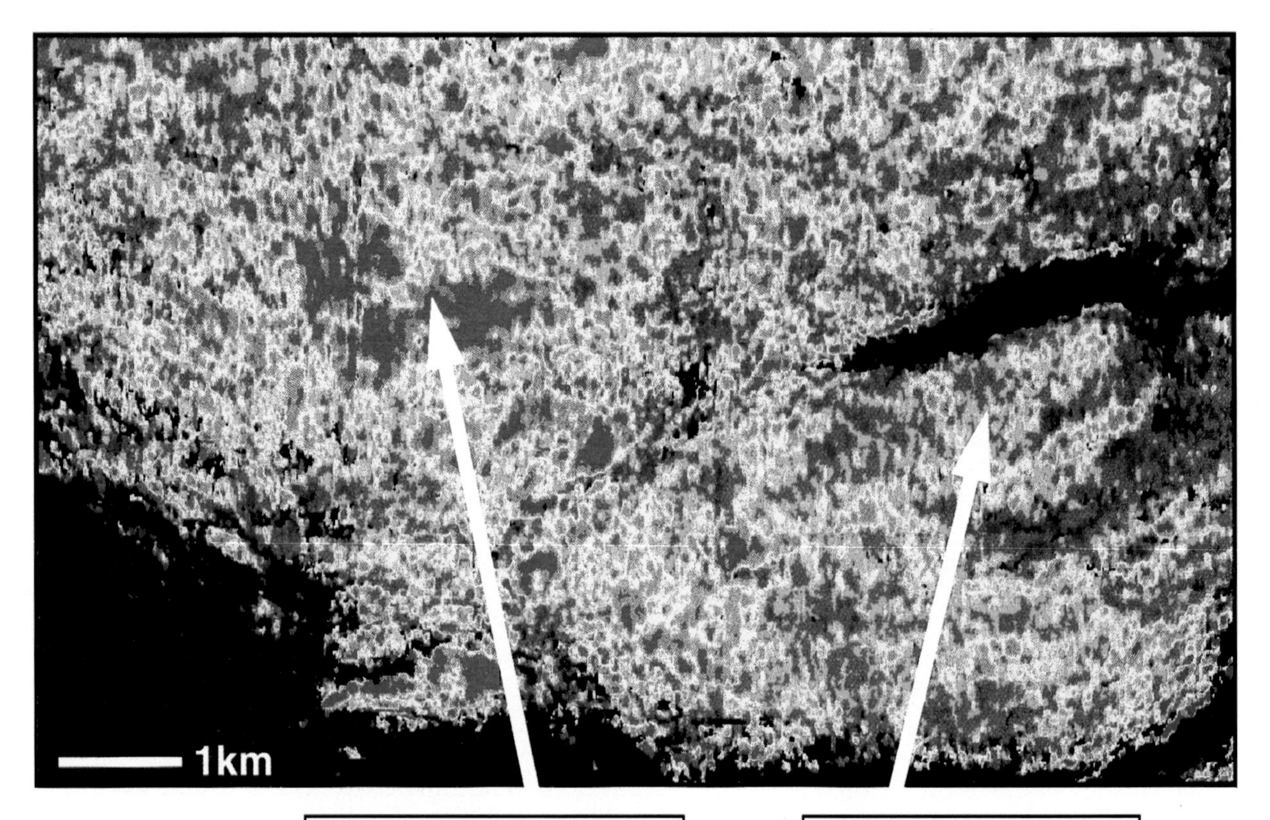
1km
Zone of maximum seismic energy
Area of lower seismic energy

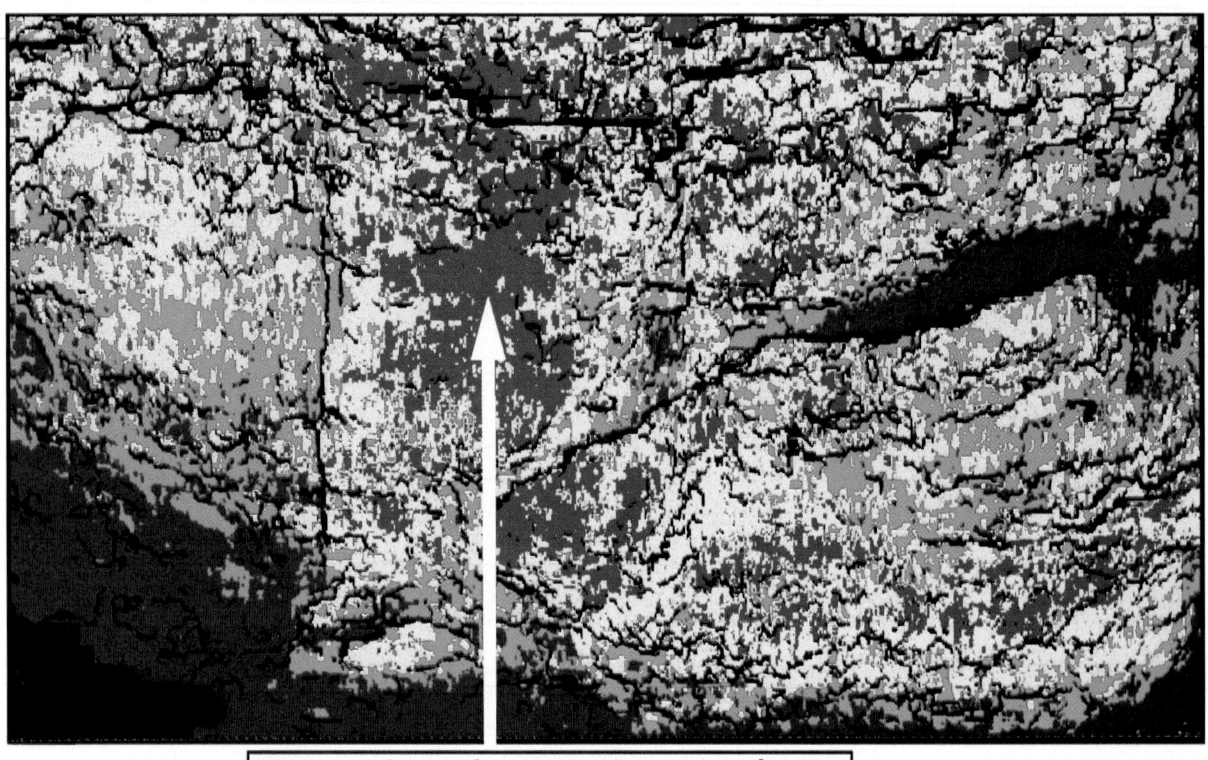
Zone of maximum zero-crossings between east-west faults

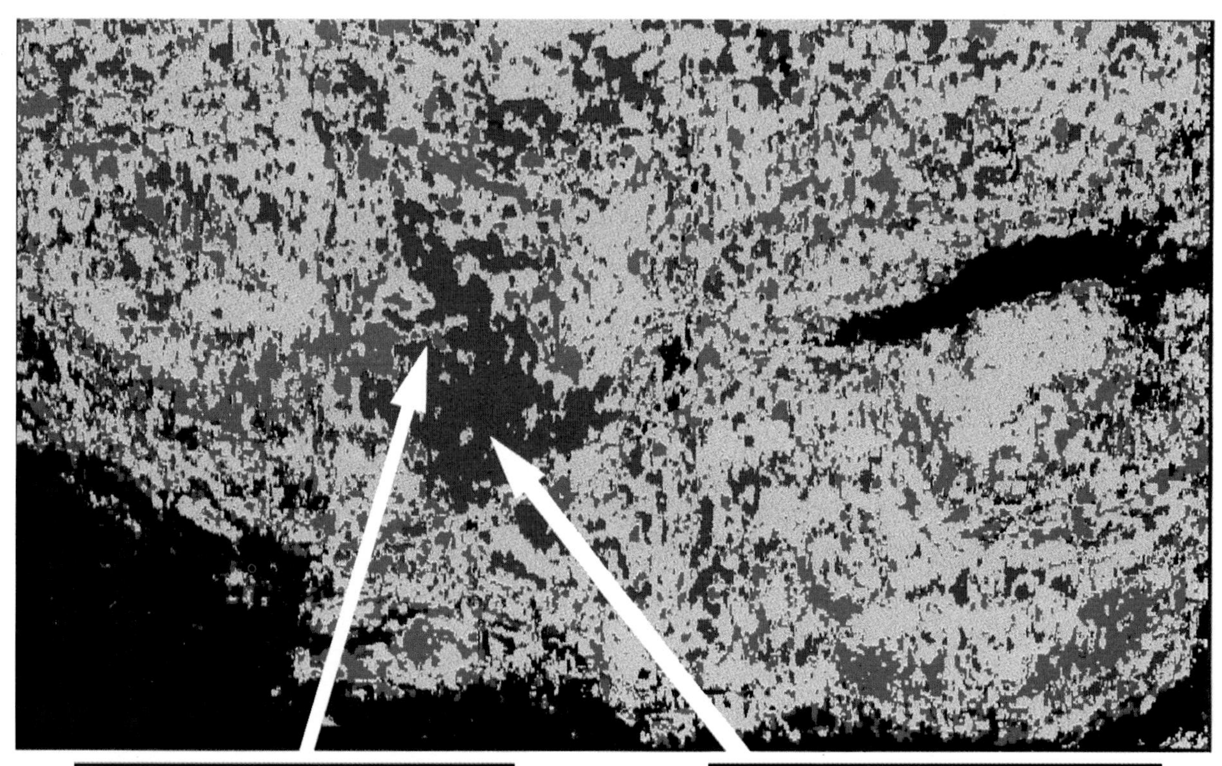

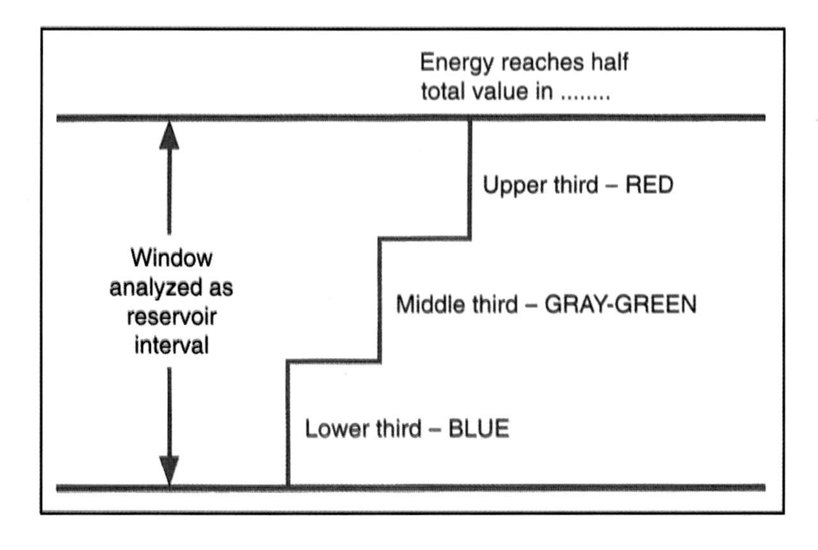

Fig. 8-38. Energy half-time map over reservoir interval from North Sea indicating vertical distribution of sands within the reservoir. (Courtesy BP Exploration and P. Mulholland.)

Fig. 8-36. (Opposite Top) Seismic energy map over reservoir interval from North Sea. High energy indicates sand in a shale-dominated environment. (Courtesy BP Exploration and P. Mulholland.)

Fig. 8-37. (Opposite Bottom) Number of zero crossings over reservoir interval from North Sea. This map indicates layering and also serves as quality control filter for horizon mispicks. (Courtesy BP Exploration and P. Mulholland.)

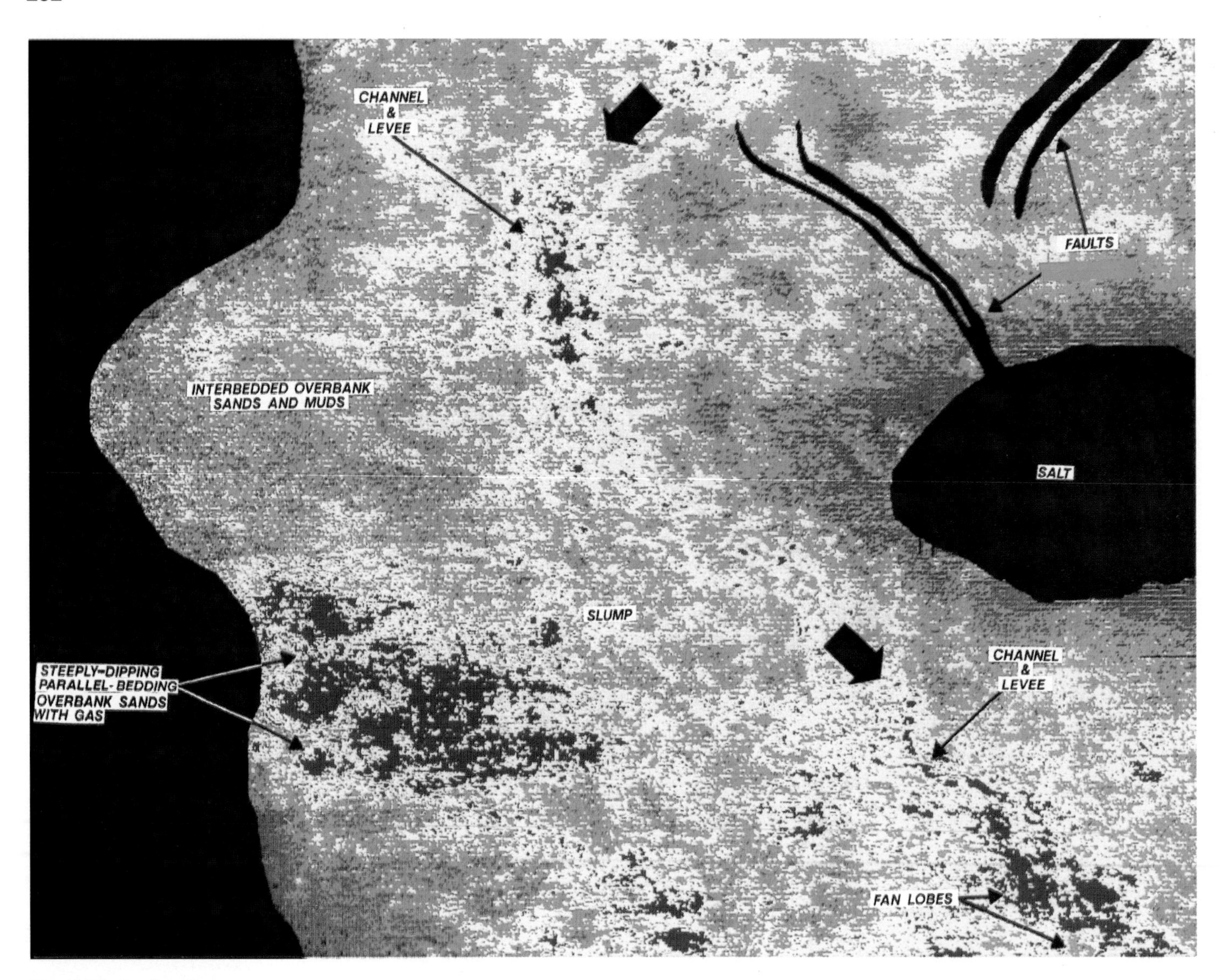

Fig. 8-39. Arc length, or reflection heterogeneity, used for mapping depositional facies in Gulf of Mexico. (Courtesy Geco-Prakla.)

Windowed amplitude is fraught with signal-to-noise problems. A time window, often hung from a structurally-interpreted horizon, may successfully capture the desired reservoir reflections but normally it captures other reflections too. This geologic noise can be very damaging unless the desired reflections are very strong. Therefore, it is often wise to use as small a time window as possible. If the reservoir reflections are readily trackable then it is normally best to do so. The approach used should always be aimed at maximizing signal-to-noise for the objective under study.

For another North Sea prospect Figure 8-36 maps the **total energy** over the reservoir interval defined as the time window between two structurally-interpreted horizons. Energy is the square of the seismic amplitude. The geologic environment here is shale-dominated so more reflection energy indicates more sand. The **number of zero crossings** mapped over the same area (Figure 8-37) indicates layering and thus should also be related to total quantity of sand. Because the top reflection here was a peak and the base a trough, the number of zero crossings between them must be an odd number. Even numbers thus indicate horizon mispicks, and the map of number of zero crossings was used as a quality control filter for mispicks based on this principle.

Energy half-time (Figure 8-38) for the same reservoir over the same area attempts to map vertical distribution of sand within the reservoir interval. Following the diagrammatic legend in Figure 8-38, energy half-time first sums energy over the interval starting from the top. Then the calculation sums a second time until half the previous total value is reached. If this point occurs above the midpoint of the interval, the sands are located primarily towards the top of the reservoir. If this point occurs below the midpoint of the interval, the sands are located primarily towards the base of the reservoir. Reds and blues, in Figure 8-38 (primarily within the high energy area of Figure 8-36) indicate areas where both these situations of internal distribution exist.

Energy half-time and other distribution attributes place even higher demands on data quality than other windowed attributes and, therefore, reliable results are difficult to obtain. **Amplitude ratio** (of top and base reflections) is a horizon-based amplitude attribute that addresses internal distribution. Relative to energy half-time it relies perhaps less on signal-to-noise but more on the zero-phaseness of the data.

Frequency-Derived Attributes

Use of instantaneous frequency to detect gas is mentioned in Chapter 5. However, it tends to be one of the less reliable approaches to the direct detection of hydrocarbons. In fact all frequency attributes tend to be noisy, and the industry has not yet really harnessed frequency for stratigraphic and reservoir studies. However, a few examples of their use do exist.

Arc length is the total length of a seismic trace over a time window. Clearly the excursion of the trace caused by increased amplitude gives more trace length, as also does oscillation caused by increased frequency. Hence arc length contains components of frequency and amplitude. It is also referred to as reflection heterogeneity. Figure 8-39 shows arc length being used to map depositional facies in the Gulf of Mexico.

Figure 8-40 shows six different frequency-derived attributes displayed for the same small area. For comparison Figure 8-41 shows five amplitude-derived attributes over the same area and also the vertical section showing the window over which all these attributes were calculated. Three of the attributes in Figure 8-41 address vertical distribution within the interval. Average absolute amplitude and RMS amplitude appear identical.

Peak spectral frequency and slope spectral frequency are both properties of the spectrum calculated over the whole window. First, second and third dominant frequency involve completely different mathematics force-fitting the spectrum to yield three maxima. First dominant frequency should primarily express the properties of the wavelet. **Second dominant frequency** should primarily express the frequency properties of the reflection coefficient series, namely the geologic layering. Thus sec-

Text continues on page 257

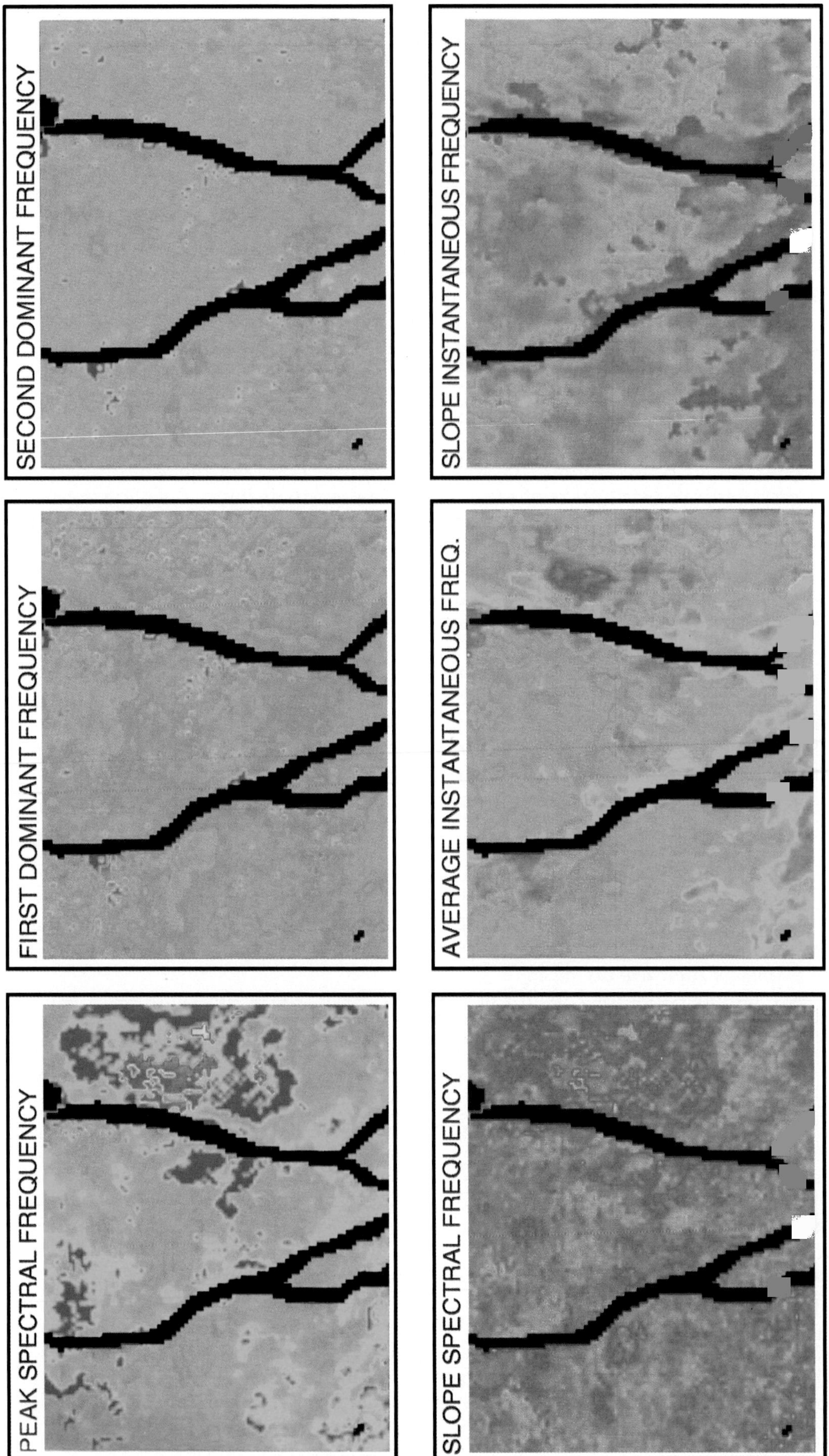

Fig. 8-40. Various frequency attributes displayed over the same area and for the same interval for comparison. (Courtesy Landmark Advance Products Group.)

PEAK SPECTRAL FREQUENCY

FIRST DOMINANT FREQUENCY

SECOND DOMINANT FREQUENCY

SLOPE SPECTRAL FREQUENCY

AVERAGE INSTANTANEOUS FREQ.

SLOPE INSTANTANEOUS FREQUENCY

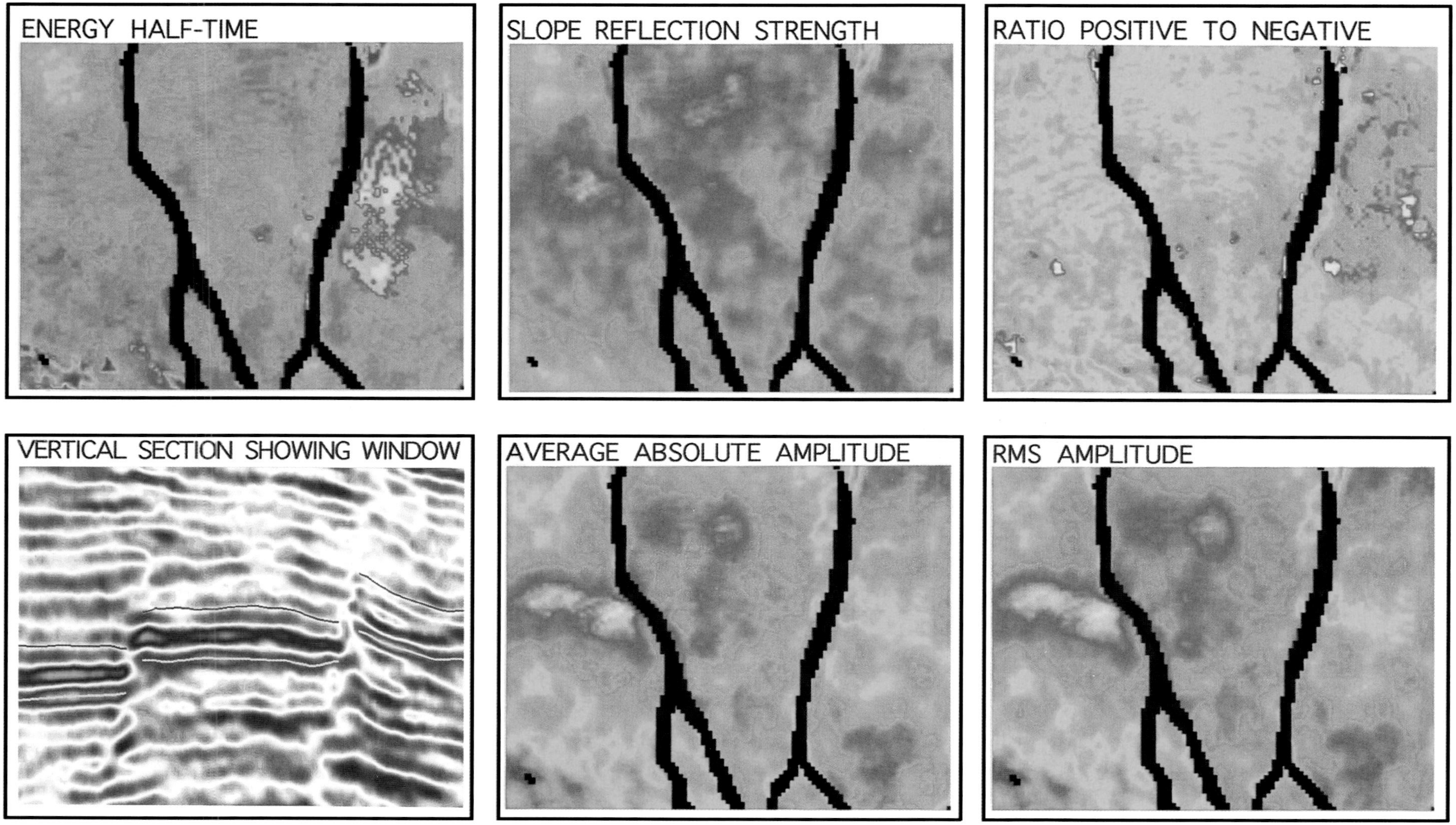

Fig. 8-41. Various distribution and amplitude attributes displayed over the same area and for the interval shown in the lower left. (Courtesy Landmark Advance Products Group.)

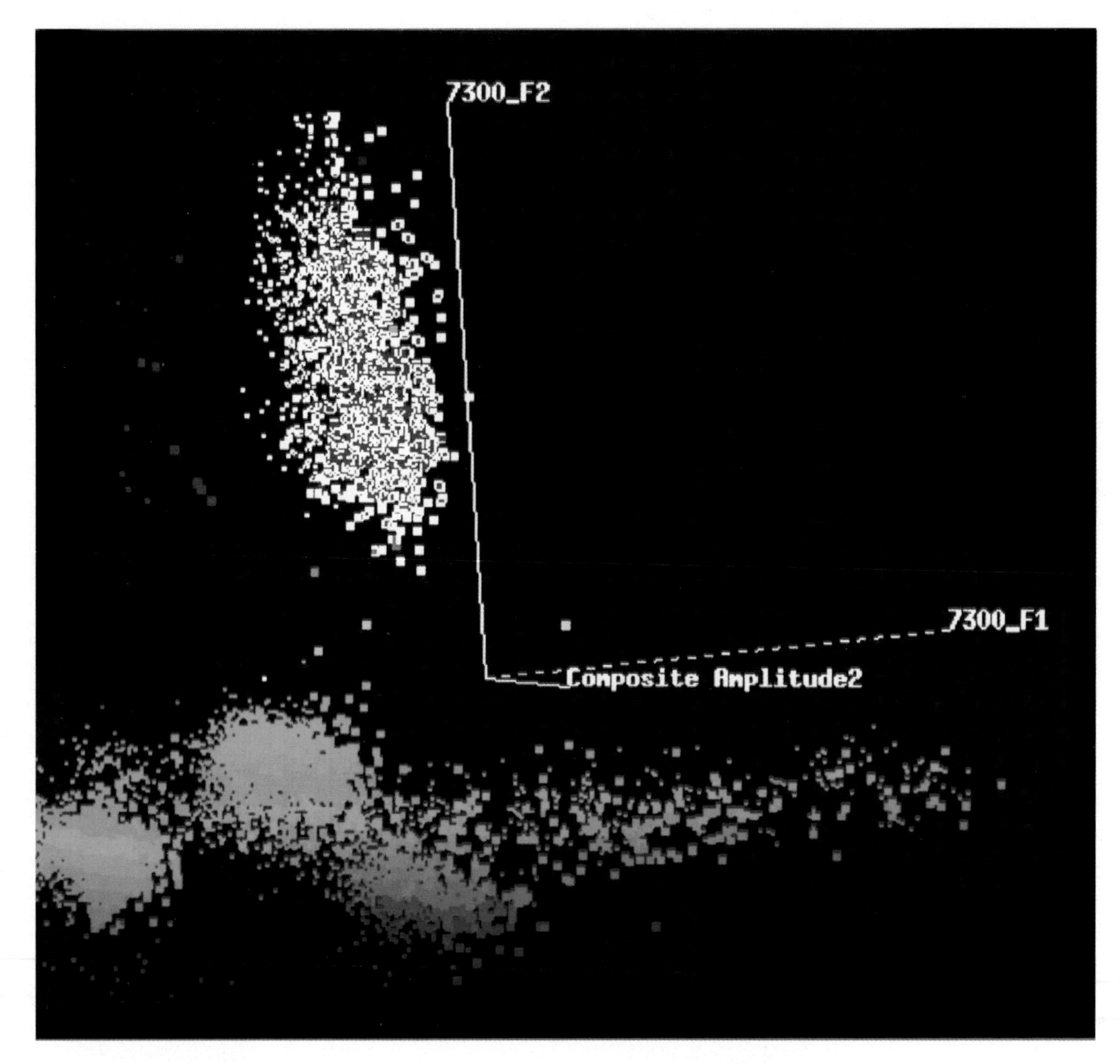

Fig. 8-42. Three-dimensional crossplot between composite amplitude, first dominant frequency and second dominant frequency used to establish statistical separation of attributes. (Courtesy Landmark Advance Products Group.)

ond dominant frequency may be the most useful frequency attribute and a limited number of studies to date suggest that this is so. Figure 8-42 shows a three-dimensional crossplot of second dominant frequency, first dominant frequency and composite amplitude. The points well separated in second dominant frequency come from one local area of the reservoir. This prompted a well which discovered some previously-bypassed reserves.

Amplitude Variation with Offset

Amplitude Variation with Offset for the identification of gas is discussed briefly in Chapter 5. AVO has become a very popular subject and is covered thoroughly by Castagna and Backus (1993). However, the amount of AVO work performed in 3-D is extremely small and Castagna and Backus barely mention the subject. Furthermore, the 3-D AVO projects which have been performed show little incremental benefit over the 3-D analysis of the post-stack amplitudes. The example presented in Figures 8-43 through 8-46 is one exception.

Figure 8-43 is the horizon slice displaying conventional stacked amplitude for a gas reservoir in the deep water Gulf of Mexico. Figures 8-44 and 8-45 show horizon slices on the same horizon from individually-migrated data volumes incorporating respectively near-offset traces only and far-offset traces only. The two are similar and both show separation into eastern and western lobes. The differences between the two do not at first appear significant.

Figure 8-46 is the arithmetic difference between the amplitudes of Figures 8-44 and 8-45; in other words Figure 8-46 is a display of the AVO attribute **far-near amplitude difference**. A channel is now clearly visible in yellow, red and green, which indicates increasing amplitude with offset. This channel, meandering across parts of both eastern and western lobes, is not visible on any normal amplitude display and is interpreted as the sand-filled channel facies of a slope fan. Furthermore, it probably indicates the areas of highest porosity and permeability within the reservoir.

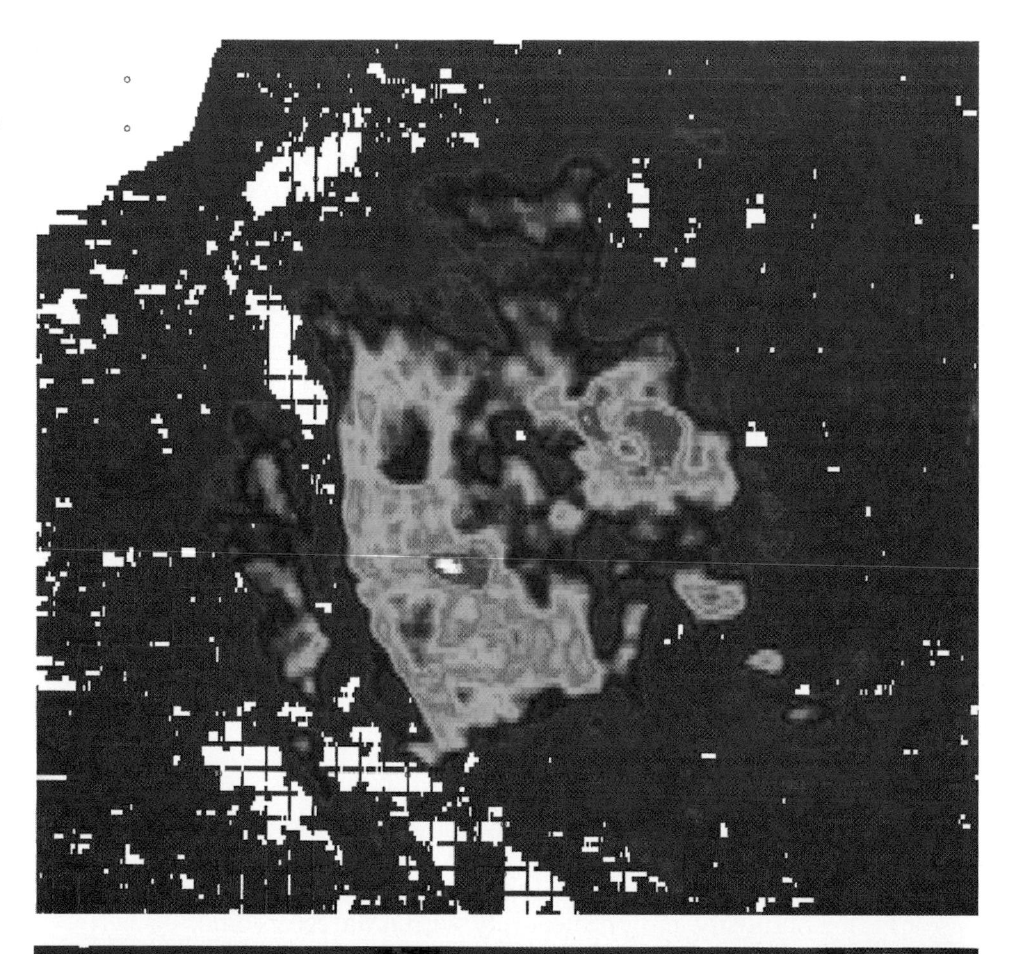

Fig. 8-44. Horizon slice in amplitude from near offsets only for a gas reservoir in the deep water Gulf of Mexico. Note the separation of the reservoir into eastern and western lobes. (Courtesy Amoco Production Company.)

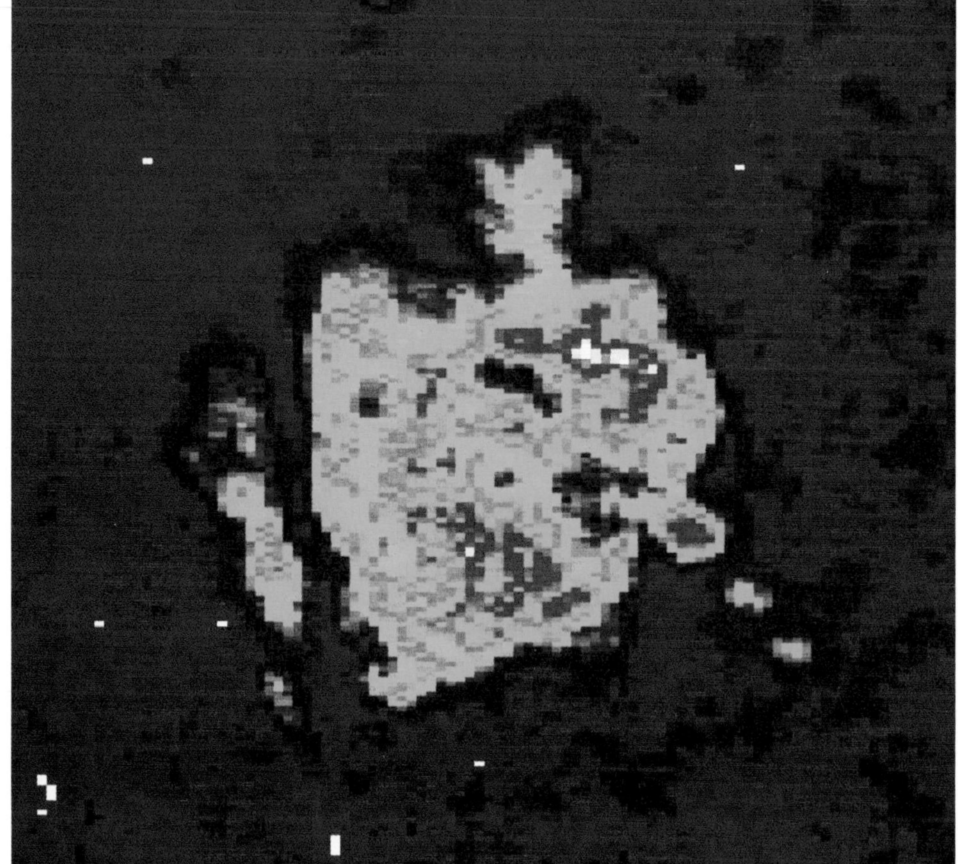

Fig. 8-43. Horizon slice in conventional stacked amplitude for a gas reservoir in the deep water Gulf of Mexico. (Courtesy Amoco Production Company.)

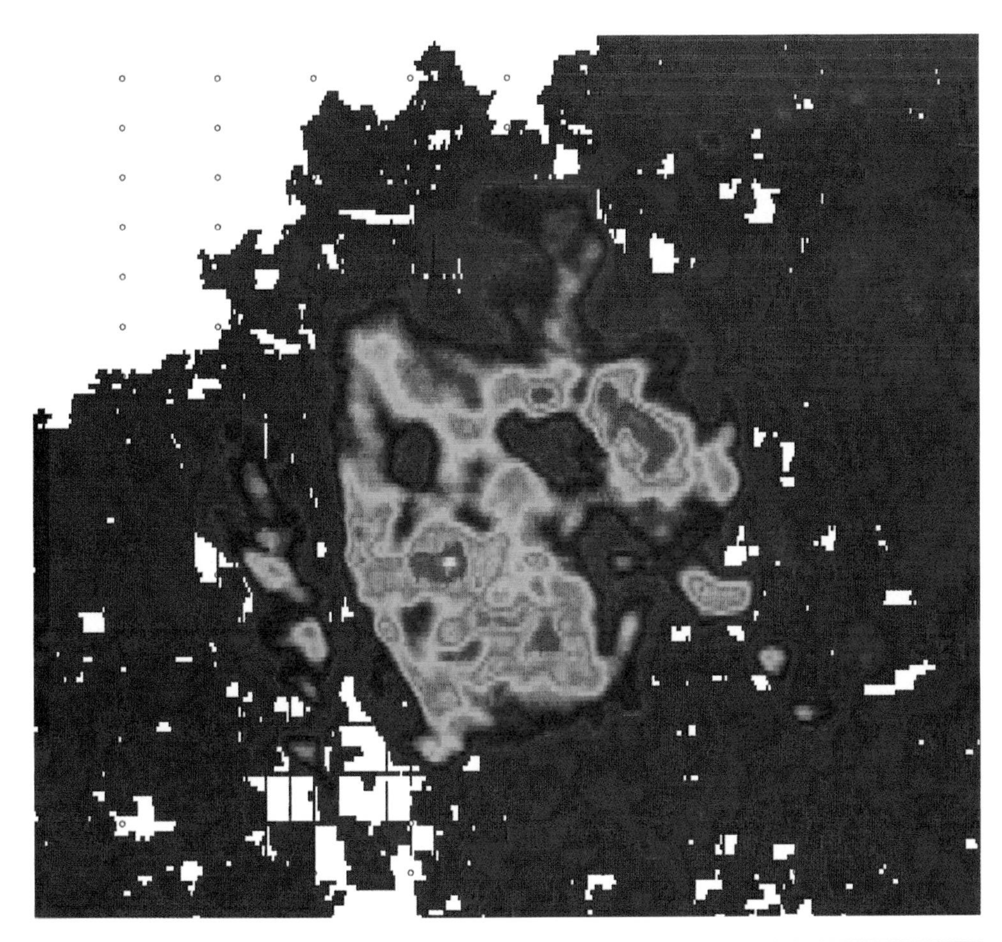

Fig. 8-45. Horizon slice in amplitude from far offsets only for comparison with Figure 8-44. (Courtesy Amoco Production Company.)

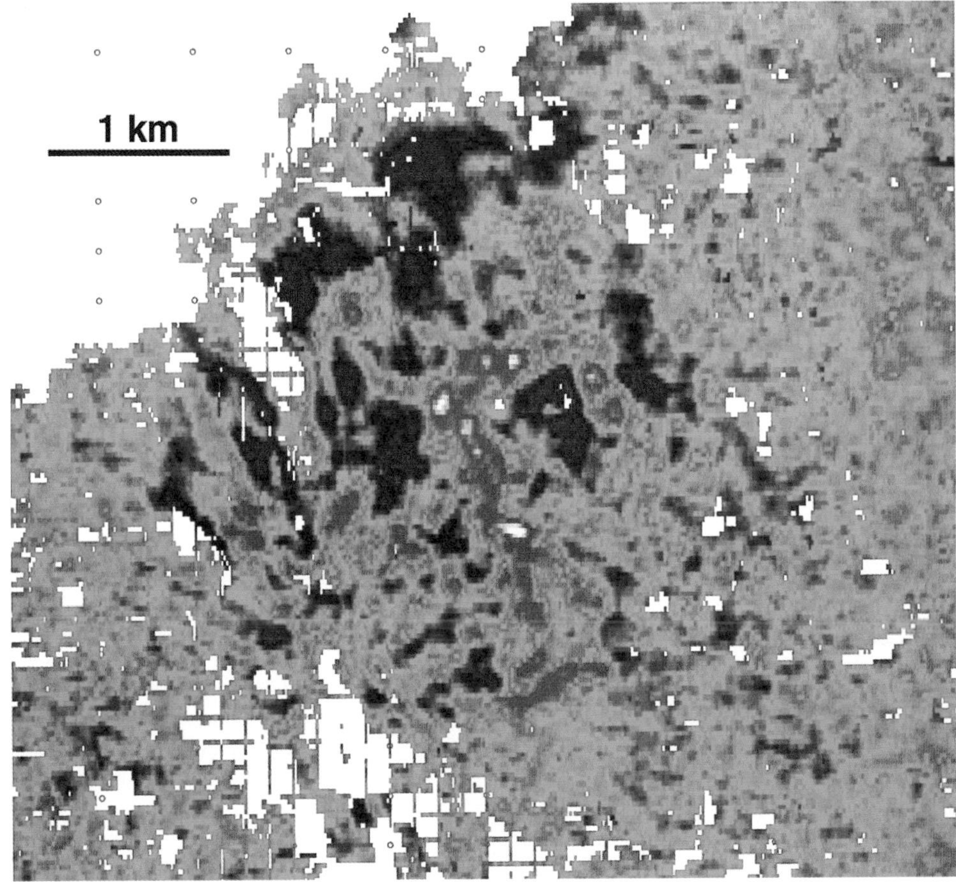

Fig. 8-46. 3-D AVO - horizon slice in far-near amplitude difference, that is the subtraction of Figure 8-44 amplitudes from Figure 8-45 amplitudes. Reds, yellows and greens indicate amplitude increasing with offset. The pattern in these colors indicates a channel not visible in normal amplitude and probably containing the highest porosities and permeabilities in the fan. (Courtesy Amoco Production Company.)

VISUALIZATION OF HORIZON ATTRIBUTES

CONTRIBUTED BY GEOFFREY A. DORN

Nature of Visualization

Visualization is the graphical presentation of data in an intuitive fashion that exposes the information in the data and provides new insight. Common examples of visualization include pie charts, bar graphs or xy plots, seismic sections, time slices, and contour maps. A structure contour map is an attempt to represent a three-dimensional surface in a two-dimensional display. Horizon attributes are typically visualized as two-dimensional map displays where the variation in the attribute value is calibrated to a particular color range. Three-dimensional visualization attempts to convey significantly more information through the representation of multi-parametric data in three dimensional surfaces and volumes.

The most common form used to represent a three-dimensional surface has for many years been the contour map. Each color on this type of map represents a range of times. The appearance of a continuous gradation can be achievable by using a larger number of individual color bins. By looking at a contour map, and understanding from the legend the relationship between color and structural highs and lows, it is possible to form a mental image of the three-dimensional surface.

This representation of the 3-D surface is limited because it requires mental integration of complex variations of data along the three dimensional structure. For example, it becomes more difficult to form a correct mental picture when the structural information must be combined with reflection amplitude along the interpreted horizon shown in Figure 8-47. The problem becomes even more complex in a practical situation when there is a need to integrate the information from structure, several horizon attributes, some well locations, and possibly culture.

These difficulties can be managed by using 3-D visualization to represent the data. In Figure 8-48, for example, the interpreted horizon is now shown a three-dimensional surface, with the reflection amplitude shown in color. It now becomes a simply a matter of observation to identify the relationship, if any, between the structure and the horizon attribute.

Perception of Three Dimensions

There are a number of cues that the mind uses to perceive and understand a three-dimensional object. These include:

- Projection
- Lighting and shading
- Depth of field
- Depth cueing
- Obscuration
- Transparency
- Stereopsis
- Parallax
- Motion

Projection refers to a method of graphically depicting three-dimensional objects and spatial relationships on a two-dimensional plane (Foley, et. al., 1990). A perspective projection is one in which parallel lines and planes converge to infinitely distant vanishing points. In a parallel projection, parallel lines and planes are made to be truly parallel in the display. This may result in an optical distortion in the perception of three dimensions, but can provide a display (e.g. an isometric orthographic parallel projection) where measured distances along each of the three axes are the same. Both projections have their application and value in the interpretive process.

When we perceive three-dimensional objects in a lighted room, or outside, part of our understanding of the three-dimensional nature of the object, and the details of its shape, are perceived from the shadows cast by light sources shining on the surface. The perception of the overall shape of the surface is significantly improved by lighting, and the shadows provide detailed information not only about major discontinuities (faults and folds) but also about portions of the surface that are relatively smoother or rougher than the surrounding surface.

Indoor lighting typically provides a number of sources of light which may shine simultaneously on a single object, so that shadows are cast in multiple directions, or very diffuse light so that shadows are greatly softened. Although some applications provide the capability of using multiple light sources and ambient or diffuse light to illuminate a surface, it is usually more intuitive and more informative to use a single light source with interactive user control of its altitude and azimuth to explore the detailed structure of a surface, and to look for different trends of discontinuity. Some applications also allow the user to control the distance between the light source and the surface and the color of the light source. A light source at an infinite distance illuminates the surface with parallel rays. A light source close to the surface illuminates the surface with diverging light rays. This affects the shadows and highlights on the surface being viewed.

When the human eye, or the camera, focuses on an object, the focus actually occurs on a plane perpendicular to the direction of viewing at a particular distance from the observer. Objects or portions of the object that are either closer to or further away from the observer are out of focus. This phenomenon is called depth of field (Foley, et. al., 1990). Although this is an additional cue to the relative distance to parts of an object, it is typically not used in technical applications that render data in three dimensions. If this technique were used, it would require re-rendering the image to clearly view different depths on the surface or in the volume which would be both time consuming and awkward. Typically the entire three-dimensional surface or volume is rendered in focus allowing the user to examine all depths in the image simultaneously with essentially the same clarity.

Depth cueing refers to a phenomenon where the more distant objects are rendered with a lower intensity than nearer objects. This can be viewed as a type of "atmospheric" attenuation. This can be important in lending realism to a computer generated image of a scene or structure. It is usually not desirable when rendering complex three-dimensional data for analysis.

Obscuration is a process where the nearer portions of a complex three-dimensional surface may obstruct or block the view of portions of the surface that are further away from the viewer. Other terms for this are visible line/surface determination and hidden line/surface removal. This is done as a matter of course in the three-dimensional rendering of opaque surfaces and solids. By allowing the interpreter control over the transparency of the rendered surfaces, it is possible to see through one horizon to the data that would be hidden behind it if it were opaque. The partial transparency of the nearer surface gives the mind very clear information about the relative position of the surfaces.

Stereopsis refers to the ability to view an object simultaneously from two slightly different directions and composite the three-dimensional image. Most people are naturally able, and accustomed, to see three-dimensional objects stereoscopically. Stereoscopic displays have seen relatively limited use to date in three-dimensional interpretation, but their use will increase over time. The increase in the amount of information that can be perceived moving from a lighted perspective display to a stereoscopic display can be quite remarkable. A stereo pair of images, one image for each eye, can be composited with the aid of a stereo viewer, a device used for a number of years in analysis of aerial photography. It is also possible to composite the images without the

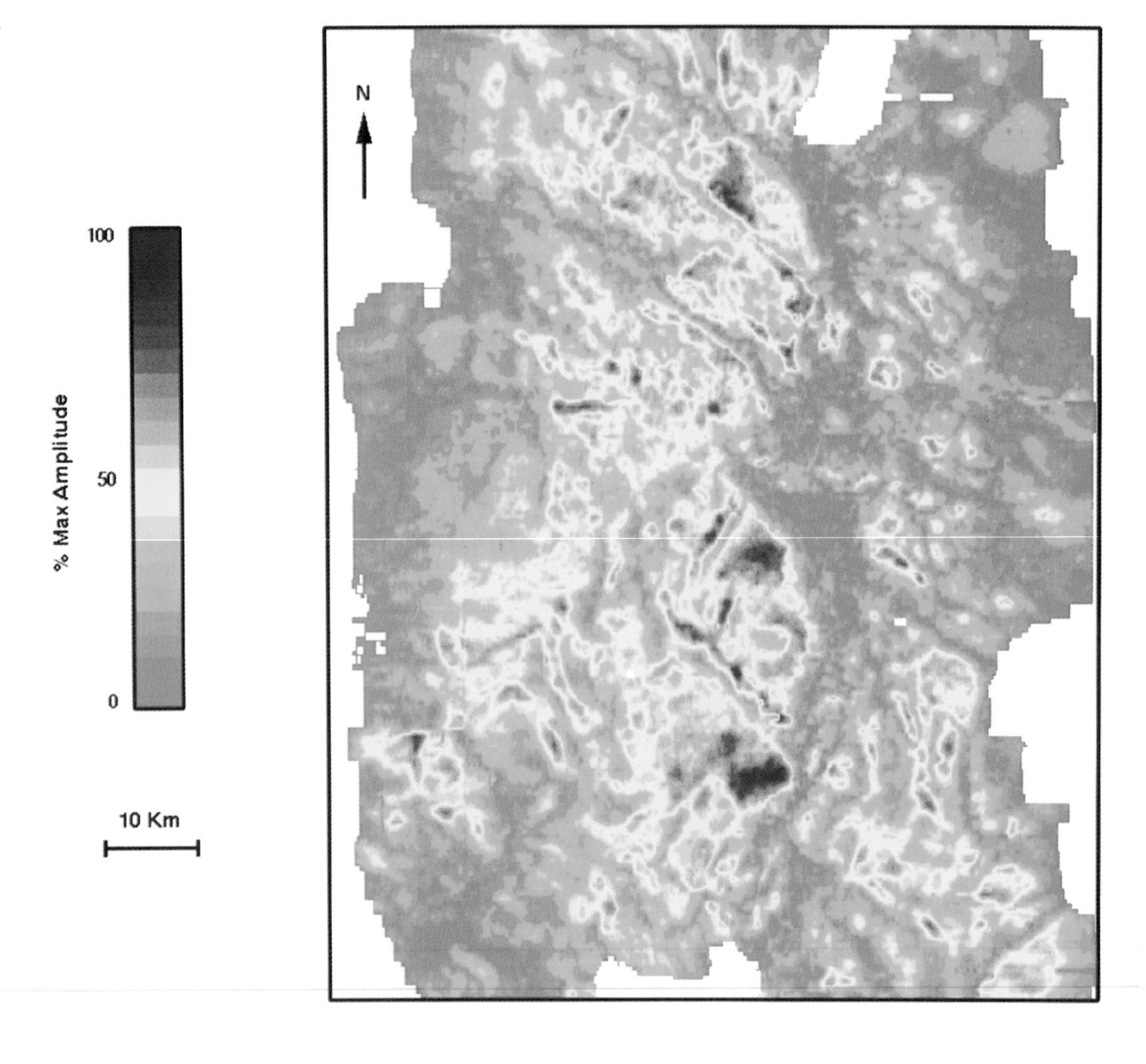

Fig.8-47. Reflection amplitude at the interpreted top reservoir reflection from a 3-D survey in the Southern North Sea Gas Basin. (Courtesy ARCO Exploration and Production Technology.)

aid of a stereo viewer, by focusing your eyes at a point beyond or behind the page. This process is very similar to that used to see the three-dimensional objects in the popular random dot stereograms (Magic Eye, 1993). Once the two images fuse into one central image, the details of the three-dimensional surface become evident.

Representation of stereo images in hardcopy has been somewhat limited. Autofusion of a stereo pair requires relatively small images, and that the images be printed with a center-to-center distance of approximately 2.5 to 3.0 inches (this is related to the inter-ocular distance between the human eyes). Somewhat larger images can be fused with the aid of a stereo viewer. Stereo representation by color separation (e.g., left eye image in blue, and right eye image in red) can be used in conjunction with red-blue filter glasses to achieve a stereo view in hardcopy. However, this severely constrains the use of color to represent the variation of data along the surface. The red-blue technique and a technique that relies on linear polarizing filters can be used to render stereoscopic images in slide or film projection (Kowalik, et. al., 1995). Hardcopy stereo viewing with a range of viewing angles is also possible through the use of a lenticular grid.

The hardware and graphics display capability to display stereo images on a workstation graphics screen have become relatively inexpensive. The most commonly used technique relies on displaying a left and right image alternately on the display screen. Each image is displayed about 60 times a second, so the composite screen refresh rate is 120 times per second. The interpreter wears a set of glasses where each lens is a liquid crystal shutter. An infrared beam is used to synchronize the left and right eye shutters to the left and right images being displayed on the computer screen. The

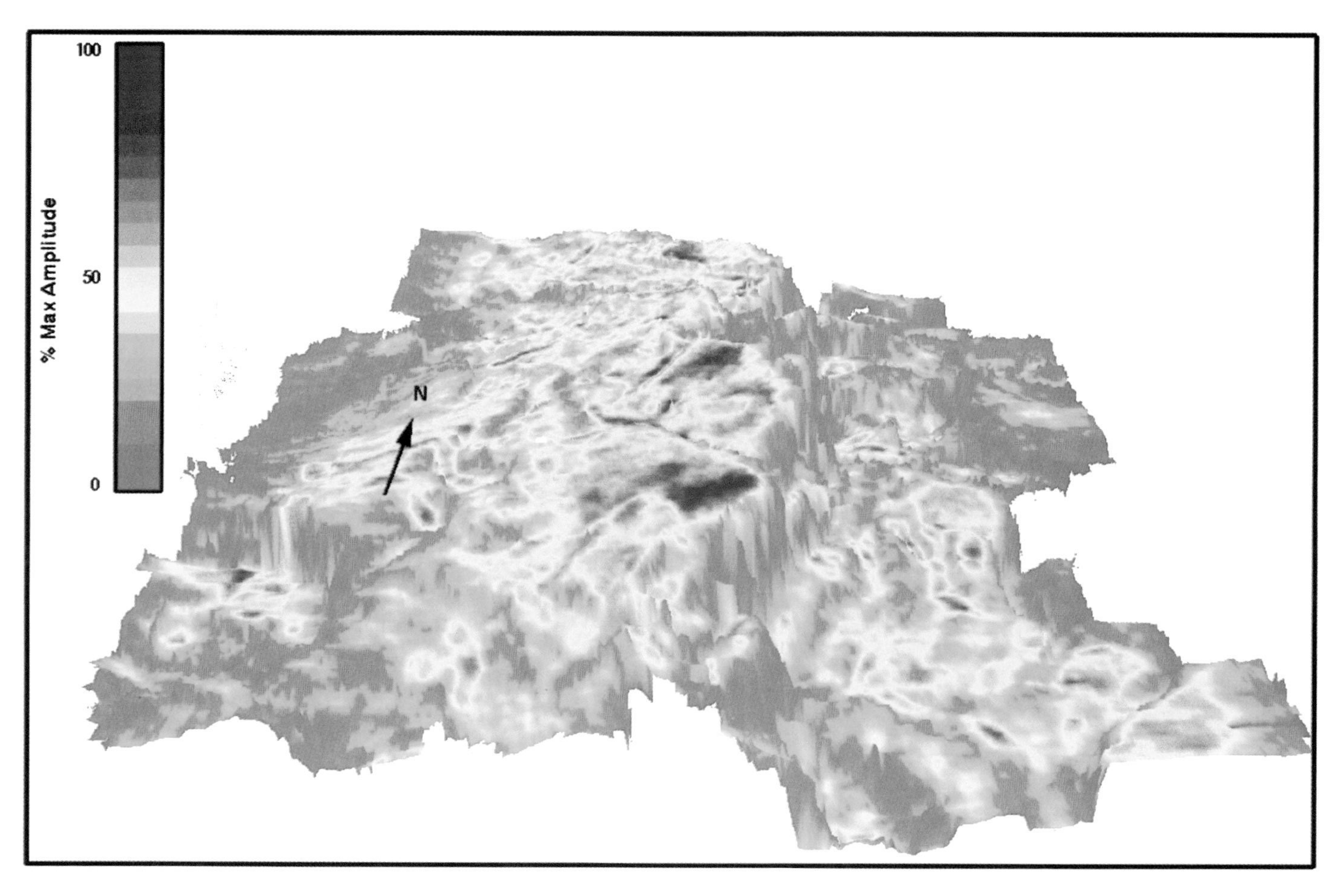

Fig.8-48. Interpreted time structure of the top reservoir reflection with reflection amplitude in color, shown as a three-dimensional surface in perspective view. (Courtesy ARCO Exploration and Production Technology.)

result is a very high quality stereo image, which on some workstations, can be rotated and moved in real-time by the user.

Work is currently underway at several research labs (including Sharp Laboratories and British Telecom) to produce auto stereoscopic displays - displays where a stereo 3-D images are perceived without any need for specialized eye wear. These displays are still a few years away from commercial application.

Parallax is the apparent displacement of closer objects with respect to farther objects when observed from two different viewpoints. The effects of parallax are most obvious when combined with motion. Driving down a highway, the objects near to the road seem to move by more quickly than those in the distance.

Motion or animation is one of the most important cues for three-dimensional perception, and one of the most difficult to describe in a medium such as a printed book. The human visual system is very sensitive to small relative changes in what is being viewed. A technique involving "flickering" between two versions of one image has been used for many years in aerial photography and in astronomy to detect small changes in a field of view over time. Moving an observer's viewpoint with respect to an interpreted horizon (distance, orientation, etc.) can greatly improve the understanding of the nature of the horizon and the attribute displayed on the horizon. Motion can be used to enhance the effects of all of the other three-dimensional perception cues previously discussed, maximizing the visual effects of parallax, stereoscopy, obscuration and changing the shadows cast by a light source.

Attribute/Structure Relationships

The interpreter frequently needs to look at the variation of an attribute of the data (for example the trace amplitude) along a horizon surface. A multitude of horizon attributes can be extracted or calculated once a horizon has been picked. Seismic

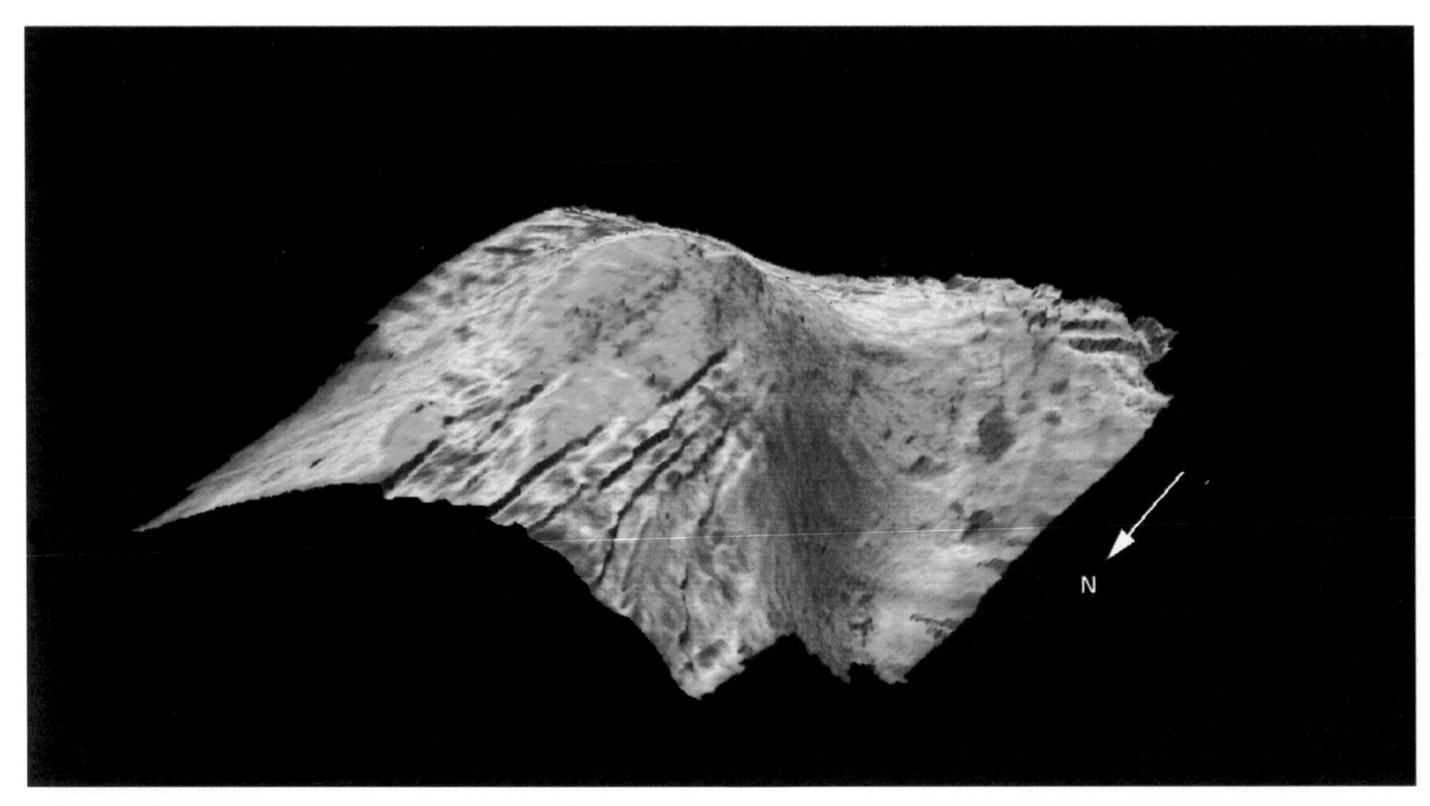

Fig.8-49. Lighted perspective display of the interpreted Cretaceous horizon structure with reflection amplitude shown in color. The view is from the northwest with a light source at 45 degrees elevation to the east. (Courtesy ARCO Exploration and Production Technology.)

Fig.8-50. (Opposite Top) Lighted perspective display of the interpreted Cretaceous horizon after the data has been flattened on a shallower upper Cretaceous unconformity. The view is from the northeast. Note the correlation of the low amplitude (green-blue) zone with the crest of the paleo-anticline. (Courtesy ARCO Exploration and Production Technology.)

Fig.8-51. (Opposite Bottom) Lighted perspective display of the residual horizon structure of the Cretaceous horizon with reflection amplitude in color. The residual structure is calculated as the time difference between the original horizon structure and a smoothed version of the horizon structure. (Courtesy ARCO Exploration and Production Technology.)

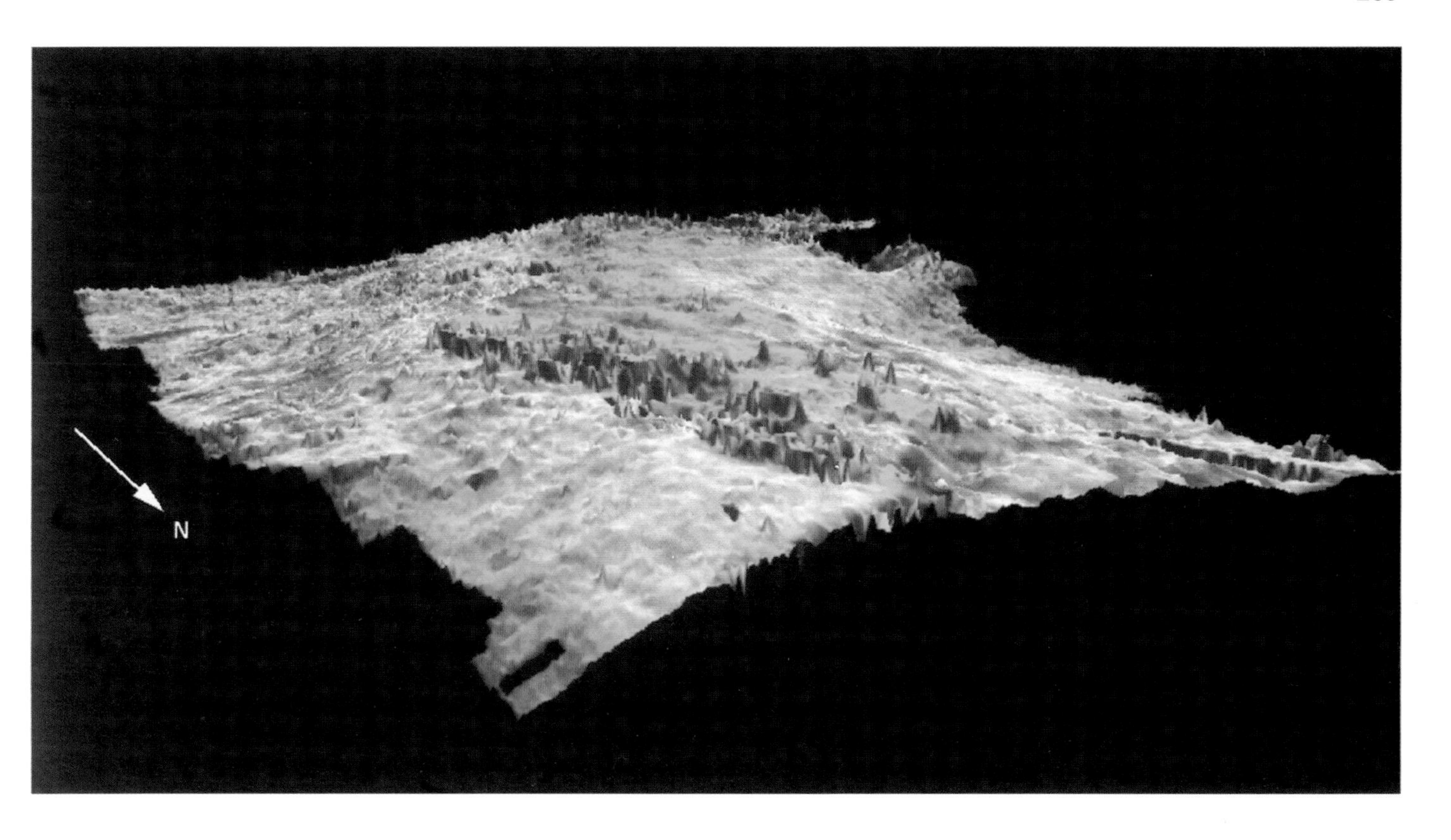
N

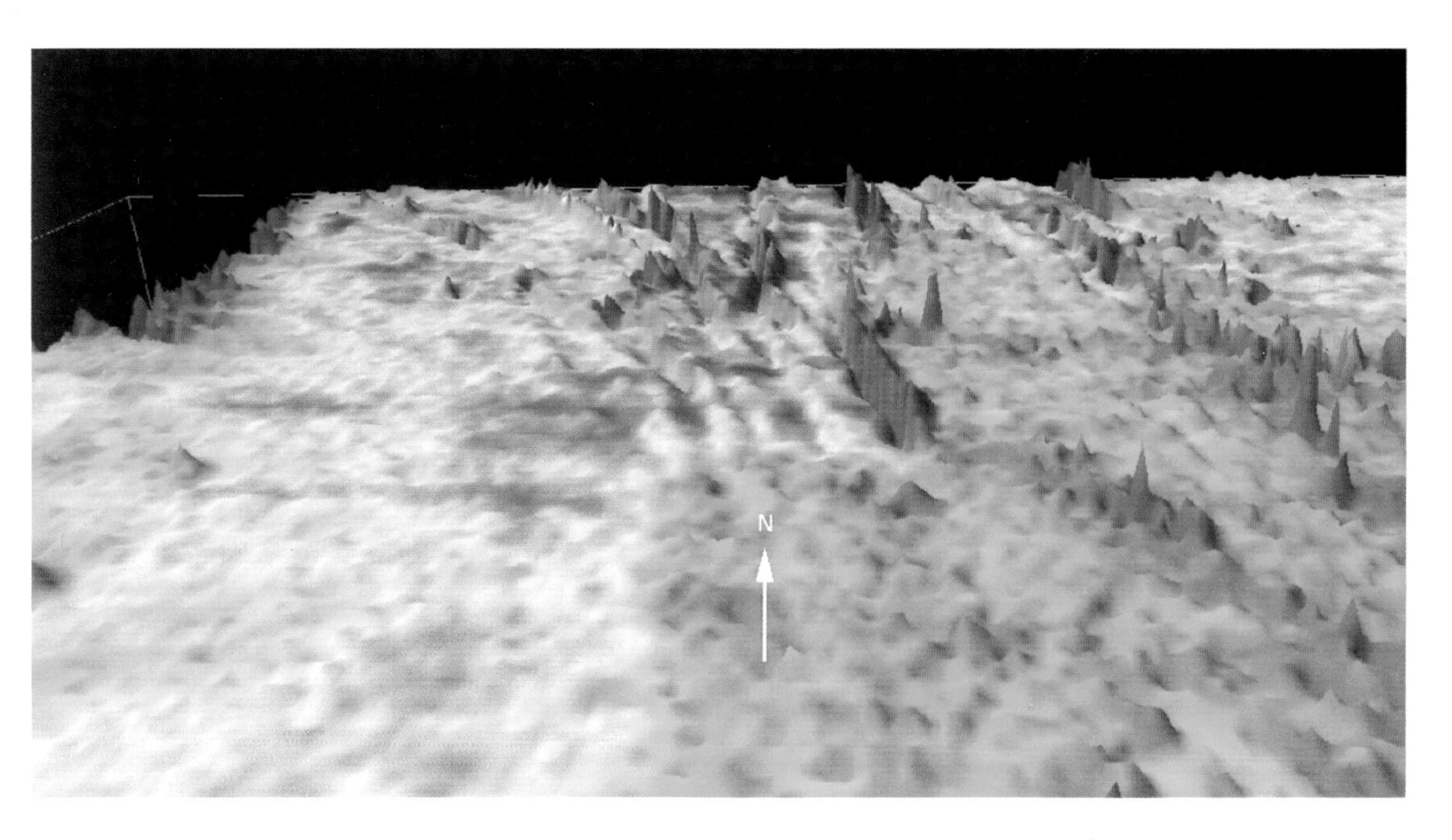
N

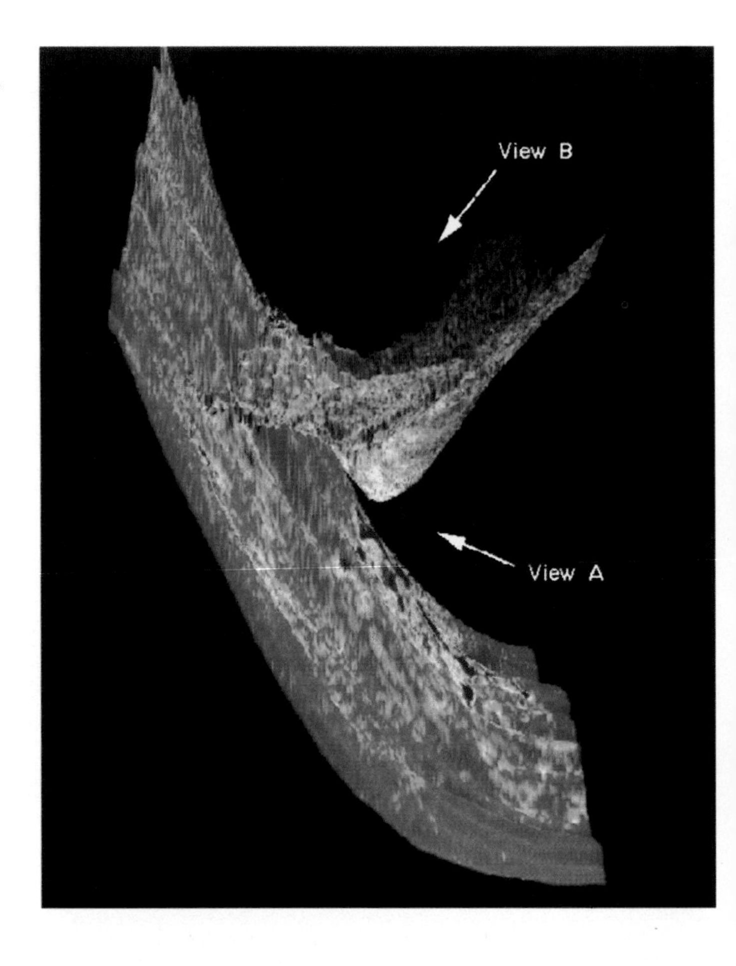

Fig.8-52. Lighted perspective display of an interpreted reservoir horizon and growth fault from a Gulf of Mexico 3-D survey viewed from the side. The color on the surfaces represents the reflection amplitude. (Courtesy ARCO Exploration and Production Technology.)

attribute analysis has been used effectively in a number of studies for fault interpretation, estimation of reservoir properties, and reservoir mapping. Denham and Nelson (1986) and Dalley, et. al., (1989), present good discussions of the use of horizon attributes for fault interpretation. Rijks and Jauffred (1991) present an overview of attribute extraction in general.

In particular, the interpreter may be interested in the relationship between the geometry of a three-dimensional interpreted horizon, and the variation of one (or many) data attributes along that horizon. An understanding of this information and its implications usually must be communicated by the interpreter to others. Visualization techniques provide an effective means of accomplishing these goals.

The ability to understand the data, gain useful insight, and communicate these with others are greatly enhanced by the use of interactive 3-D visualization. Figure 8-49 is a plot of an interpreted horizon viewed from the northwest with a single "sun" or light source off to the east (left). The shape of the surface is defined by the time structure of the interpreted horizon, and trace amplitude is mapped on the horizon in color. Several details are now very obvious. First there are a number of small throw faults trending roughly north to south. Most of these faults are downthrown to the west (the right). The faults are emphasized by a combination of shadows and low amplitude (blue) lineaments. The broad low amplitude zone clearly does not correlate with current structure in the area (it trends roughly northeast - southwest), whereas the strike of the axis of the anticline is northwest - southeast.

These displays can also be used to study the relationship between interpreted horizons. The horizon in Figure 8-49 lies slightly below an unconformity. When the unconformity is picked, the data can be flattened on the unconformity. The resulting structure is an approximation of the structure as it existed at the time deposition occurred on the unconformity surface (Figure 8-50). In this image the horizon is being viewed from the northeast. This highlights the correlation between that low ampli-

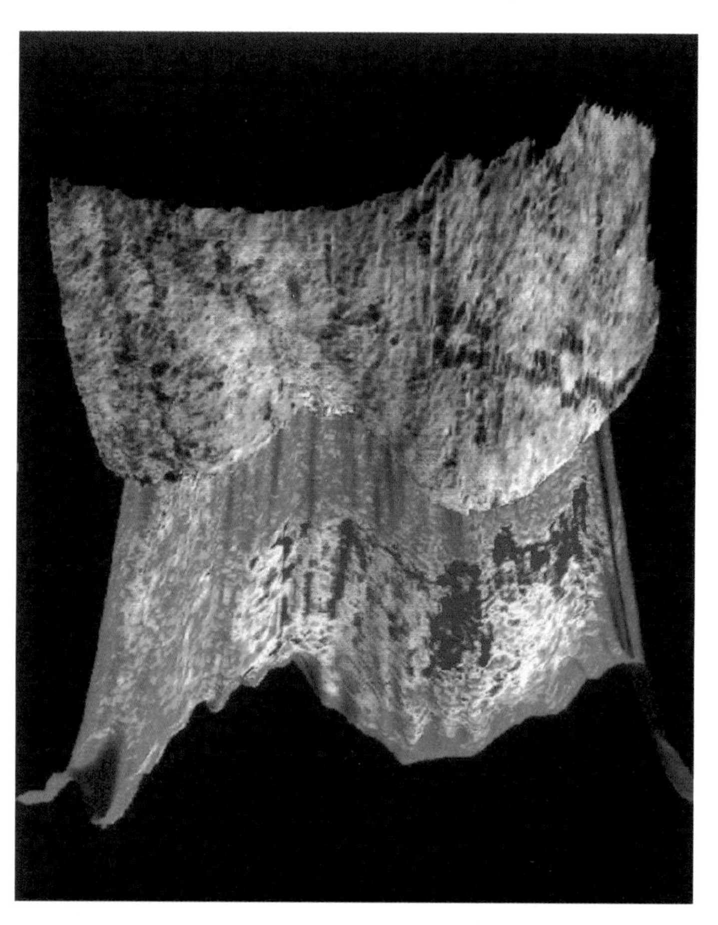

Fig.8-53. Reservoir horizon and growth fault from a Gulf of Mexico 3-D survey viewed from below, looking up the growth fault surface at the under side of the interpreted reservoir horizon. This viewpoint is shown as "View A" in Figure 8-52. (Courtesy ARCO Exploration and Production Technology.)

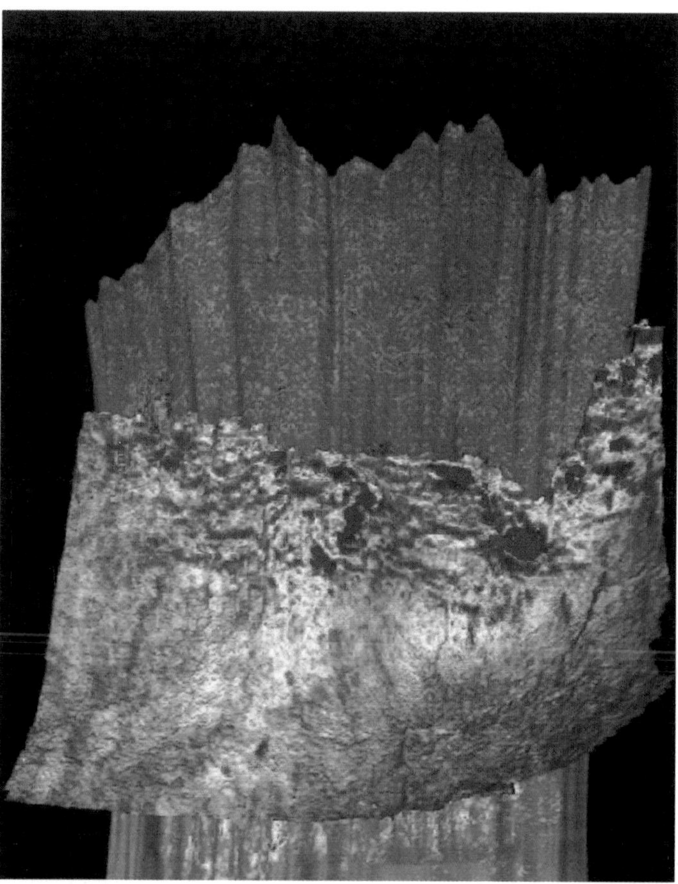

Fig.8-54. Reservoir horizon and growth fault from a Gulf of Mexico 3-D survey viewed from above, looking down at the growth fault surface and at the top of the interpreted reservoir horizon. This viewpoint is shown as "View B" in Figure 8-52. (Courtesy ARCO Exploration and Production Technology.)

tude zone and the crest of the Upper Cretaceous paleo-anticline. At the time of the unconformity, the structural trend was northeast-southwest. The low amplitude zone lies along the crest of the paleo- anticline. The fact that the north-south faults form sharp-edged boundaries to the low amplitude zone suggests that these faults existed prior to the formation of the unconformity.

Attribute/Attribute Relationships

In Figures 8-49 and 8-50, the time-structure of horizon is used to define the shape of the three-dimensional surface. It is possible to use any attribute (not necessarily time) to define the shape of the three-dimensional surface while a second attribute is represented by the color on the surface. In Figure 8-51, the regional or low spatial frequency component of the horizon has been removed. The surface is defined by the high pass filtered horizon structure - the residual horizon, as in Figures 8-3, 8-6, 8-7 and 8-8. The residual horizon will typically have its greatest non-zero values at abrupt discontinuities such as faults. In this image we are looking at the northern half of the horizon from the south. The ridges in the residual horizon structure associated with north-south trending small throw faults are clearly visible. The color in Figure 8-51 still represents reflection amplitude.

Complex Structural Relationships

Another example shows the relationship between an interpreted reservoir horizon and an associated growth fault in the Gulf of Mexico. In Figure 8-52 the horizon and fault surface are viewed from the side. Trace amplitude is shown on the surfaces in color. From the side view it is clear that the fault surface itself has a very complex structure. The interpreted horizon appears high near the fault, has a large depression and then rises again farther from the fault surface.

Figure 8-53 was created by moving the viewpoint around to look up the growth fault from below the horizon. This view highlights the structural relationship between the horizon and the fault surface. The fault surface has a very complex shape, with two major ridges with large recessed regions between and around them. The structure of the horizon shows clear control by the shape of the growth fault, with large bowl shaped depressions and ridges in the horizon corresponding to the depressions and ridges in the fault surface. Over-printed on the large scale structure of the growth fault is a finer structure of fault grooves that likely trend parallel to the direction of movement.

In Figure 8-54, the viewpoint has been moved to look down on the growth fault. The complex structure of the fault blocks along the ridge and in front of the growth fault surface are quite evident. Production of gas occurs in several of the high reflection amplitude, isolated fault blocks adjacent to the growth fault.

Relationships between Structure and Stratigraphy

In 3-D data the explorationist is provided with a unique opportunity to explore the stratigraphy of the subsurface in detail. Time slices and horizon slices allow the interpreter to see features and detail in the data that would otherwise be missed. Three-dimensional visualization provides a means of integrating the structural and stratigraphic aspects of interpretation in detail.

Figure 4-21 is a time slice at about 750 ms from a 3-D survey in the Gulf of Mexico. Obvious features of the time slice include two major faults and a large meandering stream. Figure 4-22 is an inline section extracted from the 3-D volume cutting perpendicular to the large stream. The stream appears in the seismic section as a narrow zone only a few traces wide of about 180° phase shift. One approach to interpreting the horizon is to pick the horizon across phase - peak outside the channel, trough within the channel. An alternate approach to picking the horizon is to pick it at constant phase - that is, pick the peak both outside and within the channel. Figure 8-55 is the inline from the 3-D volume, showing the horizon picked across phase (Figure 8-

Text continues on page 273

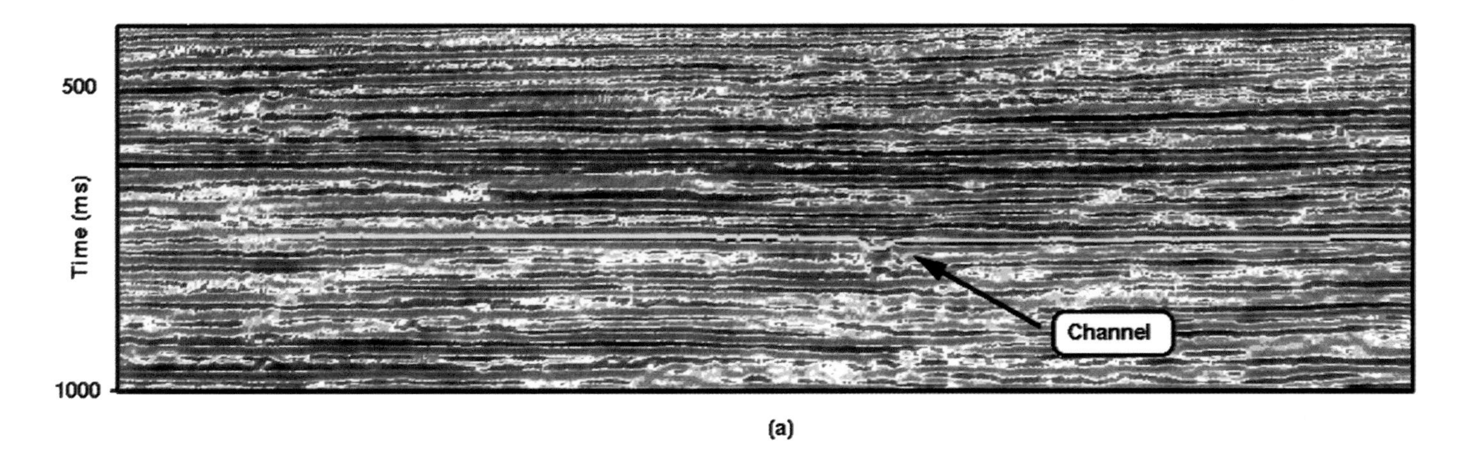

(a)

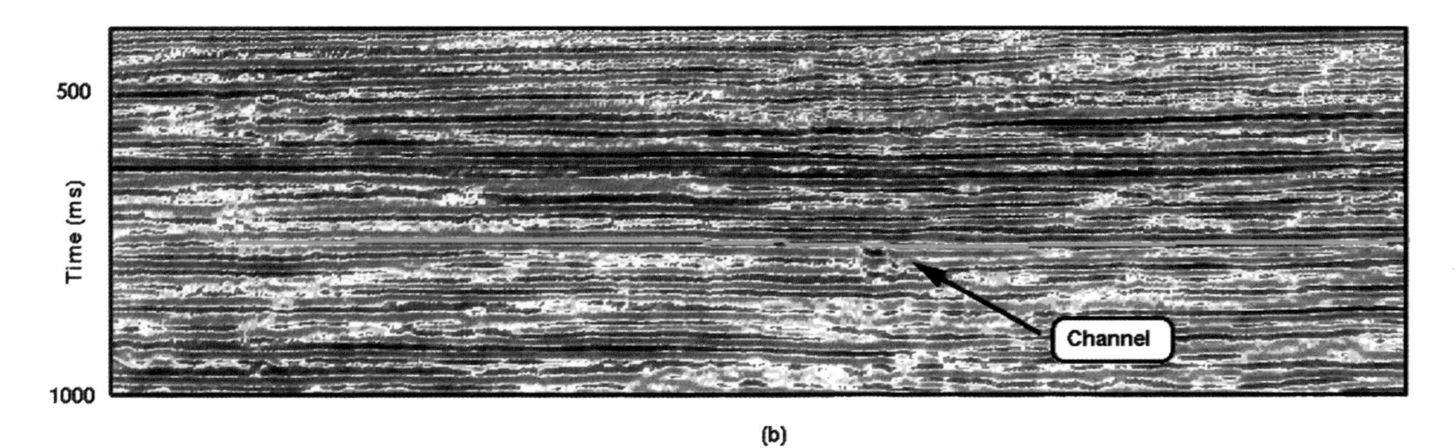

(b)

Fig.8-55. Inline section from a 3-D survey in the Gulf of Mexico showing alternate interpretations of the horizon containing a meandering stream. In figure 8-55a the horizon has been picked on constant phase (a trough) across the channel. In Figure 8-55b the horizon has been picked across phase at the channel. (Courtesy ARCO Exploration and Production Technology.)

Fig.8-56. Reflection amplitude maps at the interpreted horizon containing the meandering stream in the 3-D survey in the Gulf of Mexico. Figure 8-56a shows the amplitude for the horizon picked on constant phase, and Figure 8-56b shows the amplitude for the horizon picked across phase. (Courtesy ARCO Exploration and Production Technology.)

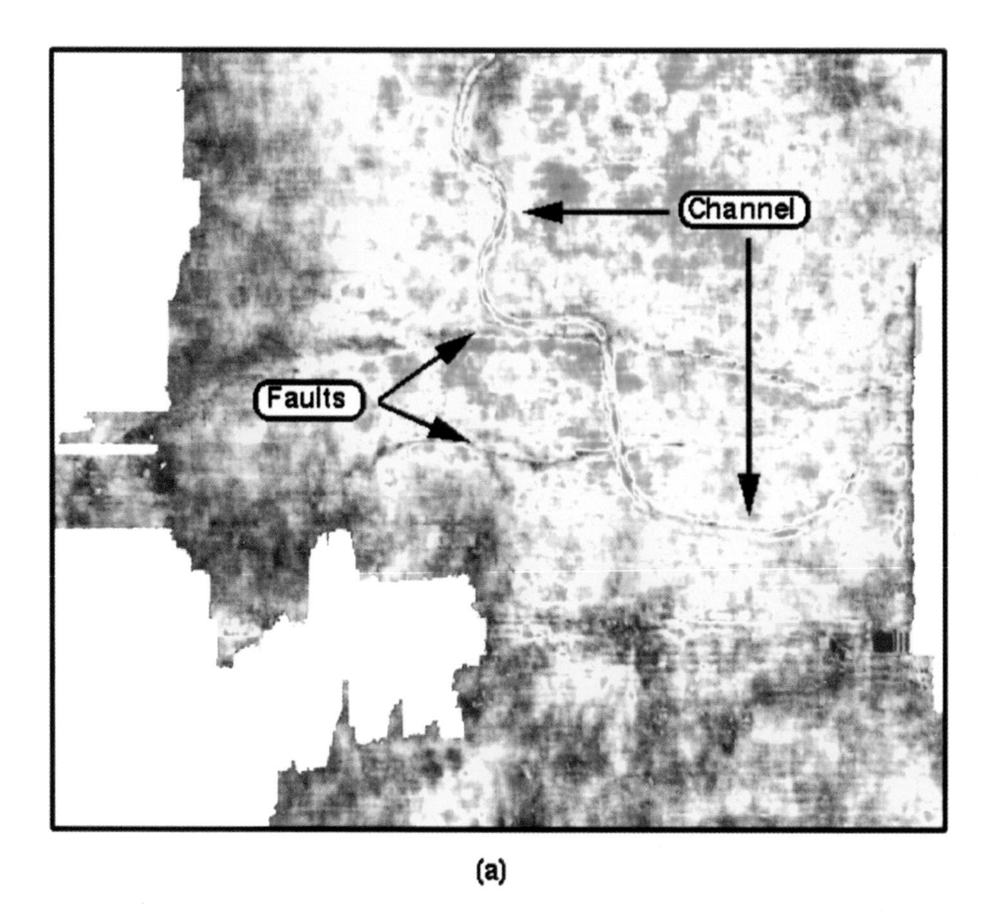

(a)

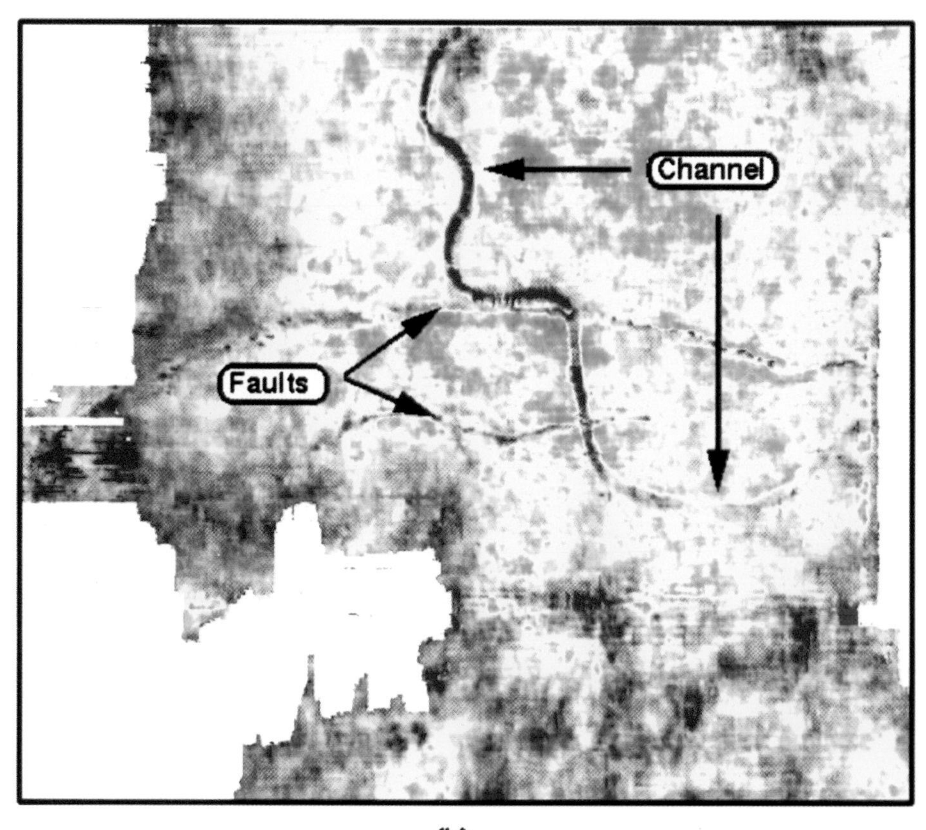

(b)

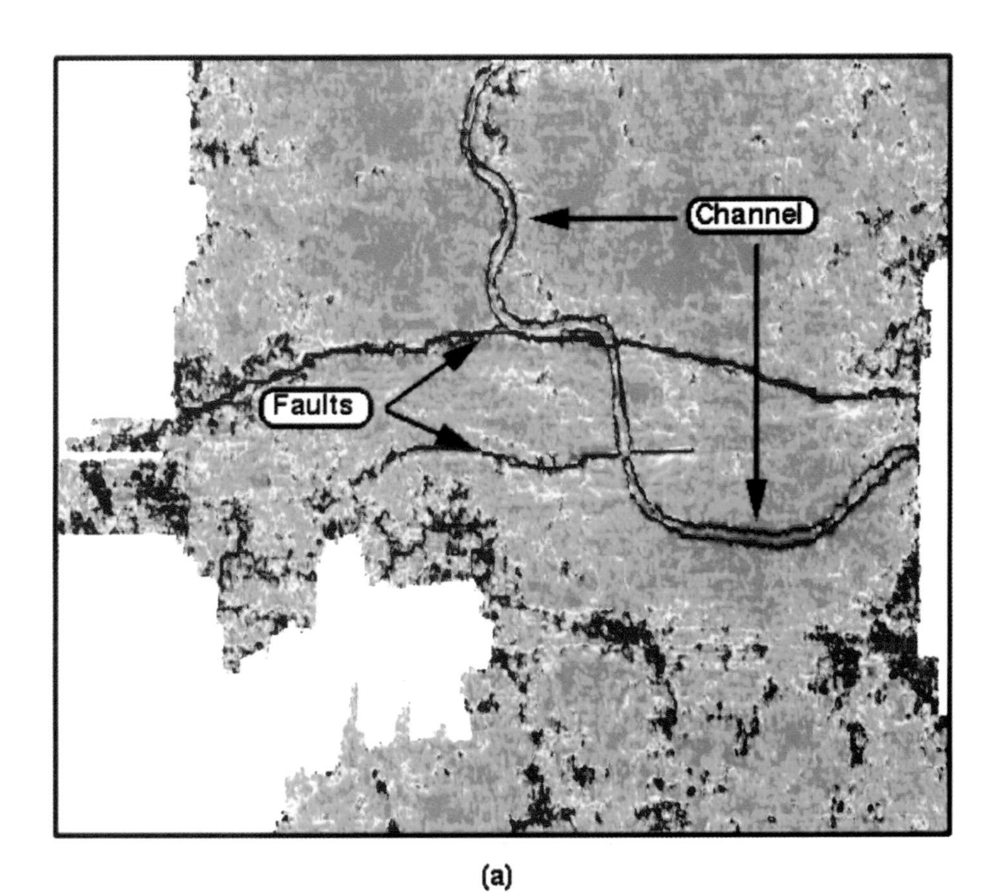

(a)

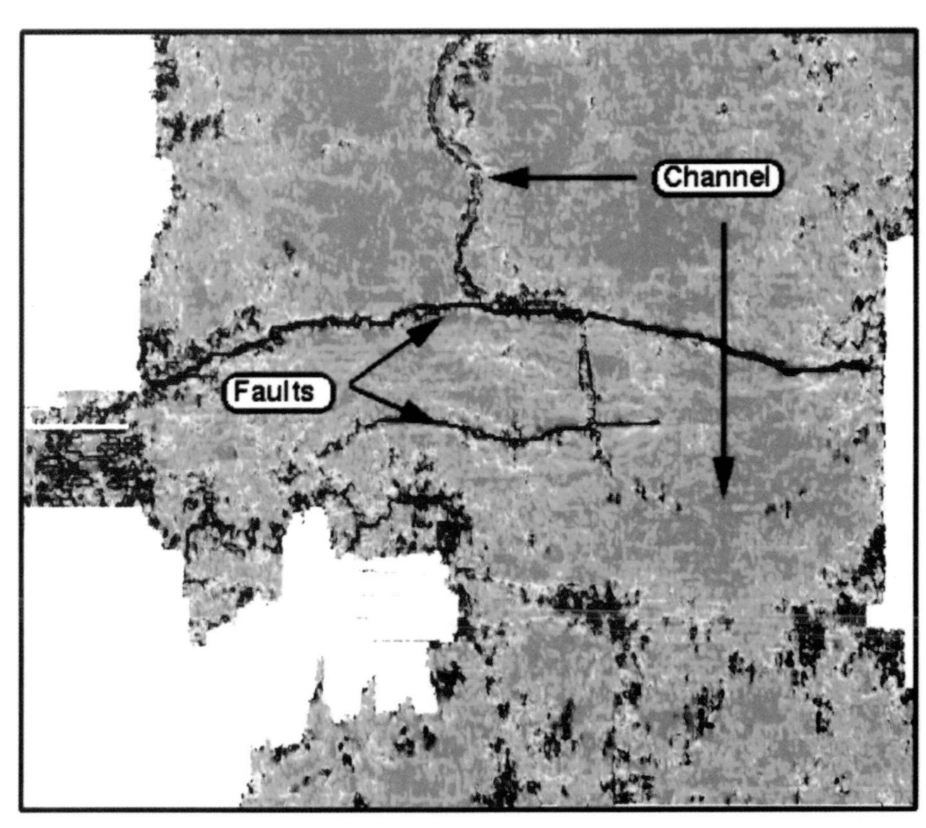

(b)

Fig.8-57. Dip maps at the interpreted horizon containing the meandering stream in the 3-D survey in the Gulf of Mexico. Figure 8-57a shows the dip for the horizon picked on constant phase, and Figure 8-57b shows the dip for the horizon picked across phase. (Courtesy ARCO Exploration and Production Technology.)

Fig.8-58. Lighted perspective display of the horizon containing the meandering stream in the 3-D survey in the Gulf of Mexico. In Figure 8-58a the surface and reflection amplitude represent the horizon picked on constant phase across the stream. In figure 8-58b the surface and reflection amplitude represent the horizon picked across phase. (Courtesy ARCO Exploration and Production Technology.)

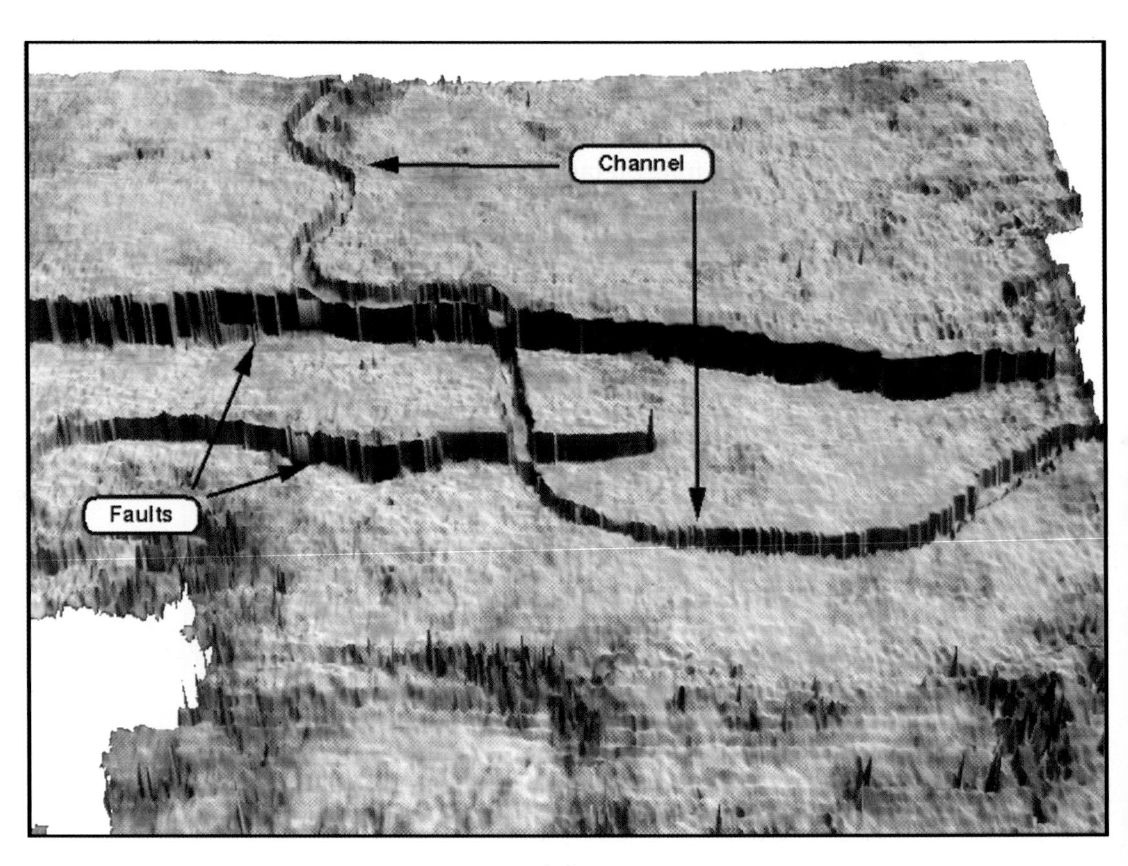

(a)

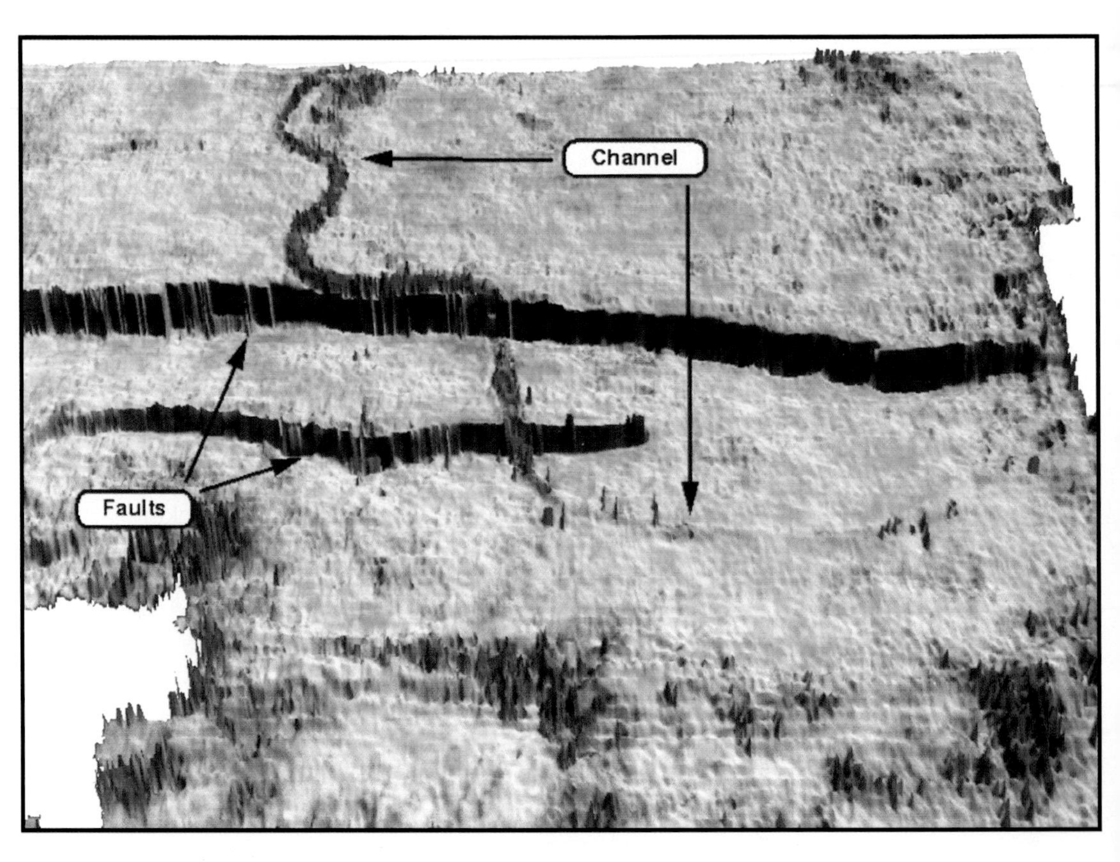

(b)

55a) and on constant phase (Figure 8-55b).

Figure 8-56 shows displays of the reflection amplitude at the interpreted horizon for the pick at constant phase (Figure 8-56a), and the pick across phase (Figure 8-56b). The dip display of the two versions of the picked horizon are shown in Figures 8-57a and b. The stream channel is interpretable in all of the displays. For the pick across phase it is most evident in the amplitude display. For the pick at constant phase, the channel is most prominent in the dip display.

There is a striking difference between the information conveyed by the two interpretations of the horizon when they are displayed in three dimensions. In Figures 8-58a and b, the horizon time surfaces are displayed with reflection amplitude superimposed in color. The constant phase interpretation (Figure 8-58a) looks like a channel. The nature of the feature is obvious at a glance because of the manner in which it was interpreted and the way in which it is visualized. The interpretation across phase (Figure 8-58b) shows the location of the channel, but it does not convey the information as well.

Integrating Information

The interpreter must often integrate more information than can be represented by a three-dimensional structure, a single mapped attribute, and a light source. For example, there may be effects caused by a shallower event (e.g., amplitude or frequency shadows). In this case a visual correlation between the shallow feature and the target horizon would provide a quick means of evaluating these effects. There are also times when a visual correlation between two attributes would be important (e.g., the area on the Top Reservoir horizon, with reflection amplitudes in color, where the contoured cumulative production exceeds a given value). There are also situations where a lighting and perspective do not provide sufficient visual information to understand the complex three-dimensional nature of a horizon.

An example combining two of these cases is shown in Figures 8-59 and 8-60. The Top Reservoir horizon structure is shown in Figure 8-59 as a 3-D surface in perspective, viewed from the north at an elevation of about 45°, with a light source shining down on the surface from above. Reflection amplitude is shown on the surface in color. The structure of the surface is difficult to comprehend from this display. Some additional visual cues are needed.

One approach to this problem, illustrated in Figure 8-60, utilizes a technique called texture mapping to map structure contours directly onto the horizon in 3-D. The contours provide the additional information necessary to understand the structure and simultaneously allow the identification of amplitude features that occur at similar depths. Texture mapping can be used to show additional data as well. In Figure 8-60 a stippling of black dots is posted on the horizon in an area where there was concern that a shallow feature might cause an amplitude shadow at the Top Reservoir reflection. The stippled zone represents the area over which the shallower feature is present. A visual assessment can now be made of the potential shadow effect on the amplitude pattern.

A second example of the use of texture mapping comes from the Pickerill Field in the U. K. waters of the southern North Sea. A geophysical reservoir characterization study (Dorn, et. al., 1996) successfully derived a relationship between corrected reflection amplitude at the Top Rotliegend (top reservoir) reflection and log-derived reservoir porosity. A seismic-guided estimated porosity map was produced (Figure 8-61). The reservoir is highly faulted, and drilling indicated that at least some of the faults, sealed by diagenesis, could act as barriers to flow within the reservoir, even if the throw along the fault was insufficient to completely offset reservoir interval. Finally, a dolomite (the Plattendolomit), deposited within the Zechstein evaporite sequence which overlies the reservoir, is rafted in the area. As the Zechstein evaporites flowed, the dolomite fractured into rafts, so that it is present over portions of the area, and not present over others. The location of the Plattendolomit rafts is important because exploration drilling had shown that overpressure might be encountered when drilling

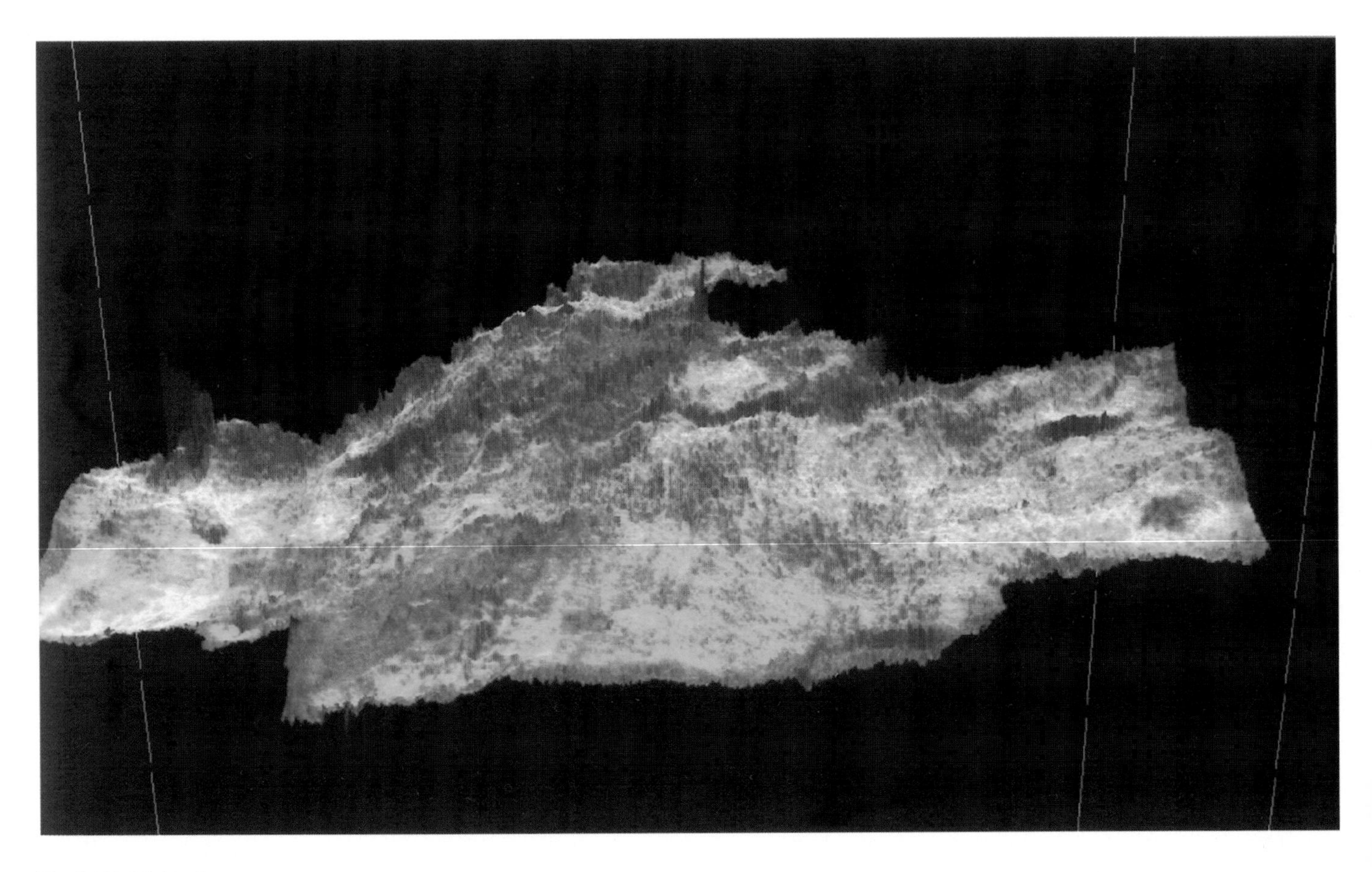

Fig.8-59. Lighted perspective display of a top reservoir horizon, with the reflection amplitude shown in color. (Courtesy ARCO Exploration and Production Technology.)

through a raft.

Figure 8-62 is a display that integrates all of the above information to aid in planning development well locations in the northwest half of the field. The structure of the surface is the interpreted Top Rotliegend time structure. The estimated porosity is shown in color on the surface. The reservoir boundary is shown in bright green, and several of the exploration well locations are shown in pink. A light source is used to cast shadows at discontinuities of the surface to highlight the location of faults. Finally, the location of the Plattendolomit rafts is highlighted by the brick pattern overlaid on the surface.

By using this type of display, potential drainage compartments can be identified. Well locations within a compartment can be optimized to avoid potential barriers to flow (small throw faults), and to encounter the highest predicted porosity. The well path can also be adjusted to avoid the Plattendolomit rafts, avoiding potential drilling problems, and helping to minimize drilling costs.

Applications of Stereopsis

One of the most important three-dimensional visual cues is stereopsis. The use of stereo displays is increasing over time because they allow the interpreter to see additional 3-D detail in the data.

Figure 8-63a is a shaded perspective display of an interpreted reef structure shown in perspective view, with a light source and reflection amplitude displayed on surface in color. This image combines the use of several 3-D cues, but the image still looks like a flat 2-D projection of a 3-D surface. Figure 8-63b is a stereo pair of the same data. Once the two images in Figures 8-63b has been properly fused into a stereo image, the horizon and reef will appear to come partially out of the page toward the viewer - it will appear as a true 3-D surface. Notice, in particular, how much more prominent the breaks in the reef appear, and the associated debris slopes.

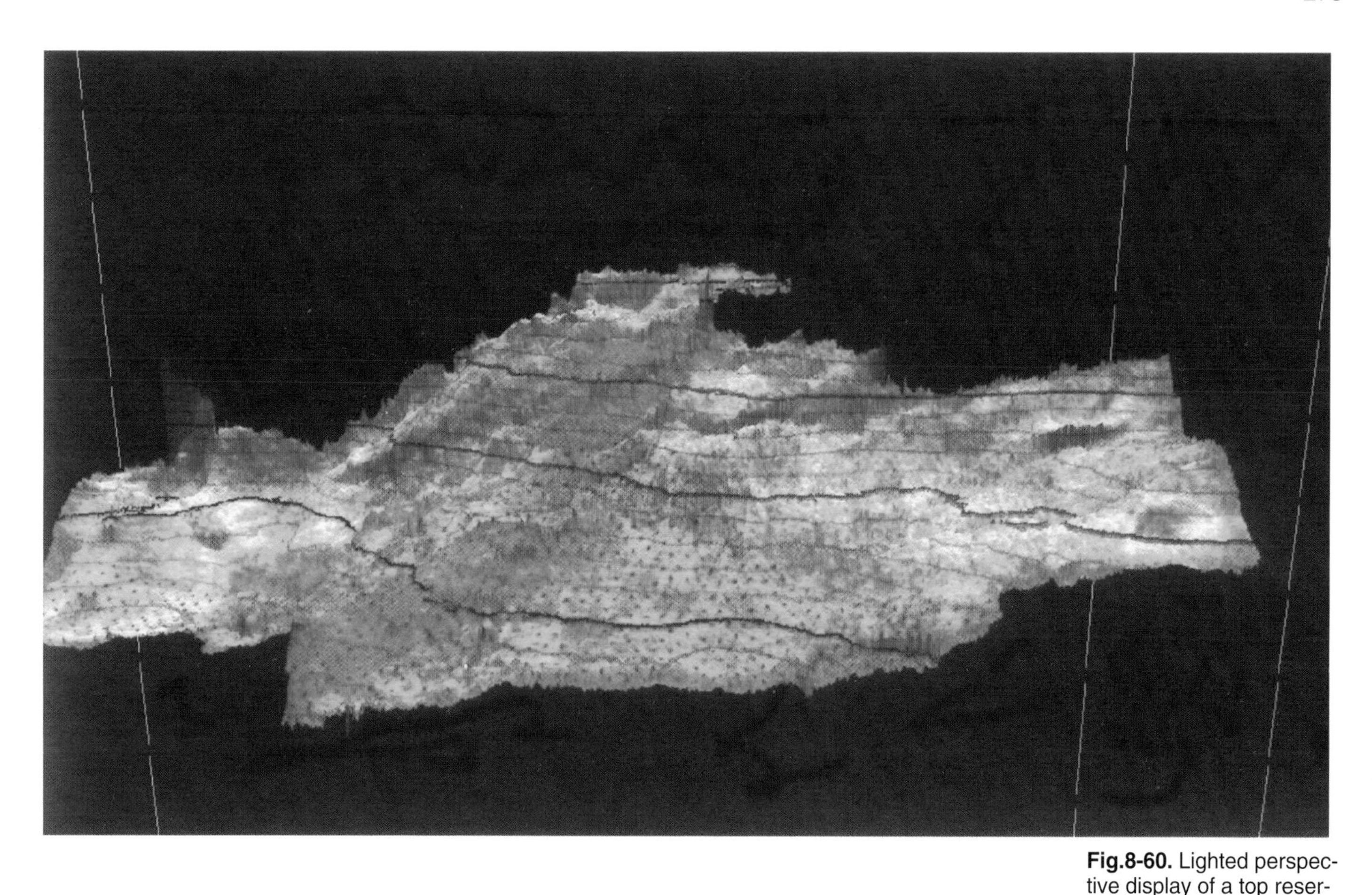

Fig.8-60. Lighted perspective display of a top reservoir horizon, with the reflection amplitude shown in color. The black contours are horizon structure contours, and the black stippled pattern represents a region where the amplitudes may be attenuated due to effects shallower in the section. (Courtesy ARCO Exploration and Production Technology.)

Figures 8-64a and b are a similar set for a reservoir horizon from a survey in the southern North Sea. Figure 8-64a shows reflection amplitude in color on the top reservoir time structure in a lighted perspective display. Figure 8-64b is a stereo pair for the horizon from the same viewpoint. The gross aspects of the structure are clearly evident in Figure 8-64a - the large tilted fault blocks, the north-south bounding fault, and the secondary northwest-southeast fault trend. The detail of the intra-block structure only becomes obvious in the stereo display. The amplitude pattern indicated by the arrow in Figure 8-64a, is clearly a small intra-block graben system when viewed in the fused stereo image.

The additional information content provided to the interpreter by stereo displays is substantially more striking with the larger stereo images that can be displayed on properly equipped workstations. Some information that can be observed in the stereo images could be interpreted from non-stereo displays, but the interpretation requires more time and effort. Other information would simply be missed without the stereo display.

Use of Motion

Animation or motion of a 3-D interpretation can also play an important role in interpretation and communication of complex structural and stratigraphic relationships. Three applications of motion are discussed below: changing the observer's viewpoint, interactive three-dimensional flattening, and interactive movement of the light source.

Motion in the form of changing the observer's viewpoint can greatly aid interpreting and understanding complex 3-D structures. Motion accentuates the effects of many of the visual cues used to perceive a 3-D object. The shifting of shadows, highlights, changing perspective, obscuration, etc., all provide very useful visual information. The creation of a movie by programmed "fly-by" or path of observation through

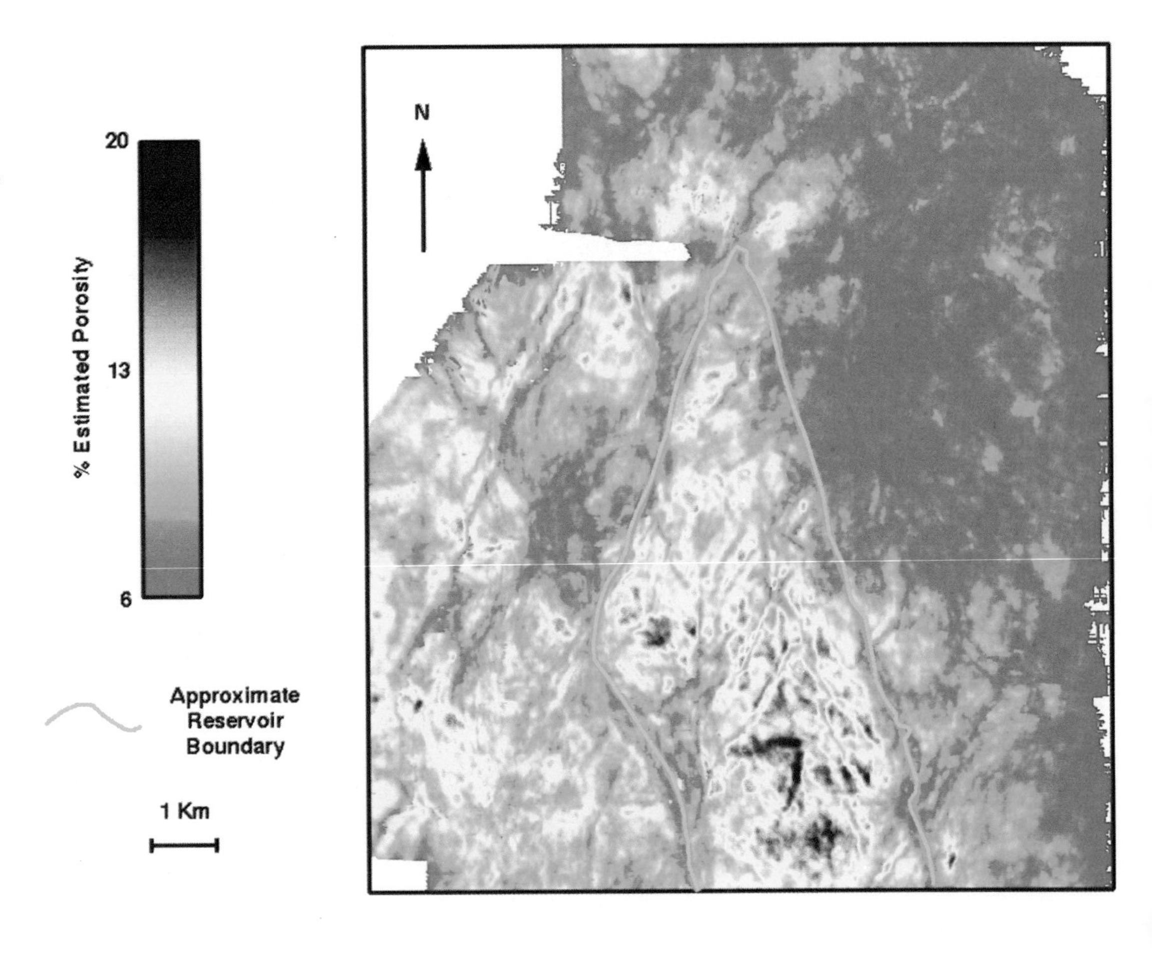

Fig.8-61. Estimated porosity, derived from an empirical reflection amplitude vs porosity relationship, for the Top Rotliegend reservoir in the northwest half of the Pickerill field, North Sea. (Courtesy ARCO Exploration and Production Technology.)

space may also be used to convey complex information to other technical personnel and to management. This technique has been used very effectively by planetary scientists for several years to explore and understand the relationships between altimetry and other remotely sensed data (e.g., view the video tape On Robot Wings - A Flight Through the Solar System, Finley-Holiday Films, 1992). The same approach can be used to convey information regarding the relationship between estimated reservoir quality and top reservoir structure, and to highlight the location of proposed or existing wells (Dorn, et. al., 1995).

Horizon flattening is a process frequently used to understand subtle geometric relationships between reflections. It can be a great aid in detecting and properly interpreting angular relationships. With interpreted 3-D horizons, flattening can be performed interactively in three dimensions (Dorn, et. al., 1995). Figure 8-65 shows a folded and faulted horizon interpreted in a 3-D survey from the North Sea. A second event, an upper Cretaceous unconformity, was also interpreted (top horizon in Figure 8-65). Figure 8-66 a-d shows a sequence of four images of the deeper horizon taken at different stages of flattening of the unconformity. By viewing the flattening as an animated movie, it is readily apparent that displacement along the faults was essentially complete prior to deposition above the unconformity. Also, the subsequent deformation did not reactivate these faults to any significant degree. The low amplitude (green - blue) zone is seen to correspond to the crest of a paleo-anticline, and it is clear that the structural trend has changed by almost 90° between Upper Cretaceous time and the present. By moving the viewpoint around while the horizon is being flattened, the interpreter can rapidly observe and understand these relationships through the survey area.

When using interactive flattening it is important to recall that this is just a flattening process - a relative vertical shifting of time and amplitude. It is not true reconstruction, and should not be used as such. It can, however, provide some very useful

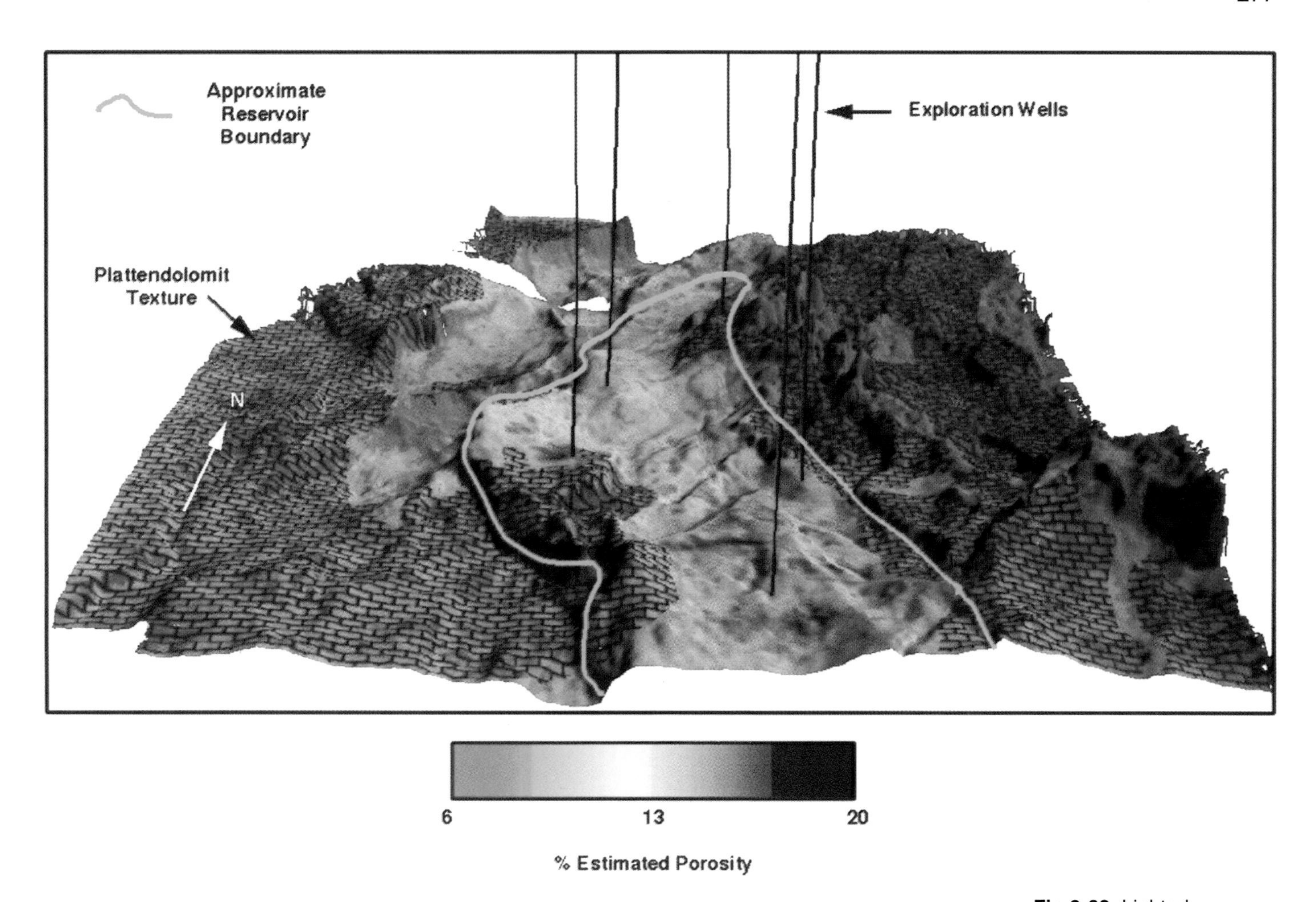

Fig.8-62. Lighted perspective display of the Top Rotliegend time structure, integrating the exploration well locations, the estimated porosity, the reservoir boundary and the areas over which there are Plattendolomit rafts. (Courtesy ARCO Exploration and Production Technology.)

geologic insight.

Figure 8-67 is a reflection amplitude map from a top reservoir horizon in a 3-D survey. Interpretation of subtle reservoir faulting through the central area of the map is important for infill development drilling. Due to a low signal-to-noise ratio, direct interpretation of fault traces from horizon attribute maps, such as Figure 8-67, is difficult.

This type of problem in fault interpretation can be addressed by the use of lighted displays of the 3-D horizon with horizon attributes draped on the surface in color. The value of this procedure is enhanced by the ability to move the light source (and, thus, change the shadows) in real time. The human eye can see differences in the lineament pattern revealed on the horizon in Figures 8-68 a-d for the light source positioned to the north, south, east and west with respect to the horizon. Human vision is even more sensitive to the subtle changes that occur as the light source is moved around and shadows change interactively. Also, by moving the light interactively, you are much less likely to miss a subtle trend than if you restrict yourself to using a few static displays.

References

Bahorich, M., and S. Farmer, 1995, 3-D seismic discontinuity for faults and stratigraphic features - the coherence cube: The Leading Edge, v.14, p. 1053-1058.

Bahorich, M., J. Lopez, N. Haskell, S. Nissen, and A. Poole, 1995, Stratigraphic and structural interpretation with 3-D coherence: SEG 65th Annual Meeting expanded abstracts, p. 97-99.

Bouvier, J. D., C. H. Kaars-Sijpesteijn, D. F. Kluesner, C. C. Onyejekwe, and R. C. van

(a)

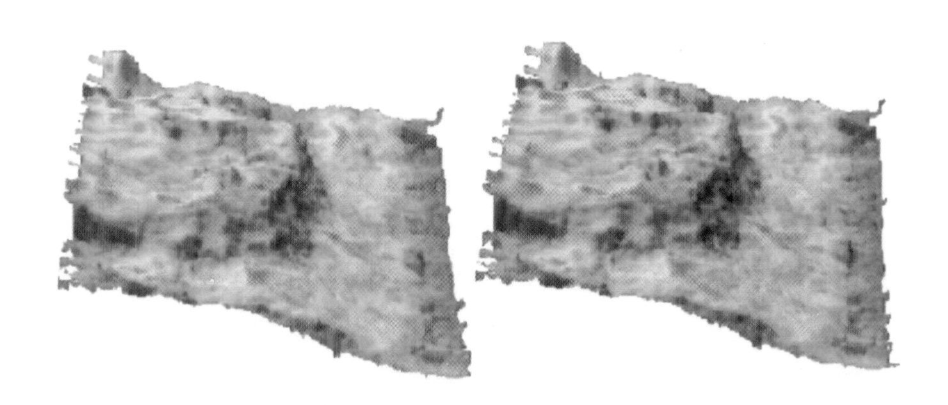

(b)

Fig.8-63. Lighted perspective display of an interpreted horizon from a 3-D survey in the Permian Basin showing a reef structure, with reflection amplitude in color (Figure 8-63a). Figure 8-63b is a stereo pair for the horizon. (Courtesy ARCO Exploration and Production Technology.)

(a)

(b)

Fig.8-64. Lighted perspective display of the interpreted top reservoir horizon from a 3-D survey in the southern North Sea Gas Basin, with reflection amplitude in color (Figure 8-64a). Figure 8-64b is a stereo pair for the horizon. (Courtesy ARCO Exploration and Production Technology.)

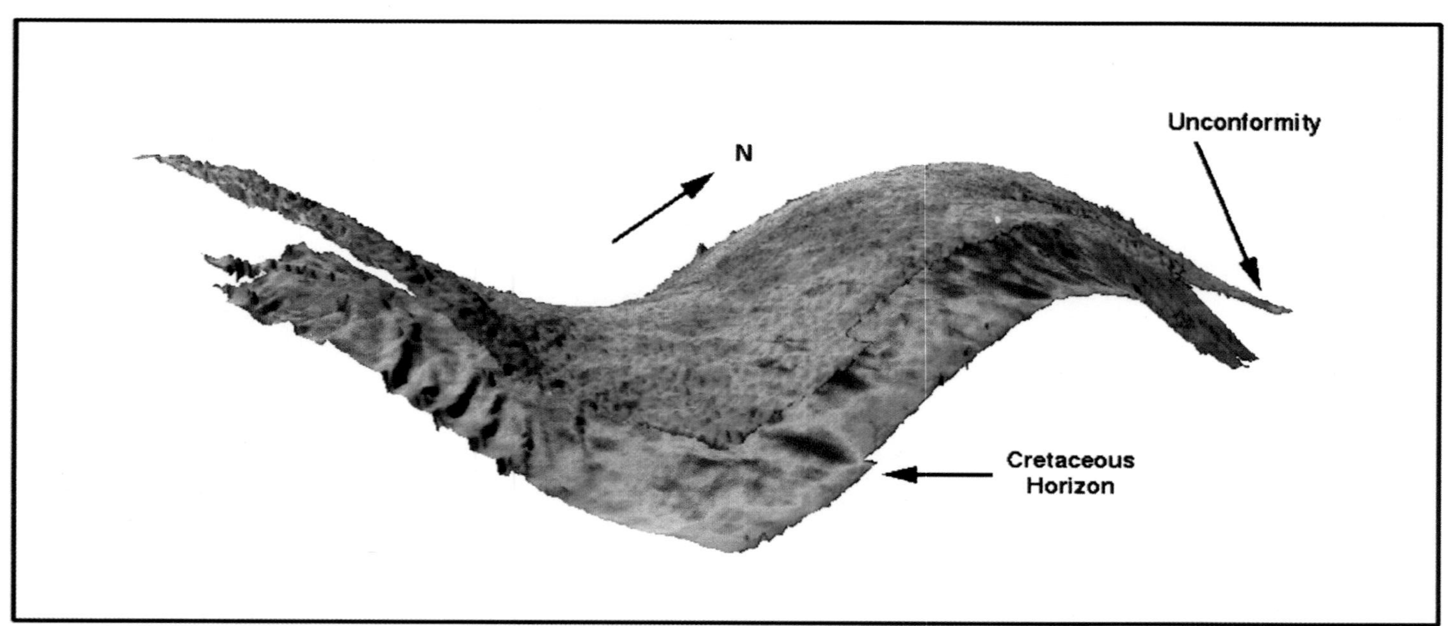

Fig.8-65. Lighted perspective display of an interpreted Cretaceous horizon and an overlying Upper Cretaceous unconformity from a 3-D survey in the southern North Sea Gas Basin. Reflection amplitude is shown on the surfaces in color. (Courtesy ARCO Exploration and Production Technology.)

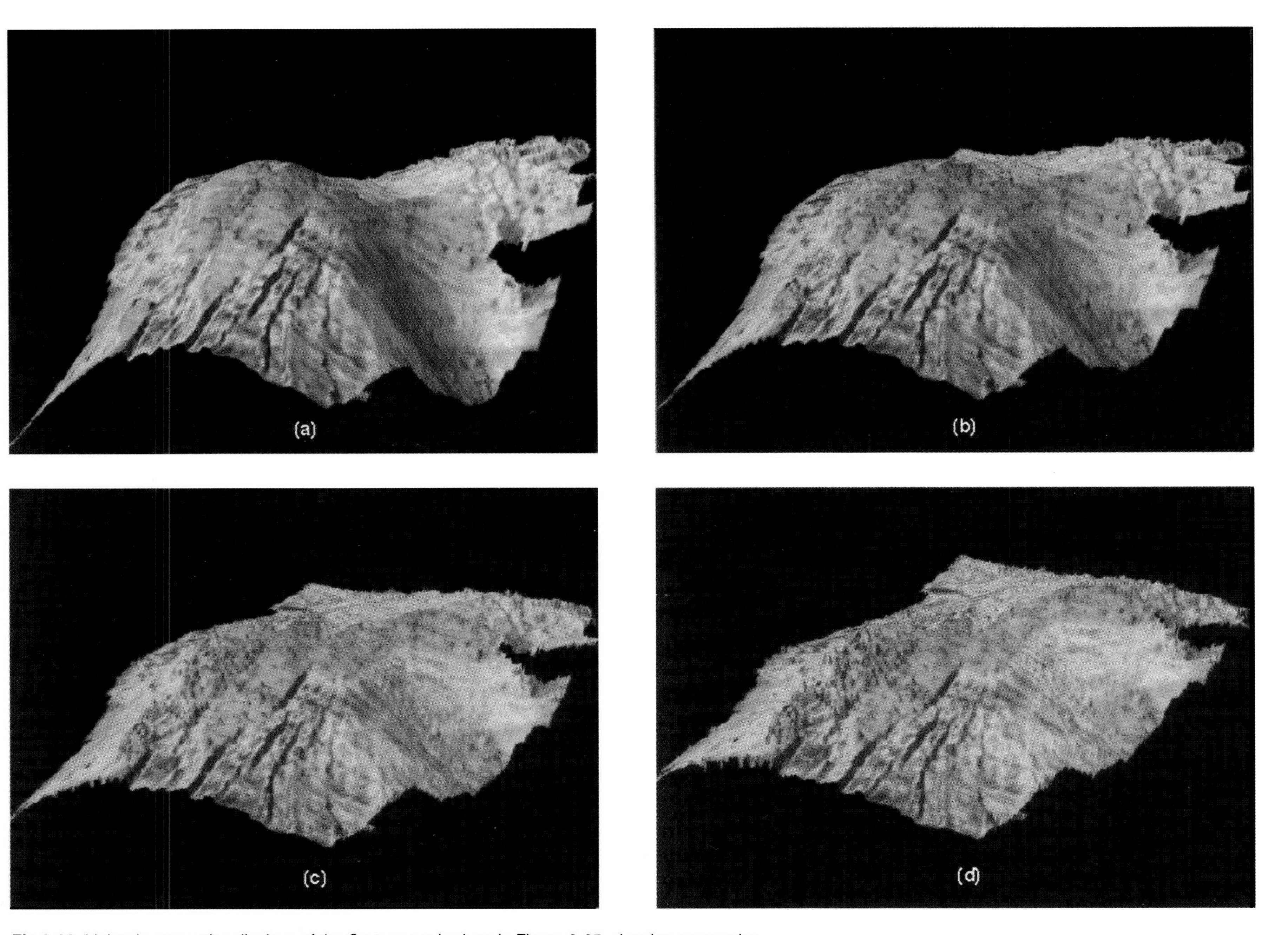

Fig.8-66. Lighted perspective displays of the Cretaceous horizon in Figure 8-65, showing successive stages of flattening of the overlying unconformity. The progression is from current structure (Figure 8-66a) to an approximation of paleo-structure at the time of deposition on the unconformity (Figure 8-66d). (Courtesy ARCO Exploration and Production Technology.)

Fig.8-67. A map of reflection amplitude at the interpreted top reservoir reflection from a 3-D survey. Notice the low amplitude lineaments associated with major east-west trending faults, and the general low amplitude zone through the center of the horizon. (Courtesy ARCO Exploration and Production Technology.)

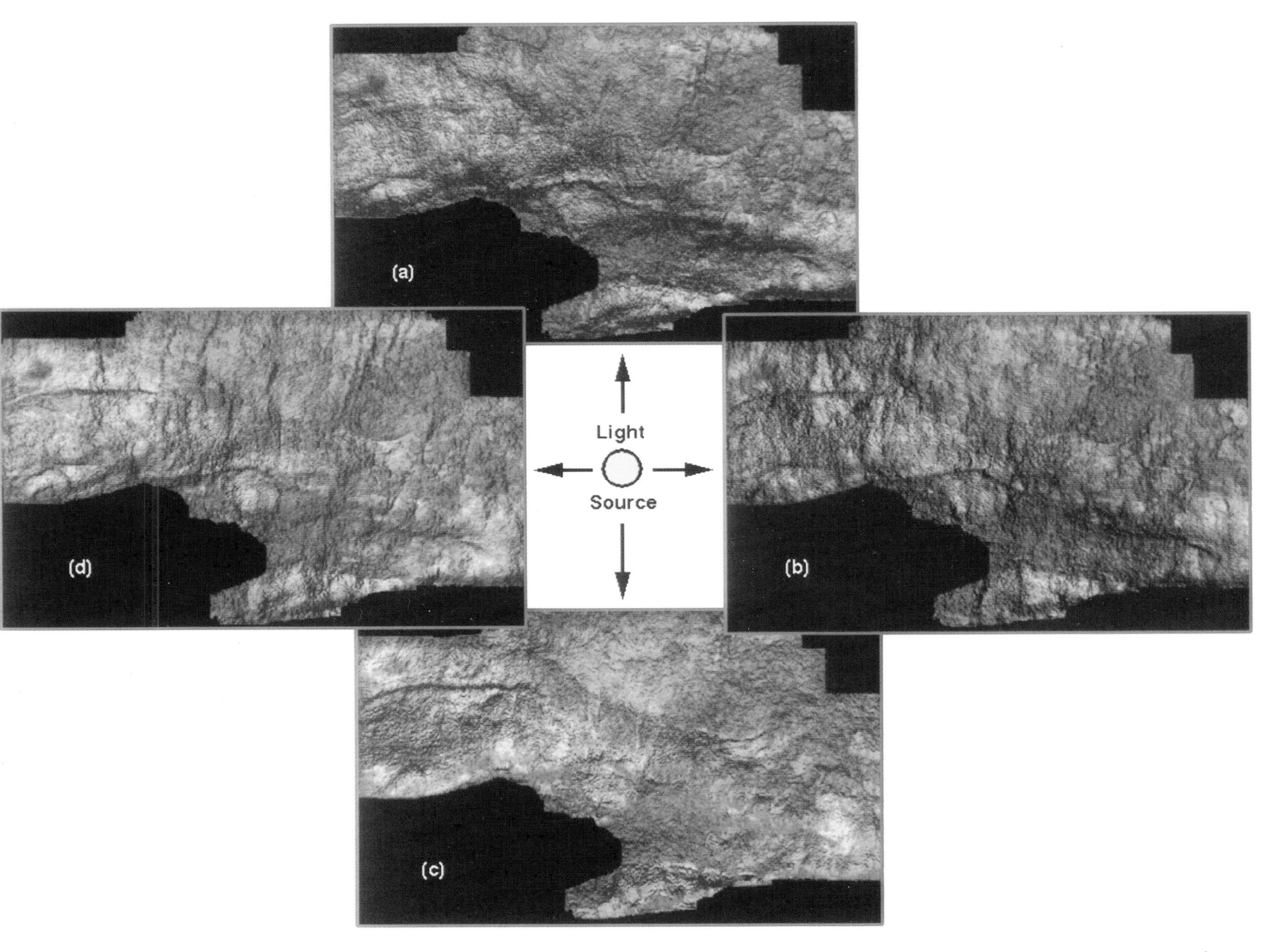

Fig.8-68. Perspective displays of reflection amplitude in color on the interpreted top reservoir time structure from a 3-D survey viewed from above. The sequence from Figure 8-68a through 8-68d shows the effects of moving the light source from the south, to the west, to the north, and to the east of the horizon surface. (Courtesy ARCO Exploration and Production Technology.)

der Pal, 1989, Three-dimensional seismic interpretation and fault sealing investigations, Nun River Field, Nigeria: AAPG Bulletin, v.73, p. 1397-1414.

Castagna, J. P., and M. M. Backus, 1993, Offset-dependent reflectivity - theory and practice of AVO analysis: SEG Investigations in Geophysics no. 8.

Dalley, R. M., E. C. A. Gevers, G. M. Stampfli, D. J. Davies, C. N. Gastaldi, P. A. Ruijtenberg, and G. J. O. Vermeer, 1989, Dip and azimuth displays for 3-D seismic interpretation: First Break, v. 7, no. 3, p. 86-95.

Denham, J. I., and H. R. Nelson, Jr., 1986, Map displays from an interactive interpretation: Geophysics, v. 51, p. 1999-2006.

Dorn, G. A., M. J. Cole and K. M. Tubman, 1995, Visualization in 3-D seismic reservoir interpretation: 5th Archie Conference, May 14-18, 1995, The Woodlands, Texas, Expanded Abstracts.

Dorn, G. A., K. Tubman, D. Cooke and R. O'Connor, 1996, Geophysical reservoir characterization of Pickerill Field, North Sea, using 3-D seismic and well data: To be published in the AAPG 3-D Seismic Atlas, edited by P. Weimer and T. Davies.

Foley, J. D., A. Van Dam, S. K. Feiner and J. F. Hughes, 1990, Computer Graphics Principles and Practice, 2nd Edition: Addison-Wesley Publishing Co.

Hoetz, H. L. J. G., and D. G. Watters, 1992, Seismic horizon attribute mapping for the Annerveen Gasfield, The Netherlands: First Break, v. 10, no. 2, p. 41-51.

Kowalik, W. S., T. M. Wissler, B. P. Dean, R. R. Brown, M. H. Hoehn, W. E. Glenn and A. Furniss, 1995, Chevron's GeoImage - enhancement and stereoscopic viewing of seismic horizons and other grids for improved data understanding and communication: 5th Archie Conference, May 14-18, 1995, The Woodlands, Texas, Expanded Abstracts.

N. E. Thiing Enterprises, 1993, The Magic Eye - A New Way of Looking at the World: Andrews and McMeel.

Oldham, A. C., and N. M. Gibbins, 1995, Lake Hope 3-D - a case study: Exploration Geophysics, v. 26, p. 383-394.

Rijks, E. J. H., and J. C. E. M. Jauffred, 1991, Attribute extraction: An important application in any detailed 3-D interpretation study: The Leading Edge, v. 10, no. 9, p. 11-19.

CASE HISTORIES OF THREE-DIMENSIONAL SEISMIC SURVEYS

This chapter presents twelve case histories demonstrating the solution of subsurface problems with 3-D seismic surveys. They have been selected on the basis of their diversity: land and marine environments, U.S. and overseas locations, structural and stratigraphic objectives, development and production arenas, authors from a variety of oil companies. Furthermore, during the evolution of this book through three editions and seven years, case histories have been added but none have been removed. Accordingly, this chapter shows a more complete record and demonstrates the progression of the technology and the increasing sophistication of the interpretation.

Case History 1

East Painter Reservoir 3-D Survey, Overthrust Belt, Wyoming

Donald G. Johnson, Chevron U.S.A., Inc.

The discovery of the East Painter Reservoir field in mid-1979 led to the initiation of the first major 3-D survey in the Wyoming Overthrust Belt. A 3-D survey was necessary because interpretation of conventional 2-D seismic data over the East Painter area did not provide a sufficiently reliable picture of the structure on the objective Triassic Nugget horizon to permit an aggressive development program. Field data for the 17 sq mi (44 sq km) East Painter 3-D survey were collected during the winter of 1979-80, and the final migrated sections were in hand by July 1980.

Interpretation of the final 3-D products resolved the previous structural ambiguities and showed the East Painter structure to be continuous and almost as large as the main Painter Reservoir feature. Information from the 3-D mapping allowed up to six development wells to be drilled at one time and helped to guide the locations of the last 13 development wells—all of them successful. The average cost per well was between $4 and $5 million. The cost of the 3-D survey was $1.6 million, which turned out to be a good value.

Introduction

The complex structures of the Wyoming Overthrust Belt in the western United States are revealed with varying degrees of clarity by the conventional 2-D seismic reflection method. In some instances, however, additional structural definition is essential for exploration and production purposes, and the 3-D seismic method can make the difference in the resolution of the structural problem.

The East Painter Reservoir 3-D survey was prompted by results of the Chevron 11-5A well, a new field discovery in 1979 located approximately 1 mile east of the eastern fault-edge of the Painter Reservoir field. This well encountered the objective Triassic Nugget horizon dipping steeply to the northwest, which verified the existence of a frontal thrust structure to the Painter Reservoir feature. Interpretation of the Nugget horizon on conventional CDP seismic data suggested that the East Painter

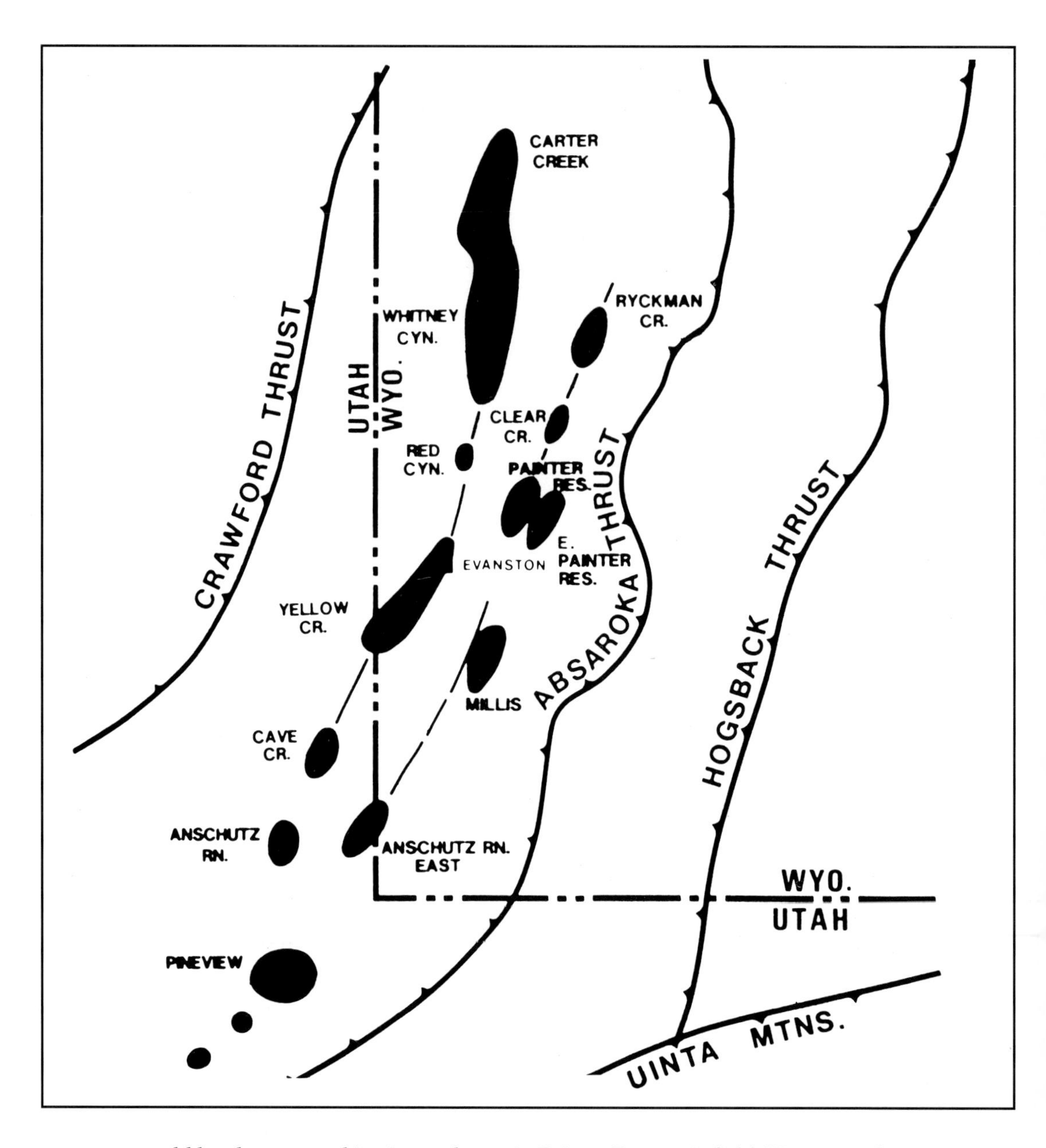

Fig. 9-1-1. Oil and gas fields of the Fossil Basin, Wyoming and Utah.

structure could be almost equal in size to the main Painter Reservoir field. However, data were very discontinuous, to the extent of nonresolution, over the central portion of the structure. The very poor data quality resulted from scattering and from destructive interference by out-of-plane energy. Because the 2-D seismic data did not provide a reliable interpretation, a 3-D survey was recommended to provide a better structural picture to facilitate development of the field.

Geology

The major geologic objectives in the Fossil basin portion of the Wyoming Overthrust Belt are located in Jurassic, Triassic, Permian, Mississippian, and Ordovician sediments which have been folded and faulted into trap position on the Absaroka thrust plate. The Cretaceous sediments lying beneath the Absaroka thrust are the key source of hydrocarbons found in the Absaroka plate structures.

In the central Fossil basin, the frontal or easternmost structural trend on the Absaroka plate includes Ryckman Creek, Clear Creek, and Painter Reservoir fields (Figure 9-1-1). Their major production is oil, condensate, and sweet natural gas out of the Triassic Nugget sandstone. The Nugget horizon is cut off by the Absaroka thrust just east of the Ryckman and Clear Creek structures, but in the Painter Reservoir area another fold-thrust trend is developed east of the Painter-Ryckman trend. It is this frontal trend that was disclosed by the East Painter Reservoir discovery (Figures 9-1-2 and 9-1-3).

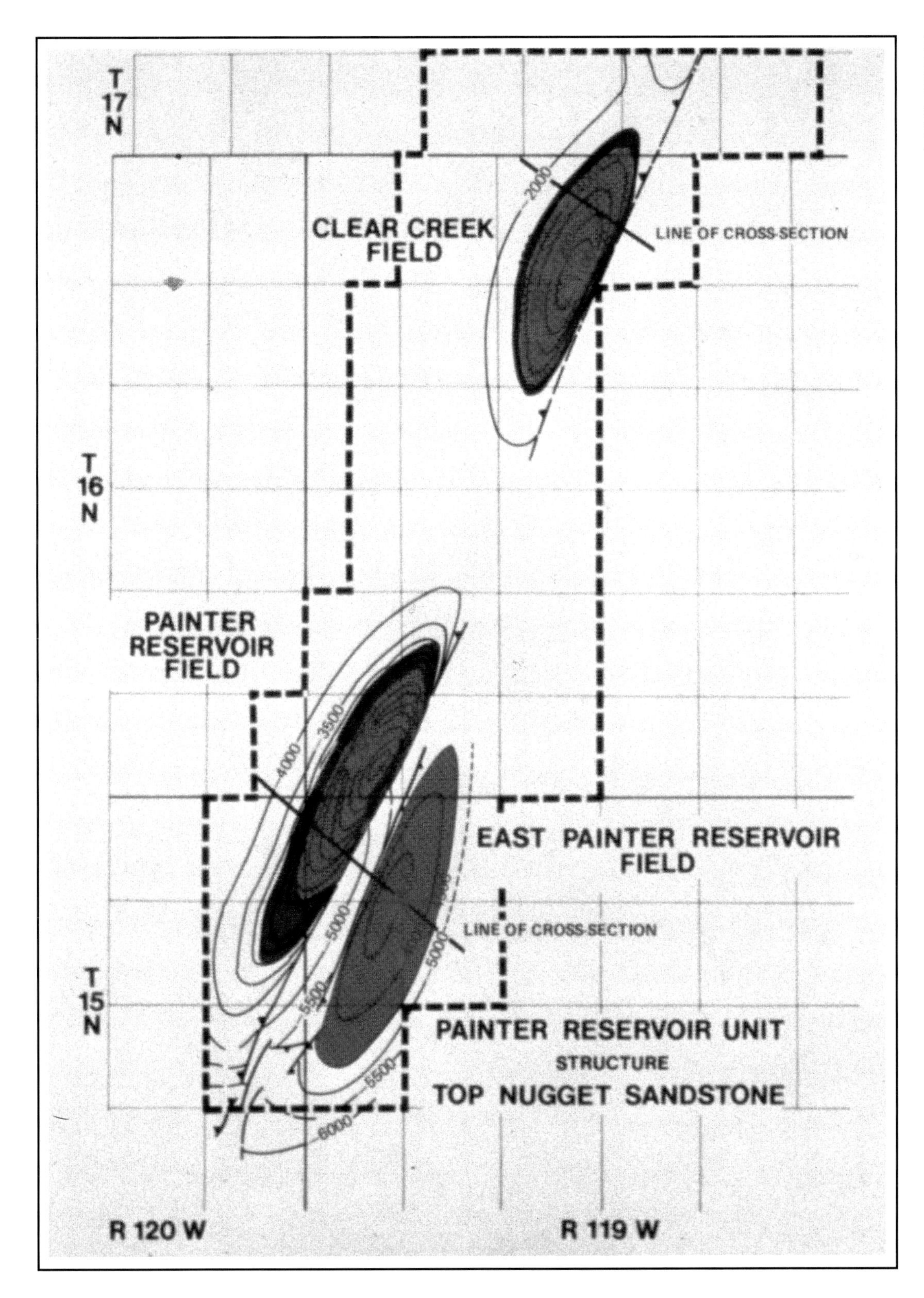

Fig. 9-1-2. Painter Reservoir and East Painter Reservoir fields, structural contour map on top of Nugget Sandstone.

Field Program

Partners in the East Painter 3-D survey were Amoco Production Co., Champlin Petroleum Co., and Chevron U.S.A., Inc. Chevron was the operator with a 50% interest in the survey. Geophysical Service Inc. (GSI) was contracted to conduct the field work and to process the data.

The survey was carefully laid out so that the entire areal time expression of the East Painter feature could be recorded. This required 17 sq mi (44 sq km) of 3-D control. The CDP sampling was designed to be twice as fine in the dip direction (100 ft; 30.5 m) as in the strike direction (200 ft; 61 m) to prevent spatial aliasing of steeply dipping data. A 4-line "swath" shooting method was used with

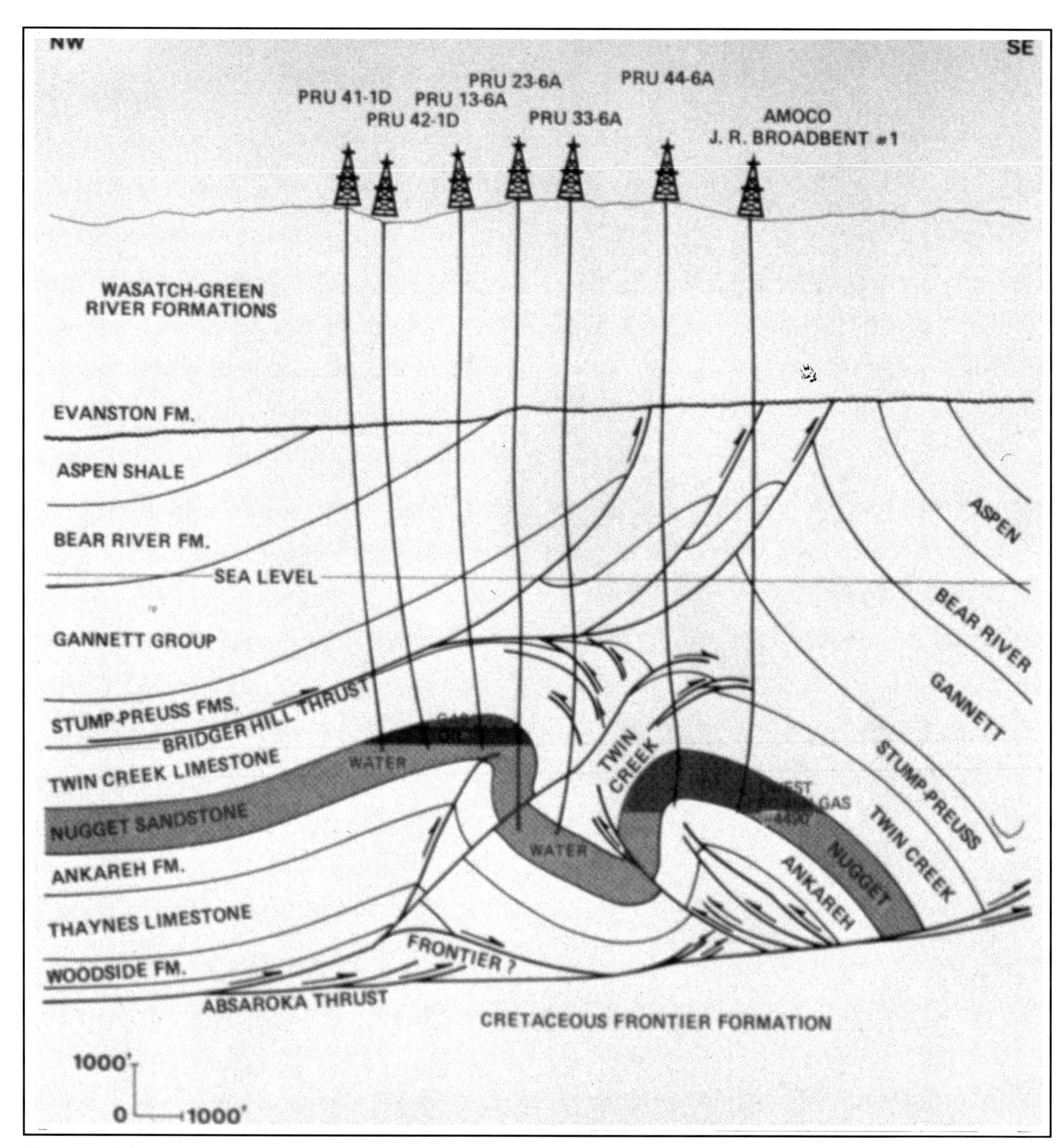

Fig. 9-1-3. Painter Reservoir and East Painter Reservoir fields, structural cross section.

dynamite in shot holes as the energy source. Where shot holes could not be located because of rough topography or because of close proximity to drilling wells and pipelines, substitute shot locations were carefully determined by Chevron and GSI to ensure adequate 3-D coverage. Shooting began in September 1979 and was completed in March 1980. The migrated products were received in early July 1980.

Interpretation and Results

Final migrated data from the 3-D survey clearly resolved the structural configuration of the East Painter feature (Figures 9-1-2 and 9-1-4). The resulting interpretation showed the structure to be as large as previously mapped and the questionable central portion of the structure to be continuous.

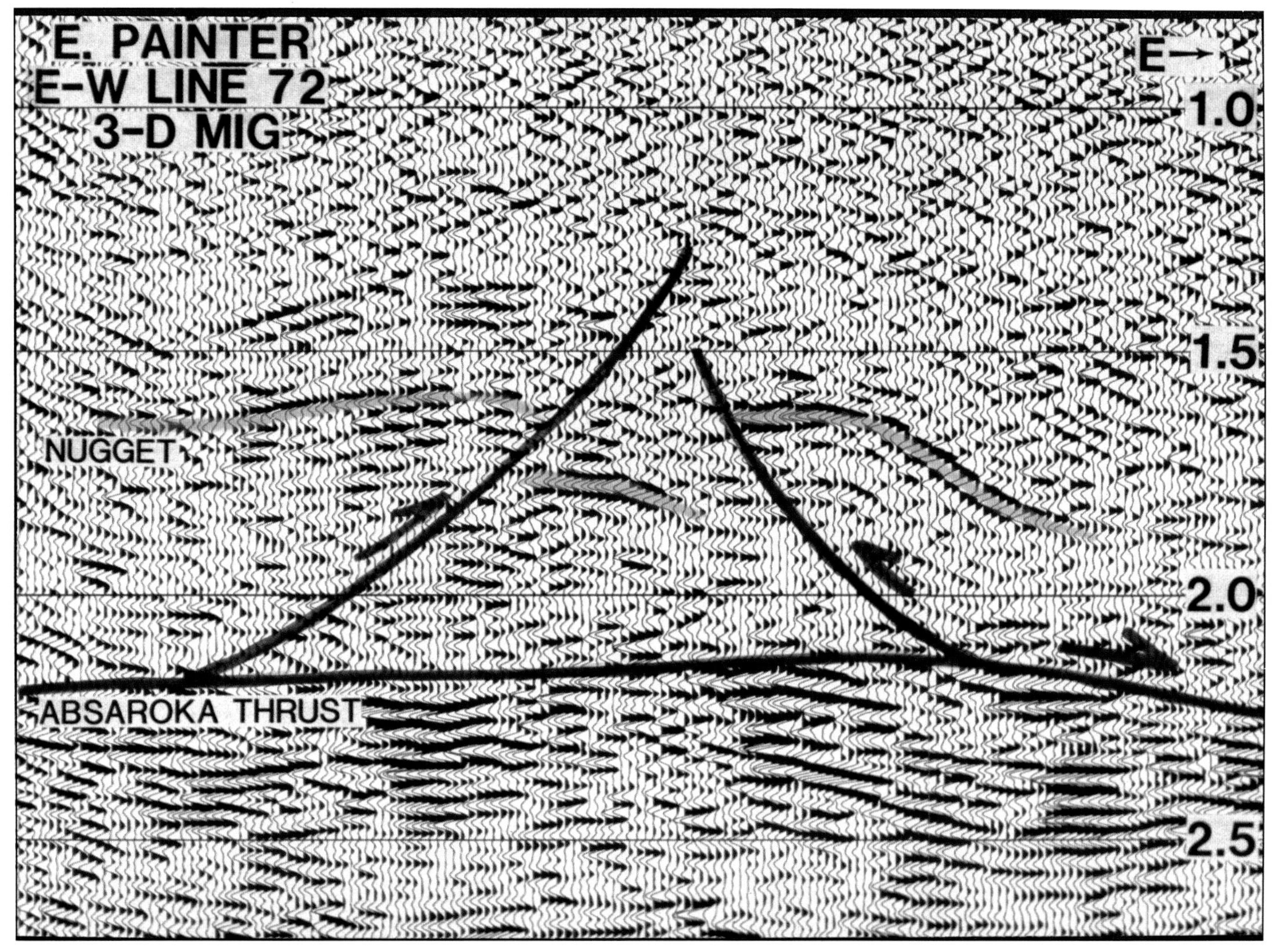

Fig. 9-1-4. East Painter 3-D migrated line 72 showing interpretation of Nugget horizon.

To date, a total of 16 wells have been drilled on the East Painter Reservoir structure. Thirteen of these were spudded after the 3-D survey was completed and their locations were guided by the 3-D mapping used in conjunction with the incoming subsurface control from the development drilling. All of the wells have been successful without any structural surprises on the Nugget horizon. The 3-D mapping allowed up to six development wells to be drilled at one time which greatly accelerated development of the field. The wells were drilled to an average depth of 12,500 ft (3,800 m) with an average cost per well of $4 to $5 million. The cost of the East Painter 3-D survey was $1.6 million—a good value!

Conclusion

Following the success of the East Painter 3-D survey, four additional thrust belt surveys were conducted during the next three years, three of which were larger than 45 sq mi (116 sq km). Without question, the 3-D seismic method is now an accepted and established exploration and development tool in the Overthrust Belt of the western U.S. As a final but necessary comment, it should be noted that the success of a 3-D survey is not automatic. Careful planning and the application of both geologic and geophysical expertise are essential to ensuring optimum results.

Statistics

Date of shooting	October 1979–March 1980
Area of coverage	17.4 sq miles
Fold	600–700%
Number of shots	1207
Collection cost	$1,214,500
Processing cost	$ 332,300
Total cost	$1,546,800
Cost per sq mile	$ 88,900

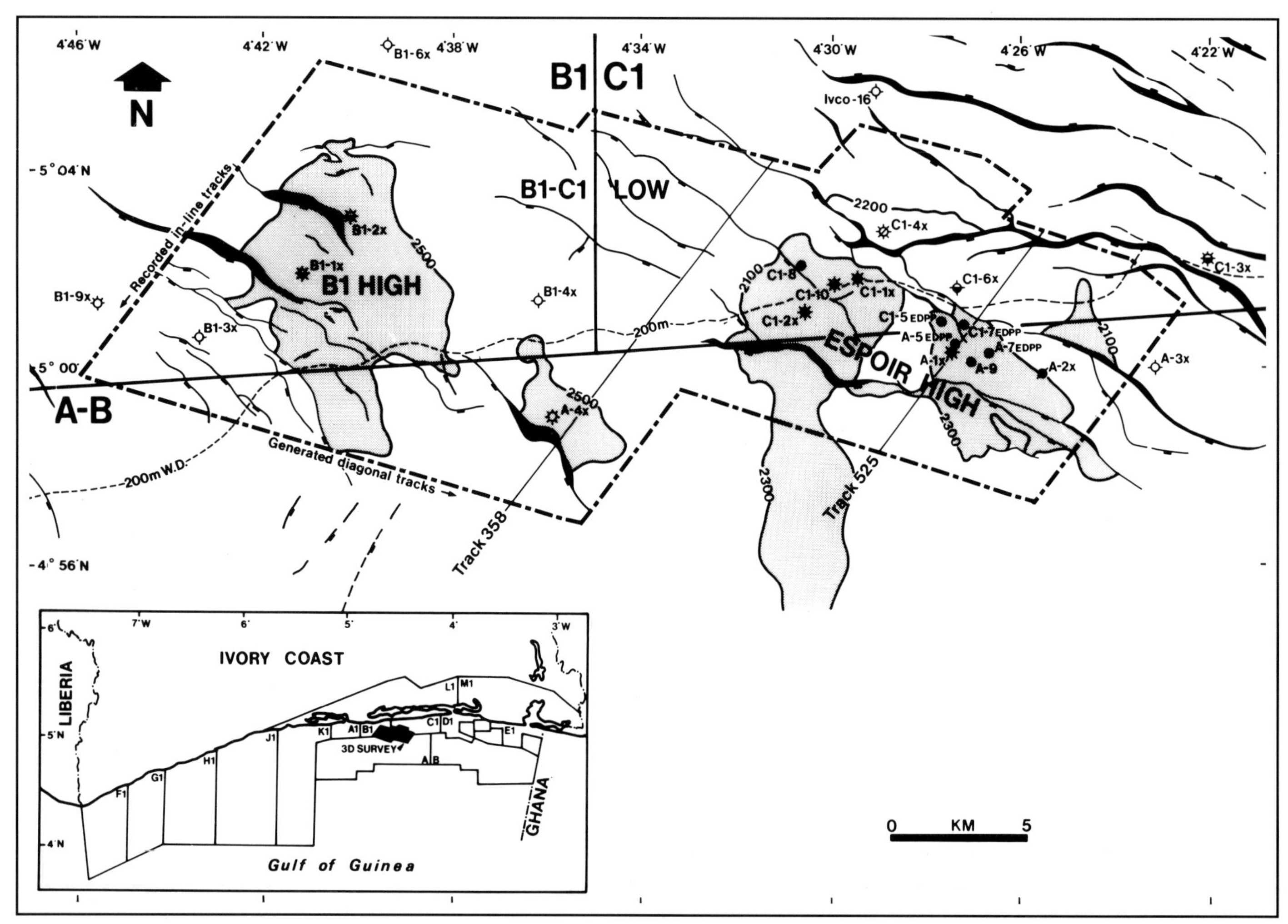

Fig. 9-2-1. Location map of 1981-82 Ivory Coast 3-D seismic survey showing structure of Albian unconformity. Contours in meters.

Case History 2

Three-Dimensional Seismic Interpretation: Espoir Field Area, Offshore Ivory Coast

L. R. Grillot, P. W. Anderton, Phillips Petroleum Co. Europe-Africa
T. M. Haselton, Consultants of Tri-D, Denmark
J. F. Dermargne, Phillips Petroleum Co. UK Ltd., England

The Espoir field, located approximately 13 km (8 mi) offshore Ivory Coast, was discovered in 1980 by a joint venture comprising Phillips Petroleum Co., AGIP, SEDCO Energy, and PETROCI. Following the discovery, a 3-D seismic survey was recorded by GSI in 1981-82 to provide detailed seismic coverage of Espoir field and adjacent features. The seismic program consisted of 7,700 line-km of data acquired in a single survey area located on the edge of the continental shelf and extending into deep water. In comparison with previous 2-D seismic surveys the 3-D data provided several improvements in interpretation and mapping including: (a) sharper definition of structural features; (b) reliable correlations of horizons and fault traces between closely-spaced tracks; (c) preparation of detailed time contour maps from time-slice sections; and (d) an improved velocity model for depth conversion. The improved mapping aided in the identification of additional well locations; the results of these wells compared favorably with the interpretation made prior to drilling.

Introduction

The discovery well, A-1X, was drilled in approximately 1,700 ft (518 m) of water to test a structural high at the Albian unconformity level (Figure 9-2-1). The well encountered

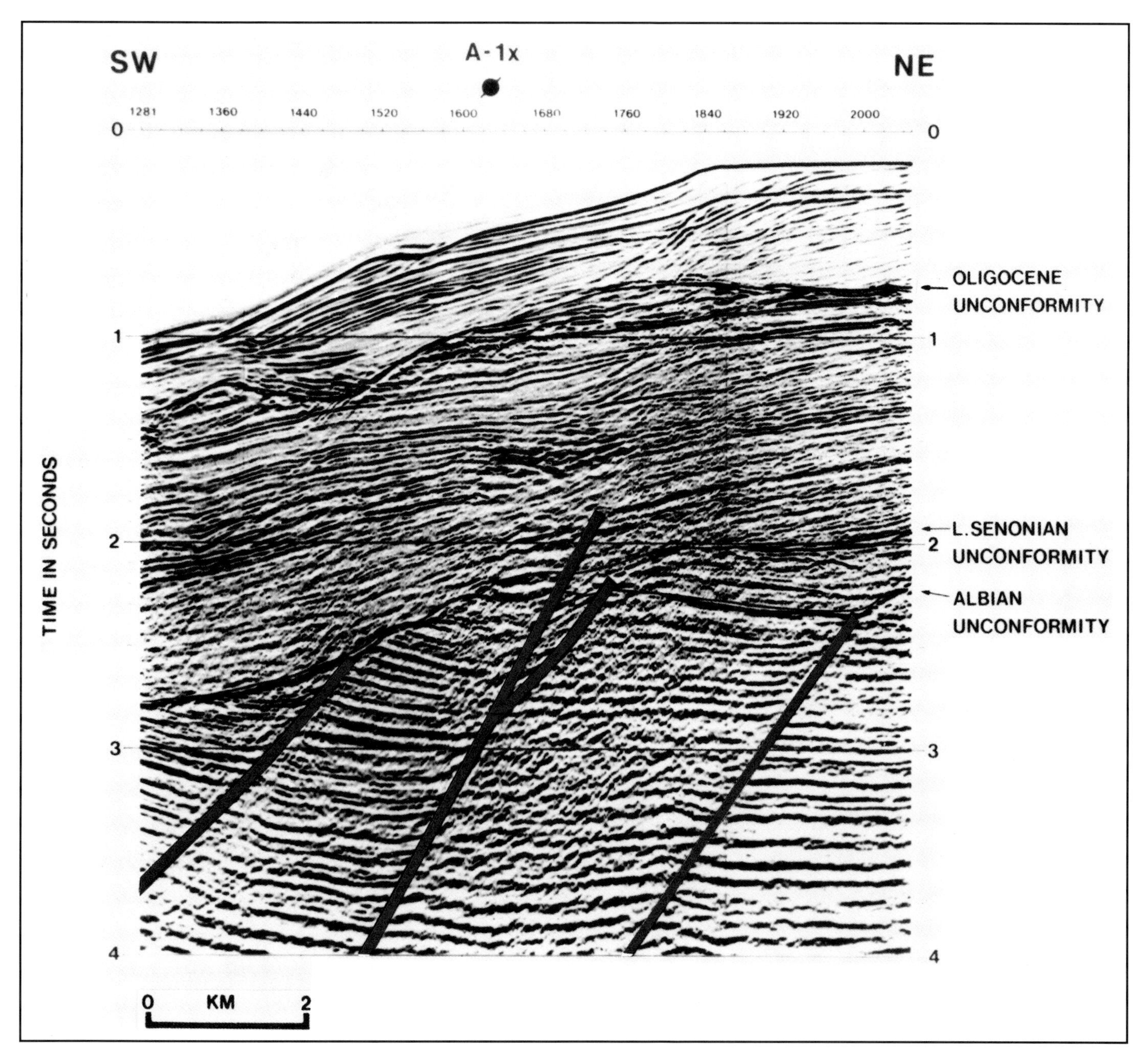

Fig. 9-2-2. In-line 525 crossing A-1X well location, Espoir field, and showing clear definition of rotated fault blocks beneath Albian unconformity.

hydrocarbon-bearing, reservoir-quality sands beneath this unconformity surface and an appraisal well, A-2X, confirmed the presence of a significant accumulation in the Espoir area. Also, additional exploration work in the adjacent B1 block revealed other features of interest associated with the Albian unconformity. On this basis, the joint venture decided to undertake a 3-D seismic program which had as major objectives the detailed mapping of the Albian structure, as well as definition of the complex faulting which appeared to be present beneath the unconformity.

The rhomboid shape of the survey area (Figure 9-2-1) was devised to include both Espoir field and adjacent structures in a single survey and to orient the recording direction perpendicular to the major faults. The survey consisted of 525 northeast-southwest-trending lines recorded during a four-month period from October 1981 to February 1982. Data were recorded using a conventional 2,400-m cable and GSI's 4,000-cu inch air gun source. Recorded line lengths ranged from 8 to 15 km (5 to 9 mi).

Results and Interpretation

The resulting in-line sections (e.g., Figure 9-2-2) clearly demonstrate the primary mapping surface (Albian unconformity) and the tilted fault blocks typical of the structural style in the area. Figure 9-2-2 also demonstrates the sloping water bottom which gives a

Fig. 9-2-3. Time slice at 2,848 ms across Espoir field showing traces of major faults beneath the Albian unconformity.

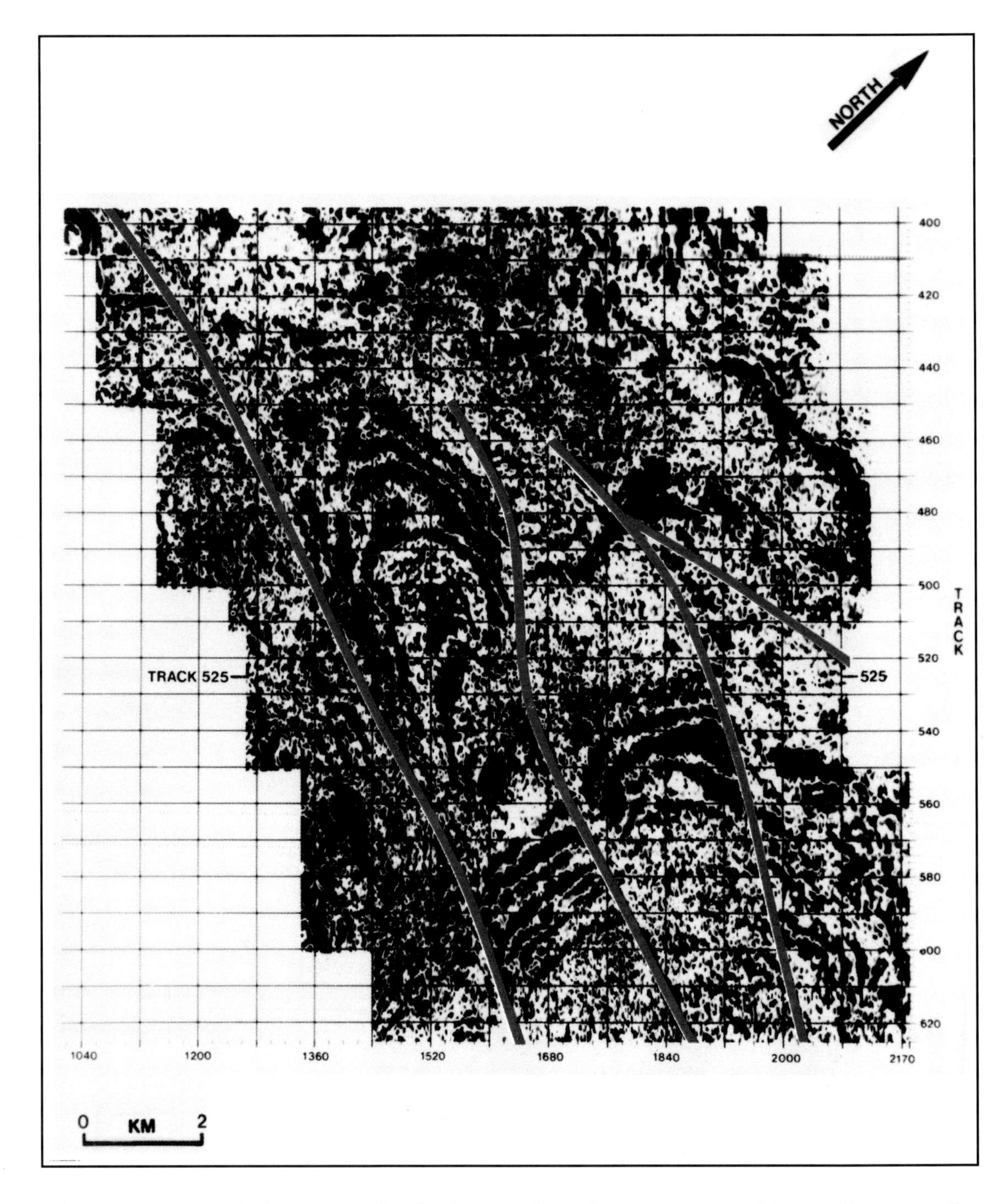

distorted structural picture on seismic time sections, both with regard to the closure at the Albian unconformity level and to pre-unconformity dips. This phenomenon was of particular concern during the velocity analysis and depth conversion stages of the mapping.

The bulk of the mapping was based on combined interpretation of vertical sections and horizontal time slices. Time slices were most useful where the reflections were distinct and not closely converging. In these areas, fault trends could be identified on time slices but generally the traces could not be mapped with the required precision. Figure 9-2-3 shows a time slice taken well beneath the Albian unconformity, which demonstrates these points. The red bands mark the traces of major faults; the individual seismic character of each fault block can be identified. However, the top part of the figure shows a zone where data quality is poorer and fault traces cannot be adequately mapped. In these areas, conventional interpretation of the closely-spaced (60 m; 200 ft) vertical sections was necessary to define fault traces, to correlate weak or complex reflections, and to map smaller depositional units.

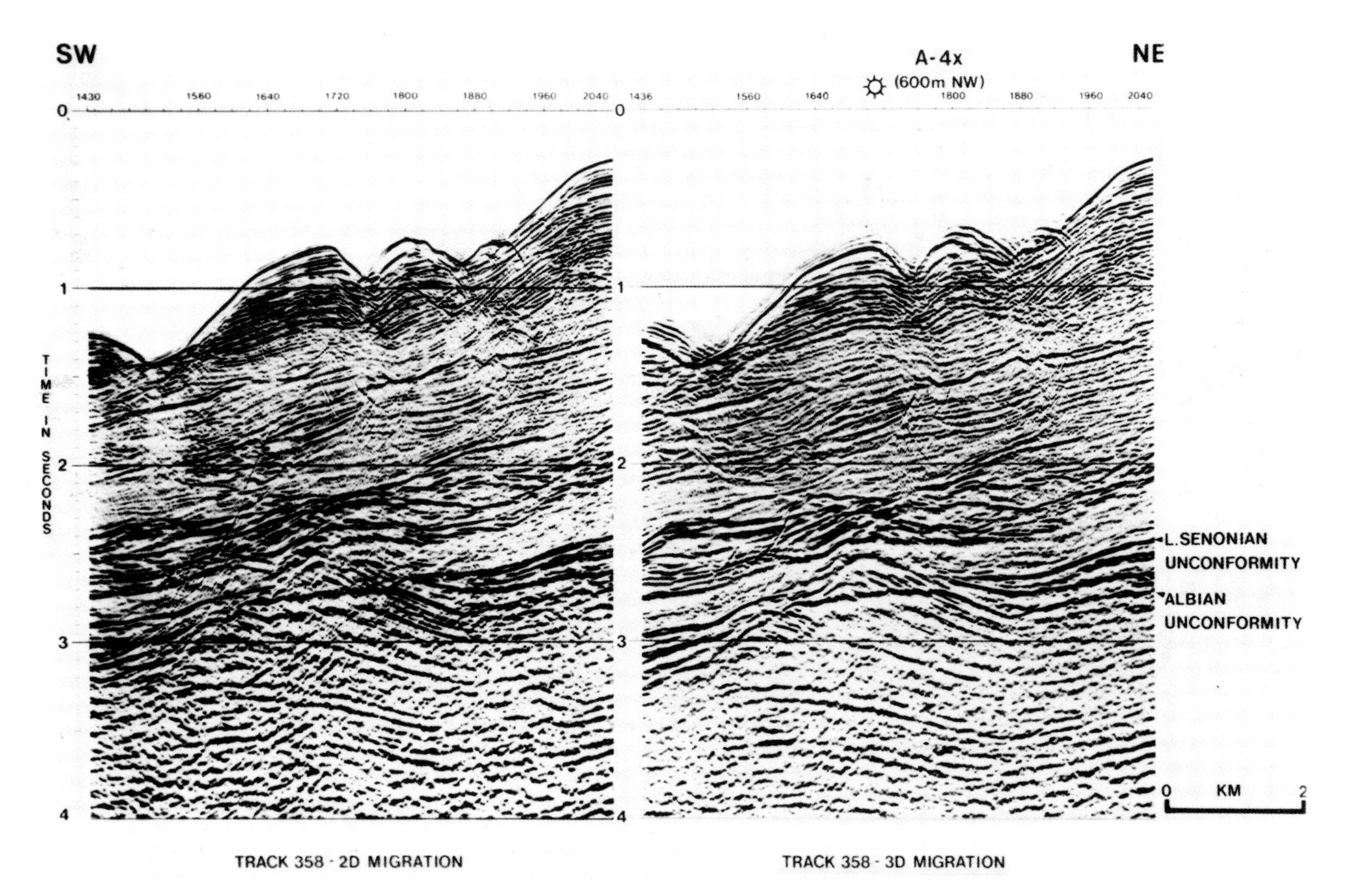

Fig. 9-2-4. Comparison of 2-D migration and 3-D migration sections across structure drilled by well A-4X showing improved definition of erosional high on Albian unconformity and fluid contact (flat spot).

In general, the 3-D data showed improved definition of the Albian unconformity surface across the entire survey area. This resulted in significant mapping revisions to the top of the reservoir interval and to changes in the extent of mapped closure over major structures in the area. Compared with previous efforts, the new maps showed increased closure at the Albian level over east Espoir together with a southward shift of the structural crest, especially in the vicinity of well A-2X.

A particularly interesting change in mapping occurred on the feature tested by well A-4X where the erosional high on the Albian unconformity is well resolved by the 3-D data. Figure 9-2-4 shows two versions of line 358 which crosses this feature near A-4X. The section on the left shows data at the intermediate 2-D migration stage. Although there is evidence of an anomaly in the center of the figure at about 2.7 seconds, the feature itself is not clear. The section on the right shows the same data after 3-D migration. The improvement in detail is noticeable and steeply-dipping intra-Albian reflections can be seen cutting through a flat spot which is close to a fluid contact defined in well A-4X. The slight tilt of the flat spot is due to the sloping water bottom. The 3-D mapping confirmed the structural isolation of the A-4X feature from the larger structure to the west.

Conclusion

In addition to the improvements in interpretation already discussed, benefits included better definition of pre-unconformity reflections, which resulted in improved mapping of intra-Albian horizons and better correlation across major faults. The ability to generate sections across well locations and individual features contributed to a better understanding and interpretation of the area. In the final stages of work, the improved velocity model derived from the closely-spaced velocity analyses aided the preparation of depth maps at the reservoir levels; this contributed to development of Espoir field and identification of further appraisal locations in the area. Overall, the 3-D survey has been a positive contribution to the evaluation of the Espoir area.

Case History 3

Field Appraisal With Three-Dimensional Seismic Surveys Offshore Trinidad

Robert M. Galbraith, Texaco Inc., Latin America/West Africa Division
Alistair R. Brown, Geophysical Service Inc.

A consortium operated by Texaco Trinidad Inc. commenced exploration in the South East Coast Consortium block offshore Trinidad in 1973. After four years of intensive exploration, a gas/condensate discovery was announced in early 1977 on the Pelican prospect. Later that year, in anticipation of the possible future need to site drilling/production platforms, a three-dimensional (3-D) seismic survey was recorded over the prospect. This survey resulted in improvements in seismic record quality, multiple attenuation, and fault resolution. A coordinated geologic-geophysical interpretation based on the 3-D seismic survey, a re-evaluation of log correlations, and the use of seismic logs differed significantly from earlier interpretations. Because of this, it is anticipated that the development of the field will need to be initiated in a different fault block from that previously envisioned.

A second 3-D survey contiguous to the Pelican survey was recorded in 1978 over the Ibis prospect. Results show significant data enhancement in the deeper part of the section and improved fault resolution relative to previous two-dimensional (2-D) control. The 3-D interpretation has revealed a much more complex fault pattern than originally mapped. Separate fault blocks will have to be individually evaluated, thus greatly increasing exploration risk.

Introduction

The republic of Trinidad and Tobago lies approximately 8 mi (13 km) off the northeast coast of Venezuela on the continental shelf of South America. The South East Coast Consortium was formed in 1973 to evaluate an offshore license obtained from the Government of Trinidad and Tobago in that year. The Consortium comprises Texaco Trinidad Inc. (operator), Trinidad and Tobago Oil Company Ltd., and Trinidad-Tesoro Petroleum Company Ltd.

The license area lies approximately 30 mi (48 km) off the southeast coast of Trinidad in the Galeota basin. This basin covers approximately 5,000 sq mi (13,000 sq km) in which thick Pleistocene to upper Miocene deltaic sandstones contain hydrocarbons in traps formed in gravity-induced structures. Closures consist of large diapiric anticlinal ridges and rollover features developed downthrown to major growth faults. To date, four major oil fields and four major gas fields have been discovered in the basin and recoverable reserves have been estimated at 1 billion bbls of oil and 13+ trillion cu ft of gas.

Exploratory drilling in the Consortium block was carried out between 1975 and 1977 with a total of nine wells drilled on four separate structures. Of this total, three were drilled on the Pelican prospect with a gas/condensate discovery declared in 1977. However, even after four years of intensive exploration, including the recording of 1,400 mi (2,250 km) of 2-D seismic data, the Consortium was still unable to determine a location for a development platform. In seeking a solution, the Consortium engaged GSI to conduct a 3-D seismic survey over the Pelican structure in 1977. Following this, the Ibis 3-D survey was recorded in 1978.

All data were recorded with 24-fold geometry along lines oriented southwest-northeast, the predominant dip direction over the block. The lines were 100 m (330 ft) apart, and the subsurface interval along each line was 33 m (108 ft). The currents in the area were commonly 6 to 8 knots at right angles to the shooting direction, so the cable drift was high. Continuously recorded streamer tracking data provided the location of

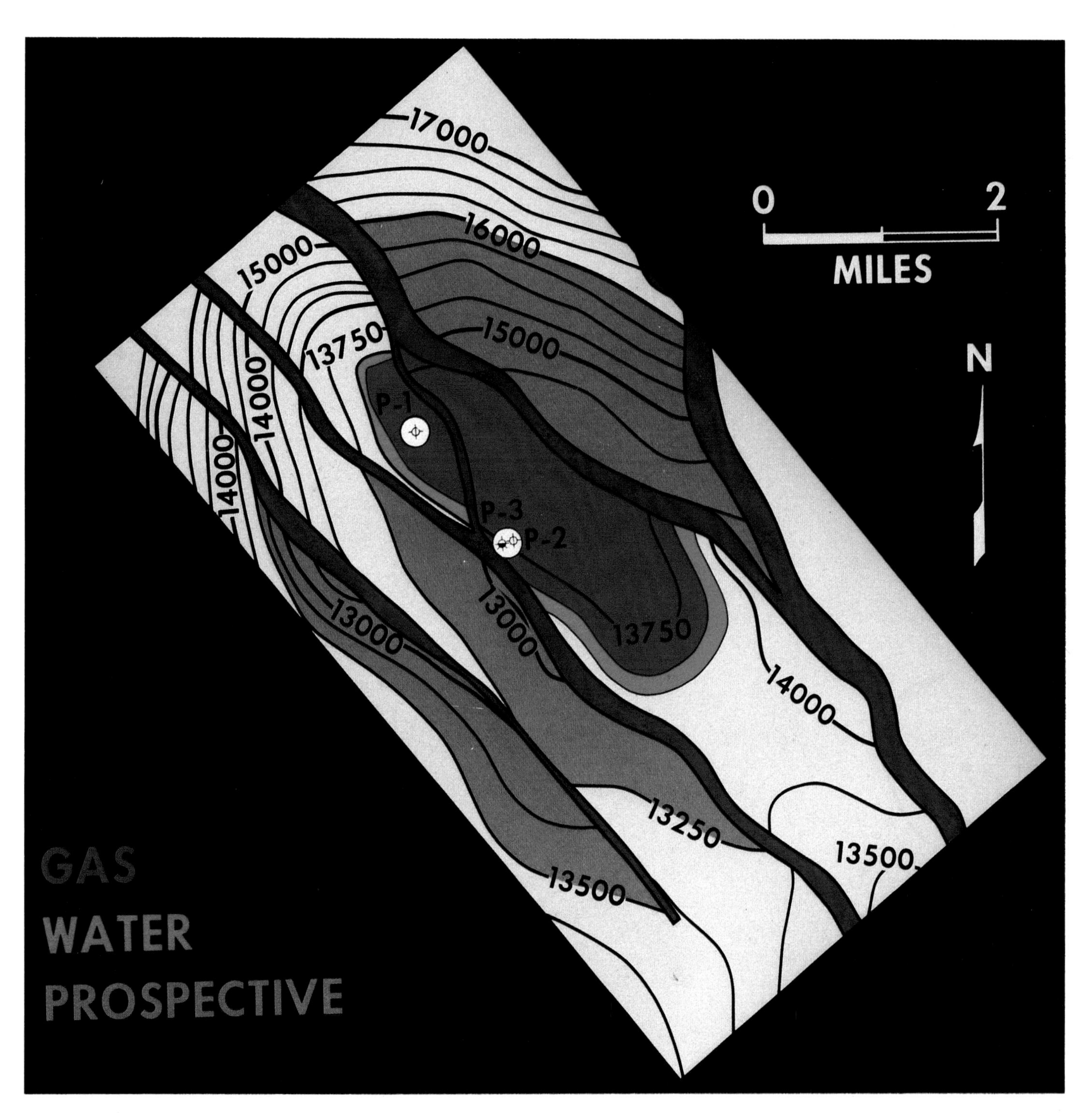

Fig. 9-3-1. Map of Pelican-3 sand, offshore Trinidad, interpreted from 2-D data. Contour interval 250 ft (76 m).

each depth point for each shot. A common-depth-point (CDP) set was then defined as those traces whose source-receiver midpoints fell within a bin 67 × 100 m. This limited the lateral subsurface smear to an acceptable level with a consequent improvement in the stack response.

Results and Interpretation

One of the reservoirs in the Pelican area occurs at the top of the Miocene. The dip at this level between the Pelican-1 well and the northwestern boundary of the 3-D survey area was mapped to be 2,000 ft (610 m) on the pre-existing 2-D data. After the primary reflections had been correctly identified using the 3-D data, less than 1,000 ft (305 m) of dip were mapped on the north flank. This decrease in dip increased the interpreted

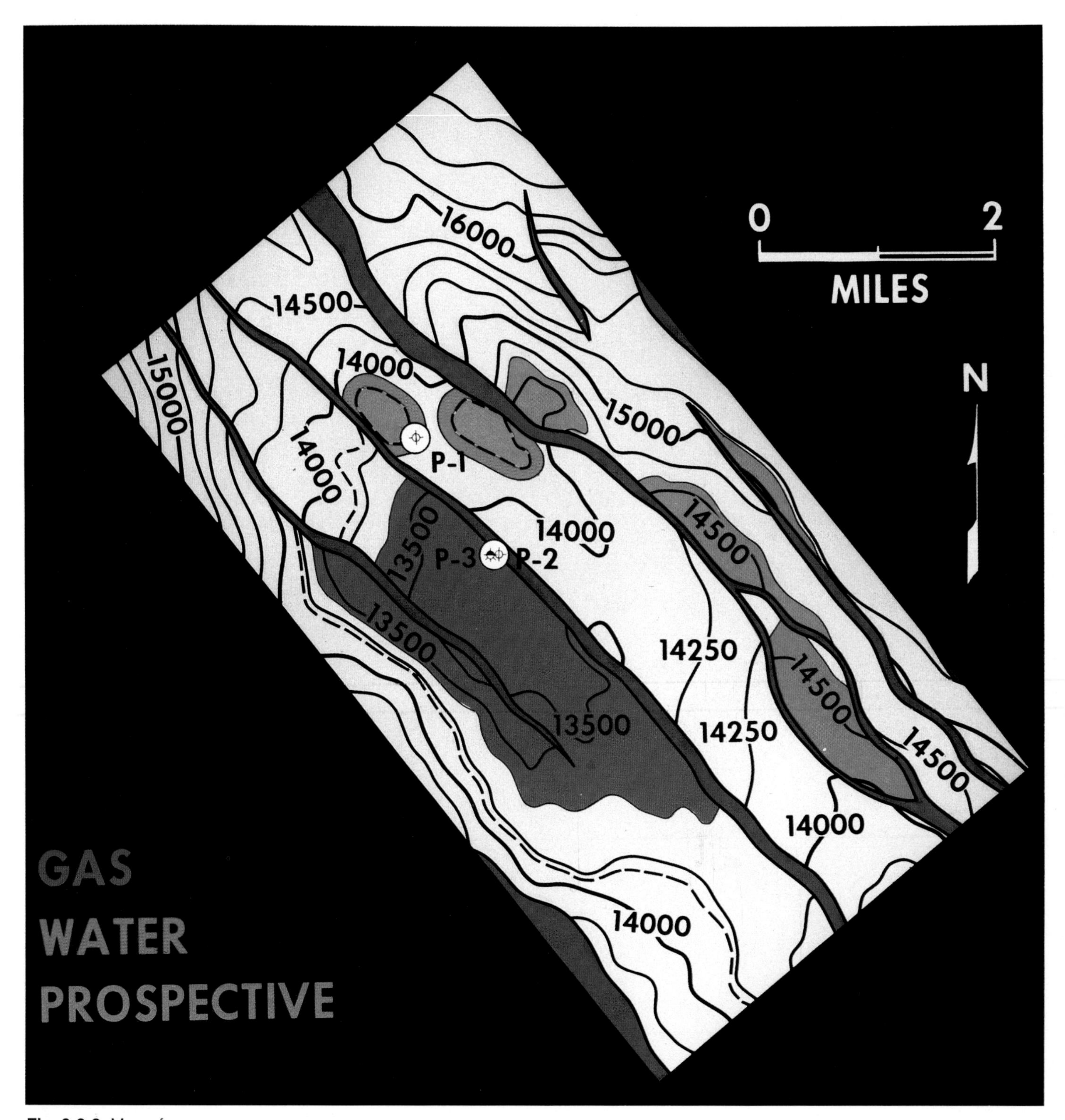

Fig. 9-3-2. Map of Pelican-3 sand interpreted from 3-D data with southeastern structural closure but not honoring water level in well. Contour interval 250 ft (76 m).

hydrocarbon-bearing area under closure by approximately 20%, thus significantly affecting reserve estimates and development economics.

The prime reservoir in the area is the Pelican-3 sand. Figure 9-3-1 shows the interpreted map at this level before the 3-D survey. Figures 9-3-2 and 9-3-3 show two interpretations made from the 3-D survey data. While a similar difference in the northwest dip exists at this level as was mapped at top Miocene, the principal difference between pre- and post-3-D interpretations concerns the faulting.

Initial interpretation of the logs from Pelican-1 and Pelican-3 wells indicated different water levels in the Pelican-3 sand. This was explained by a cross-fault separating the two

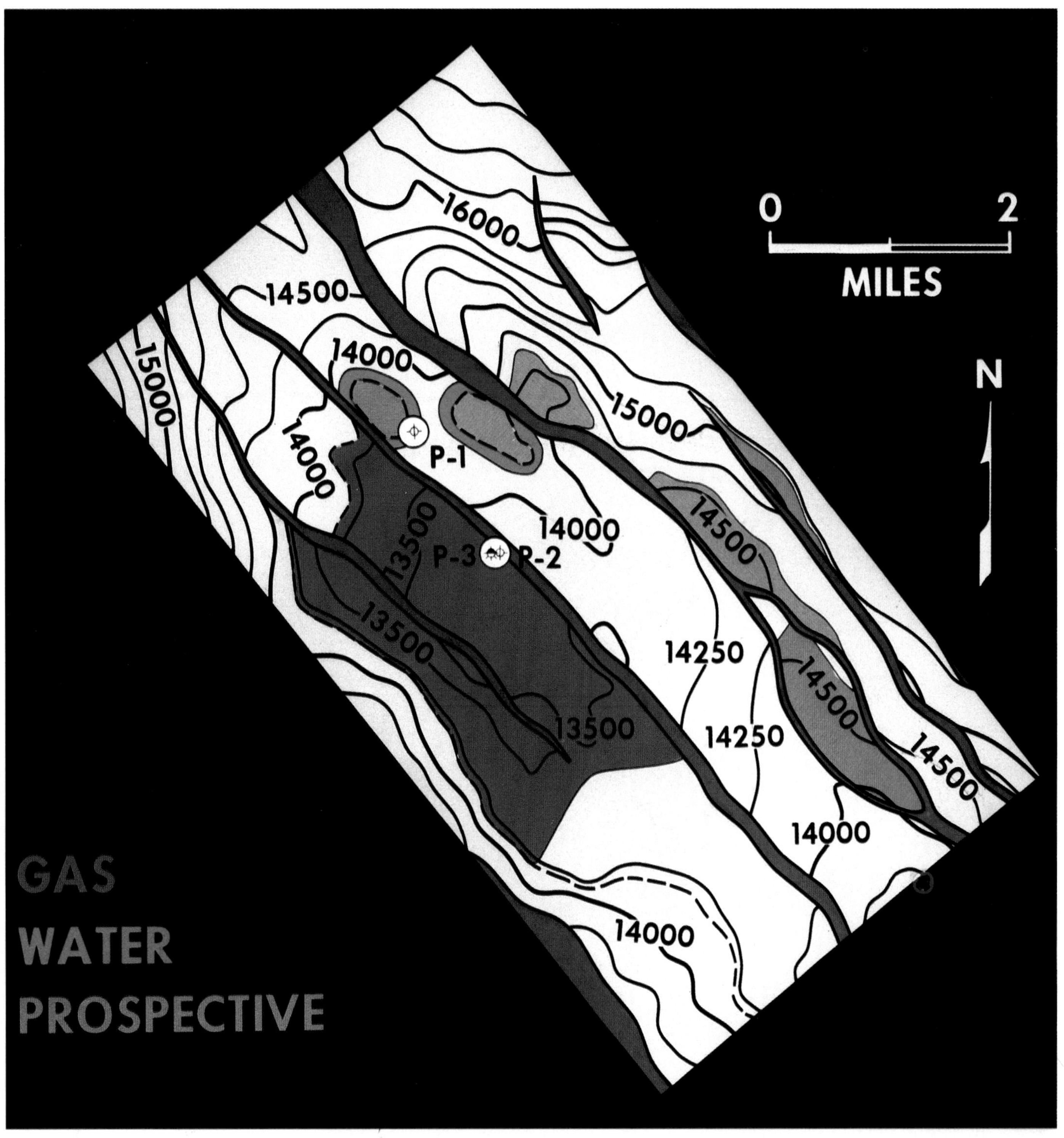

Fig. 9-3-3. Map of Pelican-3 sand interpreted from 3-D data with southeastern stratigraphic boundary and honoring water level in well. Contour interval 250 ft (76 m).

wells (Figure 9-3-1). The 3-D data precluded the possibility of this cross-fault. Instead, the growth fault has been interpreted farther northeast, thus separating the two wells at the Pelican-3 sand. The impact of this on the interpreted position of the reserves is shown in Figure 9-3-2. The recommendation based on the 3-D interpretation was therefore to initiate development drilling in a different fault block from the one proposed prior to the acquisition of 3-D control. This change in interpretation has probably saved the South East Coast Consortium the expense of at least one dry hole and possibly the cost of mislocation of a development platform.

The water level in the Pelican-3 sand in the Pelican-3 well is near 13,800 ft (4,210 m).

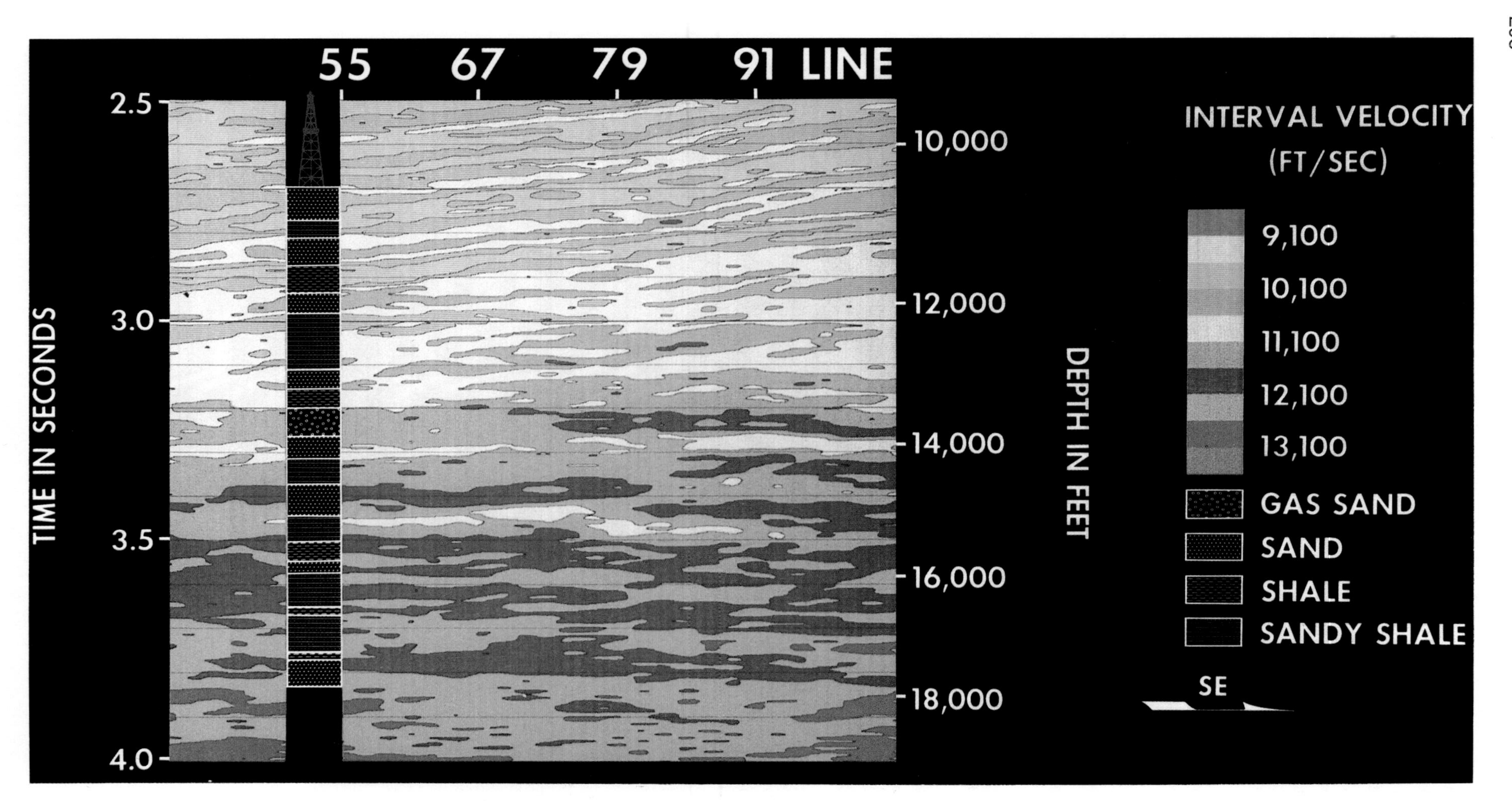

Fig. 9-3-4. G-LOG velocity section along crossline 87 through Pelican-3 well showing lateral velocity transition across inferred southeastern reservoir boundary.

The contour at this level is shown by a dashed line in Figure 9-3-2. This is 200 ft (60 m) deeper than the structural spill point of 13,600 ft (4,150 m) which, on the basis of structural closure alone, would control the downdip extent of the gas. An alternative interpretation which honors the water level in the well is shown in Figure 9-3-3. This invokes a stratigraphic reservoir boundary on the southeast.

The seismic section along crossline 87, northwest-southeast through Pelican-3 well, shows a very marked character change at the Pelican-3 reservoir level southeast of the well. This probably indicates the position of the stratigraphic boundary. This character change is evident on seven crosslines which intersect the boundary, and also on several Seiscrop sections, from which its position was mapped (Figure 9-3-3).

The G-LOG* process of seismic inversion was applied to crossline 87 through Pelican-3 well in an attempt to study the nature and validity of the stratigraphic boundary. The resulting G-LOG section in color is shown in Figure 9-3-4. Generally, the higher velocities correspond to the sands and the lower velocities to the shales. Cyclical sand-shale deposition is evident above 3.0 seconds.

The simplified lithology in the well shows the Pelican-3 gas sand between 3.20 and 3.26 seconds (Figure 9-3-4). There is no low velocity expression of this interval on the G-LOG section. However, away from the well to the southeast, the correlative interval shows an abrupt lateral increase in velocity; this is interpreted as the stratigraphic reservoir boundary. Close examination of the transition suggests layering which is also observed in the well; in the upper portion of the reservoir the transition occurs at line 70, in the next layer at line 79, and in the lower half of the reservoir at line 73. The magnitude of the velocity contrast across the boundary is approximately 600 ft/sec (180 m/sec). It is concluded that this lateral change from low to high velocity indicates the change from a porous gas-filled sand to a tight sand, in which the pores are filled with cement which is probably clay.

Conclusions

Data quality has been improved. Processing took into account cable drift, a major problem offshore Trinidad, thus limiting subsurface smear during stack. Some deep primary events have been observed for the first time. Because of increased data density, fault definition is excellent. Structural interpretations are more reliable with removal of energy from outside the plane of the section. The flexibility which permits an interpreter to generate lines in any direction is a significant benefit. The probable containment of the principal Pelican reserves by a stratigraphic reservoir boundary to the southeast has been substantially validated after a detailed study of its nature.

The 3-D results have caused major changes in the Pelican field development plans. The interpreted area under closure has been increased. The possibilities of drilling an initial dry hole and mislocating a development platform have been reduced due to improved reliability of the coordinated geologic-geophysical interpretation based on the 3-D seismic survey and a re-evaluation of log correlations. This has had a positive effect on development economics.

The 3-D seismic method has proved to be a useful tool for field appraisal in this area offshore Trinidad and will be considered over other prospects prior to commitment to expensive offshore development programs.

---

*Trademark of Geophysical Service Inc.

Case History 4

Three-Dimensional Seismic Monitoring of An Enhanced Oil Recovery Process

Robert J. Greaves, Terrance J. Fulp, ARCO Oil and Gas Company

Seismic reflection data were used to monitor the progress of an in-situ combustion, enhanced oil recovery process. Three sets of three-dimensional (3-D) data were collected during a one-year period in order to map the extent and directions of propagation in time. Acquisition and processing parameters were identical for each survey so that direct one-to-one comparison of traces could be made. Seismic attributes were calculated for each common-depth-point data set, and in a unique application of seismic reflection data, the preburn attributes were subtracted from the midburn and postburn attributes. The resulting "difference volumes" of 3-D seismic data showed anomalies which were the basis for the interpretation shown in this case study.

Profiles and horizon slices from the data sets clearly show the initiation and development of a bright spot in the reflection from the top of the reservoir and a dim spot in the reflection from a limestone below it. Interpretation of these anomalies is supported by information from postburn coring. The bright spot was caused by increased gas saturation along the top-of-reservoir boundary. From postburn core data, a map of burn volume distribution was made. In comparison, the bright spot covered a greater area, and it was concluded that combustion and injection gases had propagated ahead of the actual combustion zone. The dim spot anomaly shows a good correlation with the burn volume in distribution and direction. Evidence from postburn logs supports the conclusion that the burn substantially decreased seismic velocity and increased seismic attenuation in the reservoir. Net burn thicknesses measured in the cores were used to calibrate the dim-spot amplitude. With this calibration, the dim-spot amplitude at each common depth point was inverted to net burn thickness and a map of estimated burn thickness was made from the seismic data.

Introduction

Improving the efficiency of reservoir production can increase proven reserves. The final stages in the production of a field are enhanced oil recovery (EOR) processes. Effective management of EOR processes requires detailed reservoir description and observations of the volume of the reservoir being swept by the process. High-resolution 3-D reflection seismic surveying can be an effective tool in obtaining reservoir description, and, as demonstrated by this case study, can in some cases actually map the EOR process as it proceeds.

In this case study, 3-D seismic reflection data were used to monitor the propagation of a pilot in-situ combustion (fire-flood) process. Three identical 3-D seismic surveys were recorded over the pilot site at preburn, midburn, and postburn times. In this way, the combustion propagation was monitored over (calendar) time.

Acquisition and computer processing of the data were identical for each set of survey data, so that a direct comparison of the individual data sets could be made. To facilitate interpretation, the attributes of the seismic traces were calculated using Hilbert transform techniques as described by Taner and Sheriff (1977). Reflection strength, in this paper referred to as "envelope amplitude," was then used in the analysis of the reflection seismic data. In a unique application of reflection seismic data, the envelope amplitude traces from the preburn data volume were subtracted from their counterpart traces in the midburn and postburn data volumes, generating "difference volumes."

The combustion process substantially increased in-situ temperature and gas saturation in those sections of the reservoir affected by the burn. Both seismic velocity and density of the reservoir were changed. Zones with altered properties were detected by anomalous amplitude responses in the reflection from the top of the reservoir and from a limestone formation directly below the reservoir. The direction of the combustion propagation and estimates of its volume were based on the interpretation of these anomalies as observed in the difference volumes. The interpretation was supported by data available from monitor wells and postburn coring.

Background

Three 3-D seismic surveys were shot over a period of 15 months. The first (preburn) survey was

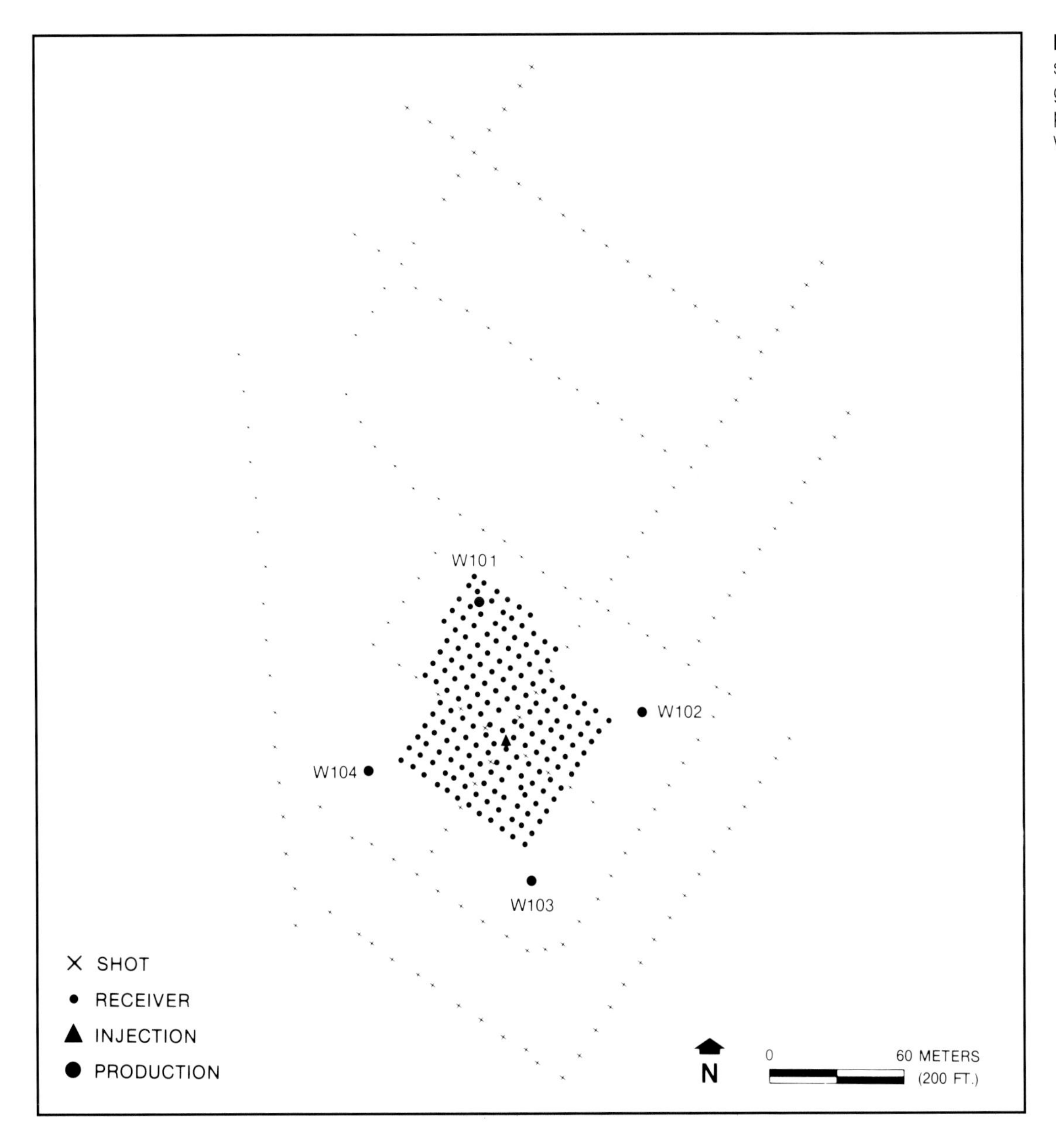

Fig. 9-4-1. 3-D seismic survey shot and receiver geometry with locations of production and injection wells.

recorded several months previous to ignition of the combustion process. The second (midburn) survey was recorded four months after ignition, and the final (postburn) survey was shot ten months after ignition.

The objectives of the seismic program were to (1) detect a change in seismic reflection character attributable to the combustion process, (2) determine the direction of burnfront propagation, and (3) determine the volume of reservoir swept by the combustion process.

The basic premise was that an increase in gas saturation in the reservoir formation would produce measurable changes in reflection amplitude. Bright spots and dim spots, caused by anomalous gas concentrations, are well-known phenomena in exploration seismology. Increased gas saturation in the parts of the reservoir reached by the combustion process was expected to create bright spots and dim spots in the shadow zone (Sheriff, 1980). The 3-D data would be used to map that progression in time.

The EOR program consisted of a five-well pilot test covering a very small portion of the Holt Field in north-central Texas. The test consisted of four production wells separated by 90 m (300 ft) with a central injection well (Figure 9-4-1). The engineering objective was to propagate the combustion process from the injection well radially outward, creating and flushing an increased oil saturation zone, the oil bank, toward the production wells. Although the concept is simple, the implementation is quite difficult and is very sensitive to the details of the reservoir geology.

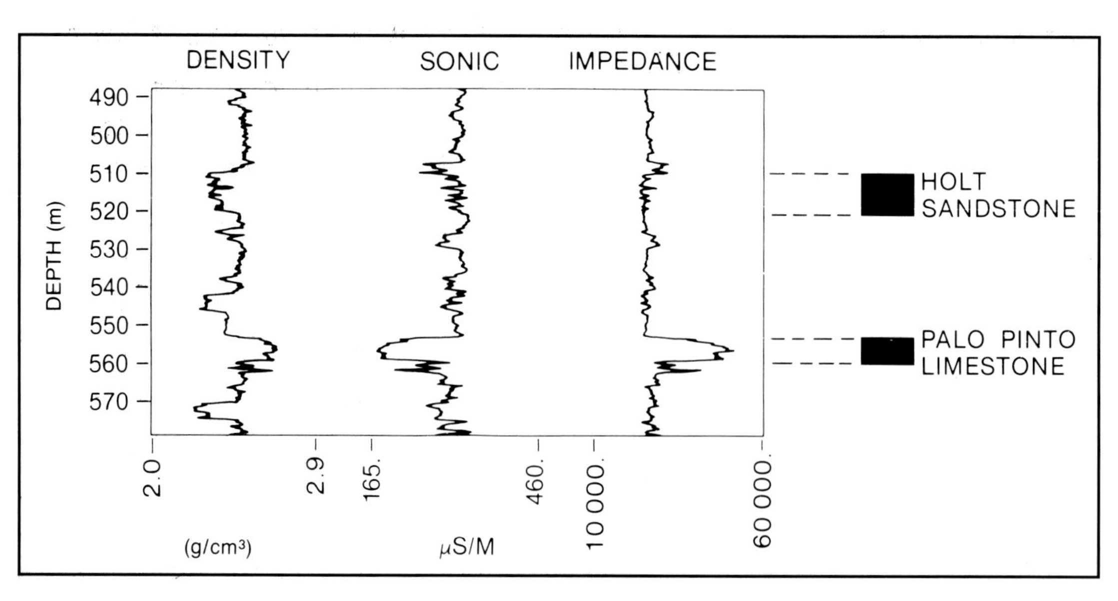

Fig. 9-4-2. Example of sonic and density logs with calculated impedance for the stratigraphic section including the Holt sandstone (reservoir) and the Palo Pinto limestone.

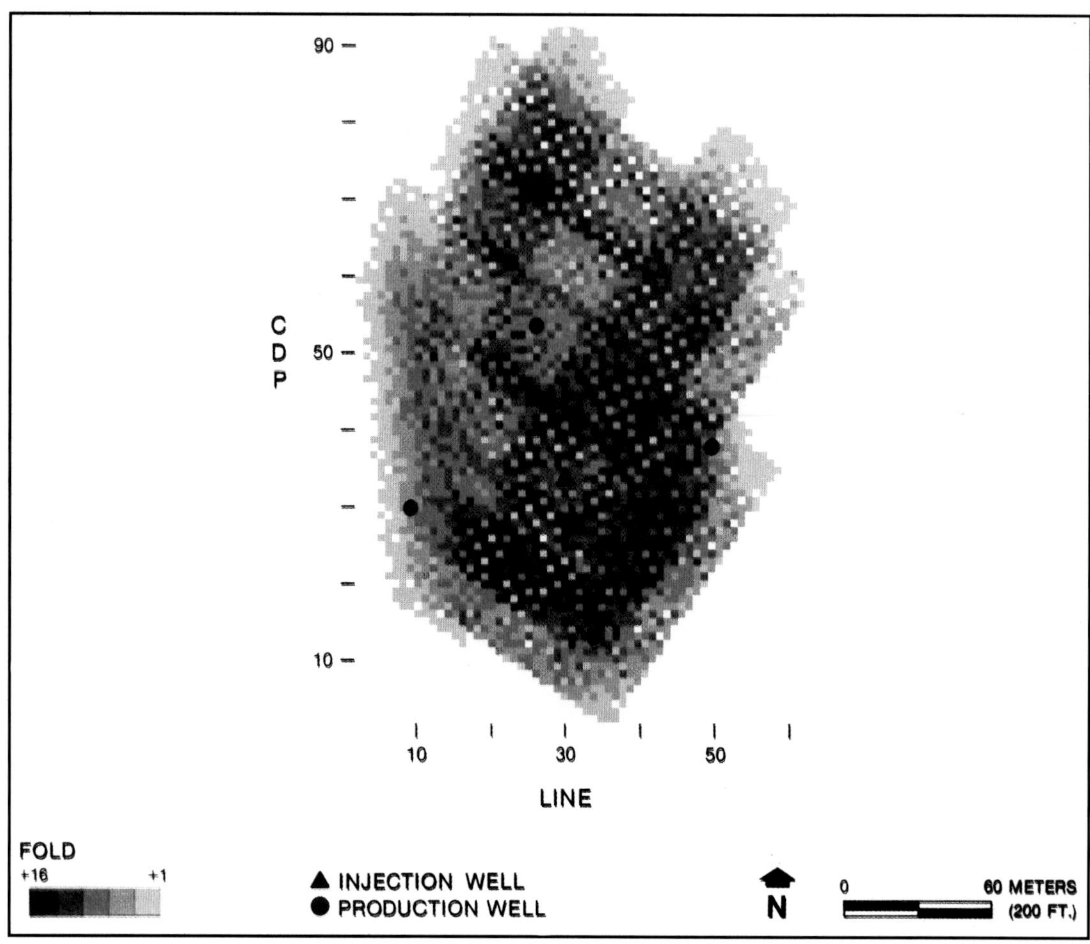

Fig. 9-4-3. CDP fold distribution of the seismic surveys. Each CDP bin covers a 3 × 3 m (10 × 10 ft) area.

The reservoir is the Holt sand, a 12 m (40 ft) thick sandstone capped by a 2.5 m (8 ft) thick limestone encased as a unit in thick shale (Figure 9-4-2). The sand is silty and laced with shale stringers and some calcite cementation zones. In this part of the field, the sand occurs at about 500 m (1650 ft) and dips to the north at 10°. A thin limestone occurs about 45 m (150 ft) below the reservoir and is identified as the Palo Pinto limestone. From extensive core analysis, the horizontal permeability of the sand was found to be several times greater than the vertical permeability. Numerous fractures were observed, and an average orientation of N27E was measured. Although a detailed model of the reservoir and the burn process was not

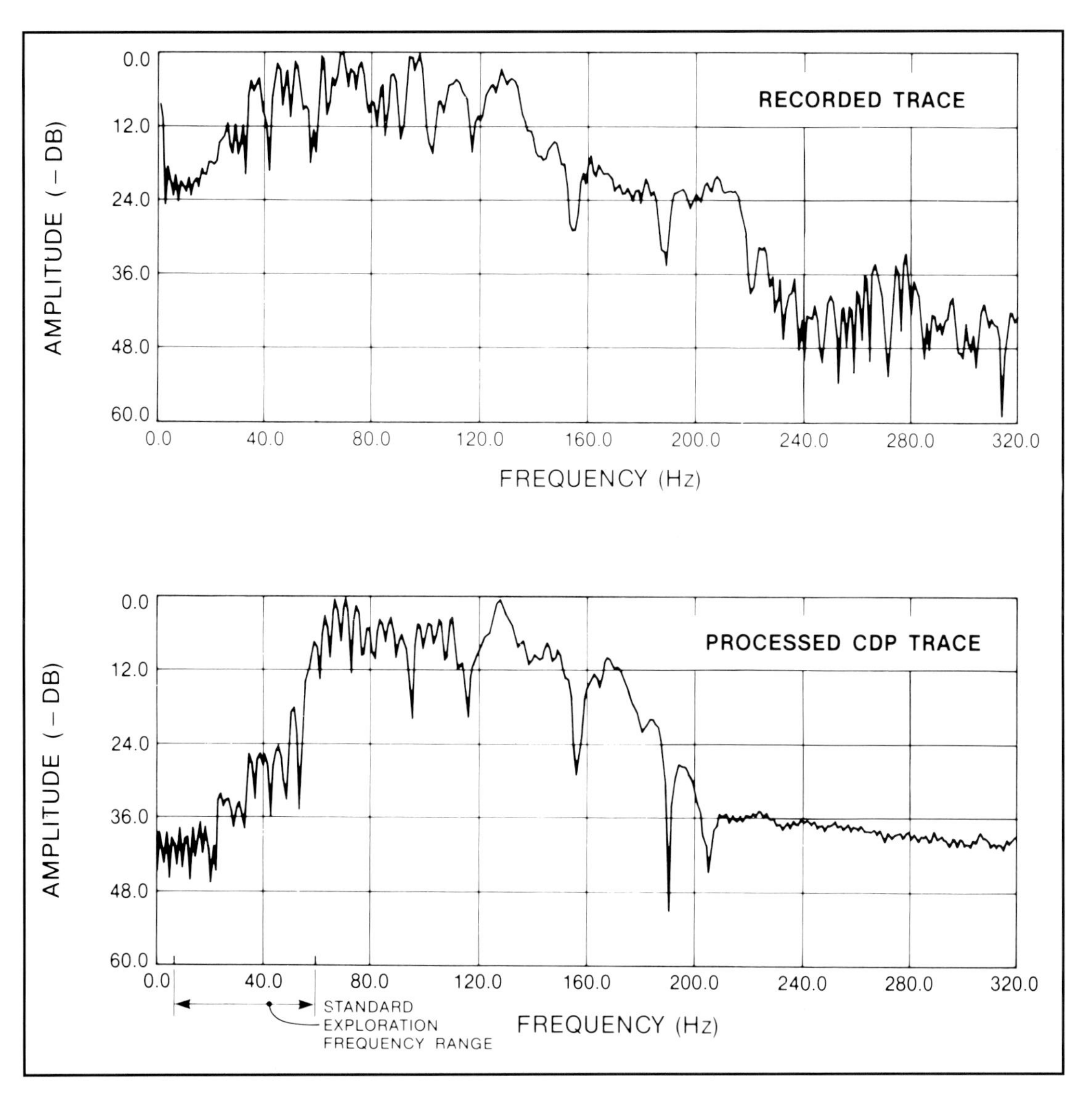

Fig. 9-4-4. An example of the power spectra of a trace before and after processing. The spectra are for the window from 0.350-0.750 s which includes the reflection sequence from the Holt sandstone to and including the Palo Pinto limestone.

constructed, some effects of the process were anticipated. The combustion process would primarily propagate updip (to the south) due to the differing fluid densities. Propagation would primarily occur laterally within the reservoir from the initiation points. Vertical propagation would be limited to fractures or other natural permeability pathways. Finally, propagation might be further guided to the southwest along fracture-induced permeability pathways.

Acquisition and Processing

Several factors guided the choices of acquisition parameters:

(1) the target area was very small (90 × 90 m);
(2) the target was relatively shallow (500 m);
(3) the data collection was to be repeated as identically as possible; and
(4) the amplitudes and spatial extents of the seismic anomalies would probably be quite small.

Simple seismic modeling based on well logs and the anticipated effect of increased gas saturation indicated that *detection* of the burnfront would be straightforward. However, very high-resolution seismic data would be required to map the lateral extent of the process and determine the net burn volume. A calculation of the resolution limit was made based on Widess (1973), and the center frequency required to resolve 7.5 m (25 ft) vertically was determined to be 100 Hz. The necessary resolution was felt to be achievable, given the shallow depth of the reservoir, if proper acquisition and processing techniques were applied.

The data collection array consisted of a modified 3-D patch geometry. Figure 9-4-1 shows the positions of shotpoints and receiver stations within the test area. This patch style of survey allowed the shot pattern to be arranged as necessary, such that the CDP data were collected with high fold over the area of primary interest (Figure 9-4-3) even though surface access was limited by buildings, wells, pipelines, etc. Furthermore, the geophones could be permanently installed at each receiver location to guarantee that the receiver array would be duplicated in each survey. A modification of the simple patch geometry was made to account for the migration of reflection points updip, by extending shot and receiver locations downdip (to the north).

The receiver group spacing was 6 m (20 ft) with a single high-frequency (40 Hz) marsh geophone comprising each group. Each of the 182 receivers was buried 6 m (20 ft) below the surface. The 165 shotpoints were distributed along crossed lines with a 12 m (40 ft) spacing between individual shots. Each shot consisted of a 2.5 kg (3 lb) dynamite charge buried at 23 m (75 ft). The recording system used was a 192-channel GUS-BUS with sampling interval of 1 ms and band-pass recording filters set a 50 Hz low-cut and 320 Hz antialias. The burial of both shots and receivers improved the signal-to-noise (S/N) ratio of the recorded data by eliminating air-wave noise and substantially reducing the amplitude of the surface-wave noise. The 50 Hz low-cut filter was chosen at this high level to eliminate surface-wave noise further and to preserve the dynamic range of the recording system for digitization of the desired high-frequency signal. For recording shallow reflection seismic data, it is especially important to eliminate surface-wave noise that can seriously degrade the quality of the shallow data window. The small, deeply buried shots, the high-frequency phones and the low-cut filter all combined to eliminate this problem. The resulting frequency range of the recorded data was substantially higher than the range of standard exploration seismic, as shown in Figure 9-4-4, and yet the range retained the minimum two-octave bandwidth considered necessary for high resolution.

The computer processing of the 3-D data sets used a standard sequence designed for 3-D CDP data. Throughout the processing, extra care was taken to retain true relative amplitude and the maximum usable frequency range. The traces were gathered into 3 × 3 m (10 × 10 ft) CDP bins. The statics and normal-movement (NMO) corrections were quite small due to the simple geologic structure and small area of the test. For more structurally complex geology, it would have been more difficult to make proper velocity adjustments because the patch geometry has the disadvantages of uneven fold and offset distribution. Three-dimensional surface-consistent statics were computed and were found to be on the order of 2 to 3 ms. Normal-moveout corrections were applied using a datumed root-mean-square (rms) velocity function derived from the well control. Standard spiking deconvolution was applied before stack. A phaseless deconvolution technique was applied to balance further the usable spectrum of each stacked trace. In the final processing step, the data were migrated using an f-k migration algorithm and the velocity function derived from the sonic logs. This approach to migration was deemed adequate, given the localized area of interest and the simple velocity structure. The frequency spectrum of a fully processed trace, windowed in the reflection zone of interest, indicates that the 40-180 Hz bandwidth of the recorded data was enhanced during processing and the center frequency of 100 Hz was obtained (Figure 9-4-4).

As postprocessing steps, the data were properly phase-corrected using well control, and the seismic attributes were calculated for each data set. To remove the geologic structure from the reflectors of interest, static adjustments were made, thereby allowing horizon views to be sliced from the 3-D data volume. Finally, in a unique step, the preburn horizon envelope amplitude at each level of interest was subtracted from the corresponding values in the midburn and postburn data volumes. The preburn data were used as the baseline seismic expression relative to which change was observed. Anomalies in the difference volumes were then interpreted directly.

Observed Anomalies—Bright Spots

Comparison of the envelope amplitudes of the reflection event at the top of the Holt sand reservoir revealed an increase in amplitude, a bright spot, which developed after the combustion process was initiated. In Figure 9-4-5, a north-south section, line 14, is shown as it appeared at preburn, midburn, and postburn times. The reflection from the top of the Holt sand is identified as a trough occurring at about 385 ms. At this horizon, the envelope amplitudes at preburn time compared to midburn time show a zone of increased amplitude near well W104, with maximum change between CDP 16 and CDP 30. By postburn time, the bright spot had increased in lateral extent from CDP 16 to CDP 36, but it had not increased in maximum amplitude relative to midburn time.

Horizon slices at the top of the Holt sand, from the envelope amplitude difference volumes, are displayed in Figure 9-4-6. The midburn difference shows a positive amplitude anomaly in the southwestern side of the data. This corresponds to the bright spot development observed in line 14 (Figure 9-4-5) at midburn time. Another, smaller bright spot is located to the southeast of the injection well at line 43, CDP 21. The difference amplitude at postburn time, Figure 9-4-6, shows that the bright spot has grown to cover most of the area within the production wells, the midburn peak to the southwest has shifted downdip toward well W104, and the maximum amplitude of the difference anomaly has increased by about 10 percent.

Observed Anomalies—Dim Spots

The strong reflection centered at 410 ms in line 14 (Figure 9-4-5) is identified as the Palo Pinto limestone. In line 14, a slight decrease in envelope amplitude occurs in the shadow of the bright spot centered around CDP 22. At midburn time, the decrease in amplitude is about 10 percent, but by postburn time, the decrease is nearly 25 percent, as marked by the change from deep orange and red shading to yellow.

A similar display of another north-south section, line 33, in Figure 9-4-7, shows a more substantial dim spot. This anomaly does not coincide with any bright spot at the Holt level at midburn time and only a modest Holt bright spot at postburn time. The dim-spot anomaly, pointed out by the arrows within the figure, is also stronger at midburn time than at postburn time. This lack of spatial coincidence (between bright and dim spots) is important in the interpretation of the results as described below.

The horizon slice difference section at the Palo Pinto reflection (Figure 9-4-8) clearly shows this anomaly. The dim spot at midburn time covers much of the pilot area with two negative-amplitude anomalies. One peak is located at the injection well, but the stronger peak lies about 30 m (100 ft) south-southwest of the injection well. The anomalies do not coincide with bright spots in the Holt reflection. A lower amplitude lobe of the dim spot extending to the southwest edge of the data does correlate with the maximum bright spot observed at midburn time. The dim spot observed at postburn time is lower in amplitude and extends over a significantly smaller portion of the pilot area. The two peaks of the midburn anomaly have merged into a ridge extending approximately southwest-northeast across the injection well with the larger area and peak of the anomaly to the southwest of the injection well.

Interpretation—Combustion Model

A simple model of the in-situ combustion process, based on combustion tube experiments, was described by Tadema (1959). The combustion process within the reservoir can be divided into various zones, with each zone defined by its relative temperature and fluid saturations. The "combustion zone" propagates through the reservoir and is defined by maximum oxidation of the heaviest, or immobile, hydrocarbons. In its wake is left the "clean-burnt sand," a hot reservoir matrix with high gas saturation. Ahead of the combustion zone are several zones at lower temperatures and with distinctive percentages of oil, water, and gases until at some distance the original reservoir temperature and fluid mixture are encountered. Of particular interest are (1) that the clean-burnt sand zone has been subjected to very high temperatures, and (2) that combustion gas, as well as some injection gas, are forced ahead of the combustion zone. If this model were expanded into three dimensions, it would consist of a series of concentric rings which propagate radially from the injection well. The model is quite simple and does not at all account for geologic complexities, but it is useful as a starting point for the interpretation of the seismic anomalies in terms of the physical process of in-situ combustion.

Generalizing this model to the Holt sand reservoir, it is important to point out that the Holt sand reservoir, at the initiation of the burn, had little to no gas saturation.

Interpretation—Detection

As stated, the first objective of the seismic program was to detect a change in reflection character attributable to the combustion process. The bright spots and dim spots are considered true combustion-caused anomalies for the following reasons. First, the changes do occur in the

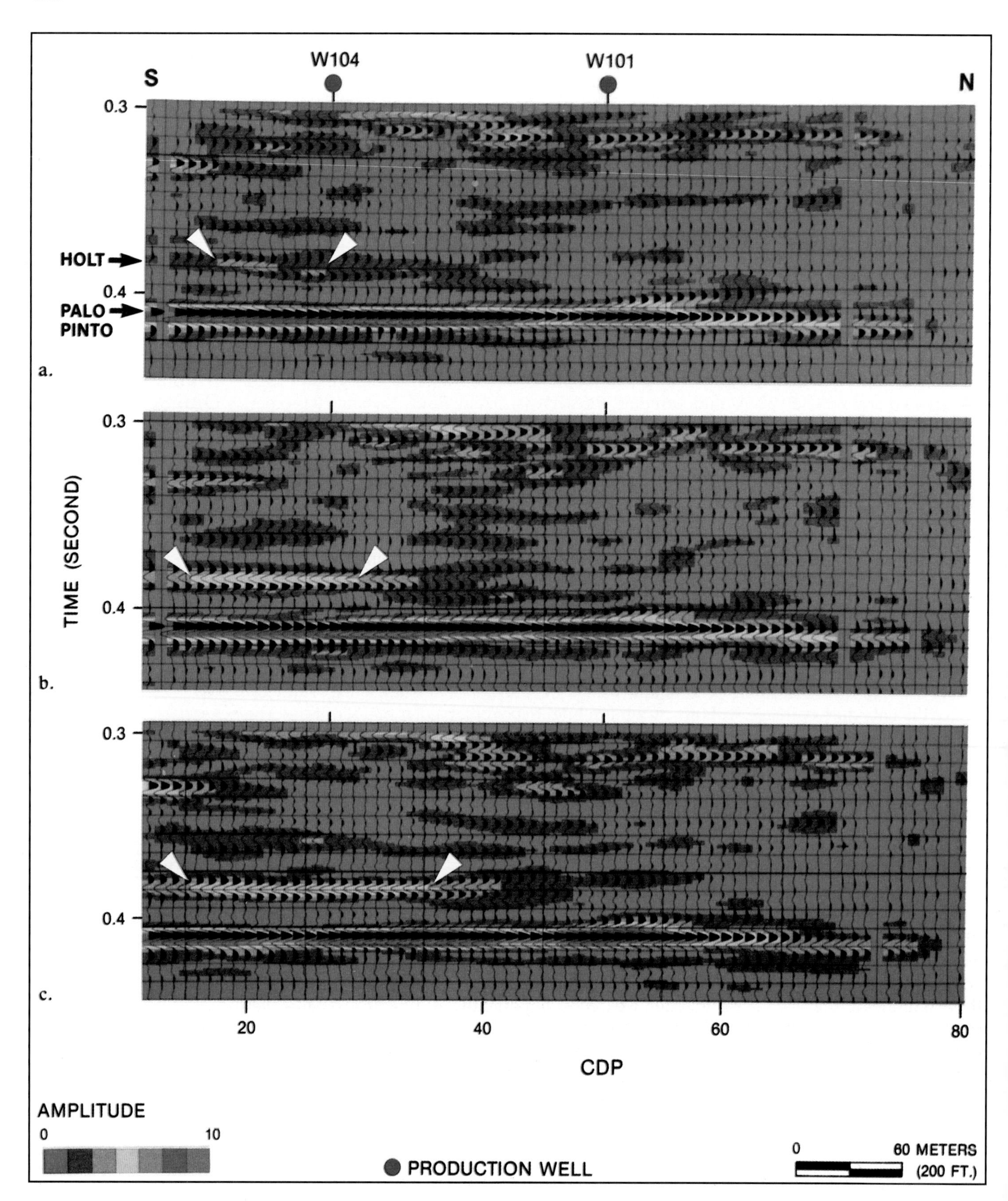

Fig. 9-4-5. Line 14, from (a) preburn, (b) midburn, and (c) postburn 3-D seismic data volumes. The reflection wiggle traces are overlain by a color scale of the calculated envelope amplitude. Dip was removed by static shifts before display. A bright spot was created (see arrows) at the top of the Holt sandstone by midburn time (b), and it increases in extent by postburn time (c). A dim spot in the reflection from the Palo Pinto limestone formed just below the peak of the bright spot.

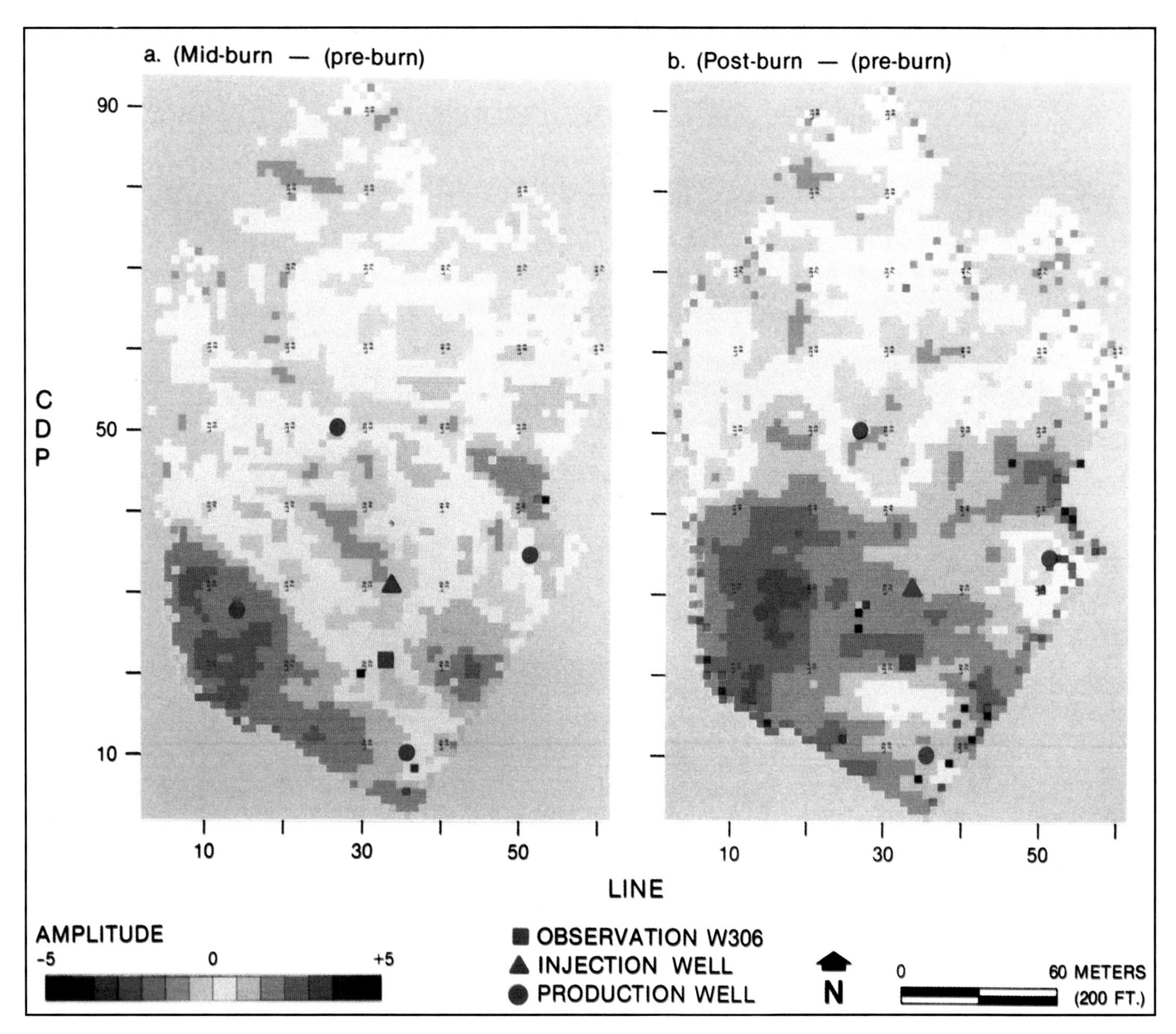

Fig. 9-4-6. The difference in envelope amplitude at the top of the Holt sandstone (0.385 s) displayed in horizon slice form. Bright spots occur as positive anomalies. Well locations are marked for position in the subsurface at the top of the Holt sandstone.

reflection from the reservoir and the reflection just below it, as expected. Searches through the difference data volumes substantially above and below the reservoir reflection showed no extended coherent anomalies. Second, the background noise level of the difference data volumes is substantially lower than the observed anomalies. Figure 9-4-6 and 9-4-8 show this by the amplitudes to the north of the injection well. Most importantly, the seismic anomalies were confirmed by well-log and core data. At the time postburn seismic data were collected, several cores and logs through the burned zone were collected.

In Figure 9-4-9, density and sonic traveltime logs showing the reservoir sand from a preburn observation well, W306, are compared to similar logs collected in two postburn boreholes. Within the zones of clean-burnt sand, the logs show substantial decreases in both density and velocity.

The combustion process was expected to increase the gas saturation with consequent changes in density and velocity. Comparison of log density values showed decreased density in burned zones averaging about 5 percent. This density decrease can be fully accounted for by a change from 100 percent fluid-filled pores to partial gas saturation. The sonic-log velocities measured in burned zones decreased 15 percent to 35 percent, averaging 25 percent. This decrease in velocity is much greater than can be accounted for by increased gas saturation in the

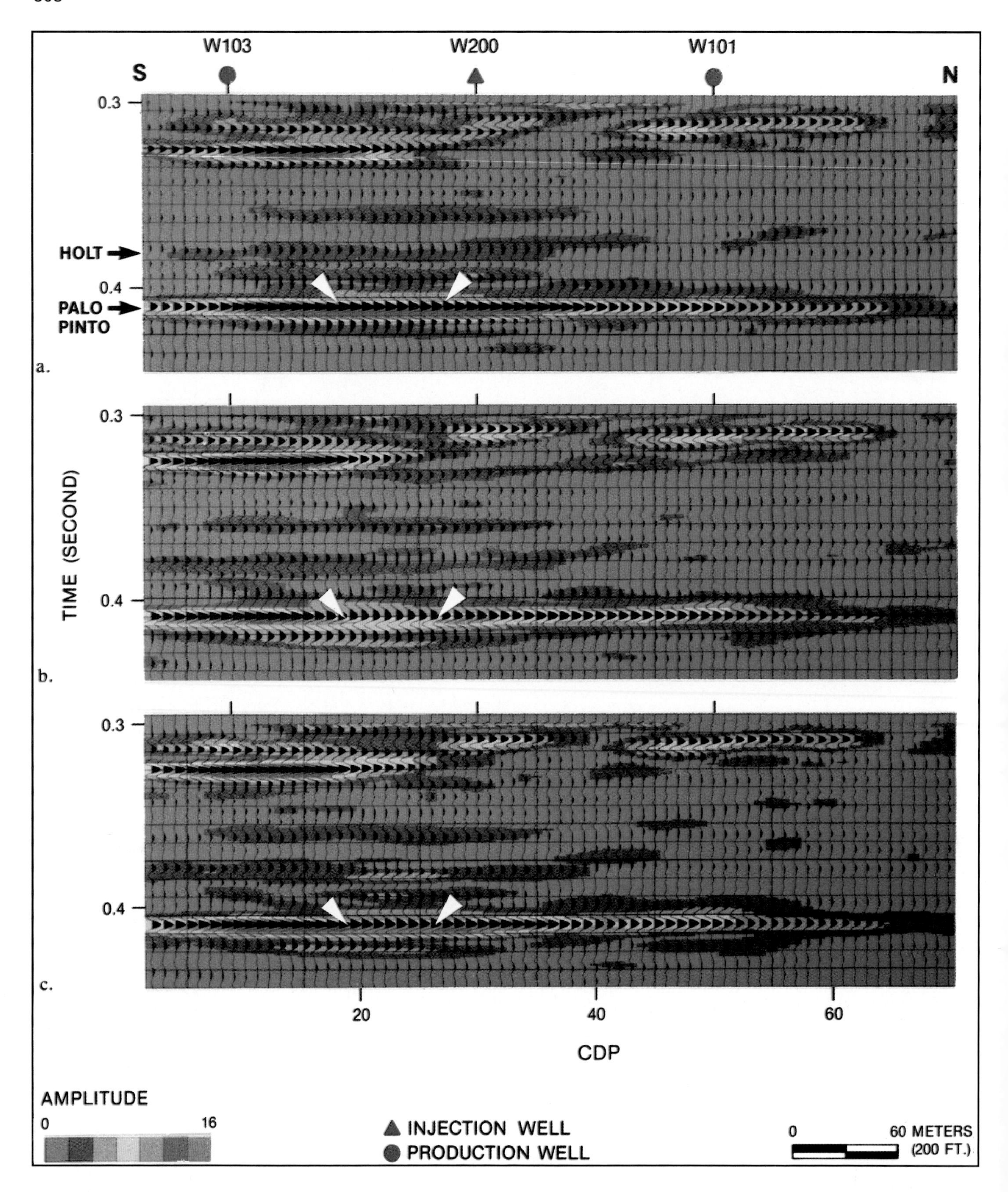

Fig. 9-4-7. Line 33, from (a) preburn, (b) midburn, and (c) postburn 3-D seismic data volumes. The reflection wiggle traces are overlain by a color scale of the calculated envelope amplitude. Dip was removed by static shifts before display. A dim spot was created (see arrows) in the Palo Pinto reflection (0.410 s) by midburn (b), but it decreases somewhat by postburn (c).

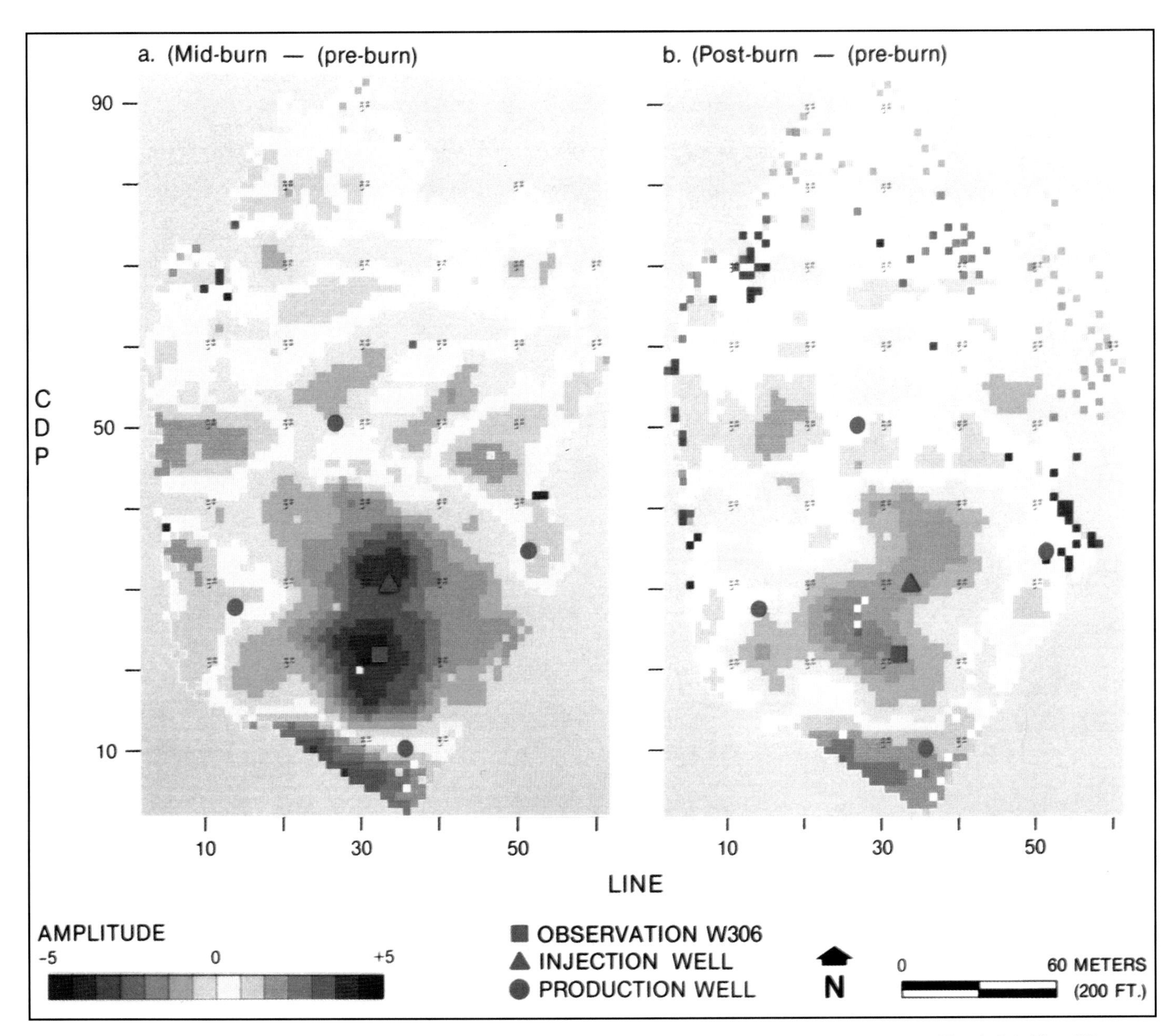

Fig. 9-4-8. The difference in envelope amplitude at the Palo Pinto reflection (0.410 s) displayed in horizon slice form. Dim spots occur as negative anomalies. Well locations are marked for position in the subsurface at the top of the Holt sandstone.

original pore space.

Ultrasonic measurements made on preburn core showed a 3 to 4 percent decrease in velocity, going from 100 percent water saturation to 100 percent gas saturation. Although this is a smaller decrease than that reported by Domenico (1976), a similar result was reported by Frisillo and Stewart (1980). In our case, the larger effect on velocity was due to permanent alteration of the rock matrix by the very high formation temperatures. Ultrasonic measurements showed a 25 percent decrease in velocity for cores heated to 700°F. The velocity decrease may be due to weakening of the rock by oxidation of organics and an alteration of clays. Therefore, the observed velocity decrease is the combined effect of changes in fluid saturation and of damage to the rock matrix.

The effect of increasing gas saturation on the seismic response is a nonlinear relation. As shown by Domenico (1974), it is the first few percentage increases in gas saturation (up to about 10 percent) that affect the seismic impedance the most. Further increases in gas saturation change the impedance very little. Therefore, if combustion gas is forced ahead of the burn zone in sufficient volume to increase gas saturation even a few percent, what is observed as a bright spot is both the clean-burnt zone and the zones ahead of the combustion point reached by steam and combustion gas.

Fig. 9-4-9. Comparison of sonic traveltime logs (above) and density logs (below) from preburn well W306 and postburn core wells W401A and W402A.

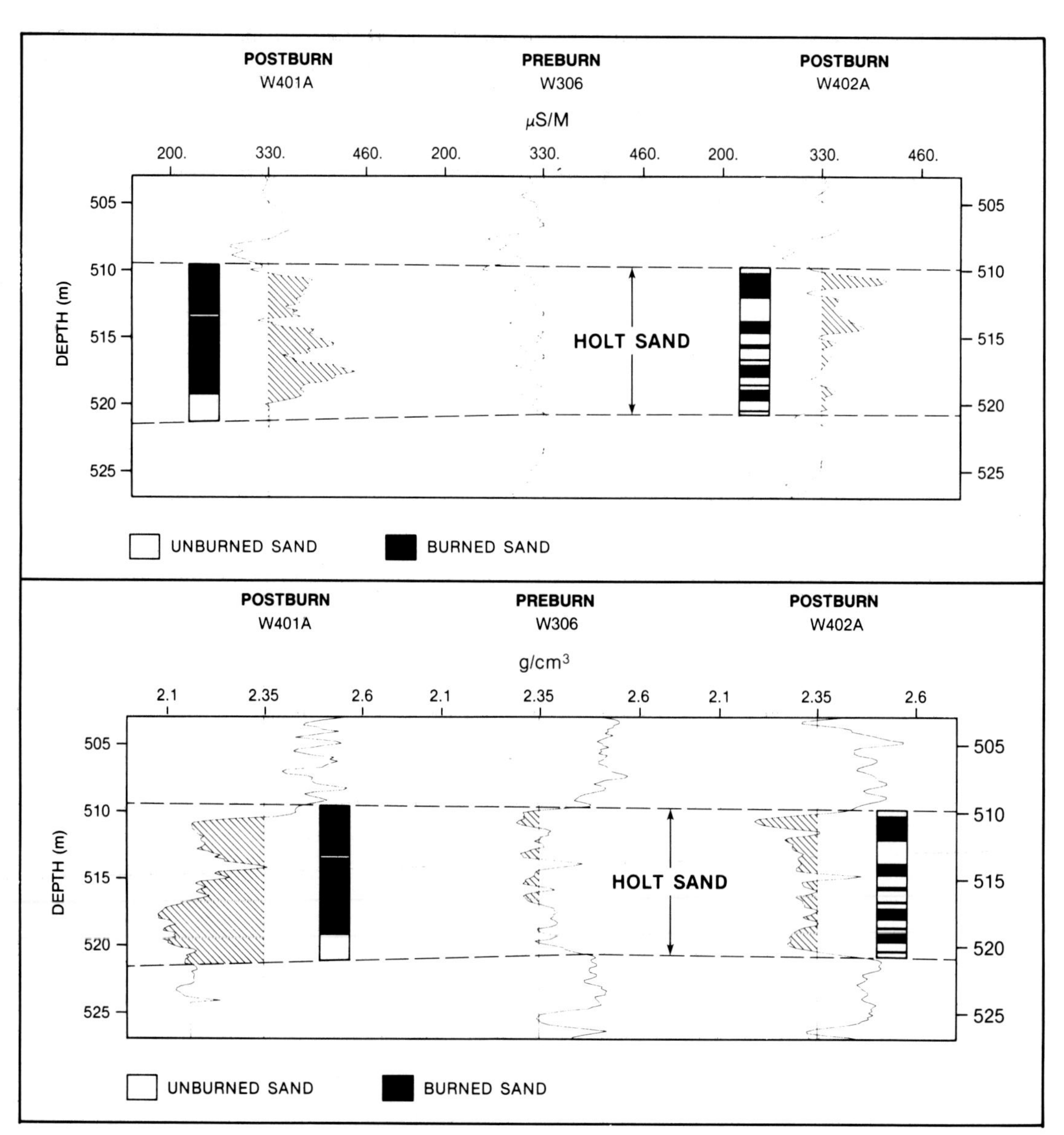

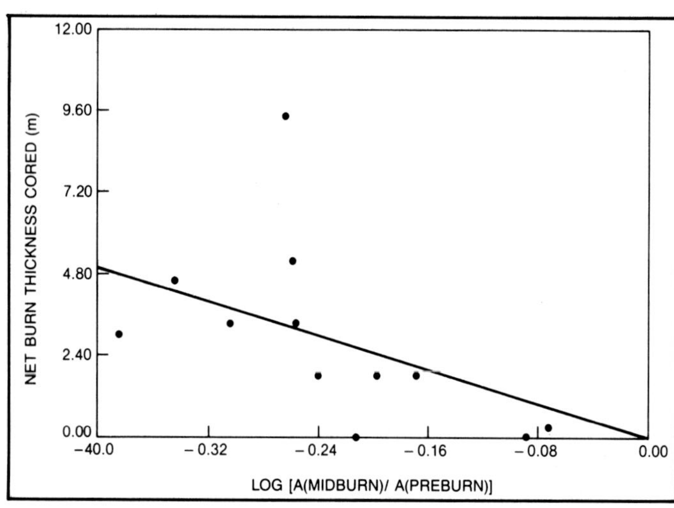

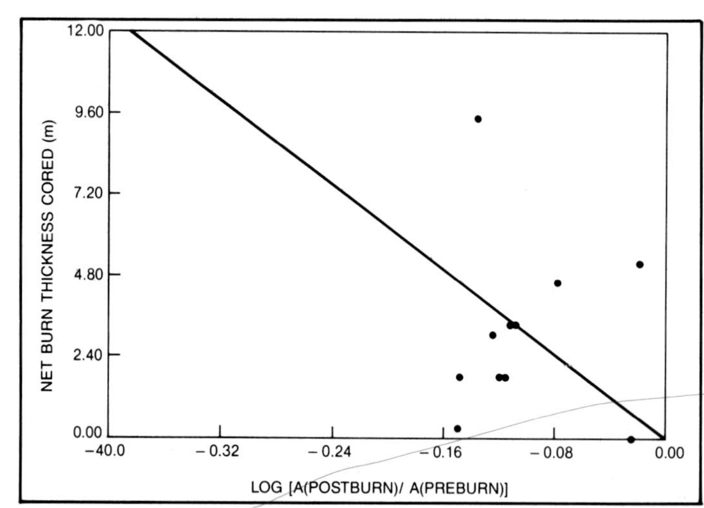

Fig. 9-4-10. Net burn thickness from postburn cores versus natural logarithm of the ratio of midburn (left) and postburn (right) to preburn dim-spot (Palo Pinto) amplitude. The line is a least-squares fit to the data points.

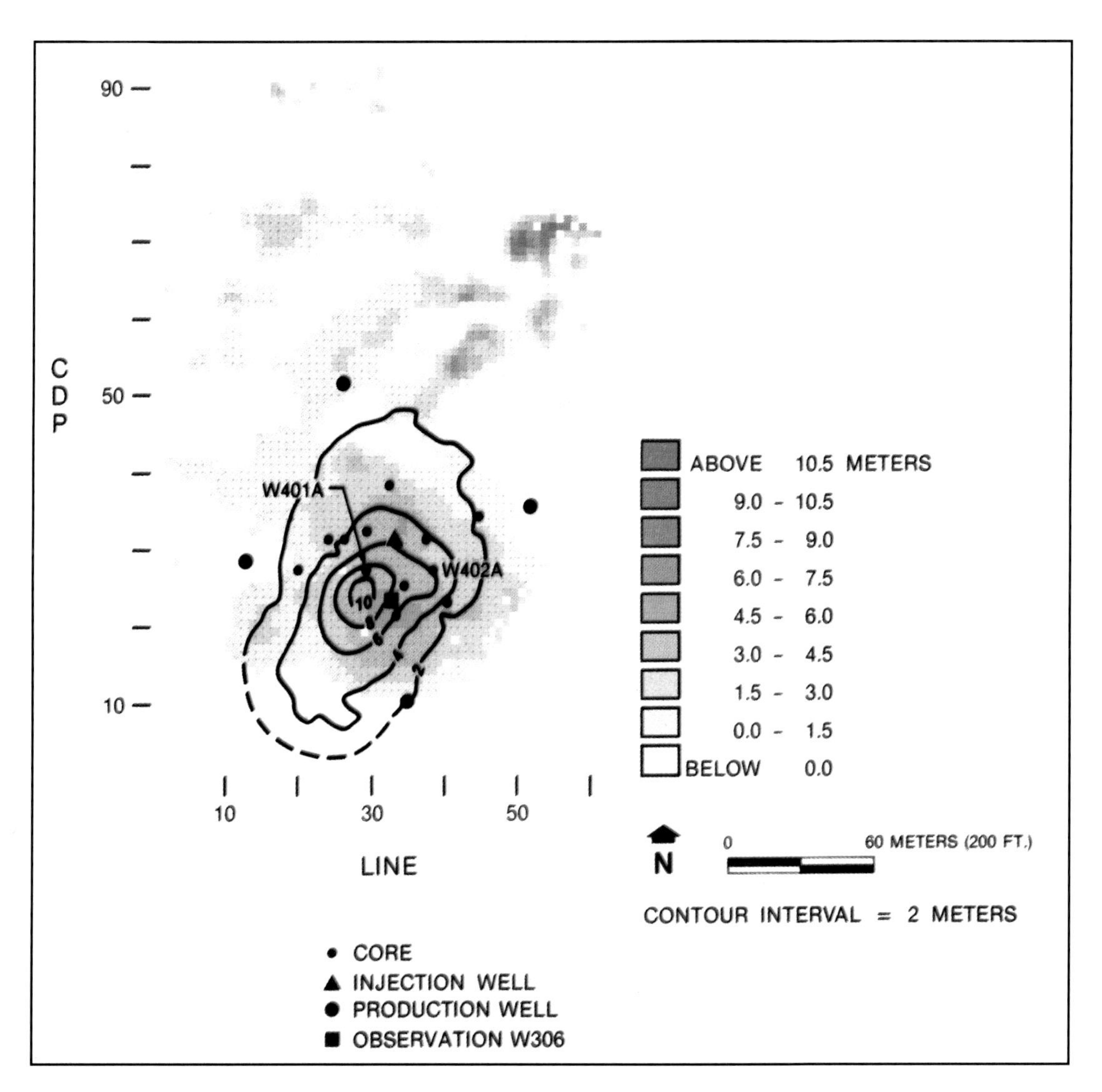

Fig. 9-4-11. Burn thickness calculated from midburn dim-spot amplitudes using equation (2) and the slope of the line in Figure 9-4-10 (left) for the calibration constant. Overlain is a line contour map of net burn thickness observed in cores.

Certainly the phenomenon of attenuation is even more complex. Similar to seismic impedance, it is the initial change from 100 percent water saturation that increases the attenuation substantially. However, unlike impedance, attenuation decreases as the rock reaches 100 percent gas saturation (Frisillo and Stewart, 1980). Therefore, in the initial stages of the process, the dim spot will reflect both the clean-burnt zone and the zone reached by combustion gases. However, as 100 percent gas saturation is reached, the dim spot will more likely be an indication of just the clean-burnt zone.

Interpretation—Propagation

Interpretation of the positions of the bright-spot and dim-spot anomalies over calendar time provides a reasonable description of the combustion propagation. First, it is quite clear from Figure 9-4-8 that the area around well W101 (to the north of the injection well) was not affected by the combustion process. This well was the only production well in which large quantities of gases were not observed. Therefore, this well was either too far downdip to be reached by the combustion process or was isolated by permeability barriers. One can also see from Figure 9-4-8 that the process did propagate to the southwest of the injection well, probably guided by the fracture system. The location of the strongest dim spot at midburn time shown in Figure 9-4-8 was verified by an observation well W306, which recorded the highest formation temperature at the time of the midburn survey.

The postburn dim spot decreased in amplitude and lateral extent compared to midburn time. It appears that the combustion process reached its maximum lateral propagation within the production area by midburn time or soon after. In a full-scale EOR project, this knowledge would be crucial in adjusting the program to sweep the reservoir more efficiently.

Although the midburn bright spot (Figure 9-4-6) also shows that the process moved to the southwest, it does not extend back to the injection well. Therefore, the combustion gas most likely propagated laterally within the reservoir until it encountered a vertical permeability pathway which allowed the gas to stream to the top of the reservoir. Once established, this pathway also allowed the burn to move to the top of the reservoir. Up to that point, the successful part of the burn was contained in the middle of the reservoir.

The postburn bright spot (Figure 9-4-6) increased in area from midburn time. A major fault (providing the southern closure to the field) is located approximately 300 m (1000 ft) to the south of the test. This fault probably blocked further southward propagation of the combustion gases, forcing them back along the top of the reservoir toward the injection well. If sustained injection of gas from midburn to postburn time had continued to fuel the combustion process, that process would have moved out beyond the production area and the area of seismic coverage.

Burn Volume

The final objective of this study was to estimate the volume of reservoir swept by the combustion process. Although the data do not have the spatial resolution to map the detailed distribution of the process, we have attempted to interpret the decreased amplitudes in the Palo Pinto reflection as estimates of burn thickness. The mechanisms of the attenuation are not separable, but several factors are certainly important: pore fluid state and interaction with the rock matrix, formation temperature, matrix velocity and density, and the increased reflectivity due to gas saturation in the reservoir.

A simple mathematical approach was chosen (after Waters, 1978) in which amplitude is expressed as

(1)
$$A = A_0 e^{-\alpha z},$$

where A_0 is the initial amplitude of the propagating wavelet, α is the attenuation parameter, and z is the propagation distance. If two seismic waves are considered identical except that one has passed through a zone Δz where the attenuation is different, then the reflection amplitude from a level past the zone of attenuation can be compared directly to find Δz. For this study, observed seismic waves are the before-burn and after-burn data traces, and Δz is the estimate of burn thickness. After taking the natural logarithms and accounting for two-way propagation, the equation becomes the linear relation

(2)
$$\Delta z = \frac{1}{2(\alpha_B - \alpha_A)} \ln\left(\frac{A_A}{A_B}\right),$$

The reservoir was cored in twelve locations within the test pattern at postburn time. The cores confirmed that the burn had occurred in somewhat vertically isolated zones within the reservoir. The net burn thickness observed in each core was compared to the logarithm of the ratio for appropriate seismic amplitudes at the CDPs corresponding to the bottom-hole location of each core. Figure 9-4-10 shows the comparison of net burn thickness to the logarithm of midburn amplitude over preburn amplitude, and also the same relationship for postburn data. Using least-squares estimation, lines were fitted through the data points and forced through the origin. The slope of either of these lines could be used in equation (2) to estimate burn thickness at all other CDPs. Since the midburn data appeared to fit the simple model better, the midburn data were used to estimate burn thickness. This reemphasizes the belief that burn propagation, at least within the production area, had ceased by midburn time. The postburn dim spot is more likely a map of formation damage only, suggesting that a more complicated model is needed to explain the attenuation due to alteration of the rock fabric.

Using the slope of the midburn line in Figure 9-4-10 and equation (2), the midburn data were converted to an estimate of net burn thickness (Figure 9-4-11). Overlain on that estimate is a computer-generated contour map based on core data. A good correlation is observed, and the correlation could have been even better if there had been core data to the southwest. This also implies that even without the core data for calibration, a good estimate of relative burn thickness could have been made using only the seismic data. The observation of seismic attenuation is a useful approach in mapping certain recovery processes. Resolution could be improved utilizing borehole-to-borehole techniques.

Conclusions

Reflection seismic surveying can be used to monitor the progress of some EOR processes. In this case study, a fireflood process was detected, its propagation direction and extent were determined, and an estimate of net burn volume was made.

The 3-D seismic data detected the burn zone and showed that the gas propagated predominately updip to the southwest. A dim spot observed in a reflector just below the reservoir level was interpreted as a map of areal extent of the burned zone. A region of maximum net burn thickness was located about 30 m (100 ft) from the initiation point of the burn. Comparison of the midburn and postburn dim spots led to the conclusion that the majority of the reservoir swept by the combustion process occurred in the first few months after ignition. The shape, orientation, and volume of the burn interpreted from the seismic data were confirmed by temperature monitor wells and postburn coring.

It was concluded that the attenuation increase, due to high-temperature alterations of the reservoir rock and pore fluid changes, was the best seismic indicator of the combustion process.

The subtraction of the baseline (preburn) data from the midburn and postburn data for interpretation of dynamic anomalies proved to be a very powerful technique. The subtraction has great potential for detecting anomalous seismic response related to active reservoir processes.

Acknowledgment

The authors would like to thank ARCO Oil and Gas Company for allowing us to publish these results. Production Engineering Research provided the financial support for this project. In Geophysical Support, L. F. Konty and D. R. Paschal were instrumental in survey planning. L. J. Hix and P. W. Wise added their expertise in data acquisition. Data processing was designed by R. Chen and S. A. Svatek. M. L. Batzle made the petrophysical measurements on core samples and provided core descriptions. J. D. Robertson contributed support and ideas during the interpretation of the data sets.

References

Domenico, S. N., 1974, Effect of water saturation on seismic reflectivity of sand reservoir encased in shale: Geophysics, v. 39, p. 759-769.

Domenico, S. N., 1976, Effect of brine-gas mixture on velocity in an unconsolidated sand reservoir: Geophysics, v. 41, p. 882-894.

Frisillo, A. L., and T. J. Stewart, 1980, Effect of partial gas/brine saturation on ultrasonic absorption in sandstone: Journal of Geophysical Research, v. 85, p. 5209-5211.

Sheriff, R. E., 1980, Seismic stratigraphy: Boston, Massachusetts, International Human Resources Development Corporation, p. 185-198.

Tadema, H. J., 1959, Mechanism of oil production by underground combustion: Proceedings, 5th World Petroleum Congress, Section 11, p. 279-287.

Taner, M. T., and R. E. Sheriff, 1977, Applications of amplitude frequency, and other attributes to stratigraphic and hydrocarbon determination: AAPG Memoir 26, p. 301-302.

Waters, K. H., 1978, Reflection seismology: John Wiley and Sons, p. 203-207.

Widess, M. B., 1973, How thin is a thin bed?: Geophysics, v. 38, p. 1176-1180.

Case History 5

Impact of 3-D Seismic on Structural Interpretation at Prospect Cougar

C. J. McCarthy, P. W. Bilinski, Shell Offshore, Inc.

Introduction

Prospect Cougar (South Timbalier 300 Field) is an oil and gas field located in 104 m (340 ft) of water, 105 km (65 mi) south of the Louisiana shore (Figure 9-5-1). Blocks S.T. 299, 300, and 301 were acquired by Shell in 1981, 1979, and 1977, respectively.

Cougar was one of Shell's first fields to be covered by a 3-D seismic survey. Its position near the beginning of our 3-D learning curve, combined with the complex structure of the field, provides an interesting illustration of the impact 3-D seismic interpretation can have on the understanding of a field's geology. The evolution of our understanding of Cougar's structure, and the impact of seismic interpretation techniques on that understanding are reviewed here.

Geology

Prospect Cougar is situated in a salt-controlled basin, downthrown to a major down-to-the-south growth fault. Hydrocarbons are trapped in an anticline formed above a salt wing.

Two sands, the B and C, account for nearly all field reserves. We will concentrate exclusively on the B Sand. The B Sand is a thinly-bedded, cyclic stack of graded sand, silt, and mudstone, deposited by dilute turbidity currents. The sand package blankets the field. Gross thickness of the package averages 50 m (160 ft). Rock properties are such that there is little or no reflection where the B Sand is wet; the presence of hydrocarbons produces a strong bright spot.

3-D Seismic

A 3-D seismic survey was acquired by Shell in 1980 (Figure 9-5-1). An area of 78 sq km (30 sq mi) was covered with an inline subsurface point spacing of 15 m (50 ft) and a line spacing of 45 m (150 ft). The total migrated area at objective depths was 26 sq km (10 sq mi).

Structural Interpretation

The structural interpretation developed for lease sale is shown in Figure 9-5-2. The map is based on a coarse, irregular grid of 2-D seismic of various vintages. The play was for turbidite sands trapped in a mildly faulted anticline. Based on this interpretation, Shell made a successful bid of $6 million for S.T. 301.

Additional 2-D seismic was acquired in support of subsequent lease sales involving blocks S.T. 299 and 300. At the time (1980), we recognized that the bright spot did not fit structure. A model of a blanket turbidite sand cut by shale-filled channels was developed to explain the amplitude pattern. The essentially unfaulted interpretation is shown in Figure 9-5-3.

After successful bids for blocks S.T. 299 ($10.8 million) and 300 ($54.8 million), the 3-D seismic survey described above was acquired to assist in developing this complex stratigraphic field. New amplitude measurements provided a detailed picture of hydrocarbon distribution. A new structural interpretation was made using paper sections on a relatively coarse (230 m or 750 ft) grid. This interpretation, combined with early development drilling, indicated that faulting was more important than previously supposed. However, the bright spots still did not fit structure (Figure 9-5-4), so stratigraphy was assumed to play a major role in controlling hydrocarbon distribution.

At that point (1983), Shell had completed development of a photographic film-based 3-D interpretation system. Although primitive by today's standards, the system allowed digitized picking from enlarged profile displays and had "movie" capabilities in both vertical section and time slice orientations. Bookkeeping features of the system finally made it feasible to interpret every line and crossline in the survey. The tight spatial grid provided by the 3-D survey enabled us to map small offsets and amplitude "glitches" which had been ignored on coarser interpretation grids.

The resulting structure map (Figure 9-5-5) shows a much more complex fault pattern. Faults were recognized as controlling hydrocarbon distribution in all but one instance. The cyclic nature of the B Sand and its high shale content make even 3-m (10-ft) faults potential seals. Because the gross thickness of the hydrocarbon column in the B Sand is generally at or below tuning thickness (35 m or 120 ft), local thinning of the hydrocarbon column associated with normal faulting produces observable amplitude decreases. These amplitude anomalies form linear patterns that often connect with actual event time offsets; in other words, they act like faults.

The fault density shown in Figure 9-5-5 was so high that it met with some disbelief. It certainly had a major impact on development strategy. Development drilling is now complete and as many as 4 years of production data are available for some wells. Production data suggested that some reservoirs were draining areas less than shown on Figure 9-5-5, prompting reexamination of the 3-D data. The 3-D seismic data have now been reprocessed and reinterpreted using a modern workstation and results from over 60 wells (including sidetracks). Our current structure map (Figure 9-5-6) shows even more faults than Figure 9-5-5. Drilling has confirmed faults picked on seismic with throws as small as 10 m (30 ft). Faults with throws of 15 m (50 ft) can be picked very reliably on seismic.

Conclusions

Three-dimensional seismic surveying has played a central role in shaping our understanding of Prospect Cougar. The field has gone from a simple, purely structural trap, to a complex stratigraphic trap, to an even more complex structural trap. The tight spatial sampling and proper imaging provided by 3-D have been the keys to unraveling the story. The tools and time available to the seismic interpreter have also had a major impact on our understanding of the field.

As a result of the detailed structural picture provided by the 3-D survey, a costly waterflood program was determined to have little chance of success and consequently was dropped. Reexamination of the seismic data after development drilling has clarified why some wells produced as expected while others declined more rapidly than anticipated, and has provided support for new drilling and workover proposals.

Acknowledgments

The authors wish to thank Shell Offshore, Inc. and Shell Oil Company for permission to publish this material. Obviously, the interpretations described here are the work of many individuals in both exploration and production, too numerous to mention here. Our passage up the learning curve with 3-D came largely as the result of a synergistic effort among geologists, geophysicists, and programmers, all trying to get a job done under both time and technology constraints.

Figures begin on page 316.

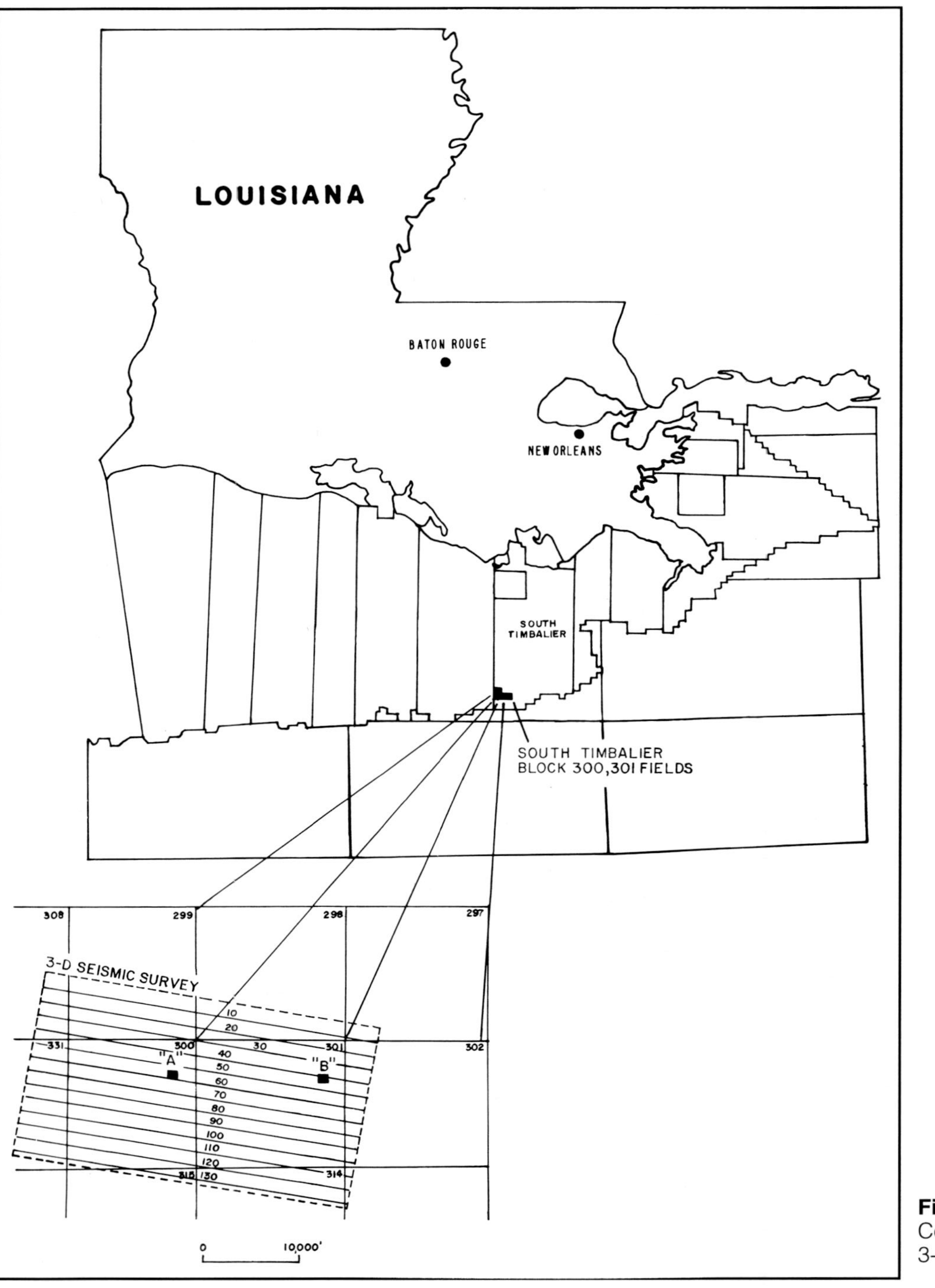

Fig. 9-5-1. Location map for Prospect Cougar (South Timbalier 300 Field). Area of 3-D seismic survey is shown.

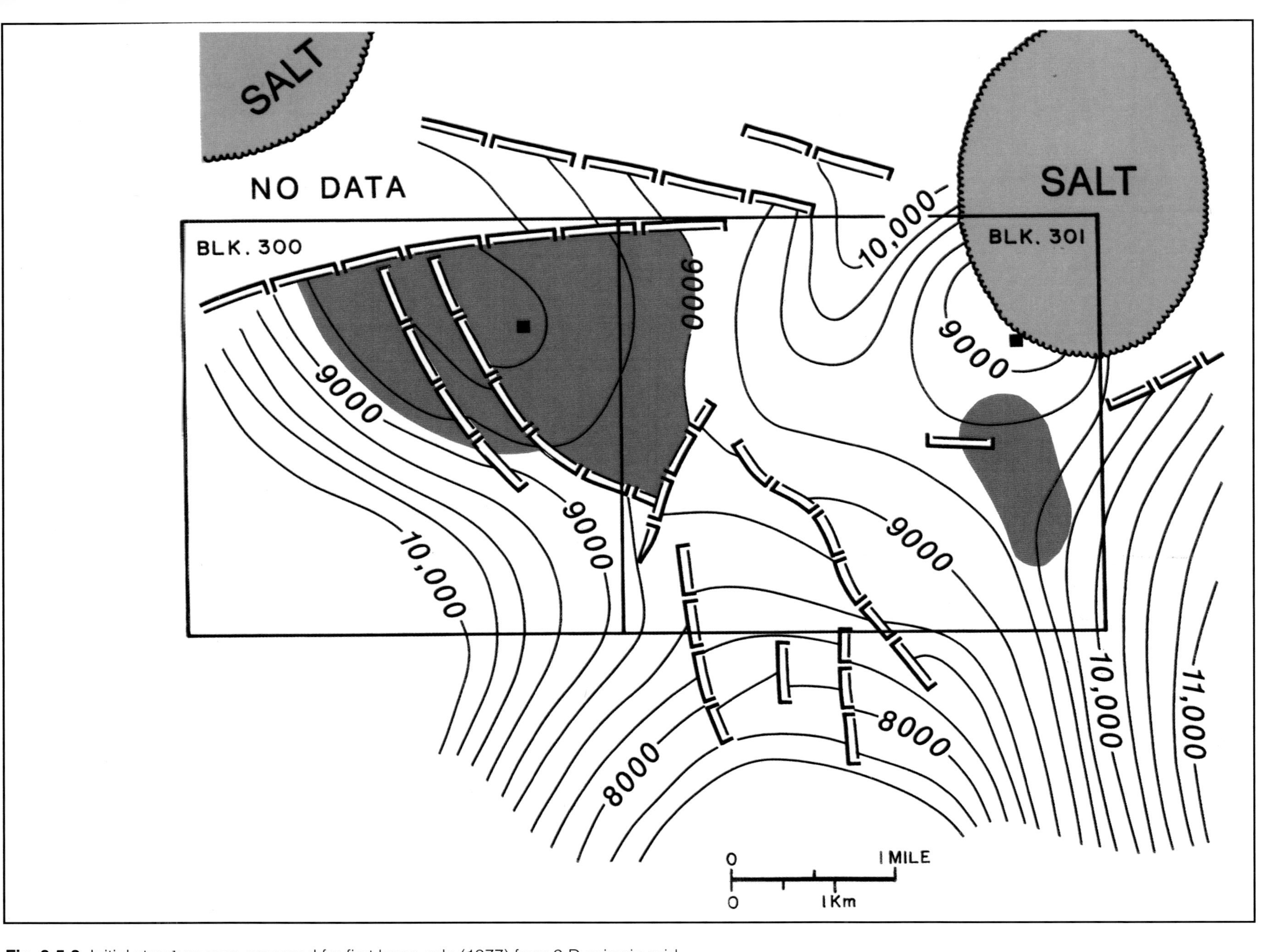

Fig. 9-5-2. Initial structure map prepared for first lease sale (1977) from 2-D seismic grid. Color indicates interpreted hydrocarbon distribution, red for gas and green for oil. Solid black squares show present location of platforms.

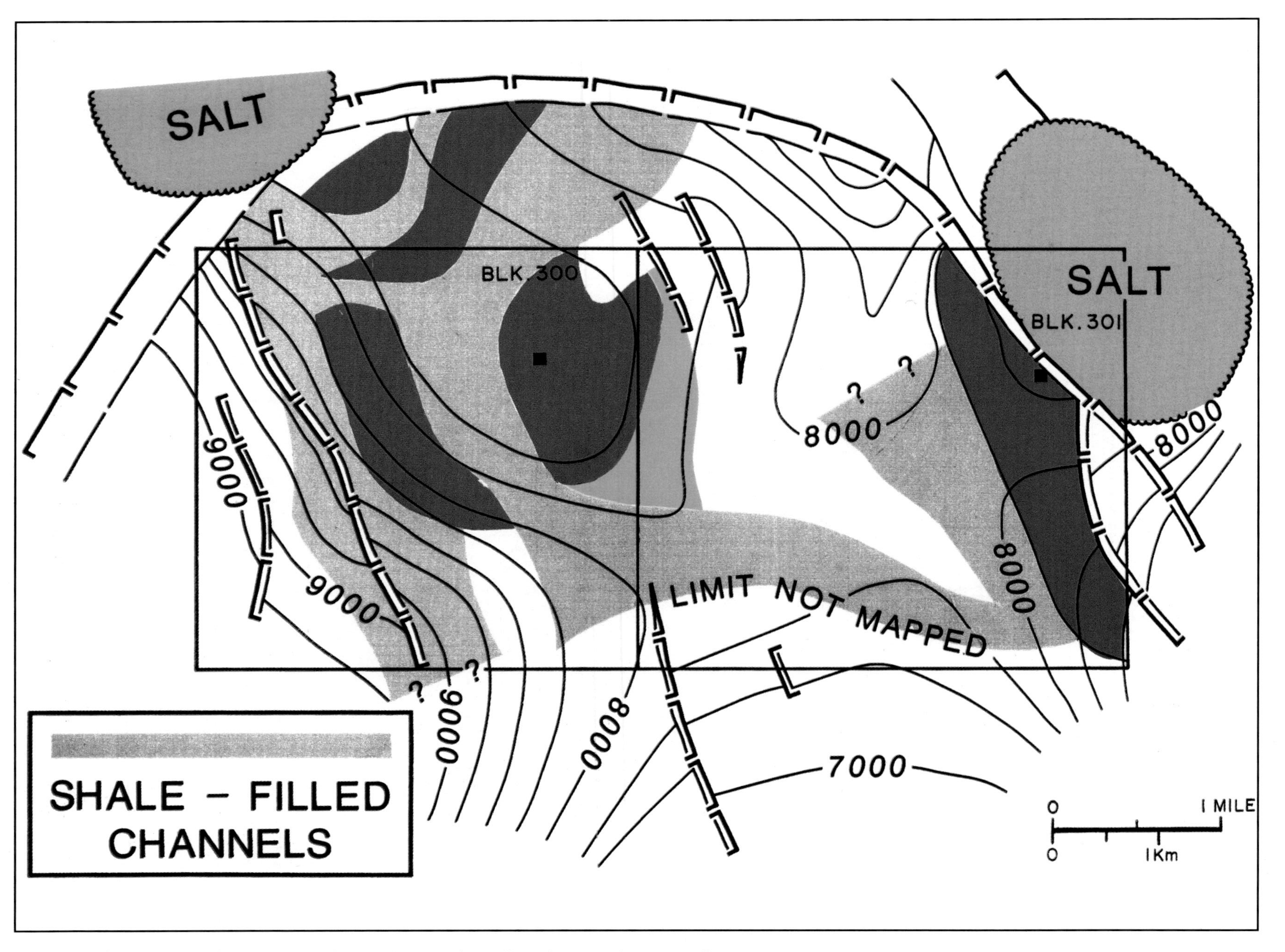

Fig. 9-5-3. Structure map from denser 2-D seismic grid after drilling three exploratory wells in S.T. 301 (1979).

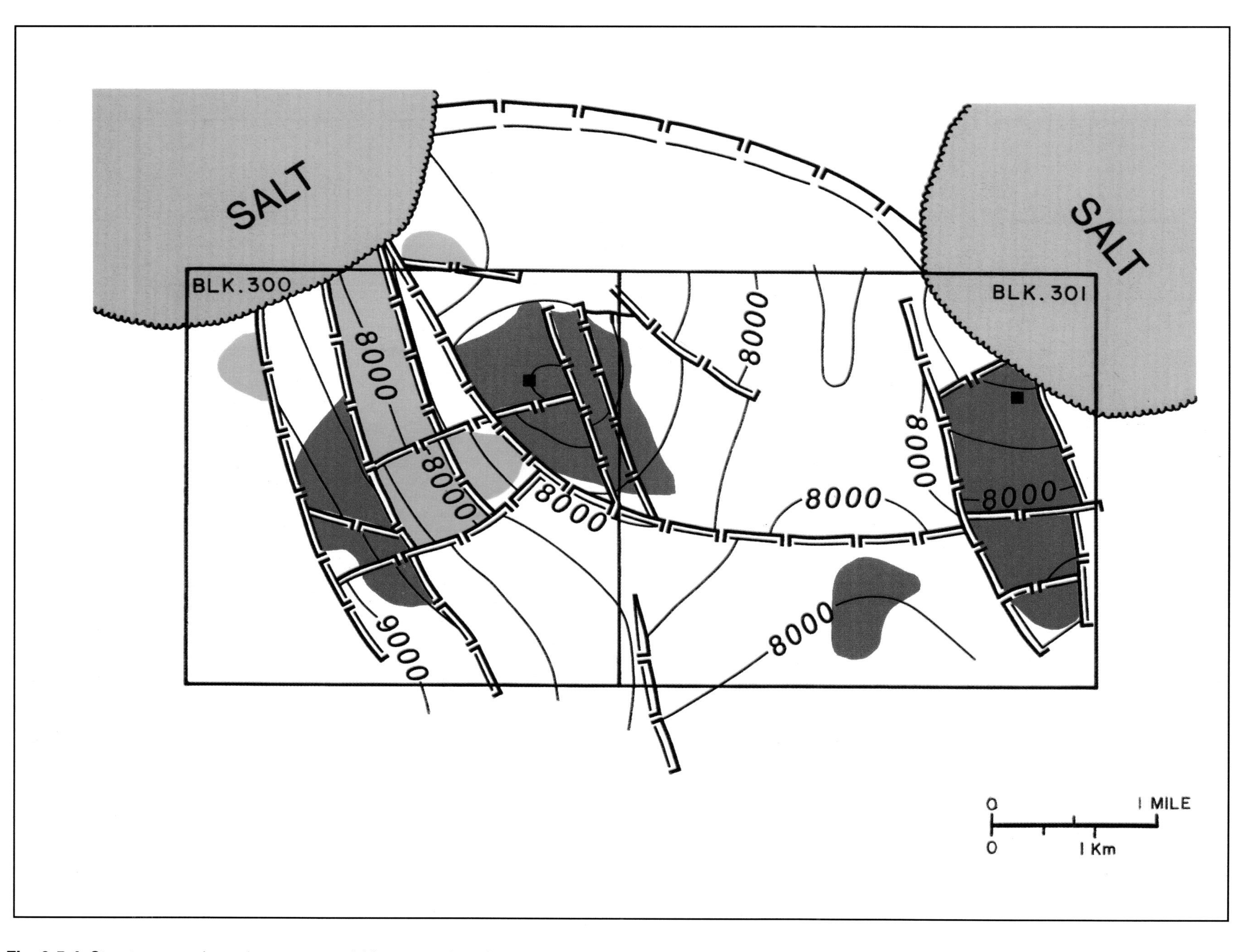

Fig. 9-5-4. Structure map based on coarse-grid interpretation of 3-D data after drilling six development wells from the S.T. 300 platform (1982).

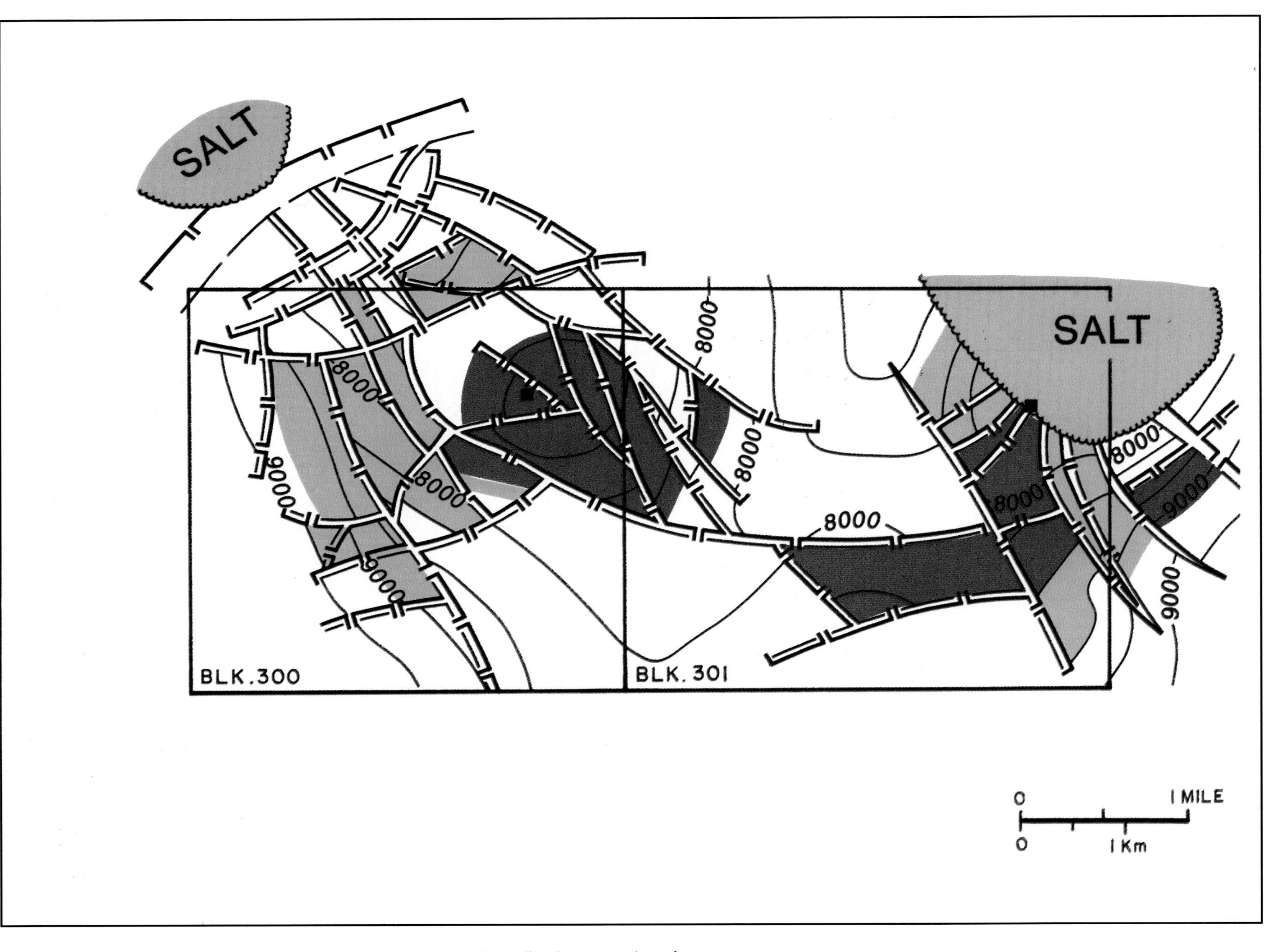

Fig. 9-5-5. Structure map based on complete line-by-line and time slice interpretation of 3-D data after 11 development wells (1983).

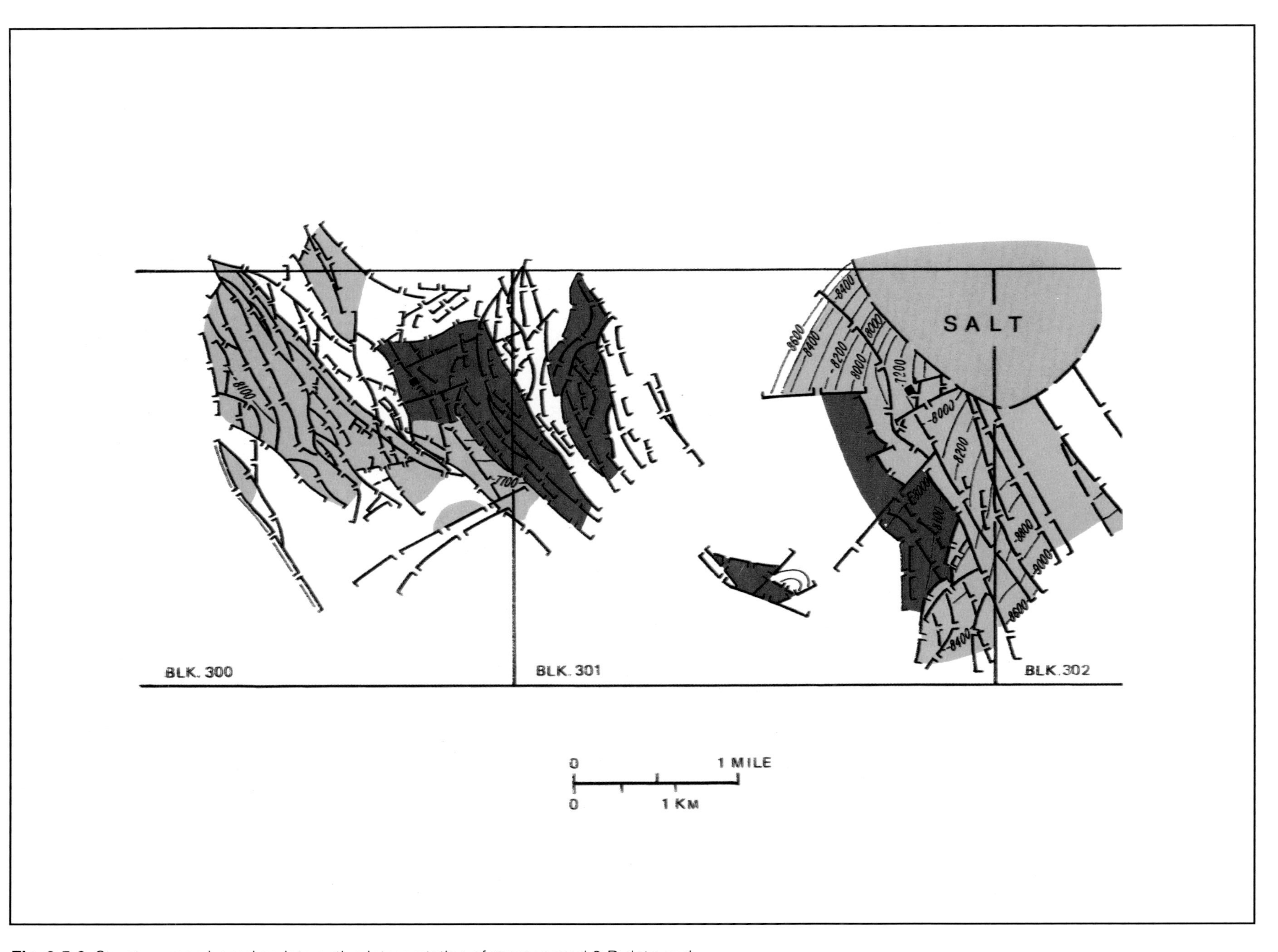

Fig. 9-5-6. Structure map based on interactive interpretation of reprocessed 3-D data and results of over 60 development wells (1988).

Case History 6

3-D Seismic Interpretation of an Upper Permian Gas Field inNorthwest Germany

H. E. C. Swanenberg, F. X. Fuehrer, BEB Erdgas und Erdoel GmbH

Introduction

In northwest Germany, sour gas is produced from Zechstein (Upper Permian) carbonates in a number of mainly medium-sized fields. A 3-D survey was acquired in 1985 to optimize drilling in the Goldenstedt field and some smaller adjacent fields (Figure 9-6-1). Recent appraisal drilling has proved significantly more potential than originally assumed in this area of complex structure and facies. Total reserves are currently estimated at 0.5 tcf of sour gas with an H_2S content varying between 7 and 26%.

Structural definition and resolution of the target horizon at 3500 to 4000 m depth (corresponding to 2.2 to 2.4 sec two-way traveltime) is severely influenced by lateral velocity variations in the overburden, which complicated processing and interpretation of the survey. Despite nominally lower coverage than regular 2-D, the superiority of 3-D data significantly enhanced structural control. Some information on the reservoir was also obtained from reflection amplitudes. The 3-D data set thus revealed details of the tectono-depositional framework not observed on previous 2-D data.

Depositional Model, Diagenesis, Reservoir Quality

Core data and modern lithologic log interpretation techniques in conjunction with dipmeter data indicate that the Zechstein carbonates in the Goldenstedt area were deposited during a transgressional cycle of the Zechstein sea in a tidal flat environment with tidal channels influenced by syndepositional faulting. In the shallow, low-energy environment, mainly mudstones together with anhydrite were deposited with hardly any porosity and permeability. In the higher-energy environment, grainstones were produced that are now found mainly within the middle and upper parts of the sequence, intercalated within mudstones. In the tidal channels, the best reservoir facies (with up to 100 m of pure grainstones) accumulated due to continuous subsidence resulting from the syndepositional faulting. Early diagenesis has altered the originally deposited limestones into dolomites. Because of the variations in primary facies and diagenetic overprinting, net thicknesses in Goldenstedt now vary between 5 and 100 m, porosities range from very low to 16%, and permeabilities reach 320 md. Fortunately, in the tight reservoir parts, fractures usually enhance permeability. The Zechstein carbonates are directly overlain by 30 to 80 m of anhydrite and several hundred meters of Zechstein salt. During the Mesozoic and Cenozoic, sedimentation in the area continued, locally governed by various tectonic events.

Structural Setting

The Goldenstedt area straddles the boundary between a relatively stable block in the north and the inverted Lower Saxony Basin (LSB), a large intracratonic inversion structure in northwest Europe (Betz et al., 1987). In this part of the LSB (the edge of which is visible at the southern end of the north-south seismic line in Figure 9-6-2), and in the similarly inverted "Goldenstedt graben," a smaller satellite structure just north of it, Lower Cretaceous sediments accumulated during an extensional phase. During a subsequent compressional phase in the Late Cretaceous, these structures were inverted. Concomitantly, thick Upper Cretaceous sequences were deposited in the adjacent rim synclines. During these tectonic phases the Zechstein salt decoupled the Mesozoic from the Paleozoic so that their respective structural styles are quite different.

*Trademark of Geophysical Service Inc.

Structural Interpretation and Time-to-Depth Conversion

Interpretation of the 3-D survey was carried out using a SIDIS* workstation on a one-pass 3-D time-migrated data set.

The seismic horizon nearest to the Zechstein carbonates is the reflection originating from the interphase between the salt and the anhydrite covering the carbonates. This "Z-reflector" is the black peak on Figure 9-6-2 immediately above the reservoir level (colored in blue). In the areas not affected by Mesozoic inversion tectonics, interpretation of this Z-reflector is straightforward. Care has to be taken only to avoid confusing the Z-reflector with strong reflections originating from anhydrites floating in the salt. Within the zones of faulted overburden (shown stippled in Figure 9-6-1) interpretability of the Zechstein is markedly reduced. The problem becomes even more serious along the margins of these areas because of the lateral velocity contrast in the overburden. Here, faults in the Zechstein can only be inferred after careful time-to-depth conversion, but mispositioning due to ray bending remains. Detailed ray tracing, on the other hand, would be highly impractical for the total 3-D area. Therefore, we followed a different scheme, based on the concept of "pseudo-average velocities," which was accurate enough and practical.

Initially, depth-dependent velocities were calculated directly from reflection-seismic traveltimes of eight prominent, identifiable reflectors and their corresponding vertical depths as encountered in the well. Subsequently, these pseudo-velocities were mapped and used as input for a time-to-depth conversion down to the Z-level. A Top Zechstein carbonate map was constructed by adding isopach values of the anhydrite cover to the depth map of the Z-horizon.

Because of complex overburden and associated velocity pull-down effects, the result needed to be thoroughly screened. This problem was solved by careful editing of the average velocity field, calculated from Z-traveltimes and top carbonate depth values at every CMP. In this way, it has proved possible to restore an uninterrupted horizon in depth, otherwise distorted in time by velocity effects.

Basically, the top carbonate map (Figure 9-6-1) describes the dissected antiform of the Goldenstedt field in the west, presumably a Late Cretaceous "trapdoor," separated from the adjacent structural highs of Woestendoellen and Quaadmoor by a fault-controlled graben.

The north-south- and east-west-oriented, apparently normal faults are interpreted to be en echelon elements of a dextral oblique slip system, active during at least the final stages of carbonate deposition in this area.

Reservoir Facies from Reflection Amplitudes

Because of highly variable reservoir quality, the production rates of wells are also quite variable. A number of positive and negative surprises were experienced in wells where reservoir quality was predicted based on geologic mapping only. Attempts therefore were made to use 3-D seismic for reservoir facies prediction. Obviously, in areas where faulting has affected the overburden this is rather ambitious.

Within the 3-D area it was possible to identify three facies types with different seismic expressions. One type (top left of Figure 9-6-3) has the most porous zone at the top of the reservoir, immediately below the overlying anhydrite. This is expressed on a black-and-white, variable area seismic section as a strong "soft kick" (trough) immediately following the "hard kick" (peak) from the top of the overlying anhydrite, the Z-reflector. This facies type has been observed in the Woestendoellen area to the northeast of the main Goldenstedt field. Comparison of a well synthetic and a seismic section shows good agreement (lower left of Figure 9-6-3). In a second facies type the most porous streaks are found in the middle part of the reservoir (top center of Figure 9-6-3). This is expressed on a seismic section as a delay of a few milliseconds of the soft kick as compared with the previous facies type. Overall, this type also appears as a lower frequency event than does the first type and is mainly found in the northern part of the Goldenstedt field. Again, the well-to-seismic match gives good correspondence. The third facies type (top right of Figure 9-6-3) is characterized by a very thin reservoir section with low porosity mudstones only. A strong soft kick following the Z-reflector cannot be found. This facies type is encountered in the southern part of the Goldenstedt field.

In order to visualize these observations over the whole 3-D area, horizon slices were produced at successive intervals below and parallel to the Z-reflector. Those for 8- and 20-ms delay are shown color coded in Figures 9-6-4 and 9-6-5, respectively. The 8-ms horizon slice (Figure 9-6-4) shows the strong soft kick characteristic of the first type as a distinct deep-blue event. By comparing Figures 9-6-4 and 9-6-1, it is apparent that this deep-blue event occurs in Woestendoellen, as well as in a number of other areas. In the Goldenstedt-N and Goldenstedt-S subareas, the reservoir reflection has not yet been reached at this particular time delay. Part of the amplitude variation over this horizon slice, however, is to be ascribed to signal deterioration and interference/absorption effects that are not related to reservoir development.

At 20 ms below the Z-reflector (Figure 9-6-5) the second facies type is clearly displayed in the Goldenstedt-N subarea, again by a distinct deep-blue event. Part of the time delay, however, is due to increased thickness of the overlying anhydrite. In the southern part of Goldenstedt a strong soft kick that would appear in dark blue colors cannot generally be found, either on the 20-ms horizon slice or on any other.

Conclusion

Using seismic data for reservoir prediction in areas of complex facies development requires detailed analysis of composite wavelet characteristics, because changes in the reflection shape cannot usually be attributed to variations of a single parameter of the reservoir only. Nevertheless, this example shows that even in rather complex cases, horizon slices can facilitate the study of the spatial distribution of wavelet characteristics. These studies can then be used to optimize well positioning in order to maximize production rates and minimize the number of wells required in field development.

Acknowledgment

Goldenstedt and surrounding fields are operated by BEB Erdgas und Erdoel, Hannover, an affiliate of Shell and Exxon, holding two-thirds of the interest in the fields. One-third is held by Mobil Oil. The authors are indebted to the managements of BEB and Mobil Oil for permission to publish this case history.

References

Betz, D., F. Fuehrer, G. Greiner, and E. Plein, 1987, Evolution of the Lower Saxony Basin: Tectonophysics, v. 137, p. 127-170.

Figures begin on page 325.

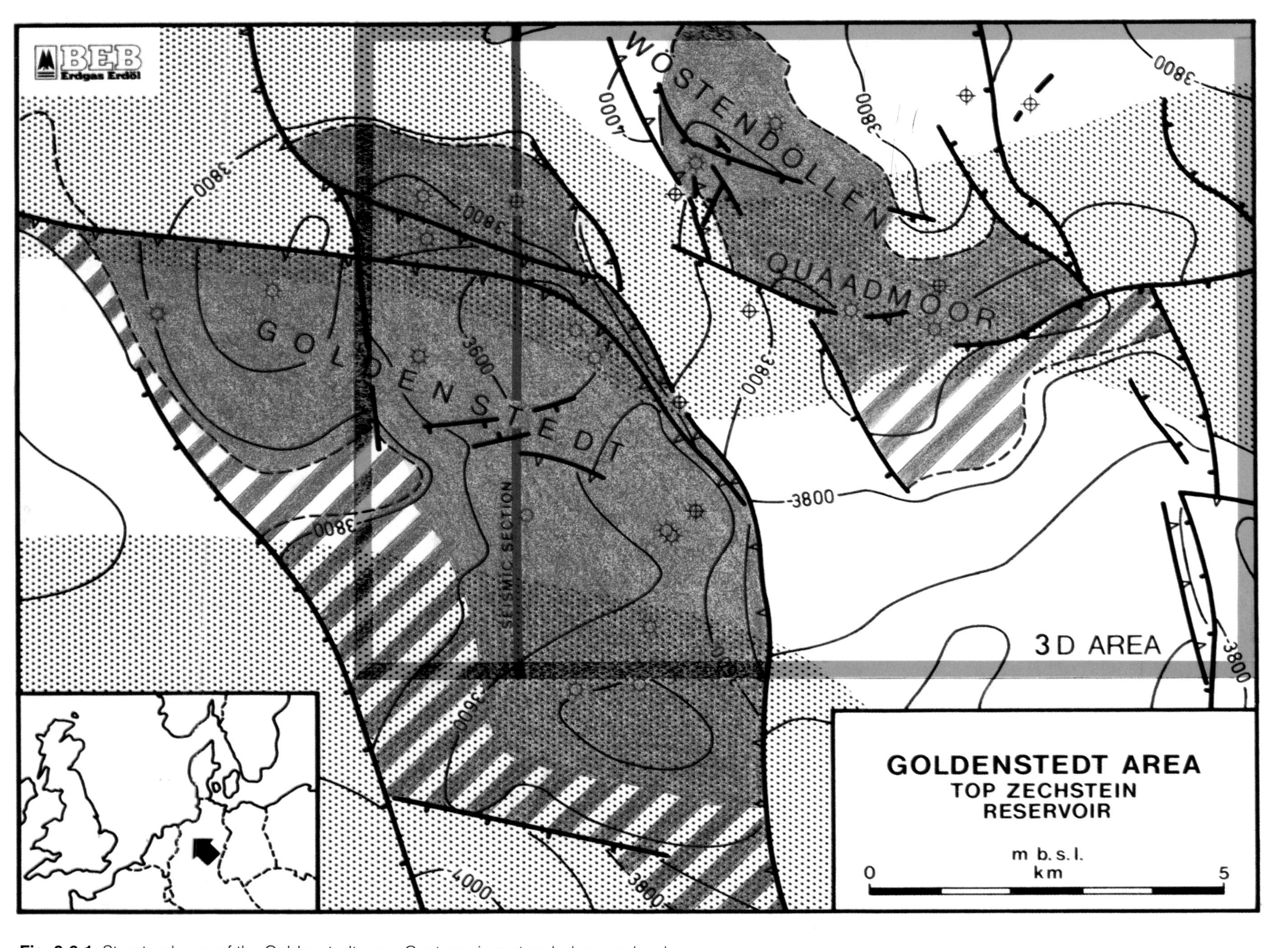

Fig. 9-6-1. Structural map of the Goldenstedt area. Contours in meters below sea level. Stippled zones indicate areas where resolution and time to depth conversion are affected by strongly tectonized overburden.

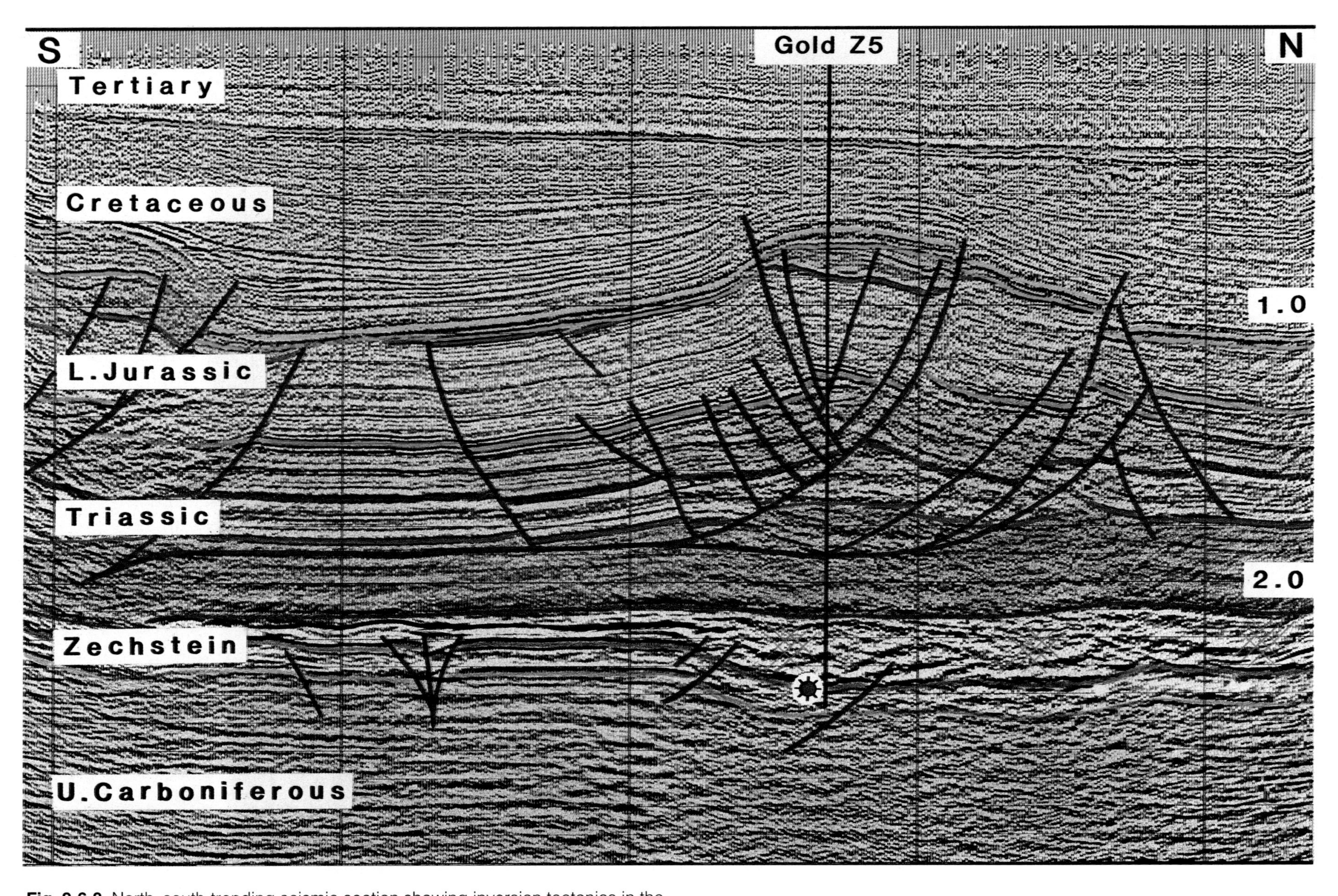

Fig. 9-6-2. North–south-trending seismic section showing inversion tectonics in the overburden. Reservoir carbonates, at 2.2 to 2.4 secs reflection time, are indicatec in blue.

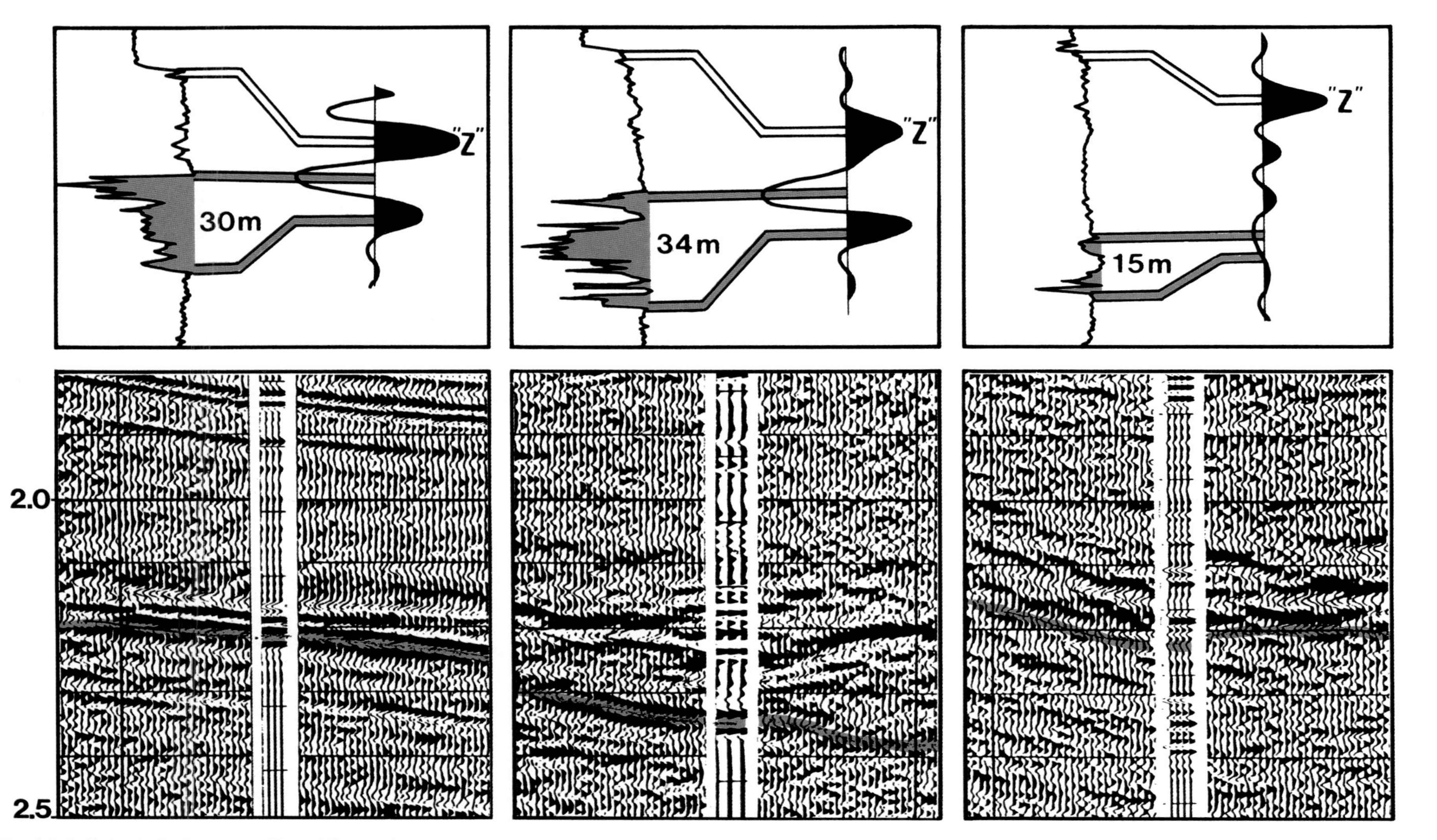

Fig. 9-6-3. Seismic facies types. Top of figure shows sonic log with the reservoir section in blue and the corresponding zero-phase synthetic. The Z-reflector originates from the top of the anhydrites overlying the reservoir. The lower part of the figure gives comparisons of synthetic seismograms and seismic section.

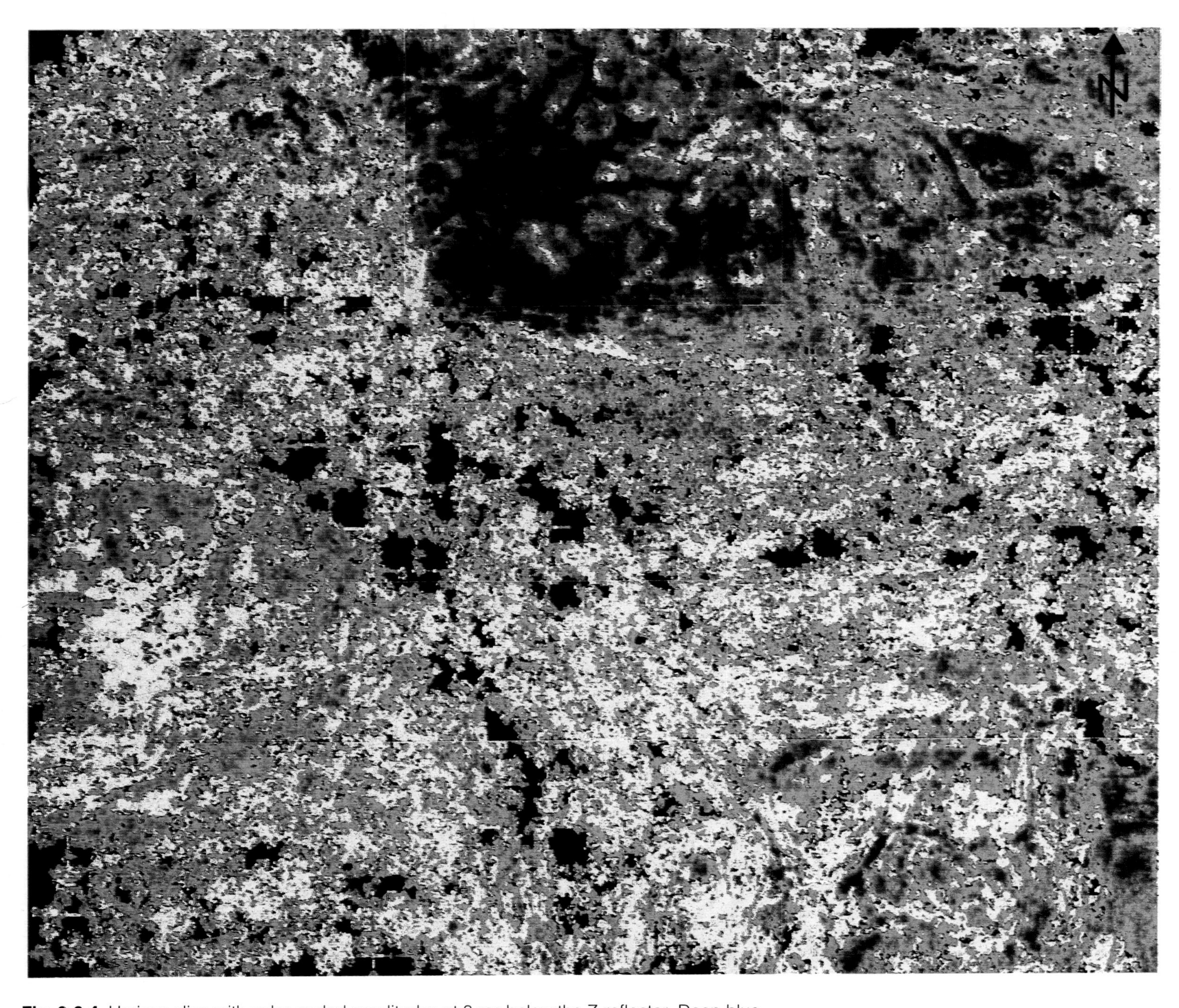

Fig. 9-6-4. Horizon slice with color-coded amplitudes at 8 ms below the Z-reflector. Deep blue represents reflection amplitudes of a strong “soft kick” (trough) indicating porosity in the upper part of the reservoir.

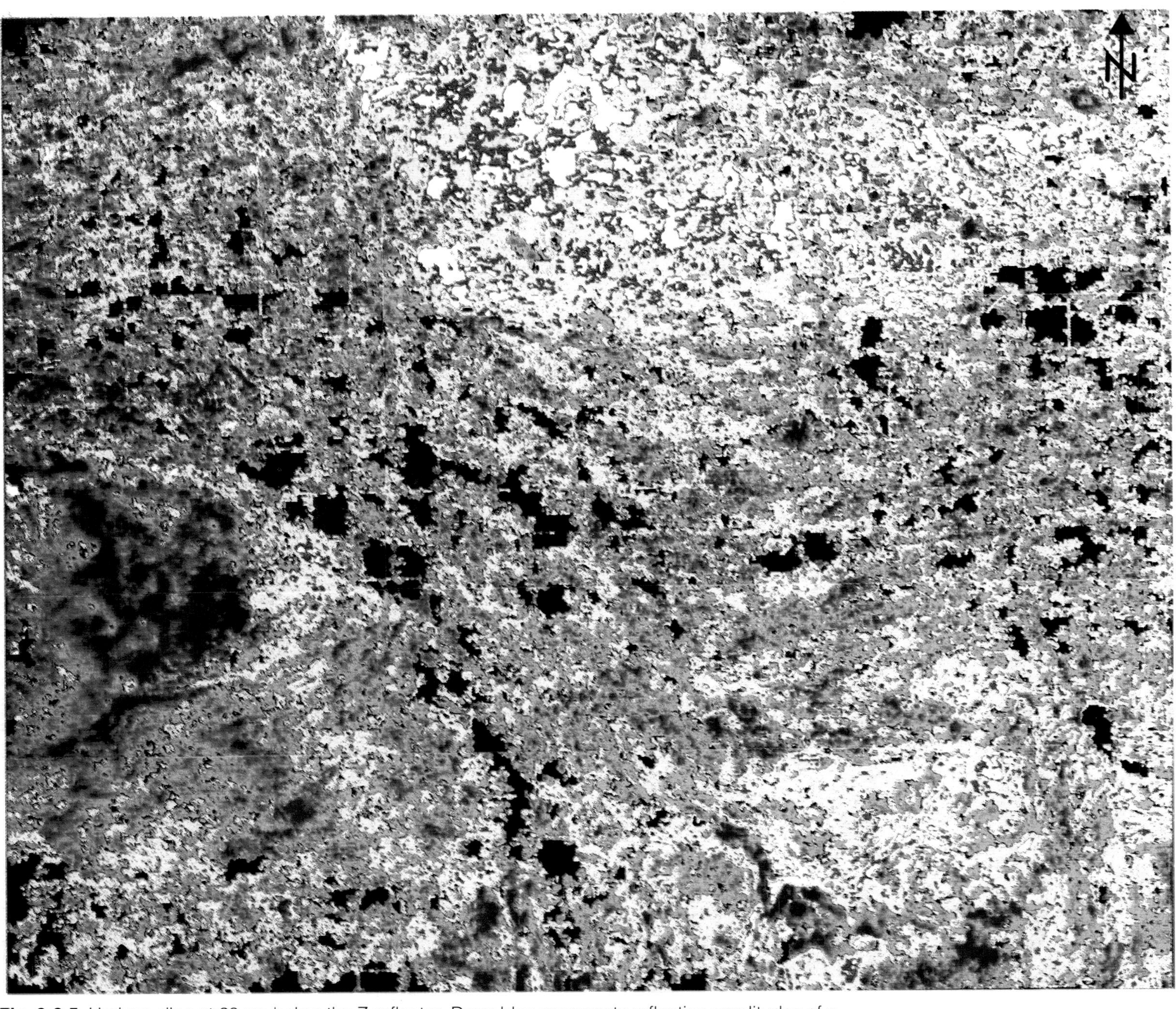

Fig. 9-6-5. Horizon slice at 20 ms below the Z-reflector. Deep blue represents reflection amplitudes of a strong "soft kick" (trough) indicating porosity in the central part of the reservoir.

Case History 7

Seismic Data Interpretation for Reservoir Boundaries, Parameters, and Characterization

William L. Abriel, Roger M. Wright, Chevron U.S.A. Inc.

Three-dimensional mapping of seismic data revealed a prospect not existing on the 1000′ 2-D grid. A bright spot was reviewed downdip of an abandoned competitor well that had apparently been faulted out. Normal 3-D mapping resulted in a new well proposal, but results from interactive workstation analysis moved the location downdip. Detailed structural and amplitude analyses revealed the presence of complex stratigraphic changes within the prospective reservoir, which were then avoided in a successful discovery well. Predictions of net pay and reservoir connectivity were confirmed. Future management of this reservoir will also be better served with combined geological, engineering, and geophysical analyses.

Introduction

An offshore area in the Gulf of Mexico was reviewed for prospects, based on a 305-m (1000-ft) 2-D seismic grid and available well data (Figure 9-7-1). No clear bright spots were observed on seismic data, and the abandoned competitor well was not encouraging (Figure 9-7-2) because of the shaly sands encountered. The area was later covered by 3-D seismic data as part of a larger survey, and the subsequent structural evaluation was quite different (Figure 9-7-3).

The 3-D survey was conducted in 1983 with an airgun source and a 3000-m streamer of 120 groups at 25-m spacing. Lines were shot at 70-m and interpolated to 35-m. 3-D processing consisted of relative amplitude scaling, zero-phase deconvolution, dip selective moveout, and finite difference 2-pass migration. Subsurface 3-D bins covered an area 12.5-m by 35-m.

Initial 3-D Mapping

A map was constructed from the 3-D data using wiggle trace variable area plots (Figure 9-7-4) and a Seiscrop table. Each line was reviewed, the faults tied, and the horizons of interest mapped. Although the area has fairly low dip, the 3-D data changed the fault pattern significantly from the 2-D map and also revealed large, undetected bright spot zones. As a result, a well proposal was considered on the 3-D line east–west through the abandoned #1 well to test the large bright spot downdip (Figure 9-7-4). The interpretation of the data suggested that the #1 well was faulted in the stratigraphic interval, and that the shaly gas sands of questionable productivity found in that well did not represent the overall stratigraphy of the downdip reservoir. Geologists and engineers reviewed the prospect with geophysicists and determined a location about 305 m (1000 ft) west of the #1 well. Gas was anticipated in a strong water-drive reservoir.

Interactive Workstation Review

Drilling could have proceeded at this point, but more time was allocated to review the 3-D data on an interactive workstation. The review showed some very important details that had bearing on the well proposal. Careful attention to color density representation of the lines (Figure 9-7-5) was possible with the workstation. The amplitudes of the troughs (negative acoustic impedance contrast) of the zero-phase data were displayed as increasing white-yellow-red, whereas the peaks (positive acoustic impedance contrast) were increasing black-green-blue. Variations in the amplitude of the bright spot could then be seen that were not visible in the amplitude-clipped black-and-white sections.

A simple review of the line containing the proposed #2 well showed it to be targeted at a slightly weaker amplitude spot (Figure 9-7-5). Past drilling experience in other areas had shown this to be a less than optimum site for development wells. The position of the #2 well on the 3-D structure map was at the north tip of a north-south fault, the throw of which could not be resolved by the 3-D data. A comprehensive review of the seismic data was in order.

Detailed Digital Structure and Amplitude

All of the lines in the bright spot area were reviewed in lines, crosslines, time slices, and in cubes (Figure 9-7-6). By picking the maximum amplitude of the first trough, a structure map was made on the top of the potential reservoir (Figure 9-7-7). The map of the bright spot reflects an updip termination against a fault, and several potential faults, within the anomaly. Structural variations along strike on the order of 15 m (50 ft) appeared to reflect potential compaction features, suggesting variations in sand distribution. After reviewing the lines in detail, it was determined that the potential reservoir was probably composed of overlapping sand units on a common water contact. Line 123 (Figure 9-7-8) showed a particularly good example of possible multiple sand lobes as opposed to a faulted uniform sand. The small faults mapped on black-and-white sections are thus more likely to be stratigraphic changes.

To better understand the distribution of the quality of the sands, the amplitudes of the reflecting interfaces were tracked, corrected for tuning effects, and represented in map form (Figure 9-7-9). In addition, a net pay map was generated (Figure 9-7-10). The procedures used were an extension of more established methods (see Chapter 7, and Brown et al., 1984). It is reasonable in this geologic area to expect the higher, detuned amplitude zones to represent thicker effective sands, and this is generally represented in the net pay map.

Revised Well Proposal

The resulting maps showed that the proposed #2 well lay in a very weak amplitude area relative to the rest of the major fault block. Net pay estimates here were not very favorable. A real danger existed that the well would be drilled in a shaly zone and would not effectively drain the potential reservoir. The connectivity of the sands appeared to be much better downdip, although considerable variation existed there as well. It was not simple to accept the movement of this potential gas well downdip. If the sands were to be uniform in distribution, the net recoverable would be less. Even so, after the seismic interpretation was considered along with other factors affecting the proposal, a new target was considered west (downdip) of the original proposed #2 (see Figure 9-7-10).

The #2 well was finally drilled after this extensive review, and the resulting well was a success (Figure 9-7-11). The #2 well was predicted to have much better sand characteristics than the #1 well and 17.7 m (58 ft) of effective pay. The #2 well actually logged 18.3 m (60 ft) of effective pay, and showed that the #1 well was in fact significantly faulted in its upper section of sands. Production tests confirm that the well is in a reasonably large and well-connected reservoir.

Future Reservoir Management

Additional wells may be considered in the future for this reservoir. The production characteristics of the #2 well may require more wells to drain all of the reserves effectively. Simple reservoir simulations are now possible using some of the data already generated from the geophysical workstation, especially the reservoir boundaries and net pay. Porosity values scaled to the #1 and #2 wells can also be generated from the amplitude data. To date, the reservoir is considered to be on a common water contact, but the apparent stratigraphic changes within the unit may potentially be permeability barriers (Figure 9-7-12).

Acknowledgments

The authors thank Chevron management for their support and permission to publish these data. Special recognition is intended for David Smith of Chevron, Eastern Region, who worked with the geologists, geophysicists, and engineers to help get the successful well drilled.

References

Brown, A. R., R. M. Wright, K. D. Burkart, and W. L. Abriel, 1984, Interactive seismic mapping of net producible gas sand in the Gulf of Mexico: Geophysics, v. 49, p. 686-714.

Figures begin on page 332.

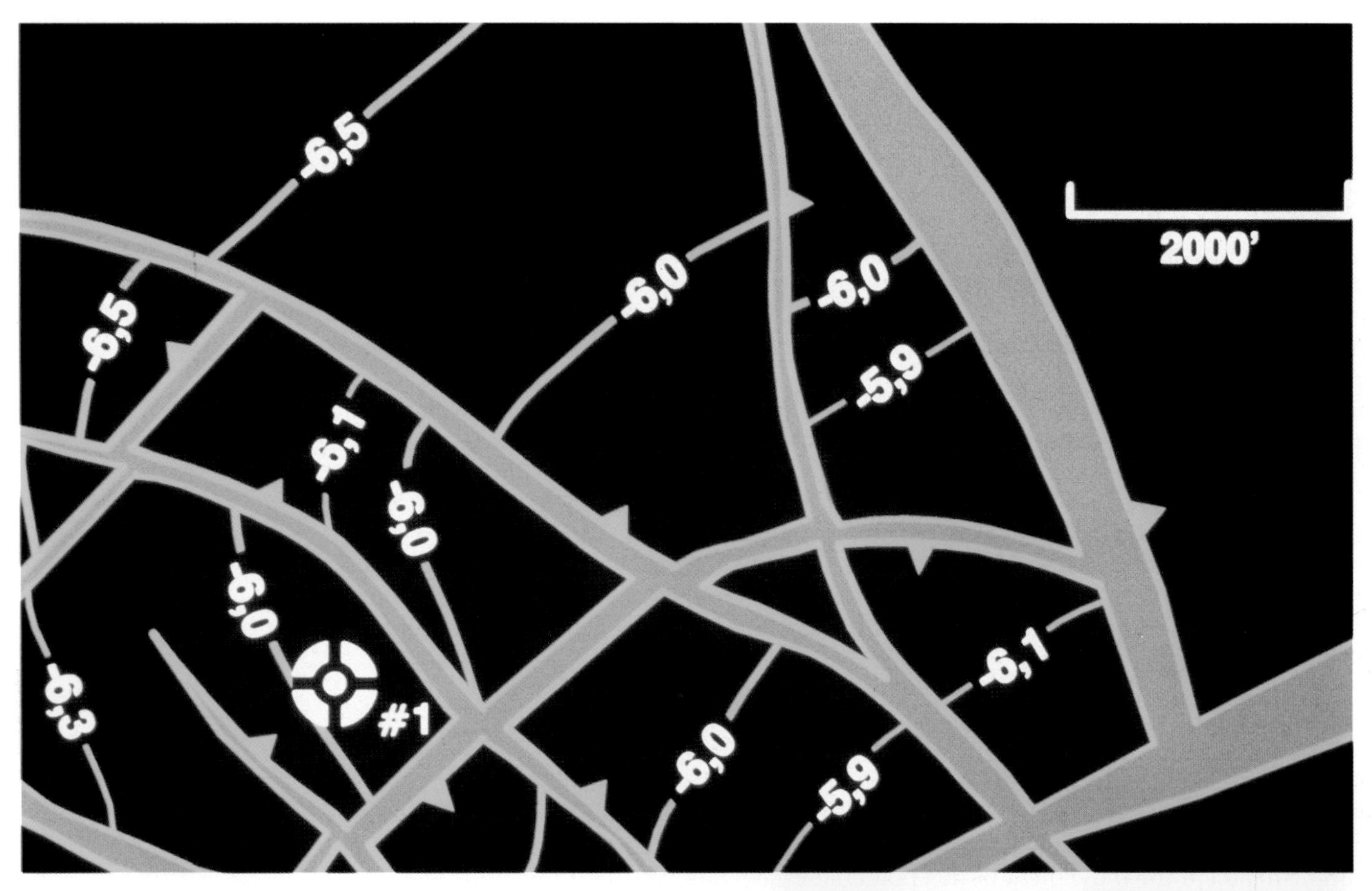

Fig. 9-7-1. Depth map of the prospect area based on 1000 ft 2-D seismic grid. Competitor well #1 was abandoned.

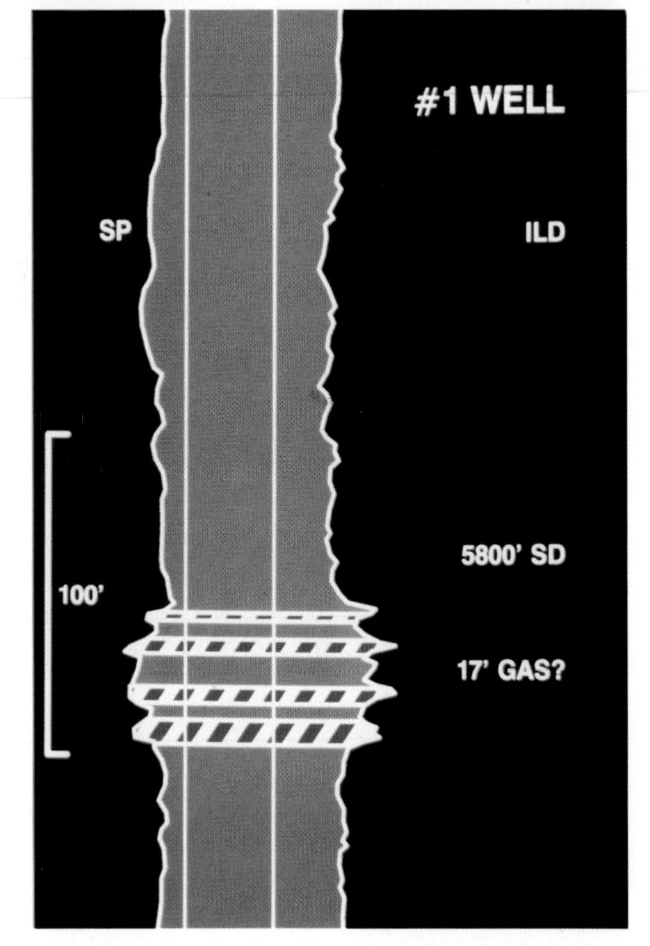

Fig. 9-7-2. E-log of abandoned competitor well #1. Sands contain gas in a few shaly stringers and are not considered economic.

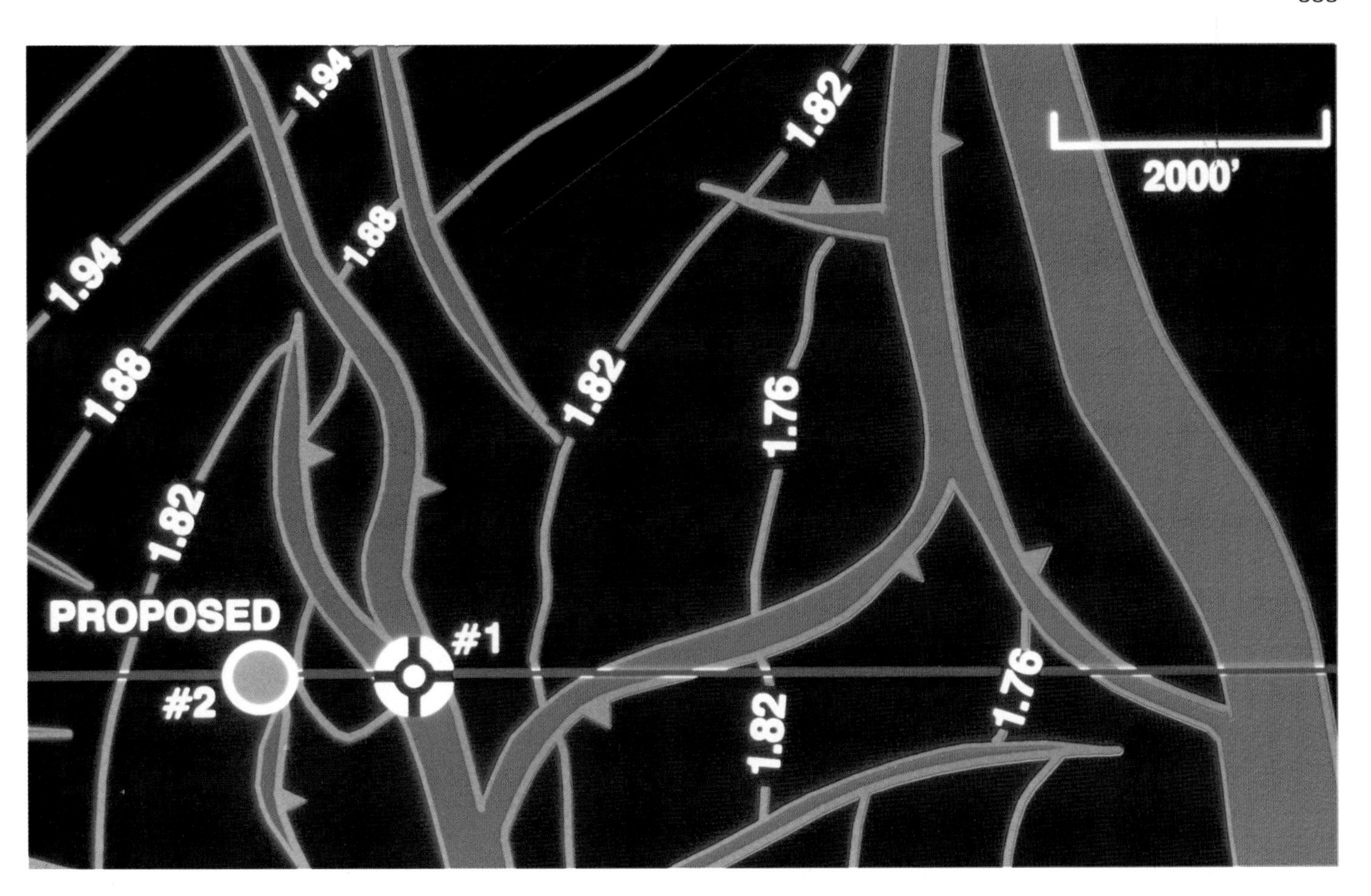

Fig. 9-7-3. Time map of the prospect area using 3-D sections and Seiscrop table. Note the significant change in fault positions, orientations, and throws as compared to 2-D map. Red line shows location of 3-D line 118 (Figure 9-7-4).

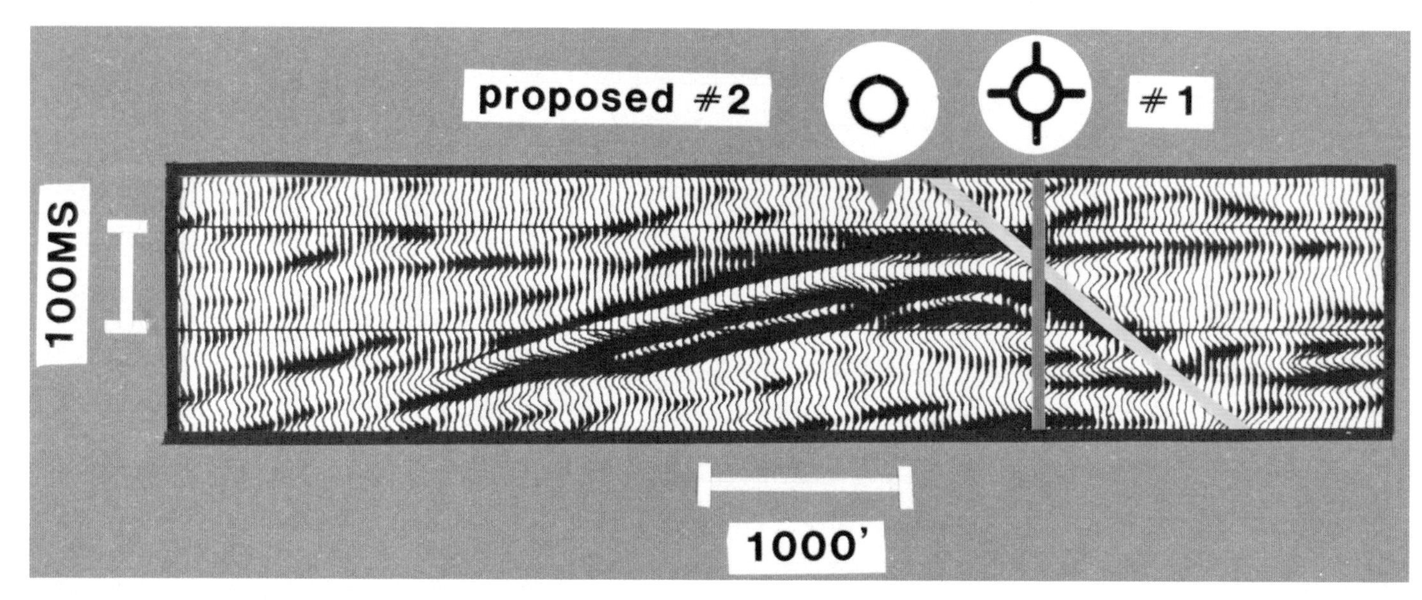

Fig. 9-7-4. 3-D seismic line 118, east-west through the #1 well. Data are zero phase with trough = negative acoustic impedance contrast. The bright spot can be drilled 1000 ft west of the faulted #1 well.

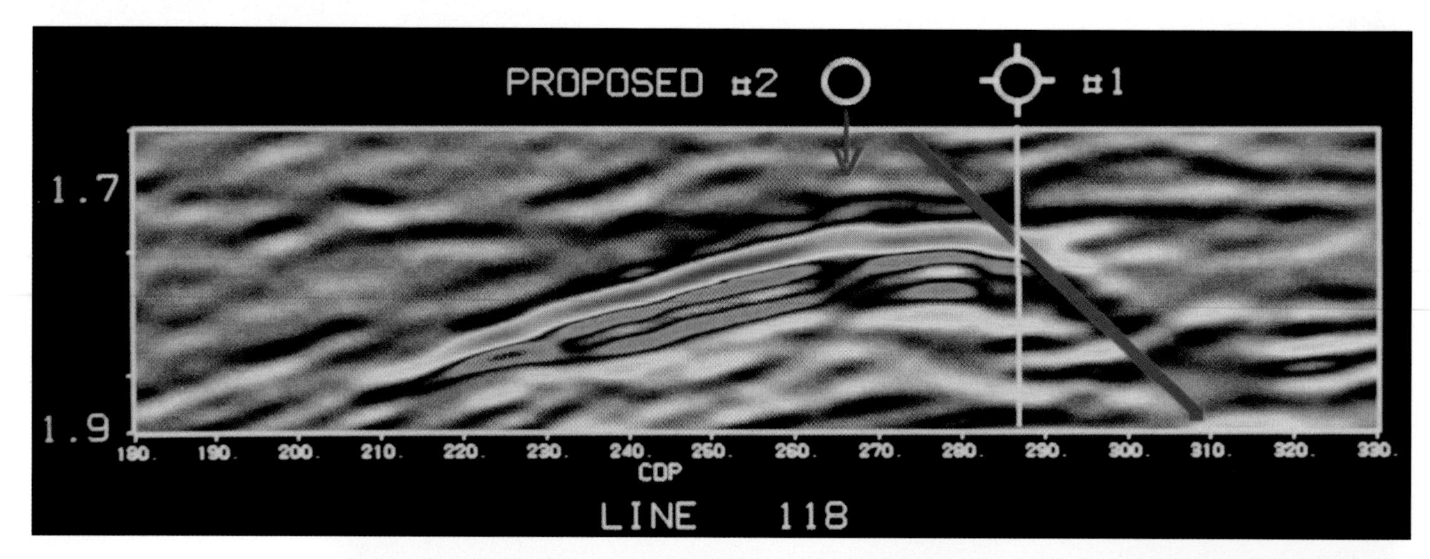

Fig. 9-7-5. Color display of line 118. Troughs are increasing amplitude white-yellow-red; peaks, black-green-blue. Note the detail not seen in clipped black-and-white data of Figure 9-7-4. Proposed #2 well is in a slightly weaker amplitude zone of the anomaly. This may not be an optimum drilling location.

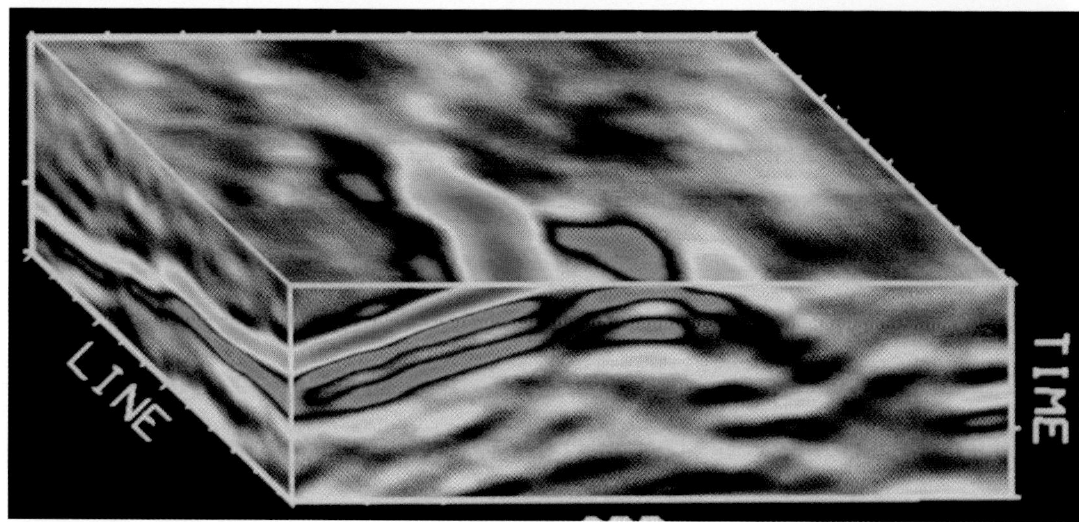

Fig. 9-7-6. Cube view of 3-D data. Front face is line 118.

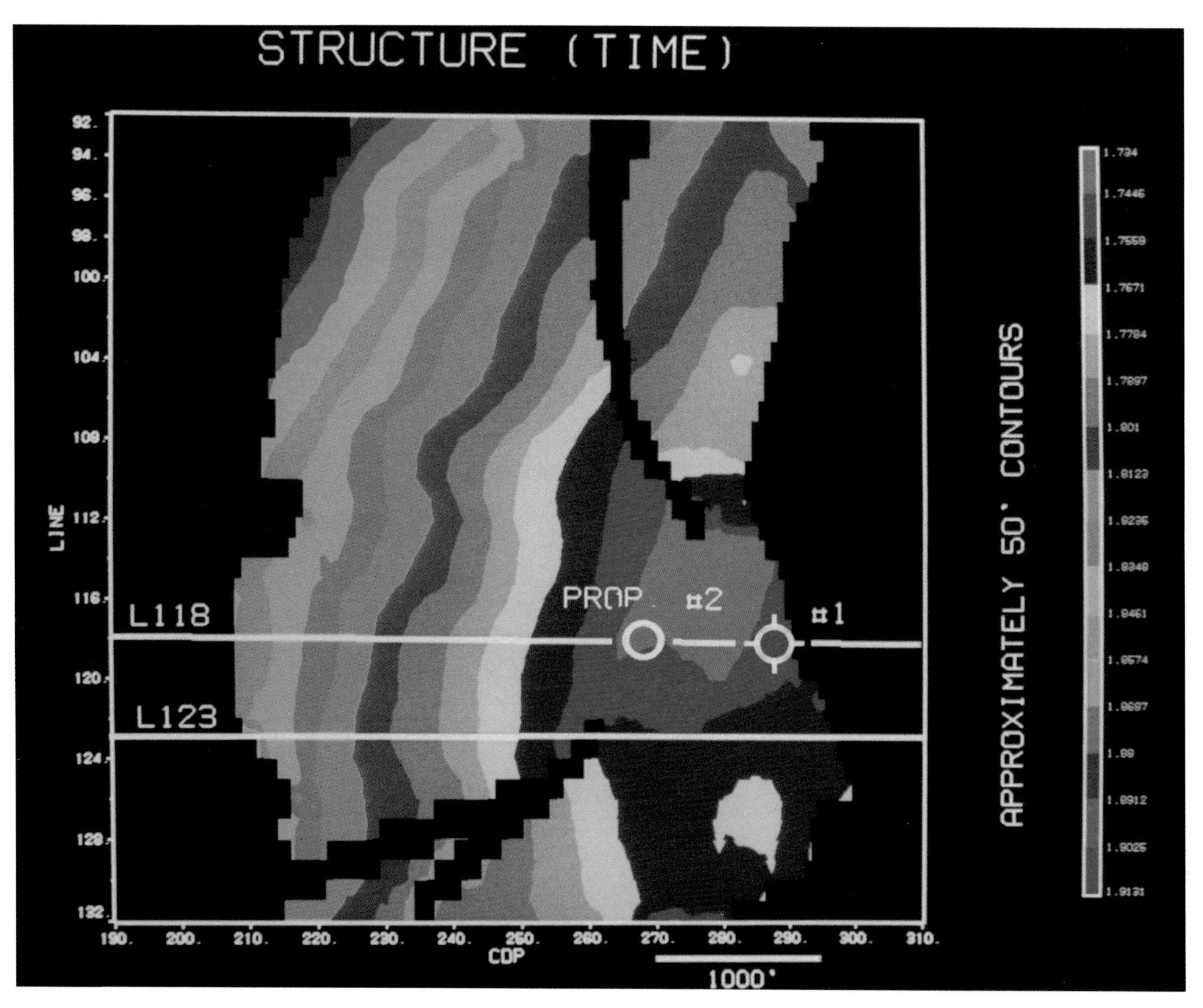

Fig. 9-7-7. Time structure map of top of bright spot based on interactive analysis, picking maximum amplitude. Color bands represent approximately 50-ft contours. Lines 118 and 123 are referenced (Figures 9-7-5 and 9-7-8). Structural breaks in the prospect map may be stratigraphic changes instead of faults. Subtle changes in the contours of up to 50 ft may be a response to differential compaction (see also Figure 9-7-12).

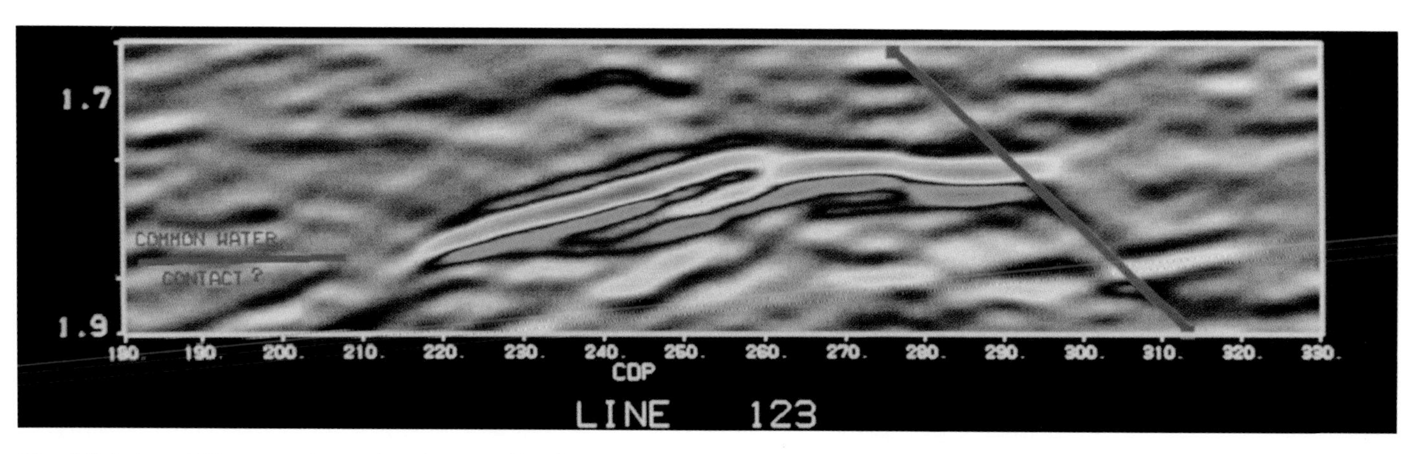

Fig. 9-7-8. Line 123 suggests overlapping sand bodies on a common water contact.

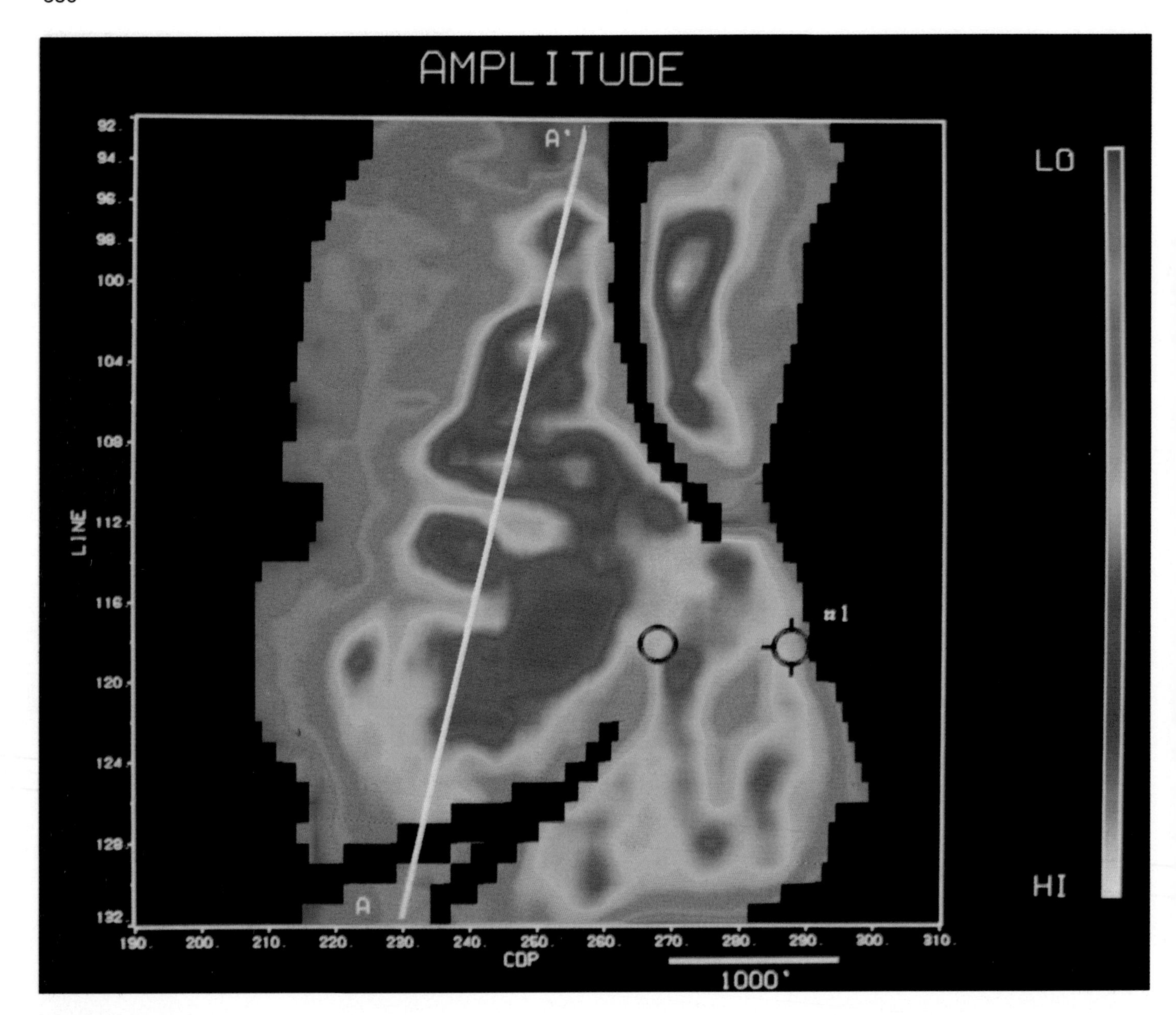

Fig. 9-7-9. Horizon slice showing detuned amplitude. Higher amplitude zones probably represent better sand development with higher porosity. Large variations suggest fast lateral stratigraphic changes. Note the #2 well proposed location is in a weak amplitude area. A well drilled farther downdip could be in a better position to produce. Cross section A-A′ is referenced to Figure 9-7-12.

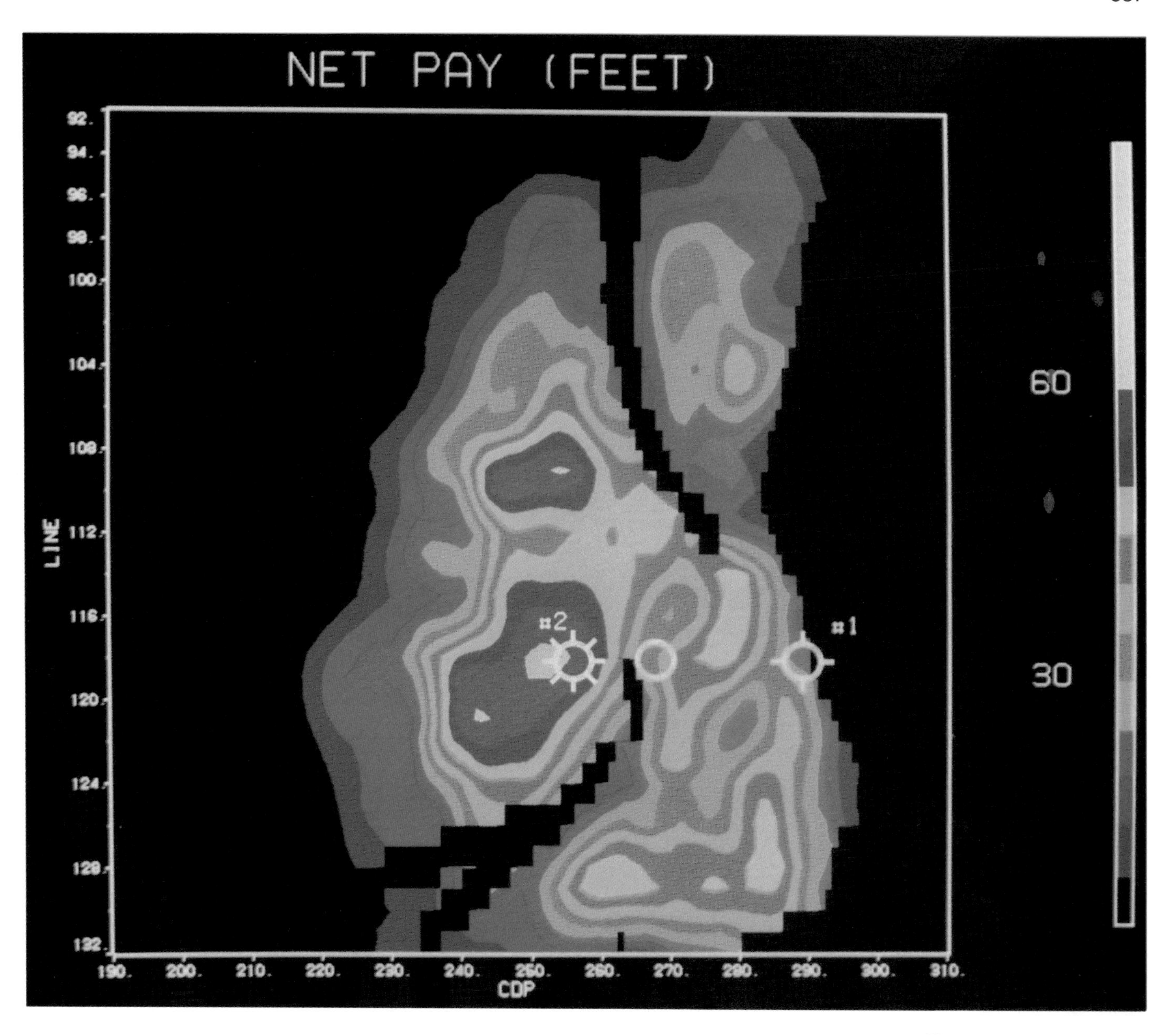

Fig. 9-7-10. Net pay map based on seismic data. Note the #2 well was drilled downdip of the original proposal. A prediction of 58 ft of net pay was confirmed in the #2 well (Figure 9-7-11), which drilled 60 ft of gas.

Fig. 9-7-11. E-log of the #2 well with 60 ft of pay. The stratigraphic interval can be correlated to the #1 well, but was mostly faulted there. Production tests show the well to be in a sizeable reservoir.

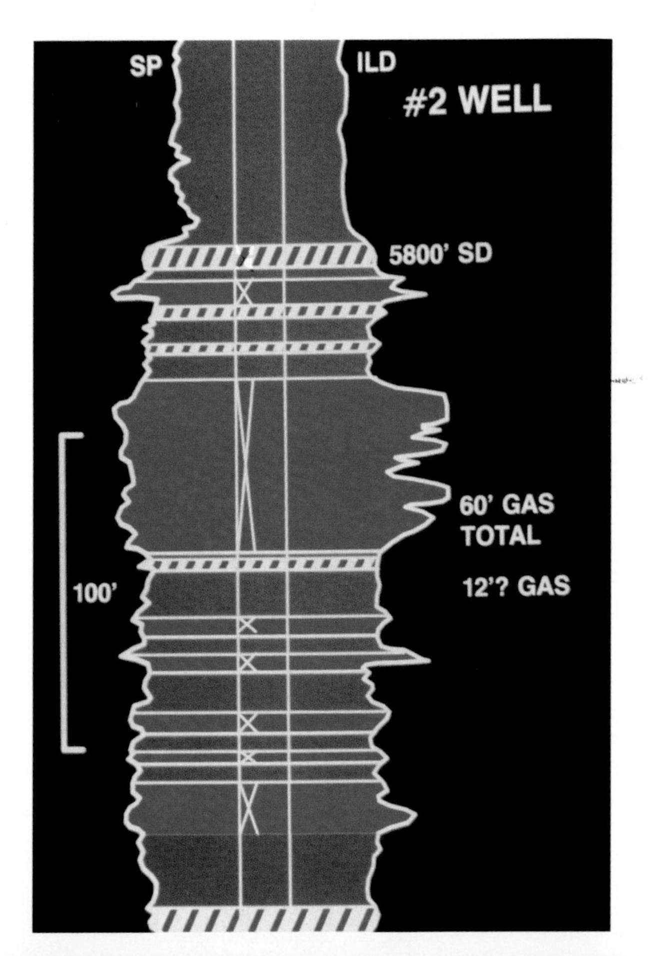

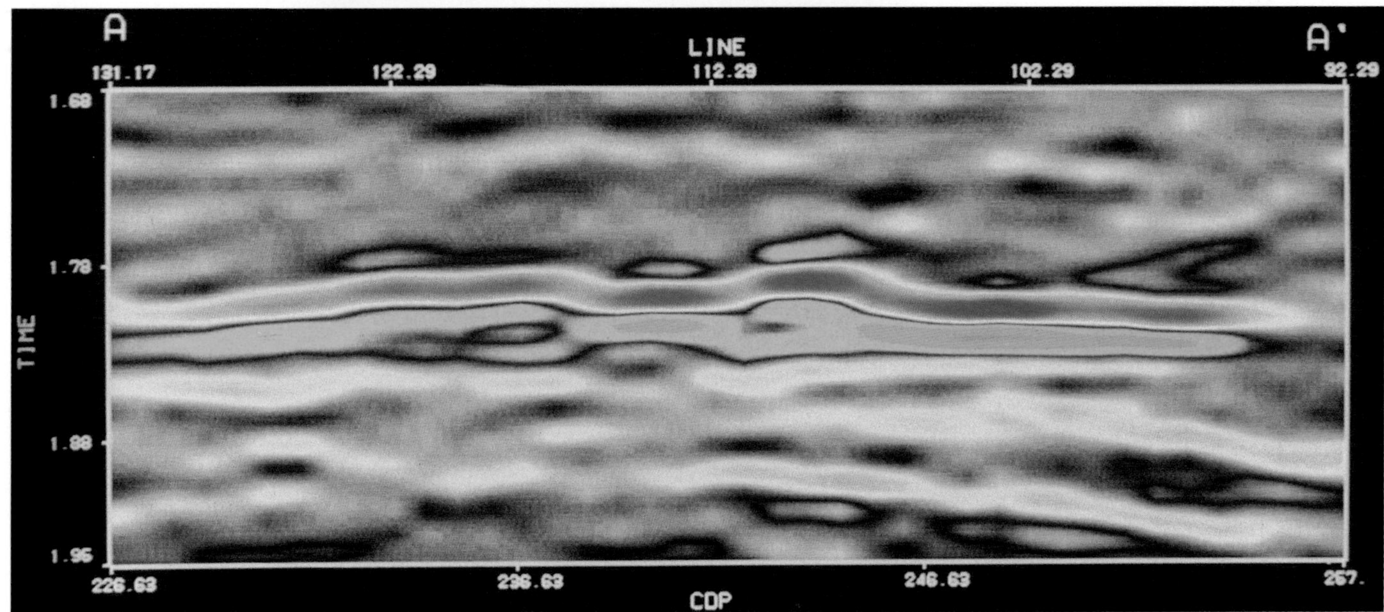

Fig. 9-7-12. Section A-A′ referenced to Figure 9-7-9. The amplitude and structure of the reservoir show rapid areal changes, but appear to be on a common water contact (note consistent flat spot). Apparent stratigraphic changes are potential production barriers.

Case History 8

A 3-D Reflection Seismic Survey Over the Dollarhide Field, Andrews County, Texas

Michael T. Reblin, Gregory G. Chapel, Unocal North American Oil & Gas Division; Steven L. Roche, Chuck Keller, Halliburton Geophysical Services

Introduction and Survey Planning

Because onshore 3-D surveys can be expensive, the results may not be considered cost-effective. This case history presents an onshore 3-D survey that was cost-effective and that shows the power of 3-D seismic versus well control and 2-D data.

Discovered in 1945, the Dollarhide field is a large faulted anticline in Andrews County which is located on the Central Basin Platform of west Texas. Production in this field is from the Permian *Clearfork*, *Devonian* Thirty-one, *Silurian* Wristen, and Ordovician *Ellenburger* formations. (The commonly used names for the reservoirs are in italics.) Well spacing is approximately 40 acres (16 ha) and the Devonian formation is currently undergoing CO_2 flooding.

In August and September 1988, a 3-D survey was acquired over a 24 mi^2 (62 km^2) area covering the Dollarhide field. The survey's primary purpose was to accurately image the location of faulting within and bounding the Devonian. This would aid in planning the CO_2 flood and possibly locate previously untested fault blocks.

Geophysicists from Unocal and Halliburton Geophysical Services (HGS) worked together to design the 3-D survey. Information (including depth of the main objective, velocity, maximum dip desired to be recorded, and reflection data quality) was compiled from previous 2-D seismic data and geologic data from well logs. Migration aperture, Vibroseis sweep bandwidth, source and receiver arrays, CMP fold, and offset geometry were all modeled and examined for optimum recording parameters. We determined that a subsurface bin size of 110 ft inline × 110 ft crossline would adequately sample the subsurface for processing through 3-D migration.

The high cost of land 3-D seismic surveys has been a deterrent to their use in both exploration and production geophysics. HGS suggested two innovations to reduce costs:

- Reduce the amount of data collected and replace it with trace interpolation prior to 3-D migration. Well control in the area reveals the general dip of the target horizon. Because the dip in the north-south direction is less steep than in the east-west direction, we could reduce the sampling in the former. We were able to use a subsurface sample interval of 110 ft (34 m) in the east-west (inline) direction and 330 ft in the north-south (crossline) direction, thus reducing by 66% the amount of data to be acquired. This also generated a further cost relief because the lessened number of receiver and vibrator lines meant fewer surface access permits to be obtained. And, economies were realized also in data processing because the number of records that had to go through CMP stack was reduced by 66%.
- Use two vibrators simultaneously to sweep two separate lines. This technique improved the productivity of the recording crews by approximately 70%. The separation of the two source signals is accomplished by upsweep-downsweep and phase rotation summing. Source separation is performed in the field during the correlation and sum processes. The isolation of the two sources using this method is on the order of 40 dB.

Data Collection and Processing

The data were acquired using a 384-channel DFS VII recording system deployed as a four-line swath. The receiver lines were spaced 1320 ft (400 m) apart with two source lines per swath (Figure 9-8-1). Each swath generated eight subsurface profiles, separated by 330 ft. After each swath, the spread was moved 2640 ft or 800 m (two cable lines) in the crossline direction. This geometry results in the subsurface swaths being adjacent as opposed to overlapping. This can be described also as "one fold crossline." With 12 swaths being recorded, the subsurface area is sampled 110 ft in the line and 330 ft in the crossline directions. The source interval averaged 440 ft (130 m). The resulting effective fold is 18-24 when source-to-receiver offsets are considered relative to the depth of interest.

Data processing techniques included: geometry description; field record quality control; surface-consistent deconvolution; preliminary stack; velocity analysis; residual static estimation;

3-D *f-k* DMO; stack; trace interpolation; and 3-D migration.

At several steps during the processing of this survey, different parameters were tested and reviewed—including the deconvolution method, benefit of DMO, migration velocity analysis, and poststack migration algorithm. A benefit was realized by including DMO in the processing sequence in that the diffracted image of the subsurface was improved. This enabled the trace interpolation algorithm to perform better in the conversion of 110 ft × 330 ft subsurface bins to 110 ft × 110 ft bins. After 3-D migration, the data volume was moved to a workstation for interactive interpretation.

Interpretation and Results

The results of this 3-D survey are impressive. Figure 9-8-2 shows the structure map of the Devonian at Dollarhide field as determined by the 40-acre-spaced well control. This map has gone through many evolutions in the 46 years since the field was discovered. Notice that the contours are relatively smooth, the anticline is cut by four simple cross-faults and bounded on the east by a fault.

The structure map from the 3-D seismic survey (Figure 9-8-3) is more complex. The contouring is more detailed and the cross-faults are not simple. The structure map shows the detail of the Devonian that the 3-D seismic has allowed us to see. This shouldn't be a surprise as our seismic data points are equivalent to a spacing of approximately four wells per acre. Considering that a seismic trace is an approximation to a synthetic seismogram from a sonic log, we indeed have a very powerful means of detail mapping subsurface structure.

There are two principal ways to look at the 3-D seismic data volume. One is the conventional seismic line display (Figure 9-8-4). On the crest of the structure, the top of the Clearfork Formation is the strong event at approximately 780 ms. The Devonian, at approximately 1000 ms on the upthrown block and 1350 ms on the downthrown block, is colored purple. The top of the Ellenburger is a high-amplitude event at approximately 1250 ms. At about 960 ms, an unconformity can be seen that helps highlight one of the more remarkable features of the data—a fault zone showing over 2000 ft (610 m) of displacement on the Devonian marker. The imaging of this fault zone demonstrates one of the shortcomings of some 3-D surveys. Due to economics, lines may not be long enough to properly image all the features (such as large faults or extremely steep dip) within the survey limits. This survey was designed to image the upthrown block so that the incomplete image east of the major fault was as expected.

The other view of the 3-D data volume, and one not available with 2-D data, is the time slice. This view allows the interpreter to see subtle features which may not be apparent or as readily interpretable on conventional seismic sections. A time slice (Figure 9-8-5) through the 3-D data volume at 1008 ms (about 4600 ft subsea or 7800 ft below surface) demonstrates this. The cross-faults are seen as northeast-southwest lineations. The previously undetected grabens not seen on Figure 9-8-2 are seen as easterly pullouts on the time-sliced peaks (blue) and troughs (red) in

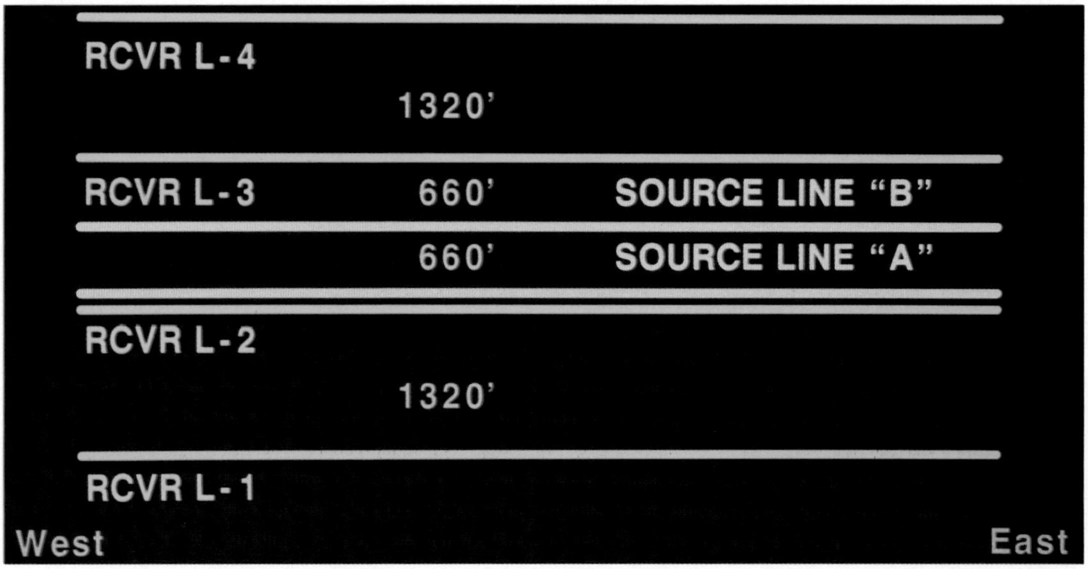

Fig. 9-8-1. Swath design of the 3-D dual source survey.

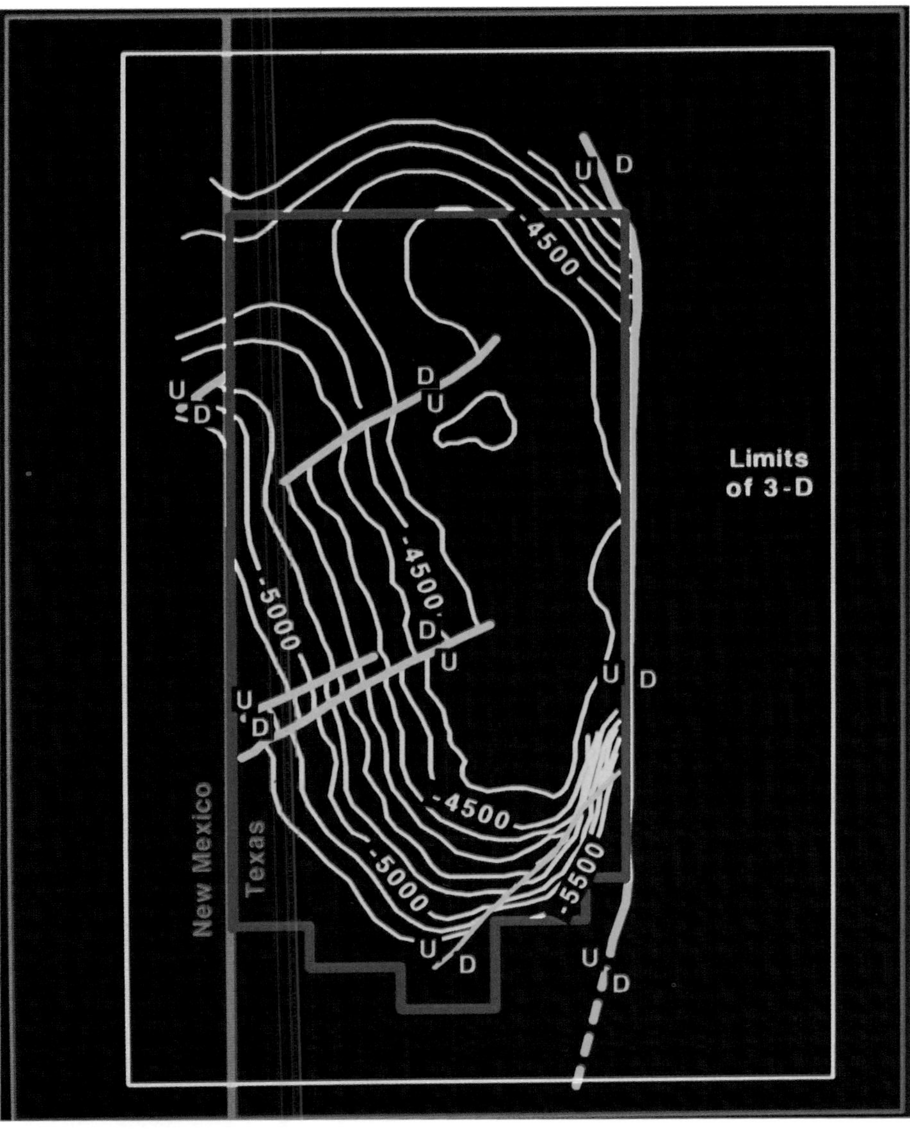

Fig. 9-8-2. Simplified Devonian structure map from 40-acre (16-ha) well control. The unit outline is red.

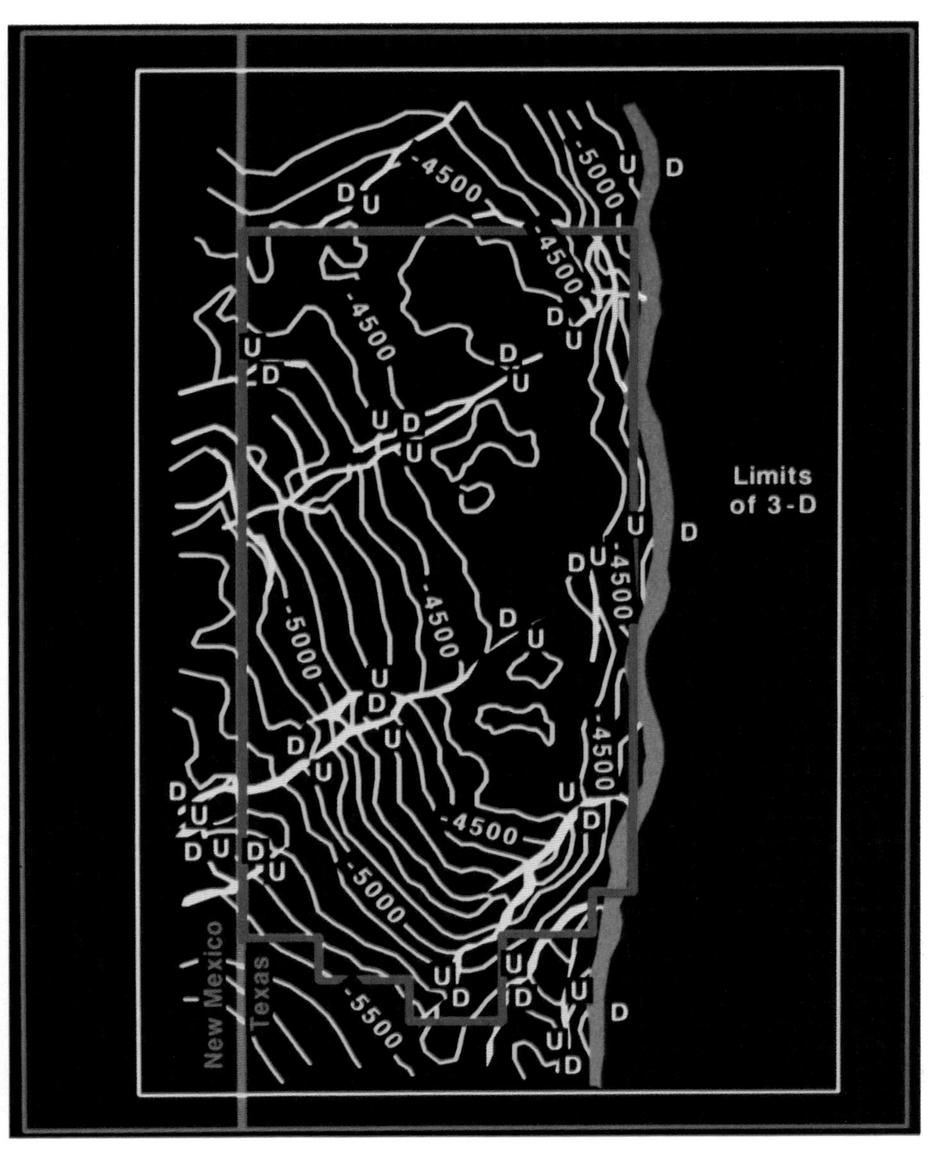

Fig. 9-8-3. Simplified Devonian structure map from 3-D seismic interpretation.

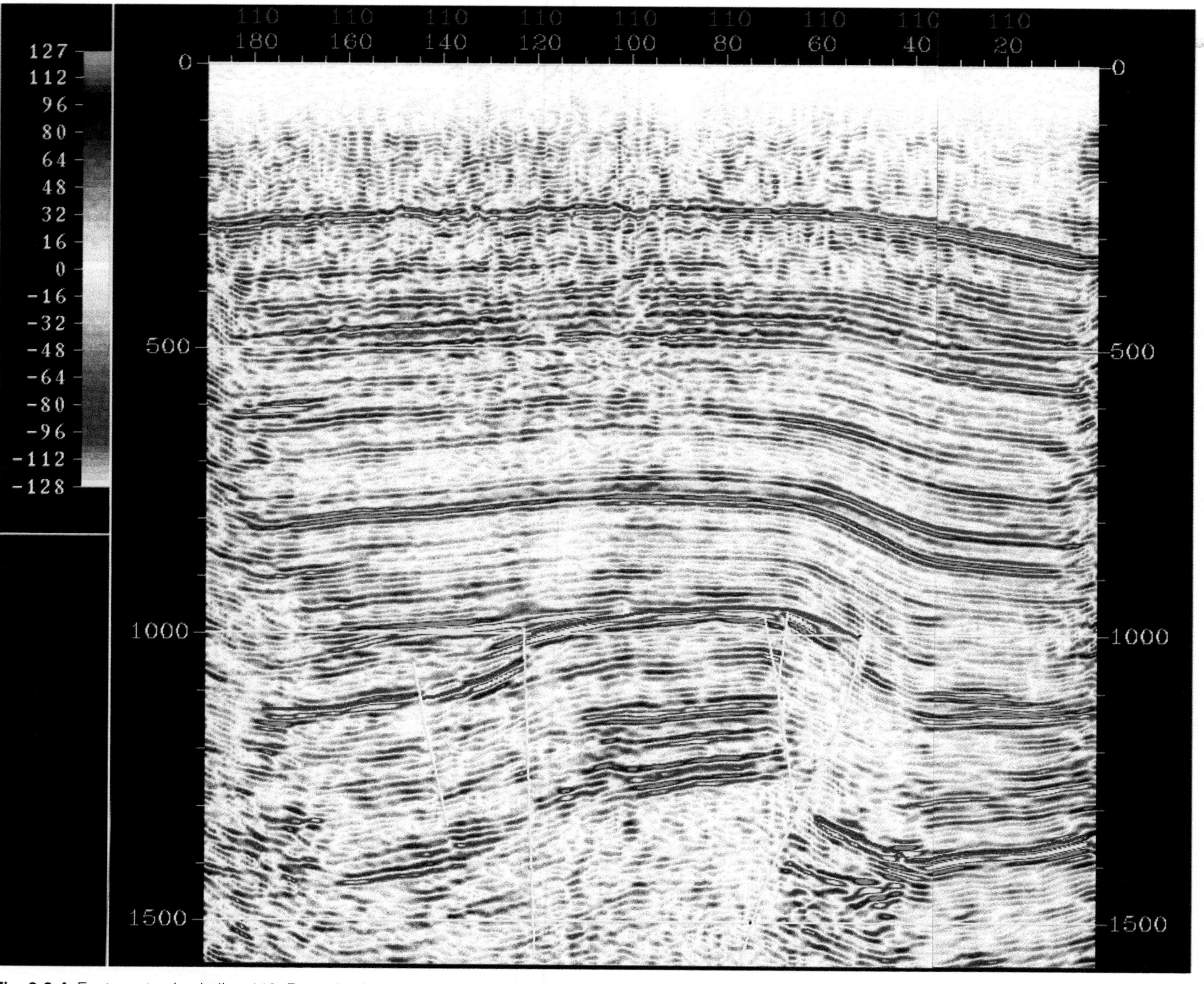

Fig. 9-8-4. East-west seismic line 110. Devonian horizon is annotated in purple; the interpreted faults are yellow.

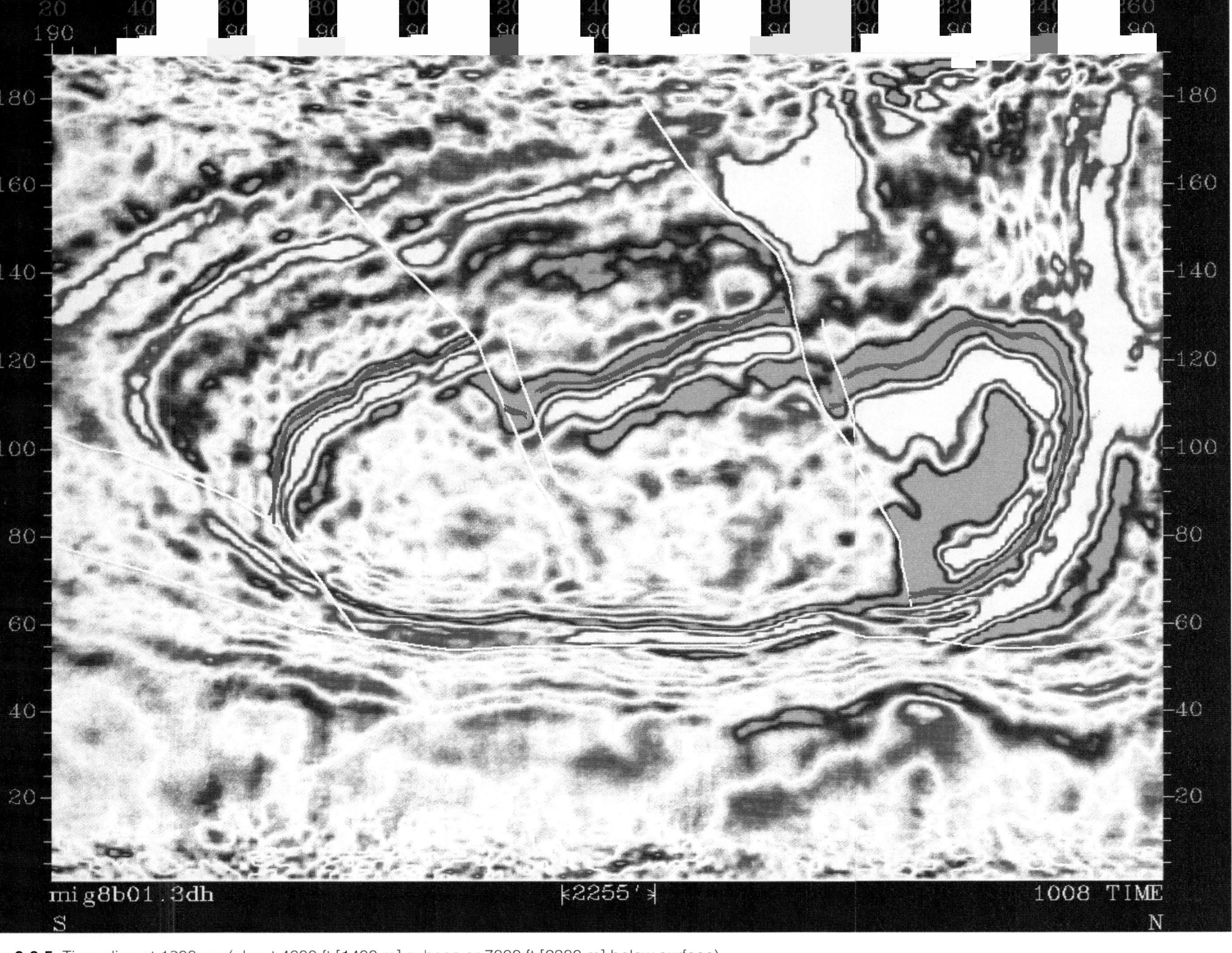

Fig. 9-8-5. Time slice at 1008 ms (about 4600 ft [1400 m] subsea or 7800 ft [2380 m] below surface). Devonian horizon is annotated in purple; the interpreted faults are yellow.

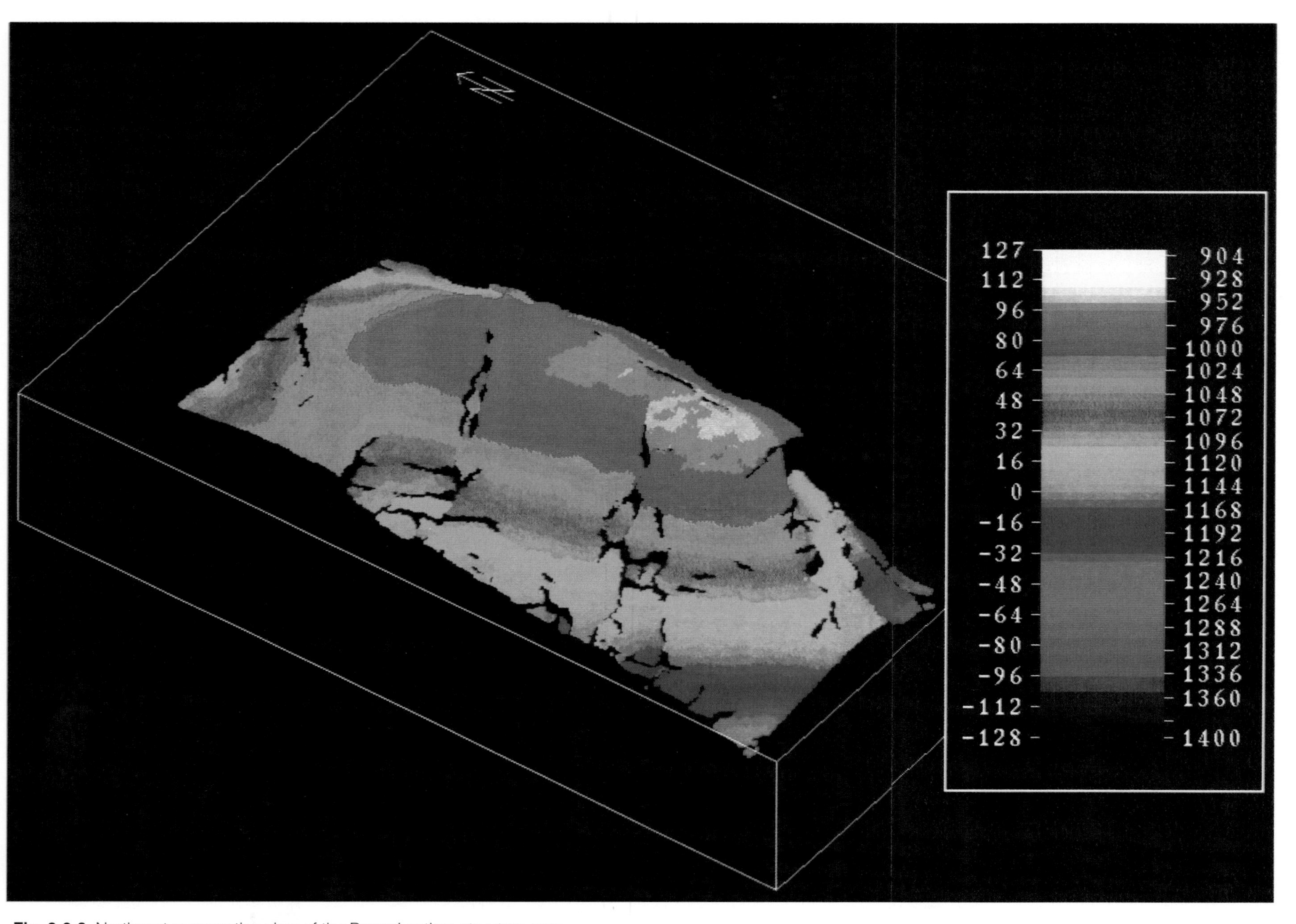

Fig. 9-8-6. Northeast perspective view of the Devonian time structure map. The color change from red to green is at the same time as the time slice in Figure 9-8-5. Note the cross-fault definition.

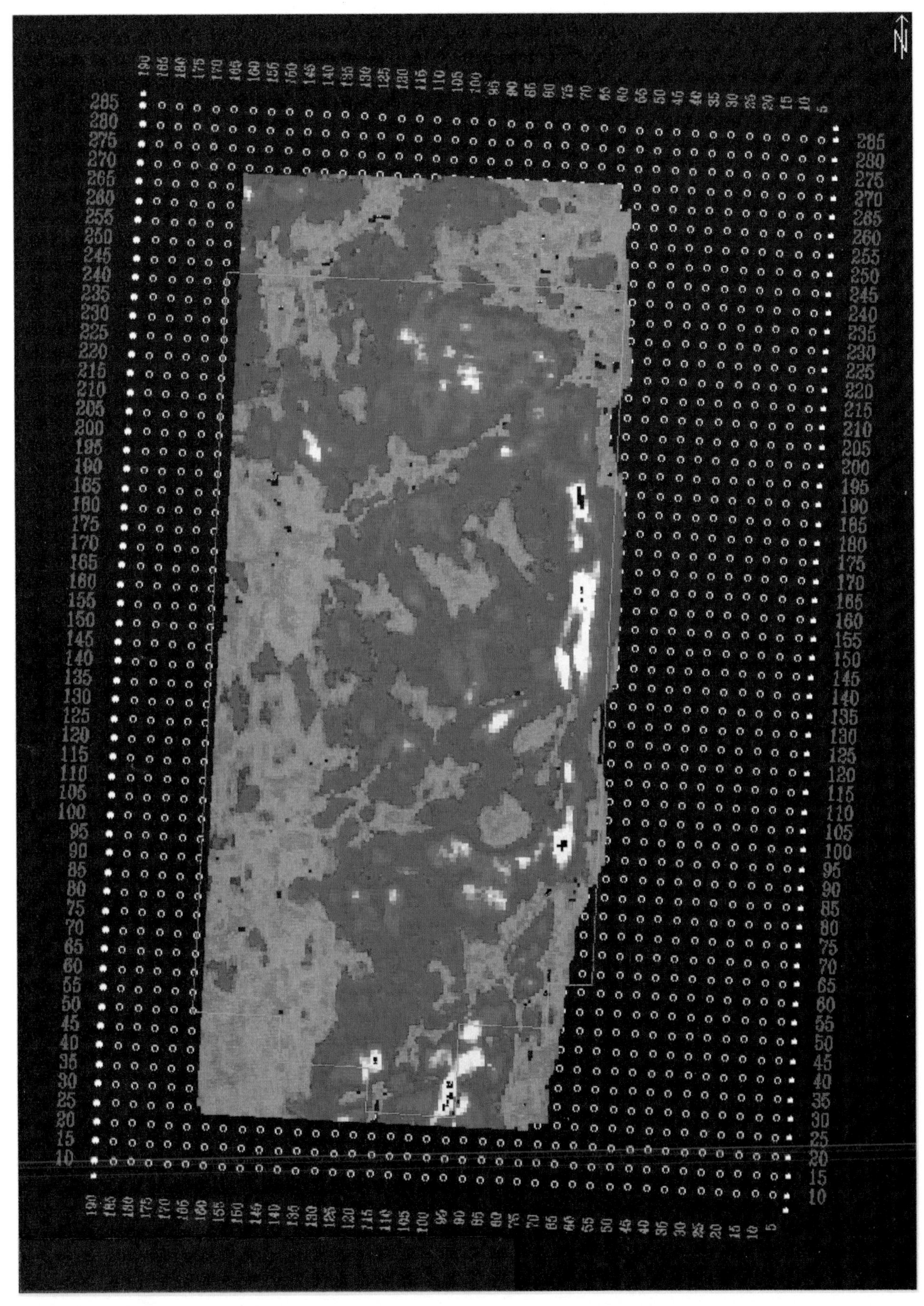

Fig. 9-8-7. Composite horizon slice (seismic amplitude map) of the producing Devonian horizon. Largest amplitudes are yellow and red and smallest amplitudes are blue and green.

the areas of the cross-faulting. Mapping of the data is now possible in both the vertical and horizontal sense. Both offer unique perspectives of the data volume.

Because the 3-D survey gives an evenly sampled volume of data, another display of the data is possible after a horizon is interpreted. In Figure 9-8-6, a perspective view of the Devonian horizon time map as viewed from the southwest is presented. It shows the northerly plunge of the anticline, which isn't readily apparent on the Devonian horizon structure map (Figure 9-8-3). The cross-faults, with their associated grabens, are quite distinct and give a real feel for the relative throw of faults. This display has helped the geologists and engineers develop a better understanding of the field shape and how the faults impact the ongoing CO_2 flood project.

Having the 3-D data volume loaded on an interactive workstation allowed the interpreters to generate various attribute displays that took us beyond the traditional time interpretation. Using the seismic peak and trough associated with the producing Devonian horizon, a composite horizon slice (amplitude map) was made (Figure 9-8-7). The hot colors, yellow and red, represent larger amplitudes and, in most cases, correspond with the better wells in the field. The cool colors, blue and green, represent lower seismic trace amplitudes along the producing Devonian. The amplitudes are interpreted to be related to the thickness of the producing zone—high amplitudes to a thick zone, low amplitudes to a thin zone. The possible exception is the linear pattern next to the north-south bounding fault where we believe the larger amplitudes are possibly related to poorly imaged steep dips. The major cross-faults are seen as northeast-southwest lineations, which divide the structure into four major fault blocks. The CO_2 flood is expected to have better results in the northern block which is denoted by the higher amplitudes. The flood was initiated in the northern block last year. The next block south has had the poorest flood results to date which seem related to the dominance of the lower amplitudes on the composite amplitude map. The third fault block was the first one flooded and has the best results to date, as could be predicted from the abundance of high amplitudes. The smallest fault block, located to the southeast, is faulted below the producing limits of the field to the north. However, a well drilled to the productive Devonian horizon in 1948 recently has been reentered and reevaluated, and could open up an extension to the field. The higher amplitudes indicate it could be a very productive block with good CO_2 flood potential.

Conclusions

Earlier in this case history, we alluded to the cost-effectiveness of this 3-D survey. One of the features that helped us sell the concept to management was the comparison of the cost to shoot this 3-D survey to a 1/2 mi grid of 2-D data and the dryhole cost of a Devonian test. The 2-D survey cost (including acquisition, surface permits, and processing) was estimated at $750,000 for 150 line-mi. The dryhole cost of a Devonian test is approximately $300,000. To date, two Devonian locations have not been drilled as the 3-D results indicated they were uneconomic. Shooting conventional swath 3-D to record 110 ft × 110 ft bins was estimated at about $1,300,000 (generating 1140 mi (1820 km) of 3-D data over the 24 mi^2). Using the 3-1 interpolation technique and simultaneous source recording, the survey actually cost $400,000. This cost is approximately half that for a 2-D survey, a third of a conventional 3-D survey, and only slightly more than a dry hole. The after-tax profit of a primary Devonian development well in the Dollarhide field is about $1,000,000. By adding one well to the field, we easily recover the cost of the survey plus give the geologists and engineers a more detailed look at a reservoir that is still being developed during the tertiary recovery stage.

The results to date are multifold. The cross-faulting of the Devonian producing horizon is much more extensive than previously mapped. This knowledge has influenced the location of several wells for the CO_2 flood and the engineers continue to use the results for future programs. Some of the newly discovered faults have generated fault traps within the field that have not been drilled and are now being evaluated to determine their potential. Evidence suggests that the fault block to the southeast may be productive, although it was drilled and abandoned over 40 years ago. Lastly, preliminary studies of the Clearfork Formation indicate the 3-D data will help in the development of the plan for the secondary recovery from that producing unit.

Land 3-D surveys can be economical and may produce results well beyond the initial goals. The two acquisition techniques discussed here are just two examples of how to shoot cost-effective land 3-D surveys. The 3-D seismic is a necessary tool to use in developing new discoveries and extending the life of old fields.

Shallow 3-D Seismic and a 3-D Borehole Profile at Ekofisk Field

John A. Dangerfield, Phillips Petroleum Company Norway

Ekofisk field in the Norwegian North Sea was discovered in 1969 but after more than 20 years of production much remains to be understood (Sulak, 1990). This case history describes some of the 3-D seismic work carried out at Ekofisk field by the Phillips Licence 018 Group of Companies: a short offset 3-D to image the very shallow layers, a normal surface seismic 3-D especially processed for shallow data, and a 3-D borehole profile, shot to image the crest of the field beneath a gas cloud.

Figure 1 shows the areas covered by the various 3-D surveys. It also shows the reservoir area that is obscured from normal seismic view by gas in the overlying sediments.

Acquisition and Processing of the Short Offset 3-D

This survey was acquired in 1987 to delineate faults close to the Ekofisk Complex that might be reactivated by the (now) 5 m subsidence of the field (Wiborg and Jewhurst, 1986). The 4 × 2 km survey was collected with flip-flop acquisition, using single water guns as the sources, and recording 1.5 s of data. The twin hydrophone cables each contained 48 five m groups. Reflection lines were 10 m apart. The processing was standard except for the extremely careful checking of recording geometry and effective time zero. One velocity function was used for the entire survey. A complete description is presented by Dangerfield (1991).

Acquisition and Processing of the Normal Surface Seismic

The acquisition in 1989 was aimed at all possible levels from sea bottom to 7 s. The 9 × 17 km area was shot using flip-flop, twin sources of 3680 in.^{3} air guns, and twin hydrophone cables, collecting reflection lines that were 25 m apart.

The data from the uppermost 2 s were processed separately in order to obtain the highest possible resolution. They were processed with particular care in muting and velocity analysis. Surprisingly, one velocity function was optimal for almost the complete area.

Acquisition and Processing of the 3-D Borehole Profile

Imaging the crestal area below the gas never has been found possible with normal seismic methods but borehole profiles generally have worked extremely well by undershooting the gas. The borehole profiles, however, have revealed many faults that can be linked in many different patterns. The advantages of 3-D data sets for fault interpretation are so strong that we took the first opportunity to acquire a 3-D borehole profile in the gas-obscured area.

In July 1989, a series of 41 walkaway lines was shot into an eight-geophone array in the deviated well 2/4 K17. Each single walkaway line thus recorded a swath of eight roughly parallel reflection lines. The geophones in the array each were separated by 15 m. The hole deviation was about 45° so the geophone array spanned about 80 m horizontally. The 41 walkaway lines were shot in a regular grid (Figure 9-9-2) with the geophone array pulled up the hole such that each walkaway line passed over the middle of the geophone array. The walkaway lines were 5 km long and perpendicular to the well. Successive lines were 40 m apart. The acquisition area was limited by the presence of the Ekofisk Complex.

The acquisition resulted in a total of 328 reflection lines. Each line was processed separately to the velocity filter stage using standard walkaway techniques; then a one-pass migration of the whole data set was run, accomplishing both NMO and 3-D migration. Unfortunately, the deepest 15 levels were so severely distorted by the gas that it was not useful to migrate these data. The input to the migration was restricted to data from geophones 3 to 6 in each array, overlapping geophones not being used. Crossline migration was performed using a sliding window of five geophone levels. The results were output as a regular 10 × 10 m grid.

Interpretation of the Short Offset Data

The excellent resolution of the data, from the sea bottom at 100 ms down to 1 km depth, showed that there were no faults with throws of more than 2 m in the region examined.

Figures 9-9-3 and 9-9-4 show two of the time slices indicating clear subglacier river deposits (Dangerfield, 1991). Figure 9-9-3 shows the presence of a break of slope, running roughly north-northeast–south-southwest where the rivers abruptly change character as shown in the following figures.

Interpretation of the Normal 3-D Data

We had expected that the results of the shallow data would be compromised severely by the areal size of the source, 17 × 20 m, and by the flip-flop acquisition, necessitating 50 m between shots in each reflection line. In practice, the results turned out extremely well. The water bottom showed little detail but 40 ms deeper, and down to 2 s, sedimentary and structural features showed clearly, setting the data from the earlier survey into a more interesting context.

Figure 9-9-5 is a time slice from almost the same time as Figure 9-9-4 but encompassing a much larger area. It shows several rivers and streams carrying glacial meltwater. At this period the ice above is believed to have been about 3 km thick. The area southwest of the Complex was fairly flat as evidenced by the complicated channeling there. The regular steps every 900 m or so in the major river north of the Complex may be due to the water channel flowing along the edges of blocks that were slightly tilted, like giant, uneven paving slabs.

The presence of two sets of nearly vertical fractures set at 90° to each other in the overburden is portrayed dramatically in the "arrowhead" appearance in Figure 9-9-6. This pattern repeats many times throughout the data, on different scales but with similar orientations, and indicates a pervasive fracture system. The orientation of one set is subparallel to that of the break of slope which controlled the river in Figure 9-9-3. Reactivation of the fracture system appears to control surface features during deposition and probably also produced the blocks suggested in Figure 9-9-5. The presence of faults displacing the "arrowhead" feature are very clearly

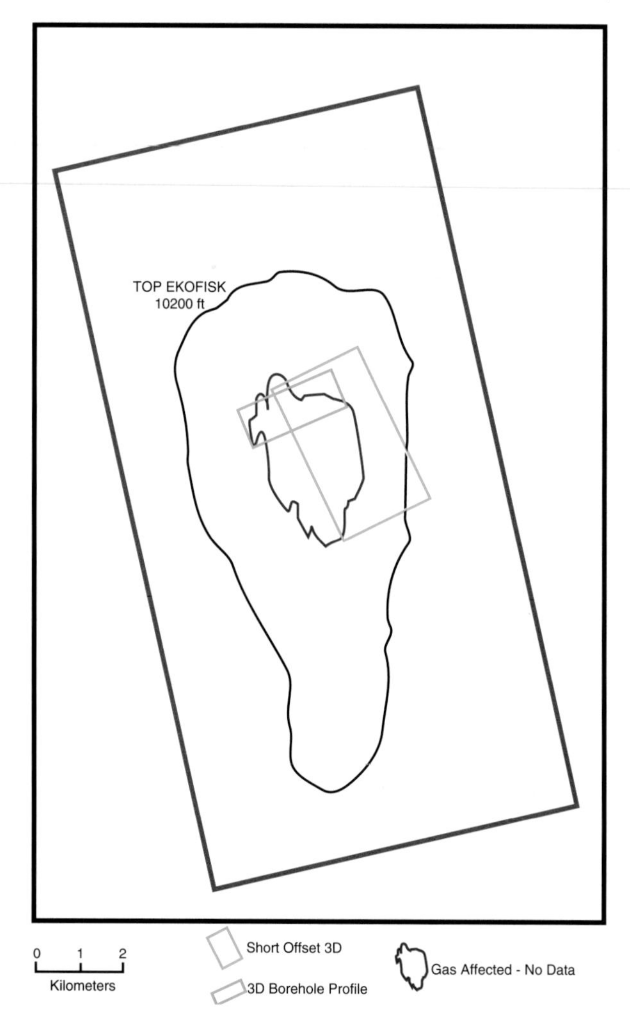

Fig. 9-9-1. 3-D seismic areas in relation to the field outline and the gas-affected area. The field is outlined in black and the full scale 3-D area in red.

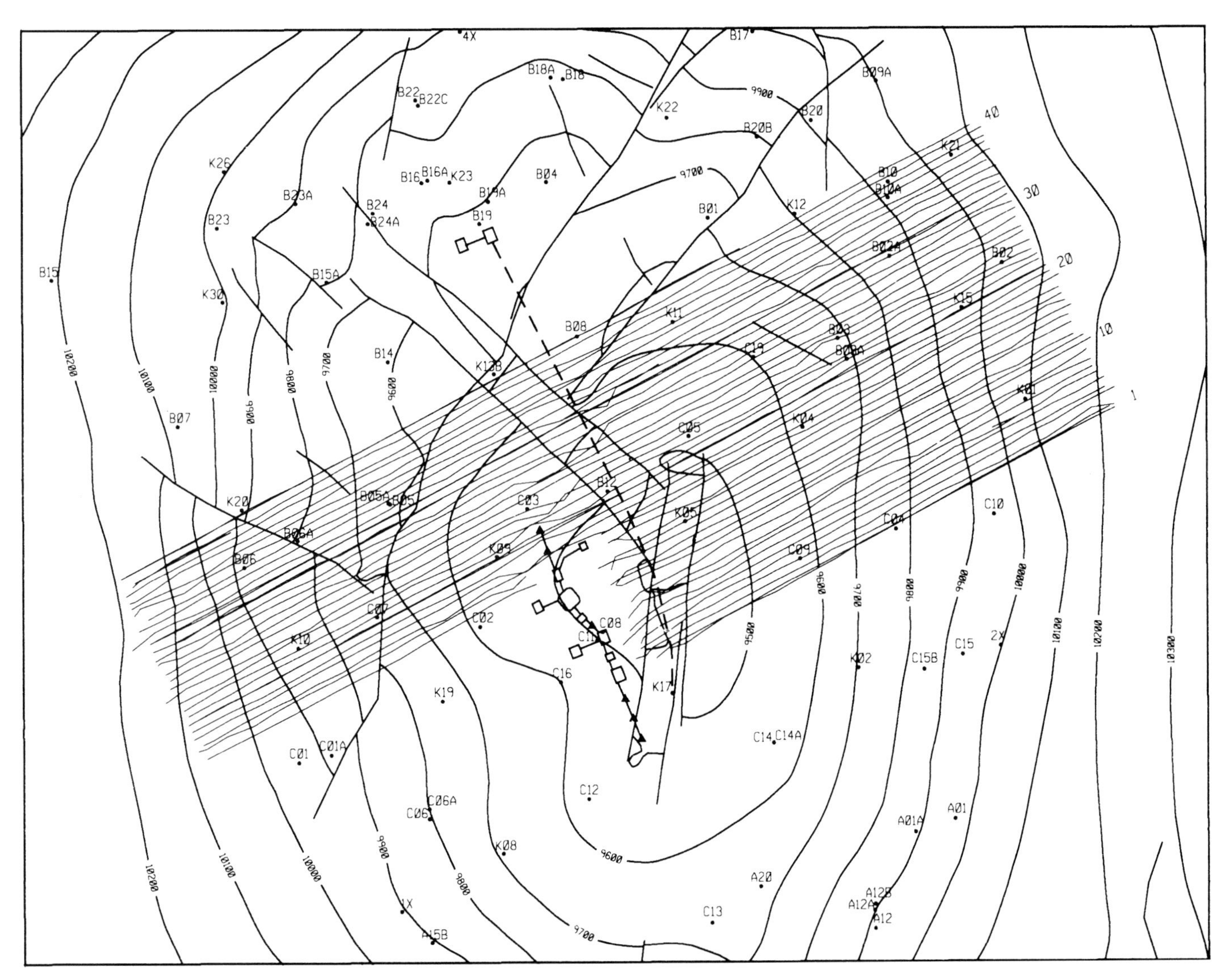

Fig. 9-9-2. 3-D borehole profile acquisition lines superimposed on the 1989 top reservoir depth map.

displayed in Figure 9-9-7. The fault displacements are of the order of 3 m.

Although only a brief selection of time slices from the shallower layers have been shown, the complete 3-D data sets form continuous series, with slight but distinct differences in successive slices, so that the evolution of the sedimentary features can be followed on a meter by meter basis.

Interpretation of the 3-D Borehole Profile

The resulting borehole profile in the northern half of the data set showed a clear image in an area where our 3-D surface seismic failed completely. Figures 9-9-8, 9-9-9, and 9-9-10 locate and compare line 40 from the 3-D borehole profile with a 3-D surface seismic line from the same place. Figure 9-9-11 shows line 16 (east-west) and crossline 150 (north-south) with the top Ekofisk Formation interpreted. The data showed the presence of many faults too small to map but with similar orientations to those of the overburden fracture system. It also showed the continuation of a crestal graben originally found in the gas-affected area by walkaway profiling (Christie and Dangerfield, 1987) and subsequently penetrated by drilling. Figure 9-9-12 shows a time slice through the reservoir with the top Ekofisk marked in green. The graben clearly displaces the Ekofisk horizon.

Conclusions

1) The normal 3-D shallow data set showed the fracture system that pervades the Tertiary and controlled some surface features during deposition.

2) High quality data close to the water bottom is readily available in the shallow section in normal 3-D data sets. This suggests that an important part of the work currently carried out by site survey vessels is accomplished in the course of a normal 3-D survey.

Fig. 9-9-3. Short offset survey time slice from 191 ms.

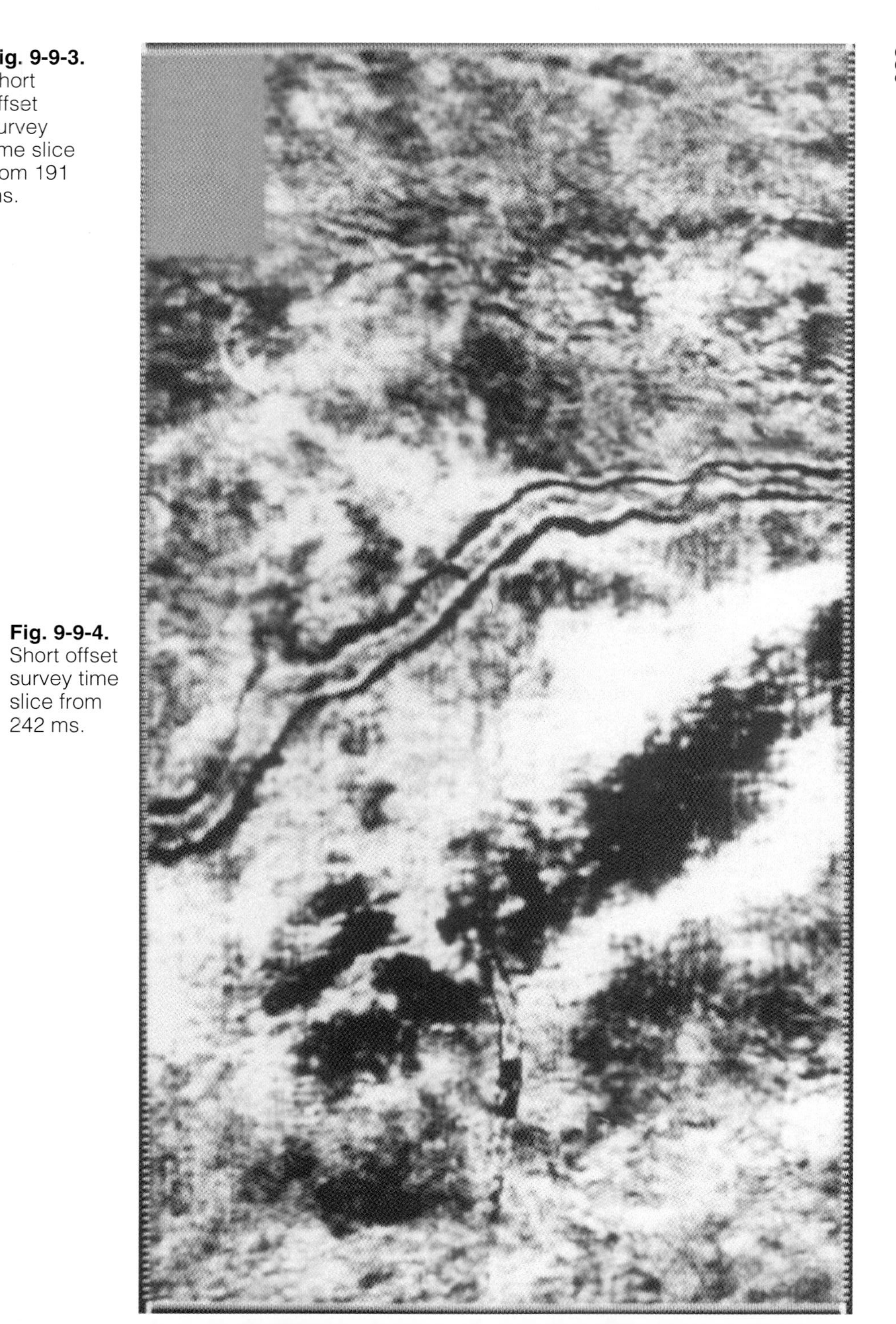

Fig. 9-9-4. Short offset survey time slice from 242 ms.

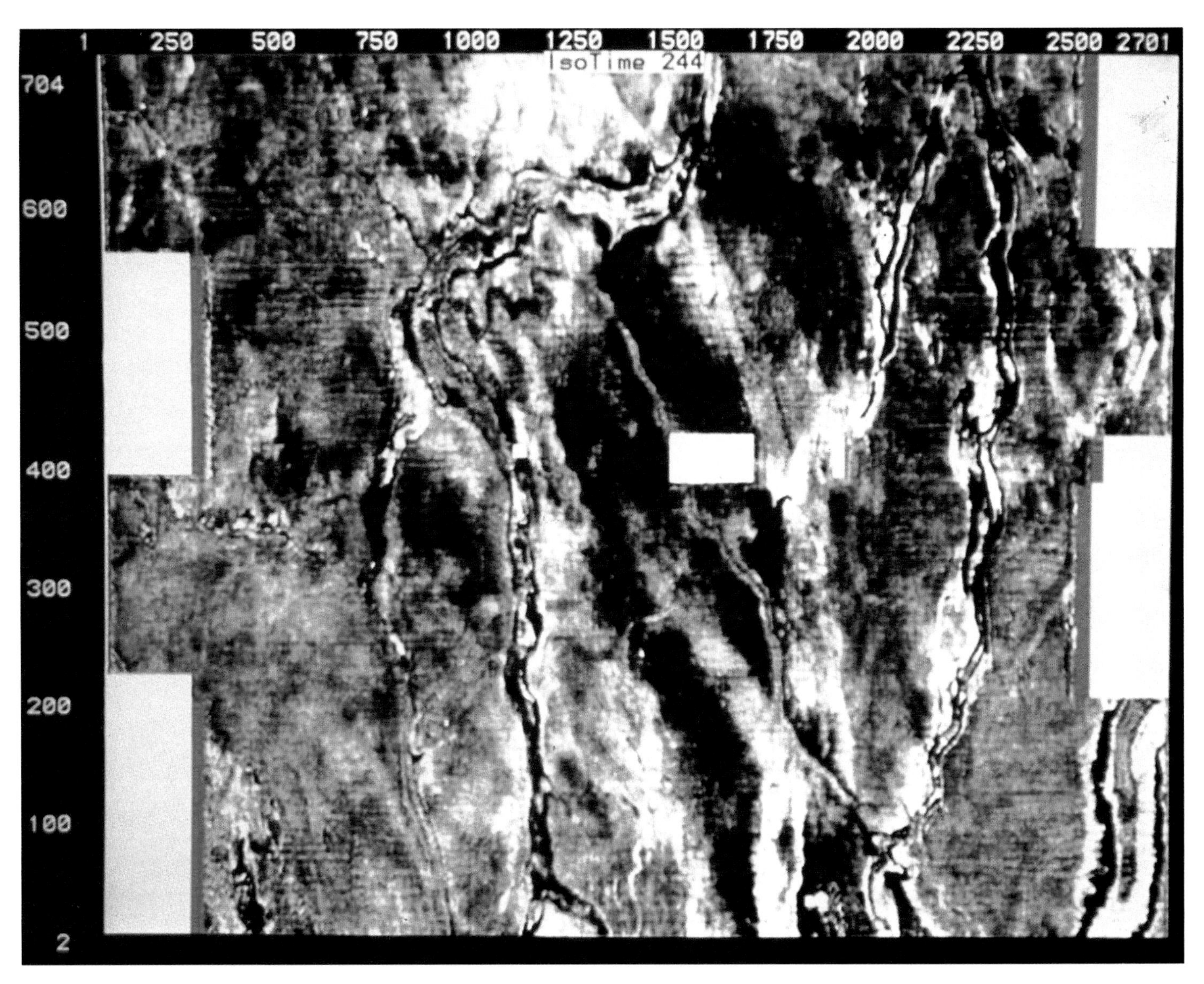

Fig. 9-9-5. Full-scale survey time slice from 244 ms.

3) The 3-D borehole profile showed the extent of the crestal grabens inside an area opaque to normal seismic methods.

4) 3-D borehole profiles should be considered as a working alternative to 2-D borehole profiles since the extra rig time and cost are surprisingly small and the benefits of 3-D are substantial. The method should be particularly suitable for time-lapse reservoir monitoring.

The statements in this case history reflect those of the author and not necessarily those of any of the Phillips Licence 018 Group of Companies. The author thanks the above Group for permission to publish the data.

References

Brewster, J., J. Dangerfield, and H. Farrell, 1986, The geology and geophysics of the Ekofisk field waterflood: J. Marine and Petroleum Geology, v. 3, p. 139-170.

Christie, P. A. F., and J. A. Dangerfield, 1987, Borehole seismic profiles in the Ekofisk field: Geophysics, v. 152, p. 1328-1345.

Dangerfield, J. A., in press, Ekofisk field—Subsidence fault analysis using 3-D seismic, *in* Reservoir Geophysics, R. E. Sheriff, A. R. Brown, D. Johnston, J. Justice, and R. Hardage: SEG/SPE/AAPG, in press.

Sulak, R. M., 1990, Ekofisk field—the first twenty years: SPE Annual Conference, New Orleans, September 1990.

Wiborg, R., and J. Jewhurst, 1986, Ekofisk subsidence detailed and solutions assessed: Oil and Gas Journal, v. 84, no. 7, p. 47-55.

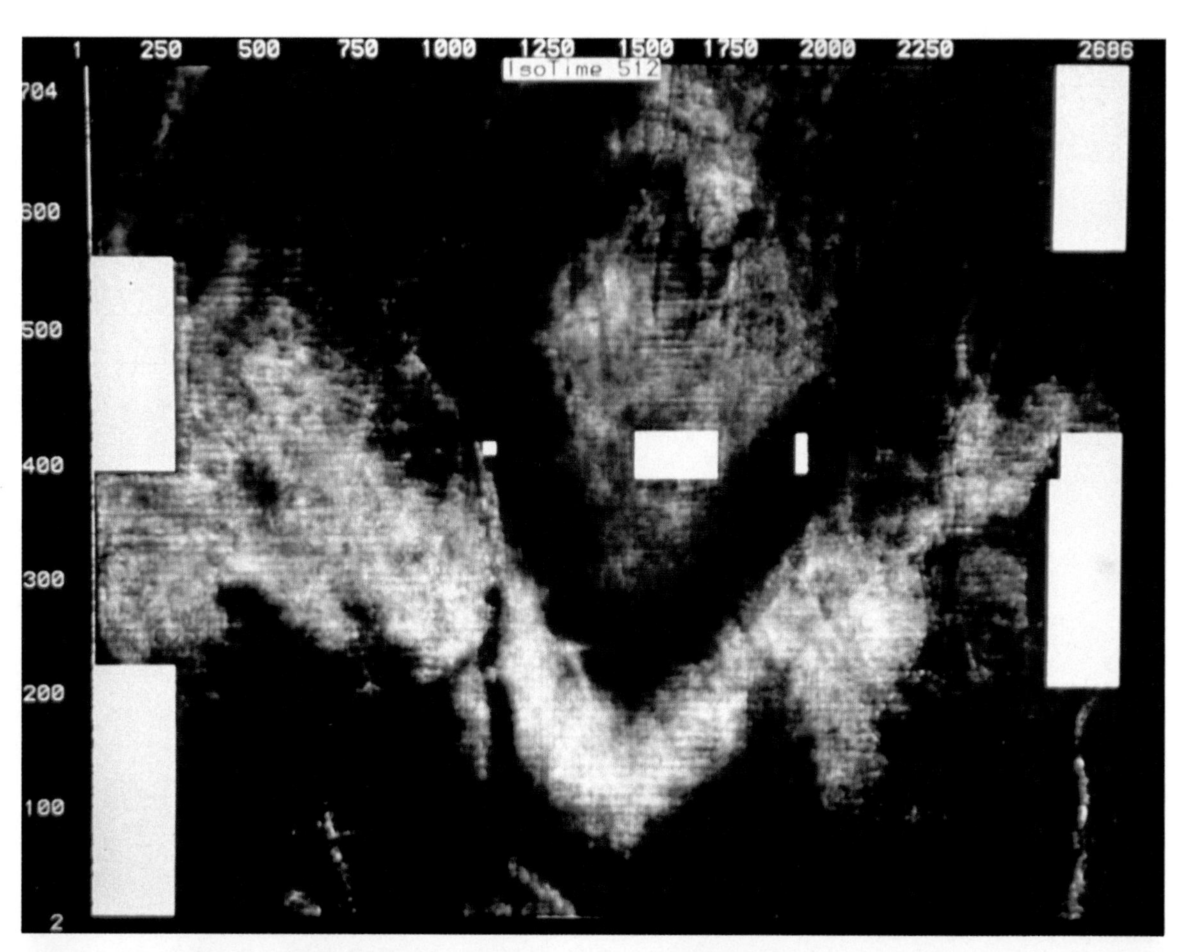

Fig. 9-9-6. Full-scale survey time slice from 512 ms.

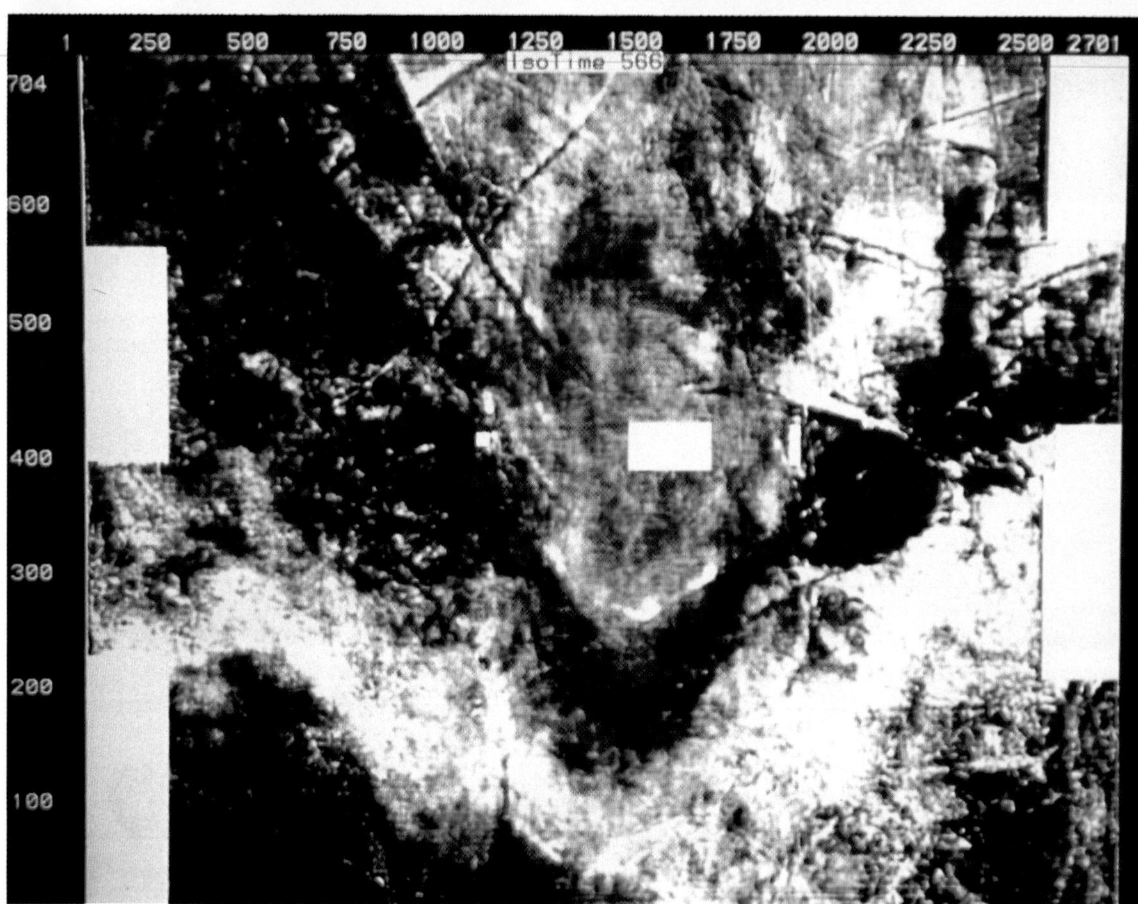

Fig. 9-9-7. Full-scale survey time slice from 566 ms.

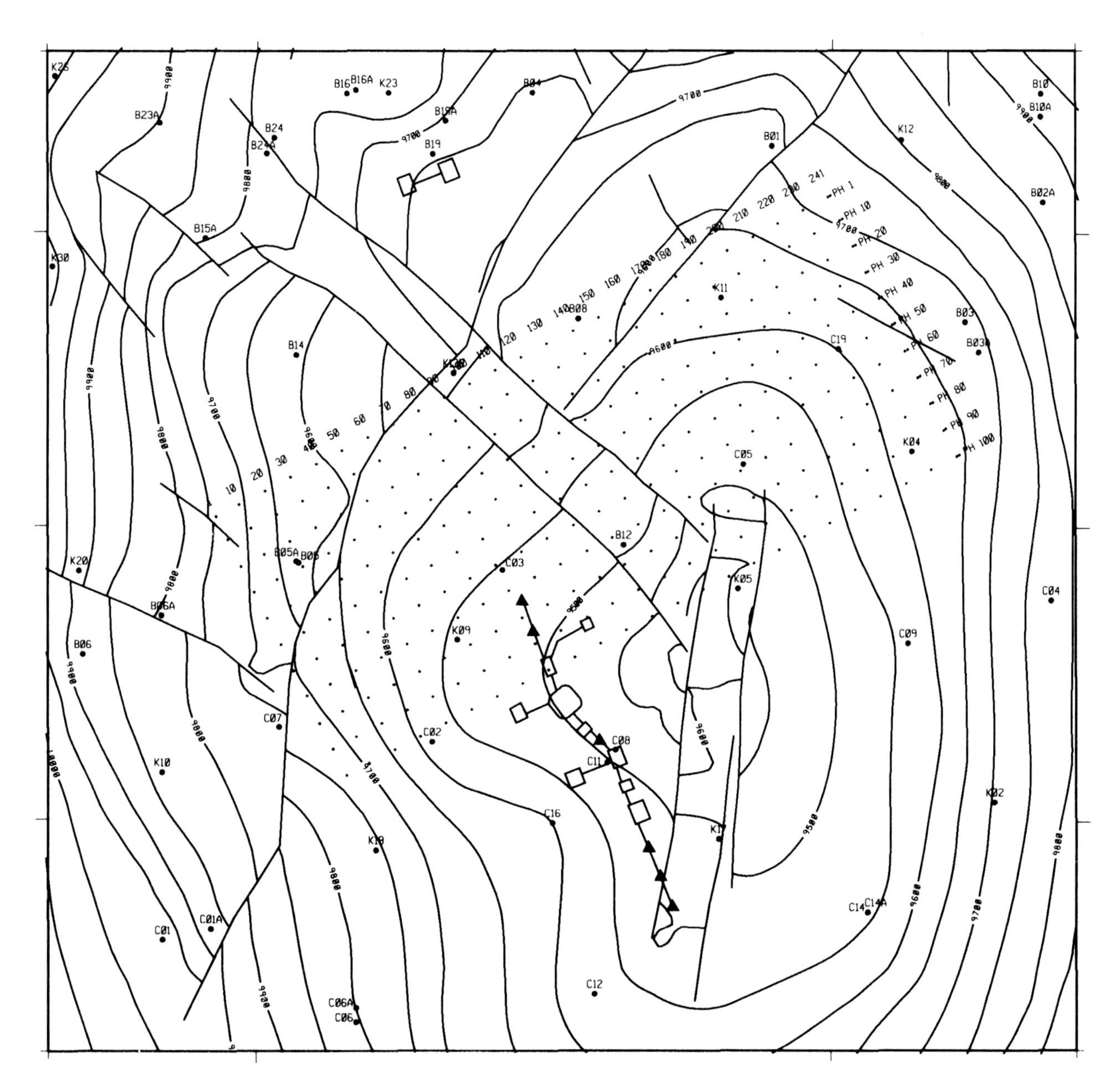

Fig. 9-9-8. Location of the 3-D borehole profile migrated data output grid superimposed on the 1991 reservoir map.

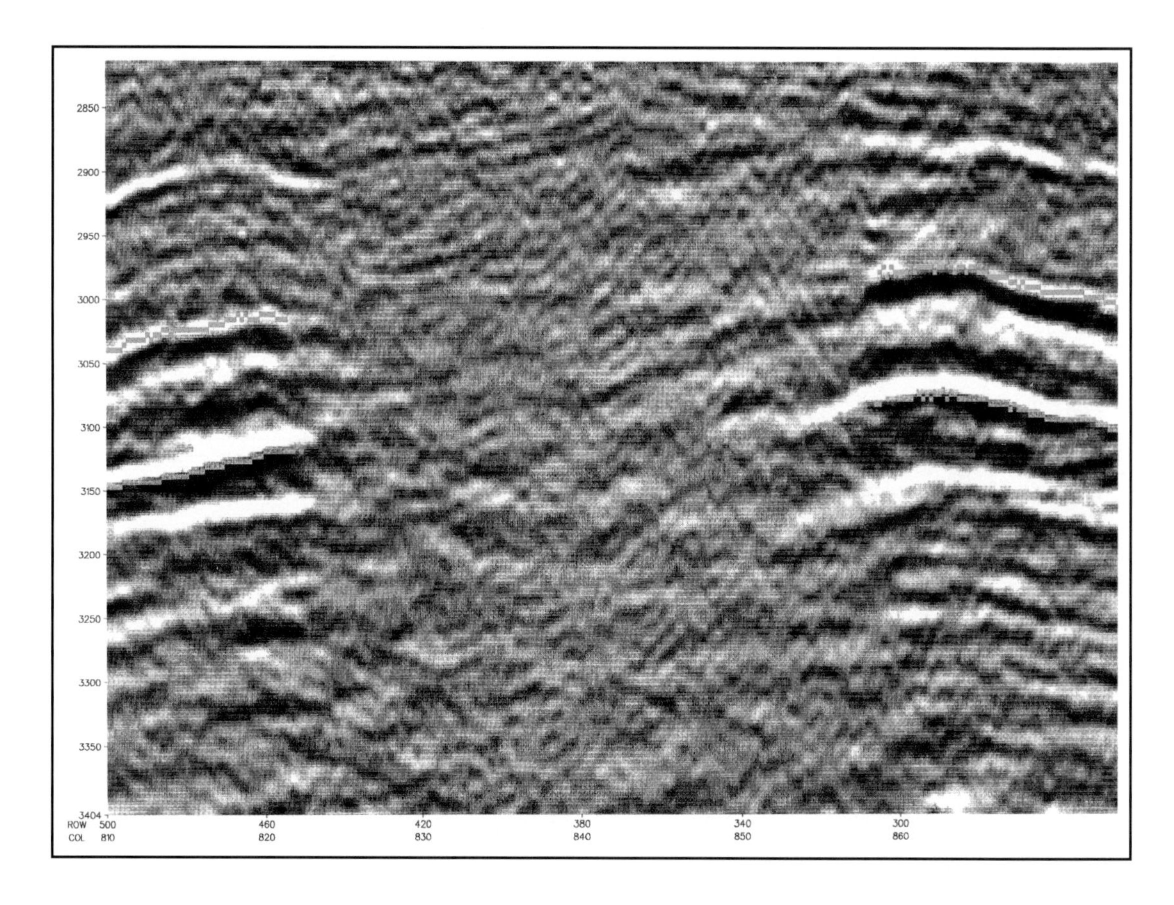

Fig. 9-9-9. Gas-affected seismic line in the borehole profile area.

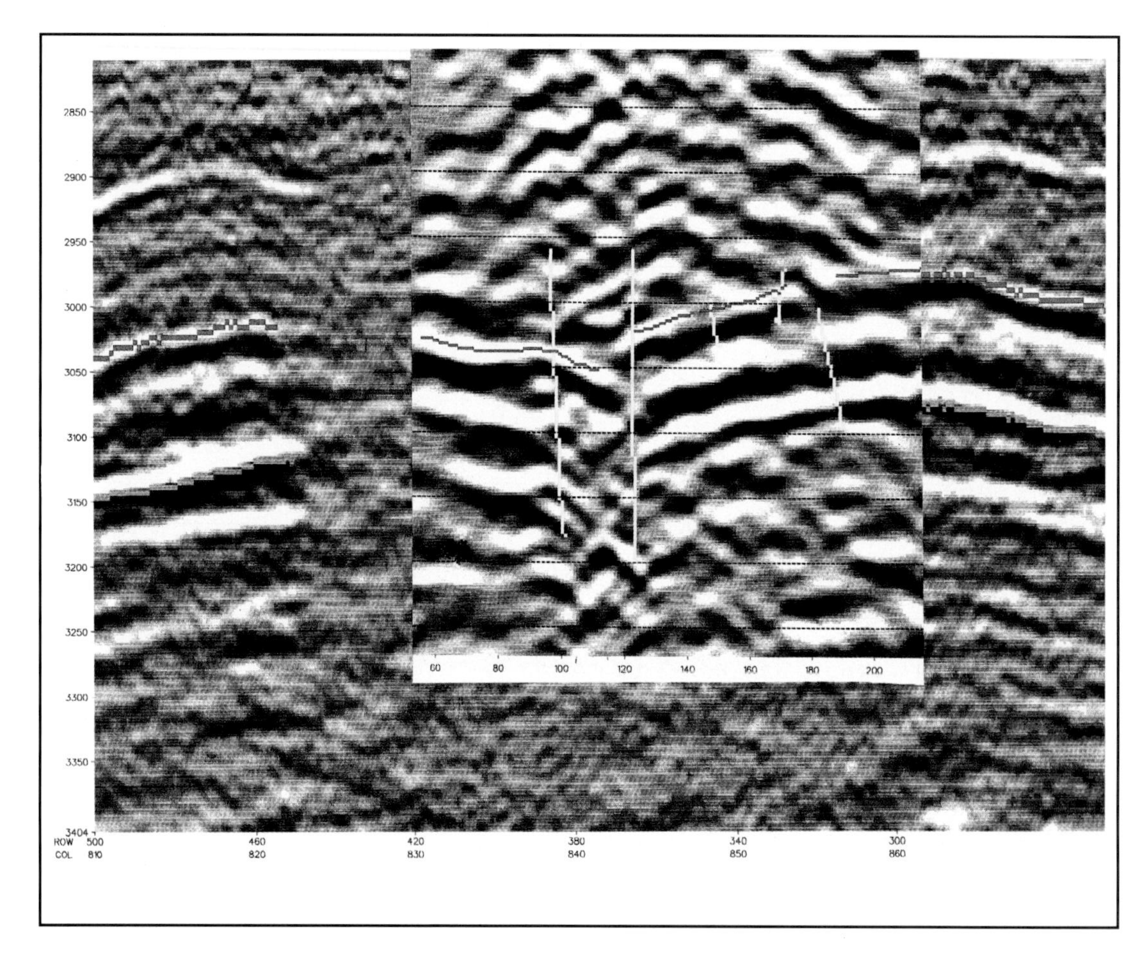

Fig. 9-9-10. Seismic line with the corresponding borehole profile line inserted.

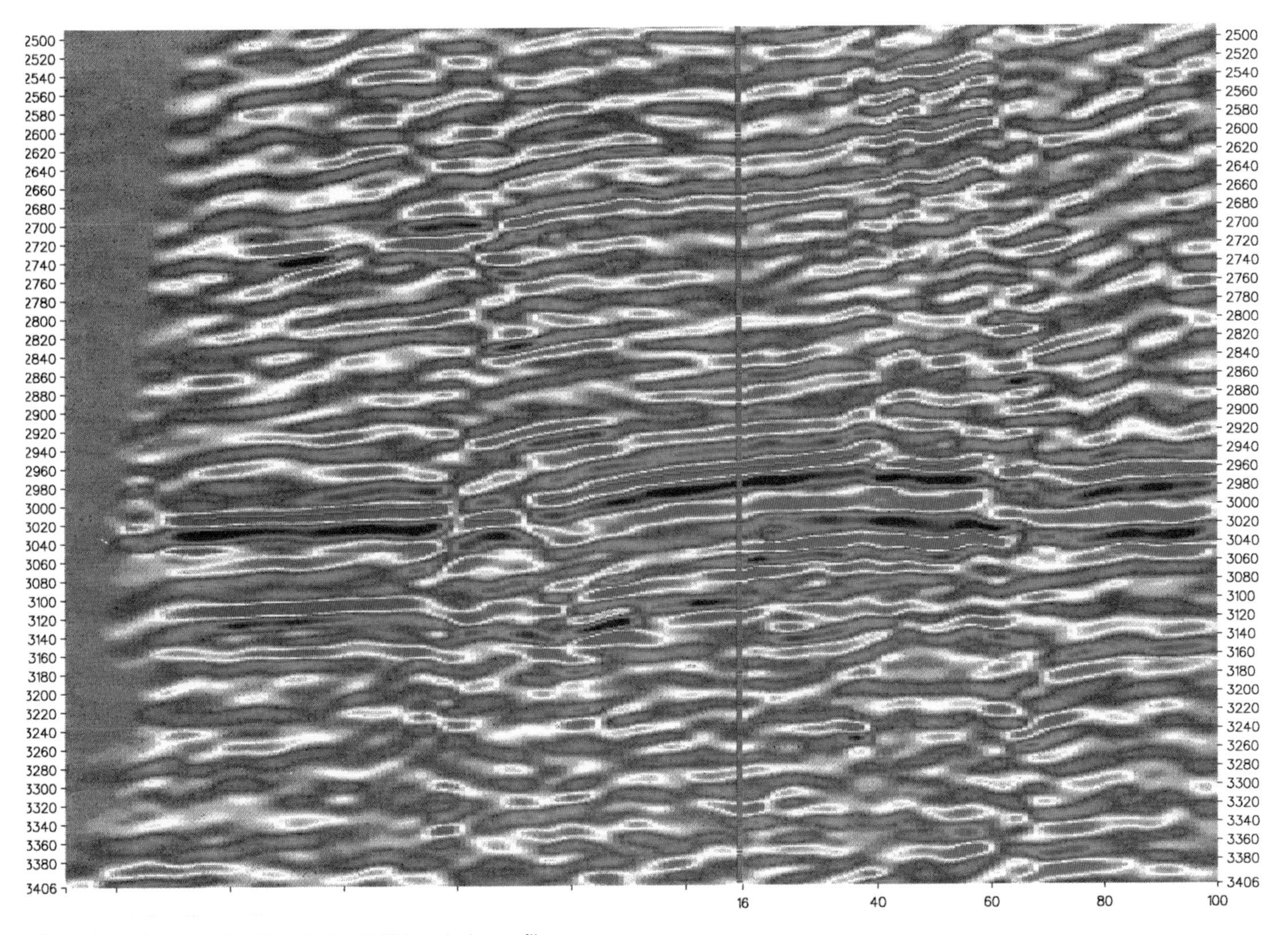

Fig. 9-9-11. Intersecting lines in the 3-D borehole profile.

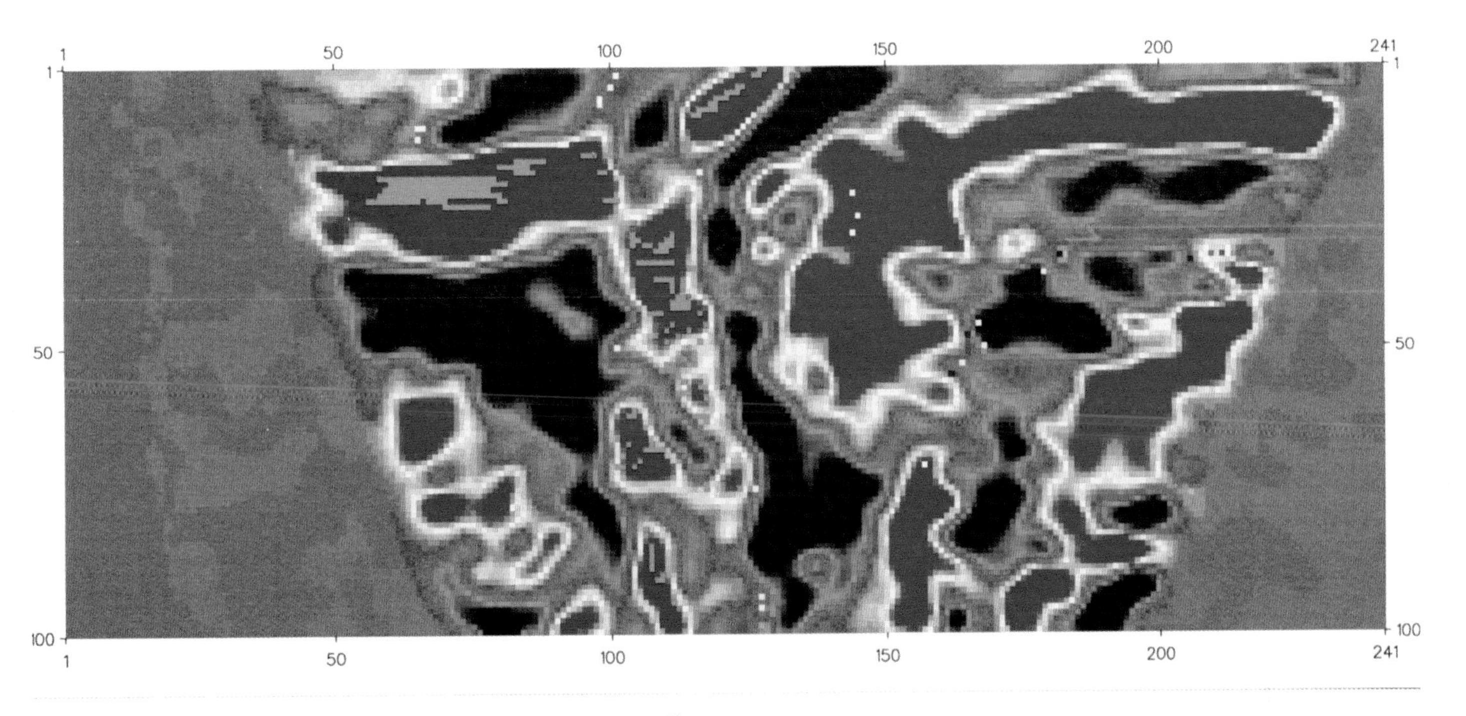

Fig. 9-9-12. Time slice at 3002 ms from the 3-D borehole profile.

Case History 10

Extending Field Life in Offshore Gulf of Mexico Using 3-D Seismic Survey

Thomas P. Bulling, Rebecca S. Olsen, ARCO Oil and Gas Company

The High Island 24L field (ARCO), located in the Texas state waters of the Gulf of Mexico, was discovered in 1967. It had produced 320 billion cubic feet (bcf) of gas and 3.0 million bbl of oil by 1986. An engineering field study completed in 1986 projected the field declining to the point of unprofitability within three years. The study found the reservoir maps had three basic problems: volumetric reserve calculations did not equal reserves produced; hydrocarbon-water contacts were inconsistent between wells thought to be in communication; and maps did not define extension opportunities. Attempts to remap the field with the existing 2-D seismic data base and well logs proved unsuccessful.

In 1986, ARCO acquired a 3-D seismic survey and, in 1987, remapped the field starting with the key producing horizons. Integration of detailed well log correlations with the dense grid of 3-D seismic allowed construction of accurate reservoir maps. These new maps helped solve the engineering problems by more accurately defining the configuration of the reservoirs, closely matching volumetrics and fluid contacts as well as defining new extension opportunities. The 3-D seismic survey and its products, along with engineering modifications and operations cost containment, resulted in the extension of the field's economic life by at least eight years. As more wells are drilled and new information integrated, additional reserves are found.

Introduction

Many fields in the Gulf of Mexico discovered with the seismic data available in the 1960s and early 1970s are facing declining production and reaching the end of their economic life. After discovery and initial development drilling, many of these fields were simply put on production and maintained. High Island 24L field is an example of such a field; only two wells were drilled between 1972 (the end of development drilling) and 1986. 3-D seismic is often needed to extend the life of these mature oil and gas fields. High Island 24L field is an excellent example of using 3-D seismic for this extension.

Background and Geology

The High Island 24L field is ten miles offshore in the Texas state waters of the Gulf of Mexico. Discovered by ARCO in 1967, the field has produced 320 bcf of gas and 3 million bbl of oil from 30 lower Miocene sands ranging from the normally pressured CM-12 (*Robulus 43*) sand to the geopressured "KI" sand (Figure 9-10-1). The field produces from anticlinal and fault traps both upthrown and downthrown to a lower Miocene growth fault system (Figure 9-10-2).

In 1986, a reservoir engineering study indicated the field would become unprofitable in two years (Figure 9-10-3, green line). The study also identified some key problems with the reservoir maps including: reservoir volumetrics did not equal production (i.e., several fault blocks produced more than maps indicated); structure maps and fluid level contacts were inconsistent (wells within the same fault block had different gas/water contacts); engineers could not reasonably forecast ultimate reserves; and maps did not define extension opportunities.

Prompted by the engineering study, ARCO attempted to remap the field to better evaluate its potential. The endeavor represented the first integration of geology, geophysics, and engineering data since development drilling ended in 1972. However, the available 2-D seismic data were inadequate and could not resolve the mapping problems. The question arose whether to abandon/sell the field or acquire more high-quality seismic data. Because the 2-D mapping proved unsuccessful, we could not place a reliable value on the property on which to base abandonment or selling criteria. Geoscientists felt chances were good that untested reservoirs would be found if accurate maps could be generated. Therefore, ARCO decided to obtain better seismic data to determine the value and potential of the field.

After considering several options, we decided not to acquire more 2-D data since 14 vintages already existed (shot between 1967 and 1985). A dense, consistent 3-D survey would greatly benefit the project in five ways: improved mapping of subtle structures; identification and evaluation of amplitude anomalies related to hydrocarbon bearing sands; resolution of deep, complicated structures; integration with data from numerous wells; and the ability to manipulate the seismic data in a variety of ways using a 3-D interactive workstation. Cost of the survey was less per mile than 2-D and less than the cost of a shallow well in this field. Geoscientists began mapping the field in June 1987 using the 3-D seismic data. Initial work focused on the most prolific reservoir sands.

Results and Examples

Significant mapping changes appeared between old 2-D and new 3-D maps. This section illustrates three examples of these changes in the "HC," "KC," and CM-12 horizon interpretations.

The "HC" (*Siphonina davisi* in age) sand at about 8000 ft (2400 m) is the most prolific reservoir. The 3-D seismic data clearly identifies subtle structures not seen on 2-D seismic lines and not found on the old maps (Figures 9-10-4a and 9-10-4b). Three wells, B-2, 9, and B-4, drilled on subtle structural highs, found the "HC" sand productive. Figure 9-10-5 is a seismic example of a subtle "HC" high tested by the successful (post-3-D) #9 well. This example shows how 3-D helped to identify untested subtle structures and allowed us to generate seismic traverses to see relationships between drilled and proposed wells.

The second example shows mapping improvements in one of the deepest producing horizons, the "KC" (*Lower Planulina* in age), at about 11,000 ft. Figures 9-10-6a and 9-10-6b illustrate the drastic differences between the pre-3-D and the post-3-D maps. These maps are similar in area and orientation. Understanding fluid contacts and reservoir juxtaposition across faults always had been a problem in the *Lower Planulina* sands. The new, post-3-D map is based on new well log interpretation integrated with the 3-D seismic. This recent interpretation explains areas where volumetrics and fluid contacts previously did not make sense.

Mapping success at "KC" is largely due to the improvement in deep seismic resolution integrated with a new well log interpretation. Figures 9-10-7a and 9-10-7b are portions of seismic lines from the 2-D and 3-D data sets. Improvement in deep resolution on 3-D line 112 (Figure 9-10-7b) is demonstrated by clearer fault trace delineation and reflection continuity.

The final example of mapping changes is from the shallowest producing horizon, CM-12 (*Robulus* 43 in age), at about 5500 ft (1680 m). Old 2-D mapping (Figure 9-10-8a) is different from the newer post-3-D map (Figure 9-10-8b). Significant to the post-3-D map is a four-way dip structure on block 90S not adequately tested by the two wells on this block. Prior to interpreting the 3-D data, we did not recognize the 90S structure and the amplitude anomaly that conforms to it (Figure 9-10-9). This anomaly is very similar to a CM-12 anomaly known to produce in wells #4 and #5. An arbitrary line known as seismic traverse 11 (Figure 9-10-10) shows the relationship between the prospective 90S block amplitude and that of the producing #4 and #5 wells. Drilled in early 1988, the ARCO 90S #1 well found hydrocarbon-bearing sands causing the CM-12 amplitude anomaly. Also, CM-12 came in very close to target depth and expected reserve size. This example illustrates the ability of the interpreter to identify amplitude anomalies in a small area. Utilizing the 3-D workstation, we can relate the untested anomalies to similar productive anomalies in the field, resulting in reduced risk and better reservoir delineation.

The post-3-D maps resulted in drilling and completing eight wells to date that will recover 40 bcf equivalent (net to ARCO). The 3-D seismic survey was an excellent investment, costing less than a shallow well in the field. The new wells, based on 3-D mapping, resulted in a reversal of the field's declining Before Federal Income Tax cash flow curve (Figure 9-10-3, red line). Other wells currently are under consideration. In total, the post-3-D maps identified 50 bcf equivalent of potential reserves (40 bcf from new wells and 10 bcf from existing wells and recompletions).

Conclusions

The 3-D seismic survey aided our interpretation of the field by providing:

- a continuous and dense grid of data across the field;
- the capability to generate traverses at any orientation;
- excellent detection of subtle structures;
- horizon slices to help define accumulations on the basis of amplitude;
- better deep resolution to help add extension opportunities.

Overall, the use of 3-D seismic data allowed for a better understanding of the stratigraphic and structural complexities of the High Island 24L field. In addition, post-3-D maps helped solve the engineering problems by more accurately defining the configuration of the reservoirs. Reservoir maps now closely match volumetrics; fluid contacts within fault blocks are consistent; and we have a better definition of the extension opportunities available. The 3-D seismic survey and its products, along with engineering modifications and operations cost containment, resulted in the extension of the economic life of the field to at least 1996.

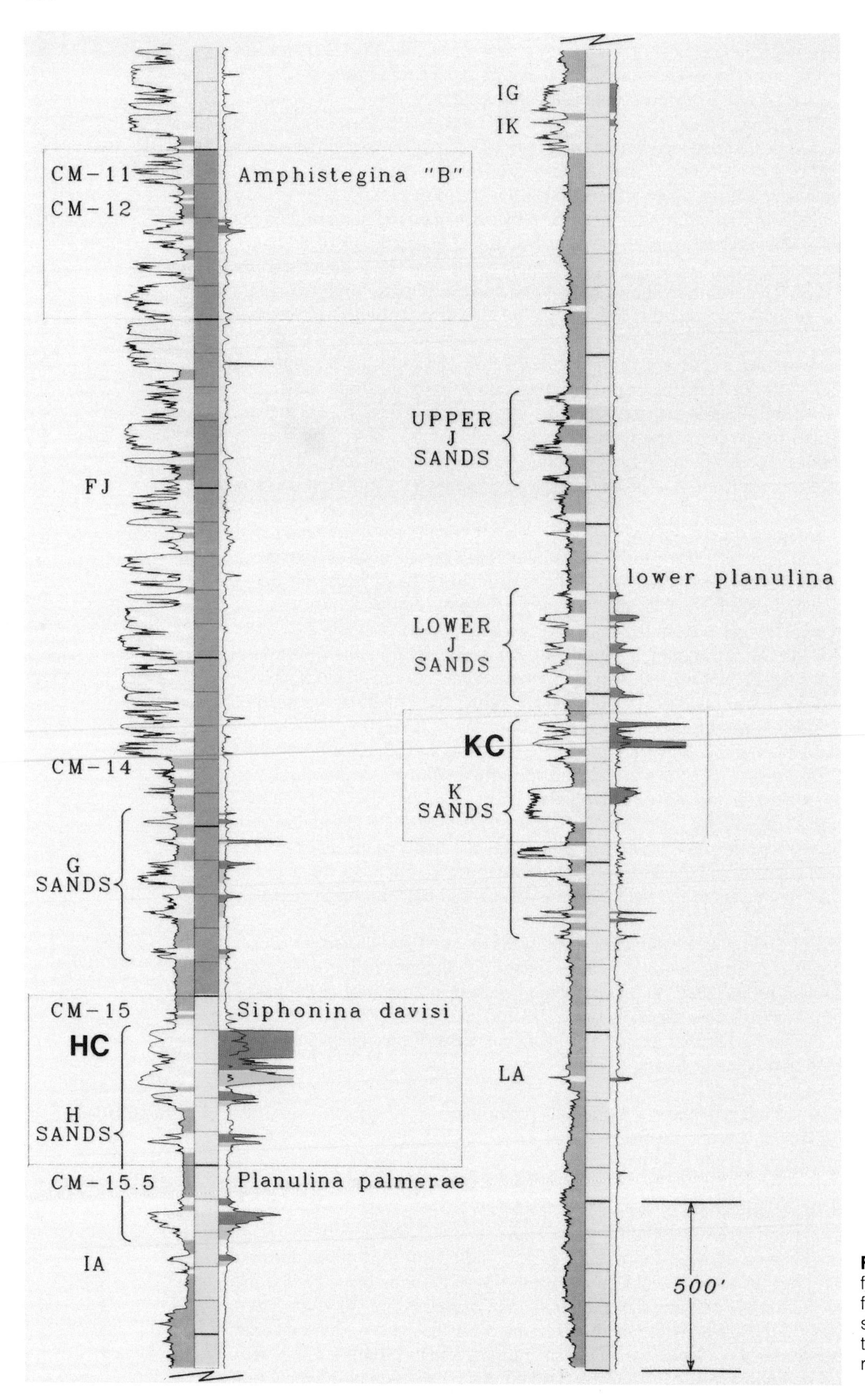

Fig. 9-10-1. The type log for the High Island 24L field illustrating the 30 pay sands. The "HC" sand is the most productive reservoir.

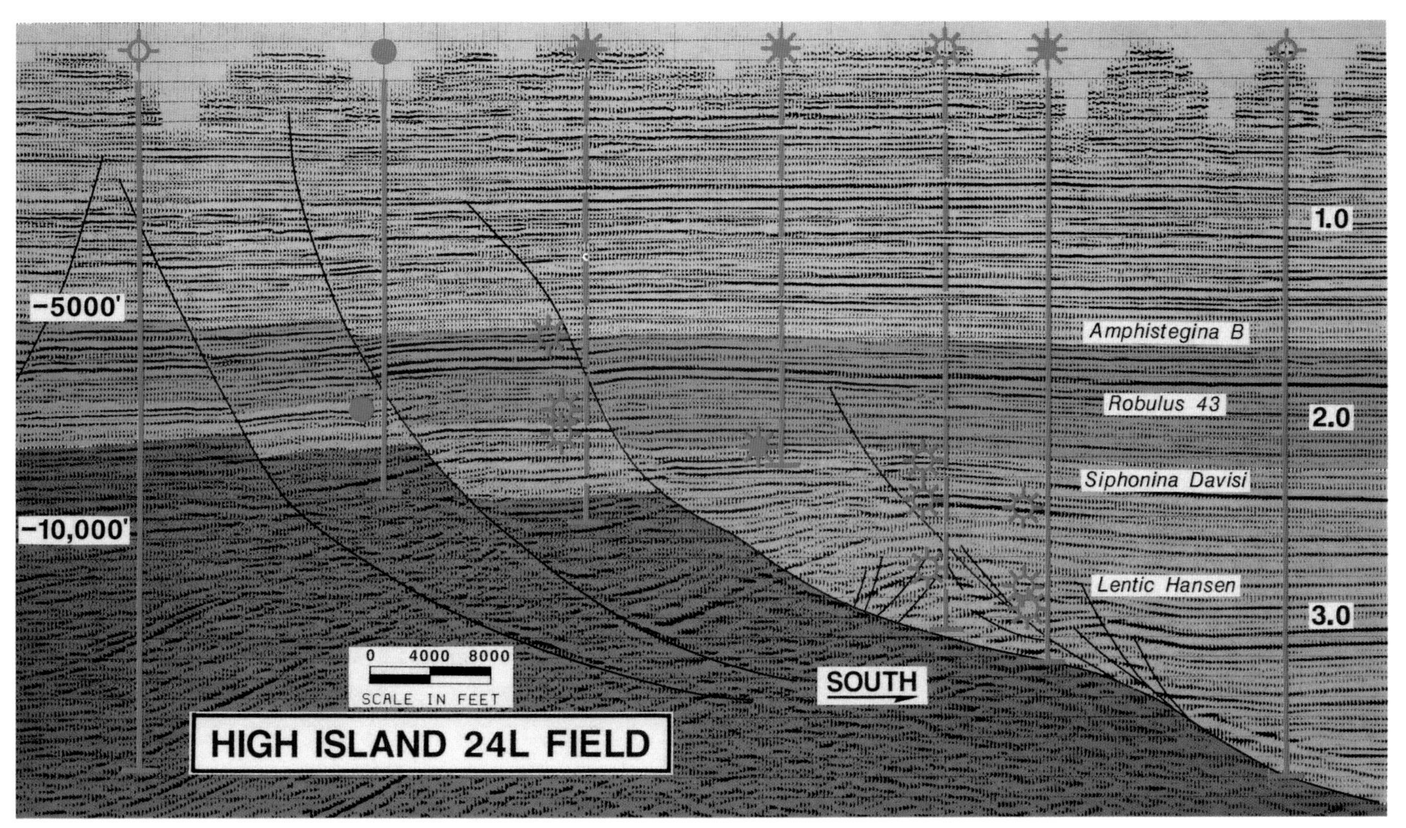

Fig. 9-10-2. Semiregional seismic traverse through the High Island 24L field illustrating typical anticlinal and fault related traps. Major producing intervals and lower Miocene growth fault system also are shown.

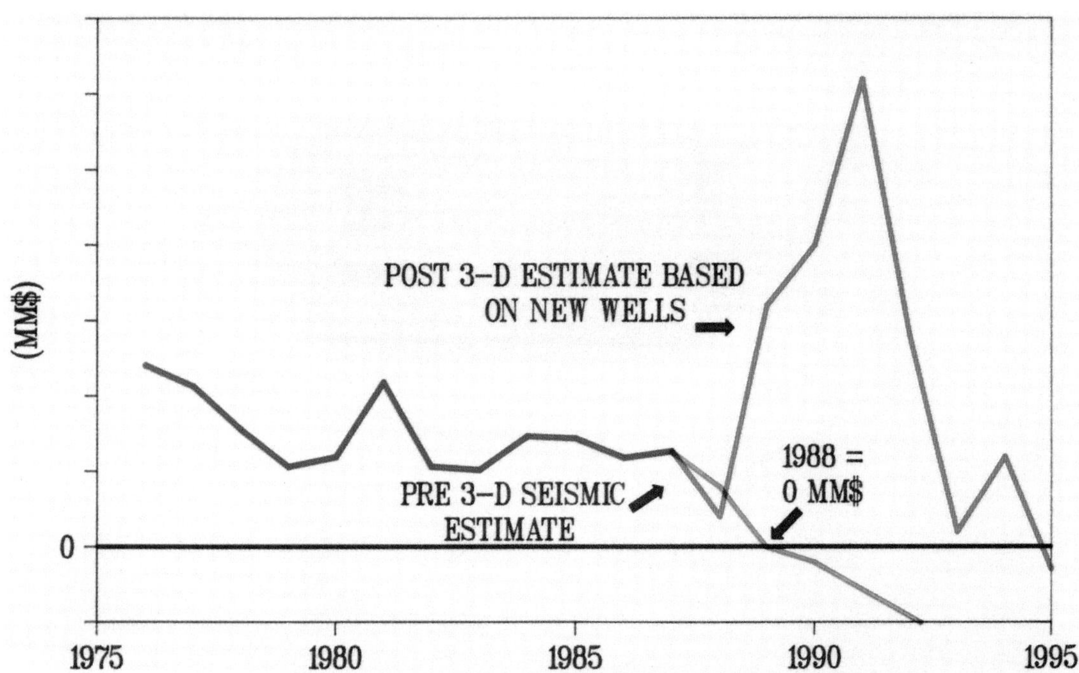

Fig. 9-10-3. Before Federal Income Tax cash flow versus time for the High Island 24L field. The green line (decline curve) represents pre-3-D seismic predictions and the red line represents the post-3-D estimate.

Acknowledgments

We extend a special thank-you to the numerous individuals who were involved in the success of this project and compilation of this case history. Thanks go to Garret Chong. Without his recommendations and support, this project may have taken a totally different tack. Also, the ARCO management team was very encouraging and supportive when budgets were tight. We appreciate ARCO's Houston drafting and reproduction for its professional and patient support. Joyce Settle typed and helped edit this manuscript and related versions. Gary Mitch helped edit this paper and advised us on its presentation. Critical to the success of extending the life of this field was the open communication and teamwork between the project geologists, geophysicists, and engineers (reservoir and drilling). Janet Miertschin, our engineering counterpart, sparked the reevaluation of this field with her 1986 field depletion study.

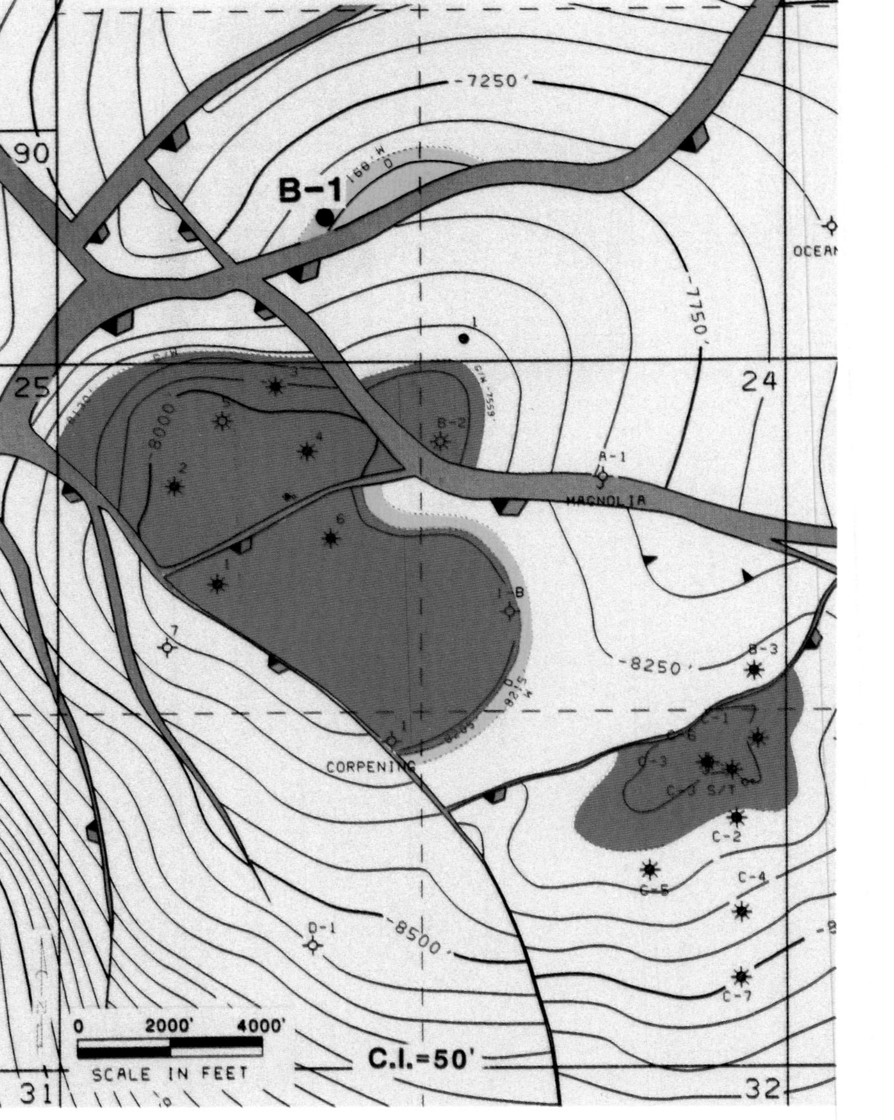

Fig. 9-10-4a. Pre-3-D seismic "HC" sand structure map.

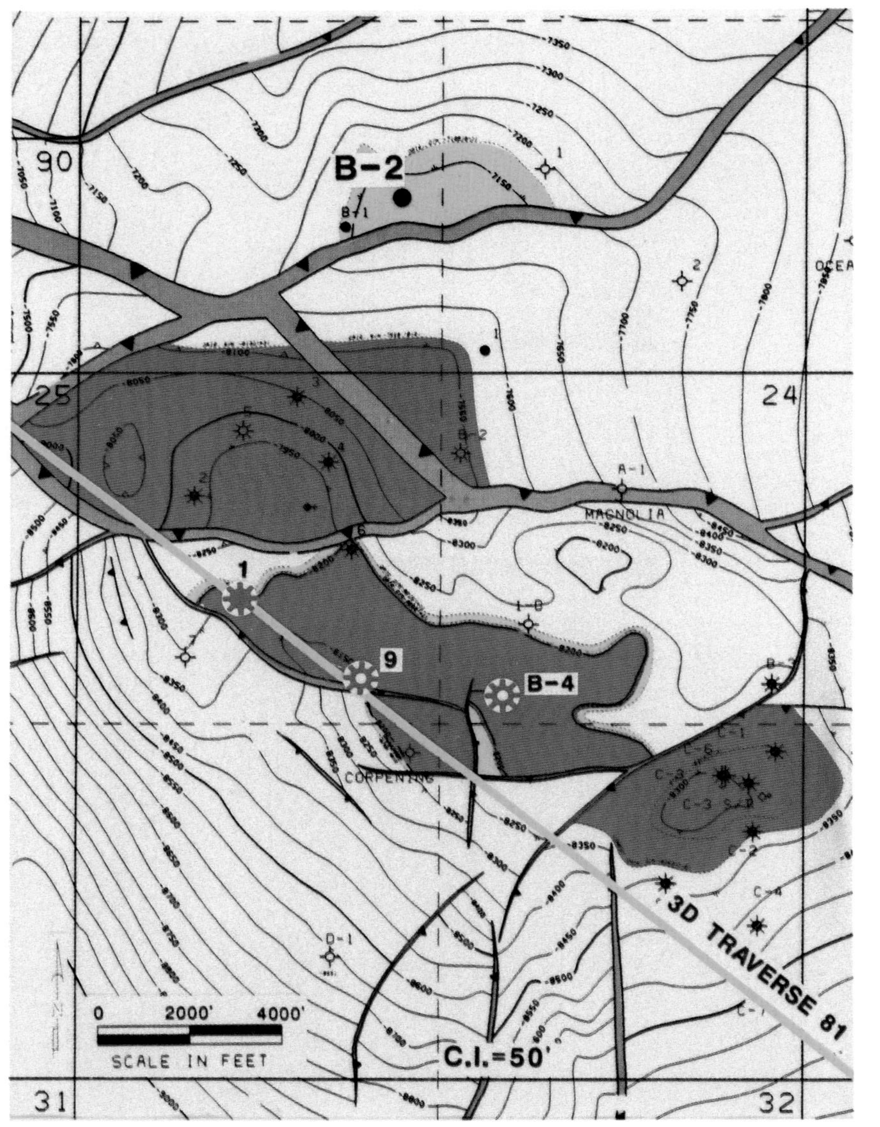

Fig. 9-10-4b. Post-3-D seismic "HC" sand structure map. Note that the subtle highs on which wells 9, B-2, and B-4 were drilled do not exist on pre-3-D structure map (Figure 9-10-4a).

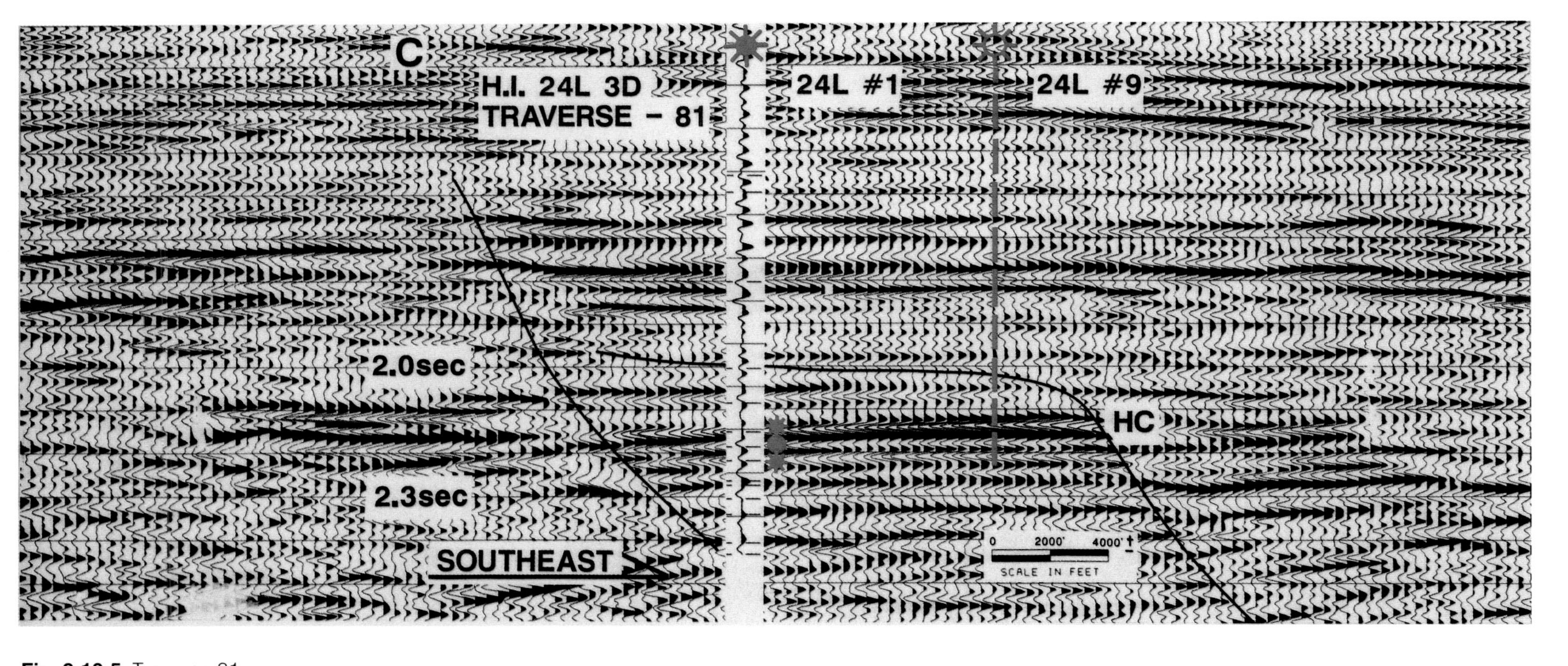

Fig. 9-10-5. Traverse 81 (arbitrary line) through a subtle "HC" high and the location of well #9. Note structural elevation and trough amplitude increase updip from 24L #1 to the proposed #9 well location. Well 24L #9 proved to be successful, close to target depth, and close to pre-drill reserve estimates.

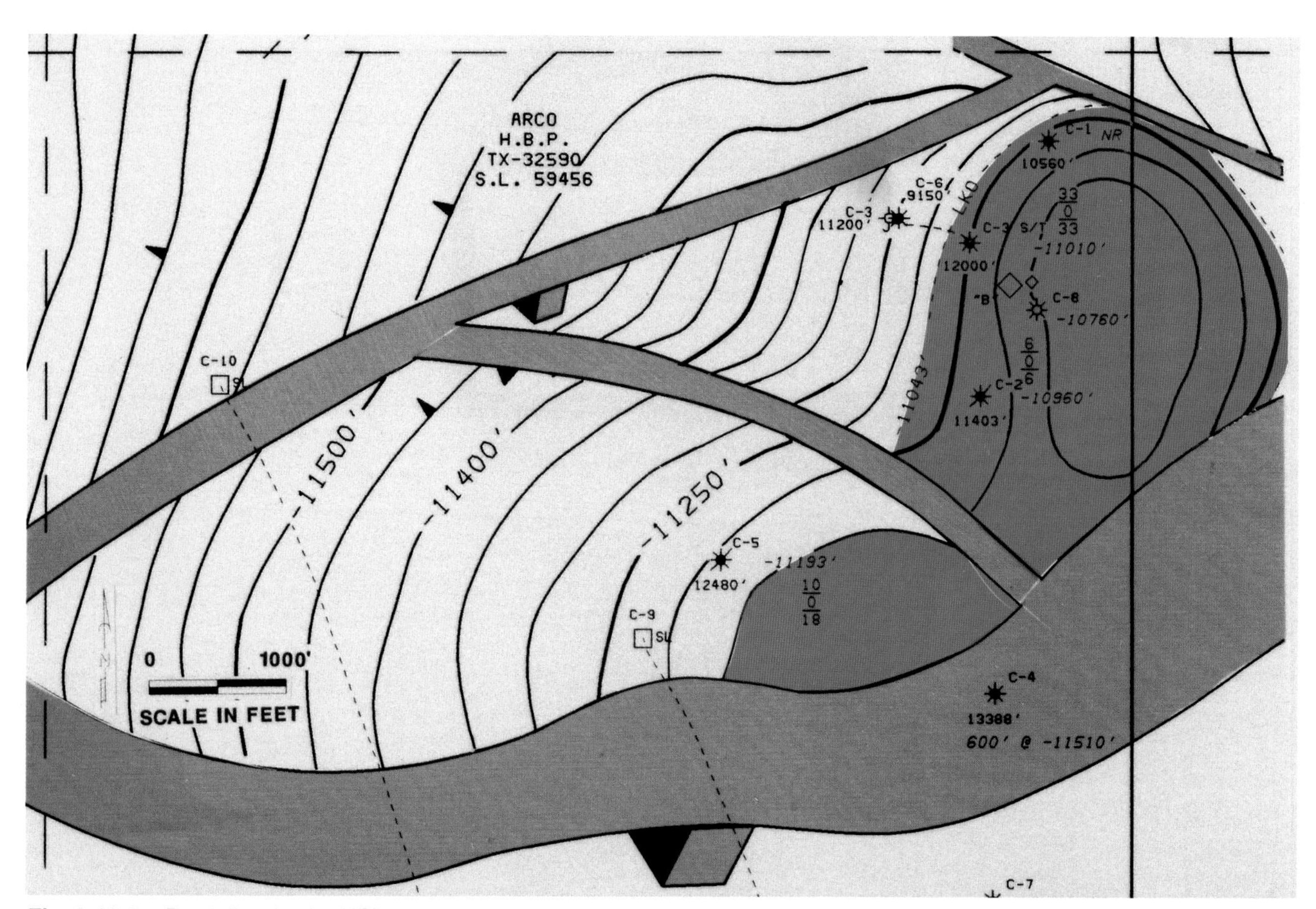

Fig. 9-10-6a. Pre-3-D seismic "KC" sand structure map.

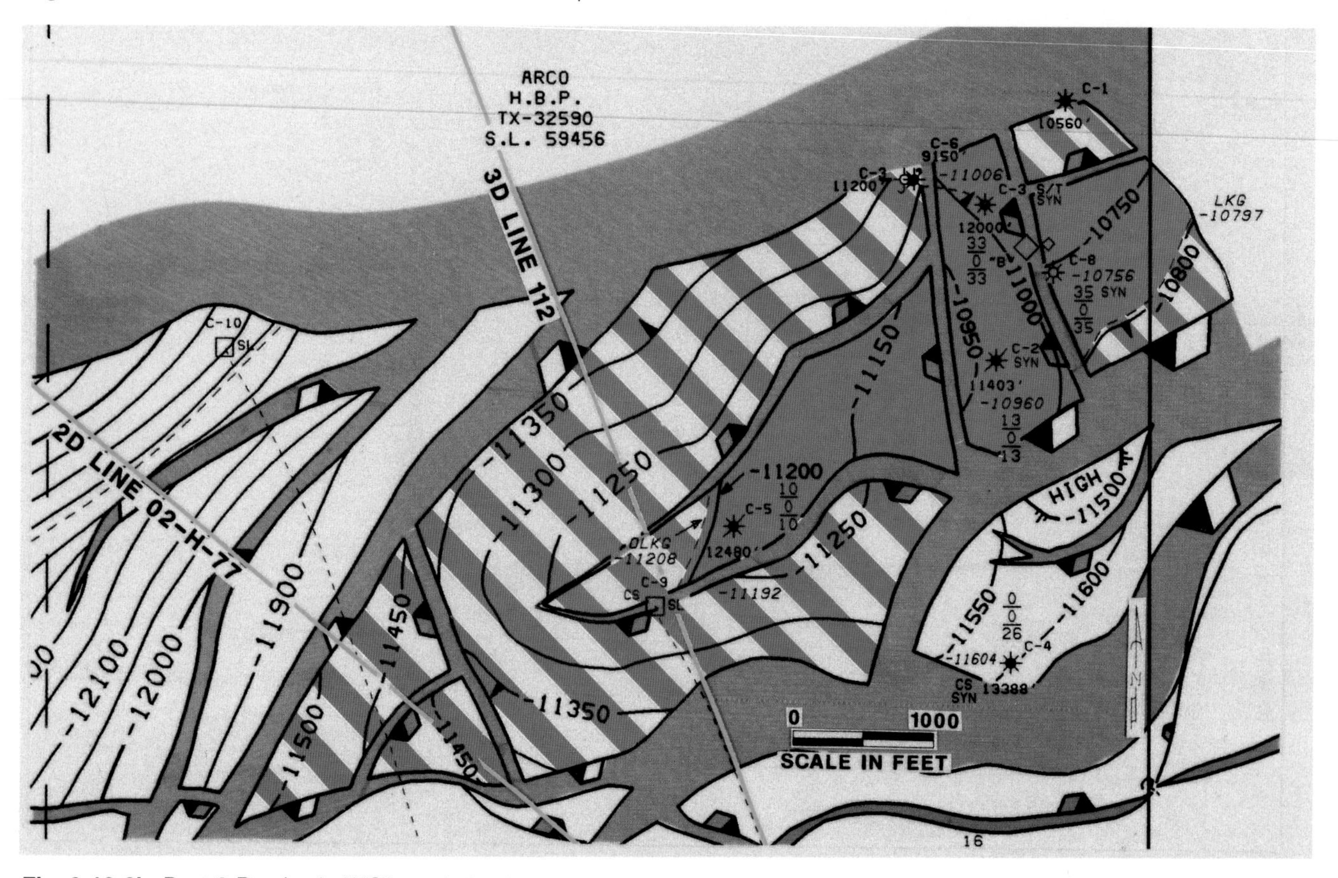

Fig. 9-10-6b. Post-3-D seismic "KC" sand structure map.

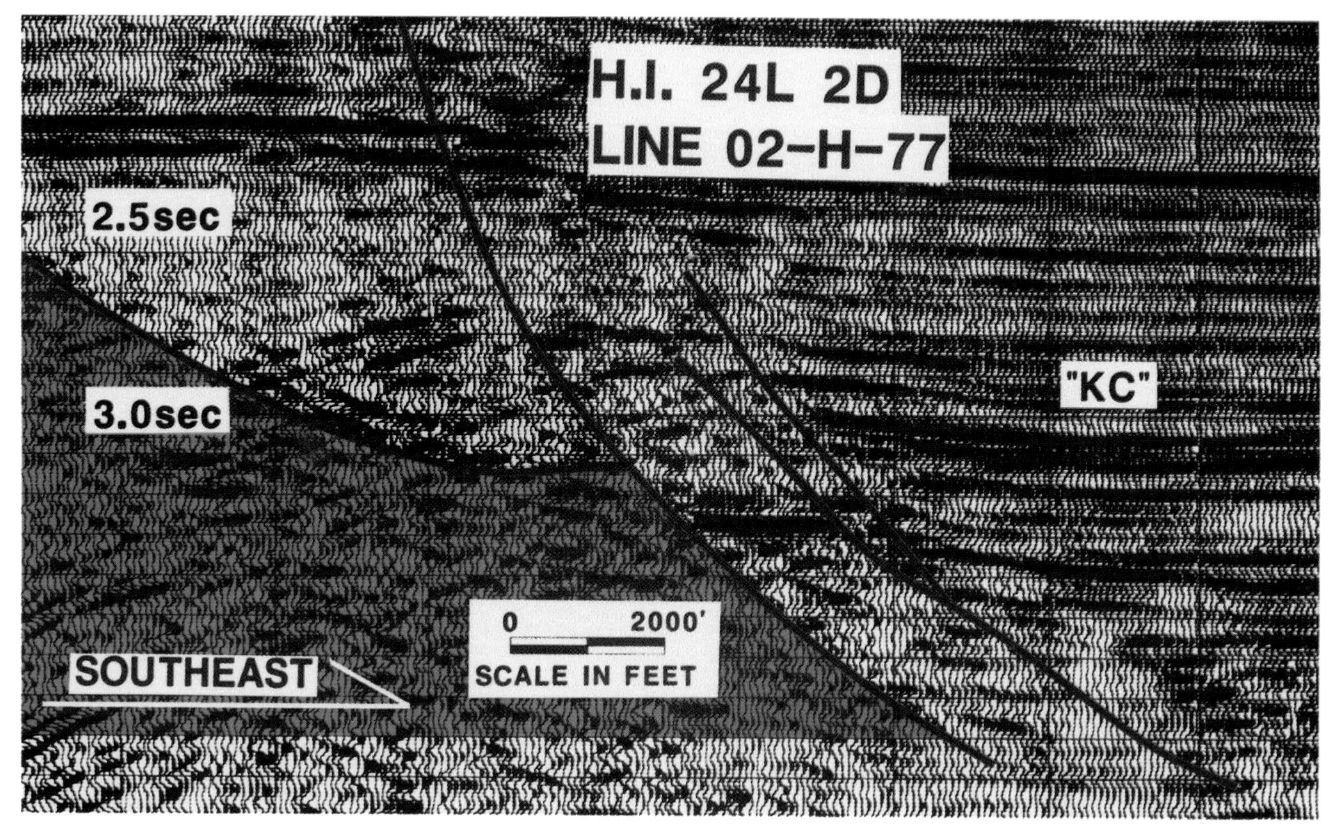

Fig. 9-10-7a. 2-D seismic line 02-H-77. Location of this line is shown on Figure 9-10-6b.

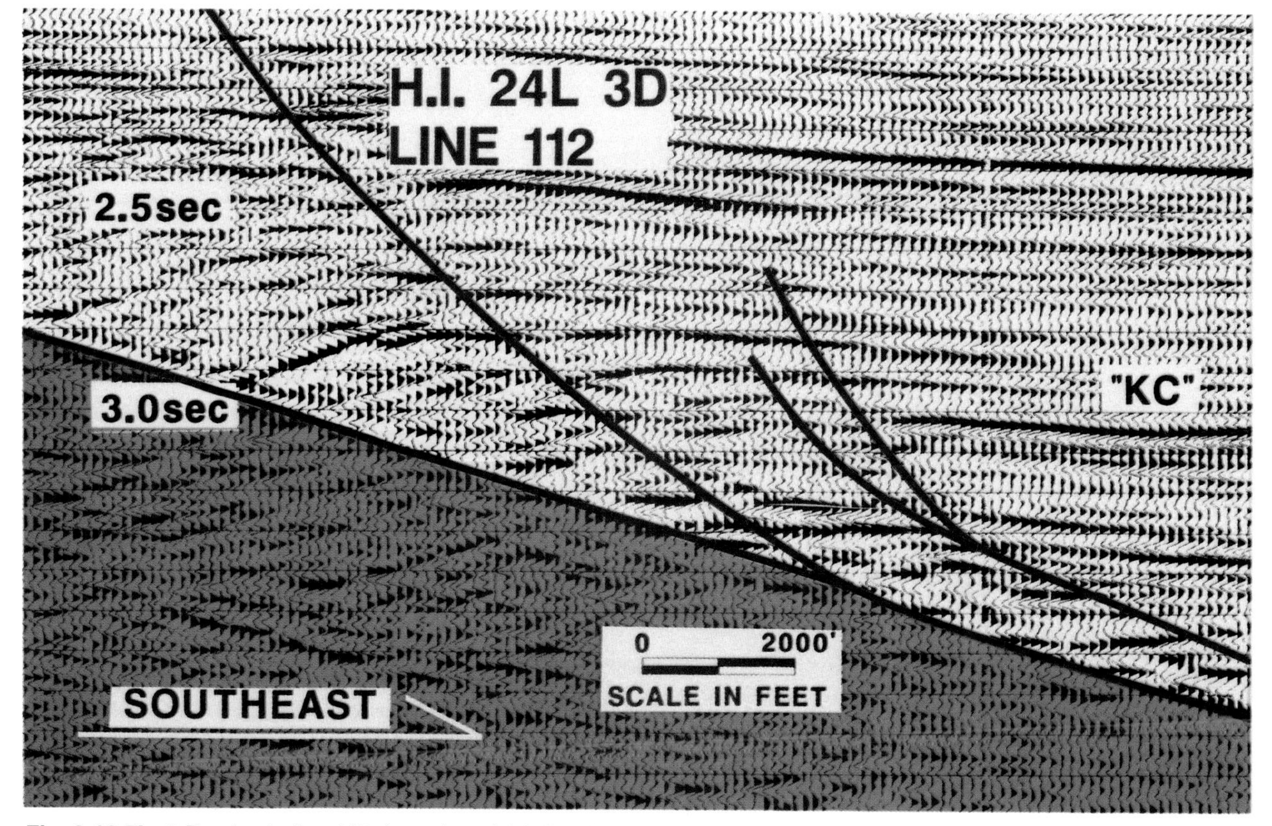

Fig. 9-10-7b. 3-D seismic line 112. Location of this line is shown on Figure 9-10-6b. Areas highlighted on Figures 9-10-7a and b are comparable. Line 112 is superior in deep resolution.

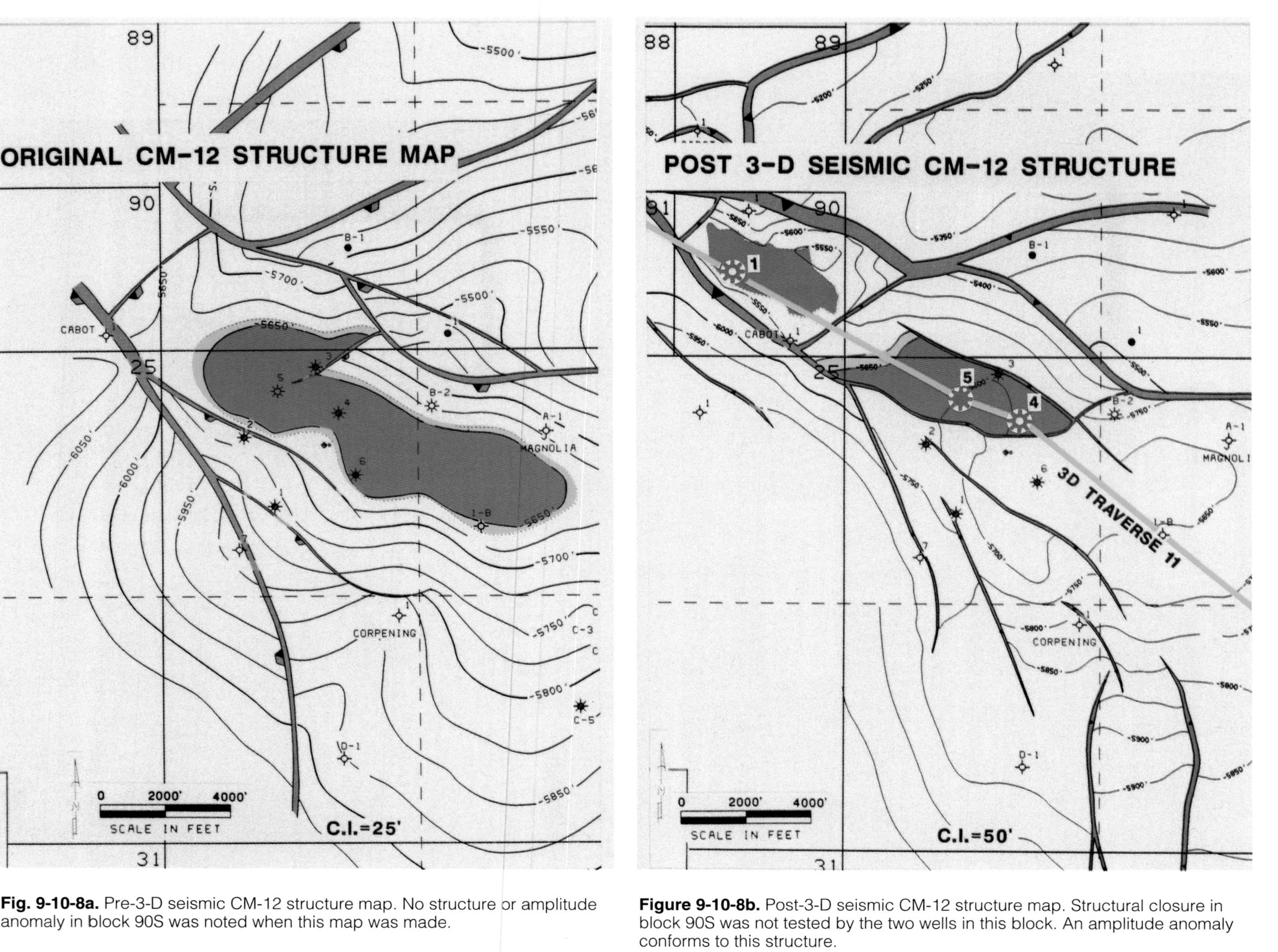

Fig. 9-10-8a. Pre-3-D seismic CM-12 structure map. No structure or amplitude anomaly in block 90S was noted when this map was made.

Figure 9-10-8b. Post-3-D seismic CM-12 structure map. Structural closure in block 90S was not tested by the two wells in this block. An amplitude anomaly conforms to this structure.

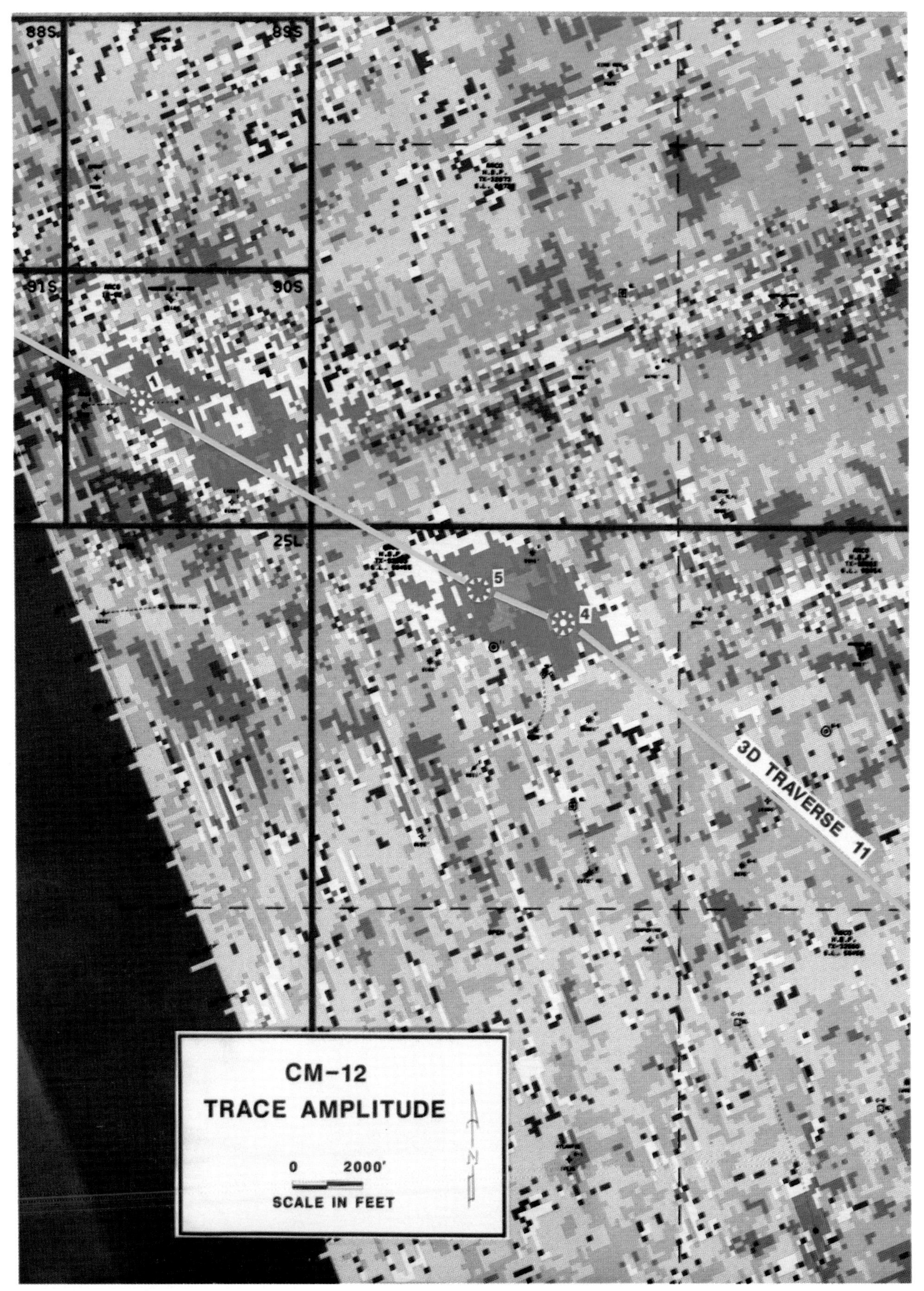

Fig. 9-10-9. CM-12 horizon slice (also termed amplitude map) illustrates strong trough amplitudes (pinks and reds) conforming to the 90S block structure (in the northwest corner of the figure). The 90S anomaly is stronger than the #4 and #5 producing anomaly to the southeast.

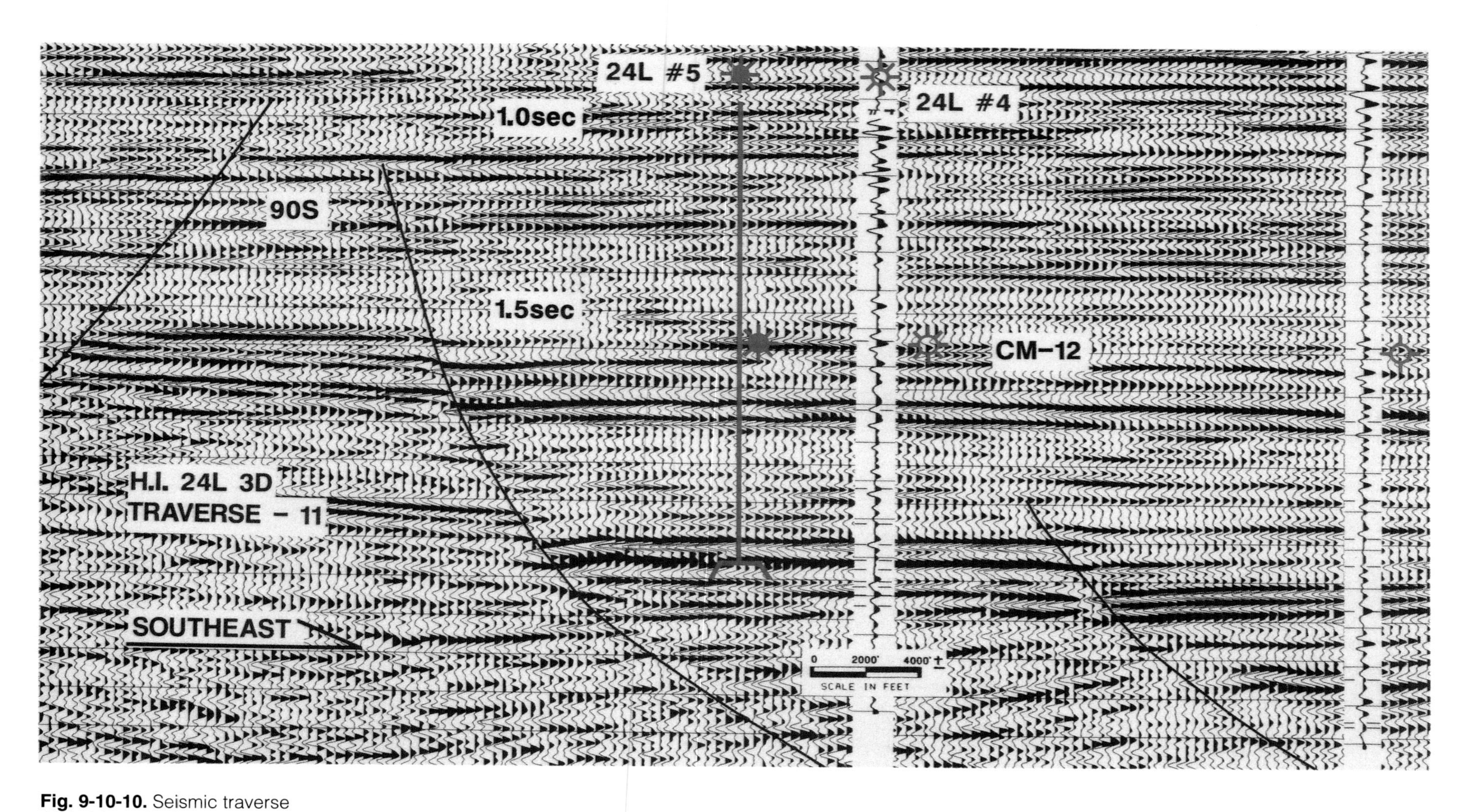

Fig. 9-10-10. Seismic traverse 11 (arbitrary line) showing the CM-12 trough amplitude anomalies marked in yellow for the 90S block and for the production in wells #4 and #5. Location of the traverse is shown on Figures 9-10-8b and 9-10-9.

Case History 11

Modern Technology in an Old Area—Bay Marchand Revisited

W. L. Abriel, P. S. Neale, J. S. Tissue, R. M. Wright, Chevron USA, Inc.

Background

Bay Marchand, a giant oil field in the Gulf of Mexico, is undergoing renewed drilling activity as the result of a recently acquired 3-D seismic survey. This classic mature field seemingly had entered its last stages of production in the mid-1980s. However, an integrated team effort by geologists, geophysicists, and engineers revitalized the area by innovative use of the new 3-D data.

Figure 9-11-1 shows the relative location of Bay Marchand to the Louisiana coastline. The field is approximately 65-70 mi (105-110 km) south of New Orleans and in state coastal waters. Water depths range from 10 to 50 ft (3 to 15 m). The field is situated across a huge salt diapir and its geology is characterized by complex faulting and stratigraphy. The latter is due to regressive marine sequences; the former is caused by both salt movement and sediment loading.

The first well, which was dry, was drilled in 1930. It then took 19 years and 11 more dry holes before the first discovery was made in 1949. This says something about the persistence of the early geologists and early management. There have been more than 800 wells drilled in the field to date.

Bay Marchand officially became a giant in 1986 when its 500 millionth bbl of oil was produced. At that time, though, it did not appear that much more could be extracted. Daily production had peaked at more than 75,000 bbl of oil/day in the late '60s and early '70s. A steady decline occurred through the remainder of the '70s and into the '80s. This decline continued in spite of engineering waterflood efforts in the '70s and acquisition of modern 2-D seismic data in the early '80s. When production dipped to 18,000 bbl of oil/day in the mid-1980s, a 3-D survey was contracted. The intent was to arrest further production decline and possibly to reverse the trend.

Objectives

The decision to shoot the 3-D survey was based on three objectives:

- Delineate new reservoirs and discover additional reserves that might be hidden in both structural and stratigraphic traps.
- Review the drilled areas and sort out the complex faulting and stratigraphy in order to determine where additional wells would facilitate production of already proved reserves.
- Assist in reservoir management via unification of the disciplines of geology, geophysics, and engineering in an attempt to obtain better models for reservoir simulation and to optimize the location of water injection wells for EOR.

Acquisition

Figure 9-11-2 shows in red the fully migrated portion of the 3-D survey. The areal extent of this area is over 60 mi^2 (150 km^2). This survey, acquired in 1988, covers most of Chevron's leased acreage (yellow line). It was recognized that a data set of exceptional quality was required to achieve the survey's objectives. To this end, the following four geophysical requirements were adopted:

- Cover all CDP bins.
- Obtain maximum resolution, both horizontal and vertical, because both are necessary to enable the isolation of many reflectors from the top and base of sand units and to evaluate stratigraphic changes.
- Obtain consistent offset and azimuth distribution.
- Obtain the best geometry—meaning that extra care had to be taken in the navigation quality control to ensure that data were obtained in the locations intended.

Achieving these objectives would require the best possible field techniques because the survey area contained many items that could significantly interfere with data acquisition. In addition to a water bottom that was nearly solid with pipelines and high voltage power

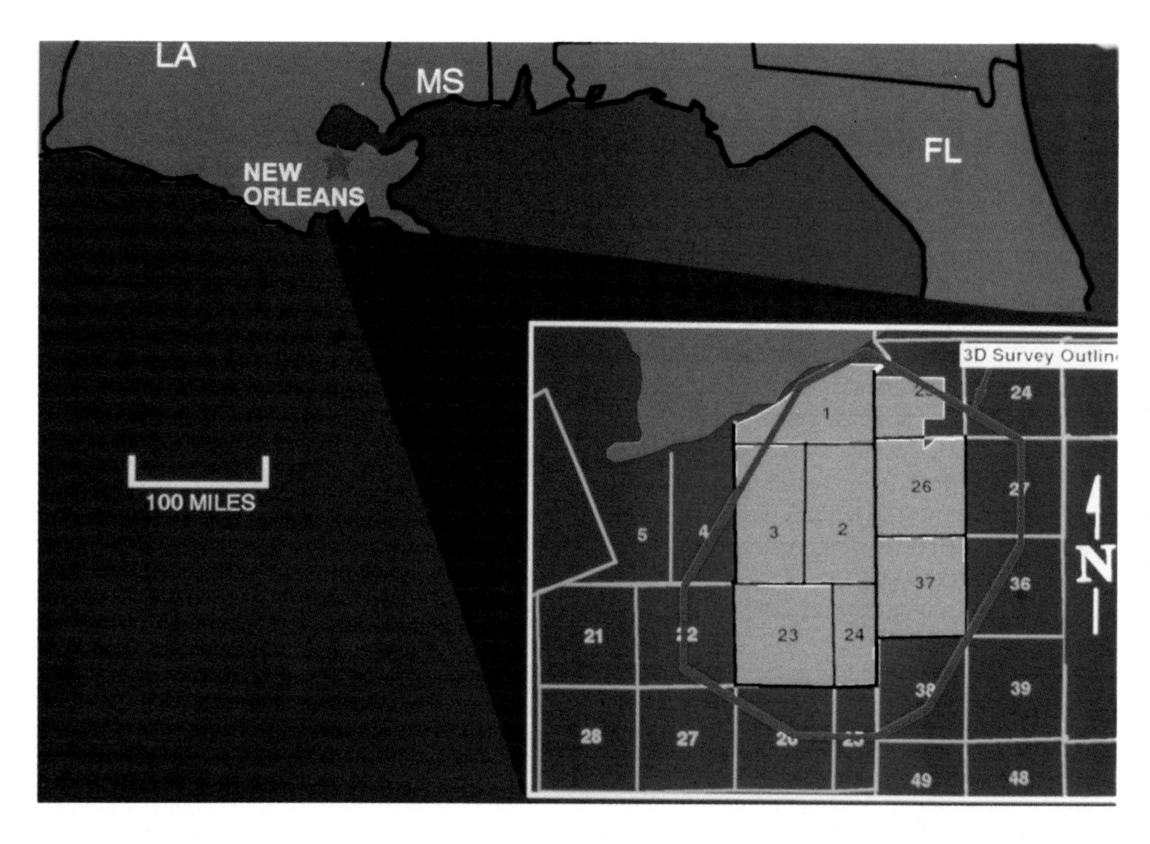

Fig. 9-11-1. Location of the Bay Marchand field offshore Louisiana.

cables, there were 114 surface obstacles in the form of single- and multi-well drilling platforms. This mandated that data be acquired via a state-of-the-art telemetry technique.

Western Geophysical's Digiseis acquisition system was employed for this survey. In this system a hydrophone is placed on the ocean floor and connected to digital electronics that are floating on the surface. All data are transmitted via antenna to the recording vessel. Since the hydrophones are not connected to each other by a long and cumbersome cable, this system makes it possible to position hydrophones in close proximity to the many surface facilities. In fact, it was sometimes necessary and possible to put the systems beneath the platforms. On most occasions, however, it was possible to obtain data at designated surface points by switching shot and receiver locations and/or undershooting the platform. The acquisition consisted of narrow swaths. The lines of receivers were laid out 880 ft (270 m) apart, and spaced at intervals of 220 ft (70 m) inline. Shots were taken every 110 ft in four rows laid out 220 ft apart and parallel to the receiver lines. This created a 3-D bin size of 55 ft × 110 ft (which was then interpolated to 55 ft × 55 ft for migration). Final coverage was 60-fold.

Due to careful acquisition techniques, the midpoint location plots differed from those encountered in a normal marine survey in two respects: (1) The bin size of 55 ft rather than the normal 82 ft was smaller, and (2) the midpoints were clustered in the center of the bins rather than scattered over the bin. It was believed these factors would significantly improve horizontal resolution in the final data.

The fold of coverage charts confirmed the excellent quality and even distribution of the data. There were no gaps due to the 114 surface structures, and no "striping" due to the hydrophones drifting to locations other than the desired sites. This is important because it increases confidence in interpretation of amplitude variations as being the result of subsurface geologic or petrophysical changes.

Structural Benefits

To date, the 3-D survey has helped the structural interpretation significantly in two key areas. First, the salt/sediment interface is better defined, and second, the complex fault geometries are better resolved.

Figure 9-11-3 is a time slice at approximately 5000 ft (1500 m). The quality is good and signal content strong. Most importantly, there is a clear definition of the salt/sediment

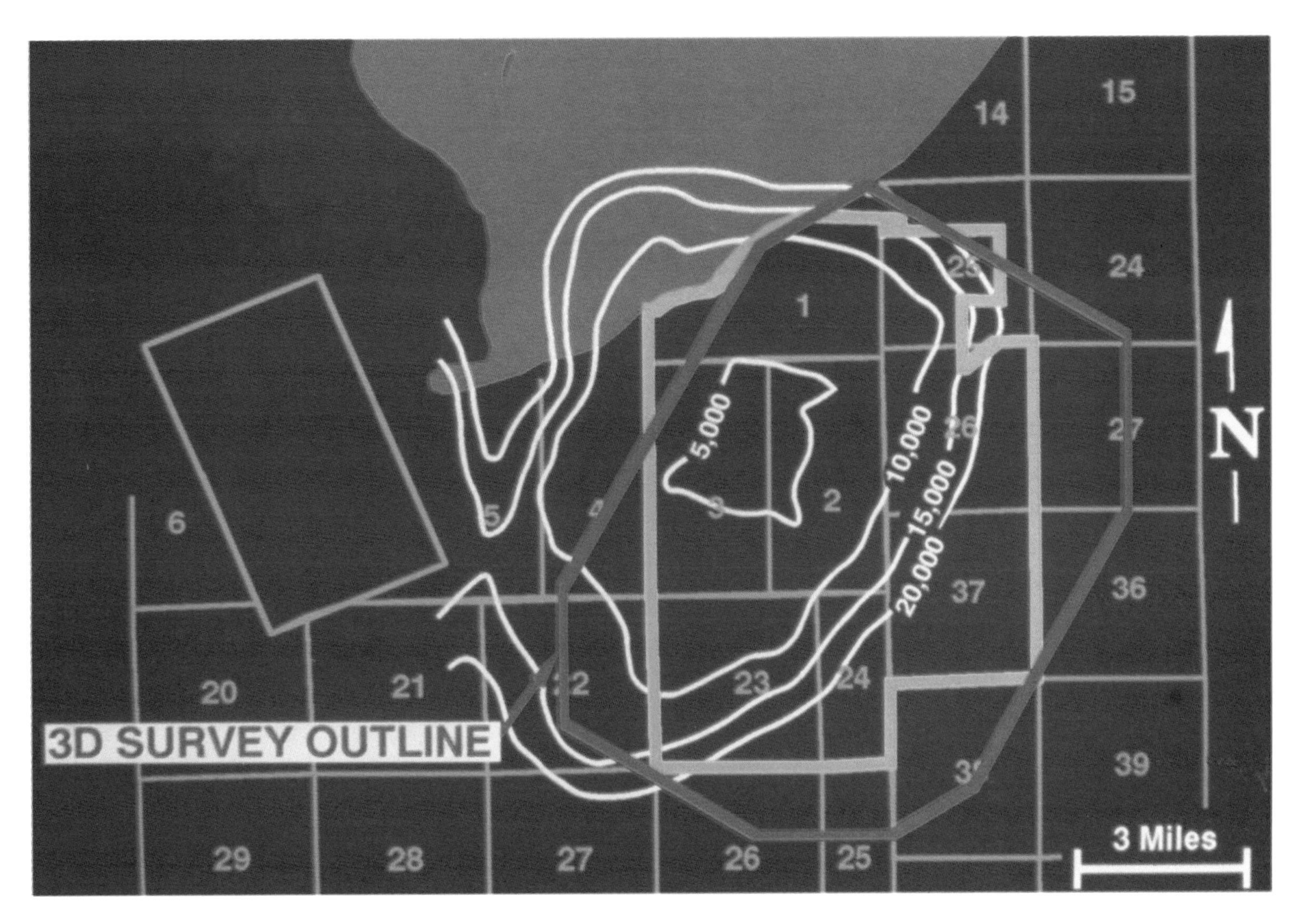

Fig. 9-11-2. Migrated 3-D survey outline superimposed on the structure of top salt.

interface in the core of the structure. This definition is very important because large amounts of hydrocarbons are trapped against this interface. Notice the irregular and unusual shape of this interface. Most geologists and management had anticipated an oval core shape and one that was much smoother. Certainly, the sharp corners of salt were not expected.

Figure 9-11-4 is a vertical section across the Bay Marchand salt dome. Again, note the sharp definition of the salt/sediment interface. Several well-defined faults also are evident. These must be accurately mapped because large reserves are trapped against them. Note, in addition, the sharp corners on the top of the salt. Some are undoubtedly enhanced by the acquisition and processing techniques employed.

Another perspective concerning the improvement in salt/sediment interface resolution is given in the next four figures. Figure 9-11-5 shows a top-salt structure map that was prepared prior to the availability of 3-D seismic control. This map, therefore, was generated using subsurface control which was quite abundant in the area. Note that there are 23 well penetration points; that is generally sufficient control over such an area for adequate top salt mapping by geologists. However, note the shape of the 5000 ft contour and compare with Figure 9-11-6.

Figure 9-11-6 is a time slice at approximately 5000 ft over the same area as Figure 9-11-5. Note the well-defined salt mass and the strong salt/sediment interface reflection. The latter indicates a large salt embayment which is not seen from the well control. This is a critical piece of information because it implies a potential reservoir, as sands may be trapped up against salt in this embayment. A usertrack along A-A′ should evaluate that potential.

Figure 9-11-7 shows the A-A′ usertrack and the concept of the prospect. The area of immediate interest is in the center of the section at 1.5 s. Note the high amplitude reflections and the good salt/sediment interface. Previous mapping had brought the highest area of salt near to the J-13 well and cut out the potential observed at the proposed location marked No.

Fig. 9-11-3. Time slice at 1.5 s showing salt-sediment interface.

1 Mile

N

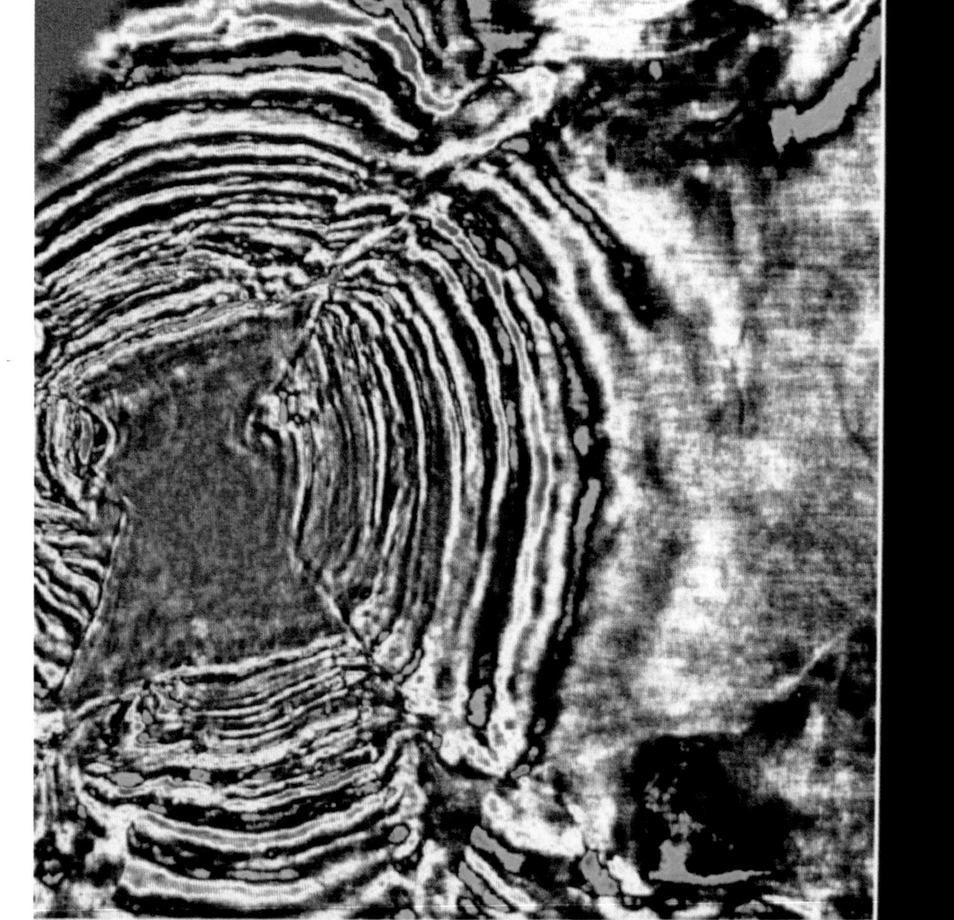

+

0

–

Fig. 9-11-4. Northeast-southwest seismic crossline from the 3-D survey showing structural detail at the salt interface.

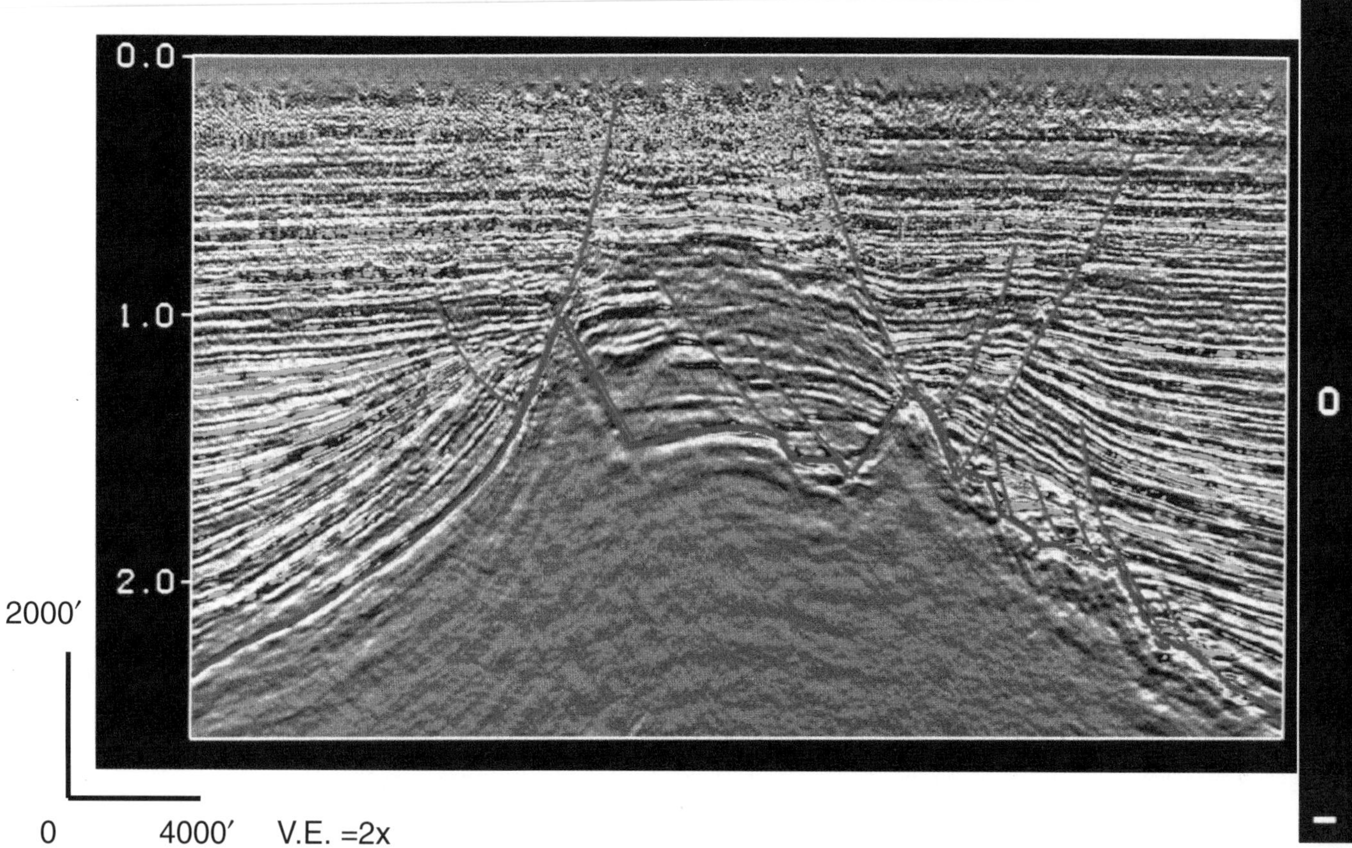

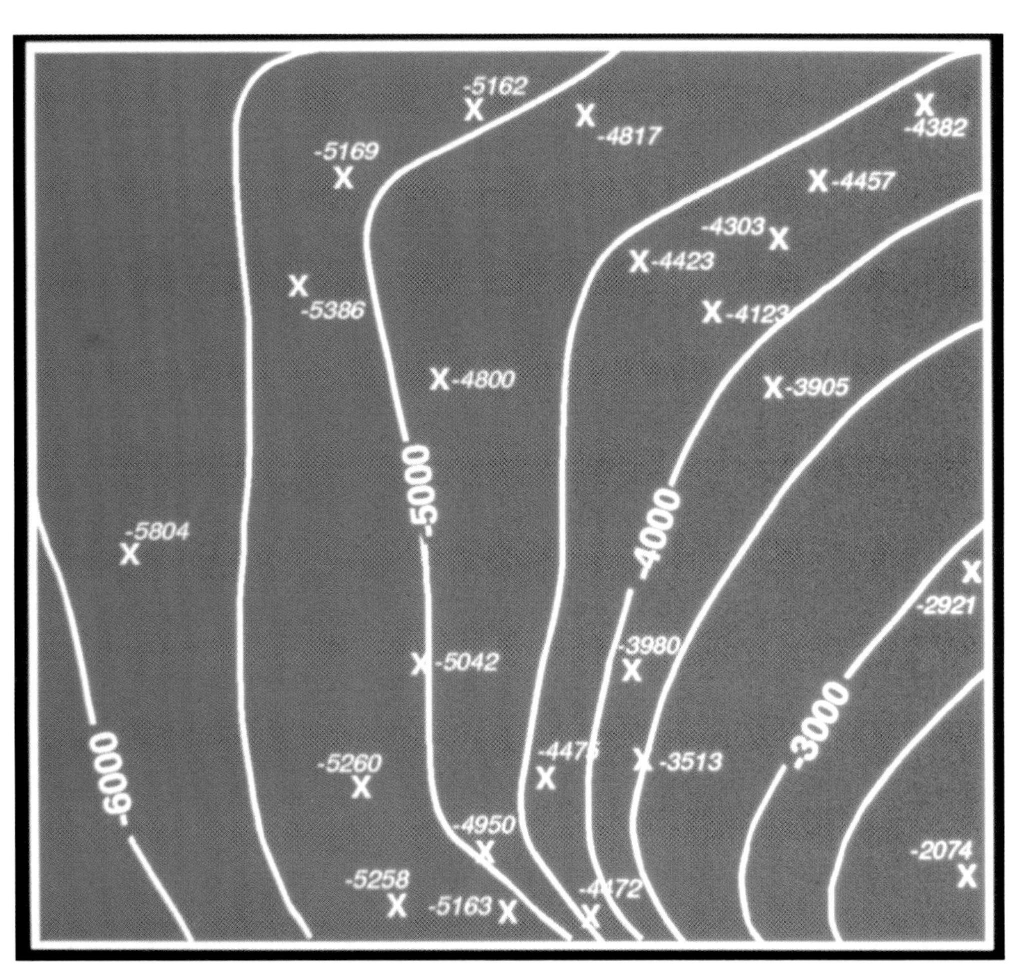

Fig. 9-11-5. Top salt structure generated from 23 wells prior to the 3-D survey.

1000′

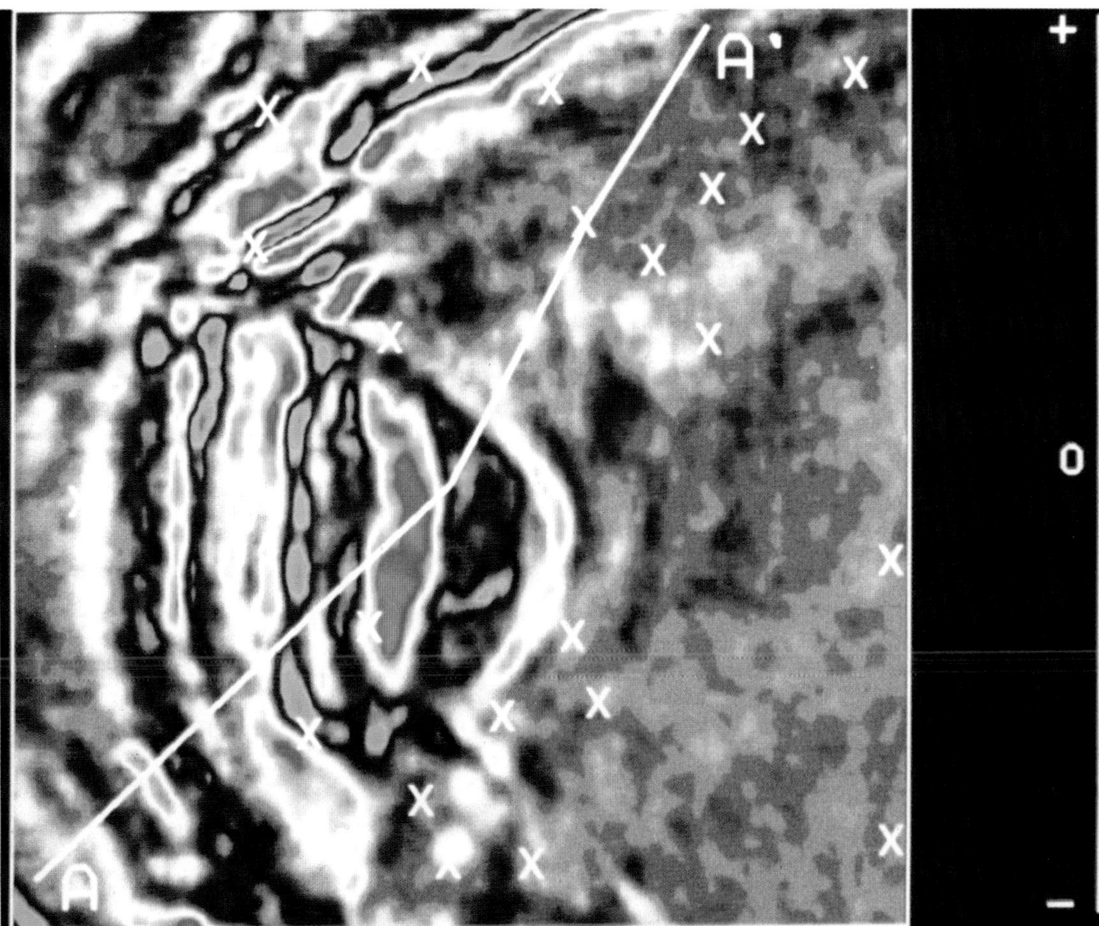

Fig. 9-11-6. Time slice at 1.5 s (detail of Figure 9-11-2) corresponding approximately to Figure 9-11-5.

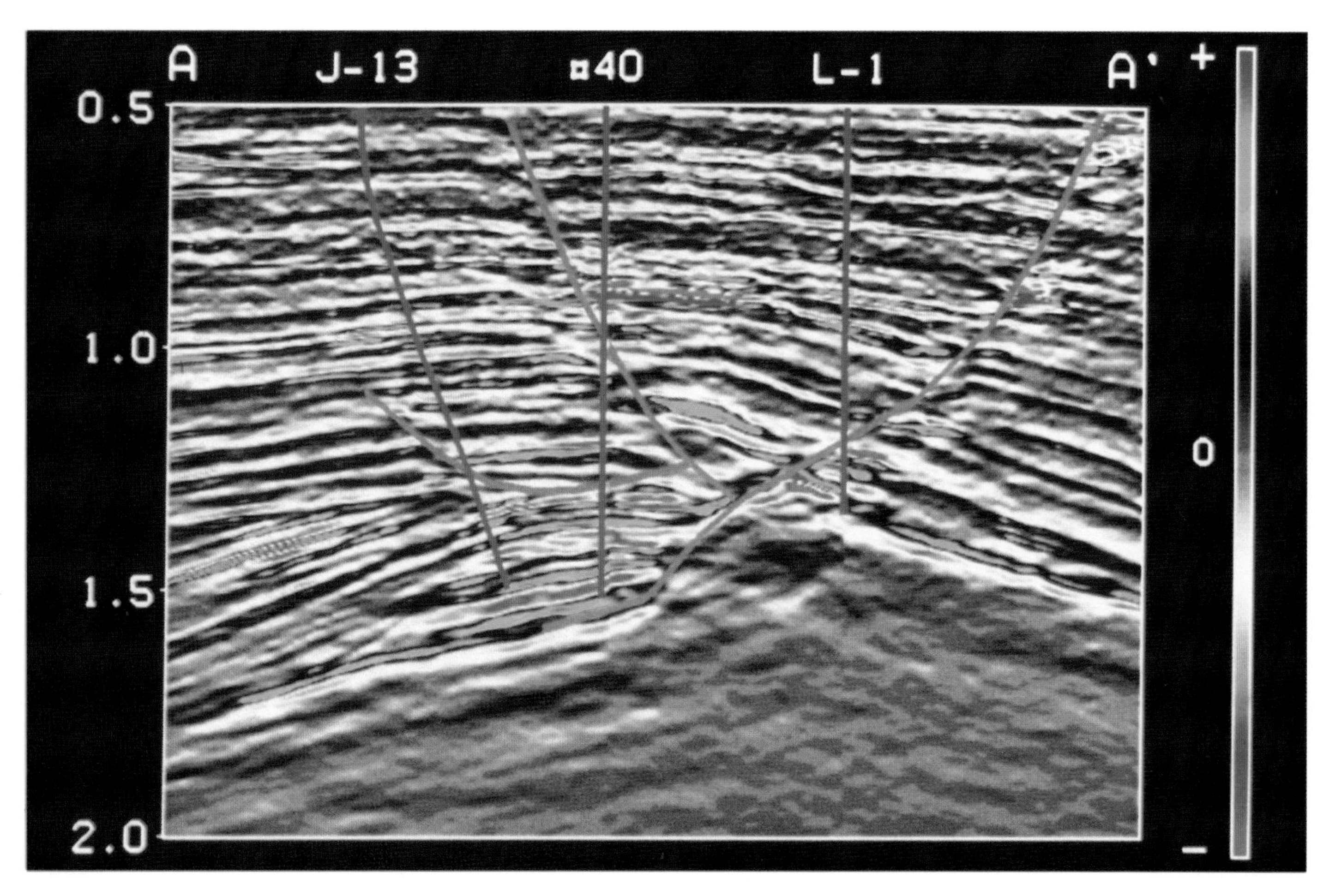

Fig. 9-11-7. Arbitrary line (or usertrack) AA′, the location of which is shown in Figure 9-11-6, illustrating new structural potential.

1000′

40. This location was drilled in 1989 to test the prospect, and found, as anticipated, thick oil and gas accumulations right up against the salt.

An example of sorting out complex faulting is discussed next. Figure 9-11-8 is a structure map of the 8200 ft Miocene sand, one of the major productive zones in Bay Marchand. This map was constructed prior to the 3-D survey and is based on subsurface control, production information, and 2-D seismic data. The solid green and red areas represent proven oil and gas reserves. The hachured red and green areas are possible and probable hydrocarbon zones. The gray areas represent shale-outs or permeability barriers interpreted from production histories. The pink lines are the 2-D grid of seismic data superimposed. This grid of data is 1/4-1/2 mile spacing. Normally this would be considered reasonable control, but it is ineffective here in fully delineating the fault blocks because several of the reservoirs are smaller than the grid size. As a result, confidence in this interpretation was very low.

Figure 9-11-9 shows the same horizon over exactly the same area as Figure 9-11-8, but it is quite different. For orientation, the yellow platforms are in the same location on each map. This revised interpretation (Figure 9-11-9) incorporates the same subsurface and production data, but the seismic control is now every 55 ft (17 m). In essence, an infinite number of seismic lines exist because of the interactive capabilities, which allow usertracks (arbitrary lines) to be generated in any direction. Therefore, more confidence can be placed in this interpretation. In fact, three wells have been drilled in new fault blocks (based on this interpretation) and all have come in essentially as mapped, establishing significant new reserves.

Stratigraphic Benefits

The 3-D data have allowed many stratigraphic insights:

- better well-to-well log correlations; the 55 ft spacing allows determination of shale-outs between wells and the tracking of horizons from one well to the next;
- improved understanding of paleo environments;

Fig. 9-11-8. Structure map of 8200 ft sand prior to the 3-D survey. Proven oil and gas are shown in green and red. 2-D seismic control is shown in pink lines. Forty well penetrations control this map.

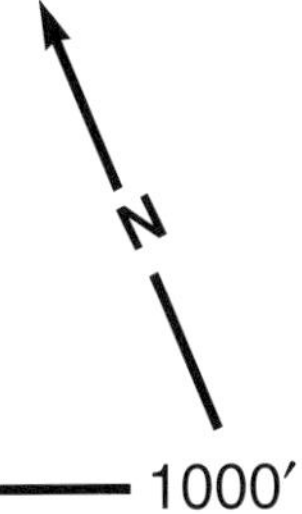

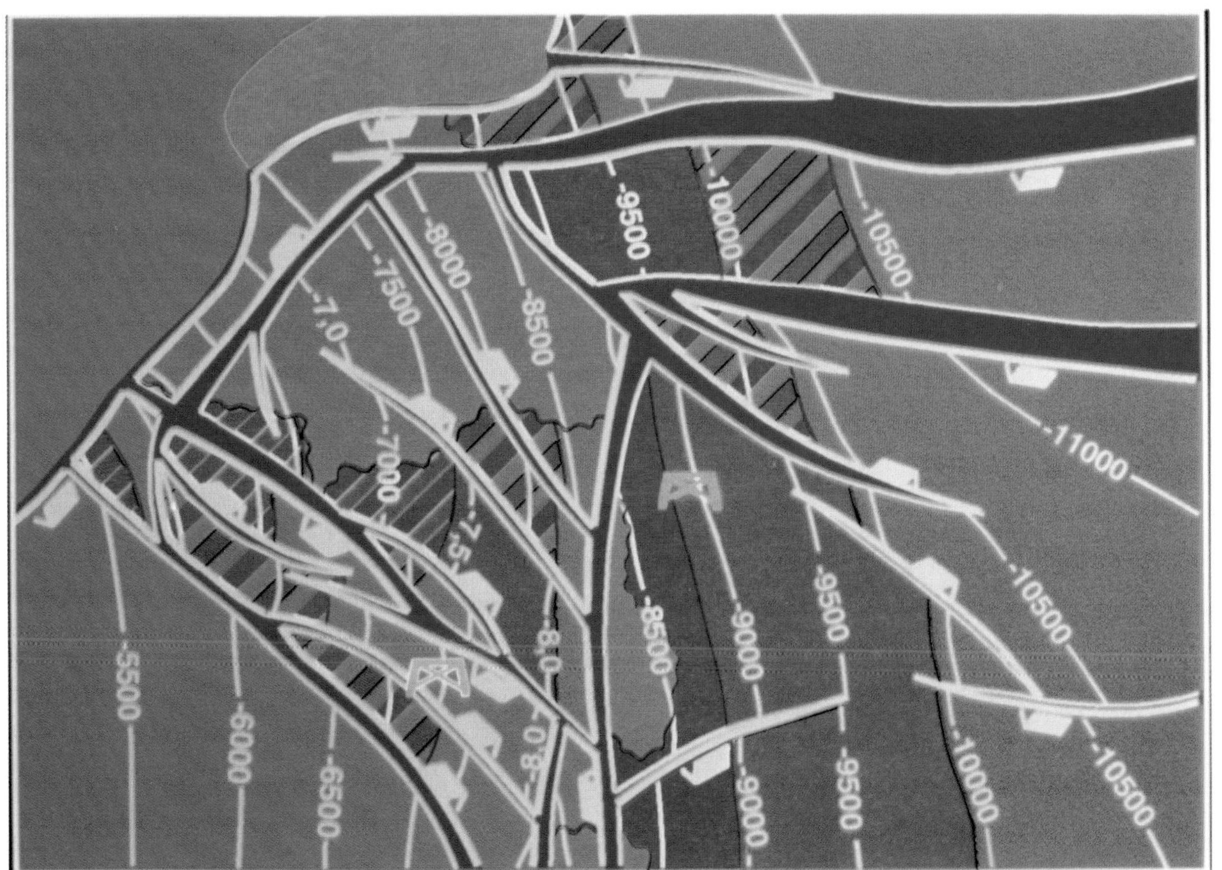

Fig. 9-11-9. Structure map of 8200 ft sand after evaluation of the 3-D survey. Note that no faults are the same as in Figure 9-11-8.

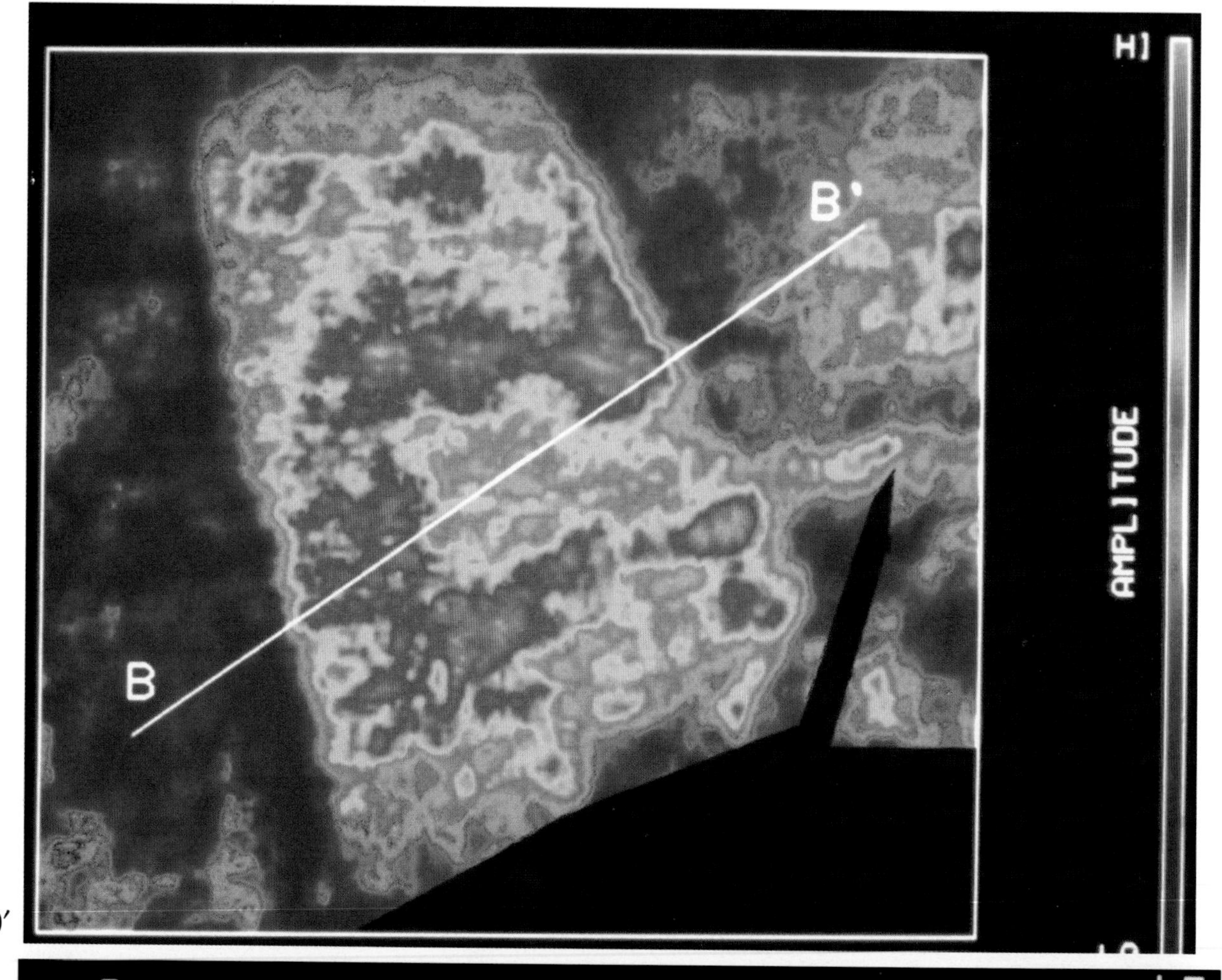

Fig. 9-11-10. Horizon slice of CP-7 sand showing potential stratigraphy associated with amplitude.

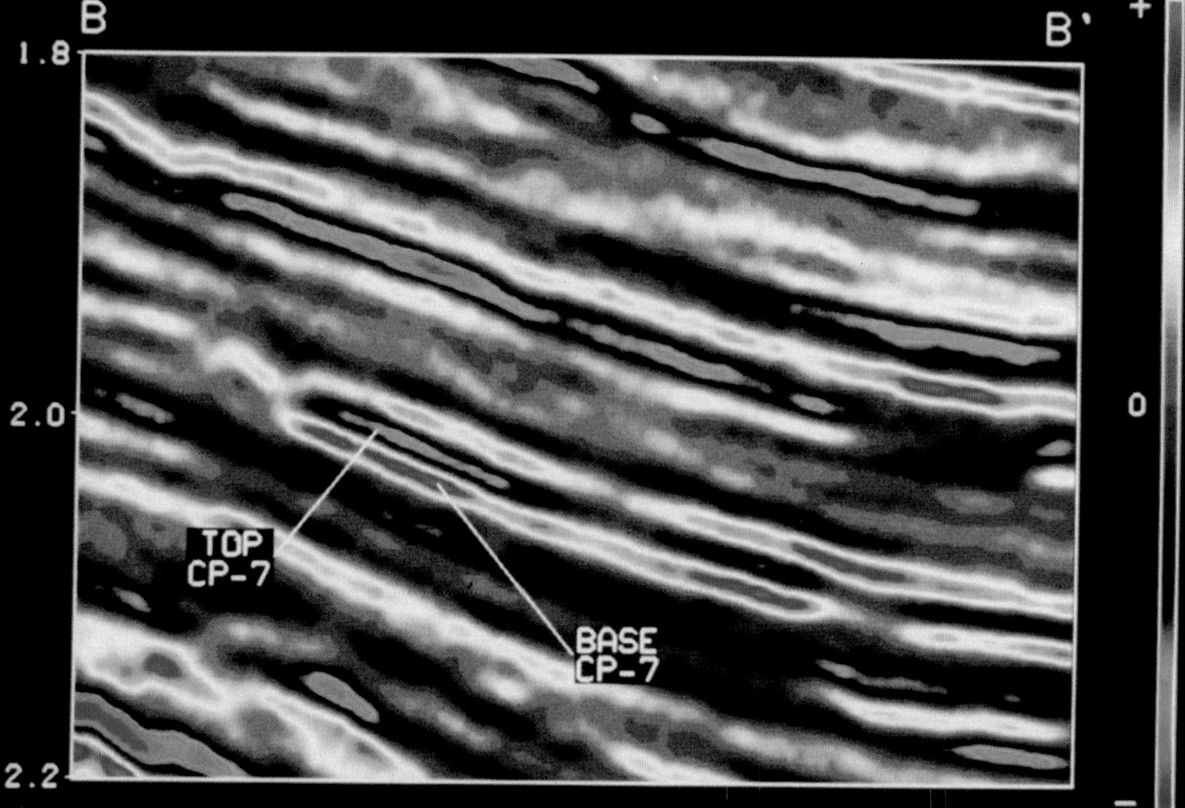

Fig. 9-11-11. Arbitrary line (or usertrack) BB′, the location of which is shown in Figure 9-11-10, showing stratigraphic terminations of CP-7 sand.

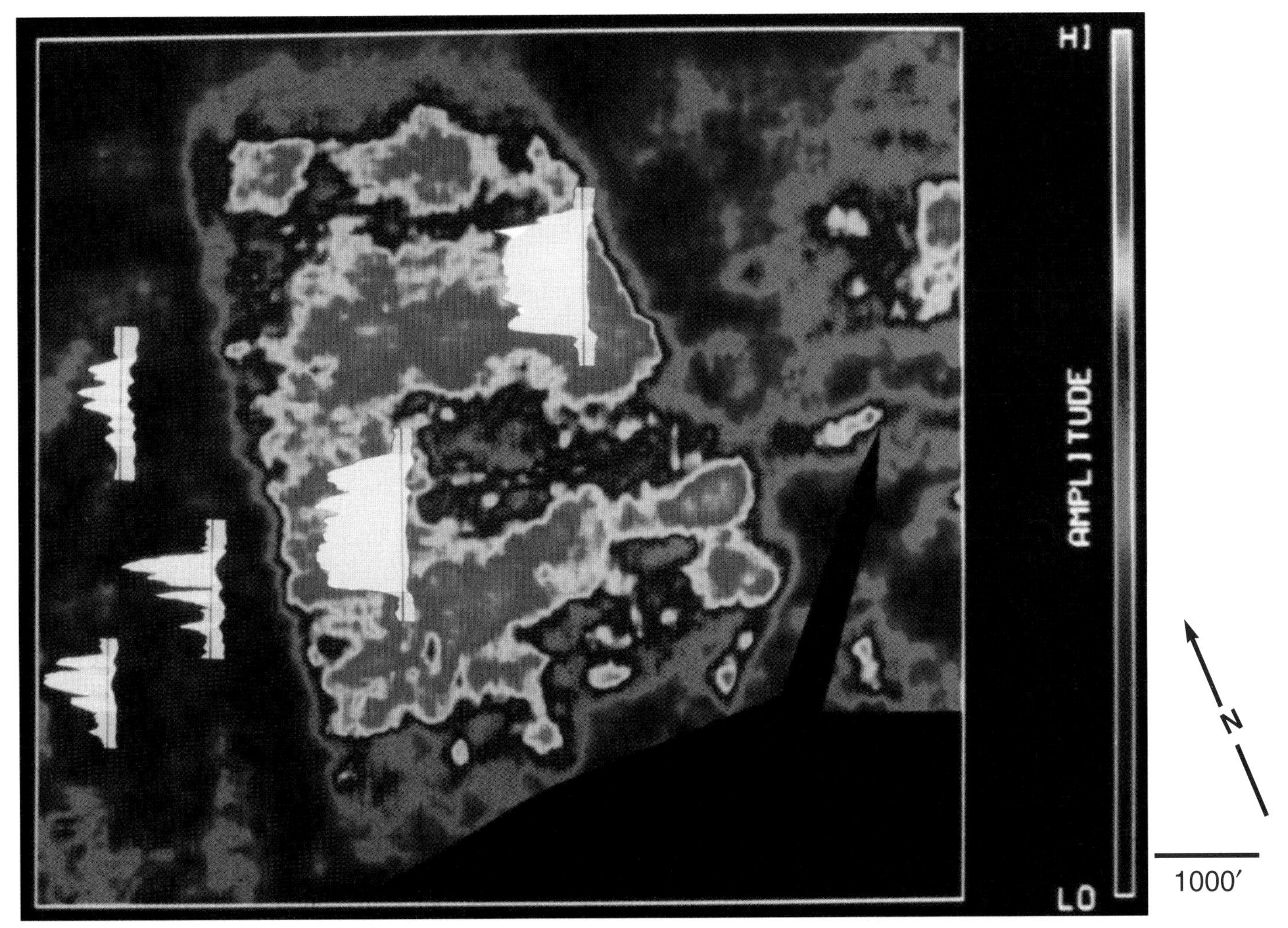

Fig. 9-11-12. Horizon slice of CP-7 sand with selected E-logs. The base of the log is the level corresponding to the horizon slice. Note that stratigraphy appears to be predictable from amplitude.

• better definition of sand distribution patterns.

The next set of figures provides an example of how a better understanding of the paleo environments has been achieved. For a 7000-ft sand horizon, the amplitudes were mapped and a horizon slice created to see what pattern was exhibited (Figure 9-11-10). Note the striking pattern of the area in red, which represents the highest amplitudes.

The amplitude distributions on horizon slices, in general, were used to help determine the location of the better developed reservoir sands. In this case, there are several things to consider. Notice how the high amplitudes terminate abruptly to the east and west. These terminations are believed to be stratigraphic boundaries. Usertracks oriented across these boundaries allowed them to be investigated. Figure 9-11-11, usertrack along the B-B′ line on Figure 9-11-10, clearly shows the strong amplitudes mapped at the CP-7 level. Again, these can be seen to terminate quite sharply at both ends of the display and without any apparent offset. These terminations correlate with the edges of high amplitude events on the horizon slice, and it is believed that the usertracks investigated substantiate the stratigraphic nature of the boundaries.

Based on the previous displays, the distinct pattern of high amplitudes in Figure 9-11-10 is believed to be associated with a meandering channel. The edge of these high amplitude events represents the edge of the meander belt. Therefore, sand within the meander belt can be expected to be better developed and sand in the overbank spill area can be expected to be less developed. Several wells penetrate this horizon and the logs from some have been superimposed on Figure 9-11-12. Logs in the meander belt do indeed show good, clean, well-developed "blocky" sand, whereas wells in the overbank spill area are poorly developed and shaly.

The next sequence of figures exhibits an attempt to define sand distribution patterns. The

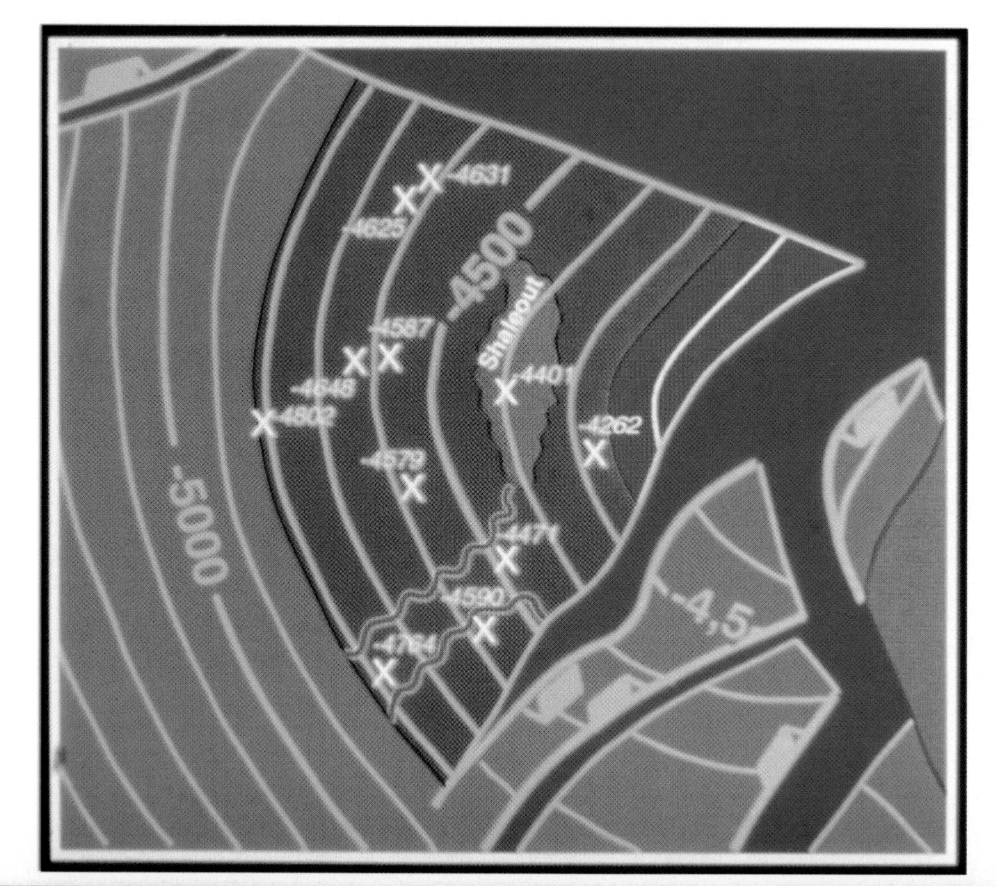

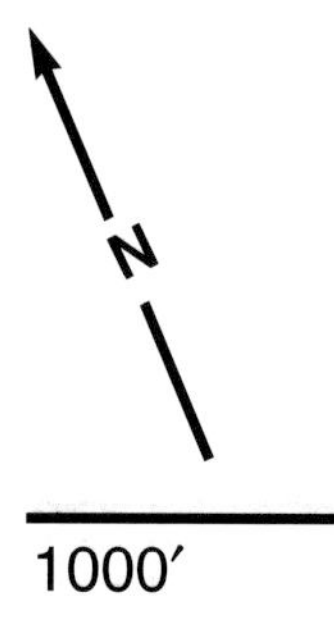

Fig. 9-11-13. Structure map of 4475 ft (1364 m) sand showing proven oil (green) and permeability barriers (red lines and gray shale-out).

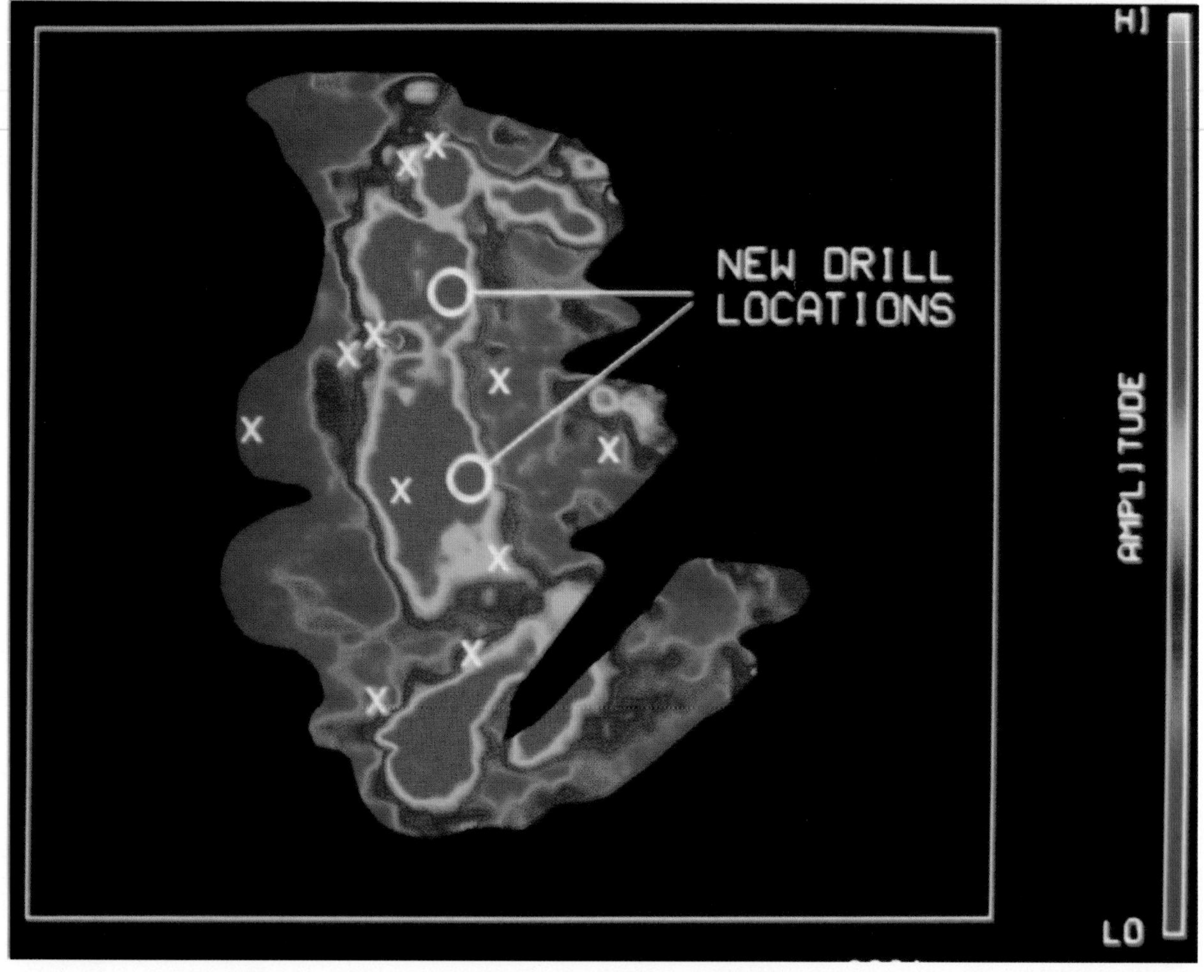

Fig. 9-11-14. Composite horizon slice after correction of tuning effects, corresponding to Figure 9-11-13 and illustrating new drilling opportunities.

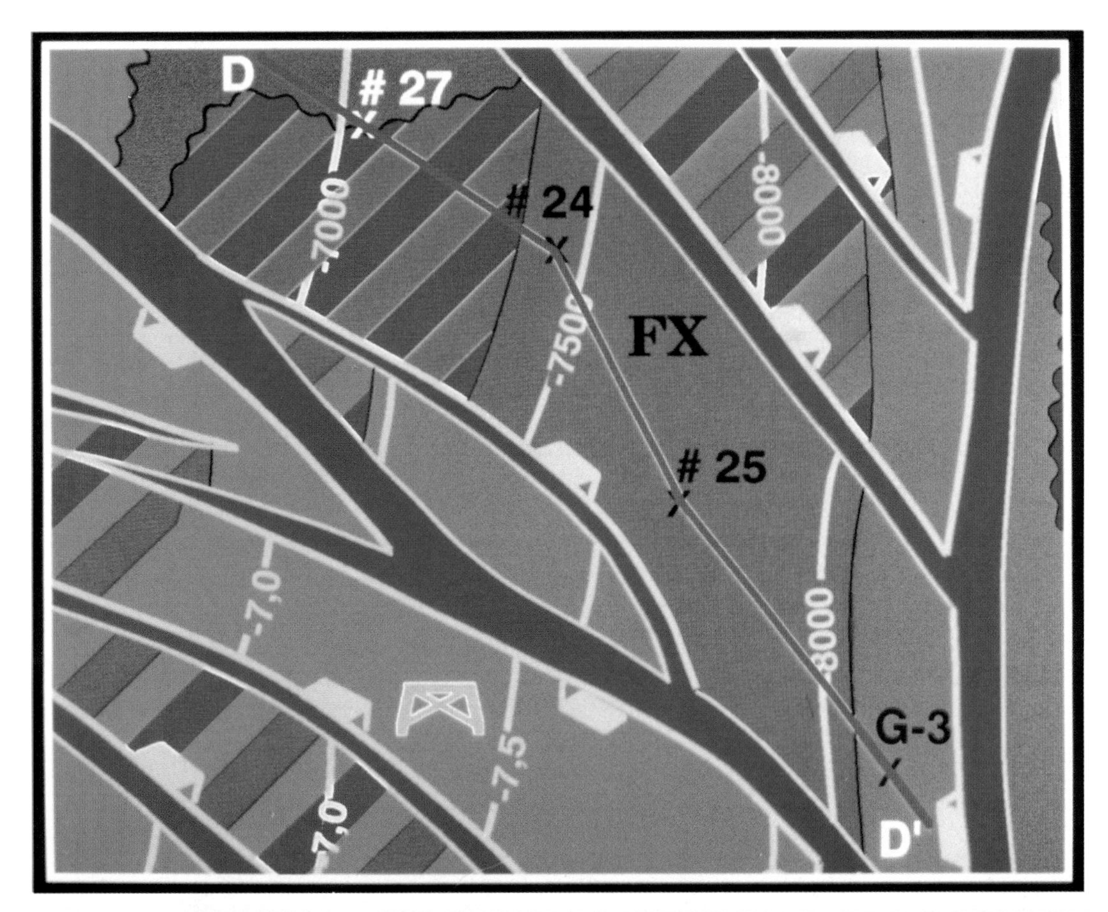

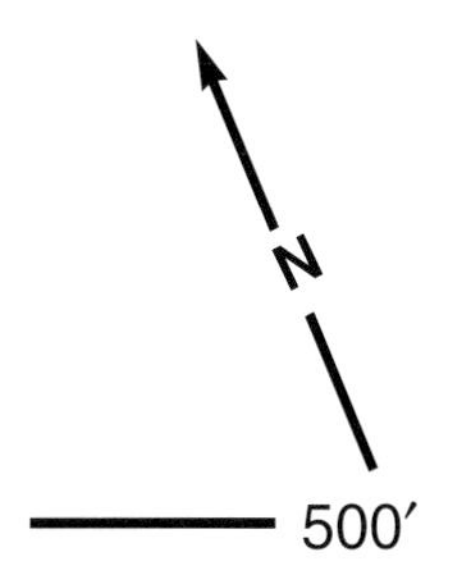

Fig. 9-11-15. Structure map of 8200 ft (2500 m) sand (detail of Figure 9-11-9) showing well penetrations in the FX block.

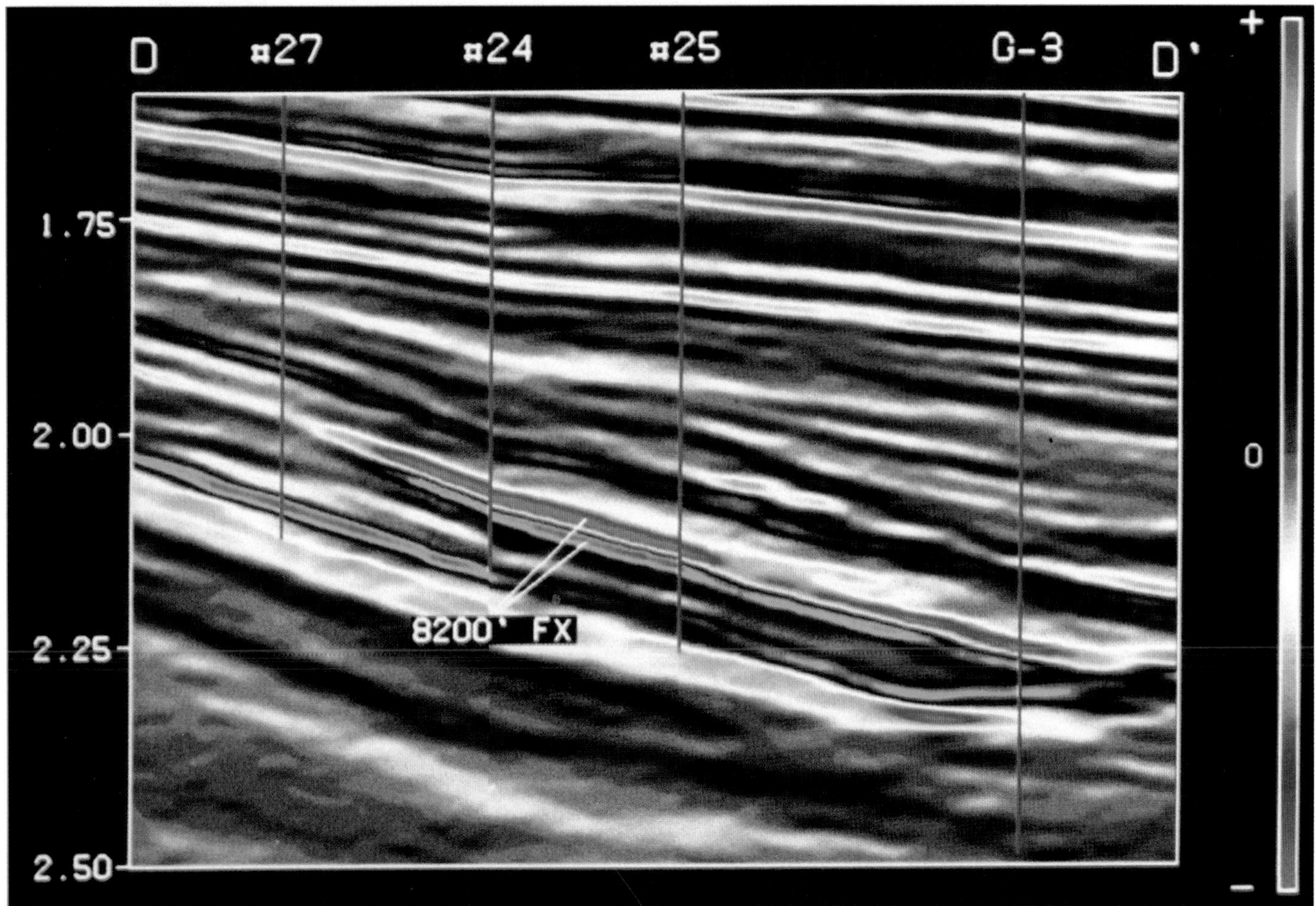

Fig. 9-11-16. Arbitrary line (or usertrack) DD′, the location of which is shown in Figure 9-11-15. Note the amplitude variations of the 8200 ft (2500 m) reservoir.

Fig. 9-11-17. Composite horizon slice after correction of tuning effects for the FX block. Note the strong variations between well control.

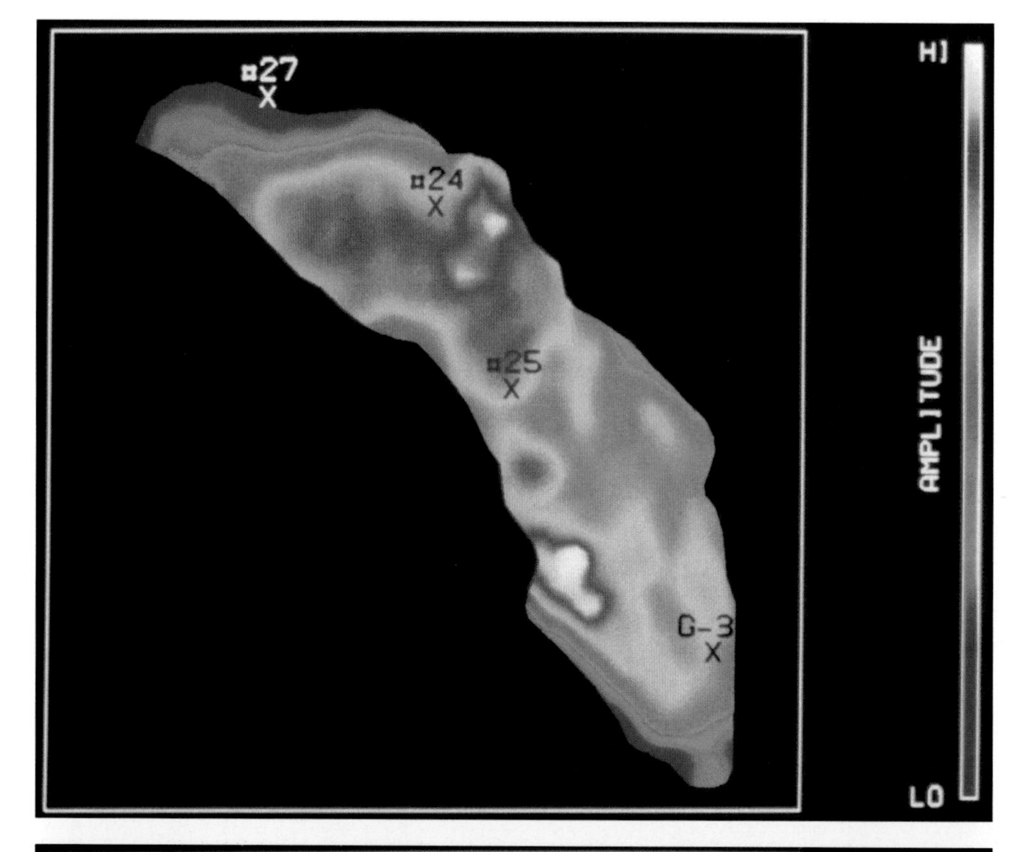

N

500′

Fig. 9-11-18. Net sand map of the FX block. Note the strong variations between wells.

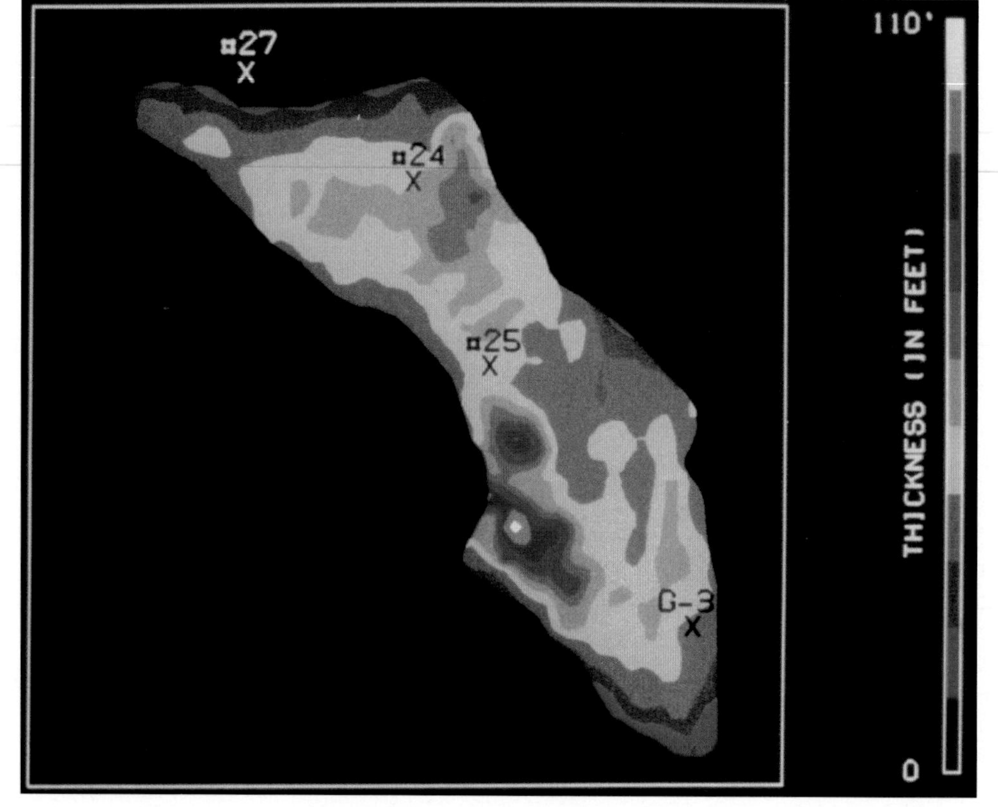

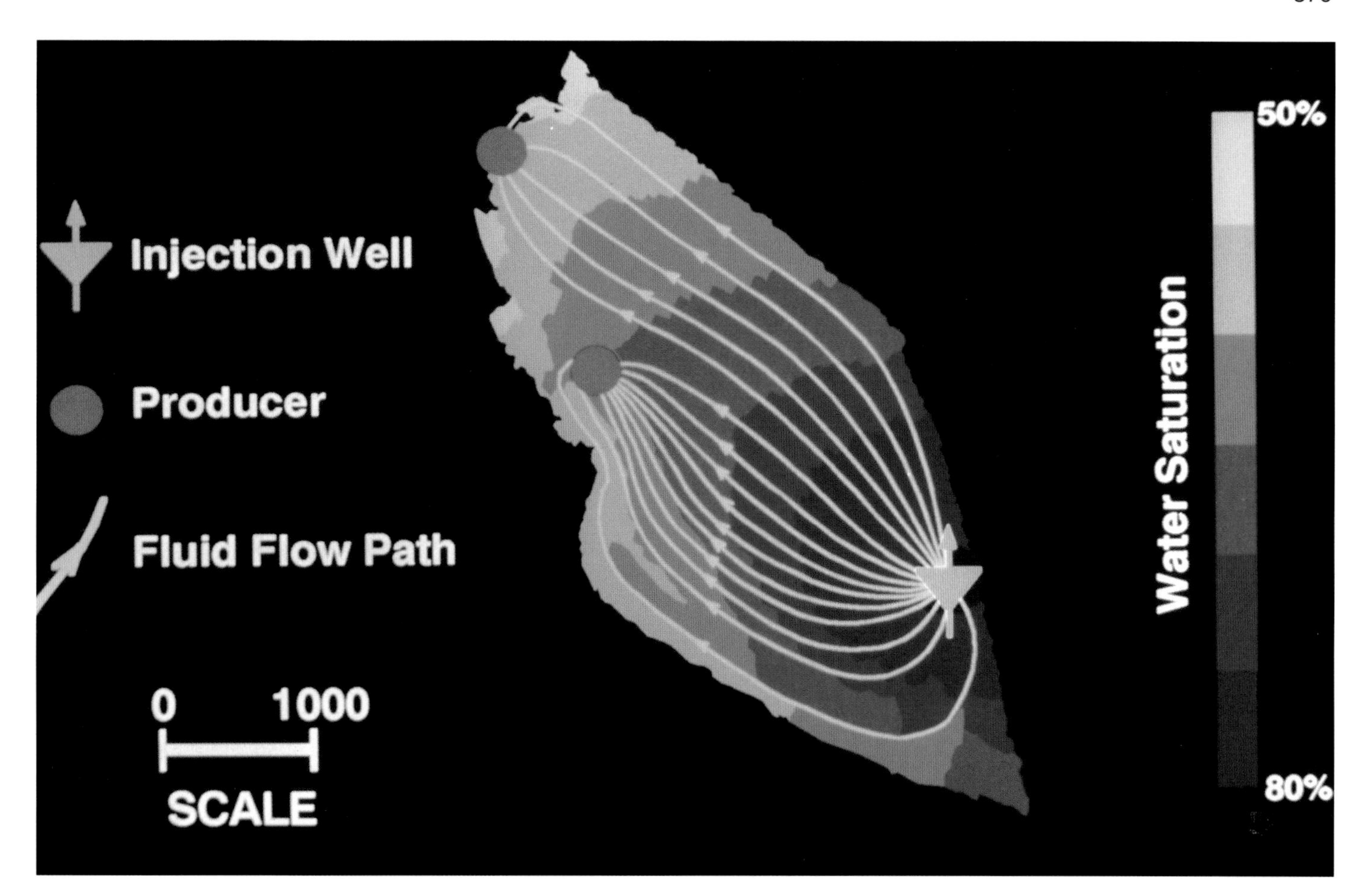

Fig. 9-11-19. Reservoir simulation of potential water injection well and takepoints at wells 24 and 25. Porosity-feet variations are based on both well and seismic control. Note that the injected water moves faster in the area to the north where the reservoir is thinner.

4475 ft (1364 m) sand in Figure 9-11-13 has produced significant reserves from several wells mostly in off-structure positions. Significant updip potential was recognized in this waterdrive reservoir, but there was also a significant stratigraphic risk. Well control indicated the 4475 ft sand thinned dramatically to the east (even shaling out in one well), while production from the wells to the south indicated a different drive mechanism and therefore the potential for permeability barriers, which are suggested by the red wavy lines.

In an effort to understand the stratigraphy of the area and to reduce the stratigraphic risk, the amplitudes associated with the 4475 ft sand were mapped and a horizon slice was created (Figure 9-11-14). This is a composite horizon slice in which amplitudes from the top of the sand were added to those from the base so that a good vertical integration was accomplished. Using the isochrons, this composite amplitude was "detuned" where thickness was at, or below, tuning. Notice the high amplitude red areas. These were interpreted to indicate the best reservoir sand. Note that the well control (white crosses) leaves the high amplitude area almost untested (only one in the better sand area). Notice also that the far updip potential no longer exists, due to probable shale-out.

Drilling of two additional wells was recommended as indicated, so that the remaining potential could be fully evaluated. The two wells were drilled in 1990 based on Figure 9-11-14. Both came in essentially as mapped, structurally and stratigraphically. Both had thick accumulations of sand at their targets, proving significant amounts of new oil.

Reservoir Management

The term "reservoir management" means different things to different disciplines and to different people. In this case history, reservoir management begins with the synthesis of information between geology, geophysics, and engineering to better "characterize the reservoir." How can 3-D data help in reservoir characterization? Again, this term means different things to different people. For the purposes of this paper, it will include:

- determination of reservoir structure;
- determination of aquifer structure;

- definition of gross interval thickness;
- location of original fluid contacts;
- calculation of porosity feet;
- calculation of net pay.

Under "optimum conditions," the six items listed above can be obtained and will assist the engineer in a more complete development of reserves and, as an obvious consequence, better reservoir management. The critical term in the previous sentence is "optimum conditions," which will be defined here as:

- having a data set with high signal-to-noise ratio;
- having confidence that the amplitude changes in the data represent geological or petrophysical changes rather than poor acquisition or processing techniques;
- having data that are zero phase and broad bandwidth.

Figures 9-11-15-19 show a mini case history of reservoir characterization that assisted an engineering waterflood project. Figure 9-11-15 shows an enlarged portion of the same structure map as Figure 9-11-9. The area to be discussed is the fault block known as the "FX" reservoir. Petroleum engineering recognized this fault block as having significant secondary recovery potential through waterflooding. This reservoir needed investigation because there had been a significant decline in pressure in the reservoir during its production history. At the present rate of decline, it was anticipated that production would cease by the mid-1990s.

What was needed to prolong production? Answer: An accurate interpretation of the structure and stratigraphy so that the remaining potential reserves could be estimated and the best position for a water-injection well determined.

There are other things to note from Figure 9-11-15. The structure is bounded on the east and west by faults, updip by a shale-out and downdip by another fault. Note that the system appears closed. Therefore it is most likely a depletion-drive reservoir. Several wells have penetrated the reservoir—two in the oil column: one updip and shaled out, the other downdip and water wet. The sands in the three southern wells were very well developed and clean.

Figure 9-11-16 is a usertrack along D-D′ in Figure 9-11-15. A good trough-over-peak (representing the top and base of the reservoir, respectively) is associated with the 8200 ft sand. Notice the clearly exhibited updip shale-out and the good downdip reservoir continuity. The high amplitudes extend down past the G-3 well which was water wet. This suggests that the amplitudes are primarily indicative of reservoir quality rather than of fluid content.

To evaluate the reservoir for waterflooding, the well log data were reviewed to determine average porosity/permeability/thickness/saturations, and the seismic data were studied to estimate these parameters between wells. Figure 9-11-17 shows the detuned composite amplitudes associated with the 8200 ft sand in the "FX" reservoir. Amplitude variations are present throughout the reservoir but the amplitude is about the same at the water-wet well G-3 and at the oil wells #24 and #25. This is yet another indication that sand quality is the primary cause of high amplitude.

Figure 9-11-18 is a net sand map, derived from Figure 9-11-17 by methods described in Chapter 7, that involve scaling the detuned composite amplitudes to well control combined with the top-to-base isochron values, and converting to depth. If only well control had been available, it would not have been possible to predict the thick area in the center, the thinning to the east, or the exact position of the updip shale-out. Using this display, appropriate volumetrics were calculated by simply adding up the values in each 55 ft^2 bin. Combining the volumetrics with engineering material balance work yielded an original oil-in-place value which justified the position of the oil/water contact as shown on the structure map.

Taking the original oil-in-place value, subtracting the oil already produced, and projecting the pressure decline curves yielded an estimate that only one-third of the original oil-in-place could be produced before pressures dropped below the point where primary recovery would be possible. Thus, the next step was to run a waterflood simulation to determine the amount of additional oil recoverable as a result of an effective sweep. Two sets of reservoir parameters were input to the simulation: one used only information from well control; the other added the variable parameters from the seismic data.

In this particular type of waterflood, streamlines are calculated to approximate fluid flow direction from the proposed injection point to the proposed take points. Figure 9-11-19 shows the flood-front at a particular stage of the waterflood. Many of these stages were calculated. When using only three points of well control, simulation using constant reservoir parameters was the best that could be done. However, the seismic data (with 55 ft spacing of control points) make a better, and possibly more accurate, solution available. In this case, it made the difference between a favorable economic forecast and an unfavorable one. With these variable parameters nearly 200,000 barrels/well of additional oil were estimated in the simulation. The additional oil also would be recovered more quickly. This obviously affected the economics of the potential waterflood.

Conclusions

The acquisition of a high-quality 3-D data set generated many improvements in both structural and stratigraphic interpretation that resulted in the delineation of new reserves in an undoubtedly mature oil field. The new stratigraphic information, in particular, led to additional economic take points in areas of already proven reserves. In addition, procedures were used to combine geophysical, geological, and engineering information to improve engineering decisions in secondary recovery projects.

The key factors in this success story came at the very beginning—the acquisition of a high-quality structural and stratigraphic 3-D survey and the proper processing of the data. These were absolutely necessary to accomplish the goals for proper management of the oil field.

As a result of the improved structural and stratigraphic mapping, the average daily production, which had reached a low of 18,000 bbl of oil/day in 1986, is now (1991) back up to 40,000 bbl of oil/day. This is a production level not seen in the past 10-15 years. It is believed that this increase will continue for several years, resulting in much greater ultimate recovery from this "granddaddy" field of the Gulf Coast.

Case History 12

Lisburne Porosity—Thickness Determination and Reservoir Management from 3-D Seismic Data

S. F. Stanulonis, H. V. Tran, ARCO Alaska, Inc.

Background

ARCO Alaska has developed a procedure to calculate total pore foot values directly from seismic amplitude, where pore foot is defined as reservoir thickness times average porosity. This was accomplished without an intermediate net pay map or average porosity map. The procedure was applied to the Lisburne Pool, a carbonate field located on the North Slope of Alaska. The upper Lisburne is Pennsylvanian in age, and stratigraphically separated from the overlying Permian-Triassic Sadlerochit formation. Structurally it lies on the crest and the southern flank of the Barrow arch. The field is bounded on the north by the major North Prudhoe Bay fault, truncated to the east by a major Cretaceous unconformity, and limited to the south and west by a variable oil-water contact. A porosity cross section through the area (Figure 9-12-1) shows the porosity is highly stratified and highly variable laterally. This makes porosity prediction difficult from wells alone.

The study was initiated in 1986 with 35 wells in the field. The area is covered by a 3-D seismic survey that was acquired in the late '70s. The survey consists of 45,000 bins with a trace every 220 ft (64 m), over an approximate 75 mi^2 (190 km^2) area. Production in 1990 was around 42,000 bbl/day from 65 wells. Production is mainly from the upper four zones (zones 7 to 4) of the reservoir, which consist of limestone and dolomite, each separated by a thin shale. The Lisburne is overlain stratigraphically by the Kavik shale, with a P-wave average velocity of 12,000 ft/s. Since this is much slower than the average Lisburne velocity of 17,000 ft/s, the horizon can be recognized as a peak on a positive polarity section. The seismic peak-to-trough amplitude of this horizon at the top of the Lisburne is plotted in Figure 9-12-2 where larger amplitudes are indicated by red, and smaller amplitudes by blue. The red, high seismic amplitude band occurring in the east, is where the Lower Cretaceous Unconformity (LCU) truncates to the top of Lisburne and replaces the overlying Kavik shale with slower Cretaceous shales. To avoid wavelet interferences and other complications in this region, the first part of this case history will be confined to the interpretation west of this truncation region.

Method West of LCU Truncation

Since there is no seismic lithology break in the upper zone of the Lisburne, the variations observed in seismic amplitude should be a function of variations in reservoir quality. One factor of reservoir quality is porosity. For a typical well in the area, the core porosity has been plotted against the sonic log velocity in Figure 9-12-3. As the core porosity increases, the sonic velocity decreases. Therefore, as the average interval porosity within the Lisburne increases, its average internal velocity will decrease, so that its contrast with the average velocity of the overlying Kavik formation will be smaller (Figure 9-12-4). Therefore, as the total pore footage increases within the Lisburne, the seismic amplitude will decrease, where 1 pore foot is defined as 1 foot of 100% porosity. Theoretically these are the results we expect to see by analyzing the surface seismic data. By plotting the total pore feet measured at each well against the seismic amplitude at that well location, that relation between seismic amplitude and total pore footage is observed in the reservoir, but within four separate regions.

Figure 9-12-5 shows that seismic amplitude decreases as the pore foot increases for each of four regions, marked R1 through R4. Not only are the regions statistically separate, they are also geographically separate (Figure 9-12-6). Therefore, the seismic amplitude at any proposed well location can be directly converted to a total pore foot value using Figure 9-12-5, since the region containing the proposed well location is known. The only problem with this method is that it will produce quantum leaps in the total pore foot map at these imaginary regional boundaries. To avoid this problem a different approach was taken.

For each well, an equation can be defined such that the seismic amplitude at that well location is equal to the slope of the regression line for that region, times the total pore foot values calculated at that well from logs, plus a constant value, "C." This C has interesting properties. Note that as the pore footage decreases, the seismic amplitude increases. At the point where the total pore foot value equals zero, the seismic amplitude equals C. Stated another way, C is the seismic amplitude at zero porosity. It is the seismic amplitude due to the impedance contrast of the Kavik/Lisburne contact, where the average velocity within the Lisburne is equal to the matrix velocity. This C value will be referred to as the seismic amplitude of the rock matrix. Just as the matrix velocity changes locally due to type of cementation, fracturing, grain-to-grain contact, etc., so will the seismic matrix amplitude (C) change locally. The parameters that cause these variations are due to the geological and geophysical imprints of the rock. This would include amplitude effects superimposed on the lithologic signature that are inadequately corrected in processing.

Four major geological imprints are responsible for the regional pattern shown in Figure 9-12-5, and are shown in Figure 9-12-7. One of these is fractures. The detailed fault patterns in Region R1 are different from those in the other three regions. It is a heavily faulted area with a different fault pattern and an assumed higher density of fractures. A smaller seismic amplitude would be expected for the same pore foot value. Region R1 is the region with the lowest seismic amplitude. The difference in seismic amplitudes between R1 and R2, for the same pore footage, may be related directly to the fracture density. Another geological imprint is the variation in thickness of the Kavik shale. Eastward, the Kavik shale is truncated by the LCU. The average velocity above the Lisburne is reduced from 12,000 ft/s to approximately 9000 ft/s where Kavik is replaced by Lower Cretaceous shales. Therefore, going from Regions R1, to R2, to R3, to R4, the amplitudes should be expected to increase for the same pore footage. Additionally, zone 7 thickens from the southern Region R2 to R4, and eastward from R1 to R4. Both of these geological imprints have a combined effect of making the pore footage appear to increase as seismic amplitude increases, if the four regions are considered as one in Figure 9-12-5. Another geologic imprint is gas. A gas cap exists in the north part of the field, above approximately –8650 ft (–2640 m) subsea (ss), and gas has a very marked affect of decreasing seismic amplitude for the same porosity. So that going northward, the seismic amplitude will decrease for the same pore footage. In Figure 9-12-7, this affect is most markedly demonstrated by the rapid transition from Regions R4 to R3 to R2 northward, in the central area of the field. Additionally, note the eastward transition of Region R1 into R2 starting at the gas cap boundary. The combination of all these factors produces the four regions that are statistically represented in Figure 9-12-5. The Lisburne is a complicated reservoir. We believe that in a simpler stratigraphic and structural setting, only one region would exist. But in spite of these complexities, the procedure gives a fairly accurate pore foot map. The procedure is as follows.

For each well the seismic amplitude, the total pore footage and the regression line for that location are known, so that C can be calculated from Figure 9-12-8. The C values are mapped in Figure 9-12-9, which defines the regression line to be used to convert seismic amplitude to total pore footage at any location. The seismic amplitude of the rock matrix (C) can be thought of as an operator. It converts seismic information to well log information. In principle, it is analogous to deriving average velocity at a well from seismic and log information. The average velocity is used to convert the recorded seismic time of a particular seismic horizon to depth, as recorded by logs in the well. Similarly, C is used to convert the recorded seismic amplitude of a particular seismic horizon to total pore footage, as recorded in the well. The procedure is different in that it only uses the second measured value recorded in the field, seismic amplitude, instead of seismic time. Thus, using the C map in Figure 9-12-9, and the recorded seismic amplitude at each trace location, a total pore foot map can be calculated.

Results West of LCU Truncation

Figure 9-12-10 is the resulting pore foot map calculated directly from seismic amplitude. Around 40,000 points have been independently calculated at a in-line spacing of 220 ft. This figure contains 20 color levels with a contour interval of 3 pore ft, equivalent to 3 ft of 100% porosity. This map can be compared to the pore foot values derived just from well control in Figure 9-12-11. Here the pore foot values are plotted on top of a color-coded structure map. The increase in horizontal resolution that seismic 3-D allows can be clearly seen by comparing Figure 9-12-10 and 9-12-11 in the seismically derived pore foot map. By summing the pore foot values over the area of concern, the total maximum geological reserves can be calculated, assuming zero water saturation.

For the last three years we have predicted pore foot values from seismic amplitude and compared them to drilled values. Results have been surprisingly accurate in spite of the complexities within the Lisburne. Seismically predicted pore foot values (from Figure 9-12-10) at these 16 new well locations were plotted against the log-measured pore foot values. The results are shown in Figure 9-12-12, which were achieved without updating Figure 9-12-10 after each of the 16 wells were logged.

Method East of LCU Truncation

The previous analysis concentrated on the area west of the truncation where a remnant of the Kavik shale still remains. The area east of the truncation is a transitional region where seismic amplitude would markedly change as a result of the loss of zones 7 through 4, and not as a result of loss of porosity. The thickness factor in the definition of pore footage is changing more rapidly than the average porosity factor. To examine the relation of seismic amplitude and pore foot in this area, a geological model was created (Figure 9-12-13). The resulting seismic amplitude of this synthetic seismic section is shown in Figure 9-12-14. The resulting peak-to-trough analysis shows that a cyclic pattern develops in the seismic amplitude as a result of the relative positioning of the shales and carbonates in the upper zones as they are being truncated. This cyclic pattern of the synthetic peak-to-trough analysis will take the shape of a banded structure in map view. The 3-D seismic amplitudes are mapped in Figure 9-12-15 and do appear banded. Again, the red regions are higher amplitudes. Also plotted are the synthetic times between this peak-to-trough event in the lower half of Figure 9-12-14. It, too, has a cyclic pattern, and the 3-D seismic peak-to-trough time differences also appear as a banded pattern in map view (Figure 9-12-16). Thus, by identifying the bands on synthetic analysis from wells with known truncated upper zones, a spatial correlation between the synthetic peak-to-trough analysis and the recorded 3-D seismic peak-to-trough amplitude and times will locate where the zones subcrop against the LCU. These interpreted subcrops are plotted in Figure 9-12-17. Additionally, for areas along these subcrops, the total zone thickness is approximately constant so that porosity thickness values could be derived as in the method west of the LCU truncation, in the first part of this case history.

Use of Porosity-Thickness Maps

Several applications are possible with a detailed seismic pore foot map. One is the construction of a truer geologic cross section. For any two wells in the area, only the porosity distribution of those wells is known. The seismically derived pore footage allows boundary conditions to be set every 220 ft laterally on the interpreted total pore foot distribution between these wells. Figure 9-12-18 shows such a porosity distribution between wells L3-08 and L3-02. The top of the Lisburne is indicated by the arrow and the red line. Notice that the peak-to-trough amplitude changes considerably between these two wells, suggesting that the total pore footage in zones 7 through 4 is changing at a faster horizontal rate than the geological cross section would indicate. The seismic cross section seems to indicate that the total pore footage is fairly constant for about 4400 ft (1340 m) from L3-08 and then decreases considerably for the next 3300 ft (1000 m). The cycle is repeated several times before L3-02 is reached. This is in contrast to the linear decrease laterally, derived from well values alone, between L3-08 to L3-02. Additionally, by studying the change in the wavelet shape through perturbing the porosity distribution within each zone separately, the zone of changing porosity can be inferred. For example, a decrease in seismic amplitude with an increase of peak-to-trough times is indicative of a zone 6 porosity enhancement. A lower zone 5 porosity enhancement has the same effect with an additional slight side lobe development. Thus, by studying the amplitude, character, and shape of this wavelet at the top of the Lisburne, a more detailed geologic porosity cross section can be constructed.

Another possible application using a detailed seismic pore foot map is the examination of the relation between faults and local pore foot anomalies. Figure 9-12-19 shows the seismically derived pore foot map with a detail fault overlay. Numerous regions are observed where the fault pattern appears to be directly related to the pore foot anomalies. In particular, locations X, O, and the area around E, A, and R. The change in pore footage around these minor faults (less than 20 ft) may be the result of preservation (or loss) of section due to this faulting, or it may be the result of porosity changes due to introduction of fluids along related fractures. Thus, local irregularities in the pore foot map can be used to study faults and their affects on porosity distribution. Conversely, examining the 3-D seismic at the boundaries of these anomalous pore foot patterns may help locate minor faults near the limits of seismic resolution, such as location F (Figure 9-12-19). Only one small lateral fault was interpreted on the east side of this pore foot anomaly. But after careful examination, numerous discontinuous fault segments were linearly mapped on the west side with throws near resolution. Additional fault cuts were found on the east side, and the lower bounding northwest-extending fault line was projected farther northward. A horst block was therefore interpreted to be at location F bounding the anomalous pore foot pattern in Figure 9-12-19.

It is well known that gas has an affect on seismic P-wave velocity, which in this environment decreases amplitude considerably for the same porosity. Figure 9-12-20 is the peak seismic amplitude map at the Lisburne horizon where blue is the low amplitude and red the high amplitude. The large blue region corresponds well to the original oil-gas contact (–8600 ft) and is within the –8650 ft contour interval. Additionally, other gas caps can be found to the south more than 150 ft deeper within the Lisburne section. If it were possible to reshoot the seismic survey with identically the same acquisition and recording parameters and ground conditions, large amplitude differences between the new and original processed amplitudes should correlate well to unswept, producible oil. Areas therefore may be mapped locating additional trapped oil.

The final application attempts to relate productivity to total pore foot values for the drainage area around a well bore, rather than just to the logged pore foot values at the well bore. Clearly, wells drain areas where the pore footage is not logged but is assumed to have a value proportional to the linearly weighed distance of the total pore foot difference between two wells. Figure 9-12-21 shows there is a relation between total pore foot and effective pore foot, where the latter term is defined as pore footage with enough permeability to contribute to production (Durfee, 1988). It is a permeability weighted pore foot value. Those wells along the upper trend have more effective porosity for the same total pore footage and are all located in the truncation areas, or within the gas cap. Figure 9-12-10 then could be multiplied by this ratio and a total effective pore foot map produced. The lower well in Figure 9-12-22 shows that production should decrease faster over time than for the more eastern well, even though both wells have the same pore footage measured along the well bore. This is due to the more porous rocks around the eastern well. Production over time at any well then can be equated to the integrated effective seismic pore footage over selected radii from the well, and a set of simultaneous equations generated with factors for fault/fracture enhancements, as well as rock type.

Conclusions

ARCO Alaska has developed a procedure to calculate porosity-thickness directly from 3-D seismic amplitude. The procedure uses observed local trends between seismic amplitude and total porosity-thickness from wells to project the seismic amplitude that should occur at that location at zero porosity, i.e., for the rock matrix. The geologic and geophysical imprints that affect this value are complicated and numerous for the Lisburne formation, and yet it was still possible to derive an accurate, detailed pore foot map. The procedure is partly an empirical technique which, although it may have wide application, must be locally calibrated to available well control. The resulting detailed pore foot map serves to enhance reservoir description and assist engineers in developing a more accurate reservoir model.

The techniques and/or conclusions are those of the authoring company and may not be shared by other Working Interest Owners.

Reference

Durfee, B. A., 1988, Matrix Characterization of the Upper Wahoo Formation, North Slope, Alaska: ARCO Internal Report, December.

Figures begin on page 386.

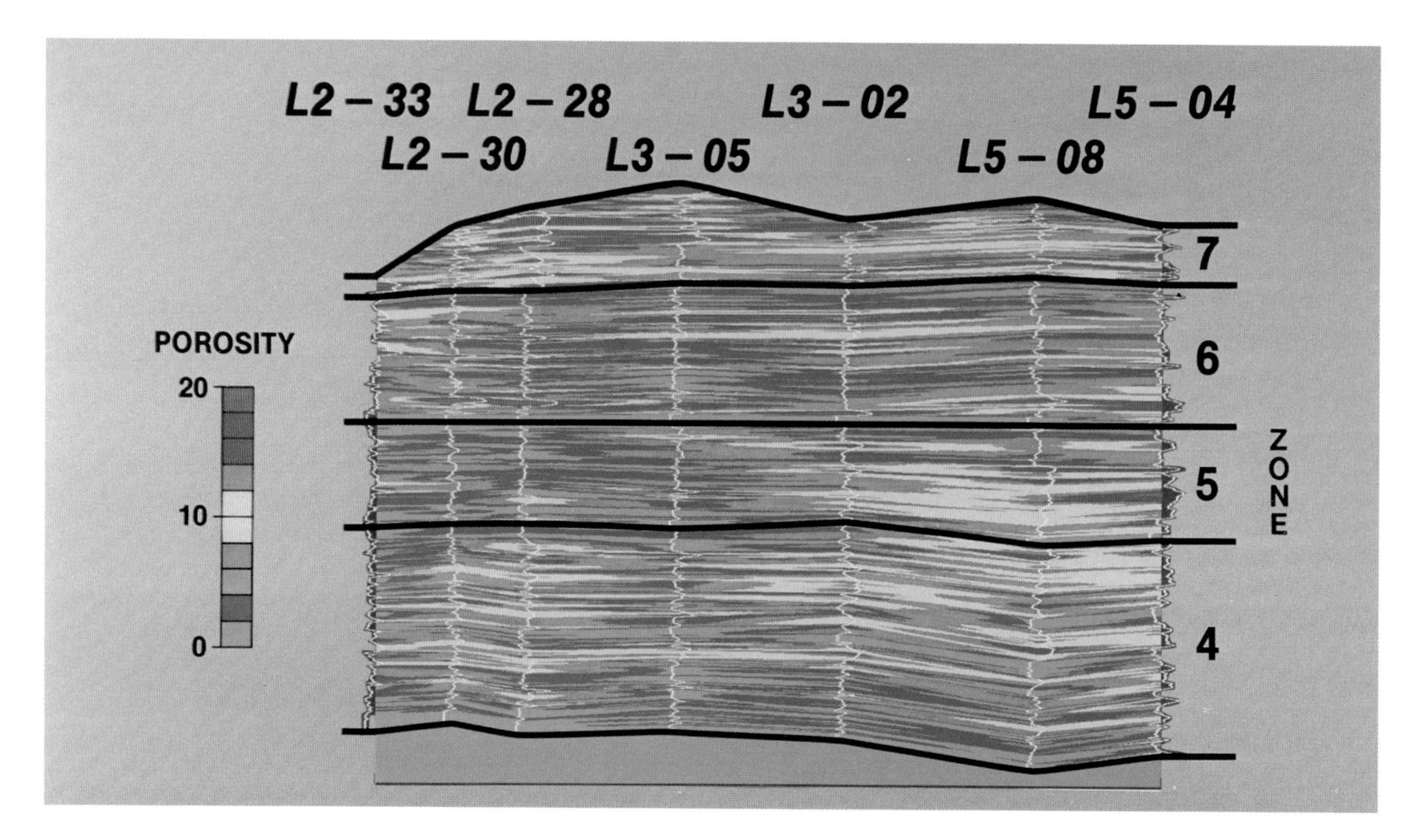

Fig. 9-12-1. Well-derived porosity cross section through Lisburne field.

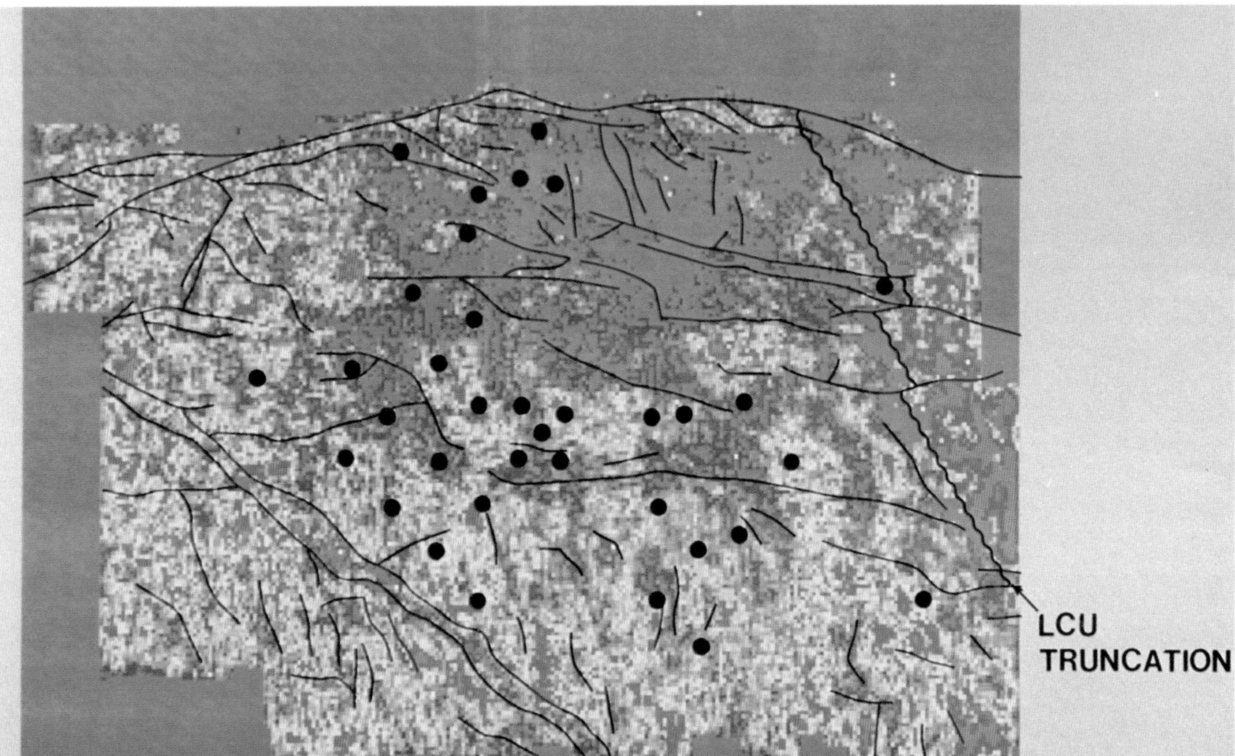

Fig. 9-12-2. Horizon slice for top Lisburne reflection displaying peak-to-trough amplitude. High amplitudes are red; low amplitudes are blue.

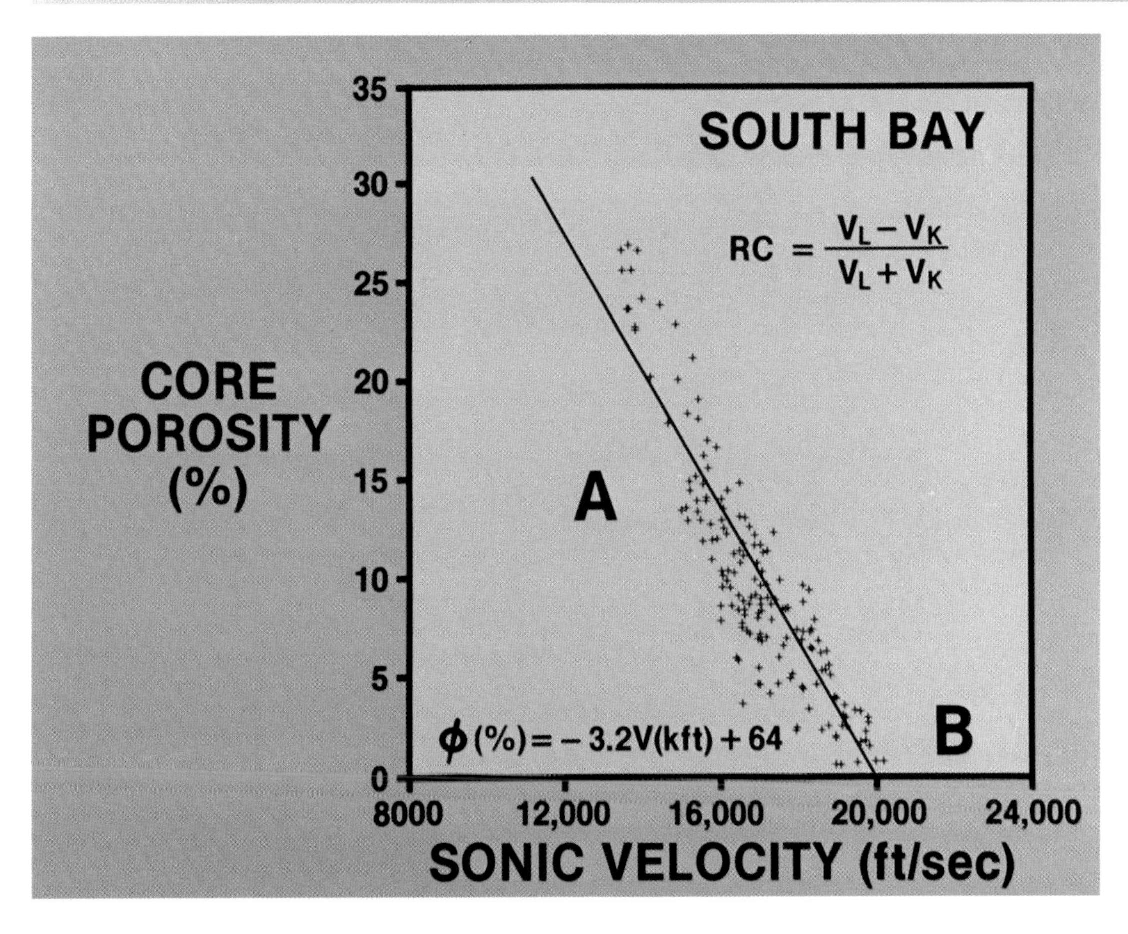

Fig. 9-12-3. Core porosity versus sonic velocity for a typical well.

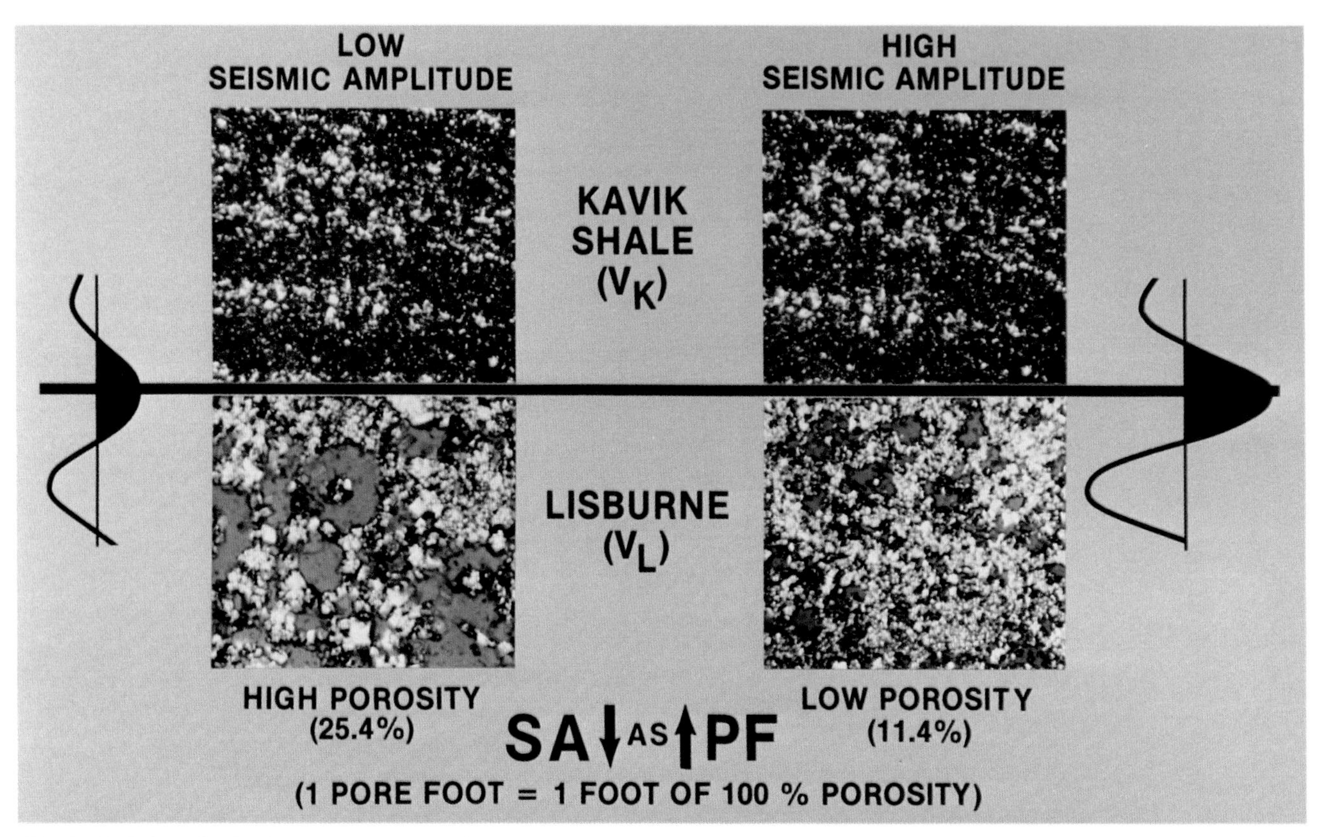

Fig. 9-12-4. Amplitude decreases as porosity increases.

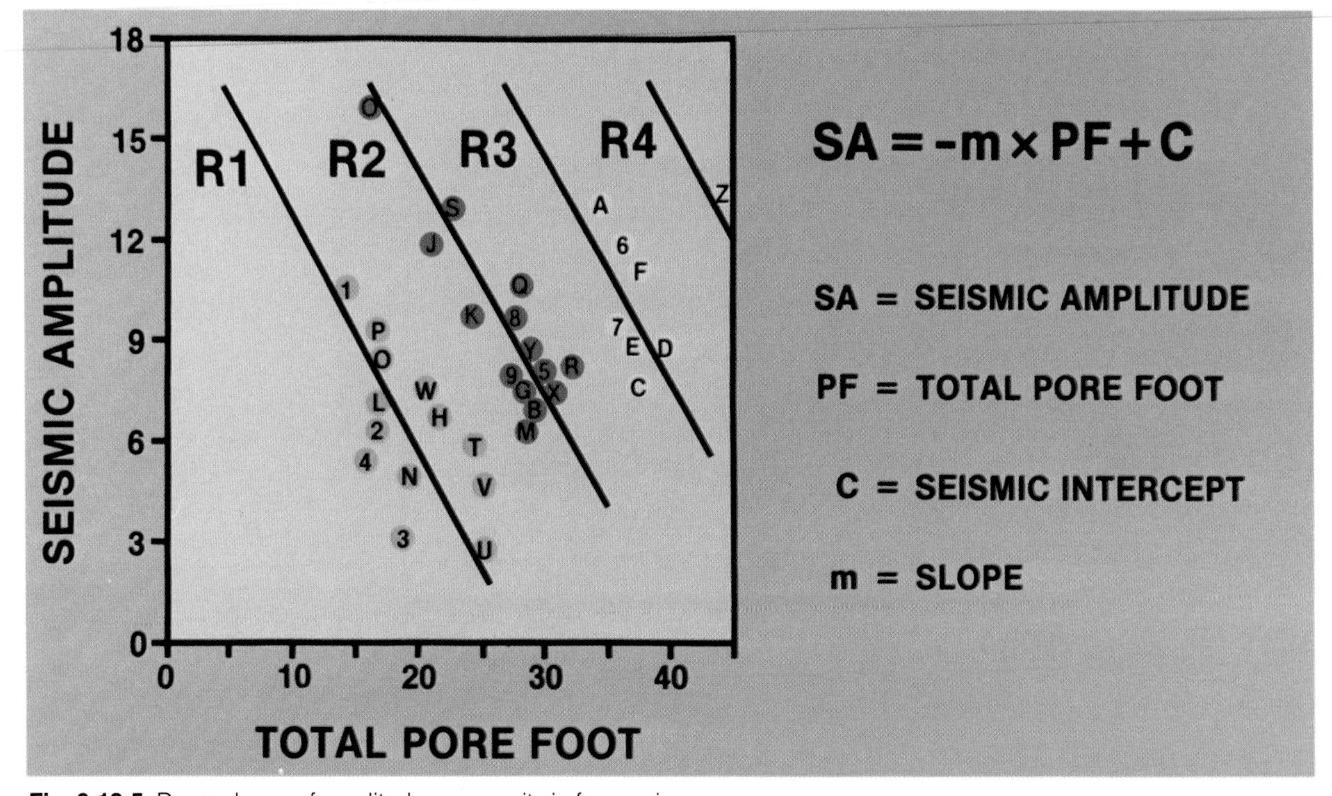

Fig. 9-12-5. Dependence of amplitude on porosity in four regions.

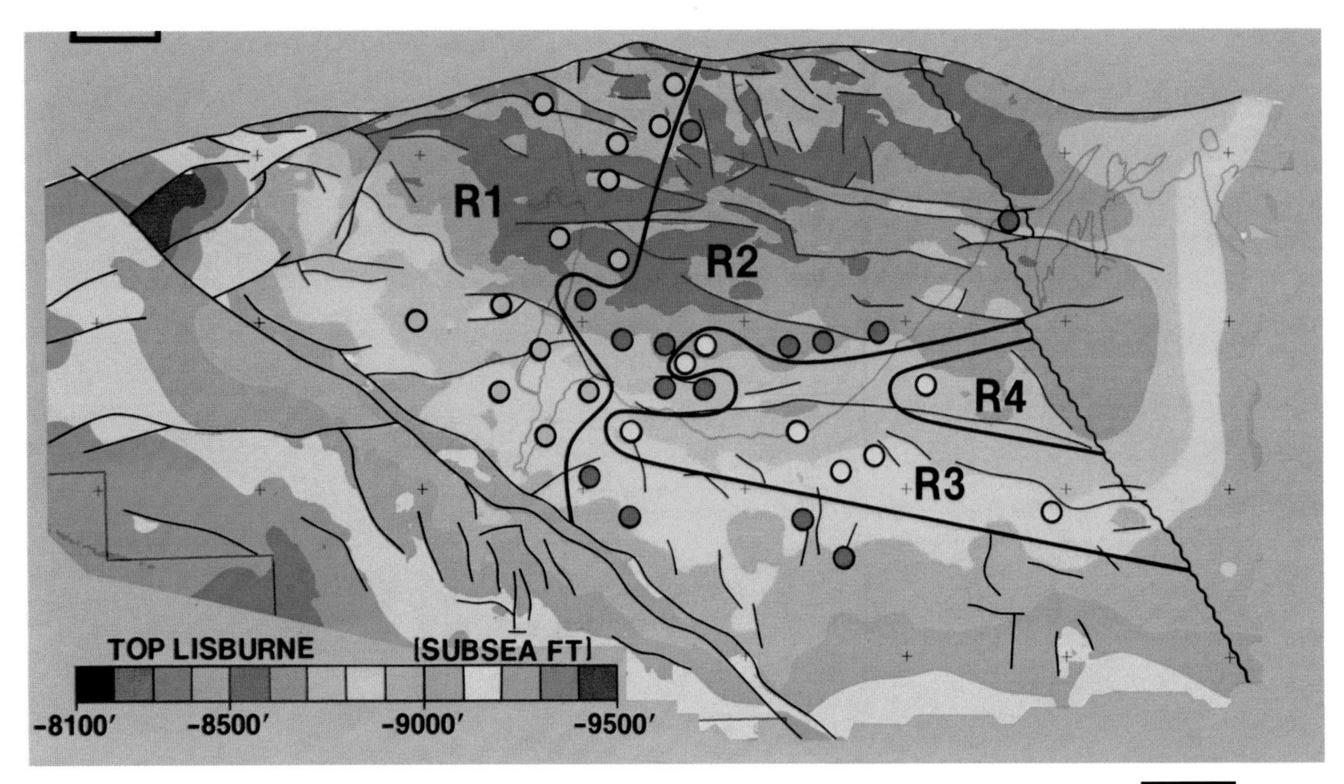

Fig. 9-12-6. Locations of seismic intercept regions.

1 Mile

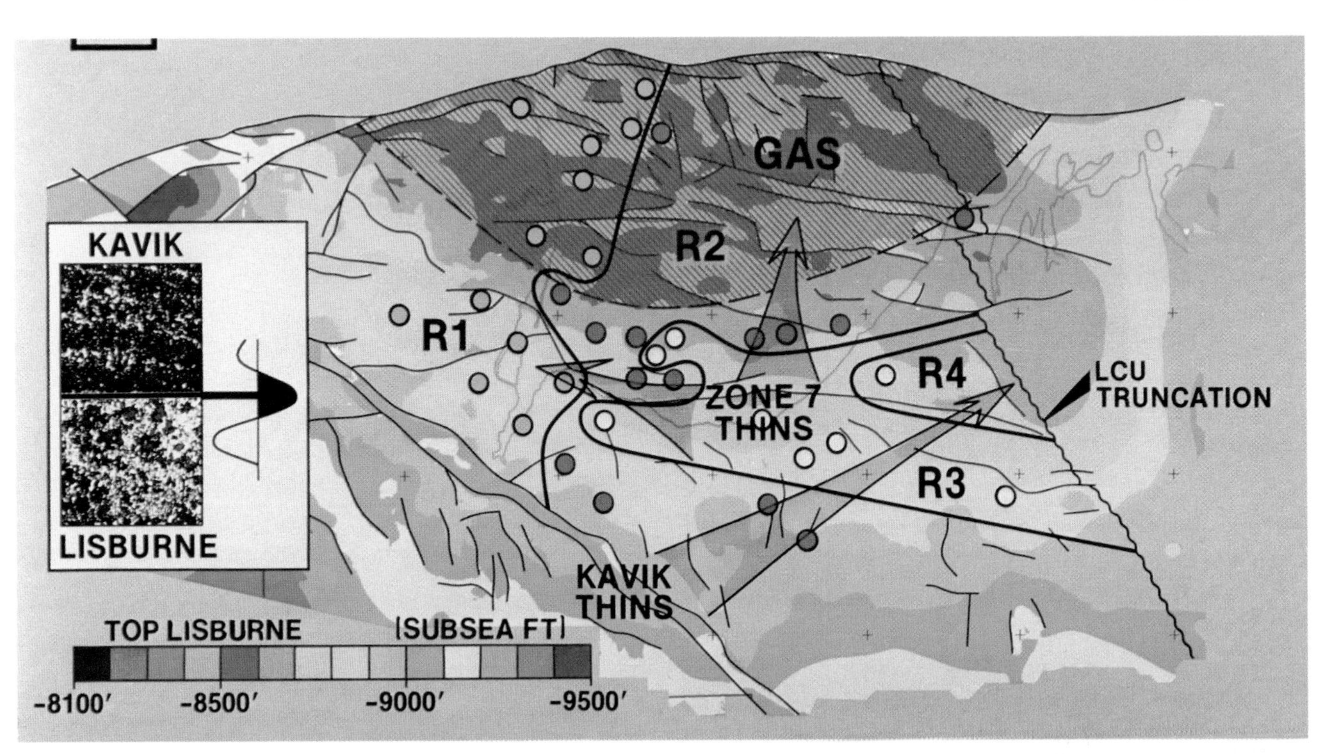

Fig. 9-12-7. Factors affecting seismic intercept regions.

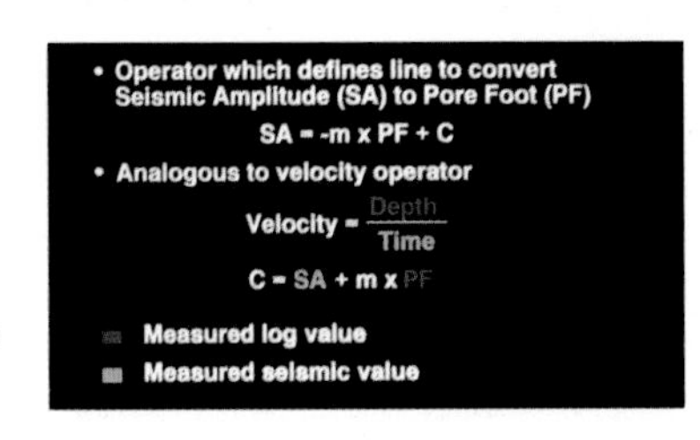

Fig. 9-12-8. Calculation of seismic intercept at each well.

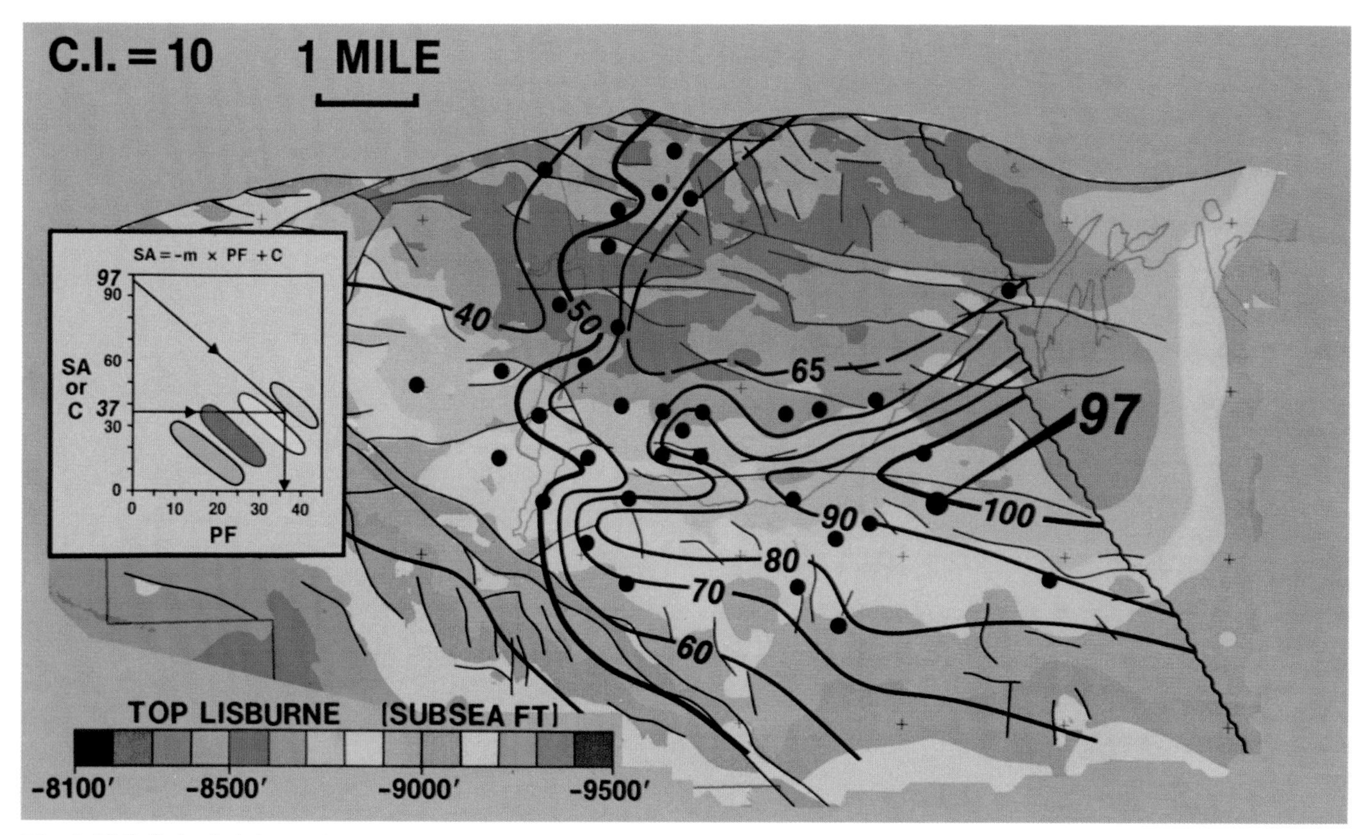

Fig. 9-12-9. Seismic intercept contour map superimposed on top Lisburne structure.

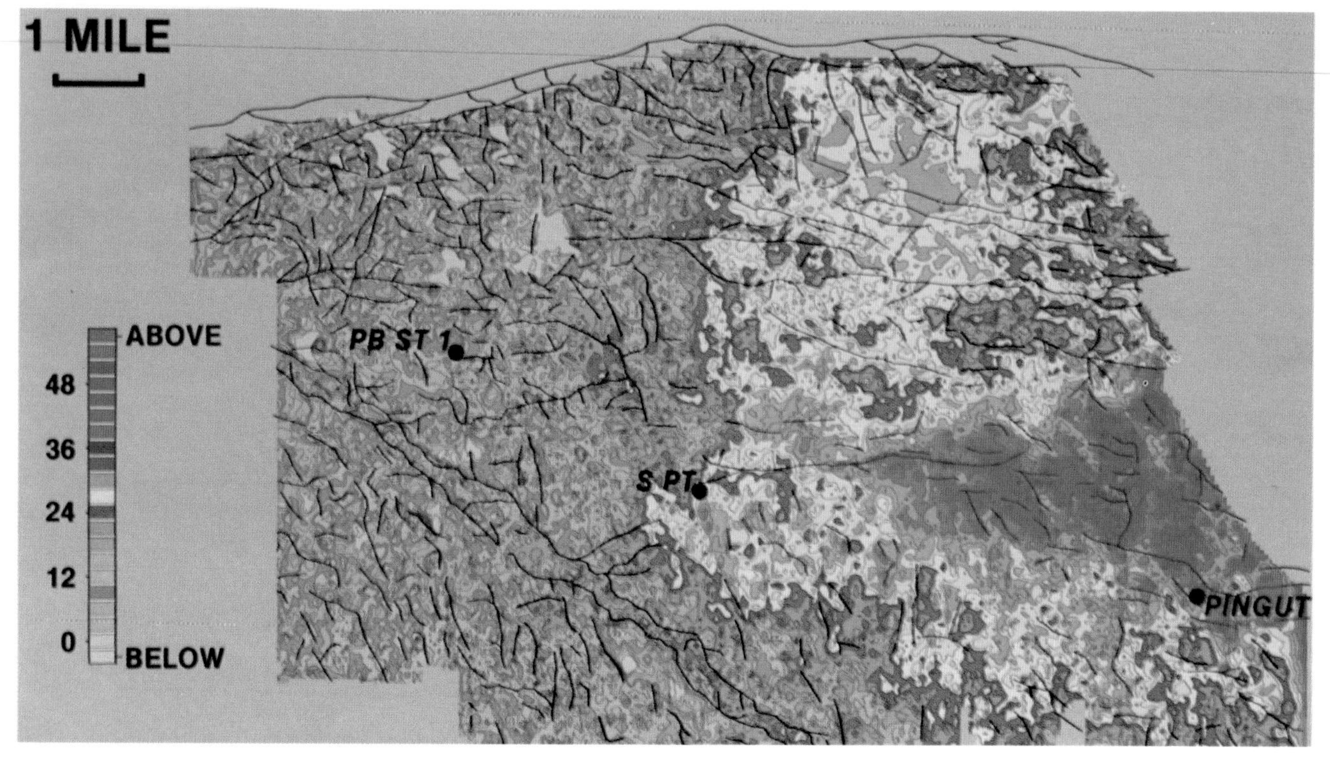

Fig. 9-12-10. Map of porosity-feet for Lisburne formation zones 7 to 4 derived from seismic amplitude.

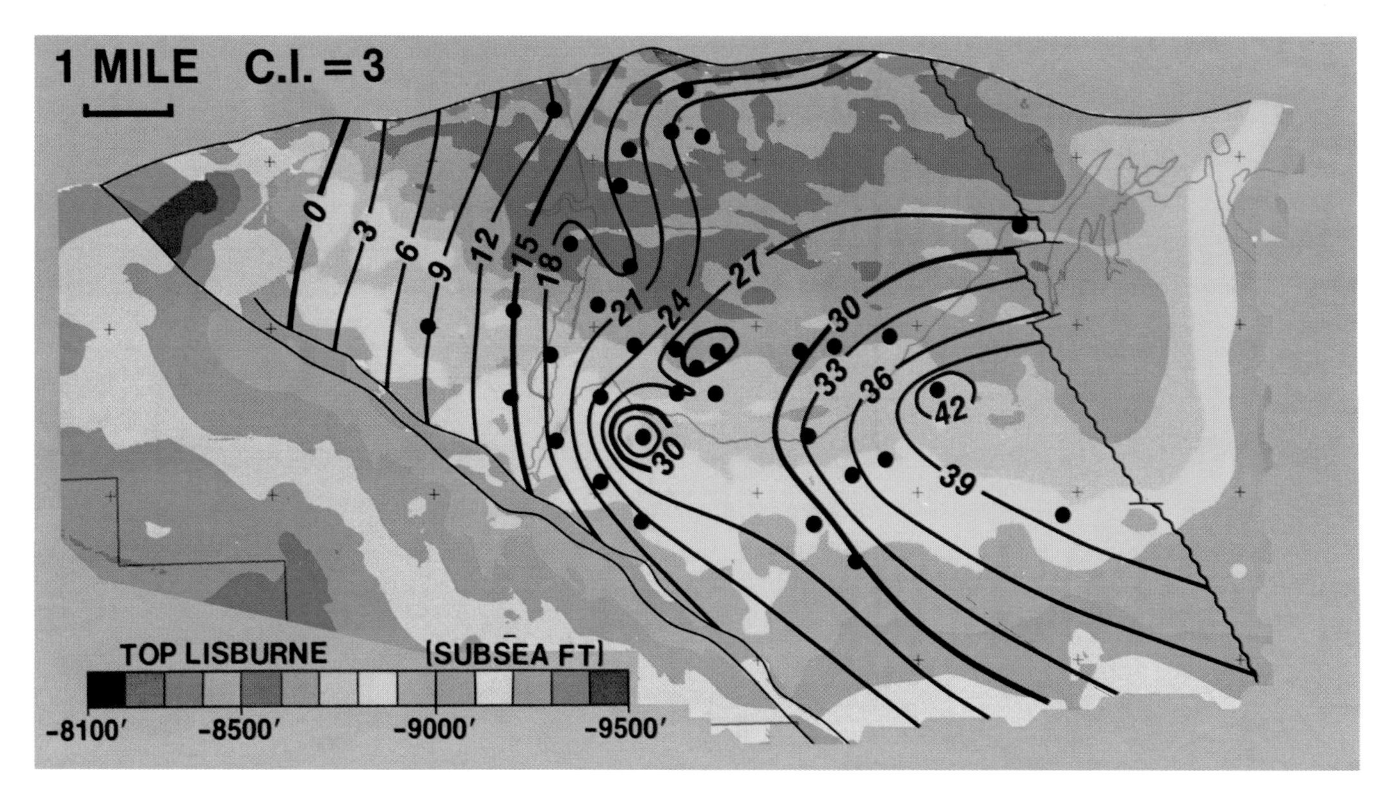

Fig. 9-12-11. Map of porosity-feet contoured from well values superimposed on top Lisburne structure.

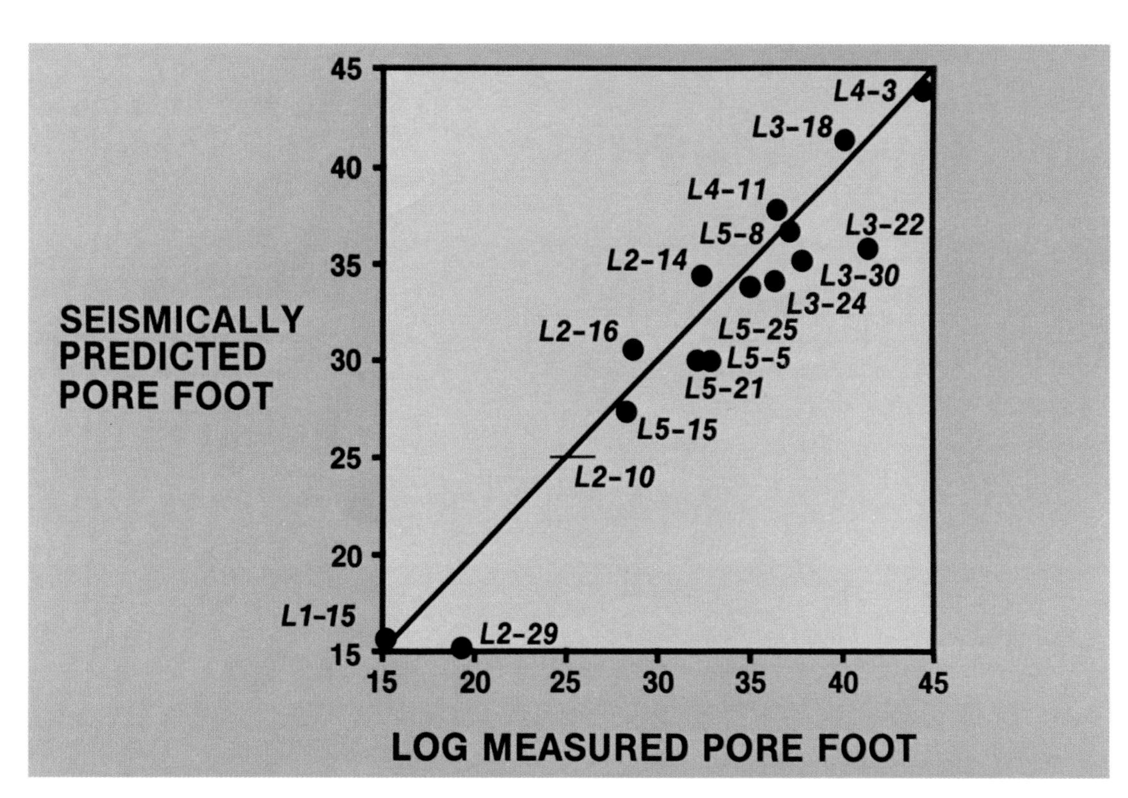

Fig. 9-12-12. Porosity-feet prediction results for 16 new wells.

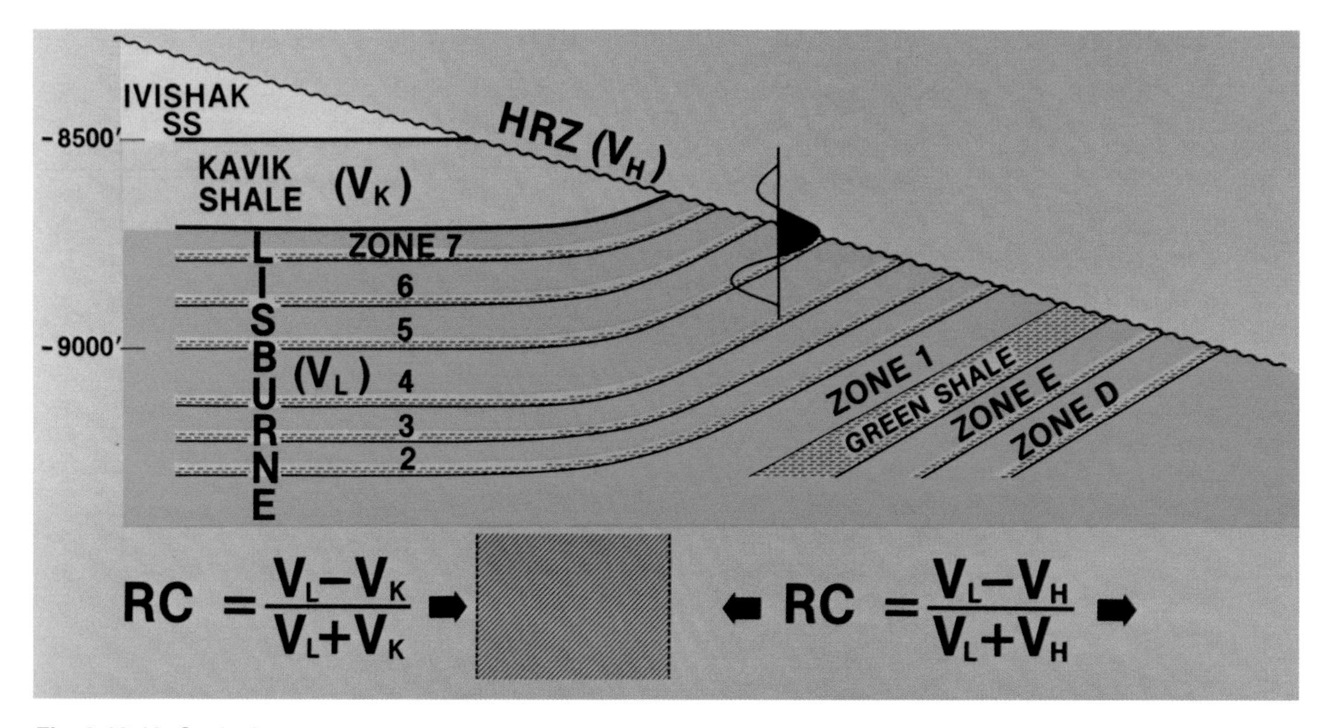

Fig. 9-12-13. Geologic cross section of Lisburne truncation.

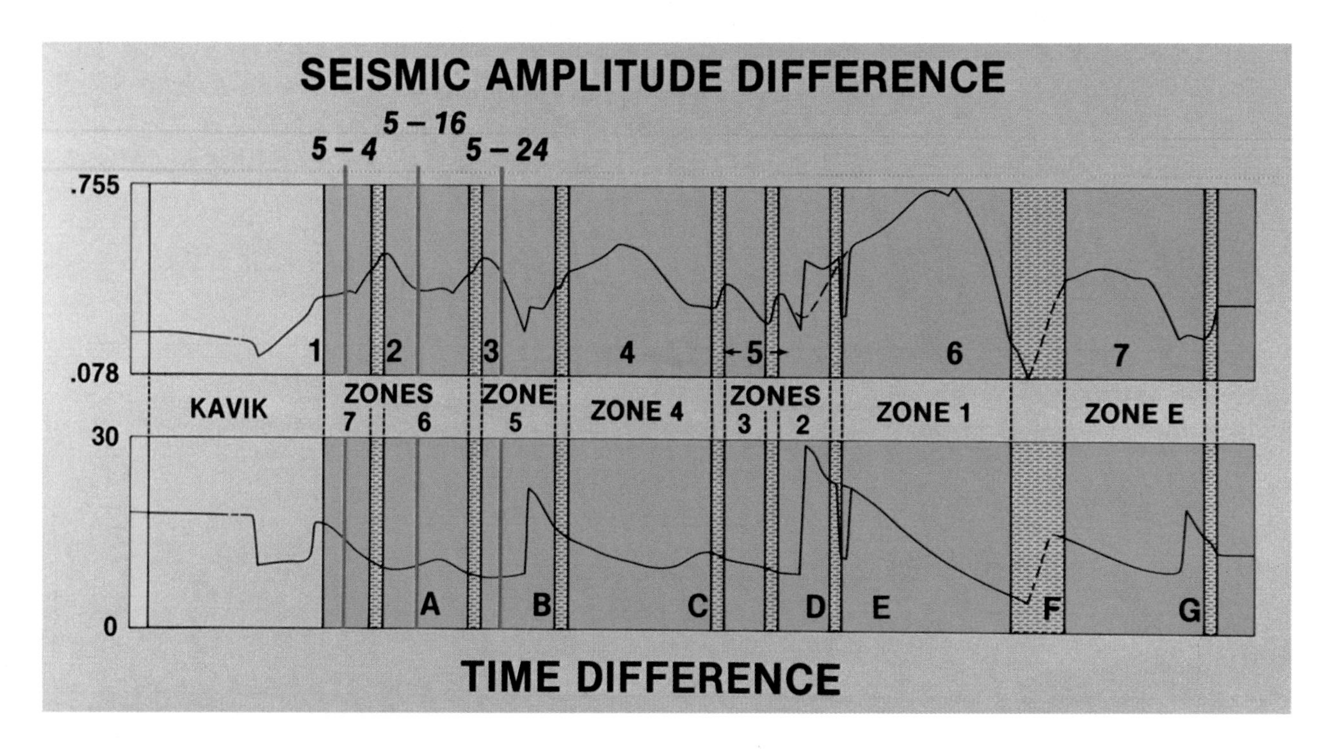

Fig. 9-12-14. Synthetic seismic amplitude and time difference along unconformity for cross section of Figure 9-12-13.

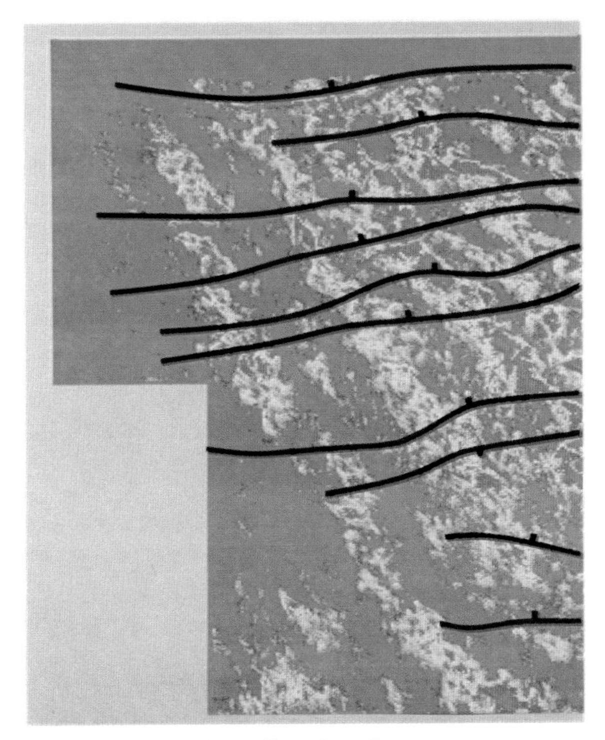

Fig. 9-12-15. Horizon slice along Lower Cretaceous Unconformity, showing banded patterns in seismic amplitude.

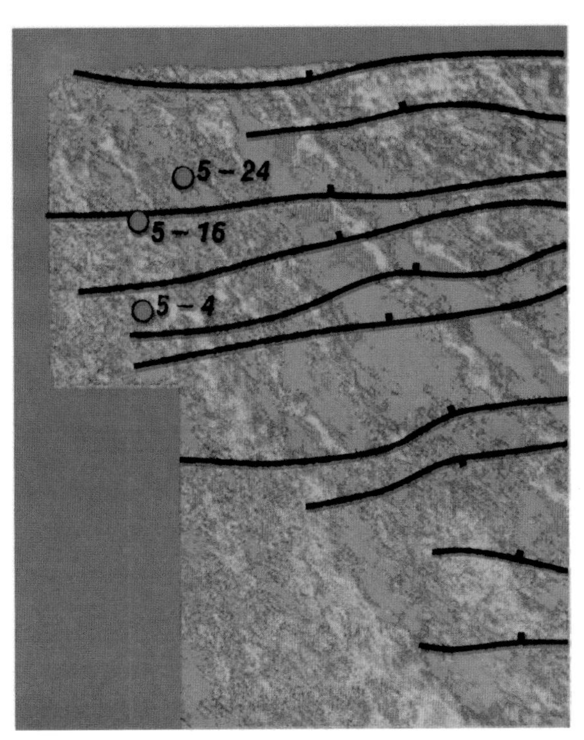

Fig. 9-12-16. Observed seismic time differences along Lower Cretaceous Unconformity, showing banded patterns.

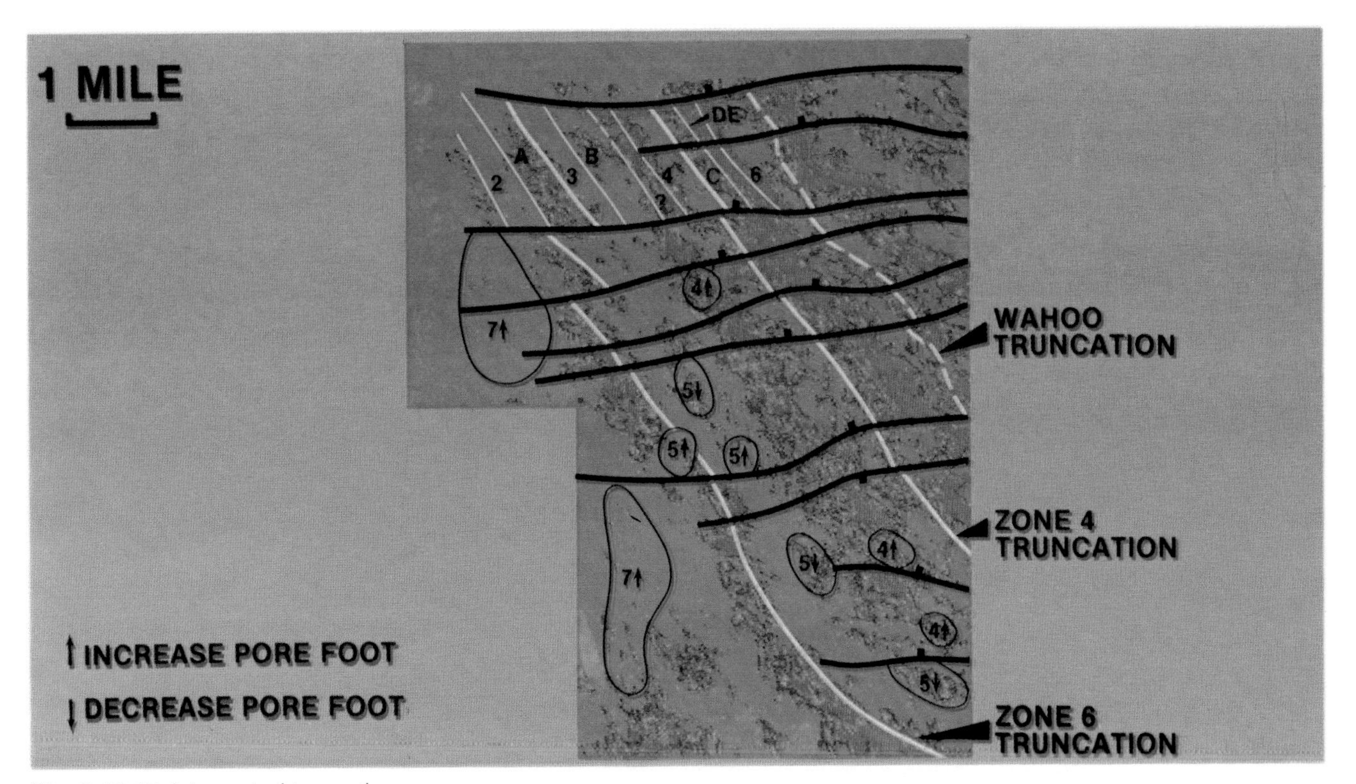

Fig. 9-12-17. Interpreted truncations superimposed on horizon slice of Figure 9-12-15.

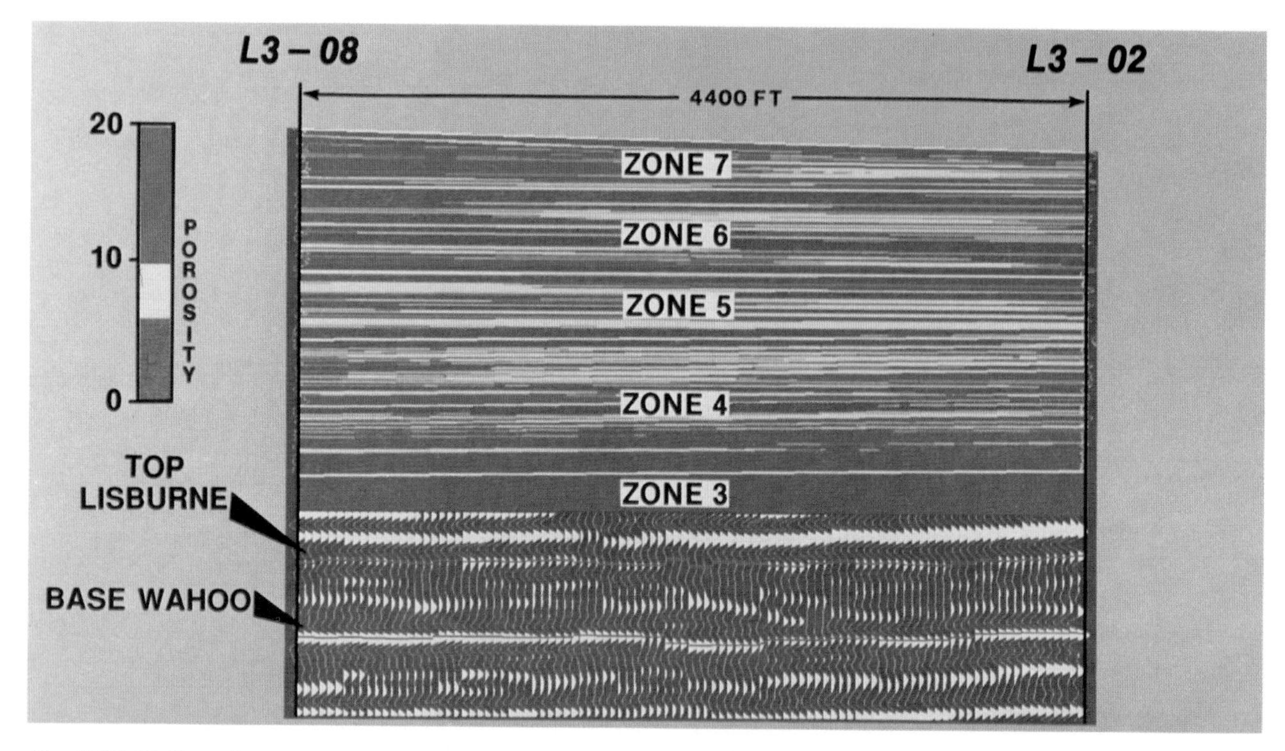

Fig. 9-12-18. Porosity cross section between wells L3-08 and L3-02.

Fig. 9-12-19. Lisburne porosity foot map and its relation to faults and fractures.

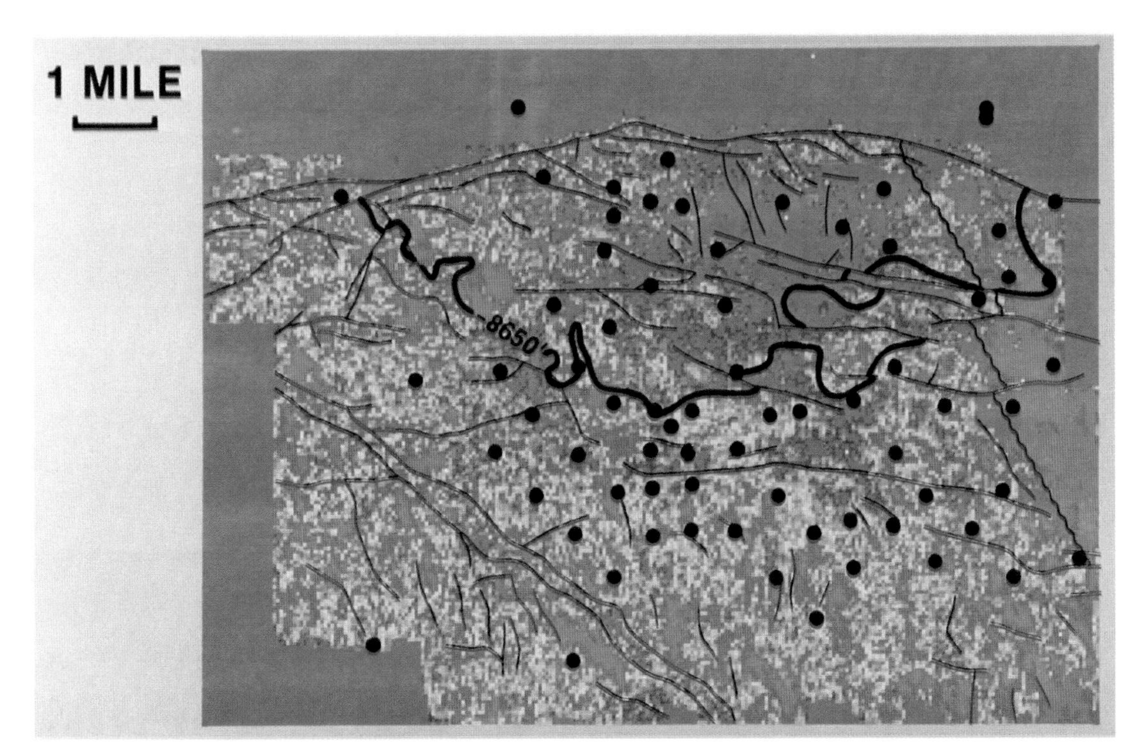

Fig. 9-12-20. Horizon slice for top Lisburne reflection showing low amplitudes (blue) corresponding to gas zones.

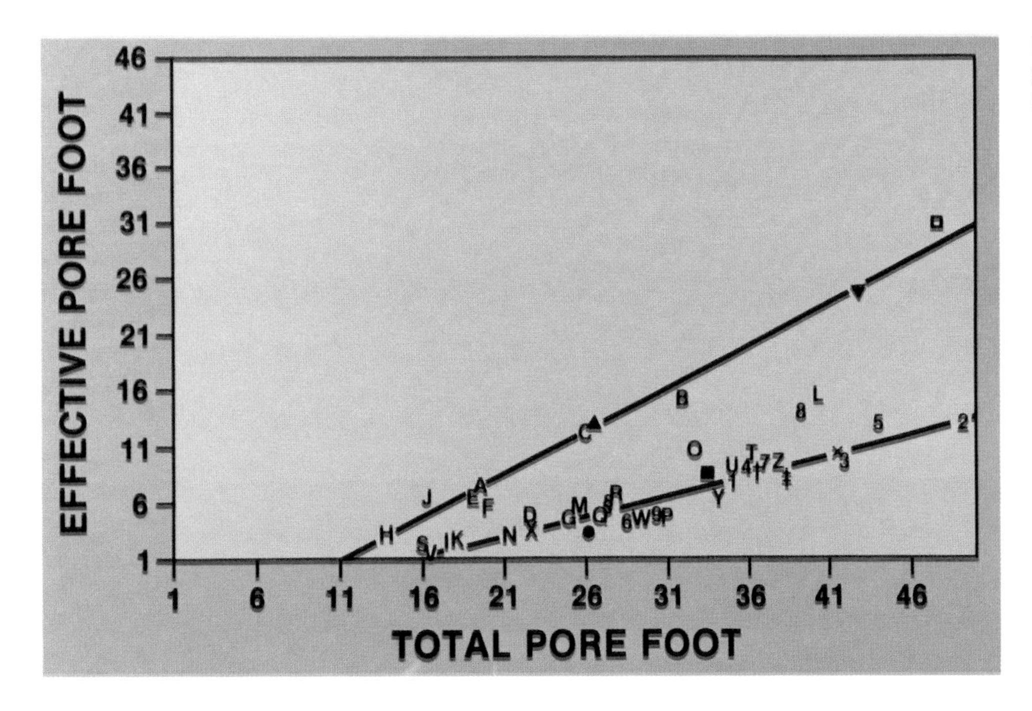

Fig. 9-12-21. Effective porosity foot versus total porosity-feet.

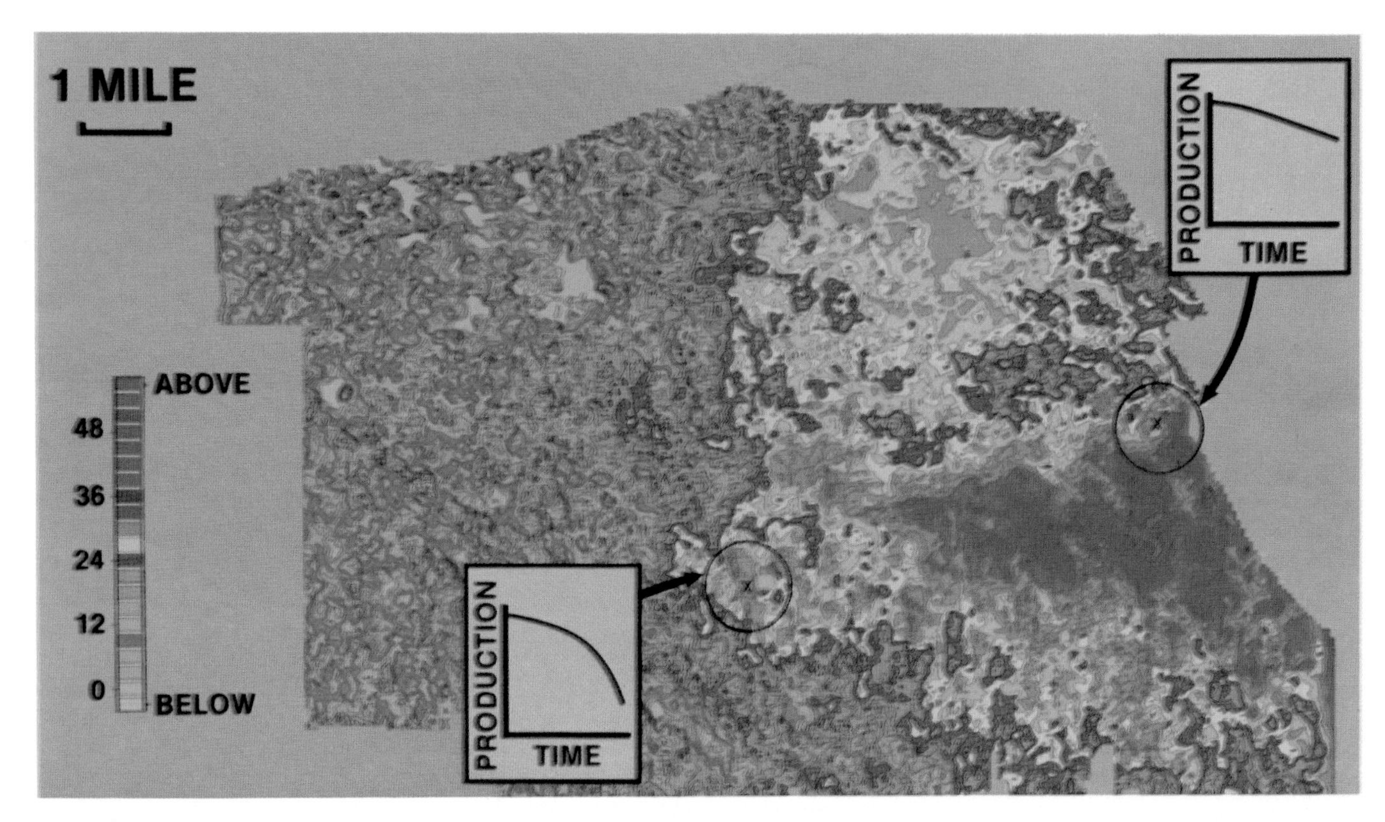

Fig. 9-12-22. Lisburne porosity foot map and prediction of production over time.

CONSIDERATIONS FOR OPTIMUM 3-D SURVEY DESIGN, ACQUISITION AND PROCESSING

Malcolm Lansley, Western Geophysical

General Issues

The optimum 3-D seismic survey would have offset and azimuth distributions which were spatially very consistent. This would necessarily mean that the fold would have to be high. Cost considerations, environmental concerns, and also the effects of facilities and other obstructions, usually result in the surface acquisition grid being quite coarsely sampled and the offset and azimuth distributions being less than perfect. These irregularities can cause difficulties in data processing and frequently result in significant spatial variations in the amplitudes and phase of the final data volume used for interpretation. Structural errors may also result. Since the variations in trace attributes follow the variations in the surface geometry used for the data acquisition, these effects cause an **acquisition footprint** or pattern of **data acquisition lineations.** Special processing algorithms may be required to minimise the effects. A careful balance between the spatial changes in trace attributes and the cost of the survey needs to be achieved during the design process.

In order to achieve a **zero-phase** wavelet for interpretation, some fundamental assumptions about the characteristics of the recorded data are frequently made in data processing. The most common of these are that the source wavelet, and the response of the recording system and sensors are minimum phase. Another common assumption is that the amplitude spectrum of the source wavelet that was introduced into the subsurface was flat. If these assumptions are not satisfied, then additional processing steps may be required in order to phase match the seismic data to the well control.

Marine Data Acquisition

There are two fundamentally different methods which are used for data acquisition in marine environments. One of these uses sensors (hydrophones) which are deployed in a streamer or streamers which are towed behind a vessel at a depth of a few meters, while the vessel is moving at a speed of four or five knots. The seismic source may be attached to the same vessel, or may be towed behind a separate vessel. In some deployments, several vessels may be synchronized to work together, each with one or more seismic sources and/or streamers. In the other technique, the recording sensors are allowed to sink to the ocean floor and are connected to a recording vessel which is not moving. These systems are called ocean bottom cable (OBS) or ocean bottom seismic (OBS). The recording sensors used may be

SM Dual Sensor is a service mark of Western Geophysical, a division of Western Atlas International Inc.

hydrophones, geophones, or both, and the last technique, using both types of sensors, is called Dual Sensor[SM] (Barr and Sanders, 1989.)

Air gun arrays used in the majority of marine 3-D surveys are normally considered to be minimum phase and are carefully tuned to optimize the amplitude spectrum. They are also configured with some redundancy of the air guns, so that if an individual gun or some small number of guns fails to operate, spare guns can be turned on to replace them and to maintain a consistent source signature. Quality control measures used to verify gun operation and timing are very comprehensive and as long as the array specifications are not violated, the minimum phase assumption in data processing can usually be made.

In conventional marine data acquisition using streamers, there are a number of ways in which the amplitude spectrum of the recorded data can be compromised, both in terms of overall bandwidth and how flat it is. The water surface represents a strong acoustic impedance contrast which can result in significant ghost reflections and corresponding spectral notches. These are present both for the airgun arrays used at the source and also at the hydrophones within the streamer. The effect on the spectral content of the data will depend upon the source and streamer depths, the water depth, and the reflection coefficient of the water bottom. In water depths up to approximately 130 meters the receiver ghost problem can be solved by the use of the Dual Sensor data acquisition and processing technique. The source ghost should theoretically exist even on data acquired with dual sensors, although the spectral effects are not normally visible to the same extent as the receiver ghost notches.

In unobstructed shallow waters, or in deeper water there may be an operational preference for conventional streamer acquisition, but this will certainly result in significant spectral notches. Shallow deployment of streamers is frequently used to try to improve the high frequency content of the data, but when the sea state worsens the noise evident on the streamer also worsens and may become the limiting factor. In order to reduce the noise, the streamer depth may have to be increased, which results in a very different spectral response, at both the high and low frequencies. It should be noted that the low frequency response is increased and the high frequencies decreased with the deeper streamer deployment.

For reservoir monitoring, any changes in streamer and source depth between the recording of successive surveys should be avoided, because of both the noise differences and the spectral differences. (S.E. Johnstad et al, 1993; S.E. Johnstad et al, 1995.)

Another consideration for marine acquisition is the use of multi-boat, multi-source and multi-streamer surveys which have been used both for obstacle avoidance and for reducing the cost of surveys. Egan et al (1991) showed that the differences in azimuth ranges when changing from an undershoot geometry back to a more conventional source-streamer geometry can cause artifacts in the processed data volume which result from the imaging processes. The current trend towards very large numbers of marine streamers with one or two source arrays towed behind a single vessel may also create problems. Beasley (1995) showed that some of these geometries can result in shadow zones with inadequate subsurface coverage which cannot be adequately resolved by the imaging processes, leaving both structural and amplitude errors in the interpretation data volume. Both of these geometries may create problems for reservoir characterization and reservoir monitoring. If the water depth is not too great, Dual Sensor ocean bottom cable techniques can resolve these problems.

The absolute positioning of marine surveys should not be a problem, with the accuracy of today's positioning systems, assuming that all of the appropriate corrections are made, and that current state-of-the-art streamer and source positioning technology is utilized. Since most marine surveys are recorded with the boat traveling in straight lines, survey orientation is still problematical. In areas with rapidly-varying velocity fields, conventional wisdom now recommends the longest source to receiver axis being aligned in the strike direction (O'Connell et al, 1993,

Manin and Hun, 1992, and Mougenot et al, 1992) in order to minimise the raypath complexity and thus make the normal moveout more hyperbolic. The problem is that in complex geology there are not necessarily dip and strike directions, and the orientation of the survey may result in imaging difficulties in the data processing stages.

For reservoir monitoring, Johnstad et al (1993) showed that absolute positioning was not a critical factor and that marine 3-D surveys could be adequately repositioned to permit subtraction of the data sets to show fluid movements in an offshore reservoir.

Land Data Acquisition

In land data acquisition, there are many factors which will affect our ability to record broad bandwidth data with a good signal-to-noise ratio. One of the more significant of these is the use of arrays: we can reduce the amount of noise recorded by increasing the size of the source or receiver arrays, but potentially we might also be attenuating the high frequency signal energy. Significant energy absorption and loss of high frequencies also occurs in the unconsolidated weathering layers near the surface. If the geophones are buried, the effects of surface noises and the absorptive losses in the near-surface can be reduced. Pullin et al, (1986) showed a dramatic improvement in signal-to-noise ratio, high frequency signal content, and static shifts on data recorded with a single geophone per recording channel buried 10 meters below the surface, when compared with data recorded with surface geophones. Although the improvement was significant, geophones are rarely buried more than a few centimeters, except for reservoir monitoring studies. For these, geophones or hydrophones are frequently deeply buried and left in place permanently so that they are available for subsequent monitoring surveys. This gives the added advantage that variations in the water table and other near-surface effects are minimized on the different surveys.

Another question that is related to the use of arrays, is when it is intended to perform pre-stack amplitude analysis (amplitude-versus-offset, or amplitude-versus-offset and azimuth) on the data. In these situations, the use of linear source and receiver arrays is quite common, the intent being to attenuate both coherent source-generated noise and random noise. The arrays also attenuate the signal differently with offset and azimuth. This signal attenuation is also frequency dependent which makes later compensation in data processing very difficult to achieve.

When dynamite is used as a land source, we can normally make the data processing assumption that the source amplitude spectrum was flat and that the phase spectrum was minimum phase. With vibroseis, however, the amplitude and phase spectra of the recorded data are frequently misunderstood. In order to optimize the data quality and minimize difficulties in data processing, it is recommended that the vibrators be set up such that the ground force signal of the vibrator is phase locked to the pilot sweep according to the SEG recommendations (Geophysics, v. 59, p. 315-322) and that the amplitude spectrum of the fundamental of the ground force signal is controlled to be flat. The use of non-linear sweeps (i.e. non-linear frequency versus time functions) will require special treatment in data processing in order to correctly handle the wavelet phase.

The source and receiver geometry used for a survey is normally a compromise between the geologic objectives, environmental and other surface constraints, and the cost. Quite often, it is possible to see effects on the finally processed data volume which relate to the geometry used to acquire the data. These effects will be reduced by acquiring a survey with as high a common-mid-point fold as possible, and with as uniform an offset and azimuth distribution as possible. Great care should be taken in the areas around surface obstructions to minimize the variations in fold, offset and azimuths.

Data Processing

In data processing the objective is to maximize the useful signal bandwidth and to provide the interpreter with a correct structural image with a known phase (usually zero-phase) wavelet, and with amplitudes that represent, as closely as possible, the spatial variations in the subsurface reflectivity. Since amplitudes are one of the fundamental attributes of seismic data and one of the easiest to analyze, correct management of amplitudes is very important. Also, since the amplitudes of data sets are differenced in reservoir monitoring studies, correct handling is essential for these. It should also be understood that when wave equation consistent processes are being used, inadequacies in the sampling, which result in amplitude variations, will also result in phase errors which are much more difficult to detect and correct.

Great care should be taken in the initial steps of **geometry assignment and data initialization**, since most subsequent processes depend upon the geometry assigned at this stage. On land surveys, special attention should be paid to areas around surface obstructions to ensure that offset source and receiver locations are correctly annotated. In marine surveys particular attention should be paid to the binning technique used, especially in areas where the source-streamer axis is in the strike direction. In this case, if the bin is made too wide in the crossline (and hence dip) direction, attenuation of high frequencies may result. Also, the use of overlapping bins should be used with care, since although the fold of the resultant stack may be more even, amplitudes will require special management in multi-trace processes such as DMO.

A space-invariant time function should be used for **spherical divergence and inelastic attenuation compensation** and velocity dependent algorithms should be avoided. Care should be taken to ensure that the amplitudes of multiples and coherent source-generated noises are not over corrected.

3-D surveys typically have high surface redundancies which make **surface consistent processes** very powerful. Surface consistent wavelet processing, amplitude compensation, and static algorithms (both refraction and residual reflection) normally provide very stable solutions for most 3-D geometries.

In the **wavelet processing**, it is important to ensure that the phase and amplitude spectra of the recorded data match the assumptions that are made in the deconvolution process. Particular attention should be paid to the phase and amplitude spectra of vibroseis data, especially when non-linear sweeps have been used and when surveys with mixed sources (for example, vibroseis and dynamite) have been acquired. Since the pilot sweep used for correlation is normally filtered with the same instrument recording filters as the data channels, the phase effects of the recording filters will be removed and the amplitude effects will be squared. Thus the observed filter effects will be very different between the vibroseis and dynamite data. Also the field vibroseis correlation algorithms are normally zero phase, and either a model-based or analytically derived inverse filter may be necessary to convert the phase and amplitude spectra.

Surface consistent amplitude compensation will correct for source and receiver coupling variations and any further data dependent amplitude scaling or balancing should be avoided.

The near-surface model used for **static correction** of land surveys is frequently derived from refraction analysis of the first arrivals of the production data. In some cases, the acquisition geometry may lead to some instability of the solution, particularly in areas with low surface redundancy around obstructions or near the edges of the survey. Rigorous quality control of the model is essential to ensure the integrity of the final structural image. Statics, although traditionally regarded as only being a problem for land data, are frequently necessary on marine surveys also. Compensation is required for the tidal effects and residual statics may also be necessary to compensate for short spatial wavelength variations in the velocities and thicknesses of the near-water bottom layers.

Frequency-wavenumber (F-K) noise attenuation algorithms work well for attenuating **coherent source-generated noise** (including multiples) as long as the data are well sampled spatially. They are not very good, however, at preserving true amplitudes, and should therefore be avoided wherever possible. With wide azimuth surveys, the characteristics of the noises may vary considerably along different azimuths because of different source and receiver array orientations. Alternative noise attenuation algorithms such as radon or F-X methods may be required to model and remove the noise. There are a wide variety of algorithms available in the industry, and the ability of any particular algorithm to correctly retain true amplitude of the signal should be verified before its use.

Dip moveout (DMO) is used to compensate for the different stacking velocities apparent on data with conflicting dips and is very sensitive to changes in both the azimuthal and offset distributions of the data being processed. Even with perfectly regular geometry, the acquisition footprint caused by the normal spatial variations in the offset distributions is frequently made worse with the application of DMO. Variations in fold and azimuths may also lead to very strong amplitude changes and artifacts which are not truly representative of changes in the subsurface. Modifications to the DMO algorithms to compensate for these variations (eg. Beasley et al., 1992) are available and should be used to minimize these effects. Unless prestack migration is going to be performed, DMO is theoretically a necessary step in the imaging process, even when the geologic dip is zero. However, it is frequently omitted from the processing sequence both as a cost saving measure and also to eliminate any problems associated with its application on data which are inadequately sampled.

The amplitudes output from a conventional **common mid-point stack** are normally corrected for variations in fold by a recovery scalar which divides the summed input trace amplitudes by the number of traces contributing to the stack. If the energy being stacked is pure signal, then the output amplitudes should be correct. When the input traces are subject to varying amounts of random and coherent noise, this recovery scalar will no longer be adequate. Although alternative schemes do exist to try to compensate for varying degrees of signal and random noise, the most consistent amplitudes will result from a survey with as uniform fold as possible.

Another consideration for CMP stacking is the range of lateral velocity variation within the offset range being included in the stack. For CMP stack to be valid, any lateral velocity variations need to be sufficiently gradual such that the normal moveout for the traces to be included in the stack is hyperbolic. Where the velocity field changes too rapidly, and non-hyperbolic normal moveout is apparent, pre-stack migration should be performed.

Although significant improvement of signal continuity is frequently apparent after the application of **post-stack random noise attenuation**, some of the energy which appears to be incoherent may, in fact, be diffraction energy which should be preserved for the migration process. It is recommended that these algorithms not be applied until after migration.

In order to complete the imaging process and accurately focus the reflection energy at its correct spatial location and time (or depth), the process of **migration** is required. There are a number of different migration algorithms and methodologies which vary in their ability to handle steep dips and also in their inherent positioning accuracy. There is a considerable amount of literature discussing the accuracies of different migration algorithms which is available elsewhere, and no discussion will be made here. The accuracy of structural information from 3-D surveys is related not only to the accuracy of the migration algorithms, but also to the accuracy of the velocity field which is used for the imaging. It should be emphasized therefore, that great care must be exercised in building an accurate velocity model for migration, particularly in areas of complex geology. Small errors in the velocity used for migration can dramatically reduce the lateral resolution in the final image.

Another issue to be considered is whether a post-stack migration will suffice, or whether a **pre-stack migration** will be necessary. As has been mentioned previously in the section on common-mid-point stack, as long as the normal moveout within a

CMP gather is hyperbolic, a post-stack migration should prove satisfactory. When non-hyperbolic normal moveout is apparent, a pre-stack migration will be necessary to correctly migrate the data. Survey orientation is sometimes carefully selected in order to minimize the effects of rapid lateral velocity variations and permit the use of more economical post-stack migration algorithms.

When selecting time or **depth migration**, the considerations are very similar. In a time migration algorithm, the migration operator is symmetric and therefore cannot correctly handle severe lateral velocity variations within the migration aperture. A depth migration operator is spatially variable and is designed to accurately position data according to the travel paths defined by the depth and velocity model provided to the algorithm. The importance of the correctness of this model cannot be overemphasized.

For reservoir monitoring studies, the velocity model used for the imaging of successive surveys should be the same or else small differences in migration distance might well be interpreted as changes in the fluid content.

References

Barr, F.J., and J.I. Sanders, 1989, Attenuation of water-column reverberations using pressure and velocity detectors in a water-bottom cable: 59th meeting, Society of Exploration Geophysicists, Expanded Abstracts, p. 653-656 .

Beasley, C. J., and R. Klotz, 1992, Equalization of DMO for irregular spatial sampling: 54th Meeting, European Association of Exploration Geophysicists, Expanded Abstracts, p. 232-233.

Beasley, C. J., and E. Mobley,1995, Spatial sampling characteristics of wide-tow marine acquisition: 57th Meeting, European Association of Geoscientists and Engineers, Expanded Abstracts.

Egan, M. S., K. Dingwall, and J. Kapoor, 1991, Shooting direction - a 3-D marine survey design issue: The Leading Edge, v. 10, no.11, p. 37-41.

Johnstad, S.E., R.C. Uden, and K.N.B.Dunlop, 1993, Seismic reservoir monitoring over the Oseberg Field: First Break, v. 11, no. 5, p.177-185.

Johnstad, S.E., R.H. Seymour, and P.J. Smith, 1995, Seismic reservoir monitoring over the Oseberg Field during the period 1989-1992: First Break, v. 13, no. 5, p.169-183.

O'Connell, J.K., M. Kohli, and S. Amos, 1993, Bullwinkle - a unique 3-D experiment: Geophysics, v. 58, p. 167-176 .

Manin, M., and F. Hun, 1992, Comparison of seismic results after dip and strike acquisition: 54th Meeting, European Association of Exploration Geophysicists, Expanded Abstracts, p. 4-5.

Mougenot, J.M., E. Robein, I. Ravat, and A. Morash, 1992, Design of 3-D land acquisition in a structurally complex environment: 54th Meeting, European Association of Exploration Geophysicists, Expanded Abstracts, p. 8-9.

Pullin, N. E., L. W. Matthews, and W. K. Hirsche, 1986, Techniques applled to obtain very high-resolution 3-D seismic imaging at an Athabasca tar sands thermal pilot: 56th Meeting, Society of Exploration Geophysicists, Expanded Abstracts, p. 494-497.

INTERPRETATION EXERCISE

Background Information

The object of this exercise is to map the structure and extent of a turbidite sand, known from well control to be visible on the 3-D seismic data as a high amplitude peak. You are provided with two vertical sections (Figures B-1 and B-2) and eight horizontal sections (Figures B-3 through B-10). The coordinates of a point at which you can identify the turbidite reflection are:

Line 539
Crossline 600
Time 1,600 ms

Structural Component of Exercise

Follow the crest of the identified blue event on each horizontal section to yield a time structure map on this horizon with a contour interval of 20 ms. Be careful to follow structural continuity regardless of lateral changes in amplitude. Faults will be seen as lateral displacements, not simply as amplitude changes. As a guide, continuity without amplitude superimposed is visible directly along the adjacent zero crossings.

Stratigraphic Component of Exercise

Using the structure map to identify the high amplitudes associated with the turbidite, outline the dark blue areas for this horizon on each horizontal section supplied. Connect these outlined areas interpretively to yield a stratigraphic map of the extent and possible flow direction of the turbidite. Because you are supplied with horizontal sections at only 20 ms intervals, there will be some gaps in coverage in the direction of dip. More interpolation and smoothing will thus be needed in the dip direction than in the strike direction.

Procedure

Take a piece of transparent paper and register it on the annotation frame of the horizontal sections. Use the vertical sections as a guide to structural continuity only. Complete the structural component before attempting the stratigraphic component.

Solution

One interpreter's map of the extent and structure of the turbidite is shown in Figure B-11. This map is based only on the data supplied for the exercise. The horizon slice and superimposed structure that were generated interactively and based on all the data are shown in Figure B-12.

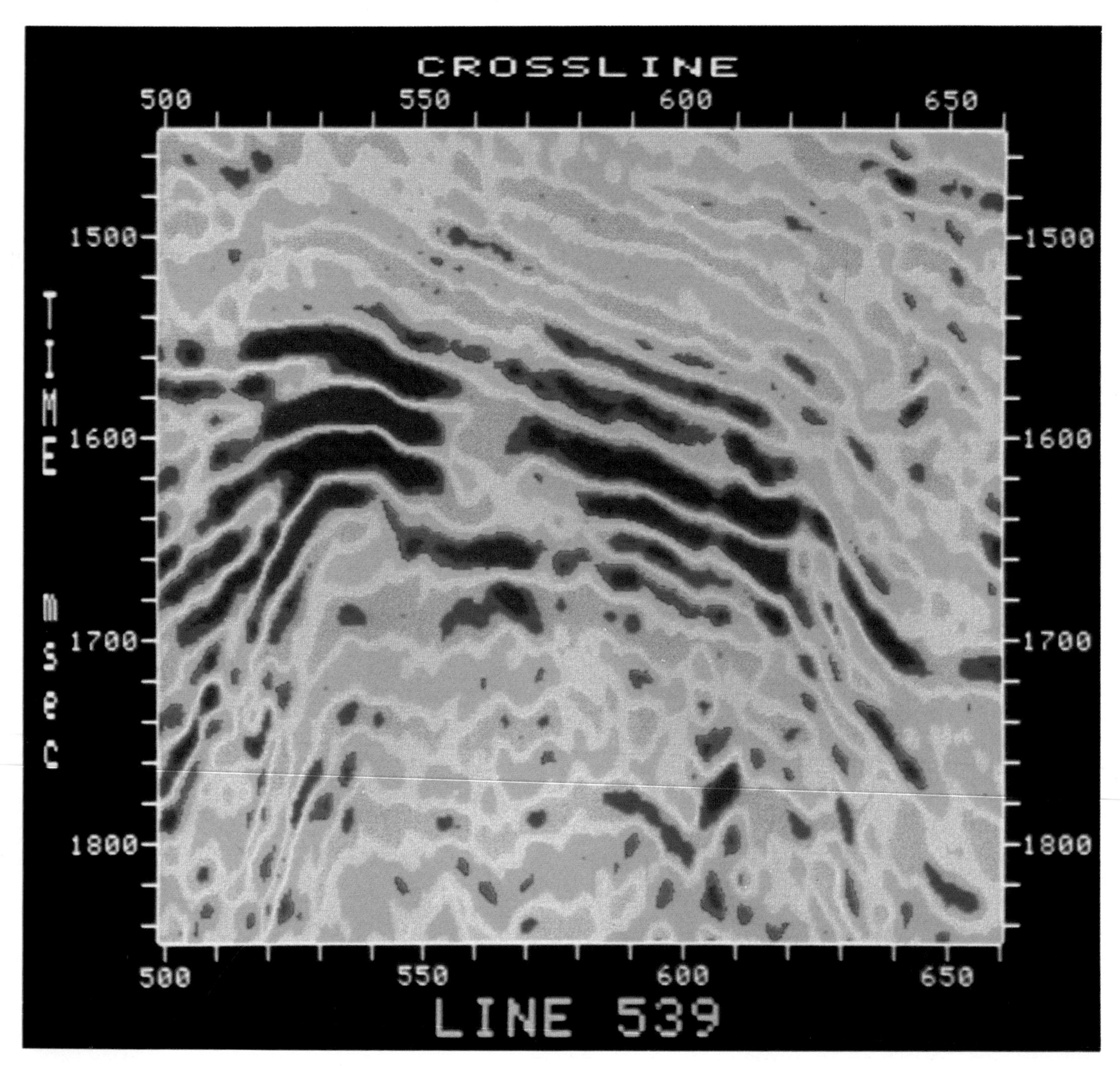
CROSSLINE
500
550
600
650
1500
1600
1700
1800
TIME msec
LINE 539

Fig. B-1.

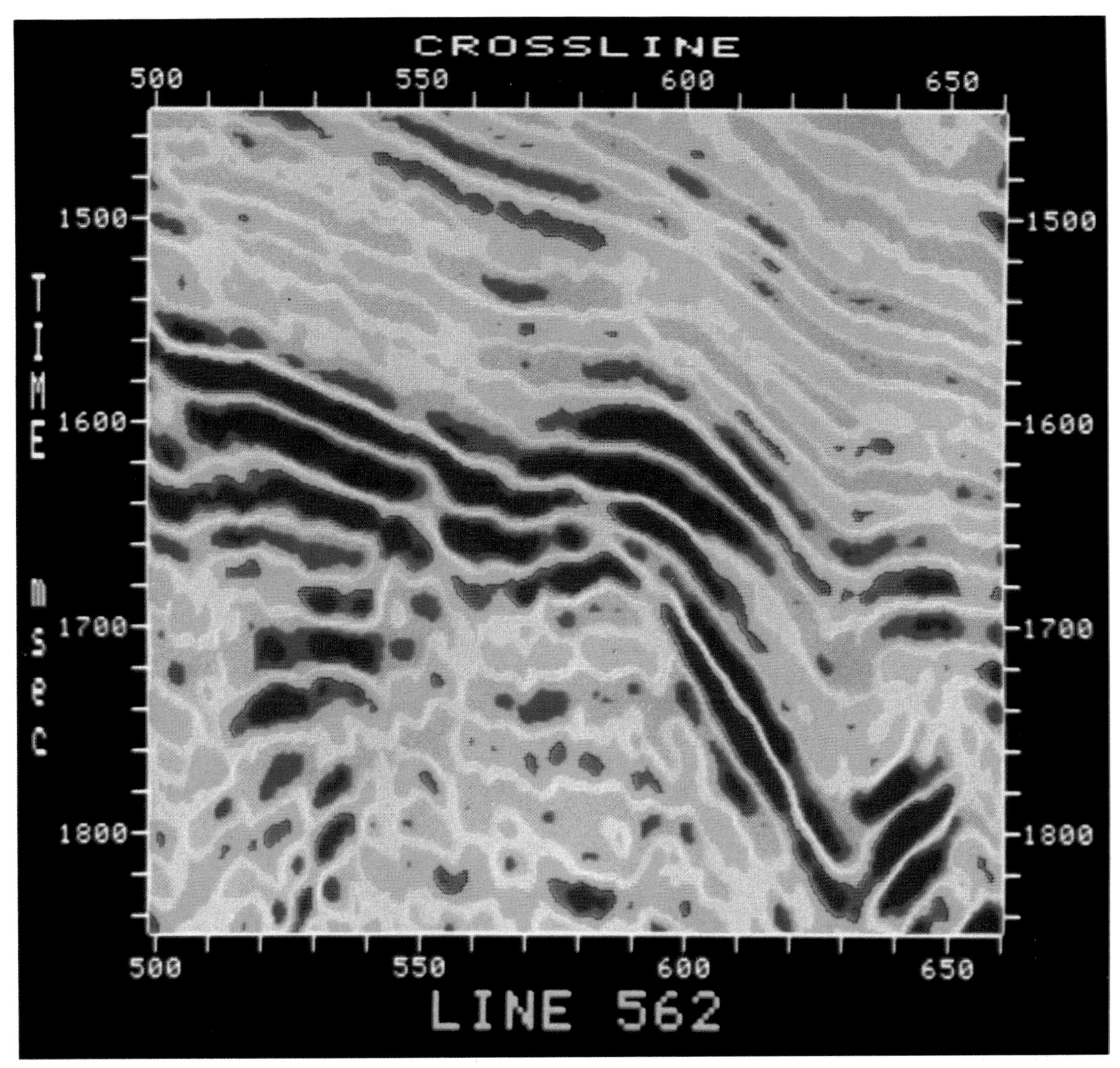
CROSSLINE
500
550
600
650
1500
1600
1700
1800
TIME msec
LINE 562

Fig. B-2.

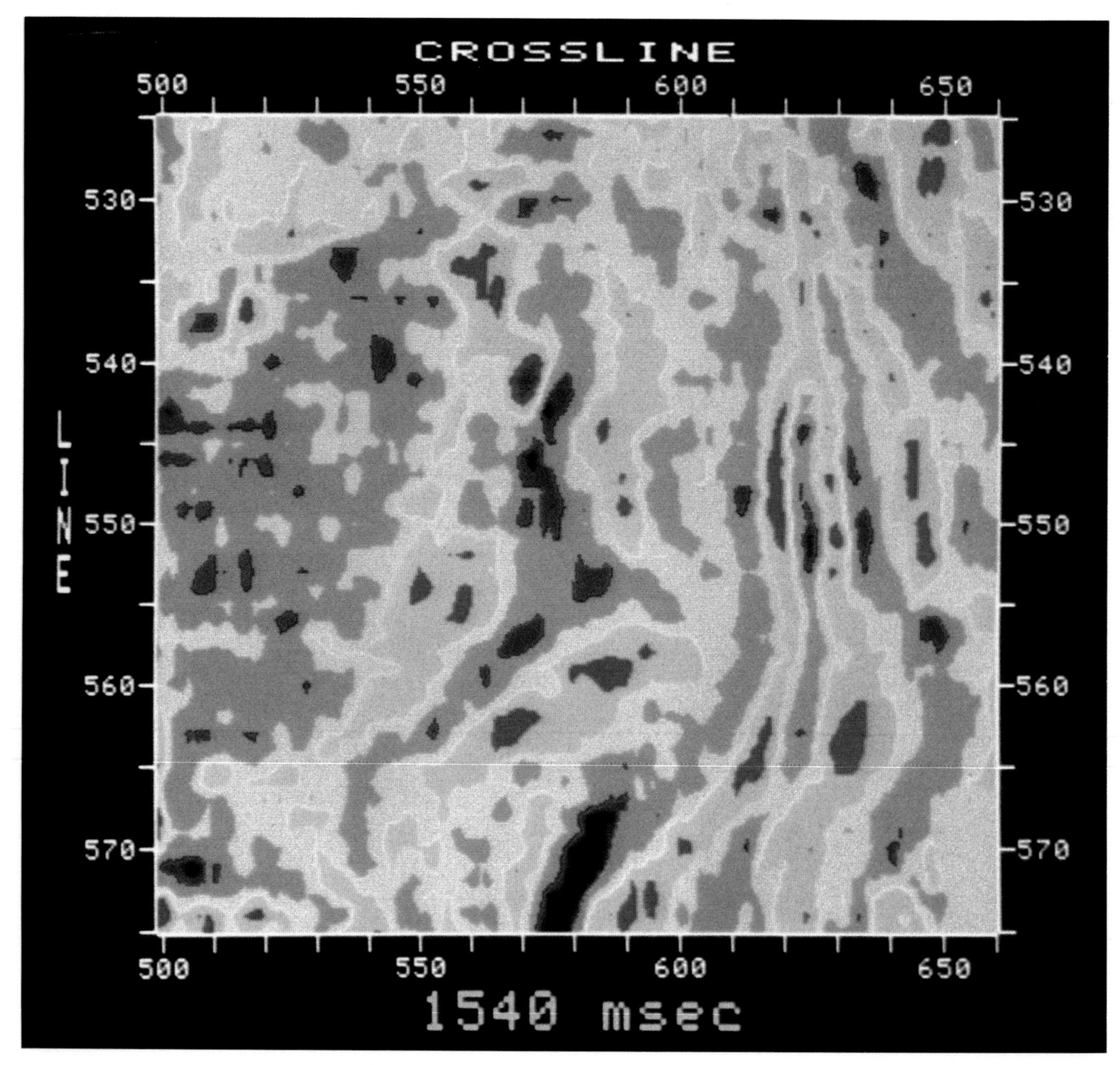

Fig. B-3.

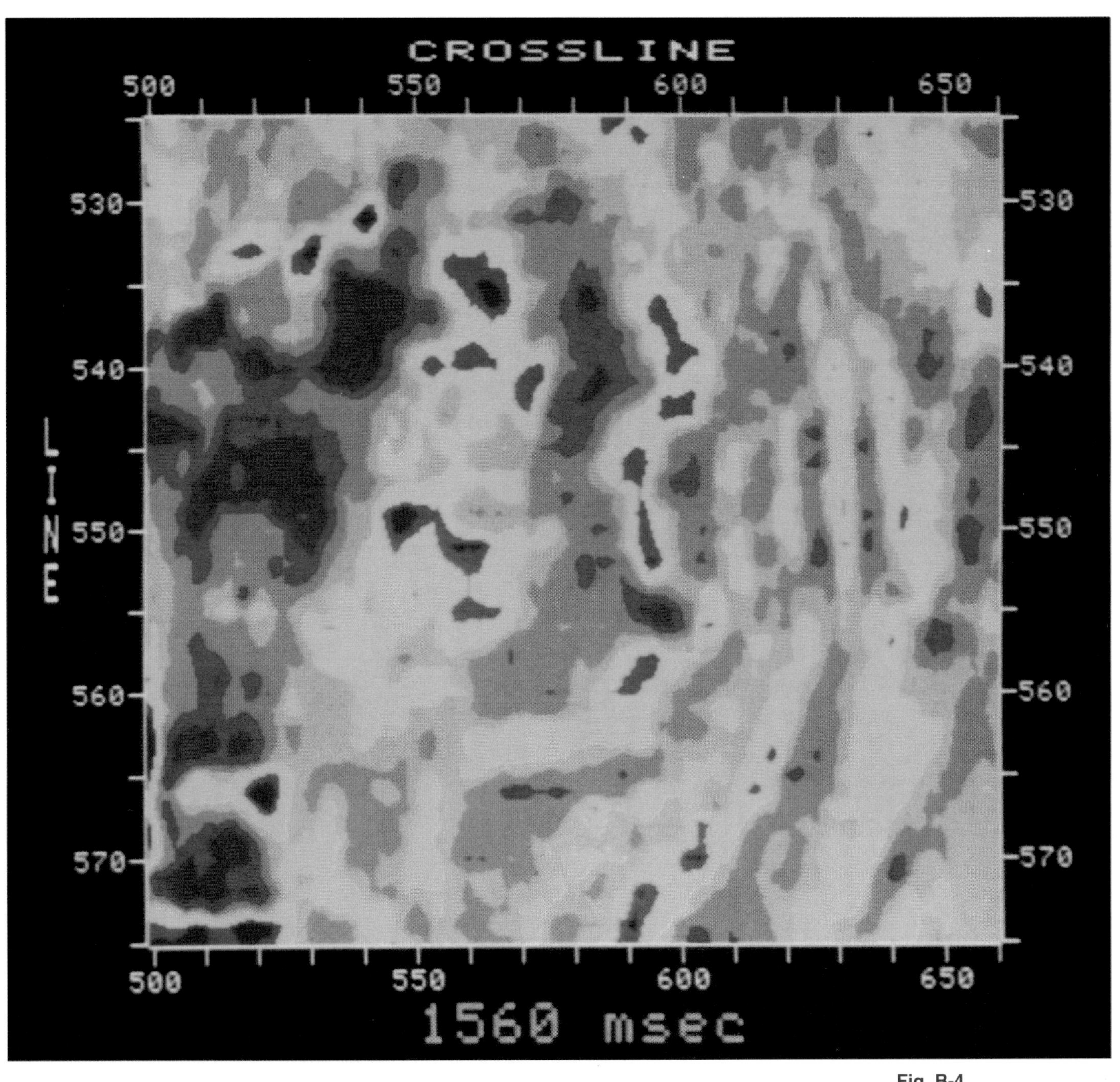
CROSSLINE
500
550
600
650
530
540
550
560
570
LINE
500
550
600
650
1560 msec

Fig. B-4.

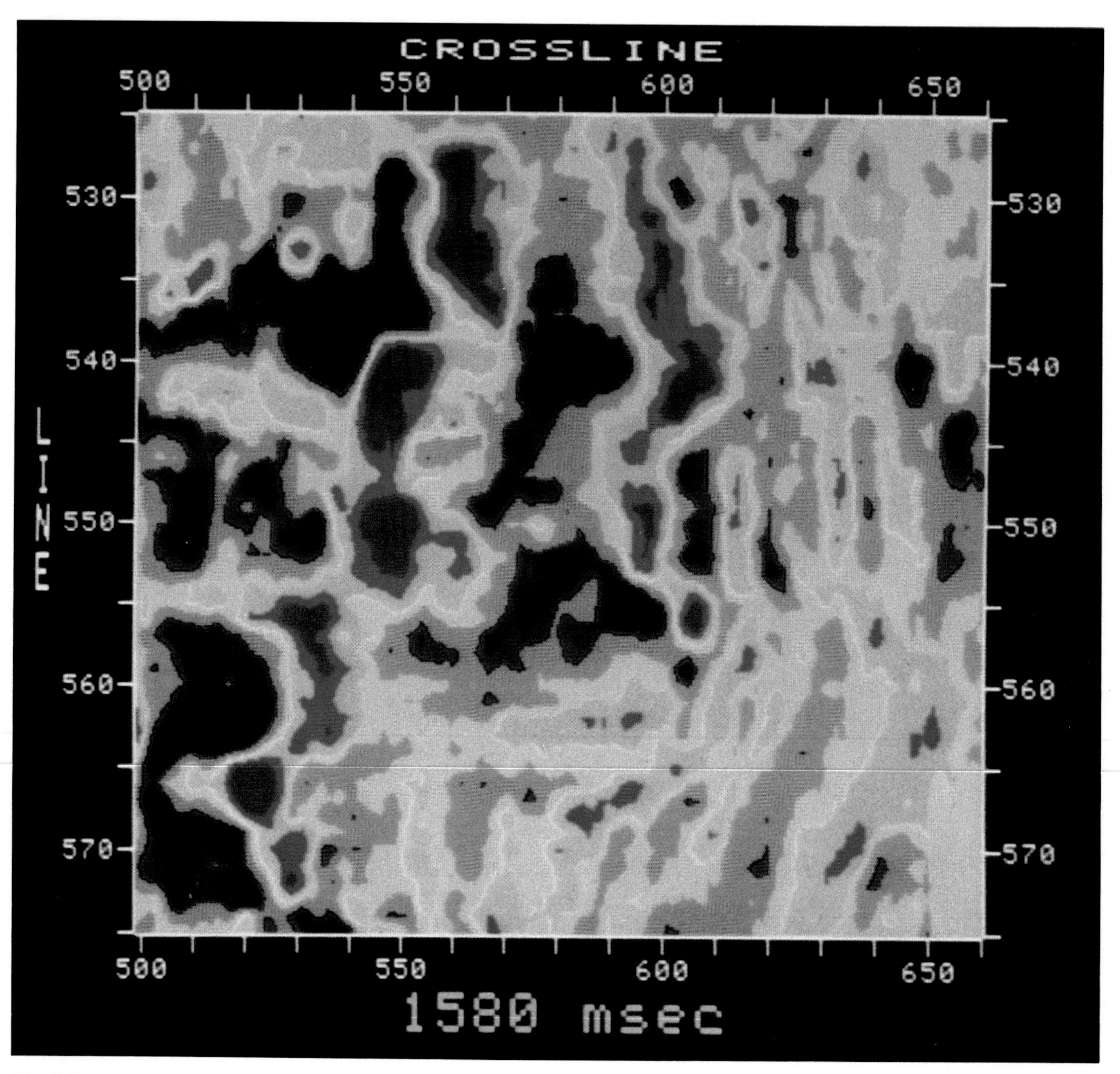

Fig. B-5.

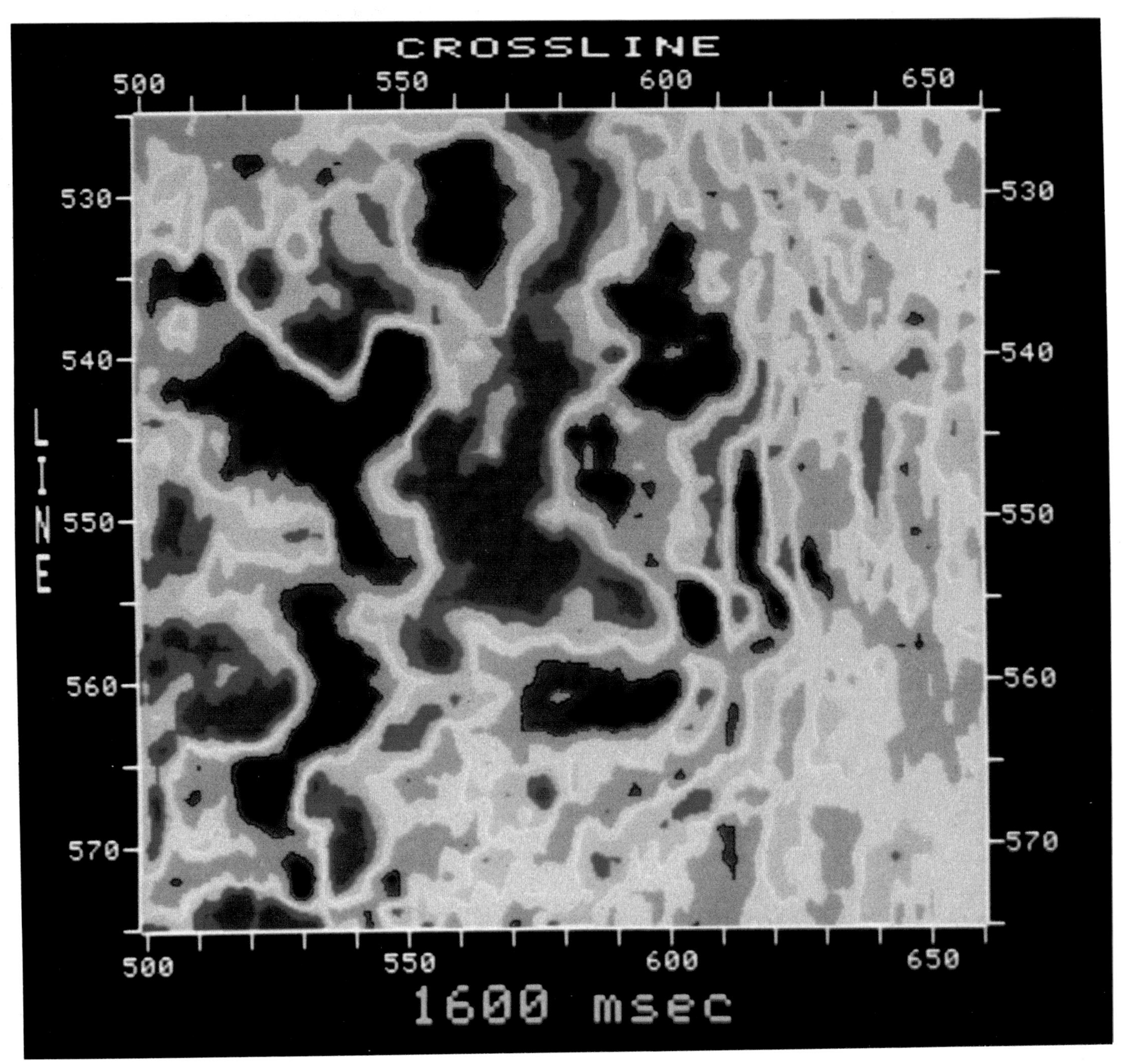

Fig. B-6.

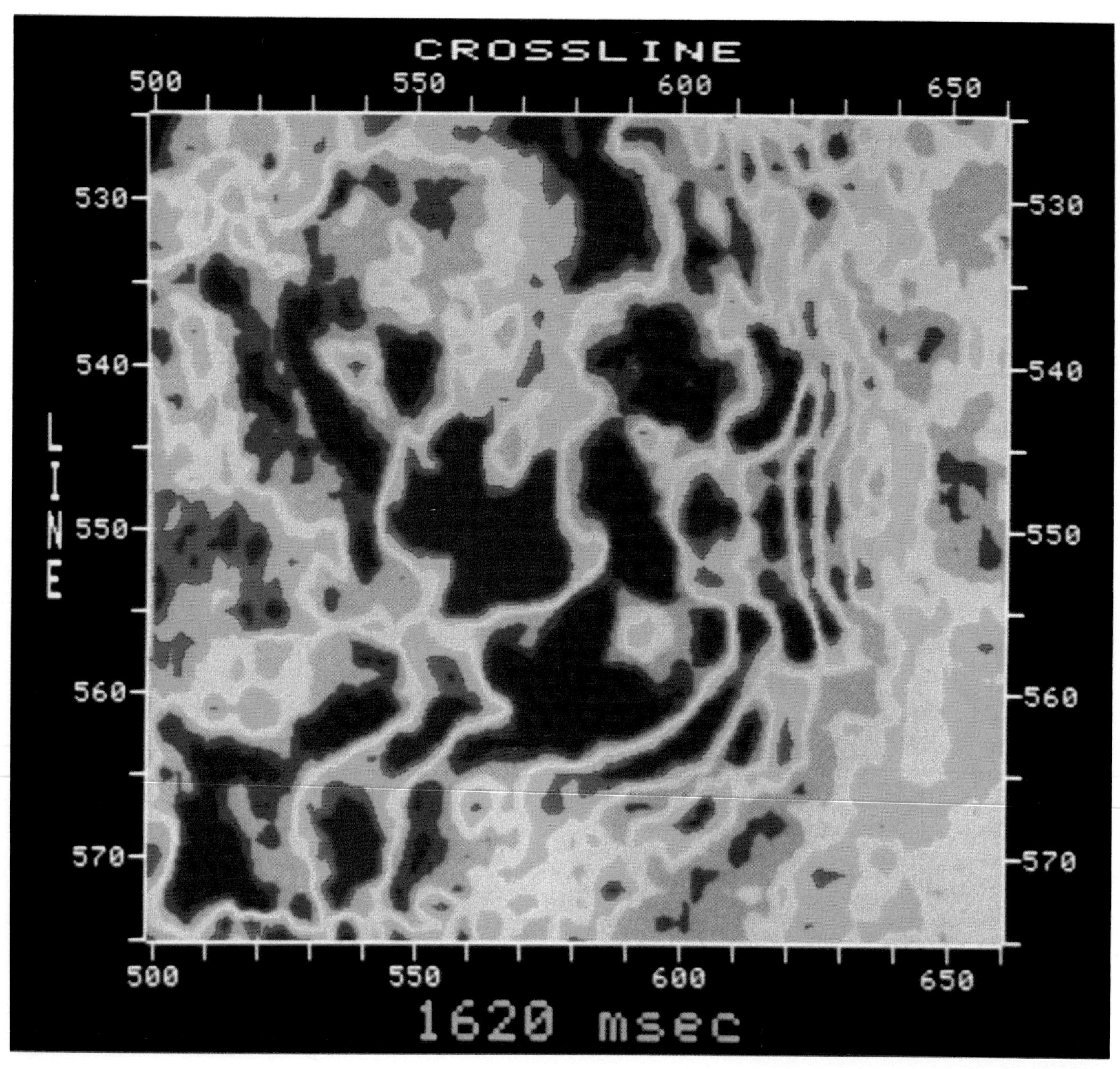

Fig. B-7.

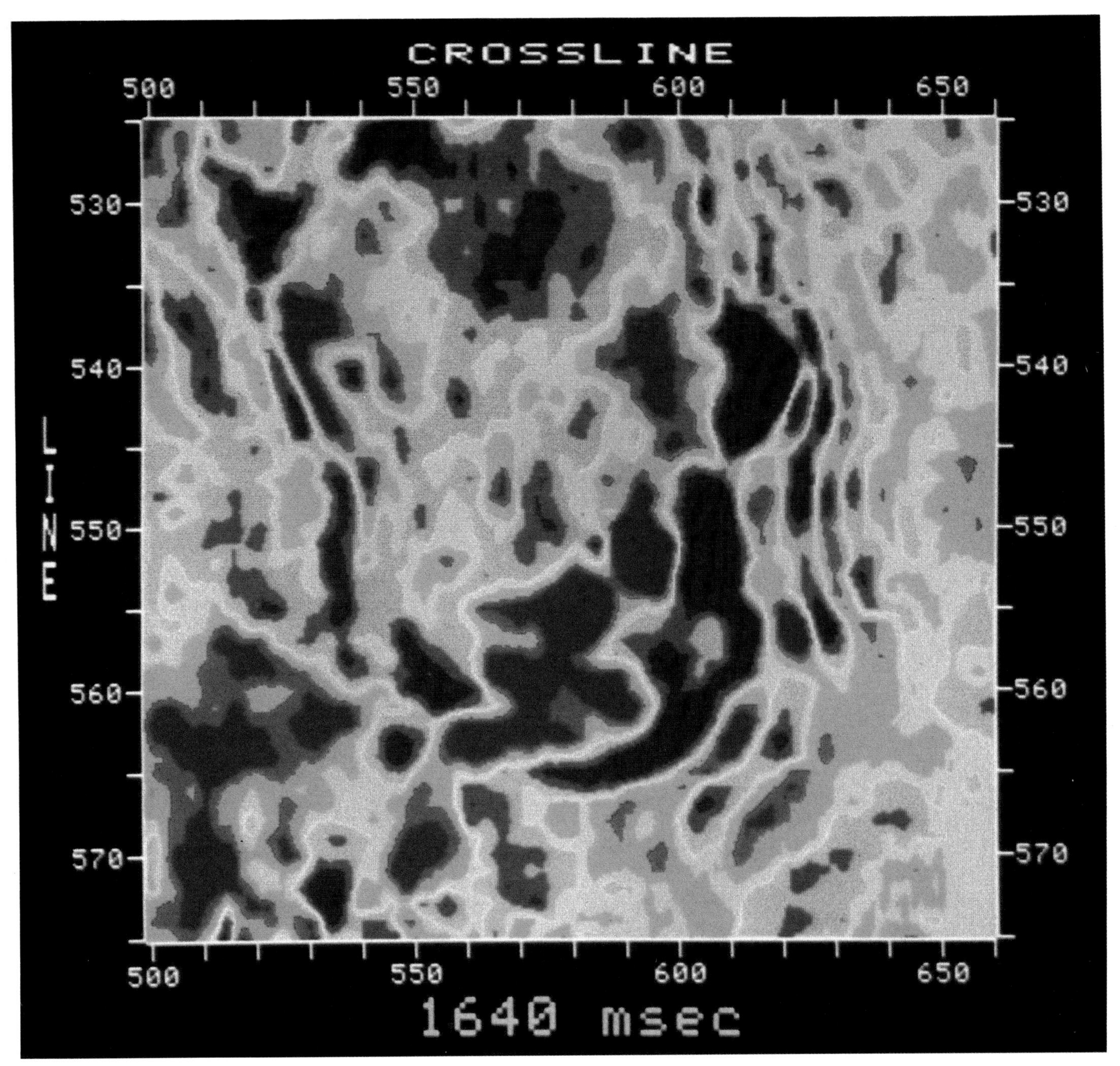
CROSSLINE
500
550
600
650
530
540
550
560
570
LINE
500
550
600
650
1640 msec

Fig. B-8.

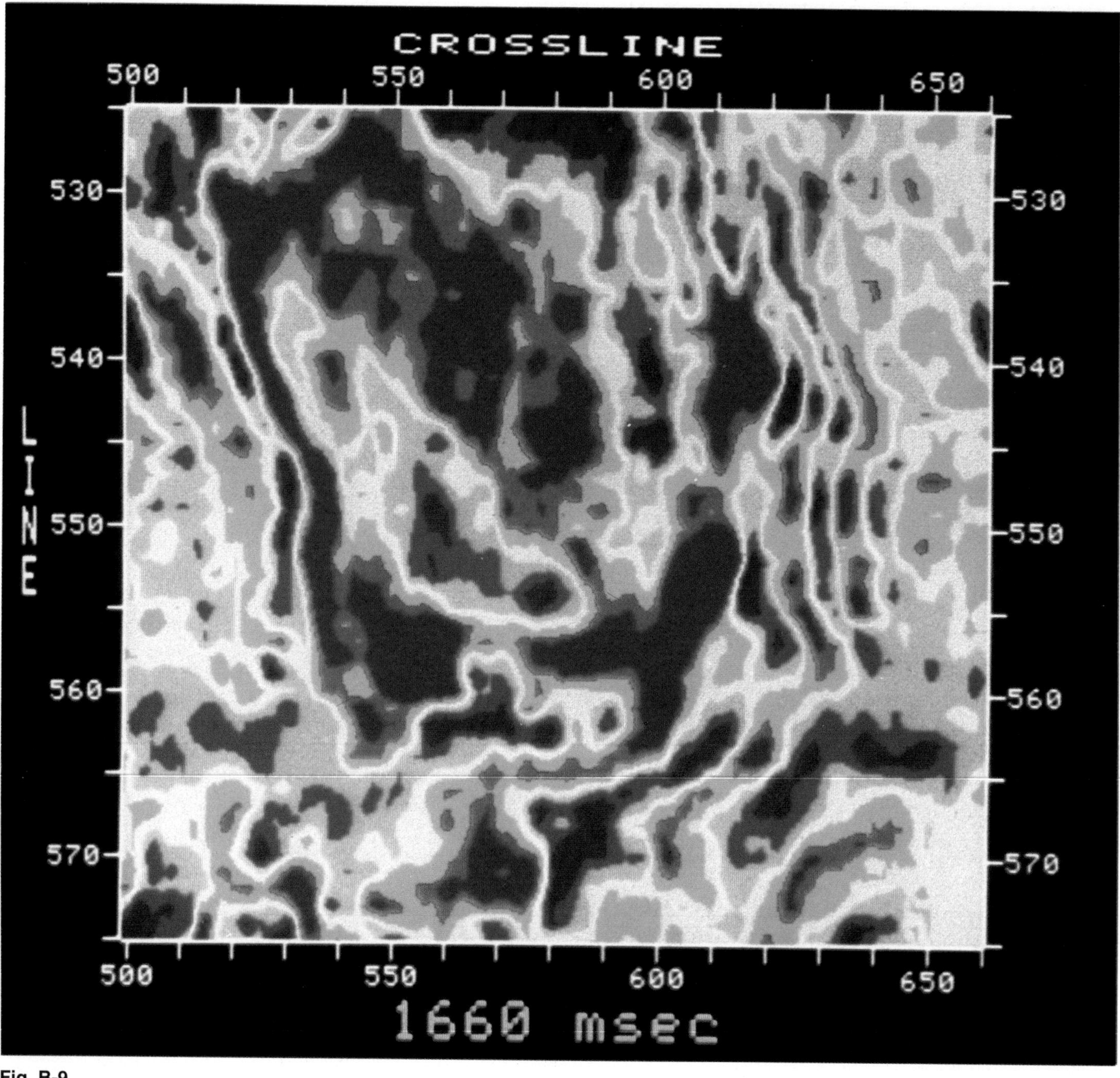

Fig. B-9.

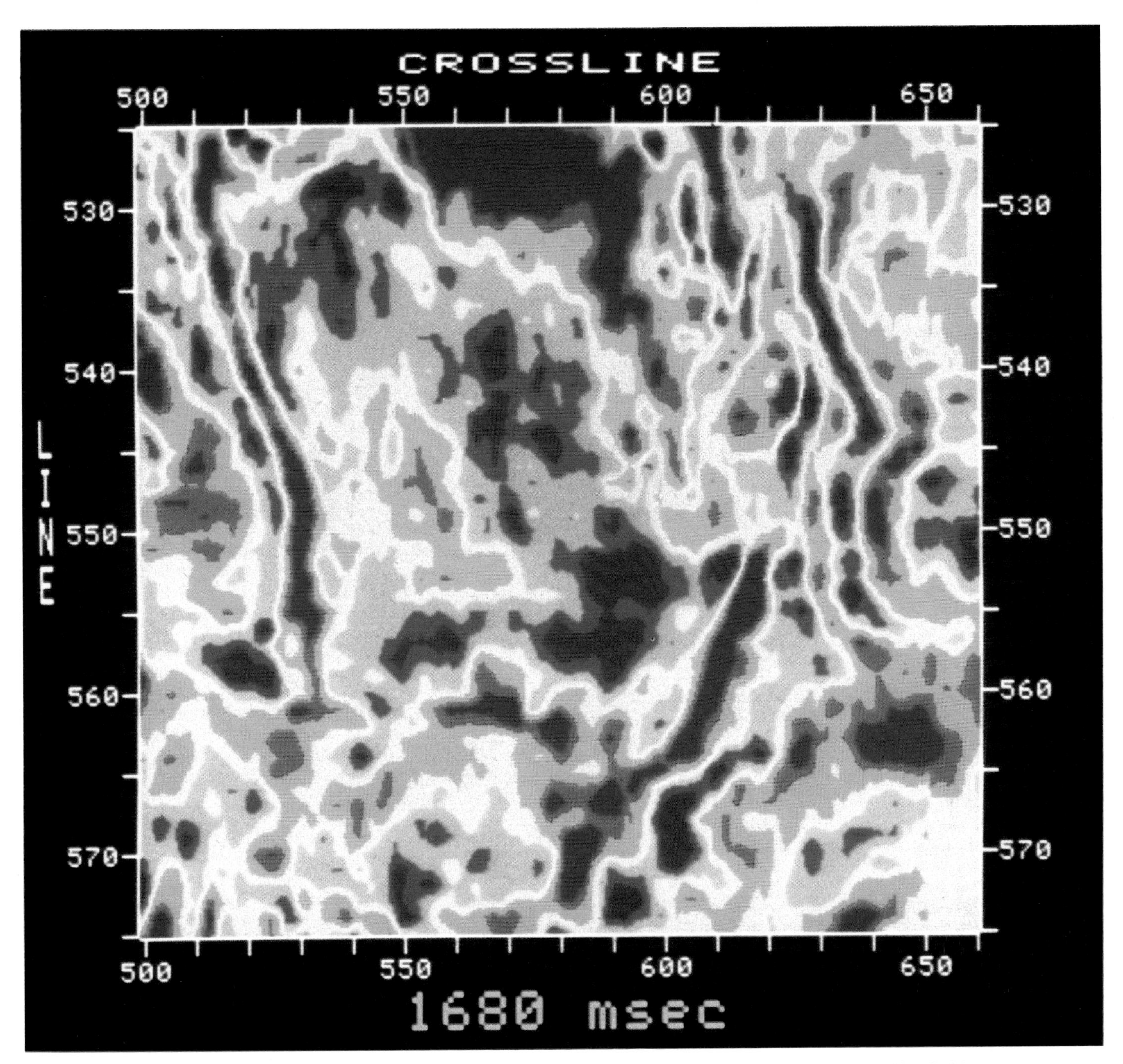

Fig. B-10.

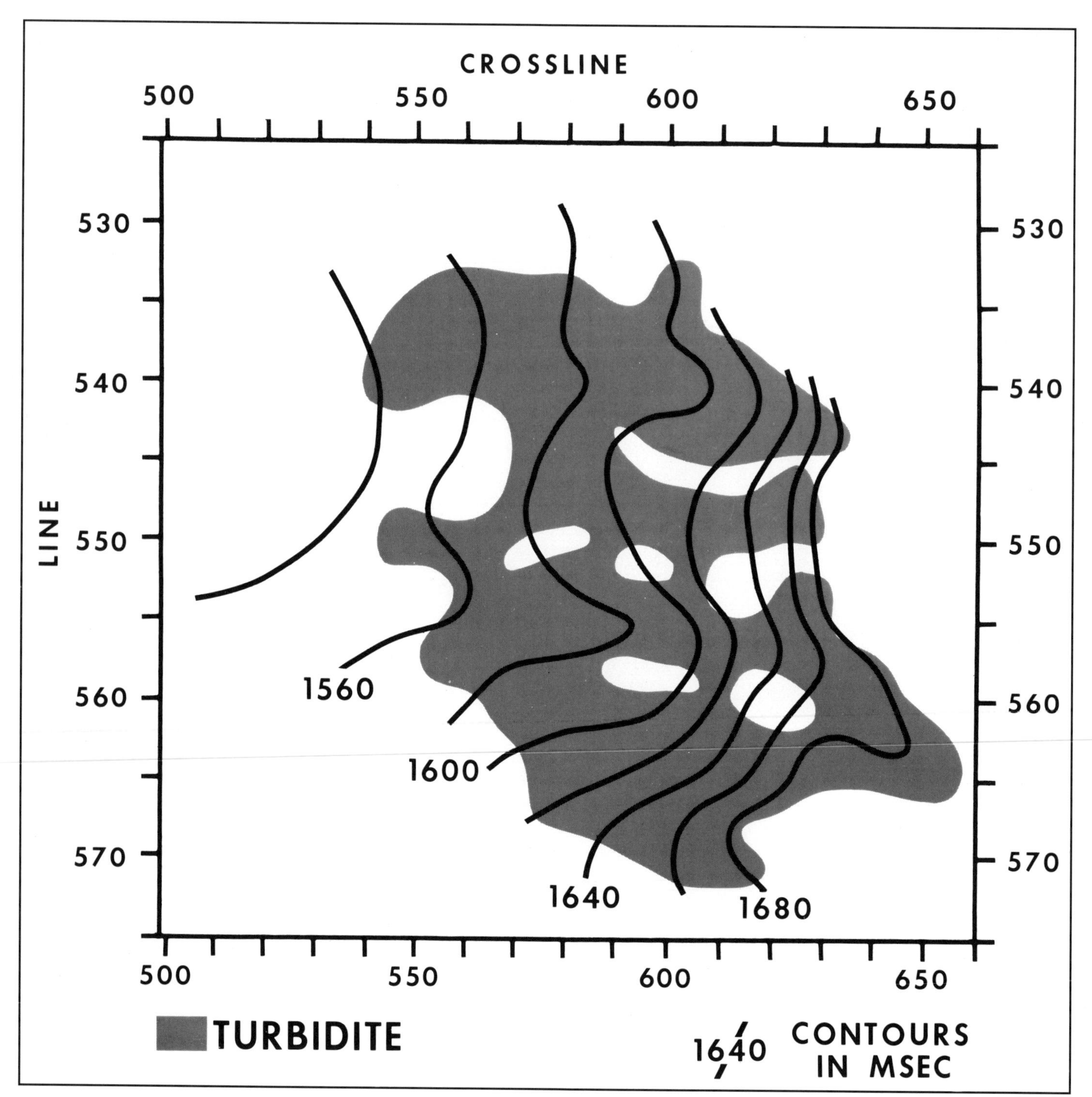

CROSSLINE
500
550
600
650
530
540
550
560
570
LINE
1560
1600
1640
1680
500
550
600
650
TURBIDITE
1640 CONTOURS IN MSEC

Fig. B-11.

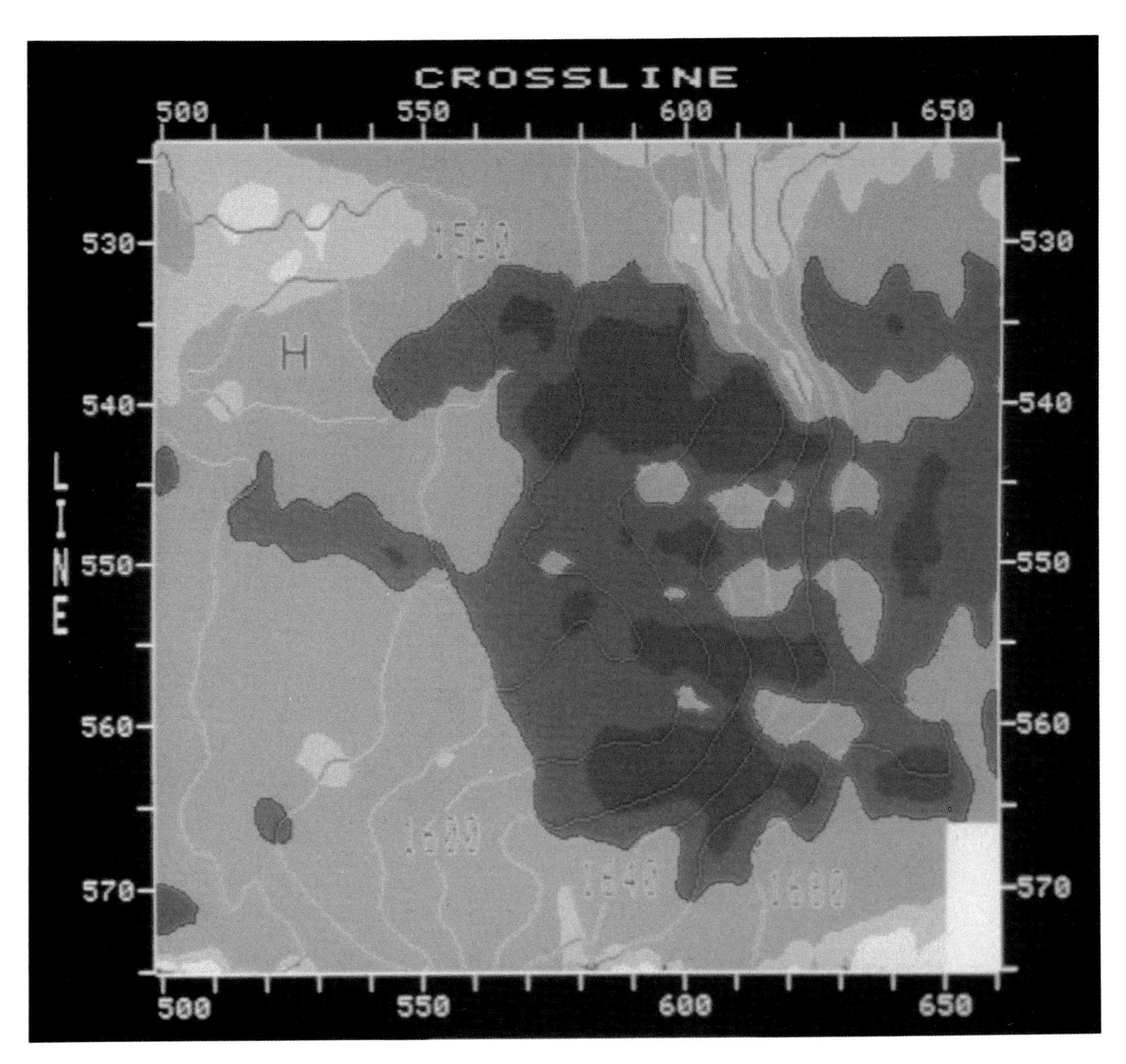
CROSSLINE
500
550
600
650
LINE
530
540
550
560
570
H
1600
1640
1680

Fig. B-12.

INDEX